Tactical Missile Warheads

Edited by
Joseph Carleone
Aerojet General Corporation
Azusa, California

Volume 155
PROGRESS IN ASTRONAUTICS AND AERONAUTICS
A Volume in the AIAA Tactical Missile Series

A. Richard Seebass, Editor-in-Chief
University of Colorado at Boulder
Boulder, Colorado

Published by the American Institute of Aeronautics and Astronautics, Inc.,
370 L'Enfant Promenade, SW, Washington, DC 20024-2518

Second Printing

ISSN 0079-6050

Preface

The design of conventional warheads for tactical missiles spans a number of different technical disciplines. Furthermore, the approaches to design take many different forms. Tactical warheads represent systems, in and of themselves, that involve primarily the mechanical and chemical branches of science. From metallurgy to structures, explosive loading to detonation physics, gasdynamics to ultra high-rate behavior of metals, this field encompasses a group of technical experts who have broad capabilities obtained through many years' experience. Although there is much literature on the subject, a text which marries the fundamentals with the more modern design approaches will assist new members of this field to learn more rapidly, as well as provide a reference text for those working actively in the field. It is to this end that the present book has been designed.

The major emphasis is on the developments spanning the period 1970 through 1990, although well-known fundamentals are reviewed where deemed appropriate. The book is a compilation of contributions from a number of authors who are experts in this field. Each author is currently very active in his field and therefore presents the fundamentals using the most modern approaches to warhead design. To set the stage, the book begins with an historical perspective that takes us from the early period, where man first develops unique methods to fend off predators, through the black powder era, the high-explosives era, and the modern era. The relationship between the level of technology in each era and the weapon systems of the time is highlighted.

The book's chapters are each self-contained articles; however, the topics are linked, and they may be divided into three groups. The first group, consisting of Chapters 1–5, provides a broad introduction as well as four technology areas that are fundamental to the physics of modern high-explosive warheads. More specifically, Chapter 2 on Explosives adds to the existing literature by focusing on explosives used and under development by the U.S. Department of Defense. This complements the U.S. Department of Energy publications. Although its goal is not to teach the new warhead designer everything he or she needs to know about explosives, it does provide a good introduction with sufficient referrals to classical works on the subject, thus giving one a place to start. It also classifies the explosives in several ways: by warhead function, by manufacturing procedure, and by warhead effects. This cross-referencing approach is extremely useful to the warhead designer.

Chapter 3 addresses an area that is key to the modern warhead designer. Full, dynamic characterization of materials, especially metals used in warheads, is critical to exploit and expand the modern design methodology. This area of dynamic behavior of materials is one of the most important contributors to the future improvement of warheads. The designer must know the limits of the materials, both warheads and targets, to do his or her job. The understanding of these limits at ultra-high strain rates is essential for the proper determination of a suitable design.

The fundamental physics underlying warhead mechanisms involves the response of metals to energy deposited upon them by explosives. Chapter 4 reviews and extends classical work on the manner in which explosives accelerate metal. Warhead designers will find it extremely useful to have the tools and understanding embodied in Chapter 4, especially for quickly developing preliminary designs, as well as for the diagnosis of observed events. The final chapter in this first group is dedicated to one of the most useful modern tools, computer simulation. Developed in the 1960s, continuum mechanics computer codes, often called hydrocodes, first came into use for conventional munitions in the early 1970s. From 1970 to the present, the use of hydrocodes has become a widespread and essential tool for the development of modern weapons. For this reason, we have included an extensive article that delves into the different types of codes from a fundamental and highly detailed viewpoint.

The second group is composed of Chapters 6–9. Chapters 6, 7, and 8 discuss three fundamental classes of warheads, namely, shaped charges, explosively formed projectiles (EFPs), and fragmentation warheads. The chapter on shaped charges addresses a generalized theory of shaped charge jet formation and jet particulation. Parametric calculations are presented to exemplify the use of such models in design. Chapter 7 gives a broad introduction into EFPs and how one uses first principles, together with hydrocodes, to synthesize an EFP design. Chapter 8 presents the broad category of fragmentation warheads, which encompasses warheads that use natural fragmentation, controlled fragmentation of various types, and aimable fragmentation. Interaction of the various types of warhead mechanisms with targets is presented in Chapter 9. The mechanics of jets, rods, and fragment penetration into various targets is reviewed in detail. Many experimental results are discussed.

The third group, Chapter 10–12, focuses upon test methodology. The most versatile and most widely used experimental tool for diagnosing warhead effects is flash radiography. Chapter 10 addresses this topic in detail and shows its use in many examples. High-speed photography, another widely used tool, is presented in Chapter 11, again with many examples. Finally, a number of special test techniques are presented in Chapter 12. This series of articles shows the methods available to acquire unique characteristics of warhead events, including the use of laser interferometry and tomographic reconstruction of flash x-ray radiographs. We will see these techniques being used more and more as the warhead mechanisms become more complex and we introduce multiple, time-phased warhead events as well as fully three-dimensional warhead designs.

By blending the technology embodied in these chapters, the warhead designer or researcher will be able to develop a unique design methodology tailored to his or her specific needs and tailored to the equipment and test ranges available.

I sincerely thank the authors who contributed an enormous amount of time and effort in constructing their chapters in accordance with the overall plan. I also thank them for their cooperation and patience with the endless changes that were imposed upon them. Thanks and sincere gratitude are due John Brinkman, who encouraged me to take on this task and reviewed and corrected the original outline. I want to thank Mary Neil for initially organizing and typing the original manuscripts as well as interfacing with many of the authors and AIAA on the early stages of this work. Thanks are also due to Cheryl Bodman for typing several chapters. Furthermore, I thank Marilyn Creed, who refined the manuscript and put it into final shape, coordinated with all of the authors, managed the final artwork, and handled the administrative details of final approvals and submission to the publisher. Finally, I express my appreciation to the staff at AIAA for their expert work in final editing and publication. Special thanks goes to Karen McQueen for her patience and endless effort in bringing this work to completion.

Joseph Carleone
June 1993

Table of Contents

Warheads: An Historical Perspective

Donald R. Kennedy*
D. R. Kennedy & Associates, Inc., Los Altos, California 94022

1. Introduction

THIS historical perspective of tactical warheads includes an overview of the antecedents of the modern warhead that extends from the dawn of prehistory of man on Earth to the modern era, which extends from World War II to the present. Much of the following, in particular the early part, is anecdotal and perhaps flavored with the author's impressions based on reading a variety of sources, most of which differed in their interpretations. More modern warhead discussions are largely based on the author's own experience as a warhead design engineer in active practice since 1950, encompassing the entire scope of non-nuclear warhead technology from underwater to aerospace applications.

For reasons obvious to those concerned with modern weaponry, the present perspective must be terminated short of a description of the most current warhead technology. However, we are fortunate in having contemporary descriptions of many warheads available in the international press, which enables this discussion to be quite current and, as a result, useful to students of the subject.

Warheads discussed here are of the type usually described as *non-nuclear* or *conventional*. Therefore, the following discussion is restricted to those warheads that either utilize the carrier's kinetic energy or employ chemical energy (e.g., high explosives) to create such effects as air blast, project kinetic-energy mechanisms such as fragments, and form and project armor defeat mechanisms such as shaped charge jets and explosively formed penetrators (EFPs). More specialized warheads, such as those for chemical or biological agent delivery, propaganda leaflets, and illuminating flares, are only briefly introduced.

As will be apparent, the target and what the user seeks to do to the target are the primary "drivers" of a warhead's design and its specific kill mechanism. However, it will also be shown that the design constraints include the delivery system's limitations, the levels of technology involved (particularly for the explosive and projected materials), and perhaps more importantly the designer's recognition that certain technologies or materials can be successfully exploited.

*President.

Finally, it will be clear that it is the designer's ingenuity that advances the state of the art and that man, as always, is still the most critical factor in modern warhead developments.

2. Prehistory and Early Historical Antecedents

Man, from the dawn of history, has sought means to exert his will and to dominate others, both man and beast. To accomplish this he has had to develop the means to deter the aggressive actions of other men. History does not record the first time that man picked up a rock or wielded a club, the true antecedents of today's tactical missile warheads.

Man's earliest targets were doubtless other men and animals: other men were competitors, and beasts were both threats and potential food. Eventually, the target became a man riding on a beast, with one or both protected by materials, such as animal skins and later chain mail and plate armor, to help protect the wearer and beast from attackers' blows.

Man has long sought the means to improve both his offensive and defensive protection capabilities. Early man learned to affix a rock to the end of a stick to increase both the hitting power and range (standoff) from his adversary. The development of shields to ward off a club made it necessary to develop a means of piercing the shield. A long, hand-held, pointed stock (i.e., the spear) was the best-known solution, one that is still used to this day. Hand-held short-range (i.e., arm-length) knives and swords were also products of the need to penetrate the defense. The hand-thrust spear evolved into the thrown spear (still in use today by primitive tribesmen). The spear, augmented by the kinetic energy of the delivery vehicle (i.e., the lance used by a knight on horseback) evolved to defeat plate and chain mail armor. The armored horse was in a sense the original main battle tank. To further project power to a greater range, the hand-projected spear evolved into the archer's bow-driven arrow and later into mechanically projected arrows (see Fig. 1). The projectors were a form of the flexible bow in which the mechanical advantage was provided by torsion springs made of horse hair and in some cases even human hair. The arrow also evolved into the crossbow and arbalest's projected bolt, which in turn evolved into today's most capable armor-piercing mechanism: the tank-gun-fired sabot, or armor-piercing, fin-stabilized, discarding sabot (APFSDS), a high-density, long-rod penetrator. The APFSDS is also known as an "arrow" projectile by some nations.

The manually wielded club probably first evolved into a projected kinetic-energy projectile in the form of the manual slingshot pellet and later the mechanical "engines of war" (see Fig. 2), the tension- and torsion-spring-driven tossing and thrusting devices variously known as catapults, ballistae, and onagers. As noted above, the latter employed such elastic materials as horse hair and animal sinews as the source of energy for the wound spring. The distinctions between the various forms of catapults are not clear-cut. An early form of the catapult was a mechanical version of the archer's flexible bow. This was to be succeeded by a more powerful torsion-spring type that could either lob or thrust its payload (i.e., early mortar or gun, respectively). The ballista (from the same root as *ballistics*) was a larger version of the catapult and, depending

Fig. 1 Sixteenth-century machine capable of launching long bolts, darts, balls, etc. (from a 1588 copper plate engraving).

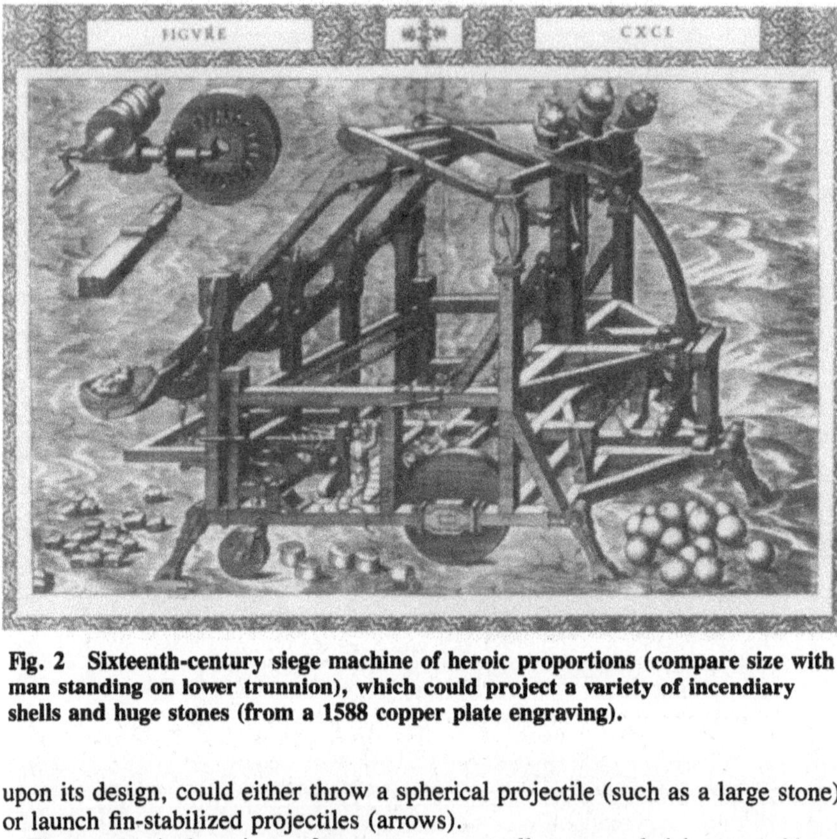

Fig. 2 Sixteenth-century siege machine of heroic proportions (compare size with man standing on lower trunnion), which could project a variety of incendiary shells and huge stones (from a 1588 copper plate engraving).

upon its design, could either throw a spherical projectile (such as a large stone) or launch fin-stabilized projectiles (arrows).

The mechanical engines of war were eventually superseded by smoothbore mortars, howitzers, and cannon, made possible by a discovery in the period 1300–1320 A.D. that black powder could propel a projectile.[1] Although the invention of black powder occurred at a time and place lost to history, it was to become the predominant material for both propelling and bursting charges for more than five centuries.

The original cannon projectiles were derivatives of the missiles fired from crossbows and are described[1] as "bolts and iron darts feathered with brass. The dart, weighing about 7 ounces, was wrapped in leather to ensure a tight fit" in the gun bore. These feeble and inaccurate projectiles were used for nearly 250 years. In a report to the British government dated March 30, 1588, Sir Frances Drake made mention of such "arrows" in his ship's stores.

Inaccuracies inherent in the aiming, firing, and fuzing of the early cannon made it necessary to extend the range of effectiveness of the projected warheads (essentially the same situation faced today). The black-powder-filled, hollow-sphere shot was an early answer. Biringuccio, writing in the late sixteenth century,[2] described the art of making cannon shot so as to break it up in a predetermined manner. These early, "controlled fragmentation" cannon shot are direct ancestors of modern guided missile warheads that use controlled fragmentation techniques.

The rifled gun, howitzer, and mortar evolved from smoothbore cannon and, because of improved accuracy of delivery, became the world standard for more than a century, including most of the present-day artillery. It is indeed ironic that today the highest-performance tank guns, those that project armor-piercing APFSDS "arrow" projectiles, have reverted to the smoothbore form.

The early incendiaries included boiling oil, lead, and the dreaded Greek Fire (the exact composition of which some claim has been lost to antiquity, whereas others believe that the general ingredients have been known for at least 12 centuries but that the exact formula is not known or "perfected"). Incendiaries were used with great effect against both men and their machines. Greek Fire was apparently last employed by both sides in the U.S. Civil War.[3] The modern-day successors to the Greek Fire and boiling oil include the magnesium incendiary bombs of WWII, which proved to be even more devastating than the nuclear "atom bomb" weapons, and the napalm of WWII and succeeding wars. Today's incendiaries include other metals (e.g., zirconium), although both the intent and end result are the same as with the earlier crude materials. In addition to incendiaries, other applications that involve burning of chemicals for light and heat include battlefield illuminants and heat sources used to decoy modern IR seeker missiles. Smoke-generating mechanisms for screening and marking purposes are also derived from the incendiaries, although some of today's smokes are generated by "cold" materials that preclude the need to combust the chemicals.

Chemical warfare as it is known today did not become an important factor until World War I. Although modern war gases are even more devastating than they were in the First World War, development of chemical agent delivery warheads (i.e., by the United States) had virtually ceased by the 1960s. Although chemical and biological warheads are probably still being developed and introduced into the inventories of such nations as Iraq and the former Soviet Union, in the United States the modern chemical warhead is more likely to employ another form of chemical mechanism, an air-blast mechanism known as a fuel-air-explosives (FAE) device. Because the FAE mechanism is employed by others, including potential adversary nations, it is discussed further under the subject of air-blast warhead technology.

Biological warfare was practiced in early siege warfare in which the diseased carcasses of men and animals were projected by catapults into the enemy's midst. Missile warheads to delivery biological agents were also developed in the late 1950s. United States development subsequently ceased because of changed national policy.

3. Warhead Technology in the Low-Explosives (Black Powder) Era (1300 A.D. Through the U.S. Civil War Period)

The German monk Berthold Schwarz is credited with the design of the first cannon in Europe in 1313. Figure 3 is a cartoon from a pre-WWII German handbook[4] purporting to show Berthold Schwarz *wie Bertold Schwarz vor zwei Sekunden des Pulvers grosse Kraft erkunden*, or as he appeared "two seconds

Fig. 3 German cartoon that reads "As the Monk Bertold Schwarz appeared two seconds after discovering the force of black powder" (ca. 1313).

after discovering the great force of (black) powder." The invention of black powder is attributed to the Chinese many centuries prior to the time of the unfortunate Herr Schwarz. It is not apparent whether the term black refers to the color (which is typically gray) or to Schwarz himself (German for "Black").

The period from the fourteenth through late nineteenth centuries saw refinements of the smoothbore cannon but continued to be based on a low explosive, black powder. The targets continued to include both man and armored man on horseback until development of the black-powder-projected small arms bullets ultimately rendered useless the armor used to protect man. Cannon fire targets also included the traditional naval and fortified land targets.

Refinements made during this centuries-long period included the invention of the modern warhead and various means for its propulsion to the target. Although the rocket was employed to a limited extent, tube artillery saw the widest use, and the targets, as in the pre–low-explosive era, continued to be man, man on horseback, naval craft, and the fortifications that were erected to protect men and their treasures from the enemy.

Tube artillery in universal use included smoothbore cannon, howitzers, and mortars. Ammunition for these projectors was historically referred to as "shot," the earliest of which were made of stone, as in the days of the catapult. Solid stone shot is known to have been used as late as the U.S. Civil War, when iron shot was unavailable to the citizens of the gold mining town of Volcano, California, who were faced with what they perceived as a possible invasion by Southern sympathizers.

Iron and lead balls had long since superseded the stone shot, with the earliest mention of iron shot in England occurring about 1350. Bronze shot appeared about the same time. Red-hot balls of clay were used in the siege of Gibraltar in 1779. Cast-iron shot was made in 12 different sizes for the British service, and on occasion they were heated to redness for use against ship targets and military stores to start fires. Steel shot was introduced into England in 1865 to pierce ship armor but was quickly replaced by chilled-iron, spherical shot by 1867.

Introduction of rifled tube artillery permitted development of projectiles with greater lengths than their diameters, or L/D ratios greater than unity. This allowed the concentration of a greater kinetic energy in a small area of impact to enhance the ability to penetrate deeply into an armor-protected target.

The U.S. Civil War also saw the introduction of iron-clad warships. Controversy raged over the choice between armor "punchers," or penetrators, and the use of massive shot to cause a "racking," or battering, of the armored target. According to an apparently well-researched source[5]:

> The theory of racking was to waste no power in punching mere holes, but to so increase the weight of the shot that the entire blow would be expended in straining, loosening, and dislocating the armor, breaking its fastenings and tearing it off, after which the vessel would be easily destroyed by shells. . . . At the same time racking could break the ribs and sides of the vessel and render her unseaworthy.

A similar controversy could conceivably take place today with respect to defeat of the complex armor arrays of the present-day land "dreadnought," the modern main battle tank.

The U.S. Civil War also saw extensive use of shot designed to produce blast and fragment effects. The shot was usually in the form of hollow cannonballs filled with black powder. As noted above, plain cased and prefragmented (i.e., controlled fragmentation) cannonball technology existed from the 1500s (see Fig. 4) and is described by Biringuccio.[2]

Armor-piercing gun projectiles of the solid shot type with elongated bodies and discarding sabots (Fig. 5) were developed in the U.S. Civil War period to constitute one of the first major breaks from traditional spherical cannon shot.

Grape and chain solid shot for such diverse targets as personnel and ship rigging were frequently employed as alternatives to both solid shot and black-powder-filled ball. Novel projectiles that evolved during the late 1800s included submissile types. A unique projectile design, consisting of interlocking metal elements, was patented in 1882 under the title "Union Combination Charge." (The writer learned of this in 1962 when he attempted to patent a modern projectile with interlocking metal elements for gun-fired defense against incoming missiles.)

Rockets have been employed in warfare for many centuries. Because of their inherent inaccuracies, they were most frequently employed as area bombardment weapons. Warheads included black-powder-filled iron cases and incendiary types.

4. Warheads in the Early High-Explosives Era (Before and After the American Civil War)

The target spectrum had by now broadened to include all previous targets plus steel-clad naval warships and land craft, including both road and rail-track-rolling stock. Personnel remained as the primary targets.

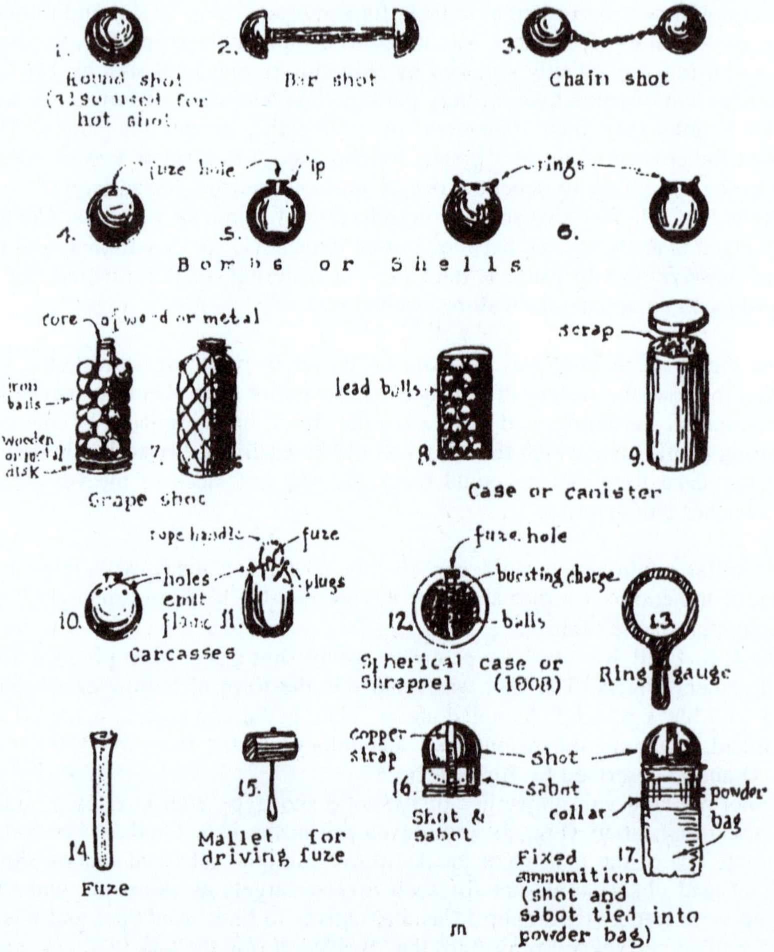

Fig. 4 Types of eighteenth-century gun projectiles.

Two important events occurring just 15 years before the Civil War marked the invention of detonating high explosives. The explosive characteristics of gun cotton (nitrocellulose) were discovered in 1846 by Schonbein and by Bottger. An Italian, Sobrero, discovered nitroglycerin the same year.

Smokeless propellants based on both nitrocellulose and nitroglycerin in France (Viele's *Poudre B* in 1884) and Sweden (Nobel's *ballistite* in 1888) finally permitted higher performance than the centuries-old black powders, even though the latter were by 1860 much refined. *Ballistite* continued in use beyond the end of WWII.[7]

An American ordnance officer, General T. J. Rodman, discovered the principle of progressive combustion of grained black powder by control of grain size and

Fig. 5 Civil War armor-piercing, long-rod, discarding sabot, subcaliber gun projectile. An invention of C. W. Stafford. Consists of chilled-iron bolt surrounded by a discardable wooden sabot. Used in limited numbers by Federal forces.

perforation. Rodman's discovery marked the beginning of the science[6] of interior ballistics.[8]

Both nitroglycerin and nitrocellulose continue to be used today in double-base propellant. Although both materials were known for some 15 years before the U.S. Civil War, neither was sufficiently developed to be employed in that war.

Experiments with compressed blocks of gun cotton resulted in the discovery in Germany (by Max von Foerster in 1883) and the United States (by Charles Monroe in 1888) of the cavity effect, the precursor to the modern shaped charge, or high-explosive antitank (HEAT) and explosively forged projectile (EFP), technologies. ["EFP" is variously defined as explosively formed penetrator, explosively forged projectile, and explosively formed projectile, all meaning the same thing. The term was derived from "SFF" (self-forged fragment). Other related terms employed since WWII include "MS" (Misznay-Schardin effect), "P-charge," "Plate-charge," and "Projectile charge," depending on the source laboratory or nation. All of these are so-called "end-effects" mechanisms and are essentially all-slug (i.e., no-jet) shaped charges.]

Picric acid was employed as a bursting charge for shell in the late 1800s but was considered excessively hazardous. Beginning about 1902, picric acid began to be replaced by trinitrotoluene, or TNT, first by Germany, and by 1912 as the standard bursting charge for U.S. Army high-explosive (HE) shells.

TNT was the most important explosive developed prior to World War I and continued to be important in both World War II and the Korean war. TNT continues in use, but its application in U.S. munitions is often as the castable ingredient of high-performance explosives such as RDX/TNT cyclotols and HMX/TNT octols. TNT is inexpensive and is still used as the primary filler of explosive ordnance by less affluent third-world nations.

Aluminum powder was first proposed as an energy enhancement additive to TNT and other explosives by Escales in 1911. This important event, although before World War II, did not contribute significantly in that war. It was first adopted by the Germans for the explosive Ammonal.

5. TNT Era: World War I to Early World War II

World War I introduced the modern era of warfare, which was to be extended and refined for the following 20 years in a series of regional wars in the 1930s and on a grand scale in World War II. WWI saw the introduction of aerial warfare, the U-boat as a major threat on the high seas, and the stalemate type of warfare exemplified by the trench warfare in France. Chemical warfare and the battle tank were also introduced in WWI. Although chemical warfare has since fallen into disrepute and its use is almost universally banned (Iraq is a notable exception), the tank and airplane became centerpieces of the wars in the 1932–1945 period.

As previously noted, this was the era of TNT, the dominant explosive for the munitions employed in land, sea, and naval warfare. The WWI warheads were basically simple, TNT-filled, plain-steel-cased gun projectiles, aerial bombs, and underwater mine and torpedo warheads. Other explosives, such as picric acid, were also employed, although on a smaller scale. Amatol, a mixture of ammonium nitrate and TNT, was frequently used as a substitute for pure TNT.

Black powder continued to be used in certain classes of ordnance, in particular the earlier hand-thrown grenades and shrapnel-filled shell bodies.

The remaining paragraphs in this section review warheads of various classes of ordnance in the period from WWI to WWII.

5.1 Gun Projectiles (High-Explosive)

The predominant warhead of World War I and World War II was the artillery-fired, plain-steel-cased, TNT-filled "common" projectile. These continue to be the standard gun-fired explosive ordnance of most nations even though much more efficient explosives and projectiles have long been available. The most important reason is their low cost as compared to other typically more complex types.

5.2 Gun Projectiles (Shrapnel-Type)

One of the common expressions heard after WWI was that a veteran had been wounded by "shrapnel." A more proper term would have been "shell fragments," inasmuch as the "shrapnel" invented by Lt. Henry Shrapnel of the Royal Artillery in 1784 was quite different from the most commonly used ammunition in the First World War. Shrapnel's invention was a spherical cased shot consisting of a thin iron shell filled with musket or carbine balls, with sufficient (black) powder inserted in the projectile to cause the shell to burst when ignited by the fuze. Until his invention, all shells depended on the bursting charge to break up and disperse the case parts (which describes the modern artillery shell rather well). It is interesting to note that true, black-powder-loaded, shrapnel-filled shells were apparently still employed as late as 1941 in one version of projectile then available for the U.S. 3-in. antiaircraft artillery. The other type was a more traditional TNT-filled steel-case projectile using a black-powder settable train time fuze.

5.3 Controlled Fragmentation Grenades: EC Blank Powder and TNT-Filled Styles

Perhaps the most common image in one's mind on hearing the expression "controlled fragmentation" is thhe mental image of a "pineapple"-style "Mills" hand grenade, widely used in WWI. Early grenades were filled with EC Blank Powder (80% NC, 8% KNO_3, 8% BNO_3) and initiated by black powder. Although the case was generally thought to break up along the grooves, research showed the case to break at the center of each platelet rather than along the groove pattern; this result is attributed to shock reflections from the grooves. The external groove pattern was apparently intended to improve a solider's grip on a muddy grenade.

5.4 Introduction of the Aerial Bomb

The aerial bomb was introduced in WWI as a natural by-product of airplanes and airships. The early aerial bombs were fairly crude by modern standards. They employed thin steel-cases, were filled with cast TNT, had fins to stabilize their flight, and used simple point detonating fuzes to detonate the bomb upon contact with the Earth. The first bombs dropped from "aeroplanes" were small and dropped by hand. As the war progressed, bombs were made larger, and both the bomb delivery methods and accuracy were improved and became more sophisticated.

5.5 Introduction of Guided Missiles in World War I

Few are aware that U.S. development of tactical guided missiles dates from World War I. Early efforts were radio-controlled pilotless aircraft and were known as "aerial torpedoes" or "flying bombs." Both the U.S. Navy and Army pursued such developments. Beginning as early as 1913, Germany had developed wire-guided glide bombs to be launched from dirigibles (zeppelins).

The invention in 1896 of wireless transmission, coupled with the invention of practical aircraft in 1903, led Elmer Sperry, whose technical expertise in the field of gyroscopes and gyrostabilizers was already well established, to develop as early as 1911 an interest in the application of radio control to aircraft. When the United States became more involved in the European war, Sperry and Curtiss undertook development of the U.S. Navy's "flying bomb" beginning in 1917. The result was a catapult-launched biplane that was as large as a manned aircraft. More than 100 trials of the Navy version took place between May 1917 and November 1918. However, signing of the Armistice before development was completed precluded the tactical use of the Navy-developed missile in the First World War.

A perhaps somewhat better known WWI "guided missile" was the "Kettering Bug" developed by the Army Signal Corps of McCook Field in Dayton, Ohio. The Bug was also a pilotless aircraft, or "flying bomb." It was rail-launched and flown on a preset course to its target. The missile was designed to fly until a preset number of propeller turns was recorded, at which time the flying bomb's wings were severed and the engine was stopped, permitting the missile's bomb warhead to fall and explode on impact. For rather obvious reasons, the accuracy of both the Curtiss-Sperry and Kettering missiles was rather poor, particularly in terms of range. Many flight trials were made, and some were successful. Although flying bombs were shipped to France and preparations made for use, as in the case discussed above, the Armistice was signed before they could be launched against the enemy.

5.6 Torpedoes: 1880s to World War I

The arrangement of the underwater torpedo, with its simple explosive-filled warhead and contact exploder mounted in the nose, remained essentially unchanged from its crude beginnings in the Revolutionary War. Until long after the U.S. Civil War, the torpedo was what is now called a mine. In some cases a "spar torpedo" was mounted on a boom extending from an underwater craft (e.g., the Confederate submarine *Hunley*) and driven so as to impact against the target. The *Hunley* successfully sank the Federal ship *Housatanic*, but the "torpedo" explosion was too close to the submarine, so it was also destroyed, with the loss of all on board.

The first "locomotive" (self-propelled) torpedo was the Austrian Luppis's surface-running torpedo in 1867. An Englishman, Robert Whitehead, working with Luppis in Austria, soon invented the first sub-surface-running torpedo operating on compressed-air propulsion.

Torpedo warheads in service in both WWI and WWII employed so-called bulk charge warheads that depended on the detonation of a large solid mass of explosive to produce the shock waves and water surges required to destroy naval targets.

In England in 1913, the so-called *H.M.S. Vernon* trials of a torpedo with a "shaped charge" warhead were performed. Apparently there is no open record of the results of those trials. Because the shaped charge of that era still featured an unlined cavity, little of the damage potential of a modern shaped charge could have been demonstrated; therefore it is possible that the *H.M.S. Vernon* trials did not sufficiently impress the Admiralty. Shaped charge torpedo warheads were not adopted in WWI.

5.7 Chemical Warheads in World War I

Although the majority of the warheads in WWI were of the chemical energy type that depended on the detonation of explosives, other warheads deployed chemicals in other useful forms such as gases, incendiaries, smokes, and illuminants. Chemical warheads were typically projected from chemical ("Stokes") mortars. Incendiaries and screening smokes consisted primarily of white phosphorous (WP). Illuminants were "star shells," flares suspended by parachutes to illuminate the battlefield at night. War gases such as chlorine and mustard gas were particularly devastating to men on both sides, with the result that conventions against their use were adopted by the Geneva Convention. Although the United States was never a party to the conventions, it supported them.

6. Preparations for the Modern Era: Beginnings in the 1920s

Much of the technology we regard as modern and contemporary in origin has roots in the period between World Wars I and II. Some have even earlier origins but were not exploited until that period. For example, the aluminum-hulled, high-mobility cross-country land combat vehicle derives from work in the 1920s and 30s by an American named Walter Christie. A Christie tank acquired by the USSR in the early 1930s evolved into the T-34 and subsequent series of tanks. The huge inventory of Soviet tanks was the impetus for a continuing worldwide effort to develop antiarmor mechanisms.

Antiarmor mechanisms have taken two forms: kinetic-energy tank-gun-fired solid projectiles and kinetic-energy, arrow-like jets generated by an explosive charge known variously as the shaped charge, hollow charge, cavity charge, and high-explosive anti-tank (HEAT). The initial discoveries of the cavity effect were made in the 1880s, but it was not until the mid 1930s in both Germany and Switzerland that the discovery was made that lining the cavity would make it possible to penetrate great thicknesses of armor plate. The metal lined cavity charge evolved into the very high performance warheads on anti-tank guided missiles, shoulder-fired rocket-propelled grenades (RPG), warhead dispersed HEAT grenades, and mines in modern inventories. Both the shaped charge and related explosively formed penetration mechanism are employed in modern antiarmor, antishipping, antisubmarine, and anti-hard-target missile warheads. However, this high performance was made possible by the development of higher-performance explosives, initiating systems, and detonation control mechanisms.

6.1 Introduction of High-Performance Explosives: RDX

RDX (Cyclotrimethylene Trinitramine), a basis of many modern high-performance explosives, was first prepared by Henning in 1899, but its explosive properties were discovered in 1920 by von Herz. The addition of aluminum to high explosives was first proposed by Escales in 1911. Aluminum was first used in such German WWI explosives as ammonal because of a scarcity of TNT, requiring use of mixtures with other, less scarce materials such as ammonium nitrate and aluminum flake with the TNT. Although the Allies had not experienced a TNT shortage in WWI, they did not exploit explosive enhancements such as aluminum and RDX until the problems of WWII made it necessary.

6.2 Transfer of Explosive Technology to and from Atomic Energy Commission Programs

The modern-day non-nuclear missile warheads derive part of their technology from the development of nuclear weapons beginning in WWII. The latter were based on two techniques of bringing together subcritical masses of fissionable materials and holding them as long as possible. The yield of the fission warhead depended on how well (i.e., simultaneously and uniformly) the subcritical mass of fissionable material was assembled and how long it was held together before it blew itself apart. One type employed a gun-type assembly technique in which propellent drove subcritical masses together in a strong, heavily confined "gun bore." The second technique employed explosives to implode a lining of subcritical materials into a critical mass at the center of the detonation. To accomplish this critical task necessitated the development of explosive lenses, detonation wave shapers, and multipoint simultaneous initiation systems to generate and control the desired uniformly spherical implosion. It was also necessary to develop more powerful and improved safety explosives materials. Appropriate information from the developing shaped charge technology was transferred to designers of the early-model atomic warheads. Their refinements were transferred back and became a part of shaped charge technology still in use.

Part of the U.S. Navy's explosive lens development for the U.S. Atomic Energy Commission (AEC) was performed at the Salt Wells Pilot Plant (SWPP) of the Naval Ordnance Test Station (NOTS) at China Lake, California. The Salt Wells facility was at one time (in the late 1940s and early 1950s) one of the more highly developed explosive process facilities in the United States if not the world. The non-nuclear warhead research of the Rockets and Explosives Department at this station in the early 1950s benefitted significantly from technology provided by a close association with the nuclear weapon developments. Much of SWPP's instrumentation, process equipment, and explosive process development technology was subsequently transferred to Picatinny Arsenal for development of Army warheads, both nuclear and non-nuclear. Technology transfers to the

U.S. non-nuclear warhead efforts included 1) the high-performance HMX-based explosives, 2) detonation wave shaping 3) high-explosive "lens" technology, and both 4) simultaneous multipoint and 5) exploding bridgewire initiation systems.

6.3 German Introduction of Lined Cavity Charge Technology

Operating quite independently (as would be expected), Germany was also involved in the advancement of non-nuclear warhead technology both prior to and during WWII. Detonation-wave-shaped, hemispherically lined shaped charges were demonstrated in 1943 to give significant improvements in penetration (see Fig. 6). The early German detonation wave shapers (i.e., *korpers* or "inert bodies") were made of *beton* (concrete). Germany's wave shaper technology may have transferred to the Soviet Union in WWII. Wave shapers are also a hallmark of all Soviet-designed shaped charges seen to date.

During WWII, German investigators also experimented with complex liner geometries, the concept of tandem shaped charges, and multiple explosive charge mechanisms, similar in some respects to the explosive lenses concurrently being developed by the United States for the atomic weapons. Figure 7 shows examples of German and U.S. WWII advanced shaped charge technology in both torpedo and tandem shaped charge warheads.

7. Missile Era: World War II to the Present

Although, as noted above, guided missiles had actually been introduced in the First World War, they were essentially forgotten and neglected in the two decades following the "War to End All Wars." However, a second technology that arose in that period and enabled the missile developments of WWII was a growing interest in and development of both solid and liquid propellant rocketry by one of the principal belligerents of the Second World War. Rocketry was of particular interest in Germany in the 1920s and 1930s. However, very few seriously pursued rocket technology in the West, with the singular exception of Dr. Robert Goddard in the United States.

In the 1930s most nations concentrated their efforts on development of the airplane for both commercial and military use. The aircraft eventually became the centerpiece weapon system of WWII. The primary armaments of aircraft were the general-purpose and incendiary bombs for the bombers and the machine cannon and rockets for the fighters and fighter/bomber aircraft. Guided missiles and rockets were developed relatively late in the war and were to become both an extension and nemesis of the aircraft weapon.

Warheads for the missiles and other armaments were developed to meet the specific target spectrums encountered in each tactical environment. In most cases targets were specific, such as tanks, ships, submarines, or aircraft. In others, the targets had a wide range of hardness, size, and degree of vulnerability. General-purpose munitions were required to have a good, although nonoptimum, effectiveness against a broad spectrum of targets.

The following sections discuss warheads developed for several of the delivery environments and target types; they include historical background where available and appropriate.

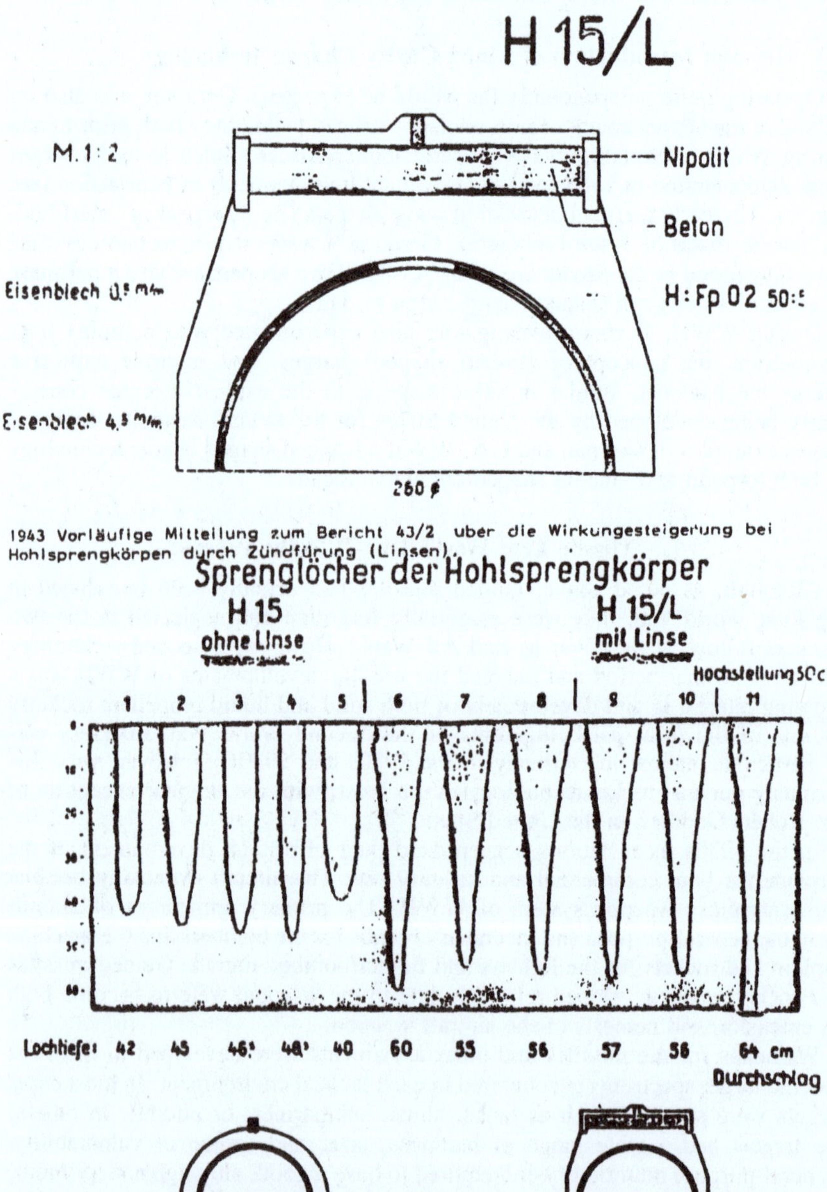

Fig. 6 German WWII research into wave shapers showing "25% increase in penetration" using *beton* (concrete) shapers.

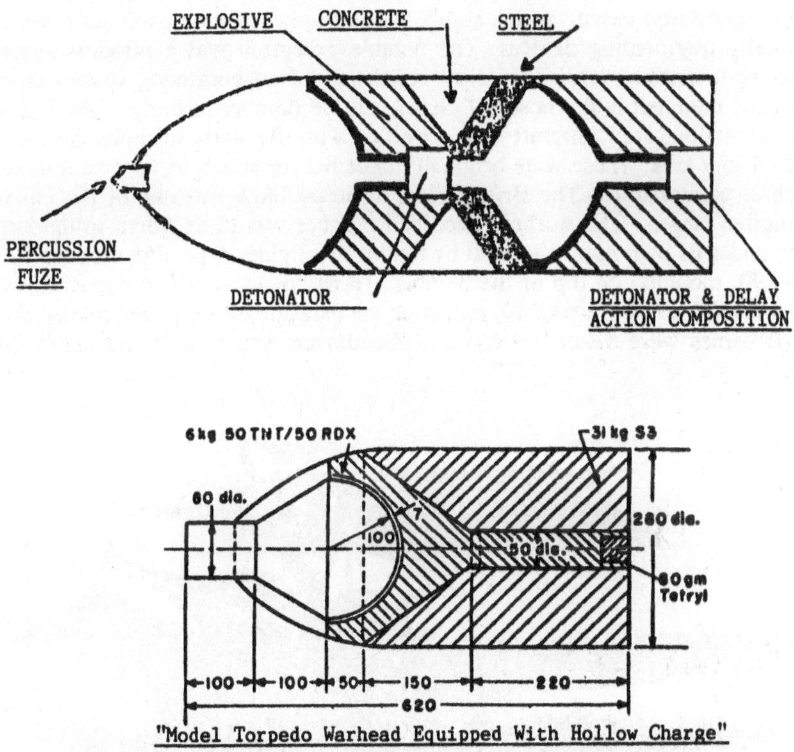

Fig. 7 Early post-WWII advanced technology shaped charged warhead concepts: (top) tandem design—Picatinny Arsenal 1947; (bottom) dual HE shaped charge—German Navy 1945. (Top figure from Eriksen, L. H., "The Shaped Charge," lecture notes, Ordance Dept., U. S. Army Picatinny Arsenal, April 4, 1947; bottom figure from "Introduction to the Theory of Underwater Explosions," Chemical-Physical Researce Institute of German Navy, Kiel, 1945; translated by David-Taylor Model Basin, AD-125-686, 1948.)

7.1 Surface Target Tactical Missile Warheads

Warheads in this category are employed in tactical missiles against a broad spectrum of land and sea surface targets. The radio-controlled pilotless aircraft developments started in WWI continued in the 1930s and contributed to the technology of the WWII-era guided missiles. The German development of pulse-jet-powered pilotless aircraft (e.g., V-1) and surface-to-surface ballistic rockets (e.g., V-2) were the best-known examples of World War II missiles but were in fact only two of many guided missiles developed for both Allied and Axis forces.

Warheads in the WWII missiles and guided bombs were typically simple, TNT-filled, iron-cased, aerial bomb or bomb-like, combination air-blast and naturally fragmenting devices. The notable exception was a pilotless-aircraft-type air-to-surface "cruise missile" consisting of an enormous shaped charge warhead mounted in the noses of German Ju-88 bomber airframes (see Fig. 8). Operationally "tired" aircraft were modified with the warhead replacing the entire cockpit area. These were originally intended for attack of very hard targets, such as capital ships. The British Fleet at Scapa Flow was one of the initially identified targets. The warhead-modified bomber was to be flown to the target area under its own power, steered by a pilot in a fighter-type aircraft, such as an Fw-190, mounted on top of the bomber. These "pick-a-pack" composites were known as MISTEL (mistletoe), reflecting the parasitic top-mounted piloted craft. Code names were *Beethoven-Geraet* ("Beethoven Apparatus") and unofficially

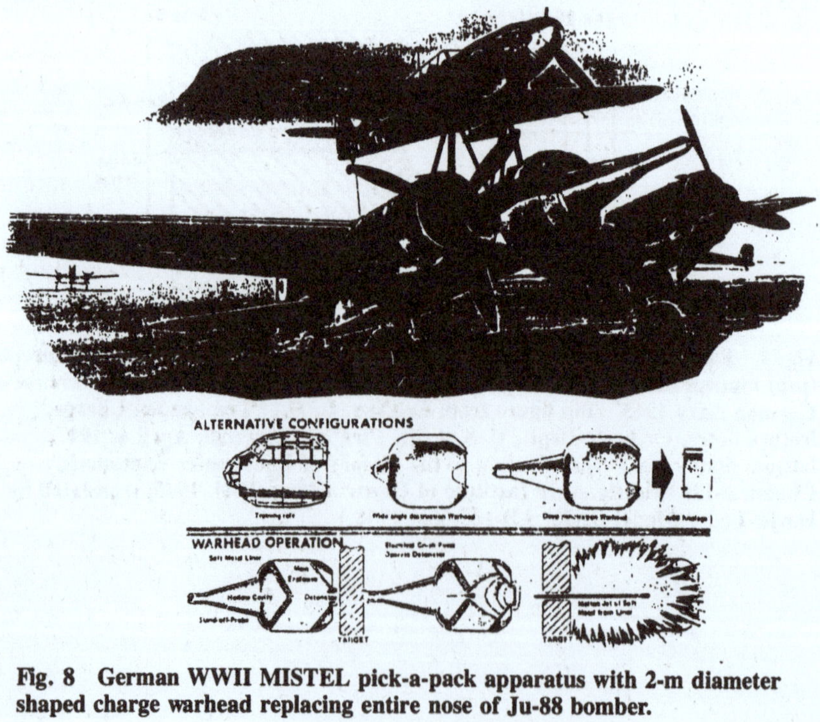

Fig. 8 German WWII MISTEL pick-a-pack apparatus with 2-m diameter shaped charge warhead replacing entire nose of Ju-88 bomber.

Vater und Sohn (by workers on the project). The composite flew to the target area using the bomber's engines and fuel. The pilot disengaged the upper aircraft and turned away while steering the large aircraft by radio command link, so that it would dive onto its target. The Germans were not as keen as their Japanese ally to fly kamikaze missions.

The MISTEL warheads remain among the most unusual non-nuclear warheads ever developed, having a wide-angled (110-deg) conical lined shaped charge approximately 2 m in diameter and weighing some 7700 lb, with 3800 lb of explosive or having a *3.8-t Gewicht hohlladung* (a 3.8-ton hollow charge), depending on the source of the information. MISTEL warheads were probably the largest shaped charges and largest non-nuclear warheads ever developed. MISTEL shaped charge liner material was identified by one source as a "soft metal layer of either copper or aluminum."[12] Although copper was already known to be a good shaped charge material, it was in critical supply in Germany and was not used in any WWII shaped charge munition. However, it is conceivable that the liner may have been made of aluminum, because MISTEL was developed by the German Luftwaffe, which apparently controlled the supply of aluminum in wartime Germany. MISTEL's shaped charge reportedly penetrated 60 ft of reinforced concrete in tests in Prussia. Tests conducted against the old French warship *Oran* in 1943 verified the warhead's capability against naval targets. Both long and short standoff nose probes were provided on various MISTEL systems, depending on whether a deep penetration or a more widespread jet effect was desired.

For comparison, the largest air-to-surface missile shaped charge warhead in the present U.S. weapons inventory was developed in the late 1960s for the U.S. Air Force AGM-65A MAVERICK missile. The warhead is 10 in. in diameter, weighs 125 lb, has 80 lb of Composition B high explosive, and carries 15 lb of lead ballast to trim the missile's center of gravity.

7.2 Ballistic (Free-Flight) Rocket Warheads

WWII also saw the modern development of free rockets for both surface-to-surface and air-to-surface employment. The warheads for the shore and surface bombardment rockets were simple metal-cased explosive-filled units, fuzed for contact bursts. The air-to-surface rockets' warheads were of two types: solid metal for kinetic-energy defeat of targets such as railroad engines [e.g., 3.5-in. aircraft rocket (AR)]; and explosive-filled, steel-cased warheads for general-purpose surface target attack. Artillery shells and bombs were also adapted as warheads for the rockets. For example, both the Navy-developed 5.0-in. AR and 5.0-in. high-velocity aircraft rocket (HVAR) used 5.0-in. common projectiles for warheads. The 11.75-in. AR ("Tiny Tim") warhead used either a 500-lb bomb body or a 12-in. Naval gun projectile. According to one source,[7] the Tiny Tim's "usual warhead, one of several, is a semi-armor piercing bomb with a total weight of 590 pounds, of which 150 pounds is an explosive charge of TNT." Tiny Tim was developed late in the war and only saw action against the Japanese at Okinawa. The Navy also developed a 14-in. AR ("Big Richard") that used a nominal 1000-lb bomb body for its warhead. The 14-in. AR was developed too late in WWII to enter the Navy's inventory.

7.3 Air-to-Surface Rockets and Missiles, Post World War II

With the onset of the war in Korea in 1950, it was realized that air-to-surface rockets required more powerful warheads to defeat such hard targets as tanks on the ground. The Navy developed both the 5.0-in. and the 6.5-in. antitank aircraft rocket (ATAR) with shaped charge warheads. The Navy's 5.0-in. Mk 25 shaped charge employed a copper liner and a Composition B explosive charge. The 6.5-in. warhead used a cast-steel liner and Composition B charge. It was developed in the remarkable time of 19 days from receipt of the urgent request from the Chief of Naval Operations in order to respond to an emergency faced by U.S. forces in Korea.

Air-to-surface guided missiles developed by the United States in the early 1950s were initially slated to use standard aircraft bombs of the Mk 80 series as their warheads. However, it became clear that the bombs could not survive the impact velocities of the new missiles (e.g., BULLPUP A). WWII records showed that most general-purpose bombs failed structurally at impact velocities much above 1100 ft/s, and the new air-to-surface missiles were to impact at velocities over 1500 ft/s. High-impact velocity-resistant semi-armor-piercing (SAP) warheads were specially developed by the Navy to cope with the high delivery velocities. The SAP warheads were characterized by thick-nosed, tapered-wall steel cases and explosive systems with internal shock mitigators. Although BULLPUP missiles are no longer in service, BULLPUP A's 250-lb Mk 19 SAP and BULLPUP B's 960-lb Mk 40 SAP warheads (see Fig. 9) continue to be employed, respectively, in the Norwegian-developed PENGUIN antiship missile and the U.S.-developed TOMAHAWK cruise missile.

Most modern antiship/anti-hard-target tactical missiles worldwide use some form of SAP warhead. Figure 10 shows the HARPOON SAP warhead. However, there are some notable exceptions. For example, the MAVERICK, with its large axial shaped charge, is now employed in an antiship role against smaller ships (see Fig. 11). The 7-in. HELLFIRE is being adopted for an antiship role. The West German KORMORAN antiship missile warhead (see Fig. 12) employs a combined SAP and radial "multi-P-charge" assembly to laterally extend the damage to the interior of its targets. The multi-P-charge warhead was possibly modeled after an experimental multiliner shaped charge device first demonstrated at China Lake in 1952 and shown to West German government representatives in 1961. The NOTS device (see Fig. 13) consisted of a TERRIER-warhead-size cylinder with 5 rows of 9 liners, each arranged in a nested, spiral array about the cylindrical case to emit a jet every 8 deg around the warhead. The liners were 90-deg apex angle aluminum cones. Each of the five rows of liners was initiated from a separate point source on the long axis, in the plane of each row of conical liner apices. It was necessary to initiate all five points simultaneously to preclude a crosswave attack on the conical liners. Figure 13 shows the result of a detonation system failure in a 1953 NOTS experiment. Note that the pattern is irregular and the jets are highly luminous, evidence of the rapid oxidation of aluminum particulates in the disturbed jets. The air-blast effect of this test was so pronounced that observers $\frac{1}{4}$-mile away in a fully enclosed above-surface, thick-walled shelter were thrust across the room (the author was the test engineer).

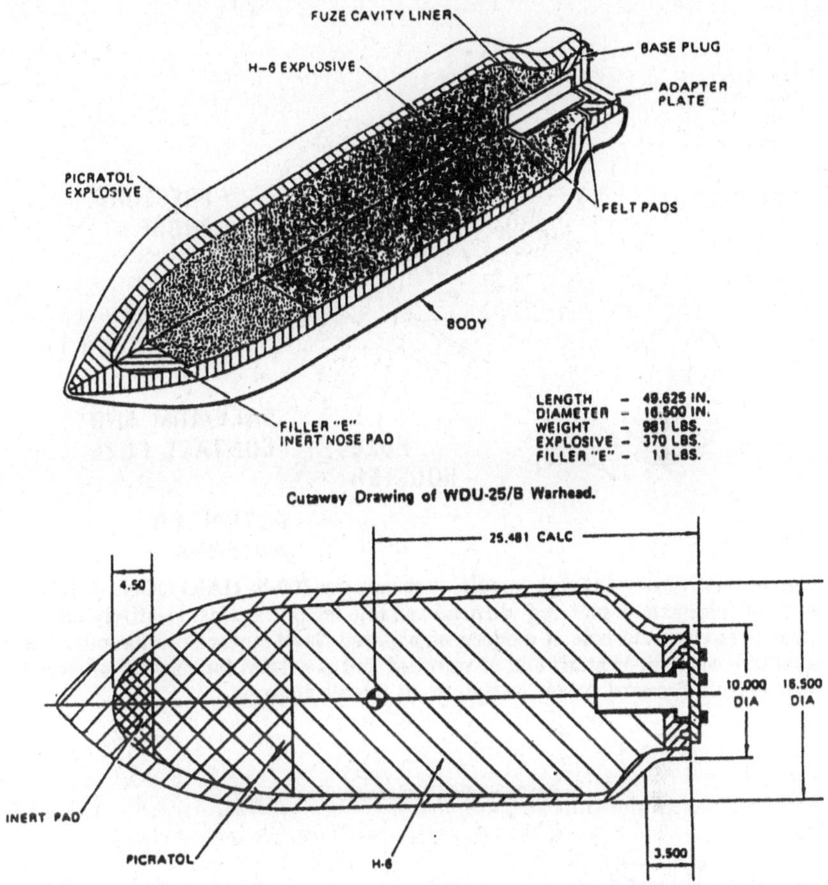

Fig. 9 U.S. Navy BULLPUP missile semi-armor-piercing (SAP) warhead developed in the 1950s.

It was later concluded that the test crew had experienced an unusually powerful fuel-air explosive (FAE) event.

The West German KORMORAN warhead's P-charge liner is similar to a wide-angle shaped charge but creates self-forged fragments rather than jets. The KORMORAN warhead also employs a simultaneous multipoint initiation system, as shown in Fig. 12. The West German-developed ROLAND anti-air-target warhead also uses a similar scheme of multiple liners (i.e., iron hemispheres) but uses a common initiation point, so that the jets are formed normal to the detonation wave's attack angle. This technique is only feasible with hemispherical liners.

The MAVERICK has been mentioned in connection with its large axial shaped charge and its utility against both land and ship targets. The shaped charge is

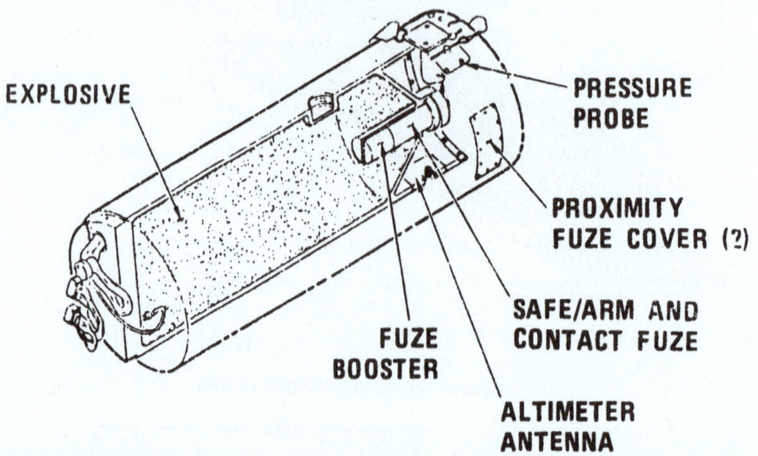

Fig. 10 U.S. Navy antiship missile warhead, the 500-lb HARPOON. A different form of semi-armor-piercing warhead suitable for use against relatively light structures characteristic of modern unarmored Naval surface combatants. Note auxiliary equipment at aft end of warhead and provision for passage of electrical signals from forward seeker section to aft missile controls.

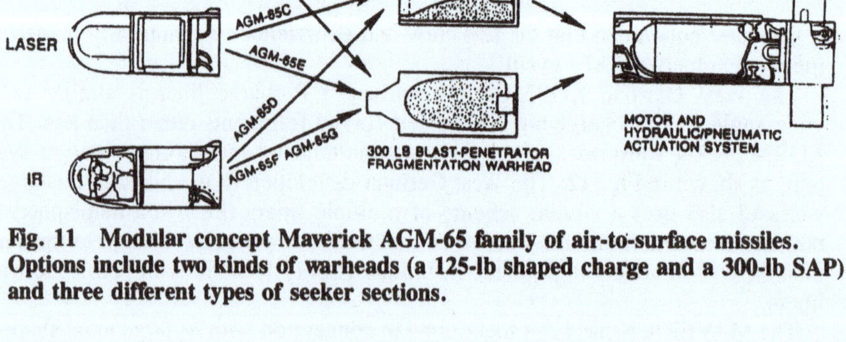

Fig. 11 Modular concept Maverick AGM-65 family of air-to-surface missiles. Options include two kinds of warheads (a 125-lb shaped charge and a 300-lb SAP) and three different types of seeker sections.

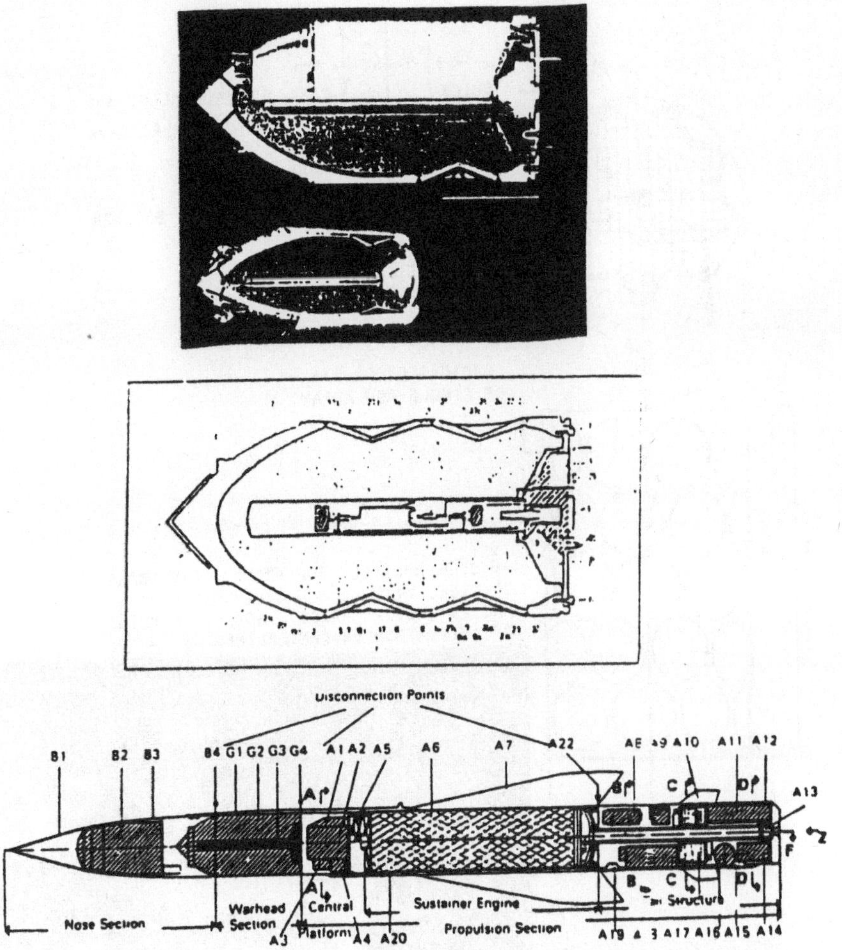

Fig. 12 West German-developed KORMORAN antiship missile. Warhead is a semi-armor-piercing type with lateral explosively formed projectile "P-charges" to extend damage when detonated inside of large ship targets. The initiation system is multipoint type with a central axis initiation near apices of each P-charge row. Perhaps the most modern antiship warhead.

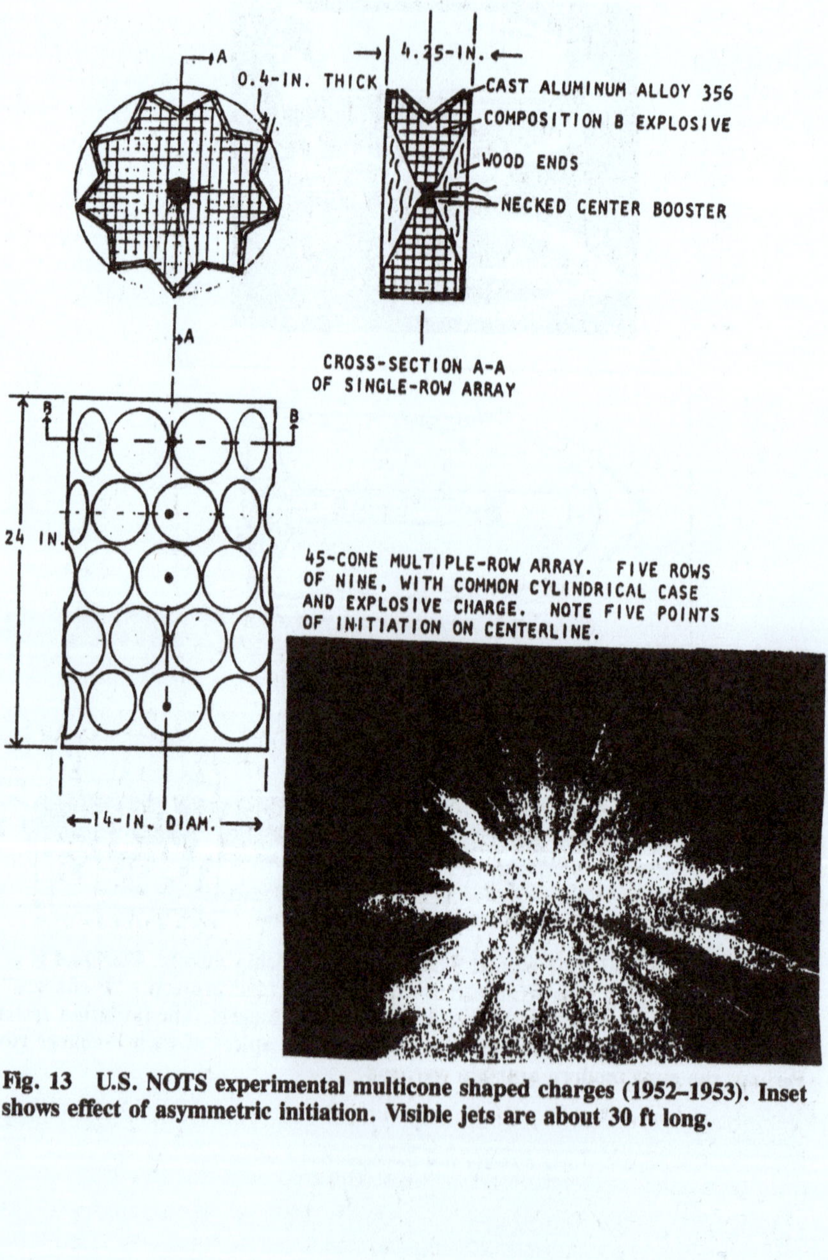

Fig. 13 U.S. NOTS experimental multicone shaped charges (1952–1953). Inset shows effect of asymmetric initiation. Visible jets are about 30 ft long.

itself unique in its use of a trumpet shaped cast-aluminum liner. However, two other MAVERICK warhead versions should also be noted. The initial version of the MAVERICK shaped charge also employed a controlled fragmentation case in addition to a two-stage axial liner. Concern expressed by a potential user that fragments might intercept the launching aircraft resulted in the fragmentation capability being deleted from the Air Force's specification. However, a subsequent "on-the-shelf" version that incorporated both a fragmentation case and metal incendiary [known as fragment-incendiary-shaped charge (FISC)] was developed, although it was never introduced into the USAF weapons inventory. Also at the time of the original shaped charge warhead development, it was suggested that the Mk 19 BULLPUP warhead would be a suitable alternative for use against a part of the expected target spectrum where a SAP type would be superior. However, because this would have required an extension of several inches of missile length, the idea was dropped. Recently, a totally new SAP-type warhead was developed for the newer versions of MAVERICK.

7.4 Antitank Missile Warheads

The antitank guided missile (ATGM) is an important member of the present-day surface-to-surface and air-to-surface tactical missile family. The armored fighting vehicle (AFV) category includes tanks, armored personnel carriers (APCs), and similar land combat vehicles. Perhaps more munitions have been developed to destroy tanks than virtually any other target on the modern battlefield. Most non-gun-fired antitank munitions employ either the shaped charge or the related Misznay Schardin end-effect mechanism. Variations of the latter include the German P-charge, the self-forged fragment (SFF), and more recently the explosively formed projectile (EFP) mechanism. Because of their more limited penetration capability, EFPs are employed primarily for top armor attack.

The shaped charge and rocket were combined in the late 1930s to provide the infantry with a means to cope with its greatest threat, the tank. The U.S. bazooka and the German *panzerfaust* are perhaps the best-known WWII antitank shoulder-fired weapons. Although different in size and concept, both employed steel conically lined shaped charge warheads and thin steel outer cases.

The WWII shaped charges were capable of defeating only two or three times their own diameter of steel armor, but the behind-armor effects of such a penetration were often sufficient to destroy a tank by disabling its crew, exploding ammunition, or igniting the fuel. As shaped charge and gun-fired, armor-piercing, kinetic-energy munitions were improved, the tank armors were made progressively heavier and more difficult to defeat. The competition of armor versus antiarmor munitions continues unabated. Modern shaped charges are capable of defeating 8 to 10 times or more their own diameter of rolled homogeneous armor, yet some of the more recently developed armors can defeat most of the shaped charges in common use (e.g., the Soviet RPG-7 system and its successors).

The end of WWII saw the development of superior-penetration shaped charge warheads based on the use of copper liners instead of the iron or steel liners of WWII. Research was undertaken in the late 1940s to better understand the physics of the shaped charge and to develop methods to overcome such problems as the effects of the spin of gun-fired projectiles, which greatly reduced the jet's penetration.

One of the earliest U.S. antitank guided missiles was known as the D-40, a rocket-jet powered and steered, wire-guided, wingless, spherical body ("guided cannonball") that carried a large, copper-lined shaped charge warhead at its center. It was a most unusual-appearing mechanism, and the development was short-lived. The first true battlefield ATGM development undertaken by the U.S. Army was the Aerophysics Development Corporation's DART ATGM, a 7-in.-diameter, land-vehicle-launched missile. Like the contemporary D-40, it employed wire guidance. The DART missile configuration was similar to the LARK antiaircraft missile, with dual tandem sets of cruciform wings, controlling flight direction by the so-called bang-bang method. Perhaps the most unique component of the DART was its shaped charge warhead (see Fig. 14) developed from 1954–1958 by the Aerojet General Ordnance Division for the U.S. Army Picatinny Arsenal. The T42E1 DART warhead introduced both new materials and new techniques to the technology of missiles and warheads. The warhead was a full 7-in. fuselage diameter (the same size as the 1970s-era HELLFIRE ATGM) and was designed to defeat 6-in. of steel armor at a 65-deg obliquity, with a requirement for producing maximum lethal effect behind the armor. Based on research performed at NOTS, China Lake, in 1952, DART's warhead introduced the following methods and materials to non-nuclear warhead technology:

- Filament-wound, epoxy-bonded, glass-fiber, minimum-lateral-hazard structural outer case
- Copper-foil strips, laid up longitudinally in the filament-wound case to provide electrical connections from the nose trigger switch to the base-mounted fuze (a precursor to modern flat wire technology)
- Low-density detonation wave shaper, initially of a foamed-in-place styrene and subsequently the first application of epoxy-bonded phenolic microballoons in a "syntactic foam" and the first U.S. shaped charge warhead to incorporate a detonation wave shaper

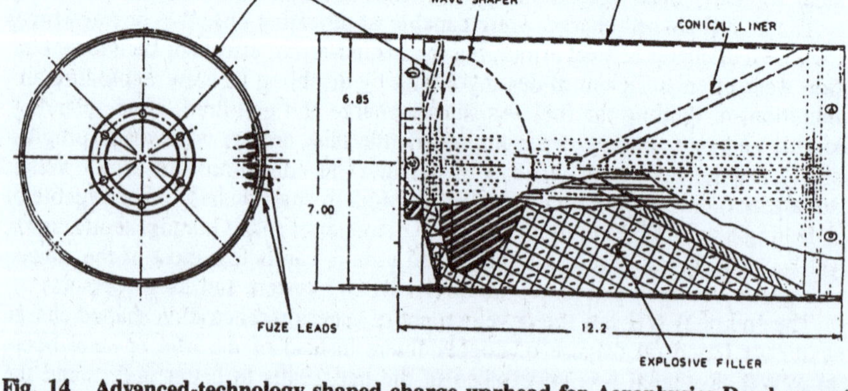

Fig. 14 Advanced-technology shaped charge warhead for Aerophysics Development Company's DART missile, the first U.S. ATGM (1954).

- Aluminum-alloy shaped charge liner, precision cast in a permanent mold, used as-cast; liner originally had a tapered thickness wall, later changed to a straight wall
- First known non-nuclear warhead application of HMX/TNT octol as an alternate filler to the Army's specified-baseline explosive, Composition B

The DART warhead technology was evaluated in several series of comparison firings against up-armored T26-E5 tank targets at the Aberdeen Proving Ground. DART was demonstrated to be superior in terms of the behind-armor side effects (i.e., exit hole diameter, quantity and spread of spall, high overpressure and thermal effects, blinding light, etc.) of a fuel-air explosion from combustion of the aluminum jet. However, the penetration, although amply meeting requirements, was $\frac{2}{3}$ that obtained with an equal-diameter copper-lined shaped charge that produced a small-diameter but much deeper penetration hole but only modest behind-armor effects as compared to the aluminum lined charge. The result was the Army's rejection of DART's advanced techniques in favor of the more conventional point-initiated, copper conical lined, shaped charge, typical of all subsequent U.S. Army-developed ATGM and rocket warheads, including the SHILLELAGH, DRAGON, and TOW ATGMs, as well as the shoulder-fired successors to the 2.36-in. and 3.5-in. M-28 bazookas, the rocket-propelled grenades (RPGs) M72 LAW (early 1950s) and VIPER (mid 1970s).

The Swedish-designed AT-4, more recently acquired by the U.S. Army, employs a mechanism to provide behind-armor effects of the same kind but with much less power than produced by the DART. The previously discussed MAVERICK shaped charge warhead is the only U.S.-inventory shaped charge warhead that takes full advantage of the enhanced behind-armor effects produced by combustion of an active metal jet behind the defeated armor.

Certain novel techniques pioneered by the 1954–1958-era DART subsequently appeared in both Soviet ATGM and West German ATGM warheads developed in the 1960s and 70s (see Fig. 15). German WWII documents revealed that German investigators had introduced 1) wave shapers using concrete inert bodies between the initiation point and the top of iron hemisphere liners, 2) tandem shaped charges, 3) zinc liners as substitutes for steel, and 4) use of high- and low-velocity explosive in an antiship torpedo shaped charge (see Fig. 7) which had high-velocity HE both surrounding and running aft along the axis of the steel-hemisphere shaped charge liner. The lower-velocity, higher underwater-efficiency, aluminized explosive surrounded the core high-velocity component, an elegant solution even today.

Techniques to increase the penetration capability of shaped charges have been the subject of extensive research since the beginning of shaped charge research. The use of two or more charges in tandem has intrigued investigators in both military and industrial (oil well completion charge) laboratories. The fact that some have performed as intended gives hope that tandem shaped charge technology problems may soon be routinely resolved. Problems that need to be resolved include the jet traffic and the effects of the first unit's detonation and jet formation on its tandem unit. Figure 7 illustrates one approach to this problem. An alternative approach is a single explosive charge with a more complex tandem liner arrangement that produces a sequence of jets, each tailored to accomplish a specific task in the target defeat process.

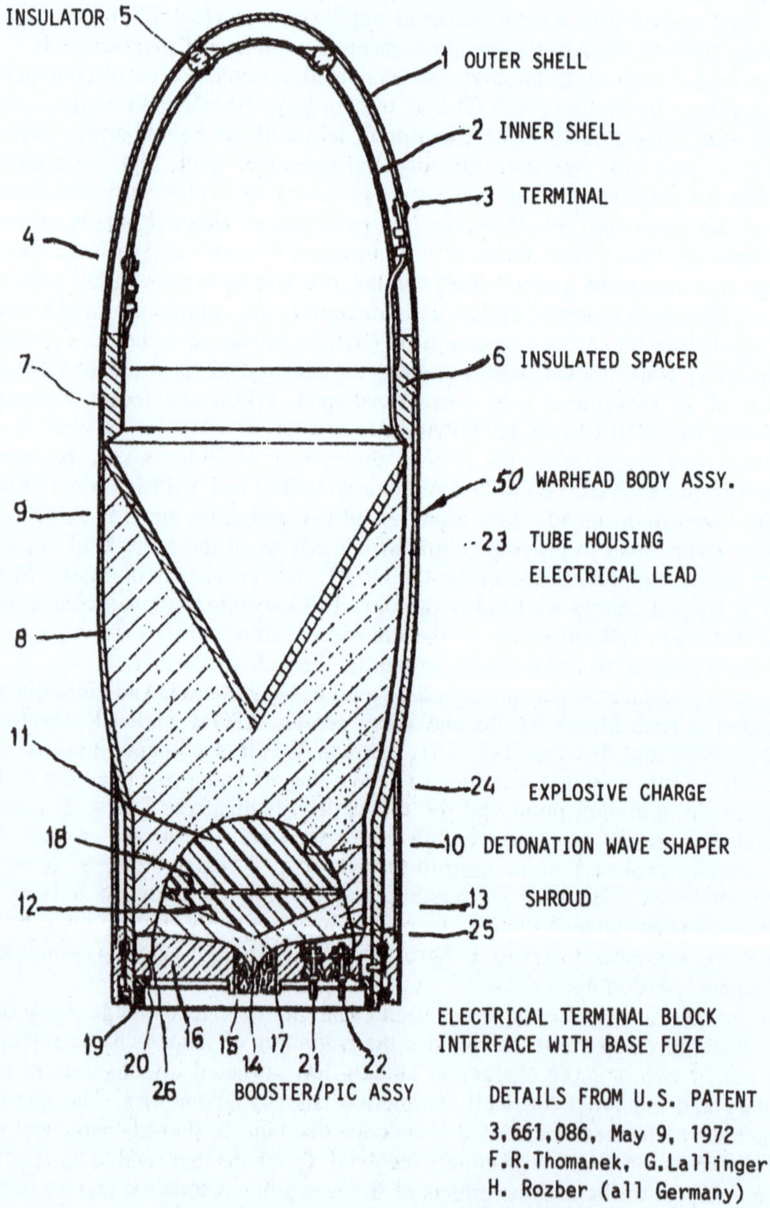

Fig. 15 MILAN warhead patent drawing.

Other methods to increase penetration capability included improvements in the quality and precision of the shaped charge assembly, improved liner geometry (e.g., variable contour and thickness), and the development of liners made of significantly higher density materials than the traditionally employed copper.

7.4.1 Second-Generation U.S. Army Shaped Charge Warheads

The second generation of U.S. Army ATGM HEAT warheads developed subsequent to the DART included the 6-in. SHILLELAGH, the 5-in. TOW (for "tube-launched, optically tracked, wire-guided"), and the 4-in. DRAGON. All of these warheads were characterized by copper conical liners, single-point initiation, cast HMX-TNT (Octol) explosive charges, and a thin-wall aluminum case with a cylindrical forebody and conic truncate aft section. The ATGM fuze systems were point-initiated, base-detonated (PIBD) types with a forward-mounted crush switch and a base-mounted safing-arming-firing (SAF) unit. The crush switch was typically two thin hemispherical shells separated by electrical insulators, as shown in Figs. 15 and 16, the 1965 proposal design for the XM-224 MAW (medium assault weapon) warhead. MAW was later known as DRAGON. Each shell is one side of the fuze electrical circuit connected to the electromechanical SAF at the warheads base (aft) end. Contact of the two shell elements at target impact completes the circuit to fire the SAF unit's electric detonator. Note the resemblance to the MILAN nose crush switch in Fig. 15 but the different approaches to the shaped charge explosive system in which MILAN employs both a detonation wave shaper and precision initiation coupler (PIC) assembly.

Both the SHILLELAGH and the basic 5-in. TOW were fitted with a 60-deg straight-walled, copper conic liner with a flat apex (see Fig. 17). DRAGON's liner was a 42-deg cone with a hemispherical apex, resembling the approximately 2.6-in. conic liner of the shoulder-fired, rocket-propelled grenade introduced in the mid 1950s and known as LAW (light antitank weapon). LAW's mid 1970s-developed, ill-fated successor was an RPG known as VIPER; VIPER's liner was the same geometry and size as the LAW, and VIPER's performance was 150% that of the LAW.

The SHILLELAGH, TOW, and DRAGON ATGMs and LAW RPG were all developed in the 1950–60 period and were thus available for use in combat in the war in Vietnam. TOWs were used effectively. LAWs were used much more extensively but had technical problems and were generally disliked by the users.

Armor/antiarmor technology had advanced significantly in WWI, stagnated in the 1920s and 30s, and accelerated spectacularly in WWII. Each advance in armor technology was eventually offset by an overmatching advancement of an antiarmor mechanism. The most significant single advance was the WWII development of the shaped charge and related end-projected technologies, such as the all-slug shaped charge, variously known as the Misznay Schardin mechanism, self-forming fragments (SFFs) and explosively formed penetrators (EFPs). Both antitank belly-attack land mines and antipersonnel mines (e.g., CLAYMORE) took advantage of the Misznay Schardin effect. The mines were used effectively by both sides in Vietnam. (Viet Cong forces reportedly captured the initial lot of CLAYMORE mines delivered to the U.S. Army.)

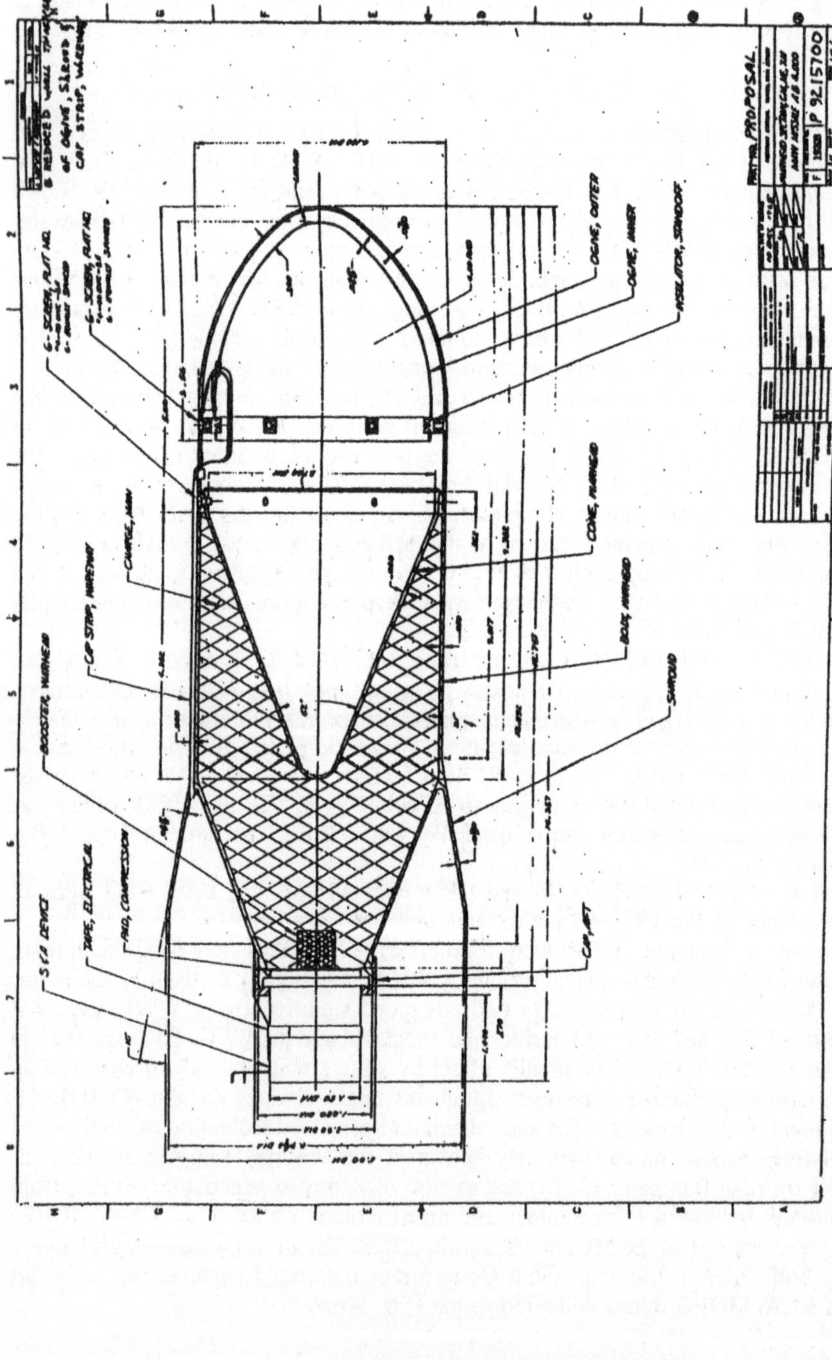

Fig. 16 Proposal design of XM-224 MAW warhead.

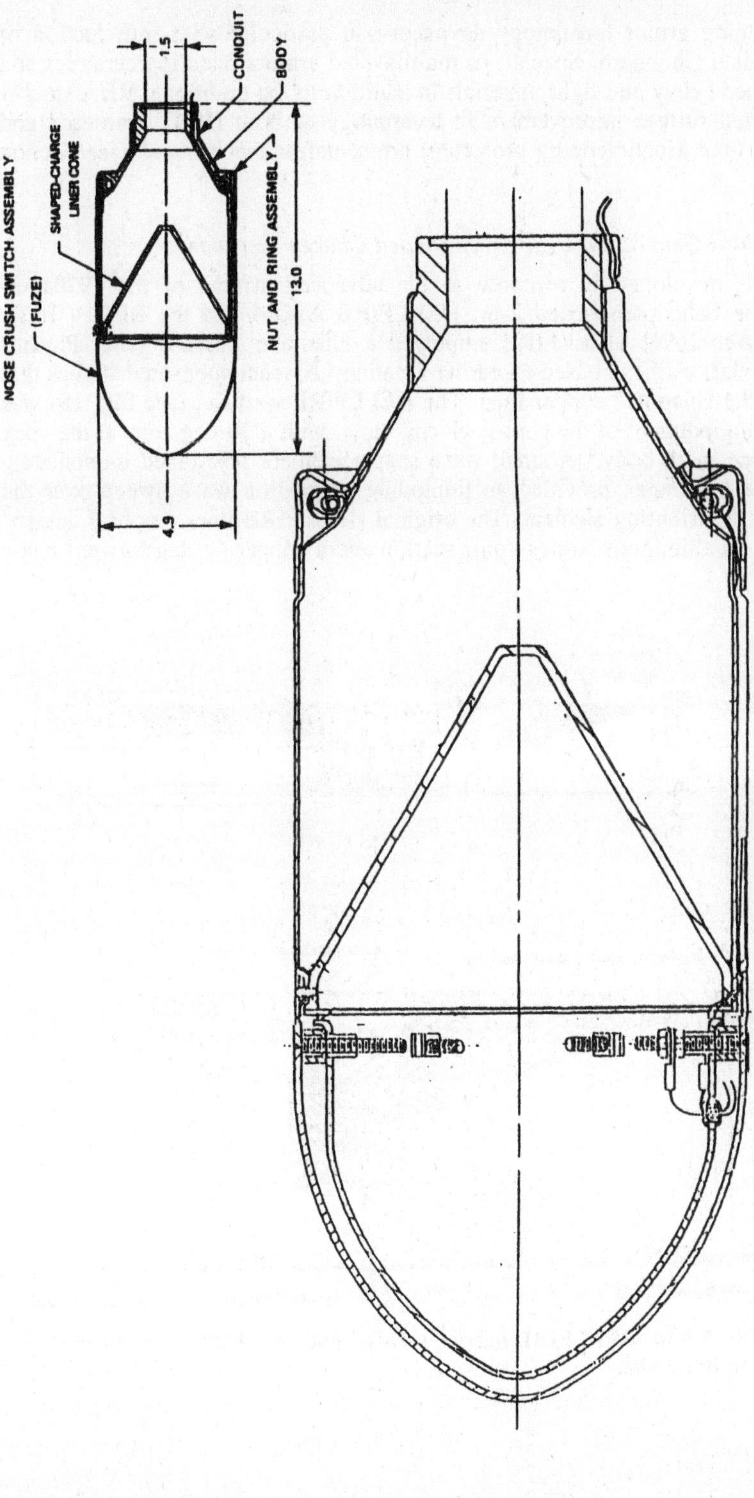

Fig. 17 Original M207 TOW warhead with 60-deg, flat-apexed, copper conical liner.

Continuing armor technology advances—in particular with introduction of the so-called Chobham, Special, or multilayered armors featuring ceramics and sandwiched heavy and light materials in addition to the traditional RHA steel—necessitated further improvement in technology of both HEAT warheads and KE (gun-fired kinetic-energy projectile) armor defeat mechanisms (see Section 7.4.5).

7.4.2 Third-Generation U.S. Army Shaped Charge Warheads

ATGMs developed in response to the advanced armors of the 1970s included the helicopter-carried 7-in. HELLFIRE ATGM and the RPG VIPER, an improved LAW. HELLFIRE employed a Firestone Defense (later Physics International) warhead based on earlier Picatinny-Arsenal-sponsored studies that evolved the "biconic" copper liner. The HELLFIRE warhead (see Fig. 18) was the first embodiment of the copper biconic liner, with a 30-deg cone at the apex and 42-deg main body.[8] A small wave shaper element positioned immediately aft of the liner apex provided an imploding detonation wave sweep over the initial jet contributing element. The original HELLFIRE liner proposal design[9] specified an aluminum 30-deg conic section and a copper 42-deg forward conic

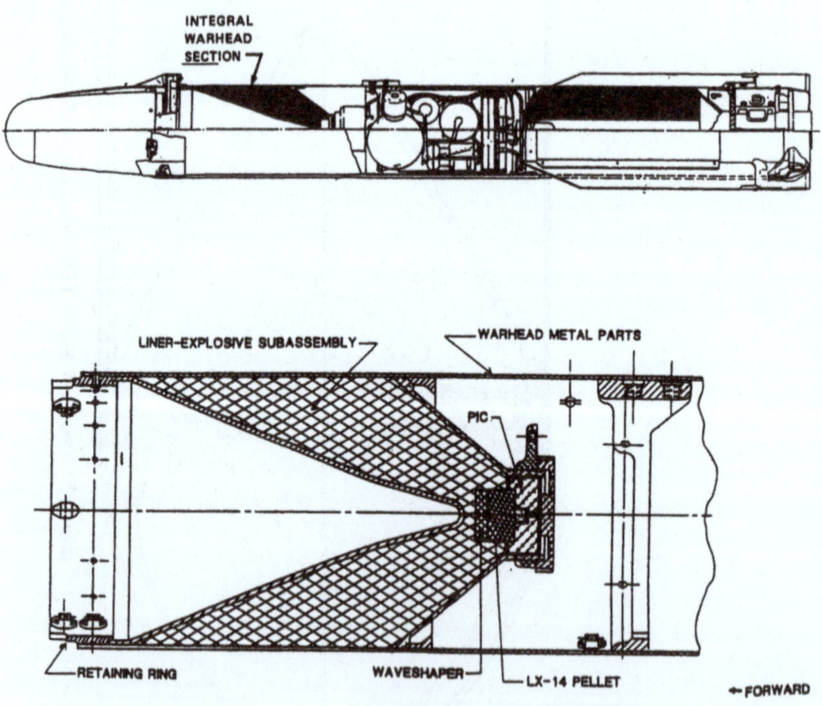

Fig. 18 7.0-in. HELLFIRE missile warhead and sketch showing warhead location in missile.

section. Firestone's X-ray flash radiographs of experimental biconic liner jets clearly showed that a too-acutely angled copper liner with an imploding wave attack would potentially produce a noncoherent jet (known in the parlance as a "poly-furcated" jet), a condition directly attributed to the hypervelocity (exceeding 10 km/s) jets produced by the 30-deg conic section. The liner collapse and jetting velocity exceeded the bulk sound velocity of copper. Aluminum was selected because it has a nearly 37% higher sound velocity than copper (5.38 km/s for Al, 3.94 km/s for Cu).

Perhaps in consideration of fabrication costs, the Army elected to make the entire liner of copper, with a noncoherent precursor jet an accepted consequence. The specified penetration performance of HELLFIRE was exceptionally high for its time of development, and, even though the early jet effects were lost, HELLFIRE met its penetration requirements.

HELLFIRE and VIPER employ a pressed, high-performance, HMX-based explosive designated LX-14, representing a significant change from the long-used melt-pour explosive technology. The Soviets used pressed explosives for approximately two decades prior to U.S. adoption of pressed charges, with the earlier Soviet ATGM warheads using pressed RDX/wax charges and the more recent versions having an HMX-based pressed explosive similar to U.S. LX-14. Cast HMX explosives continue to be used by many U.S. and most European shaped charge warhead manufacturers.

7.4.3 Upgrading of the Second-Generation Warheads

Because of the continuing improvement and adoption of superior armor, it became imperative for both the TOW and DRAGON warheads to also receive performance enhancements. Revised metal parts and explosives were introduced. DRAGON warhead improvements included a high-precision, trumpet-shaped liner that almost doubled penetration of the earlier version. Both the 5-in. "Improved TOW" (ITOW) and the subsequently developed 6-in. TOW-2 warhead adopted a downscaled biconic liner and explosive configuration of the 7-in. HELLFIRE. Extendable nose crush switch assemblies that increased the standoff to 3 cone diameters were also substituted for the short crush switches such as that shown in Fig. 17. Figure 19 compares the HELLFIRE, ITOW, TOW-2, and VIPER warhead metal parts assemblies, including the wave shaper and booster elements. Figure 20 illustrates the TOW-2 extendable nose probe in both its stowed and extended positions, and the TOW-2A version with a precursor charge in the tip of the probe.

Another major advancement in armor was introduced in the Israeli incursions into Lebanon and Syria in 1981 in the form of a 1970 West German invention known as explosive reactive armor (ERA). Although the explosive armor concept originated in Australia in 1944 and was the subject of U.S. research and development in the 1950s, it was refined to a useful form by Dr. M. Held of West Germany. ERA created a major perturbation in the world armor/antiarmor community because it made it possible for a lightweight applique to significantly disrupt the jet of even a major-caliber HEAT warhead. Although ERA appliques were quickly adopted by the major potential belligerent forces, the inherent

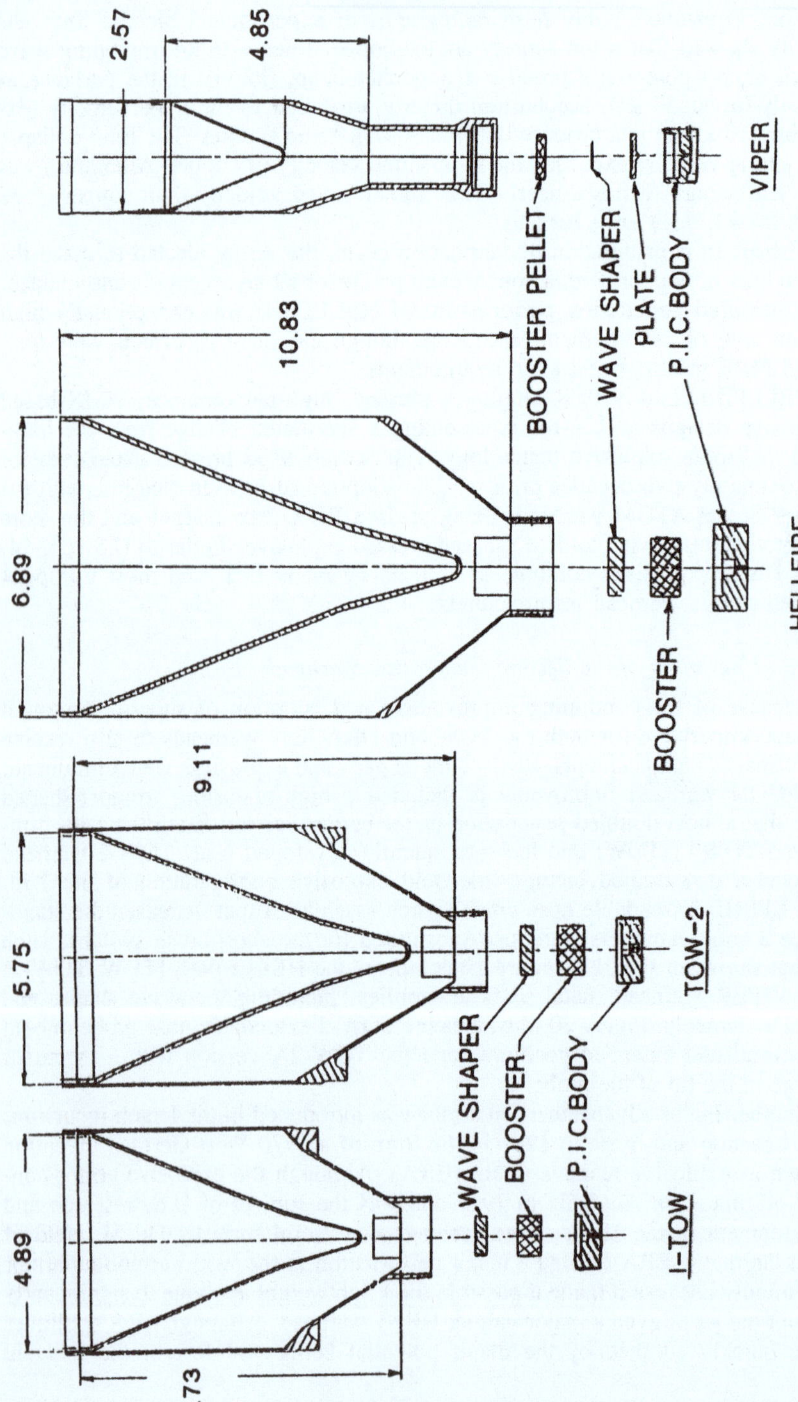

Fig. 19 Comparison of metal parts and explosive geometries of TOW, ITOW, TOW-2, and VIPER warheads. (Sketch courtesy Mason & Hanger.)

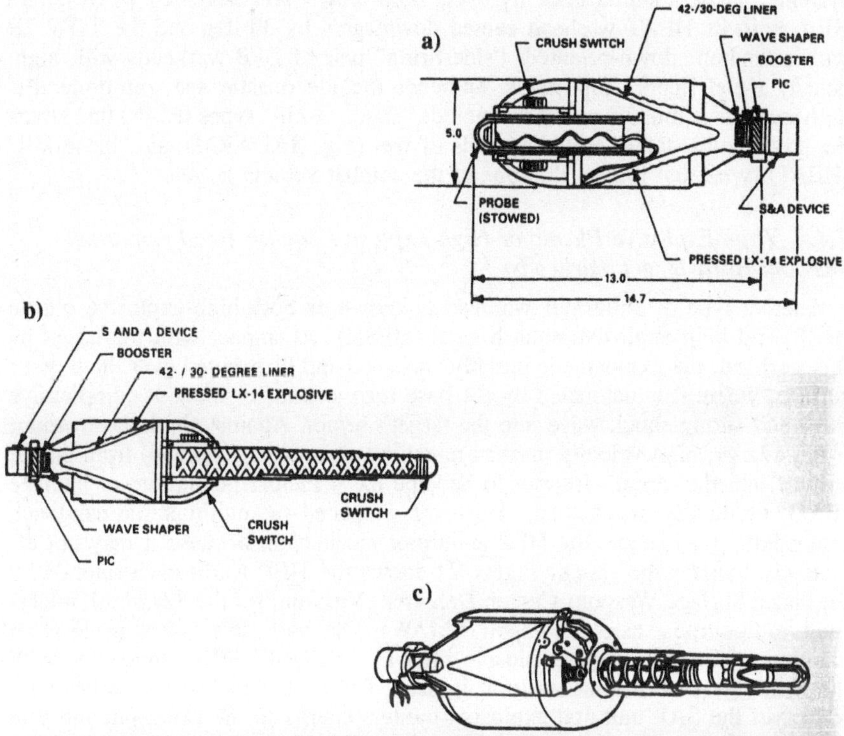

Fig. 20 Warhead probes: a) M207E1 warhead (ITOW missile) with extendable probe stowed; b) M207-E2 warhead (TOW-2 missile) with probe extended (TOW-2 has 6-in. diameter versus 5-in. ITOW); c) TOW-2A with precursor shaped charge in forward probe element.

hazard of ERA to friendly troops caused some potential users to be wary of its adoption. Israel was the first to use and the first to abandon ERA. The USSR's adoption of ERA sent a shock wave through NATO's political entities, in particular the U.S. Congress, with defense press pronouncements that all currently fielded antitank missiles were unable to defeat an ERA-equipped tank target.

One response to ERA was the tandem shaped charge, another patented invention (patent applied for in 1974) of M. Held of West Germany. The forward HEAT charge is to cause reactive armor function prior to detonation of the larger main HEAT charge warhead. The 6-in. TOW-2 was upgraded to the TOW-2A version by incorporation of a smaller 38-mm shaped charge in the tip of its extendable nose probe. Other ERA responses included the research and development of very high density metal liners producing jets that are less affected by ERA appliques, including multiple-unit arrays on Soviet tanks "inadvertently" seen by members of the U.S. Congress' Armed Forces Committee in a 1989 visit to the USSR.

Other solutions to the high-efficiency frontal-array armors include top attack systems. These include both fly-over, shoot-down ATGMs such as Sweden's BILL with its HEAT warhead canted downwards by 30 deg and the TOW-2B with a tandem, down-oriented, "side-firing" pair of EFP warheads with high-density metal liners. Top attack warheads include missile and gun-projectile-delivered smart munitions. These include "shoot-to-kill" types (EFPs) that attack the target from standoffs of hundreds of feet (e.g, SADARM) and "hit-to-kill" (HEAT) types that impact the tops of the combat vehicle targets.

7.4.4 High-Explosive Plastic or High-Explosive Squash Head Antiarmor and Anti-Hard-Target Warheads

Another type of antiarmor warhead is known as both high-explosive plastic (HEP) and high-explosive squash head (HESH). At impact with the target by this warhead, the explosive is initially squashed and distributed over the target's surface, where it is detonated by the base fuze to transfer a major compressive force and strong shock wave into the target's armor. Against single-layer armor plate, a large, high-velocity mass of metal is scabbed (i.e., spalled) from the far surface into the target's interior to become the principal behind-armor damage (BAD) production mechanism. However, a spaced or multimaterial sandwich armor array may defeat the HEP antiarmor mechanism because it may not effectively transfer the shock. Figure 21 shows the HEP warhead developed by the Naval Surface Weapons Center, Dahlgren, Virginia, for the 82-mm shoulder-fired, multipurpose assault weapon (SMAW). The warhead's case is made of an aluminum alloy. The high-explosive filler is a pressed RDX composition with the ability to deform plastically at impact. The rubber pad at the forward interface of the SAF unit and explosive tranfers energy to the firing pin and stab detonator during the explosive deformation. The HEP warhead is particularly effective against both log and timber and sand bagged bunkers, reinforced concrete structures, buildings, and light armored vehicles such as APCs and infantry fighting vehicles.

An alternative, advanced configuration high-performance, shaped charge known as the high-explosive antiarmor (HEAA) warhead was subsequently developed for the SMAW to provide a defeat mechanism for use against HEP-resistant multimedia targets. The latter might include the heavier classes of armored vehicles with deliberate or incidental (e.g., external stowage bins) spaced armor arrays, and mobile and fixed targets equipped with effective spall-suppression liners.

7.4.5 Kinetic-Energy Antiarmor Warheads

The most effective antitank defeat mechanisms have traditionally been, and continue to be, kinetic-energy projectiles fired from high-pressure cannons, the main armament of all modern-day main battle tanks. A descendant of the solid shot of ancient cannon, the typical modern kinetic-energy projectile is a subcaliber, discarding sabot-launched, high-L/D (length-to-diameter), high-density-metal, fin-stabilized, long-rod or arrow projectile, typically referred to as APFSDS (for armor-piercing, fin-stabilized, discarding sabot). Metals employed for the penetrators may include steel, tungsten, tungsten carbines, and depleted uranium. These may be used either as the main penetrator or as a composite with

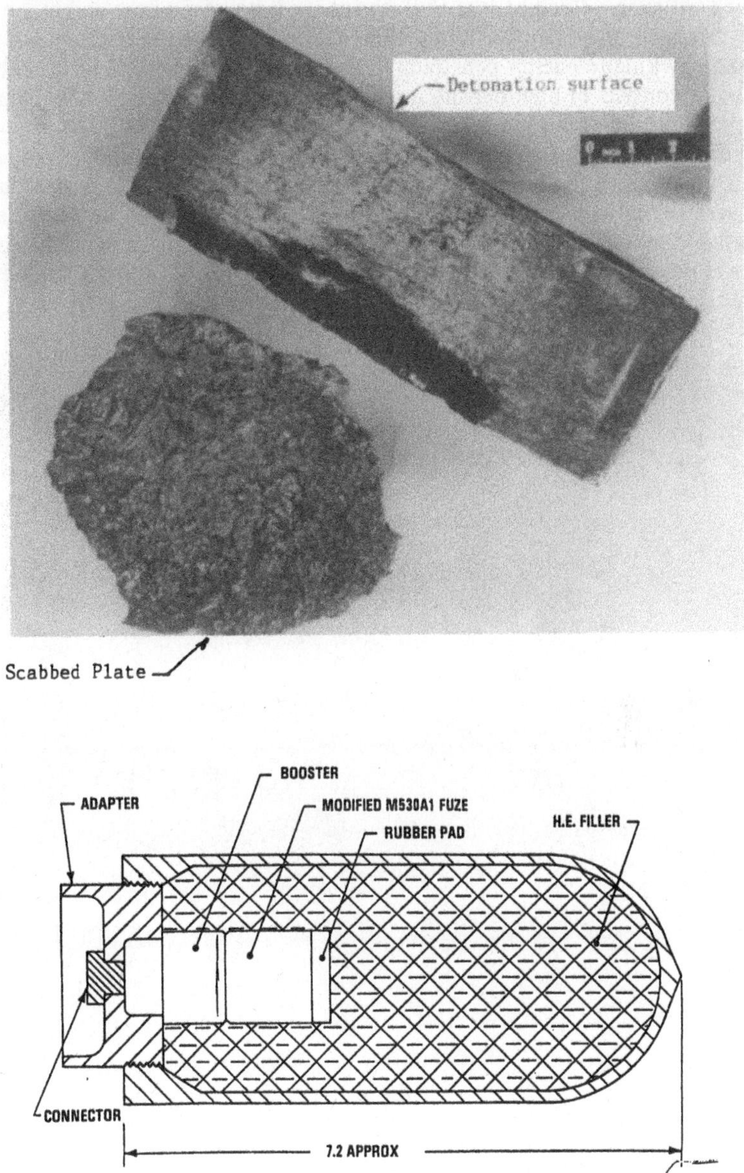

Fig. 21 82-mm SMAW HEP or HESH and typical HEP scabbing effect on steel armor plate.

another metal (e.g., steel) housing the high-density penetrator. The projectiles are typically driven to velocities in excess of 1.5 km/s by the cannon propellant charges. At such velocities, the trajectories are flat and times-to-target are very short. Massive and complex armor arrays are required to defend against such high-energy kinetic-energy penetrators. The behind-armor effect of the residual penetrator consists of high-density metals from both the projectile and the armor spall, ricocheting about the target interior.

Figure 22 illustrates a typical modern-day gun-cannon-fired KE projectile. In this example, a high-density tungsten penetrator is carried in a steel carrier that has external ribs to interface with and lock into a three-piece, multisector, discardable sabot fabricated from light metals such as magnesium or aluminum. The KE projectile shown in Fig. 22 was designed for a 120-mm smoothbore cannon.

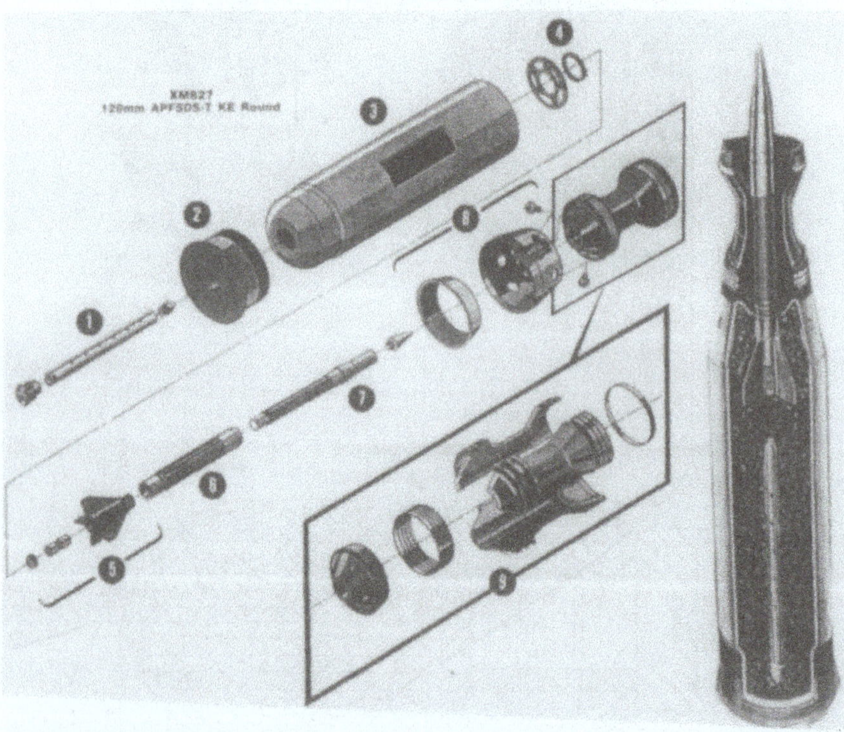

1- Primer Assembly
2- Base Case
3- Combustable Case and Propellant
4-Spring Disc and Retaining Ring
5- Fin and Tracer Assembly
6- Sheath
7- Penetrator Assembly
8- Case Cover and Cap Assembly
9- 3-Segment Sabot Assembly

Courtesy of Rheinmetall/Honeywell

Fig. 22 Components of APFSDS-T kinetic-energy, 120-mm tank-gun-fired round.

7.5 Anti-Air-Target Gun and Tactical Missile Warheads

The aircraft became a major threat in tactical warfare beginning in WWI. Considerable effort has since been devoted to their defeat from both the ground and from other aircraft. The early anti-aircraft techniques included the use of aircraft machine guns in aerial combat, and ground fire in the form of rifle, machine gun, and small- to medium-caliber gun projectiles. All continue to be important anti-air-target mechanisms.

At the outset of WWII, the U.S. Army's "triple-A" (i.e., anti-aircraft-artillery, or "ack-ack") capability comprised 3-in. 50-caliber mobile and 105-mm fixed guns firing powder-train-type time-fuzed, TNT-filled, and older-model ball "shrapnel"-filled projectiles. The Navy used both twin 3-in., 50-caliber gun mounts and 5-in. guns on their ships. These were augmented by water-cooled .30- and .50-caliber machine guns to engage close low-flying threats. The attack on Pearl Harbor served as a reminder that the aircraft problem was getting beyond the capability of the then-existing gun systems. High-firing-rate cannon of 20- and 40-mm caliber with HE-filled point-detonating fuzes were introduced. The 3-in. guns were replaced by the higher-altitude-capable 90-mm AA gun by the United States, and the famed 88-mm gun was introduced by the German Army. It was reported that the average number of rounds expended per aircraft brought down by ground fire was around 10,000 rounds per target. This changed significantly with introduction of the variable-time (VT) radio-proximity fuze. The VT fuze was one of the more important developments of the Second World War.

The introduction in WWII of rocket-propelled missiles, both ballistic and guided, significantly added to the capability to destroy aircraft targets in both surface-to-air and air-to-air combat. The air-to-air ballistic rocket (57-mm R4M) was introduced by Germany late in the war. It used a shaped charge warhead (Fig. 23). Its direct descendants are employed today by the former Soviet forces.

Development of surface-to-air and air-to-air guided missiles was vigorously pursued by Germany's scientists, although few such missiles became available for tactical use in WWII. Interest in the United States began late in the war but increased rapidly in the post-war period in the Navy Bumblebee project and projects for such missiles as the Army and Navy LARK, the Air Force BOMARC, and the Army NIKE AJAX, NIKE HERCULES, and homing all-the-way killer (HAWK) surface-to-air missile systems. The air-to-air missile developments included the Navy SPARROW and SIDEWINDER and the Air Force FALCON. The Navy Bumblebee project evolved into the "Three-T" missiles: TERRIER, TARTER, and TALOS.

U.S.-developed air-to-air unguided rockets included the 2.75-in. AAFFAR (the "Mighty Mouse"—based on the WWII German-developed R4M air-to-air rocket) and the 5.0-in. HPAA (Zuni). These were both developed at NOTS, China Lake, in the 1948–1955 period. NOTS also developed but never introduced into inventory a 2-in. air-to-air rocket known as GIMLET. The Air Force BIRDDOG 8-in. air-to-air ballistic rocket project was developed in the mid 1950s and was originally provided with a 150-lb controlled fragmentation warhead (the first to employ discrete precut fragments in a composite epoxy-fiberglass matrix). Even though two live firing air-to-air tests were highly

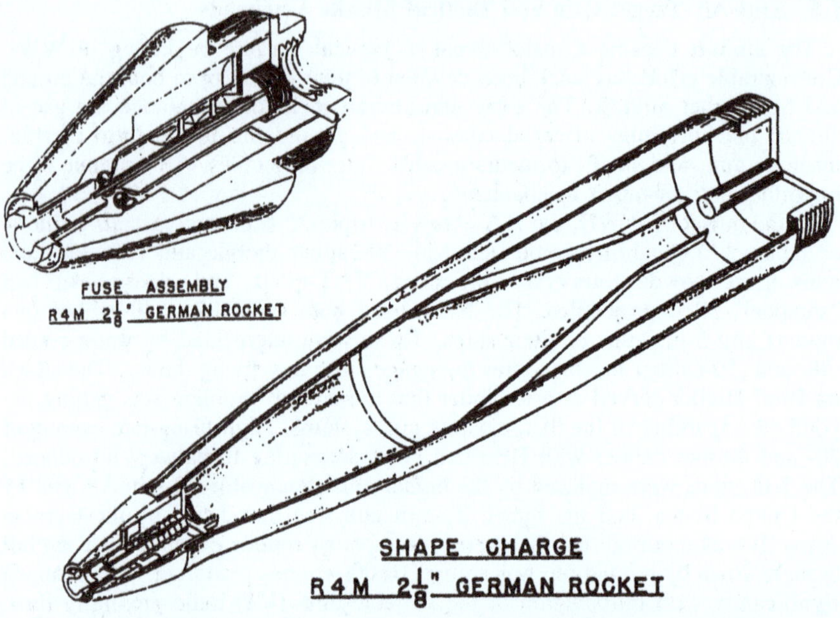

Fig. 23 Shaped charge warhead for German R4M air-to-air rocket.

successful and two droned B-17 bombers were brought down by unplanned wide misses, the Air Force reconfigured the system to become the MB-1 GENIE nuclear-tipped free ballistic rocket. A GENIE with a 1.5-kiloton nuclear warhead was test-fired in the air over the Nevada test facility in 1957.

The non-nuclear, or so-called conventional, warheads for the early missiles were of two basic types: 1) those with natural or controlled fragmentation and 2) pure blast. Pure blast warheads were used on the surface-to-air REDEYE and guided aircraft rocket (GAR-4) FALCON, both of which were designed to penetrate and detonate inside the aircraft skin. Most "first-generation" anti-air-target warheads employed some form of fragment size, shape, and weight control.

The operations analysts, having realized in WWII that the combination of proximity fuzing, good guidance, and fragment size and distribution control would pay off in terms of a higher probability of aircraft target kill, began to exert considerable influence on the warhead development community. Therefore, much of the attention of the warhead research and development community was to be devoted to antiaircraft kill mechanisms in the period following the Second World War. The principal kill mechanisms that were to be studied are discussed below, with typical applications noted:

First-generation systems:

- Internal blast only (AIM-4 FALCON, REDEYE)
- External blast-fragmentation, both natural and controlled (NIKE AJAX, NIKE HERCULES, SIDEWINDER, SPARROW, TERRIER, HAWK, PATRIOT)

Second-generation systems:

• Shaped charge, unitary and multiple jet (e.g., French-German ROLAND)

• Submissile (NIKE HERCULES, NIKE ZEUS)

• Rod warheads: continuous rod (e.g., SIDEWINDER, SPARROW III, TALOS) and discrete rod (e.g., AIM-9L, PHOENIX)

Third-generation systems:

• Mechanically and electronically aimable, azimuthally biasable, enhanced, "balanced" fragment and blast effects (no inventory examples, as yet)

7.5.1 *Internal Blast Warheads*

As the name implies, this type of warhead must penetrate into the target before it is detonated so that the maximum transfer of destructive energy can be provided from a small warhead. The Navy's 2.75-in. air-to-air folding-fin aircraft rocket (AAFAR), known as "Mighty Mouse," is an air-to-air rocket developed during the late 1940s and early 1950s. At the time of the 2.75-in. AAFFAR development, its warhead was required to penetrate an aircraft target, typically a large bomber. The warhead design engineer responsible for the 2.75-in. rocket warhead later joined Hughes Aircraft Company to be the principal engineer for development of the Air Force FALCON guided missile warhead system, which also employed an internal burst warhead. The IR-version of the missile was reputed to typically fly into the target aircraft's jet engine tail pipe. However, as time passed and targets became fighter aircraft rather than large, slow bombers, it was apparent that the missile would no longer attack in such a mode; instead, it typically just missed its target in and around the engine area. A retrofit fragmentation warhead development was undertaken in 1970 (AIM-4H). FALCON systems were outmoded by that time, so warhead development was terminated. However, technology of the warhead's tungsten fragment pack was transferred to other applications (e.g., PATRIOT).

REDEYE and its successor, the shoulder-fired antiair missile STINGER, are other examples of anti-air-target missiles with small, internal burst warheads. They have sufficient external effects that even an external contact or near miss may be adequate to inflict useful damage to certain targets, such as helicopters and similar high-vulnerability aircraft.

7.5.2 *Externally Burst, Blast-Fragmentation Antiair Warheads*

The most commonly used anti-air-target warhead in the world today is the simple but highly effective type directly derived from the WWII period. The early antiaircraft gun projectiles and warheads used plain cases and allowed the HE to fragment the casings in a random manner. The random sizes of fragments ranged from useless dust to large shards with a wide range of fragment sizes and shapes in between. Operations analysis studies revealed that the probability of kill of an aircraft could be much enhanced if the size, shape, and distribution of the fragments could be controlled. The only problem to be resolved, and one that continues to be in a state of flux, is how heavy the fragments should be. The desired fragment mass has varied from 30 grains (at 7000 grains/lb, or about 2 g) to hundreds of grams. (The Swedish BOFORS L-70 40-mm gun-fired controlled fragmentation round discharges tiny 3.8-grain tungsten balls!) Both the NIKE AJAX and early SIDEWINDER warheads developed during the late 1940s and early 1950s used 30-grain fragments. Yet the NIKE HERCULES developed in 1954 was designed to discharge nearly 18,000 cubical steel fragments weighing 140-grains each. The 1100-lb, oblate spheroid warhead had a 3-pi fragment distribution, and the 650-lb H-6 explosive charge imparted an initial velocity in excess of 8800 ft/s exclusive of the HERCULES missile's velocity. HERCULES exploited a highly refined fragment size control method that became a standard practice. The fragments were made as cubes and were laid up on a "GRP" (fiberglass in epoxy matrix). Thus the ultimate fragment size, shape, and distribution control were achieved in a minimum parasitic weight structure. The HAWK and the present PATRIOT warheads employ essentially the same construction technique but with different fragment sizes. The PATRIOT is in effect a smaller version of the HERCULES, with different sizes of fragments and a change of inner shell structure material being the principal differences from its "parent."

Many fragment control methods have been developed. The choice of method depends on the application, the operational environment, the cost-effectiveness requirement, and very often the ingenuity of the designer. Methods to achieve fragment size and shape control, in addition to the discrete fragment-in-matrix method discussed above, include, but are not limited to, the following:

- Notched square steel wire in continuous coils (e.g., later-model hand grenades) or brazed single rings (early AA missile warheads)
- Machined internal grooves (e.g., Pearson grooves)
- Cast patterns, internal or external (e.g., early hand grenades, gun-fired projectiles)
- Embossed patterns (e.g., ICM grenades)
- Metallurgical control (heat treatment, special alloys)
- Internal sleeve with embossed pattern (e.g., SIDEWINDER 1A)

7.5.3 Shaped Charge Warheads: Single and Multiliner

The first generation of anti-air-target missile warheads were the controlled fragmentation types employed on such systems as the NIKE AJAX, TERRIER, SPARROW, and SIDEWINDER missiles. The second-generation warhead studies took place in the 1950s. These were to examine three distinctly different mechanisms: 1) warheads employing the shaped charge, 2) warheads deploying submissiles, and 3) warheads producing discrete rods and expanding continuous rods. The shaped charge studies were performed at the U.S. Naval Ordnance Test Station, China Lake (later known as the Naval Weapons Center and now as the Naval Air Warfare Station).

The shaped charge mechanism was explored as a part of the studies of the "vaporifics" effect. The effect was first observed in 1947 at the New Mexico School of Mines by Dr. John Rinehart. The effect was noted as an unexpectedly great, blast-like damage accompanied by a brilliant flash of light when aircraft structures (e.g., aluminum surfaces) were struck by very high velocity fragments. In the NOTS experiments, both single and cluster experimental devices (multicharge and multiliner, as shown in Fig. 13) were fired at very long standoffs against aircraft structure targets. Liner materials and geometries were found to be critical, with aluminum-alloy, wide-angled, thick liners demonstrating extraordinary damage effects at standoffs up to 240 charge diameters at low altitude. Altitude chamber studies showed that damage effects could be produced at much greater standoffs because the jet's particles were less severely decelerated and ablatively burned away, as had been observed at the local surface altitude of 2400 ft.

The concept of the single shaped charge was that it could be aimed from the interceptor missile to hit its target, even in a high-velocity miss encounter. Realization that the necessary aiming speed and accuracy capability was not yet possible within the extant state of the art, multiple-liner warhead devices with a large number of liners implanted about the surface of a cylinder were explored and found to be a useful alternative. Hypervelocity fragment beam warhead mechanisms were found to be a significant improvement over the first-generation warheads. However, success with the continuous rod warheads in New Mexico soon attracted the attention of the Navy, and so emphasis was placed on the research and development of the latter type of devices as discussed below.

Two multijet shaped charge warheads were subsequently developed in West Germany. These are the ROLAND antiaircraft missile and the previously mentioned KOMORAN antiship missile warheads. The latter possibly derived from a technology transfer in visits to the United States in the early 1960s by Professor Hubert Schardin and his protégé, Dr. Franz Thomanek, who is credited as an original discoverer of the value of the lined shaped charge in the mid 1930s. It is noted that the latter claim is disputed by the Swiss, Henry Mohaupt. Although, Mohaupt claims prior discovery, he has yet to present contemporary evidence to substantiate such a claim. Thomanek's work resulted in many of the German shaped charge weapons in WWII. A company he established produced nearly half the total shaped charge weapons procured by the German Government in WWII. He later founded and directed MBB's Shrobenhausen works. Dr. Mohaupt is credited with the introduction to the United States of the shaped charge, from which the Army was to develop the ubiquitous bazooka of WWII fame.

7.5.4 Submissile Warheads

The submissile types of anti-air-target warheads were under the research and development cognizance of the U.S. Army, with research centered at the Army Ballistic Research Laboratories, Aberdeen Proving Ground in Maryland. Development was both performed and sponsored by the Picatinny Arsenal.

A submissile warhead consists of multiples of individual explosive-loaded and fuzed spheres or blocks, clustered about a central dispersion system. The latter is typically a solid propellant that propels the submissiles radially into a pattern of sufficient size and density to intercept the potential target aircraft or missile. Other dispersion methods include 1) spinning the dispenser to use centrifugal force to distribute the submissile units, 2) using an expanding bladder to distribute the payload radially, 3) using aerodynamically self-deploying units, or 4) using an individual propellant charge in each unit.

Antiaircraft/antimissile submissile warheads were developed for the NIKE HERCULES (see Fig. 24) and NIKE ZEUS missiles in the late 1950s and early 1960s. They were successfully developed but both so complex and costly compared to simpler and equally effective warheads that none were introduced into the anti-air-target missile inventory. However, submissile warheads and dispensers were subsequently developed for use against surface targets, and many such systems are in use today.

7.5.5 Rod Warheads, Continuous and Discrete Rod

Rod warheads were studied by the U.S. Navy Bureau of Ordnance, principally at the New Mexico Institute of Mining Technology (NMIMT) in Soccorro, New Mexico. Work was also carried on at the Naval laboratories at the U.S. Naval Ordnance Test Station in China Lake, California; the Naval Ordnance Laboratory (NOL) in White Oak, Maryland; the Naval Proving Ground (NPG) in Dahlgren, Virginia; and by contract to Johns Hopkins University at the Applied Physics Laboratory (JHU/APL) in Silver Spring, Maryland.

According to pioneer researcher M. L. Kempton at the New Mexico Institute of Mining and Technology,[10] the concept for use of rods to achieve structural kills of aircraft was first proposed by Dr. John Workman, the laboratory director. The initial studies were performed by Kempton in the late 1940s and involved the packaging of multiples of 12-in. lengths of square steel rods in Navy 5-in. 38-caliber gun-fired projectiles. The rods were expelled from the rear of the projectile by a propellant charge and dispersed into a conically expanding, tumbling pattern by the centrifugal force and gyroscopic effect imparted by the rapidly spinning projectile. The 12-in. rods were determined to be insufficient for the desired structural kills. Kempton then conceived of the idea of using U-shaped rods with twice the length of the 12-in. rods. The centrifugal force opened the rods into lengths twice as long and with higher probabilities of achieving structural kills. Work with folded rods in the artillery shells led Kempton to the concept of the explosively expanded, continuously connected rods for antiaircraft missile warhead applications.

According to Kempton, the discovery of the continuous rod principle was "somewhere between 1951 and 1953." (The writer believes it to be 1952 because in his visit to NMIMT in the summer of 1953, the continuous rod studies were well underway). A claim was made in the 1953 visit by Dr. C. R. Cassity, then attached to the NMIMT staff, that Dr. Workman's idea for the long rod may have come from the cables hung from balloons that protected London in WWII. The cables were intended to destroy aircraft that flew through the protective screen. Figure 25 shows a proposed screen of cables for the defense of London against an anticipated future aerial attack on England by Nazi Germany. It is noteworthy that the *LIFE* magazine article containing this diagram was published two and a half years prior to the outbreak of WWII in September 1939.

Work at China Lake took place in 1952. Discrete lengths of 0.5-in. square steel rods were packed around a central explosive charge. Detonation drove the rods away at high velocity but also severely bent the rods. The rod warhead was considered for a 5-in. rocket application, but the effort was stopped to permit

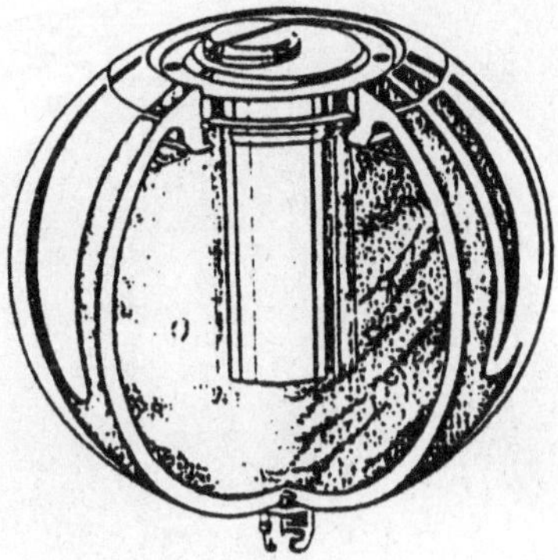

Blast Submissile

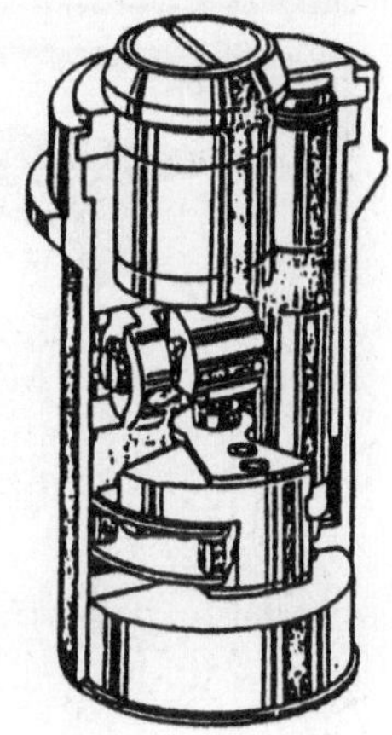

T1409E1 Fuze for Submissile

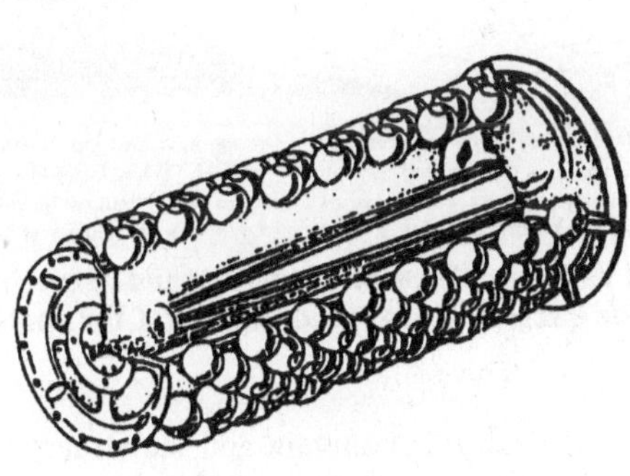

Assembled Warhead

Fig. 24 T46E1 NIKE HERCULES submissile antiaircraft warhead developed in the late 1950s. Carried 205 individually fuzed submissiles. Each 4.5-lb submissile contained 1.6 lb of H-6 explosive. Center core included 23 lb of mortar ignition powder, arranged in conic form so as to produce uniform submissile distribution. Warhead weight was 1023 lb.

London's six zones of defense against a hypothetical fleet of enemy bombers are indicated in an exaggerated form in this drawing, sketched as though looking from a point southwest of London toward the English Channel in right foreground and the North Sea and Holland in the background. What looks like a picket fence spitted with sausages is the "balloon barrage" shown on the opposite page. London is the cluster of buildings on the near side of the

Fig. 25 Proposed balloon defense of London against expected future air attacks, using cables to sever aircraft wings (predecessors of the continuous rod warhead?).

the ordnance engineer team (G. Fountain and the writer) time to concentrate on the higher-priority development of a controlled fragmentation warhead for the initial version of the SIDEWINDER missile.

Continuous rods warheads were characterized by a double layer of 3/16-in. to 1/4-in. square steel rods welded at opposite ends to form a large expanding hoop when the central high-explosive charge was detonated. The principal problem to be resolved was to deliver a sufficiently uniform impulse that the nest could open without breaking rods or tearing the welds. To accomplish this, appropriately contoured inert attenuators were placed between the explosive and the rod nest of the earlier designs. However, the low-density mass (typically magnesium) also reduced the launch velocity. Therefore, in later designs, the explosive charge was shaped at its hollow central core to ensure a uniformly delivered impulse. A performance problem of the continuous rod warhead was its rather low velocity, on the order of 4000 ft/s. One of the higher-velocity continuous rod warheads was developed in the early 1960s by Aerojet Ordnance for the Hughes HM-55 missile. Originally built for Sweden, it was later made for the U.S. Air Force as an advanced-version FALCON. The HM-55 rod expansion velocity was approximately 5400 ft/s, a performance achieved in part by design of the initiation system's booster to avoid the creation of shocks that could break the rods. The novel booster was transferred to other applications, including MAVERICK and various non-nuclear products of the Lawrence Livermore National Laboratories.

Warhead development for Navy antiaircraft missiles SPARROW, SIDEWINDER, TERRIER, TARTER, and TALOS was sponsored by the Navy, with the Applied Physics Laboratories, Johns Hopkins University (APL/JHU) in Baltimore, Maryland as the contract operations analysis and project management organization. Industrial contractors such as Rheem (later to be a part of Aerojet General Ordnance), Bendix Aviation, Chicago Midway, and others participated in the development and manufacture of specific warheads. Aerojet Ordnance was also a developer and manufacturer of the safe/arming/firing (SAF) mechanisms under contract to the Naval Ordnance Laboratory facility, then located at Corona, California.

The continuous rod warhead became the Navy's favored method by the late 1950s, and virtually all Navy missile warheads were to replace their original fragmentation warheads with the "CR" types. The Army followed suit for a time but eventually reverted to the controlled fragment types that continue in use today.

The USSR apparently did not embrace the concept of the continuous rod warhead, even though they were doubtless well aware of the U.S. technology. The Soviet designers favored use of large controlled fragmentation warheads with very heavy fragments. The U.S. Navy also reverted to the use of discrete lengths of rods, with emphasis on high velocity, high mass, and a controlled tumble rate.

7.5.6 *Aimable Effects Controlled-Fragmentation Warheads*

The U.S. Armed Services have sponsored several studies of aimable-effects non-nuclear warheads. The earliest known studies were for special antiballistic missile warheads and were undertaken in 1944 under the impetus of the German V-2 rocket attacks on London. Early concepts considered physically aiming "headlight"-like, end-projection-effects (i.e., Misznay Schardin) controlled fragmentation warheads to project a cloud of fragments into the anticipated zone of intercept.

Both antimissile and antiaircraft aimable warheads have been the subject of research and development activity since the early 1950s to the present Strategic Defense Initiative (SDI) program. Many novel techniques have been explored, including warheads that

- Unfold to expose all fragments in the target direction
- Slough off a quadrant of the warhead so that the remainder is aimed at the target

• Have external explosive charges to compress the desired firing face so as to concentrate the effects toward the target

Problems with the mechanically aimed warheads include their complexity, high parasitic weight, and an excessive function time that affects accuracy. In the early 1970s, research and development were sponsored by the U.S. Air Force of "electronically aimed" warhead mechanisms (see Fig. 26) that could respond to bias the fragment energy (velocity) and, to a lesser degree, the total mass of fragments toward the expected target. Aimable cylindrical warhead (ACW)

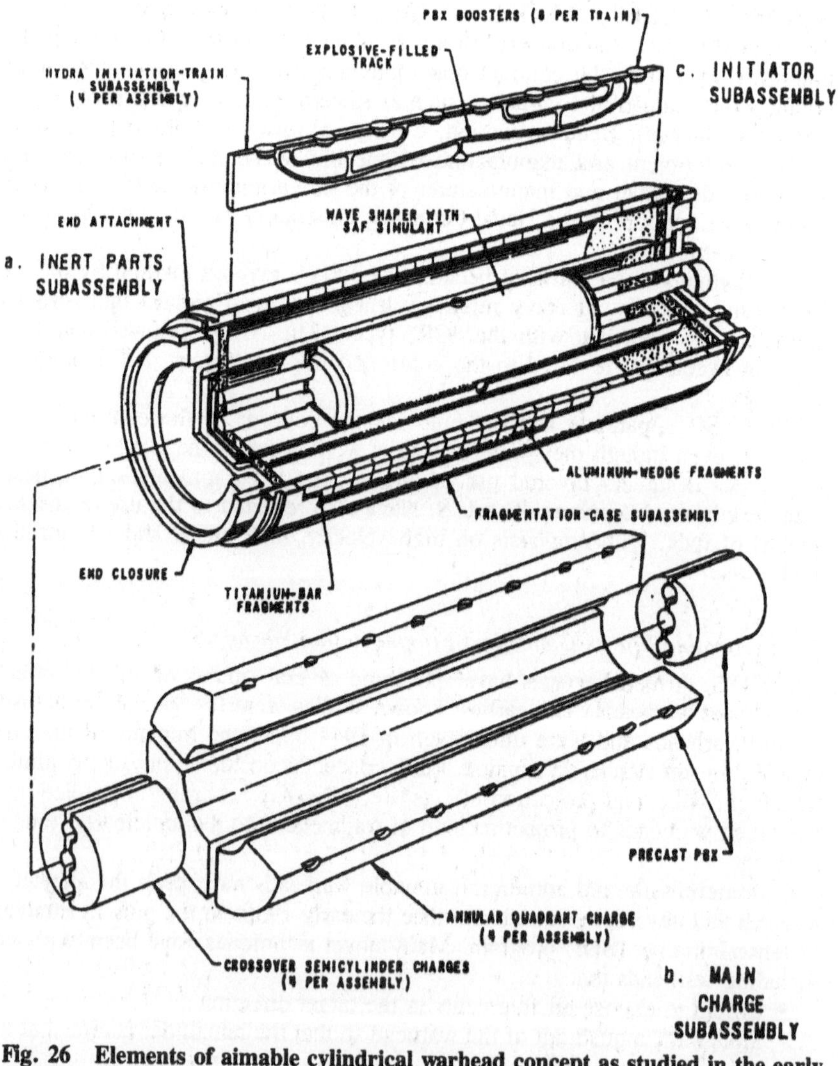

Fig. 26 Elements of aimable cylindrical warhead concept as studied in the early 1970s for possible antiaircraft missile applications.

technology was enabled by development of target detector devices (TDDs) that provide target azimuthal location information as the missile carrying the warhead passes its target. Track testing demonstrated this to be feasible even at passing or crossing velocities exceeding 2000 ft/s. The operating principle is a series of multipoint initiation lines disposed about the periphery of the warhead center at the explosive-to-fragment case interface. A cylindrical wave shaper at the center causes the detonation to implode onto the surface opposite the line (or two lines) of initiation. The effect is a significant fragment velocity increase over that achieved with either single-point or dual-end initiation. The effect also increases the concentration of the fragments on the aim side but is also such that an aiming miss is not too serious. A target on the warhead's "aim" side sees a significantly enhanced performance. Fragment initial velocities of 11,000 ft/s were achieved on the aim side at C/M (explosive charge to case mass) ratios of only 1.4. Research was performed in both AIM-9 SIDEWINDER and AIM-7 SPARROW sizes. Variations of such aiming techniques continue to be under consideration for possible future applications.

8. Modern Torpedo Warheads

Virtually all torpedo warheads were until very recently of the so-called bulk charge type. Such warheads are basically cylinders of high explosives with the explosive selected for maximum underwater effectiveness against both surface and submerged targets. High explosives containing aluminum in a finely divided (powdered) form have been the principal types employed from WWII to date for bulk charge applications. Typical U.S. Navy torpedo applications include the 21-in. Mk 48 heavyweight (3480-lb) torpedo and the 12.75-in. Mk 46 lightweight (570-lb) torpedo, both of which have bulk charge warheads containing aluminized explosives. The heavyweight MKA8 torpedo warhead explosive payload is nearly seven times greater than the lightweight Mk 46. Torpedo warheads of the former USSR are also of the bulk charge type, but the charges are typically much larger[11] than those in Western practice, as is typical of Soviet warheads in general.

As discussed in an earlier section, torpedo warheads incorporating shaped charge mechanisms were first considered in 1913 and again in WWII. However, it was not until recent years that antisubmarine torpedoes have adopted the shaped charge as a primary kill mechanism. The applications are the smaller, lightweight torpedoes designed for the attack of submarine targets. Lightweight torpedoes seldom weigh more than a few hundred pounds and are modest in size (typically a 12.75-in. diameter and 100 to 111 in. long). Warheads are thus constrained to weigh less than 200 lb, with a fair fraction of that weight devoted to the case or containment structure, which must be very strong to allow the aerially sown lightweight torpedoes to enter water at high impact speeds.

The last U.S. lightweight torpedo with a small bulk charge warhead was the Mk 46. Its development began in 1959, but by the mid 1960s the warhead designers recognized that the submarines deployed by the Soviet Navy employed double-hull structures, a practice apparently adopted from captured German WWII late-war developments. The problem with the double hull was that the effectiveness of the bulk charge is rapidly diminished with standoff from the primary target, the pressure hull. It was apparent that if double-hull construction was to dominate the threat submarines, the hull spacing might render small bulk charge warheads incapable of destroying their intended targets.

Underwater shaped charge studies, first conducted in WWII, were resumed in 1952 by researchers at the U.S. NOTS who explored the effects of various metal liner materials. The Navy (assisted by the U.S. Army BRL) performed a series of full-scale underwater experiments of large, copper-lined shaped charges in the mid 1950s. With a realization that the shaped charge might be the only solution to the small non-nuclear torpedo warhead, serious research was resumed under Naval Surface Weapons Center (NSWC) White Oak Laboratory sponsorship in the mid 1970s. The U.S. Navy's Mk 50 BARRACUDA, the British STINGRAY, and the French MURENE lightweight torpedoes have all adopted the shaped charge kill mechanism. The choice of liner material is a key parameter and may be different in each country's application.

Enhancement of the effectiveness of the lightweight torpedo warhead continues to be a high-priority goal as the target submarines are made stronger to resist such attacks. Increased hardness of the target submarine is achieved in several ways: 1) increased spaces between the inner pressure hull and outer shell, 2) increased materials strengths and thicknesses, and 3) increased use of compartmentalization to minimize the effects on buoyancy of a flooded compartment. Lightweight torpedo applications are perhaps the most challenging of the non-nuclear warhead problems.

9. Modern-Day Artillery Warheads

The most widely used tubed artillery warheads continue to be of the simple blast/steel-cased fragmentation types similar to those employed in both World Wars. Principal differences in today's projectiles are greatly increased ranges through the use of better-designed guns, propellants, auxiliary thrust devices, and improved aerodynamic design. Both preset time and proximity fuzes continue to be the types used to provide the higher kill probabilities.

Post-WWII artillery shells also include shaped charge (HEAT) mechanisms in addition to fragmenting cases. They are known to some as HEAT-Multi-Purpose (HEAT-MP) because of their multiple kill mechanisms of fragmentation and blast, in addition to a shaped charge jet. These are typically the alternate load in the modern tank, with the kinetic-energy, long-rod penetrator the first choice round for antiarmor application. As noted in Section 2, the APFSDS high-density, high length-to-diameter "arrow" round (known only as "sabot" to the modern tanker) is the principal antiarmor mechanism in the modern land combat vehicle. The kinetic-energy shot has always been the preferred ammunition for gun-fired applications, whereas shaped charge are preferred for low-velocity or lightweight applications such as the shoulder-fired antiarmor weapon.

Perhaps the greatest advancement in effectiveness of gun-fired artillery in recent decades was the introduction of the so-called dual-purpose/improved conventional munitions (DP/ICM). These are but one of several kinds of payloads that can be delivered by "cargo round" ammunition. Cargo rounds can carry any of a wide variety of payloads and disburse the contents at specified times or points along the trajectory by payload expulsion, typically from the projectile's base. Lateral payload distribution is provided by the centrifugal force of the spinning projectile.

Cargo round payloads may be any of a wide variety, including the following:

- Flechettes (small arrows) in so-called BEEHIVE ammunition, for which the payload expulsion is usually forward to minimize velocity loss
- Propaganda leaflet payloads
- Pyrotechnics (parachute flares)
- Top attack munitions [e.g., SADARM (Sense And Destroy ARMor) smart munitions or "dumb" DP/ICM such as M42 high-explosive dual-purpose (HEDP) grenades]

The most commonly produced special artillery round in NATO today is the DP-ICM type. The U.S. 155-mm M483 howitzer projectile (see Fig. 27) exemplifies this type of ammunition. It is adopted directly or in variations for both gun and mortar applications by many other nations including Great Britain, West Germany, Israel, Spain, Greece, and others. The DP/ICM version deploys 1.5-in. diameter, 0.45-lb combined shaped charge and controlled fragmentation grenades (see Fig. 28). The U.S. 155-mm M483A1 projectile carries a payload of 88 of the basic M42 and M46 (higher-strength case variant) grenades with axial shaped charge and lateral fragmentation capability, as shown in Fig. 28. The grenades are stacked 11 units high in 8 tandem "sticks." At the desired point along the trajectory, a preset time fuze causes a propellant charge in the nose to eject the grenade payload rearward. The aft closure's constraints are broken by the force of the moving payload, and the sticks of grenades are expelled aft where the grenades separate and disburse by the combined spin and air loads. The grenades deploy ribbon-tape stabilizers that also act to arm the grenade fuzes. The grenades are dispersed into fairly uniform patterns and impact on the upper surfaces of targets such as combat vehicles or on the ground. Although nominal performance is quoted at a modest 2 in. plus, the grenades were demonstrated by one producer to defeat 100 mm of armor plate and to produce a very high-velocity radial fragmentation effect. Some of the more recently developed grenades in Europe are claimed by their developers to have a significantly greater penetration ability.

A multiple launch rocket system (MLRS) is an artillery rocket system that fires large, unguided, barrage rockets for distances to 30 km or more. The basic payload of the MLRS is 644 of the above-described DP/ICM grenades. The latter (designated M77) are a variant on the M42 HEDP, having tape stabilizers designed to accommodate the different delivery environment.

10. Special Warheads

This category includes such special purposes as delivering propaganda leaflets; delivering incendiaries and chemical and biological agents; firing signals and illuminants into the sky; and delivering mines, special ground target attack mechanisms, and fuel-air explosives. It is a very broad and important area of concern to the warhead systems engineer. Some were briefly introduced in the preceding discussion of modern artillery cargo projectiles. Key types are briefly discussed below.

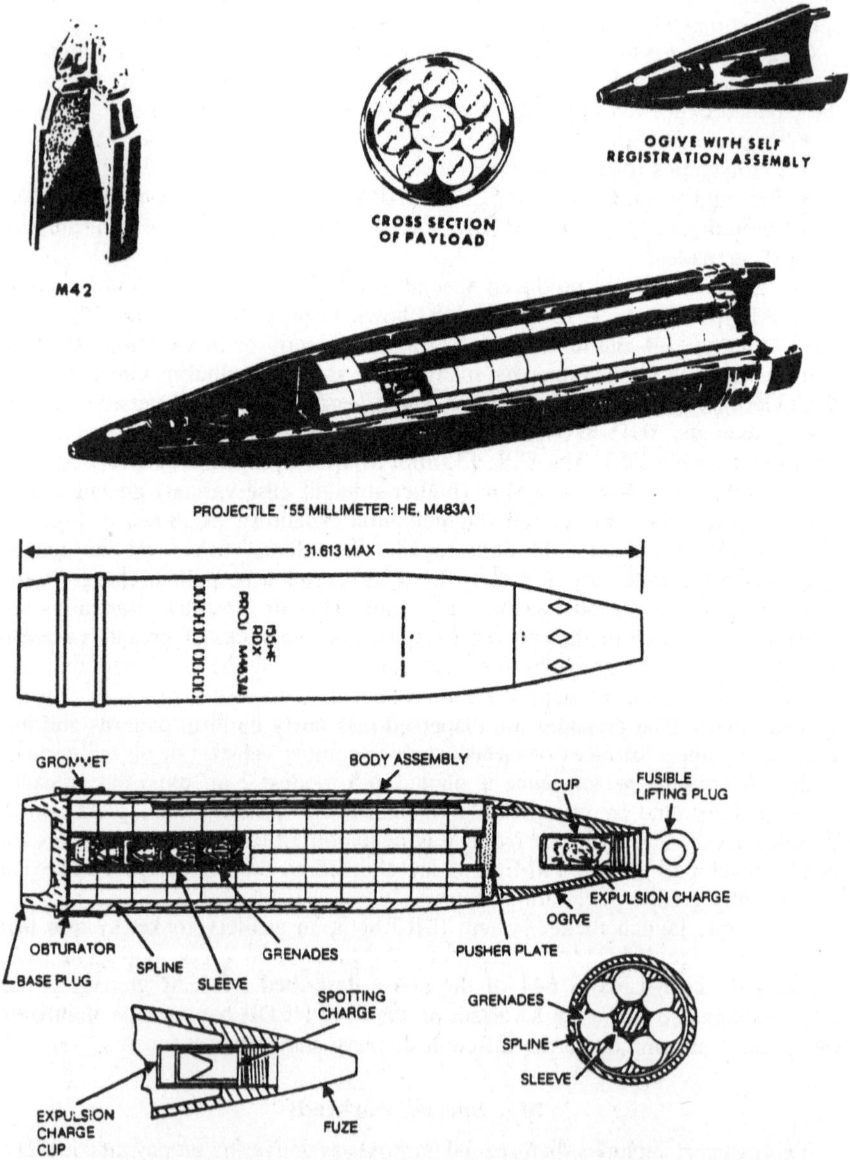

Fig. 27 Modern U.S. Army "cargo" projectile. Shown is 155-mm, HE, M483A1 improved conventional munition. Employs preset M577MT (mechanical time) or M742ET (electronic time) fuzes to release payload at selected point along trajectory.

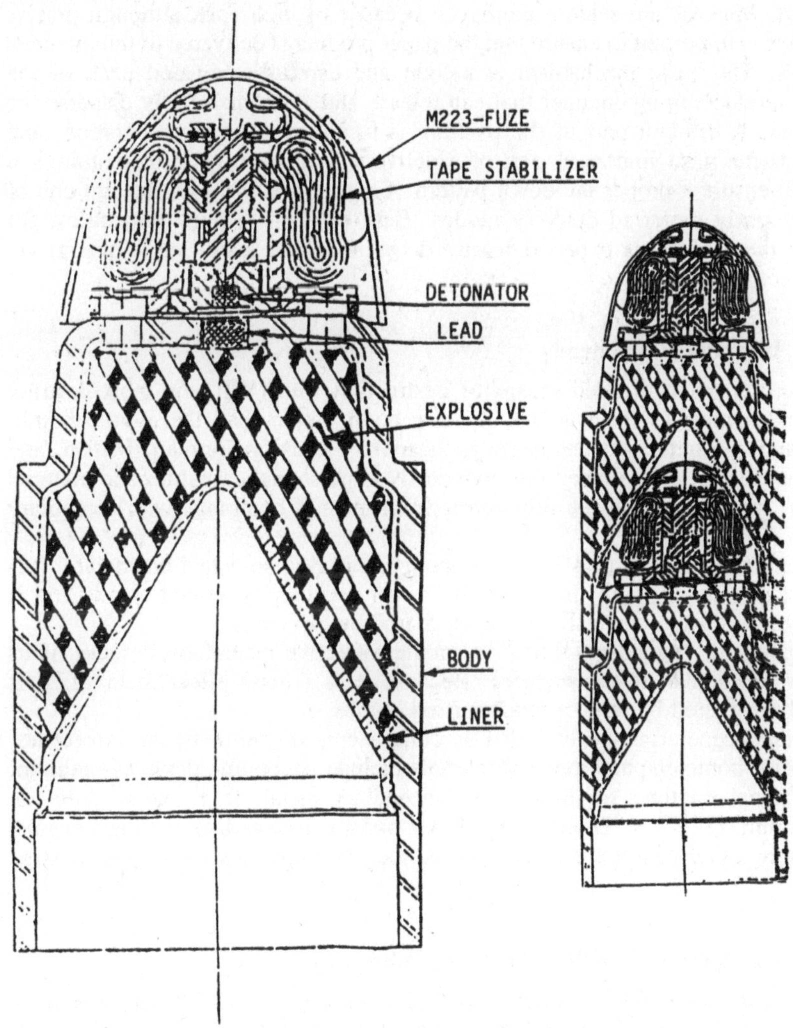

Fig. 28 Cross section of M42/M46/M77-type HEDP ICM grenades as employed in cargo projectiles (Fig. 27) and MLRS basic warhead. Illustrates full nesting to maximize packing fraction and support payload to withstand longitudinal gun-launch accelerations. Grenade diameter is 38.5 mm. Differences between three grenade types involve minor case and stabilizer variations.

10.1 Propaganda Warheads

Modern warfare depends as much on intimidation and swaying of minds of the enemy as it does on destruction. The special warheads employed to deliver such psychological products may be artillery shells, bombs, or dispensers in aircraft. Missiles are seldom employed because of high cost, although precise delivery is important to ensure that the paper product is delivered to the intended readers. The basic mechanism is a tight and carefully prepared pack of the paper products in a container that can release and aerodynamically disperse the payload. A difficult part of the packing is to ensure a proper separation, and perhaps the most important part of a delivery is to have good meteorological data to ensure a proper lay-down pattern. Cargo artillery projectiles are one of the currently preferred delivery modes. Base ejection "cargo" projectiles, for which the payload is expelled rearward by propellant action on a piston, are most commonly used.

10.2 Incendiary Warheads

Fire is a much dreaded means of destruction. In WWII, the most destructive mechanism in the war was the fire bomb rather than the atomic bomb. Although the latter was devastating, incendiary raids on German, British, and Japanese urban areas proved far more costly in personnel casualties and material losses, largely through the fire storms from massed bombings with incendiary magnesium bombs.

White phosphorous (WP) is a commonly employed so-called incendiary. It is in fact a relatively poor fire starter but is an excellent screening smoke and is both very effective against, and much feared by, personnel.

Napalm, or "gelled gasoline," is another effective incendiary because of its ability to stick to target surfaces. Because it is a mass effect material, it is usually delivered by large aircraft-carried bombs.

Other incendiaries are included as components of multipurpose projectiles, grenades, bomblets, and mines. Materials include zirconium, depleted uranium, and special mixtures of plastics and incendiary metals [e.g., Quasi-Alloy of Zirconium (QAZ™)]. These materials are usually arranged so as to accompany or follow other destructive mechanisms such as fragments and shaped charge jets.

10.3 Chemical and Biological Agent Warheads

The expression *NBC warfare* stands for nuclear, biological, and chemical warfare. Although official U.S. policy has for several years suspended the development and production of chemical and biological warheads, the continued and expanded use of chemical warfare, particularly in the Middle East, makes it an imperative to continuously assess such mechanisms from both potential application and defense positions.

Chemical and biological delivery warheads are classed as "weapons of mass destruction" and are referred to as "the poor man's nuclear weapon." For this reason alone, NBC constitutes a major threat to the world population.

Chemical agents can be delivered in many different ways, including rocket or gun artillery, missiles, bombs, grenades, aerial sprayers, surface sprayers, hand-thrown canisters, etc. The need is to deliver the chemical agent where it will accomplish the desired goal. Neither great accuracy nor speed are necessary requirements. Chemicals include lethal nerve gases, lachrymators (tear gases), lung and skin irritants, foliage defoliants (e.g., "Agent Orange" of Vietnam War fame), or (as in recent use in the Middle East) mixtures of these and other agents.

Biological agents are more often employed for strategic rather than tactical applications and may include both viruses and bacteria for use against both man and vegetation (e.g., food crops).

Until recent years, most chemical warheads were of a bulk filled type with the agent dispersed by a high-explosive burster charge along the central long axis. These warheads combined the hazards of both high explosives and chemical agents in their most potent forms. Major problems in both manufacture and the handling and storage of such double-hazard munitions have long been hallmarks of chemical munitions. The problem is particularly encountered in long-term storage for which the deterioration of munition components led to leakage and degradation of safety. It is difficult to transport and dispose of such munitions. Perhaps the most significant event in recent times (since the mid 1960s) was the development of multicomponent munitions that carry innocent components and mix them into a high-potency form only while en route to the target from the launcher. The Navy-developed binary in-flight generation (BIGEYE) "eyeball"-delivered free-fall bomb is such a case in point. The bomb carries two nonlethal components in separate compartments that are combined and mixed by internal equipment while en route to the target to create the nerve agent "VX." The bomb design permits separate storage of the components so that no accidental mixing can take place.

Development of biological agent warheads has been much less publicized. In the late 1950s, a biological warhead was developed for a U.S. Army theater ballistic missile. It was never produced. The warhead's payload consisted of spherical, plastic-cased, self-dispersing submunitions, similar to but much larger than the ubiquitous self-dispersing antimaterial bomblets such as the BLU-26 and its many successors. Because of the characteristics of live agents, it is necessary to store the warheads in refrigerated containers. The latter can be quite complex and costly. No other overt applications are known.

10.4 Illuminants and Signals

Battlefield illumination and colored signals (e.g., star shells, flares) are essential elements of modern warfare. Artillery projectiles of the base expulsion cargo type are, as noted above, the currently preferred delivery method. The illuminant may be a single parachute-suspended "candle," or, if the projectile is large, multiple units may be carried.

10.5 Mine and C^3 Device Delivery Warheads

As introduced above in the discussion of cargo-type artillery projectiles, the delivery of a wide variety of payloads has been made possible. Other applications include delivery of top attack antiarmor submissiles and smart munitions. The latter include such devices as the SADARM mechanisms, which are carried in tandem mounted groups of three mines that are expelled from the cargo round's base. Each unit deploys a stabilizer that both retards and causes the mine to rotate so that its downward-looking sensor describes an ever-decreasing circle on the ground below the descending mine. If a target is located in the field of search, the mine is functioned to form an explosively forged projectile that defeats the vehicle's top armor.

Another useful cargo round application is delivery of electronic devices capable of several functions such as locating targets, jamming signals, etc.

10.6 Fuel-Air Explosive Warheads

The modern fuel-air explosive mechanism is another of the Soviet-designated "weapons of mass destruction." It is therefore an important mechanism that may well be one of the more widely used future mechanisms.

The FAE mechanism is an ancient and often-encountered mechanism in everyday life. The combustion in the cylinder of the auto engine is an FAE event, and so is the explosion of coal dust in a mine or wheat dust in a grain storage silo. The mechanism has been explored since at least WWII for warhead applications. The British studied "Algas," a dispersion of aluminum powder in gasoline, which was demonstrated to be highly efficient for internal blasts.

There is evidence that the German Army may have employed a form of FAE on the Russian defenders in the WWII siege of Sebastapol. German 28-cm and 32-cm NEBELWERFER barrage rockets were employed in massed arrays, with groups of 32-cm-diameter liquid incendiary rockets fired first so as to disperse their contents of gasoline, kerosene, or naphtha as vapors, followed seconds later by the 28-cm high-explosive rockets that possibly detonated the fuel-air mixtures. The Soviets noted widespread personnel casualties "without a mark on them," a typical comment concerning a victim of blast injury. Reports in the popular press suggest that similar mechanisms may have been employed by the Soviets against personnel targets in Afghanistan.

U.S. development of FAE was pioneered at the Naval Weapons Center, China Lake, in the 1950s. FAE weapons were employed in Vietnam. Both the Air Force and Sandia Corporation developed large FAE devices, primarily for large-area forest destruction to permit vertical assault landings in the jungles in Vietnam. The Navy-developed weapons were free-fall, aircraft-dispenser-delivered canisters. The fuel used was ethylene oxide, which was dispersed radially by a high-explosive burster as the canister impacted on the Earth's surface. Second event detonators were then projected into the expanding cloud and detonated at a time when a good mixture of fuel and air was present. Overpressures on the order of 300 psi were attributed to such weapons.

Fuel-air mechanisms research in the United States was suspended for several years by orders of the Carter Administration. As a result, the development of the second-generation weapons was delayed, although developments apparently continued by the Warsaw Pact. The successful results of the USSR research were apparently used, as noted above, in the war in Afghanistan, which was basically an antipersonnel war.

FAE research and development continues on a worldwide basis today, seeking higher-energy source materials, possibly exploiting the energies available in certain solid materials such as reactive metals.

One potential FAE application is its possible use against a field of shallow buried land mines. When the idea was first proposed by the writer in the 1960s, the interest was to use the FAE-explosion-produced overpressure to react mines with respect to the soil cover by the difference in their intertial response. In a cleared area, such an action could expose a mine's location permits it and disposal by another means such as gunfire. The FAE mine clearance systems developers apparently did not understand the original intent and sought instead to detonate pressure-sensitive mines. Perhaps it had not occurred to them that the future mine layers could simply change to a fuze that required a multiple pulse or a higher-pressure signature.

11. Conclusions

The above discussion of the various types of non-nuclear warheads is by no means complete or inclusive of the many variations conceived of by man over the many centuries of periodic warfare.

However, a sufficient diversity of type and function is shown to illustrate the fact that man's ingenuity knows few bounds when it comes to devising methods to inflict harm on man, his family, possessions, and property in times of war between nations.

The fundamentals of both piercing and energy transfer mechanisms seem to remain constant. The means, the materials, the efficiencies, and the velocities and accuracy continually improve with time. Few "new" ideas are completely novel. Most are adaptations and combinations of prior mechanisms.

The developing enhanced efficiency and capability of modern non-nuclear warhead mechanisms, coupled with ever-increasing accuracy and range of delivery against selected targets, permits non-nuclear warfare to have a very high lethal capability, enough so that the need to resort to the use of the universally-dreaded nuclear warfare may be rendered unnecessary in future combat.

Today's warheads are enhanced by the availability of superior materials, computerized computational models, and advanced, high-precision warhead delivery mechanisms. Yet, in spite of such superior tools, materials, and manufacturing technology, it still requires the ingenuity of a few talented individuals to conceive of new solutions to both old and new problems to advance the state of the art of non-nuclear warhead technology.

References

[1]Cornish, C. W., "History of Ammunition," Royal Ordnance Factories Consultancy and Advisory Service, Ministry of Defence (Procurement Executive), Central Administration, Kingsmeadow Road, Nottingham NG2-1EQ (no date or report number but of a recent origin, cites Telex #37-531).

[2]Biringuccio, V., *The Pirotechnia of Vannoccio Biringuccio*, first published 1540, translated from the Italian by Cyril S. Smith and Martha T. Gnudi in 1943, American Institute of Mining and Metallurgical Engineers, Cambridge, MIT Press Paperback Edition, 1966.

[3]Sharp, A., "Greek Fire," *Civil War Times Illustrated*, Vol. 23, No. 7, Nov. 1984, pp. 32–37.

[4]*Taschenbuch für den Artilleristen* Rheinmetall-Borsig, Düsseldorf, Germany, 1940.

[5]Dickey, T. B., "The Armor Punchers," *Civil War Times Illustrated*, Vol. 18, No. 5, Aug. 1979, pp. 24–31.

[6]Davis, Lt. Col. M. H., "Important Developments in Ammunition History," U.S. Army, George Washington University, School of Engineering, Defense Training Course, "Course in Metal Components of Ammunition," Sept. 19, 1941.

[7]*U.S. Rocket Ordnance Development and Use in World War II*, Office of Scientific Research and Development, War Department, Navy Department, U.S. Government Printing Office, 1946.

[8]Clark, E. L. *Development of a Warhead for the HELLFIRE Missile System*, Firestone Tire and Rubber Co., Defense Research and Products Division, Summary Report, Sept. 74–Dec. 75, Rept. No. DRD-68, Feb. 1976.

[9]HELLFIRE Proposal design for Rockwell International was jointly prepared by J. E. Backofen of the Battelle Columbus Laboratories and D. R. Kennedy of the Shock Hydrodynamics Division, Whittaker Corp., May 6, 1976.

[10]Kempton, M. L., personal telephone communication, NMIMT, Soccorro, New Mexico, Dec. 1988.

[11]Friedman, N., "The Naval Institute Guide to World Naval Weapons Systems," Naval Institute Press, Annapolis, Maryland, 1989.

[12]Coles, B., and Rickson, P. L., "Mistletoe—The Deadly Parasite," *Air Classics Quarterly Review*, Vol. 4, No. 3, Fall 1977, pp. 38–47 and 86–99.

Bibliography

Books

Biringuccio, V., *The Pirotechnia of Vannoccio Biringuccio,* Trans. by Cyril Stanley Smith and Martha Teach Gnudi, The M.I.T. Press, The Massachusetts Institute of Technology, American Institute of Mining and Metallurgical Engineers, 1942.

Manucy, A., *Artillery Through the Ages*, National Park Service, Interpretive Series History No. 3, U.S. Department of the Interior, U.S. Government Printing Office, Washington, DC, 1949.

The Howell Torpedo Honeywell Inc., No. 0-22113-10-18, Jan. 1967.

U.S. Rocket Ordnance Development and Use in World War II, U.S. Joint Board on Scientific Information Policy for Office of Scientific Research and Development, War Department, Navy Department, U.S. Government Printing Office, 1946.

Weapons, An International Encyclopedia from 5000 B.C. to 2000 A.D., St. Martin's Press, New York, 1980.

Filley, B., *Junkers JU 88 in Action* Part 1, Aircraft No. 85, Squadron/Signal Publications, Inc., 1988. (Discusses the "Mistel" apparatus.)

Military Explosives, Department of the Army, No. TM9-1300-214, Nov. 1967. (Histories of explosives.)

Pearson, L., "Developing the Flying Bomb," *Naval Aviation in World War I*, edited by A. O. Van Wyen, Chief of Naval Operations, U.S. Government Printing Office, Washington, DC, 1969, p. 70.

Caruana, A. B. *British Artillery Ammunition 1780* (Only available information concerning source.)

Taschenbuch für den Artilleristen Rheinmetall-Borsig, Dusseldorf, Germany, 1940.

Nagele, H., and Schaap, D., *Geen oorlog'geen minitie*, Meijer Pers bv, Amsterdam, Netherlands, 1979. (300-year history of military production in Dutch language, gift from Euromissile.)

Canby, C., *A History of Weaponry*, Leisure Arts Limited, London, The New Illustrated Library of Science and Invention, no date.

Gunsten, B., *Rockets and Missiles—The Illustrated Encyclopedia of the World's*, Crescent Books, New York, 1979.

Rogers, Col. H. C. B., *A History of Artillery*, The Citadel Press, Secaucus, NJ, 1975.

O'Hart, Maj. T. C., *Elements of Ammunition*, Ordnance Department, Army of the United States, Wiley, New York, 1946.

Gander, T., and Chamberlain, P., *Weapons of the Third Reich*, Doubleday, New York, 1979.

"Engines of War," *Encyclopedia Britannica*, 1966, pp. 399–400.

Weeks, J., *Men Against Tanks—A History of Anti-tank Warfare*, Mason/Charter Publishers, New York, 1975.

Ford, B. J., *Allied Secret Weapons—The War of Science*, Ballantine's Illustrated History of World War II, Weapons Book No. 19, Ballantine, New York, 1971.

Ford, B. J., *German Secret Weapons—Blueprint for Mars*, Ballantine's Illustrated History of World War II, Weapons Book No. 5, Ballantine, New York, 1969.

Pawle, G., *Secret Weapons of World War II*, Ballantine Books, New York, 1957.

Halbritter, K., *Halbritter's Arms Through the Ages—An Introduction to the Secret Weapons of History*, Viking Press, New York, 1979.

Magazine Articles

Coles, B., and Rickson, P. L., "Mistletoe—The Deadly Parasite," *Air Classics Quarterly Review*, Vol. 4, No. 3, Fall 1977, pp. 38–47 and 86–99.

Held, M., "Air Target Warheads," *International Defence Review*, Vol. 8., No. 5, Oct. 1975, pp. 719–724.

Soedel, W., and Foley, V., "Ancient Catapults," *Scientific American*, Vol. 240, No. 3, 1979, pp. 150–160.

"Critical Points: Metal Improves Even Explosives," *Metal Progress*, April 1947, p. 604. (A discussion of aluminum additives to explosives.)

Holder, P., "Roman Artillery (2)," *Military History, Past and Present*, Dec. 1986–Jan. 1987, p. 30.

Snyder, D., "Torpedoes for the Confederacy," *Civil War Times Illustrated*, Vol. 24, No. 1, March 1985, p. 40.

Anderson, W., "Abhorrent to Civilization—The Explosive Bullet in the Civil War," *Civil War Times Illustrated*, Vol. 24, No. 5, Sept. 1985, p. 36.

"The Camera Overseas: London Awaits the Enemy's Bombers," *LIFE Magazine*, Vol. 2, No. 13, March 29, 1937, p. 56 (Balloon defense).

Di Salvatore, B., "A Reporter at Large—Vehement Fire I," *New Yorker*, April 27, 1987, p. 42; and May 4, 1987, p. 38.

"100 years ago, August 1881: re interesting torpedo experiments at Naval Torpedo Station," *Scientific American*, Aug. 1981, p. 12.

Sharp, A., "Greek Fire—The Spirit of Christian Warfare?" *Civil War Times Illustrated*, Vol. 23, No. 7, Nov. 1984, p. 32.

Dickey, T. B., "The Armor Punchers," *Civil War Times Illustrated*, Vol. 28, No. 5, Aug. 1979, p. 24.

Ineichen, H., "Die Schweiserischen Raketenbatterien im 19.Jahrhundert," *Schweizer Soldat*, Nov. 1978, 11/78, p. 33.

Kennedy, D. R., "Dart," *Armor Magazine*, Sept.–Oct. 1981, p. 22.

Papers/Course Notes/Miscellaneous

Held, M., "Trends in Warhead Design," paper presented at Third International Seminar on Defence Technology sponsored by Pakistan Ordnance Factories Board Wah Cantt and Monch Publishing Group (circa 1985).

Tolch, N. A. (a consultant to the Aerojet General Corp.), "Lectures on Fragmentation" (circa 1960).

"A Bit of Undersea Warfare History—Whitehead," (torpedo) *Naval Undersea Warfare Museum Foundation Newsletter*, No. 8 (circa 1987).

Chapter 1

Introduction to Tactical Warheads

Joseph Carleone*
Aerojet General Corporation, Azusa, California 91702

A MISSILE leaves its launcher targeted for an enemy aircraft. Early in its flight, the missile fuze senses the launch and arms the firing train, which leads to the high-explosive controlled-fragmentation warhead. After a few seconds of flight, the proximity sensor aboard the missile detects the aircraft and sends the firing pulse to the initiator that detonates the warhead. As the explosive reaction releases the stored chemical energy, the high-pressure gases accelerate and expand the steel casing. The casing walls stretch and reach a velocity of 1500 m/s or more. As the casing reaches its limit and fractures, an expanding cloud of over one hundred fragments forms. A section of the cloud intercepts the target. Fragments perforate the target, entering into fuel and equipment. The aircraft plummets to earth in flames.

Soldiers await the approach of a heavily armored enemy tank. Their portable missile launcher tracks the huge vehicle as it comes within effective range. The missile is launched, senses the environment, and arms the warhead. The missile is guided to the target by signals from the tracking system. Upon impact with the tank, a crush switch on the missile closes the circuit that fires the shaped-charge warhead. The high-pressure explosive gases collapse the thin copper shell that lines the hollow cavity within the explosive charge. This collapsing process forms a long, cylindrical metal jet about 3 mm in diameter that is traveling at speeds up to 9600 m/s. At such velocities, when the jet impacts the hard, heavy armor, enormous dynamic pressures of megabar proportions are achieved, causing even the hardest armor to flow. As the target material is locally deformed and moved away, the successive portions of the jet continue to impinge on this localized area, increasing the depth of the hole. Eventually, the jet erodes its way through the target. As the jet emerges into the vehicle, the rear surface of the target erupts. (The target material leaves the rear surface of the target by a complicated mechanical process that will be discussed in more detail in another section.) An expanding cloud of fine, high-speed particles is formed. The remainder of the jet goes on to damage some critical component of the tank's fire control system. The fragment cloud from the target damages hydraulic lines

*Vice-president, Armament Programs, Electronic Systems Division.

and other equipment. The jet also deposits thermal energy, causing an increase in temperature and pressure within the vehicle. The tank is rendered helpless. Frequently, the emerging or residual jet will impact one or more rounds of ammunition, causing explosions and fire that catastrophically damage the tank beyond repair.

An aircraft delivers a dispenser full of smaller devices called submunitions over a group of enemy tanks and other vehicles. The dispenser releases the submunitions, which sense the release and arm the explosively formed projectile (EFP) warhead. During their controlled fall, the sensors detect the targets. Immediately upon sensing a target, the firing signal detonates the warhead. The thin contoured metal disc embedded in the explosive surface is accelerated by the action of the explosive gases. During this acceleration, up to 2500 m/s in 60 μs, the disc is also shaped into a flared rod configuration. This EFP rod, drag-stabilized by its flared shape, travels to the target. It impacts the top of the target, where the armor is relatively thin. The projectile begins to penetrate in a manner similar to the shaped charge. But, because of its overall lower velocity and shorter length, it creates a different kind of target eruption, which is composed of larger fragments. The fragments seriously damage the interior components of the tank. The tank is defeated.

The events just described demonstrate in three realistic scenarios how warheads make use of and control explosive energy to achieve a specific mission. This text will focus on the events that occur between initiation of the high explosive and the damage of the target. The fundamentals of the technologies associated with these events will be presented at a basic level to provide the background necessary to study each warhead type.

The text is intended to serve as pedagogical aid as well as a reference for some of the most recent developments in warhead technology. It will not devote a lot of attention to early work on the subject and will not be an all-encompassing review of work done to date. The reader is referred to the series of design handbooks[1–4] for details on much of the work through the 1960s. It is our intention to provide the background and understanding necessary to pursue research and development of warheads in the future. Therefore, choices on which topics to include or exclude, as well as where to place the primary emphasis, had to be made. We believe that a reasonable and practical choice has been made to serve the primary goals of the text.

1. Classification of Warhead Types

Generally speaking, a warhead may be defined as the specific device or part of an armament system that damages a desired target and renders it incapable of performing its intended function. Within this definition there are two very broad categories of such devices: 1) those in which the damage is caused by the action of an explosive that accelerates metal or produces blast effects, and 2) those in which propellants accelerate metal penetrators to lethal velocities. In a more traditional sense, the former (high-explosive) devices are normally referred to as warheads, or sometimes *chemical energy* warheads, and the latter as *kinetic energy* penetrators. This terminology does not truly reflect the mechanisms involved. The so-called chemical energy warheads, for the most part, damage the target by accelerating metal; that is, the resultant kinetic energy causes the damage. Therefore, we will use the term *high-explosive* warhead, or simply, warhead throughout this text. The kinetic energy penetrators that are accelerated by gun or rocket propellant will not be included in this text.

The family of high-explosive, conventional warheads can be conveniently divided into two categories, directed energy and omnidirectional, as shown in Fig. 1. Directed energy warheads "focus" the explosive energy by the use of a cavity lined with metal, normally called a liner. The liner mass is accelerated in a specific direction and the amount of focused kinetic energy is about 15–20% of the total explosive chemical energy released. Directed energy warheads can be divided further into shaped charge, hemispherical charge, and explosively formed penetrator, as depicted in Fig. 2, although there is not a clear dividing line among these types of warheads. In fact, a continuum of directed energy warhead configurations exists with the shaped charge at one extreme, which produces a very high-speed metal jet (tip speeds of 7–10 km/s) of low mass, and the explosively formed penetrator at the other end, which produces a high-mass rod or fragment of moderate velocity (typically 2–3 km/s). It will be convenient for discussion purposes, however, to make the distinction among the various directed energy warhead types.

Omnidirectional warheads are effective in all directions surrounding the warhead. They may be divided into fragmentation and blast warhead types. Fragmentation warheads use the explosive energy to expand a metal casing into a number of small projectiles or fragments. These fragments are formed by three major approaches: natural fragmentation, in which the fragment size is controlled only by the natural fracture characteristics of the case material; controlled fragmentation, in which the case design is biased to produce a selected fragment size; and preformed fragments, in which fragments are fabricated and then embedded into a casing matrix material or bonded onto a thin surface skin around the explosive.

Several other omnidirectional warheads are actually variations of the fragmentation warhead. Two examples are the discrete rod warhead and the continuous

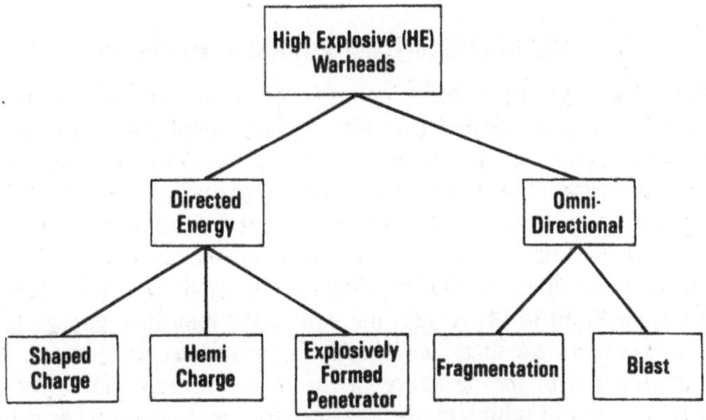

Fig. 1 Family of high-explosive warheads.

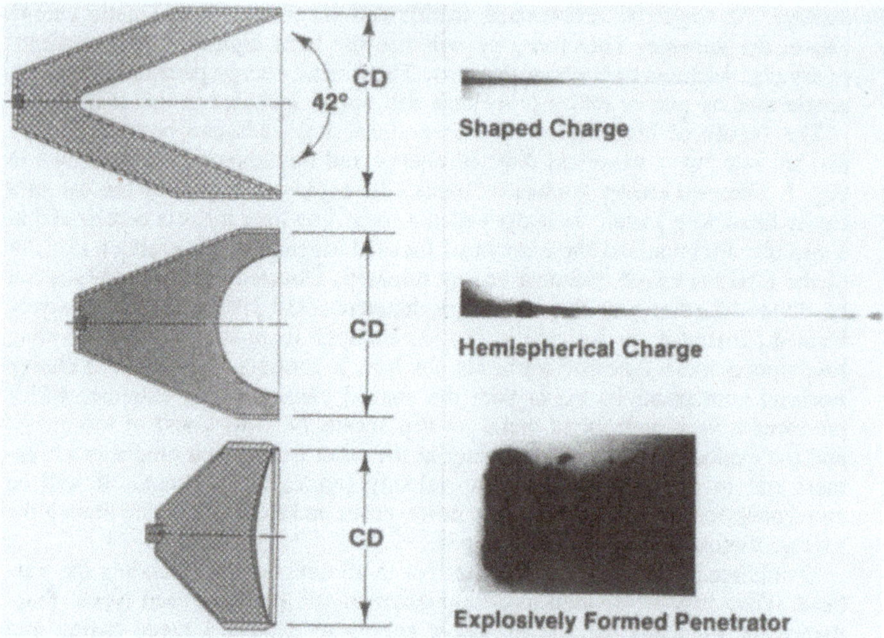

Fig. 2 Three types of directed energy high-explosive warheads.

rod warhead. Details of these warheads will not be reviewed in this Introduction (see Ref. 1).

Blast warheads use the explosive air shock and resulting overpressure to cause damage to the target. Blast warheads usually have some fragmentation capability.

Of course, these are generic types of warheads. Many combined effects warheads or multipurpose warheads have evolved. For the most part, however, the principles and functions of warheads can be traced back to these generic types. Therefore, we will focus our attention upon the basic types.

2. Mechanics of High-Explosive Warheads

The defeat of a target by a high-explosive warhead may be described by the energy conversion process depicted in Fig. 3. The stored chemical energy in the high explosive is converted to kinetic energy by the warhead mechanism. Depending upon the type of warhead, this kinetic energy may reside in a directed jet or projectile of metal or in a number of metal fragments. For simplicity, let us call the formed metal output of the warhead the *penetrator.* Some of the energy is also transmitted by the motion of the gases and corresponding air shock. During its flight to the target, the penetrator may lose energy to the surroundings by means of air drag, ablation, or combustion. In general, the metal penetrator interacts with the structure of the target vehicle. This interaction is a complicated process in which the target structure, and in some cases protective

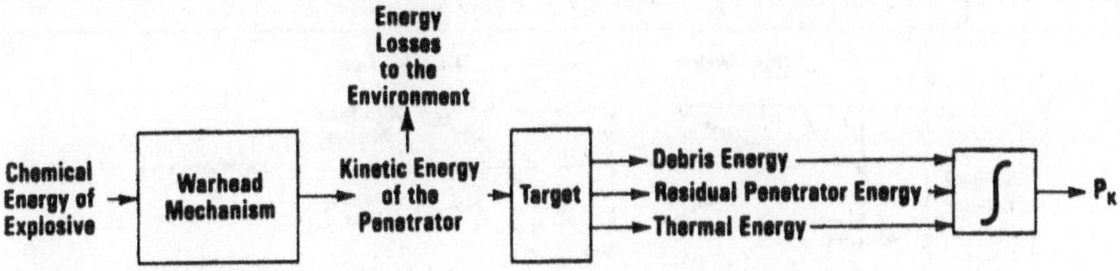

Fig. 3 Energy transfer process that occurs in the functioning of a high-explosive warhead.

armor, absorbs a portion of the penetrator's kinetic energy. During the penetration process, various types of fracture mechanisms occur in both the target and the penetrator. When the outer structure and any protective armor is completely penetrated, the resultant energy delivered within the vehicle renders the damage to "kill" the target. Somehow, we must integrate this delivered energy to estimate the probability of damaging the vehicle sufficiently to render it ineffective, or rather, the probability of kill, P_k.

In this text we will focus upon explosive-metal interaction, warhead mechanism formation, and target interaction. Determination of P_k is beyond the scope of this work.

Five general events occur during the functioning of a high-explosive warhead. These events involve chemical, thermal, and mechanical effects and are listed as follows:

- Detonation
- Metal acceleration
- Defeat mechanism formation
- Flight to the target
- Terminal ballistics

Detonation refers to the exothermic chemical reaction that takes place in a high explosive. This reaction propagates through the unreacted explosive with such rapidity that its rate exceeds the speed of sound in the unreacted explosive. Therefore, the advancing reaction zone is preceded by a shock wave, which is often called the *detonation wave*, or *detonation wavefront*. The rate of advance of this reaction zone is called the *detonation velocity*. In a high-order detonation, the detonation velocity will continue without decay through the unreacted material and is termed stable.

This reaction produces pressure pulses of very short duration with extremely high peak pressures on the order of 300–400 kbar. These pressures decay rapidly as the explosive products expand. This high pressure expansion provides the impulse to accelerate the metal casings or liners that are part of the warhead design and gives rise to the blast waves and air shocks that are present when a warhead detonates.

This *metal acceleration* event (which is not strictly applicable to blast warheads) essentially converts the chemical energy to kinetic energy with the metal moving at a very high final velocity. During the acceleration phase, the metal may also

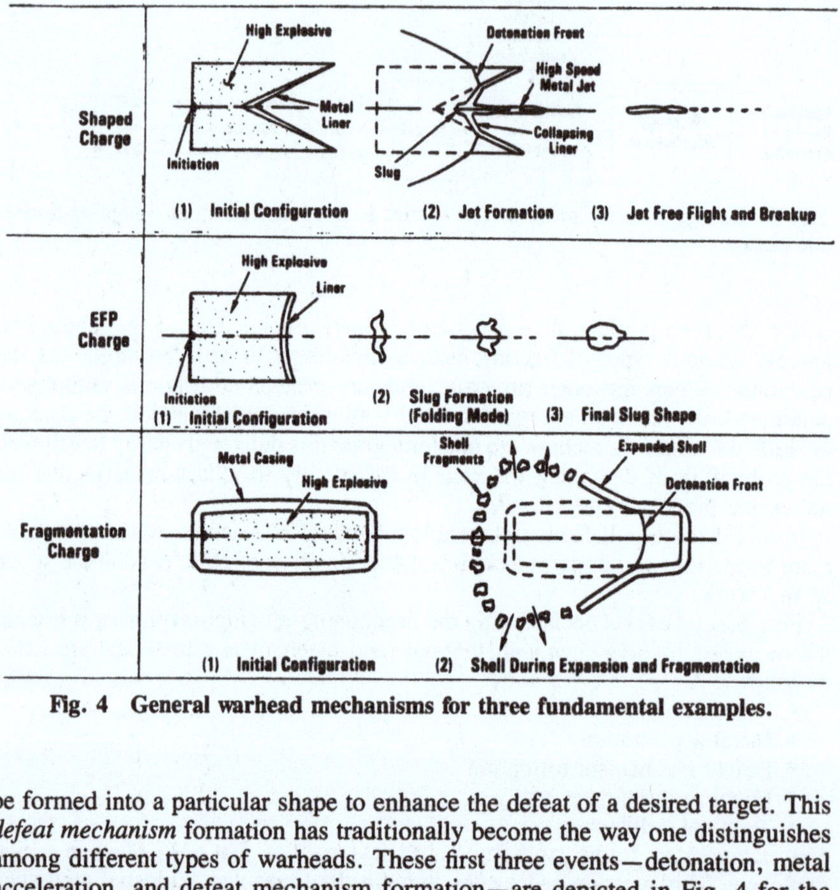

Fig. 4 General warhead mechanisms for three fundamental examples.

be formed into a particular shape to enhance the defeat of a desired target. This *defeat mechanism* formation has traditionally become the way one distinguishes among different types of warheads. These first three events—detonation, metal acceleration, and defeat mechanism formation—are depicted in Fig. 4 for the examples of a shaped charge, explosively formed penetrator (EFP), and a fragmentation charge.

2.1 Directed Energy Warheads

In the shaped charge, the detonation wave sweeps over the conical metal liner, causing it to collapse axisymmetrically. As it collapses, liner ring elements are squeezed closer and closer to the axis. By forcing the mass radially inward, high pressures and stresses are created. The material heats up and flows plastically. The geometry is such that a momentum exchange occurs on the axis of the charge, forming a high-speed metal jet (tip speed in excess of 8 km/s) that constitutes about 15% of the liner mass. The formation and flow of the jet seem to occur in a superplastic taffy-like fashion. The remainder of the liner mass forms the slower-moving slug (sometimes called the carrot because of its shape). The apex portion of the liner is driven more intensely than the base portion because of the larger ratio of explosive to metal in the apex region. This causes the tip of the jet to be faster than the tail. Tip velocities can vary between 7 and 10 km/s for typical shaped charges, depending on the specific design. The effective jet tail velocity is about 2 km/s. The jet, therefore, stretches during its flight to the target and eventually breaks into a number of segments or particles. As long as the jet segments remain aligned, the jet still penetrates deeply; however, when the shaped charge is placed at long distances from the target, the penetration is reduced because imperfections in the jet cause the segments to become misaligned. Therefore, the shaped charge is typically used in direct-hit weapons, which place the warhead very close to the target.

In these applications, almost no jet energy is lost to the environment. In Fig. 5 is shown the penetration of a typical shaped charge into armor steel as a function of distance from the front of the warhead to the target (this distance is called *standoff*). The unit of measure used in Fig. 5 is the explosive charge diameter, abbreviated as CD. This is a very important measure for warhead designers, as you will observe in the following chapters, since to a first-order approximation many warhead events scale in some fashion with the charge diameter. A photograph of a sectioned stack of armor plate that has been penetrated by a shaped charge jet is shown in Fig. 6.

More advanced shaped-charge liner designs have evolved. Varying the liner angle along the length achieves greater jet stretching. Furthermore, wall thickness variations along the length have been developed to optimize performance. This faster stretching gives greater jet length during the penetration process at

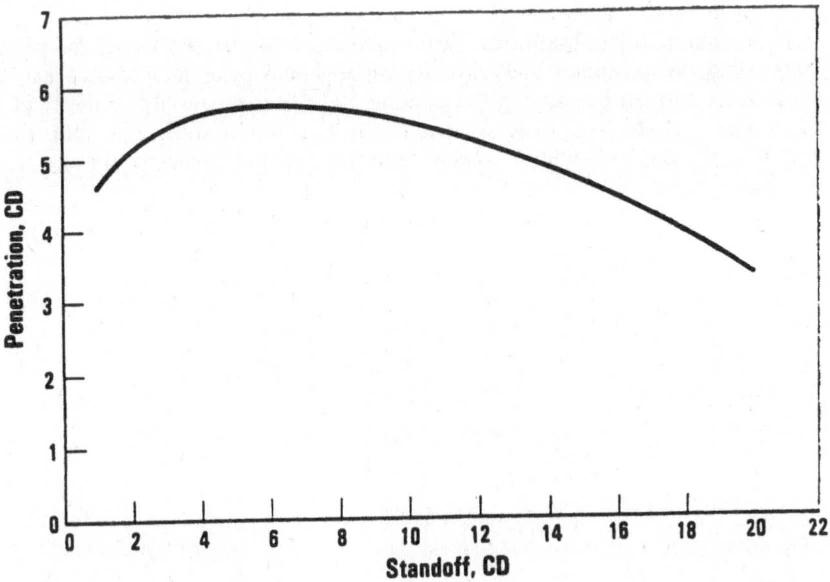

Fig. 5 Typical penetration performance of a precision, conical shaped-charge warhead into armor steel (not an optimized design).

short standoff and hence gives greater penetration. Details of the mechanics of penetration will be discussed in a subsequent chapter. Suffice it to say here that for shaped charge velocities, penetration is governed primarily by total effective jet length and jet density.

In contrast to a shaped charge, an explosively formed projectile (EFP, known also by the names explosively formed penetrator, self-forging fragment, Misznay-Schardin charge, P-charge, and mass focus warhead) deforms a dish-shaped metal liner into a coherent projectile while accelerating it to velocities up to 3000 m/s. Nearly all of the liner mass remains in the EFP. The primary distinction between EFPs and shaped-charge warheads is that EFPs can be used over very long standoff distances of 150 m or more, although there are applications where the EFP is used at moderate standoffs (0.5–10 m) because of its behind-armor effectiveness when the targets are relatively thin.

In fact, the traditional shaped charge and the EFP are two extremes of the spectrum of metal-lined cavity charges. For example, as the cone angle of a shaped charge is increased beyond 150 deg the liner no longer splits into a jet and slug, but a coherent mass or primitive EFP is formed. This effect has been clearly shown by Held,[5] whose results are reproduced in Fig. 7. We divide them into categories such as shaped charge, hemi charge, and EFP for convenience because the kill mechanisms in each case are formed differently and the problems associated with each are different. There are grey areas, however, between the distinct categories. For example, by design, one can collapse an EFP liner hard enough to cause some jetting and elongation in which a long stretching rod is formed that eventually breaks into segments (see Ref. 6). Such a device is neither a classically shaped charge nor an EFP. These hybrids are common and are used in various applications.

Within the family of traditional EFP warheads, however, work over the period 1975–1985 demonstrated that virtually all practical penetrator shapes can be explosively formed by varying the contour and thickness profile of the liner in the proper way. In fact, many shapes, such as compact spheroids, long rods, hollow caps, and rods with a conical flare have been explosively formed from thin metal discs.

Fig. 6 Comparison of hole profiles for a shaped-charge jet and an EFP warhead.

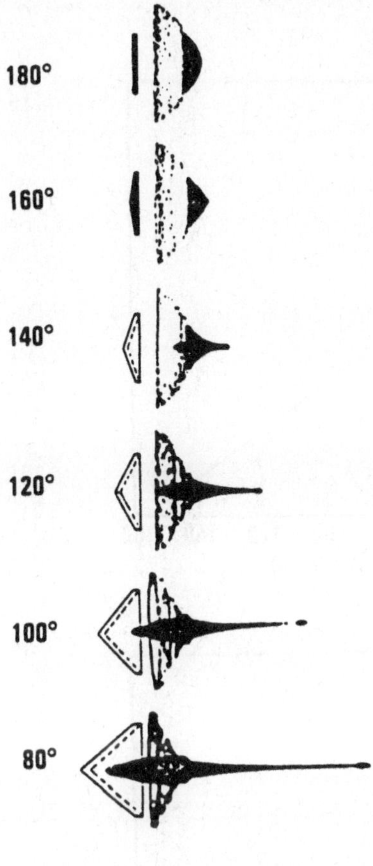

Fig. 7 Radiographs of the collapse of liners with varying cone angles, showing the continuous spectrum of lined cavity charges between shaped charges and EFPs. (Reprinted from Ref. 5 by permission of American Society for Nondestructive Testing, Inc., Columbus, Ohio.)

The penetration capability of the EFP is directly related to its shape, orientation, and velocity at impact. For compact shapes, orientation is not a problem. As it travels, the air drag will reduce its speed and performance will decay. For rod-shaped penetrators, however, orientation is critical. If the rod has no aft flare or other means of stabilization (fins, spin, etc.), then it will tumble shortly beyond 8–10 m of flight. Such tumbling renders the rod ineffective for long standoff applications. When the rod is flight-stabilized by an aft drag cone that is formed from the proper liner design, it will traverse long distances and impact the target essentially end-on to achieve maximum penetration performance. Therefore, the primary reduction in penetration with distance will be due to loss of velocity from air drag. The effect of drag on a typical EFP is shown in Fig. 8. A comparison of the penetration of a rod-shaped EFP and shaped charge is shown in Fig. 6.

The difference between the penetration performance of a shaped charge and an EFP is clearly exhibited in Fig. 6; however, an even more dramatic difference exists between the after-armor effects caused by each type of warhead. In

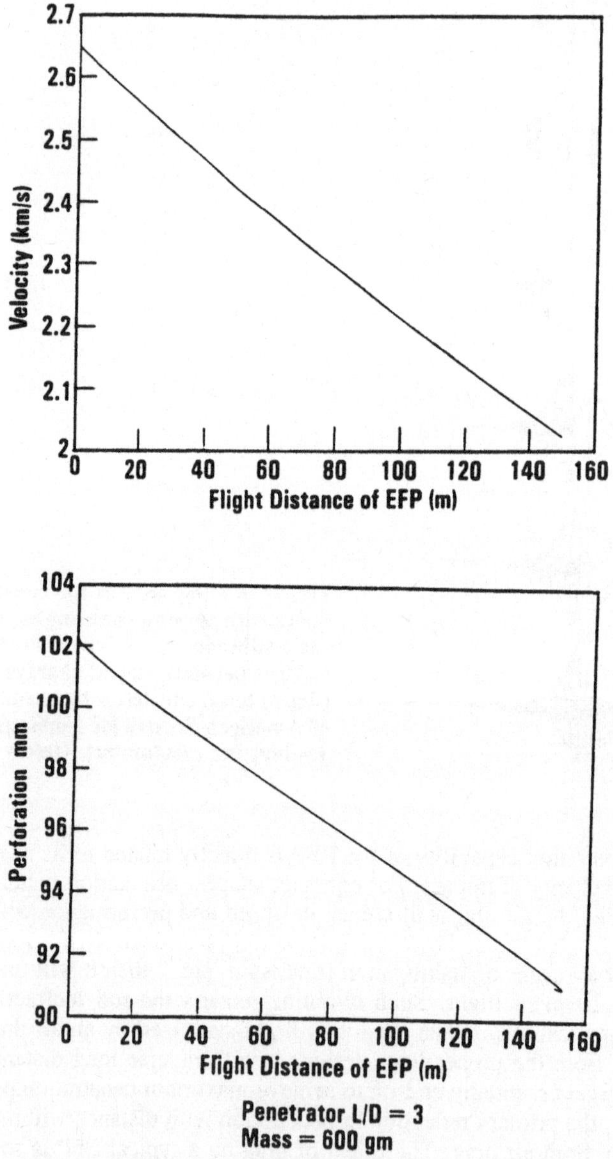

Fig. 8 Effect of drag on performance of a typical EFP.

Fig. 9 is shown flash radiographs of a jet and an EFP as they exit an armor target. The jet produces an expanding debris cloud of very fine target particles (of several grams each) surrounding the residual jet. The debris particles have very little penetrating power and are stopped by a 16-gauge steel plate. The primary behind-armor damage is created by the residual jet. On the other hand, very large armor debris fragments are created by the EFP as it perforates the armor. These fragments have significant penetration capability.

2.2 Omnidirectional Warheads

In contrast to the shaped charge and EFP warhead, fragmentation warheads possess omnidirectional capability. As depicted in Fig. 4, upon detonation, the explosive rapidly expands the warhead casing and propels a large number of fragments in a radially expanding cloud. Such fragments are accelerated to speeds of up to 1700 m/s depending on warhead design. There are two major objectives of fragmenting warhead designs: 1) control of the size of the fragments and 2) control of the velocity and pattern of the fragments. The first objective involves the selection of casing material and method of fragment control (notched casing, preformed fragment, etc.) The second objective involves the charge design (explosive mass-to-metal mass ratio, overall case configuration, etc.) and explosive initiation techniques (single point, line, multipoint, etc.). The different types of fragmenting warheads are distinguished by the method of

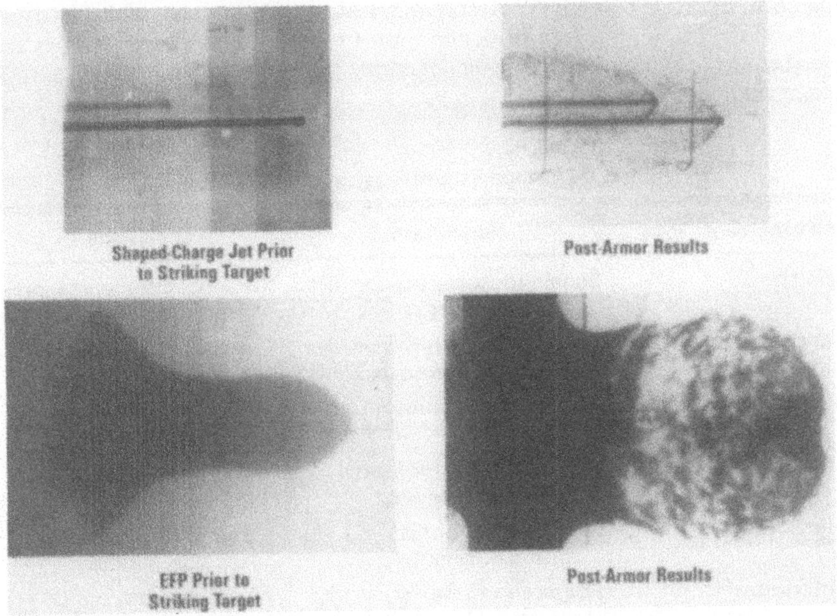

Fig. 9 Flash radiographs showing behind-armor effects caused by shaped charge and EFP perforation of armor targets.

fragment control and will be briefly reviewed here. The control of velocity and pattern involves specific designs and is beyond the scope of this introduction. Held gives an interesting overview of omnidirectional warheads.[7]

A simple metal tube or pipe filled with a high explosive will fracture into many fragments when the explosive is detonated. The distribution of fragment sizes and shapes will be a strong function of the material characteristics. If the material is brittle, very tiny fragments will be formed. If the material is very ductile, long strip fragments will be generated along with some smaller ones. This so-called *natural fragmentation* usually produces a mix of fragments that is not suitable for most applications. Therefore, various methods to control fragment size have evolved. The more common methods are listed in Table 1.

One of the most obvious approaches is to take our natural fragmentation case and put grooves in the wall to provide stress risers that cause the case to fracture at predetermined locations, thus controlling the size of the fragments. Both external and internal grooves have been investigated, with experimental data indicating that internal grooves are far superior in the ability to produce controlled fragment sizes. Also the geometrical configuration of the notch has been shown to be very important (see Ref. 8). A photograph of an internally grooved fragmentation casing is shown in Fig. 10.

Another approach is to use preformed fragments. In fact, the use of preformed fragments in the shape of balls, cubes, blocks and rods provides considerable latitude in the choice of the number and size of fragments. In practice, preformed fragments are either embedded in a matrix material such as plastic or sandwiched between two thin-walled shells. Both of these techniques provide lower structural rigidity. Because of early explosive gas leakage in the case of a matrix or parasitic weight in the case of the thin-walled tubes, lower fragment velocities are obtained for preformed fragment concepts compared with natural or grooved fragmenting casings.

Table 1 Comparison of fragmentation warheads

Fragmentation type	Advantages	Disadvantages
Natural	Simple structure, low cost, high structural rigidity	Non-optimum mix of large and small fragments
Internally grooved	Fragments of relatively similar size, good structural rigidity	Fragment size dependent on wall thickness
Metallurgical control	Same as natural but with higher cost	Complex manufacturing process
Preformed in matrix	Precise control of fragment masses and numbers	Weak structure, high cost
Insert control	Simple assembly, high structural rigidity	Fragment size dependent on wall thickness
Notched wire	Fragments of relatively similar size, suitable for automatic assembly	Weak structure, costly production technique

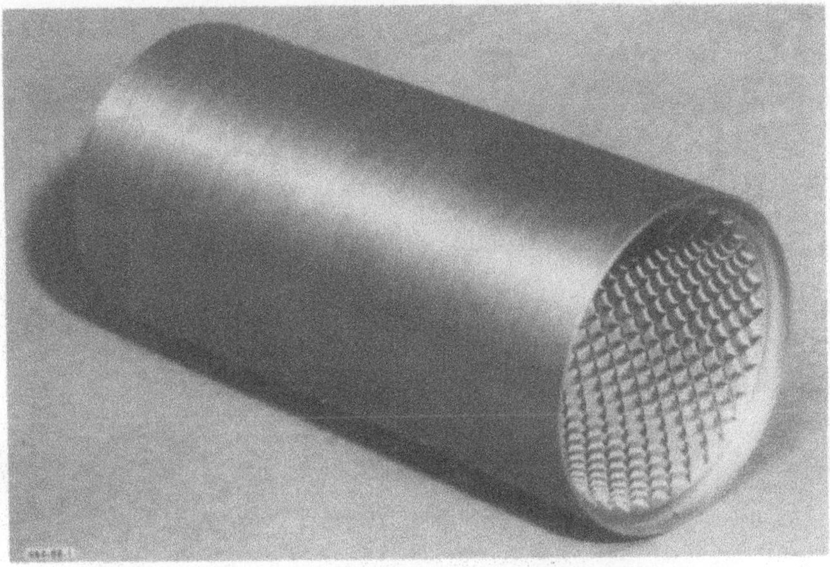

Fig. 10 Example of an internally grooved fragmentation casing.

Another technique for achieving controlled fragmentation is to use inserts in the explosive at the warhead casing-explosive interface. Two types of inserts have been investigated: individual, such as triangular or roof-shaped strips of material, and continuous, such as wire mesh. This technique is especially suitable where the strength of the fragmentation casing is important and is also applicable to materials that are unsuitable for grooving techniques.

Notched-wire techniques have also been used to make warhead cases. With one technique, a mandrel is spirally wrapped with a continuous length of notched wire, then brazed, to form a rigid structure. Another technique is to form rings or hoops of notched wire. Brazing of the individual rings forms a solid casing.

Of course, the various types of warhead defeat mechanisms can be easily combined. A fragmentation case on a shaped charge is one of the most obvious combined warheads. In Fig. 11 is shown such a design, the Combined Effects Bomblet (BLU-97/B), which is the payload for the Combined Effects Munition, an air-delivered dispenser weapon.

2.3 Modern Approach

The basic warhead mechanisms already discussed are certainly not new. Since 1970, however, the following three areas of research have dramatically affected the course of warhead technology development: 1) the development and use of computer simulation techniques for explosive events and terminal ballistics, 2) the characterization of materials in response to dynamic loads and the recognition of the importance of controlled metallurgical properties and precision

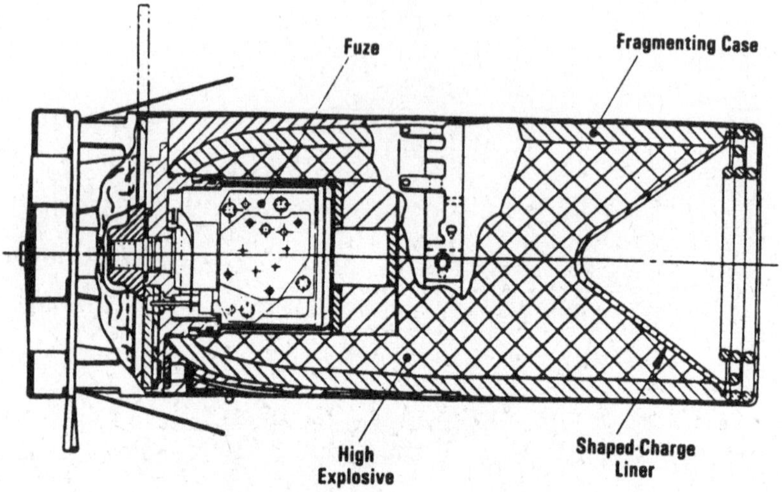

Fig. 11 Combined effects bomblet.

hardware, and 3) the use of higher energy and more homogeneous explosive compositions.

The use of two- and three-dimensional continuum mechanics computer codes, often called hydrocodes, has significantly influenced the development of modern warheads. These codes permit the simulation of warhead events that involve extreme pressures (on the order of 50 GPa) during an extremely short period of time (5–100 μs). In fact, hydrocodes have proven to be indispensable in warhead design (see Refs. 9 and 10). As an example, in Fig. 12 is shown a computer simulation of the formation of an EFP using the EPIC code. (The EPIC code was developed by G. Johnson;[11] however, various versions exist. This particular version is the one in use at Aerojet.) In addition to simulating the shape of the EFP, the code yields estimates of stress, strain, strain rate, and so on within the EFP as well as for the warhead case during formation. It also provides the pressure distribution within the explosive as a function of time and estimates how momentum is transferred from the expanding gases to the metal. Various codes are used for different aspects of the problem. In Fig. 13 is shown the sequence of an EFP perforating two plates. This simulation was conducted using the HULL code.[12] Of course, hydrocodes are applicable to the other warhead types as well, and many examples will be presented in the following chapters.

To use these computer-aided modeling techniques effectively requires an understanding of the response of the liners and casing under such highly dynamic rates. Moreover, the design process requires such an understanding to choose the proper grade of material for use in the warhead. For example, dynamic behavior of copper requires high purity and fine grain size to delay the breakup of stretching jets. These important aspects, both for computer design and materials

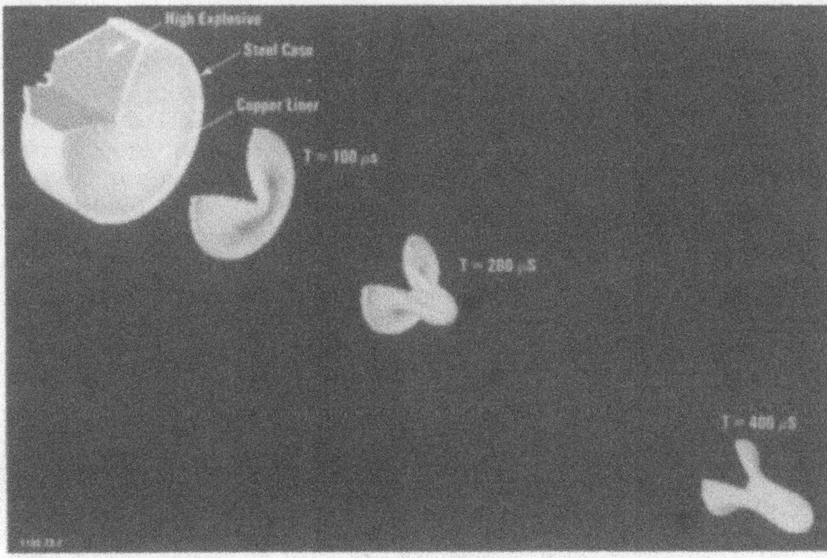

Fig. 12 Computer simulation of the formation of an explosively formed penetrator.

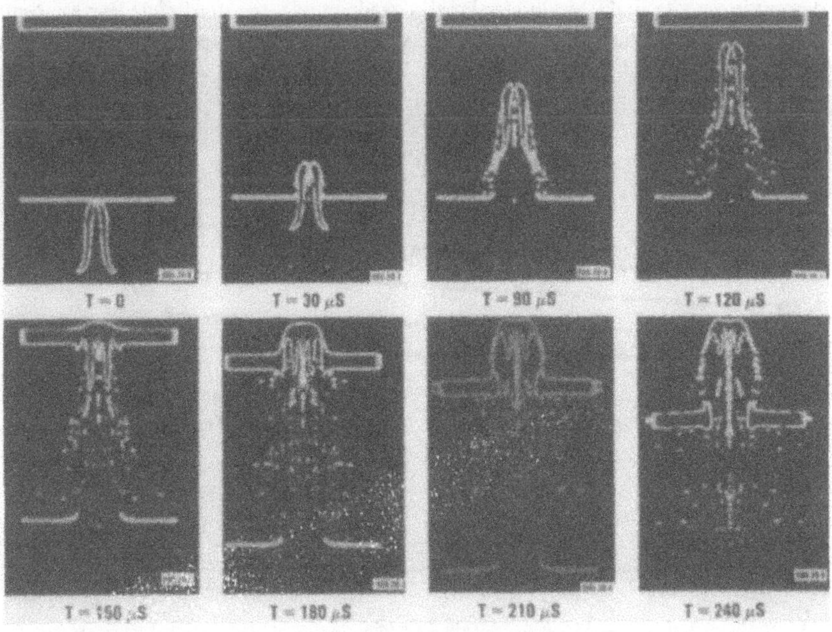

Fig. 13 Calculations of terminal ballistic effects of an EFP warhead.

selection, cannot be assessed by standard laboratory materials characterization tests. Since the variation in properties with rate of loading can be large, as evidenced by Fig. 14, which shows a sudden jump in flow stress for strain rates greater than 10^3 per second, several tests have evolved to measure dynamic materials response. These are summarized in Table 2. Data from these tests are used to develop constitutive materials models for use in hydrocodes. Until the onset of material failure, the current understanding is adequate. More research needs to be done to provide a satisfactory understanding of dynamic failure mechanisms, although significant inroads have been made in this area.

High quality control of materials and precision of manufacture are requirements that result from an understanding of the dynamic processes involved during warhead functioning. Whether the concern is the axially symmetric collapse of a shaped charge liner or the uniform fracture of a fragmenting case, imperfections in materials or fabrication become extremely amplified when compared to quasistatic applications. In more traditional applications, imperfections can be masked by other regions of the material since there is time for the system to come to equilibrium and statistical variances to balance out.

Such critical quality issues naturally carry over to the explosive. The use of TNT-based explosives is tending to be replaced by pressed or cast plastic bonded explosives for many reasons. Performance and reliability are two key reasons. For high-precision warheads, such explosive performance considerations as energy density, detonation pressure, and detailed structure at the front of the detonation wave all come into play. A critical variable for a smooth detonation wave is the grain size of the explosive in the warhead. For example, the median weight-average diameter of the HMX in 75% octol is approximately

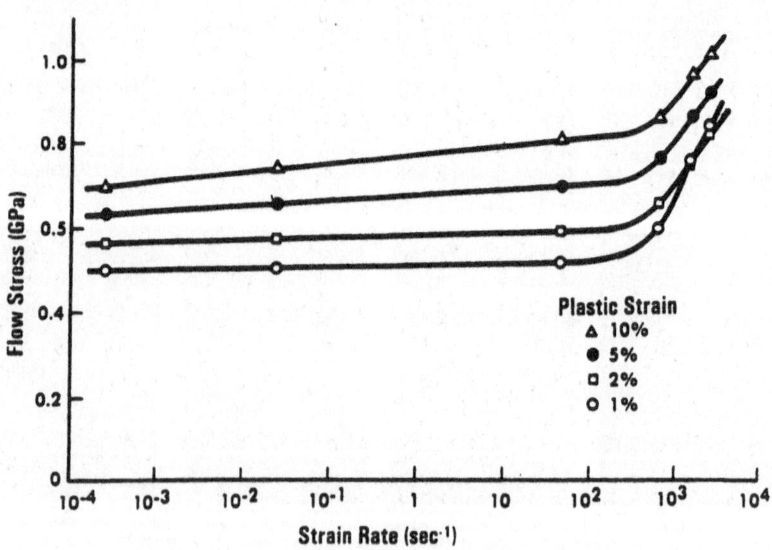

Fig. 14 Variation of flow stress with strain rate.

Table 2 High-strain-rate test techniques

Dynamic test	Strain	Strain rate	Pressure	Disadvantage
Hopkinson bar	Moderate	Moderate	Low	Unknown friction can induce errors
Torsion bar	Large	Low	Zero	
Taylor anvil	Large	Moderate	Moderate	Nonuniform loading in space and time
Plate impact	Small	Very high	High	Since elastic and plastic strains are comparable, it is necessary to know elastic response
Expanding ring	Moderate	High	Low	Sample material may be shock hardened

500 μm. Attention must also be paid to homogeneity of the explosives, and much work has been done to improve processing techniques. The trend toward insensitive munitions places further restrictions on high-performance warheads.

These three areas—simulation modeling, precision materials, and high-energy explosives—have had the most impact on modern warhead technology and in many cases will be the focus of this text. References to older work will of course be necessary to provide the foundation, but the goal is to present the latest improvements and directions for the future.

3. Applications

The selection of the type of warhead is governed by the desired target and the weapon system. For tactical weapons, the targets can be conveniently divided into three categories: heavy (primarily tanks), light (armored personnel carrier, armored artillery, etc.) and soft (aircraft, antiaircraft missile sites, trucks, etc.).

In Fig. 15 the typical warheads used for each target category are shown. Note that we have introduced another warhead type called *multilens*. This warhead is a hybrid of the fragmentation concept. However, instead of simple fragments, the surface of the warhead has either small shaped charge (or hemi charge) liners or EFP liners embedded in it. Thus, a spray or shotgun of jets or EFPs can be produced.

Application of various warhead types is indicated in Fig. 16. For each major type of warhead, specific examples of their use in weapon systems are given. Many more obviously exist, but only a few were selected. In addition, the overall warhead shape (*L*/*D*) is indicated for each type. Of course, these warheads may be scaled into almost any size. For comparison purposes, however, the more common sizes (not necessarily the size of the system examples stated) are listed along with the hardest applicable target for that size. Some of these systems are in production whereas others are in development. Details of the operation of each system are clearly beyond the scope of this introduction, and the interested reader is referred to the specific military agency for more details.

	Target		
Warhead	Heavy Armor	Light Armor	Soft
Shaped charge	X		
EFP	X	X	
Multilens		X	X
Fragmentation			X

Fig. 15 Warhead/target matrix.

	Warhead Type	Example(s)	Size/Shape: L/D	Size/Shape: Diameter, D (in.)	Hardest Applicable Target	Operational Standoff (CD)
SHAPED CHARGE	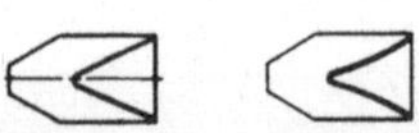 CONICAL LINER, TRUMPET LINER	CEB (BLU-97/B) WASP Maverick Rockeye TOW Dragon LAW (M72)	1.5 - 2.0	2 - 2.5 2.5 - 4 4 - 7	Heavy top Heavy side Heavy frontal	0 - 10
EFP		SFW SADARM	0.5 - 0.9	4 - 6 6 - 8	Heavy top Heavy side	10 - 1000
MULTILENS	MULTISLUG 	AMIS	1.4	5.4	Light	
MULTILENS	MULTIJET 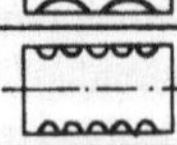	Roland BLU-85 Walleye Condor	~1.3	Various	Soft	Various
MULTILENS	FOCUSED MULTISLUG 	Defense Suppression Submunition	0.5 - 0.9	4 - 6	Soft	
CONTROLLED FRAGMENTATION	SCORED CASING 	CEB (cylinder) M-74 (sphere)	Various		Soft	Various
CONTROLLED FRAGMENTATION	PREFORMED FRAGMENTS  OTHER: NOTCHED WIRE MESH INSERTS	M-250 Chaparral	Various		Soft	

Fig. 16 Examples of warhead application.

References

[1]"Warheads—General," *Engineering Design Handbook,* Army Materiel Command, AMCP 706-290, AD 501 329, July 1964.

[2]"Design for Terminal Effects," Army Materiel Command, Ammunition Series, Section 2, AMCP 706-245, AD 389 304, July 1964.

[3]"Elements of Terminal Ballistics, Part One, Introduction, Kill Mechanisms and Vulnerability," Army Materiel Command, AMCP 706-160, AD 389 219, Nov. 1962.

[4]"Elements of Terminal Ballistics, Part Two, Collection and Analysis of Data Concerning Targets," Army Materiel Command, AMCP 706-161, Nov. 1962, AD 389 318.

[5]Held, M., "Flash X-Radiography in Ballistics," *Materials Evaluation,* Vol. 43, 1985, pp. 1104–1109.

[6]Carleone, J., and Johnson, G. R., "Multi-Slug Self-Forging Fragment Warheads," Proceedings of the First Self-Forging Fragment Conference, U.S. Army Ballistics Research Laboratories, Aberdeen Proving Ground, MD, Oct. 13–14, 1982.

[7]Held, M., "Air Target Warheads," *International Defense Review,* Vol. 5, No. 5, Oct. 1975, pp. 719–724.

[8]Pearson, J., and Finnegan, S. A., "The Shear-Control Method of Warhead Fragmentation," *Proceedings of Fourth International Symposium on Ballistics*, Monterey, CA, Oct. 17–19, 1978.

[9]Carleone, J., Chou, P. C., and Simpson, R., "A Synthesis Method for Modern Warhead Design," *Proceedings of Fourth International Symposium on Ballistics*, Monterey, CA, Oct. 17–19, 1978.

[10]Carleone, J., "Computer Simulation of Explosive Devices," *Aerojet Technology,* Vol. 2, No. 1, 1984, pp. 4–11.

[11]Johnson, G. R., "Analysis of Elastic-Plastic Impact Involving Severe Distortions," *Journal of Applied Mechanics, Transactions ASME,* Vol. 98, No. 3, Sept. 1976, pp. 439–444.

[12]Matuska, D. A., and Osborn, J. J., "HULL Documentation," Vols. I and II, Orlando Technology Incorporated, Shalimar, FL, Rev. Oct. 1987.

Chapter 2

Explosives

Erwin Anderson*
Naval Surface Weapons Center, White Oak, Silver Spring, Maryland 20903

Nomenclature

D = detonation velocity
D_i = detonation velocity of the ith component of a mixture
D_∞ = detonation velocity at infinite diameter
$D(d)$ = detonation velocity as a function of diameter, d
D(TMD) = detonation velocity at TMD conditions
$D(\rho)$ = detonation velocity as a function of density, ρ
d = charge diameter
d_c = critical diameter
d_f = failure diameter
l = length
P = detonation pressure
$P(\rho)$ = detonation pressure as a function of density, ρ
Q_D = detonation energy
Q_c = total chemical energy of decomposition
R = correlation coefficient of least-squares fit
r = charge radius
t = wall thickness
γ = adiabatic exponent; see Eq. (7)
ρ = density
ϕ = parameter in Kamlet-Jacobs equations; see Eq. (13)
ϕ_i = volume fraction of the ith component of a mixture

1. Explosive Materials

THIS section provides a review of explosive detonating materials with emphasis placed on Department of Defense (DOD) conventional explosives. Explosive materials developed by the Department of Energy (DOE) at either the Los Alamos National Laboratory (LANL) or the Lawrence Livermore National

* Research Physicist.

Laboratory (LLNL) are described in other works.[1–3] Furthermore, additional emphasis is placed on presenting information on newer explosive compositions developed in the last two decades.

The development of an explosive takes a minimum of three to five years from the initial concept to a point where the material has been sufficiently characterized to be considered, with minimal risk, for munitions use. Currently, a large number of high-explosive material concepts are being evaluated in basic research and exploratory development. Past experience suggests that only a small fraction of these material concepts will eventually be shown to possess all the characteristics required of military high explosives. The elapsed time from the selection of an explosive for use in a munition to the time of actual production of this explosive in this munition may be as long as three to eight years or more. The explosive materials considered in this section are confined to those materials that have been approved for military use or have reached a level of development where they have been selected for use in a weapon or have been assessed to be viable materials for future weapons. The widespread use of plastic-bonded explosives (PBXs), for example, cannot be fully appreciated without considering the numerous weapons systems, now under development, for which these materials have been selected.

1.1 High-Explosive Components of Explosives

1.1.1 Introduction

Most military explosives consist of mixtures of materials, at least one of which is a high explosive. These mixtures, or formulations, are used instead of pure high-explosive material to achieve one or more desirable properties that are not present in the pure high explosive. The most desirable property is the desensitization of the high-explosive component.

1.1.2 High-Explosive Molecules

The pure, single-compound, high-explosive materials that are important for military use are listed in Table 1, along with their theoretical maximum (or crystalline) density (TMD) and their chemical formulas, molecular weights, and heats of formation. The three most important materials on this list are TNT, RDX, and HMX; these materials are the most extensively used high explosives and are found in practically all U.S. munitions. TNT (MIL-T-248A) is available in chip or solid block form, and RDX (MIL-R-398C) and HMX (MIL-H-45444B) are manufactured as powders. (Note that material specifications will be given, if available, for the materials referred to in this chapter: MIL = Military specification; WS = Weapon specification; JAN = Joint Army-Navy.) These three materials are manufactured by the Army, the last two by the Holston Army Ammunitions Plant in Tennessee. RDX and HMX are produced in six and eight classes, respectively, according to their particle size distributions. HMX is produced in two grades depending upon the maximum allowable amounts of RDX as an impurity: Grade A, 7% RDX; Grade B, 2% RDX. Only Grade B HMX is used in PBXs.

TNT is the lowest-performing explosive of the three in terms of detonation velocity, or pressure. It is used in mass-produced weapons where low cost is the major selection criterion. Because it has a reasonably low melting point (80.8° to 80.9°C) that is well separated from its decomposition point (250° to 300°C), it can be safely melted and poured into weapons.[4] TNT undergoes a 10.5% loss in volume upon cooling from the liquid melt state (ρ = density = 1.48 g/cm^3) to the solid state (ρ = 1.654 g/cm^3). Special care must be taken to avoid the formation of cavities within TNT-based explosive charges during cooling. On a manufacturing scale, it is difficult to produce charges close to the theoretical maximum density of TNT. Typically, charges are produced at densities between 91% and 97% of their TMD.

When higher performance is required, either RDX or HMX is chosen. The performance of RDX or HMX is significantly higher than TNT, as is their cost. During the early to mid-1980s, TNT cost approximately $1.00/lb, RDX cost from $2.00 to $4.00/lb, and HMX cost from $7.00 to $13.00/lb, depending upon the class of the material. Although these three materials are produced at the multimillion-pounds-per-year level, there are significant differences in their production capacities. TNT is plentiful; RDX is readily available, except during the highest usage rates in wartime, when demand quite often exceeds the U.S. production capacity. During peacetime, the selection and use of HMX- containing explosives are reviewed to ensure that wartime production capacity will not be exceeded by the total of all HMX weapon needs. The use of RDX-containing explosives is encouraged, unless a definable need exists for higher explosive performance. RDX melts at 205°C with a violent, exothermic decomposition. The melting point of HMX is inconveniently high (285°C) for easy manufacture. More importantly, the HMX specification allows as much as 7% RDX in the HMX as a contaminant. Before the HMX melting point can be reached, the material will pass through the melting point and violent decomposition of the RDX contaminant. Also, both RDX and HMX are too sensitive to be used in the pure state as main charge or booster explosives without desensitization. For these reasons, RDX and HMX are formulated into mixtures with TNT or plastic binders, for example, to provide both desensitization and a material that is practical for production.

The other high-explosive molecules listed in Table 1 that have found military application in the past are DATB and Explosive D. DATB[10] has found application in a single main charge pressed explosive, PBXN-4, for use in missile warheads. Explosive D (JAN A-166A), or ammonium picrate, has been used since the early 1900s as a projectile fill.[11] It is an insensitive but low-performing explosive that is currently used only in the Navy's 16-inch projectile and in the picratol explosive formulation for the Bullpup warhead. DATB and Explosive D will probably continue to see very limited application in the future.

Nitroguanidine (NQ), TATB, and NTO are insensitive high-explosive molecules. NQ has been evaluated for some time in explosive formulations without much success.[12] It is extremely insensitive and usually requires the addition of a more sensitive material, such as RDX, in order to provide a reliably detonating material. Its relatively low explosive performance in explosive formulations has in the past eliminated NQ formulations for consideration in weapon applications. During the 1980s this material again underwent evaluation as a possible component of insensitive explosives. TATB (WS23158) is a high-density, extremely insensitive explosive with excellent cook-off properties.[13] Because of its extreme insensitivity, it has been called the "wooden explosive." It has found application in one DOE explosive, PBX-9502 (TATB/Kel-F 800 = 95/5), and one Navy booster explosive, PBXN-7. For most military applications, it requires addition of a sensitive material, such as RDX or HMX, to achieve a reliably detonating material. Its high cost ($34.00+/lb) limits its application to low-volume specialty explosives. NTO is a more recently developed insensitive molecule that underwent extensive evaluation in the late 1980s.[14] Its insensitivity, coupled with a projected raw material cost lying between that of RDX and HMX, makes it an attractive potential alternative to TATB.

PETN, tetryl, and HNS are booster explosive molecules. Their sensitivities are such that they usually are not used as components in main charge explosives. PETN (MIL-P-387A) is used alone in detonators, primers, and explosive fittings, and it has been formulated with appropriate desensitizing binders into flexible demolition explosives and into extruded explosives, such as PBXN-301. Tetryl (MIL-T-00339B) has been used, without desensitization, as a booster explosive. By 1970, it was the booster material used in most Navy weapons. During the early 1970s, the Army decided to discontinue tetryl production because it was a major source of environmental pollution, and continued production would have required a costly modernization program.[15] Other booster materials were then available at one-half to one-third the raw material cost. Tetryl has been phased out from consideration for use in all new weapon programs and has been replaced by the pressed PBX booster materials, such as PBXN-5, PBXN-6, and PBXN-7. HNS (WS-5003C) is a heat-resistant explosive with a sensitivity comparable to that of tetryl.[16,17] It is available in two classes: HNS-I, which is a superfine-particle-size material, and HNS-II, which possesses a relatively large particle size. It is used as a mild detonating fuze[18] and is applied in the crew module of the F-111 aircraft and in the Gemini and Apollo space systems.[19]

In addition to the above noted molecules that are components of main charge explosives, Table 1 also lists three primary explosive molecules. These materials are extremely sensitive to friction, electrostatic discharge, and impact. Lead azide (MIL-L-3055A) and lead styphnate (MIL-L-16355 and -17186) have been standard primary explosive materials in the past. Lead azide had been the primary explosive of choice and was used extensively in primers and detonators. Lead azide has the undesirable property of forming copper azide when in contact with copper or copper alloys in the presence of moisture.[20] Copper azide is much more sensitive than lead azide. As a result of a demilitarizing accident during demilitarizing (demilling) of lead azide–containing projectiles at McAlester AAP in 1971, which was attributed to initiation of copper azide that had formed within the charges,[21] lead azide was prohibited by the Navy to be in contact with copper and its alloys.[22] This policy led to the phasing out of lead azide use and the eventual development of DXW-1 (WS18967), which is the code name for mercuric 5-nitrotetrazole.[23] It is the current primary explosive of choice for future naval munitions and is expected to see fleet use within a few years.

Table 1 Explosive molecules

A. High-explosive molecules

Explosive	TMD g/cc	Chemical formula C	H	N	O	Molecular weight, g	ΔHf, kcal/mole	Ref.
DATB	1.837	6	5	5	6	243.1	−23.6	5,1
Explosive D	1.717	6	6	4	7	246	−94	5,1
HMX	1.905	4	8	8	8	296.2	17.93	5,1
HNS	1.740	14	6	6	12	450.3	18.7	5,1
NQ	1.810	1	4	4	2	104.1	−23.6	5,1
NTO	1.930	2	2	4	3	130	−14.3	5,6,7
PETN	1.780	5	8	4	12	316.2	−128.7	5,1
RDX	1.806	3	6	6	6	222.1	14.71	5,1
TATB	1.938	6	6	6	6	258.2	−36.85	5,1
Tetryl	1.730	7	5	5	8	287	4.67	5,1
TNT	1.654	7	5	3	6	227.1	−15	5,1

B. Primary explosive molecules

Primary explosive	TMD g/cc	Chemical formula C	H	N	O	Pb	Hg	Molecular weight, g	ΔHf, kcal/mole	Ref.
Lead azide	4.76/4.8	–	–	6	–	1	–	291	112	5,1
Lead styphnate	3.06	6	3	3	9	1	–	468	92.3	1
DXW-1	>3.15	2	–	10	4	–	1	428.6	–	5,8

C. Other molecular components of explosives

Primary explosive	TMD g/cc	Chemical formula H	N	O	Cl	Al	K	Molecular weight, g	ΔHf, kcal/mole	Ref.
Aluminum	2.702	–	–	–	–	1	–	26.98	0.00	8
Ammonium nitrate	1.725	4	2	3	–	–	–	80.05	−87.27	1
Ammonium perchlorate	1.95	4	1	4	1	–	–	117.49	−70.69	8
Potassium nitrate	2.109	–	1	3	–	–	1	101.10	−117.76	8

Other component molecules used in explosives are listed in part C of Table 1. Aluminum powder is used as a metal fuel to increase the total energy of the explosive. This use of aluminum in explosives was first proposed in the nineteenth century (see Table 2 for other explosive development dates). Aluminum powder is available in a number of grades and particle sizes ranging from a flake "bomb" grade (MIL-A-512) to small (< 10 μm) spherical particles (MIL-A-23950). PBXs use spherical or near-spherical grades of aluminum powder with particle sizes usually below 25 μm. The use of ammonium perchlorate (AP) is limited to formulations that have been optimized for underwater applications. It is used as an oxidizer in conjunction with aluminum as a metal fuel. Ammonium

Table 2 Milestones in explosives development

Date	Milestone
1841	Ammonium picrate (Explosive D) first prepared by Marchand
1863	TNT first prepared
1877	Tetryl first prepared by Mertens
1888	Ammonium picrate patented as an explosive by Nobel
1899	RDX first prepared by Henning
	Use of aluminium in explosives to increase power proposed by Escales (German patent—1900; U.S. patent–1911)
1901	Explosive D used by U.S. Army as an explosive fill for projectiles
1902	Germans adopt TNT as standard filling for projectiles
1907	Explosive D adopted by U.S. Navy as projectile fill (dual charge of Explosive D/black powder)
1914	Tetryl first used as an explosive
1918	German studies indicate 15% to 20% aluminum addition to explosives increases underwater power
ca 1920	von Herz discovers RDX value as an explosive
1920s to 1930s	Composition B developed by British
1936	Germans adopt Trialene (RDX/Al/TNT) explosives
1939	RDX first used on large scale
early 1940s	Composition B used by U.S.
early 1940s	Composition A (RDX/beeswax = 91/9) used by British
1941	British use of Torpex (RDX/TNT/aluminum)
1942	Torpex-II in use by U.S. Navy
1943–1944	Nine accidents involving Torpex-II
1945	Composition A-3 adopted by U.S. Navy for use in 3-in./50-cal projectiles
1945	Tritonal (TNT/aluminum-=-80/20) used in U.S.
June 1945	HBX-1 first loaded into production weapons
late 1940s	HBXs (RDX/TNT/aluminum/wax) developed
1952	Early PBX (RDX/polystyrene/DOP) developed
1955	H-6 adopted for use in Terrier missile warhead
1956	H-6 adopted by Navy as fill for Mk 80 series bombs

nitrate (AN) can also be considered an oxidizer, but principal use in military explosives has been as a low-cost, readily available, but lower-performance substitute for RDX. It possesses the undesirable property of undergoing irreversible growth under temperature cycling, which reduces AN-containing explosives to applications where very short shelf lives are acceptable. Fertilizer-grade ammonium nitrate has been involved in some of the largest peacetime accidental explosions.[28,29] Potassium nitrate (KN) has been used in mixtures with AN to reduce the severity of the AN irreversible growth.

These 18 molecules, almost all of which are used in powdered form, provide either the detonating material or additional sources of energy for practically all U.S. military explosives. Their performance, sensitivity, and other properties, along with their use in compositions, are discussed in the following sections.

1.2 Explosive Compositions

With a few exceptions, most of the 18 explosive molecules, especially the important ones like RDX or HMX, cannot be used directly in powdered form as booster or main charge explosives, principally because many of the powders, even when compacted to high density, are too sensitive for military use. A secondary reason is that those powders having acceptable sensitivity levels when compacted do not always possess the physical characteristics that would permit them to be pressed into stable, high-quality charges that maintain their physical integrity over long periods of time.

TNT, Explosive D, HNS, and tetryl are materials that, on the other hand, have all the properties that permit them to be used directly. TNT, in powdered or chip form, can be melted and resolidified upon cooling into a continuous crystalline solid charge that is relatively insensitive for many applications and possesses sufficient physical integrity that it can be used in warheads. Explosive D provides an insensitive, high-quality charge when pressed into projectiles. Both HNS and tetryl can be pressed into charges that have acceptable sensitivity and physical properties to be used as booster materials. The primary explosives lead styphnate and DXW-1 can be used in the pure form when pressed.

The other high-explosive powders must be utilized in conjunction with another material, which provides desensitization and/or provides the combined material with physical properties that permit the mixture to form a solid explosive charge. These mixtures are called explosive compositions or formulations. This section discusses the types of explosive formulations, their nomenclature and composition, and their use.

1.2.1 Types of Explosive Formulations

There are numerous ways to categorize explosive formulations:

1) Types by general warhead function
2) Types by manufacturing procedure
3) Types by warhead effects
4) Other classifications

These categories are not mutually exclusive. Most explosive formulations can be categorized by all four categories. For example, PBXN-103 is a nonideal (#4), castable (#2), main charge (#1), underwater (#3) explosive.

Types of explosive formulations based upon their general warhead function. All explosives can be divided into two very general categories: 1) primary explosives and 2) secondary explosives. Military secondary explosives can be further subdivided into other classes: 1) booster explosives and 2) main charge explosives. Primary explosives are extremely sensitive within well-defined limits and are selected to be so. The secondary explosives are substantially less sensitive than primary explosives.

Primary explosives are sensitive formulations or compounds, such as lead azide or lead styphnate, that are used to initiate detonation in high explosives. They are sensitive to heat, impact, and friction and undergo a rapid reaction upon initiation. These sensitive explosives are separated from the booster explosive by the interrupter of the warhead's fuze, exploder, or safety and arming device.[20] Primary explosives are used in detonators and in explosive bolts, valves, and cutters.

A booster explosive is an explosive acceptable for fuze components whose detonation would normally be communicated to the main charge explosives of a fuzed weapon when the fuze is in either the armed or unarmed condition. This class includes explosives used in leads, relays, detonating cords, boosters, and other components on the main charge side of the interrupter.[30] Booster materials are substantially less sensitive to shock, impact, friction, and electrostatics than are primary explosives.

Main charge explosives are compounds or formulations such as TNT, Composition B, or PBXN-106, which are used as the final charge in any explosive application. These explosives, because of their insensitivity, ordinarily require initiation by a booster explosive.[30] They are less sensitive to shock, impact, friction, and electrostatics than are booster explosives. The main charge explosive is the explosive that causes the damage from weapons such as bombs, rockets, projectiles, torpedoes, and mines.

The interrelationship of these three types of explosives is presented in a very practical and simplified fashion as follows. A properly designed armed warhead will detonate when it perceives it is on target. The fuze system of the warhead determines the presence of a target by some predesigned means, such as by simple target impact or by a sophisticated electromagnetic or sonar target determination. This perception of being on target must be rapidly, safely, and reliably transmitted to the main charge for it to detonate and cause target damage. The fuze system signals the initiation system to initiate the detonator, for example, by capacitor discharge through a wire. The prior heating of the wire causes the primary explosive to initiate and build to a detonation. The output shock from the detonator is then transmitted to a booster explosive that is able to take the relatively weak output shock and build it to a stronger shock sufficient to initiate the main charge. This transmission of the shock signal from the detonator to the main charge may be accomplished, depending on the design and sophistication of the warhead, by a booster material in the form of a simple right cylindrical pellet, or a more sophisticated design requiring channels of booster explosives, called leads, or even a complex explosive logic circuit.

The output shock of the booster is then transmitted to the main charge. If it is of sufficient strength, this shock will initiate the detonation within the main charge, and thus the warhead will function as designed. The resultant damage from the warhead is caused by the extremely rapid release of energy and gas from the main charge. For safety reasons, the sensitive primary and booster explosive components of the warhead are kept to the minimum weight that allows the component to function. Table 3 lists the three explosive types, their weight range as found in U.S. Navy weapons, and the more typical range of weight values found in modern weapons. In general, the weights of the components differ by three orders of magnitude, from primary to booster to main charge.

Types of explosive formulations based upon their method of manufacture. Explosives are quite often categorized by the method of manufacture. There are four major methods of manufacture:

1) Pressing
2) Casting
3) Extruding
4) Injection molding

In terms of volume of production, the majority of explosives are cast (or castable) explosives, followed by pressed explosives. Combined, these two types constitute almost all explosive production. Extrudable and injection moldable explosives are specialty explosives that have a very limited production volume.

Types of explosive formulations based upon the warhead effect they produce. Another way to label main charge explosives is to categorize them according to the major warhead effects that they produce or, almost equivalently, according to the type of warhead in which they are primarily used. Most warhead explosives are designed or optimized to produce the maximum output for a specific warhead effect. Each warhead or explosion effect has its own specific design criteria for producing maximum output.

The major warhead or explosion effects are as follows:

1) Metal-accelerating or brisant
   a) Fragmentation
   b) Shaped charge
2) Air-blast
3) General purpose (GP)
4) Underwater
5) Internal blast
6) Cratering

A brisant or metal-accelerating explosive is one that has been optimized to produce a high detonation pressure. When in intimate contact with metal, these explosive materials, upon detonation, will accelerate metal to high velocities. They are applied in warheads, with fragmenting missile designs; they are also

Table 3 Approximate weights of warhead charges

Explosive type	Weight range	Typical range
Primary	100 mg–4.2 g	100–500 mg
Booster	5 g–14.5 kg	50–200 g
Main charge	0.1 kg–600 kg	0.1–600 kg

applied in shaped charge warheads, which require an extremely rapid collapse of the shaped charge liner to produce maximum target penetration. Air-blast explosives are optimized to produce the maximum peak pressure and impulse in air over the largest possible distance from the explosive charge. Air-blast capability is of considerable interest for bomb-type weapons but is rarely the sole mechanism used for producing warhead damage. It is usually coupled with another damage mechanism, such as fragmentation capability, to produce a more general purpose (GP), multiple damage mechanism warhead. The major use of GP explosives is in the Mk 80 series bombs. In this application, the damage is caused by the high-velocity fragments produced by the breakup of a thick steel wall bombcase and the subsequent peak pressure and impulse in air produced by the rapid expansion of the hot explosive decomposition gas products. Achieving a general purpose explosive formulation usually requires a compromise in the optimization of the various damage mechanisms. Underwater explosives are compositions that are optimized to produce the maximum underwater shockwave and bubble energies. They are used in torpedoes, underwater mines, and antimine and depth charges. Although they have been under consideration during the last two decades as damage mechanisms, there have been no well-developed explosives to date that are optimized for internal blast or cratering. The internal blast effect refers to the generation of relatively long-duration overpressures in confined spaces, such as bunkers or ship compartments, produced by the detonation of an explosive charge and the subsequent combustion, in the air of the chamber, of the explosive's detonation products and combustible warhead materials. Cratering refers to an explosive's ability to produce large-diameter depth or volume craters in soil and rock.

Other classifications. Along with the above ways of classifying explosives, other classification systems are occasionally employed, such as describing materials by their detonation or sensitivity properties. For example, based on their detonation properties, explosives can be classified as either ideal or nonideal. Price[31] has classified compositions within a limited number of explosive types depending upon their shock sensitivity characteristics.

1.2.2 Explosives Nomenclature

A. TNT-based and other classical explosives. A number of names are used to describe classes of the explosives materials. The cyclotols are mixtures of RDX and TNT in various ratios and are usually expressed in the form X/Y cyclotol, where X is the weight percent of RDX and Y is the weight percent of TNT ($X + Y = 100\%$). An example is 75/25 cyclotol, which consists of 75% RDX and 25% TNT by weight. Composition B (Comp B) is a desensitized (by 1% wax) 60/40 cyclotol. In a similar fashion, the octols are mixtures of HMX and TNT. An example is 75/25 octol, which is a mixture of 75% HMX and 25% TNT by weight. Torpexes are mixtures of RDX, TNT, and aluminum, or equivalently, they are aluminized cyclotols. Use of torpexes is forbidden within the U.S. Navy because of the accidents that occurred during World War II (Table 2). For this reason, the torpexes were modified by the addition of approximately 5 parts of mixture of wax and Composition D-2 desensitizer to 100 parts torpex explosive.[27,32] These desensitized explosives are known as the HBXs (high blast explosives). The three HBXs approved for military use are HBX-1, HBX-3, and H-6. In the United Kingdom, the "torpex" terminology is still used and applies to both the undesensitized and desensitized aluminized cyclotols. Minols are mixtures of TNT, aluminum, and ammonium nitrate.

Since the early part of World War II, the United States and UK adopted the nomenclature "Composition X" to describe a series of important explosives:

1) Composition A—pressed mixtures of RDX and wax or wax substitutes: examples are Comp A-3 (RDX/wax = 91/9) and Comp A-5 (RDX/stearic acid = 99/1).

2) Composition B—RDX/TNT/wax = 60/40/1.

3) Composition C—RDX/binder; examples are Comp C-3 and Comp C-4.

4) Composition D—inert desensitizers for explosives; the only currently used example is Composition D-2 (wax/nitrocellulose/lecithin = 84/14/2).

Informal usage permits the shortening of the above terminology; Composition A-3, for example, is usually denoted Comp A-3 or sometimes just A-3.

B. PBX nomenclature. Each of the major developers of new explosive compositions has its own system for naming or identifying explosive compositions. The purposes for these systems are 1) security reasons and 2) the establishment of a formal, recognizable method of identifying the status of an explosive. Many of these systems involve a code-word method of naming explosives. The intention of the these code words within the DOD was to reduce access to recently developed explosive technology. These code-word systems can be used to establish within an organization a method for readily identifying the status of the explosive— experimental, qualified, production, and so on. A number of organizations, including the U.S. Navy, the Lawrence Livermore National Laboratory (LLNL), and the Los Alamos National Scientific Laboratory (LANSL), use this approach to distinguish between experimental explosives and explosives that have been approved for production or weapon use.

The U.S. Navy has formally adopted the following nomenclature for plastic bonded explosives (PBXs): the capital letters PBX are followed by a single capital letter designating the status of the explosive, whether experimental or approved for fleet use, and then a dash, followed by a number between 1 and 399. If the material is experimental, the capital letter after PBX identifies the organization responsible for its development and is either W for White Oak, C for China Lake, or AF for the Air Force. Furthermore, if the experimental explosive has received preliminary safety approval, called "Qualification," for use outside the explosive research facilities at White Oak and China Lake, then the number is followed by the letter Q in parentheses. If the PBX has been formally approved for Navy use within a specific weapon, then the letter following PBX is changed to the letter N. A specific number is assigned to each PBX depending on its method of manufacture:

1 to 99 = pressed PBXs

100 to 199 = castable PBXs

200 to 299 = extrudable PBXs

300 to 399 = injection moldable PBXs

These nomenclature rules are summarized in Table 4.

Table 4 U.S. Navy PBX nomenclature

P	=	plastic
B	=	bonded
X	=	explosive
C	=	experimental; China Lake NWC/CL developed explosive
W	=	experimental; White Oak NSWC/WO developed explosive
AF	=	Air Force developed explosive
(Q)	=	qualified
(I)	=	interim qualified; now called "Qualified" (Q); obsolete terminology
N	=	final qualified in Navy weapon
1–99	=	pressable
100–199	=	castable
200–299	=	extrudable
300–399	=	injection moldable

The application of these rules is illustrated by the following explosive designations.

1) PBXW-113
2) PBXW-113 (Q)
3) PBXN-110
4) PBXW-100
5) PBXN-103
6) PBXC-117 (Q)
7) PBXW-7
8) PBXN-5
9) PBX(AF)-108

PBXW-113 is the fourteenth castable PBX developed at NSWC, White Oak. PBXW-113(Q) is the designation for the same explosive after it received qualification status from the Navy. PBXN-110 is the designation for the same explosive after it received approval for fleet use in a Navy warhead. Similarly, PBXW-100 was the first castable PBX developed at White Oak, which later received the PBXN-103 designation when approved for fleet use. When an experimental PBX reaches a certain level of development, it is assigned the next available number for its manufacturing category at either White Oak or China Lake, which maintain independent lists of PBX numbers. If that PBX is later approved for fleet use, it receives the next available PBX number, in that manufacturing category, from the Navy's PBX number list. Thus, in most cases, because of the independent PBX number lists, the number assigned to a given PBX will change when it receives approval for fleet use. This is illustrated by number changes for the PBXW-113/PBXN-110 and PBXW-100/PBXN-103 explosives. PBXC-117(Q) is a qualified experimental castable PBX developed at China Lake. PBXW-7(Q) is a qualified experimental pressed PBX developed at White Oak. PBXN-5 is the fifth pressed PBX approved for fleet use. PBX(AF)-108 is an Air Force–developed, experimental PBX.

The LANSL uses an explosives nomenclature system in which experimental explosives are given a designation consisting of the letter X, followed by a dash and a four-digit number (e.g., X-0008, X-0280). Production explosives are given a designation consisting of the letters PBX, followed by a dash and then a four-digit number. The first two digits refer to the percent mass of the major high-explosive ingredient. One example is PBX-9404, which consists of 94% of the high-explosive ingredient HMX.

At the LLNL, a number code is applied to all experimental research explosive formulations produced in amounts of more than 50 grams. The code consists of the letters RX followed by a dash, then two digits, another dash, and two capital alphabetic characters (e.g., RX-01-AA and RX-13-XD). The two digits (01 and 13 in the examples), assigned in arbitrary sequence, define a general class of explosives. For example, RX-01 refers to nitromethane liquid formulations; RX-02 refers to PETN extrudable formulations. The final two letters, assigned in arbitrary sequence (AA, AB, etc.), refer to a specific formulation within the general class. For those explosives under consideration for weapon application and production, their codes change to an LX number when 1) a material specification is written, 2) the material's chemical, physical, and explosive property evaluation is complete, and 3) the material has some definite application. The LX number code consists of the two letters LX followed by a dash and the two digits, a second dash, and a single digit, e.g., LX-04-0 and LX-04-1. The first pair of digits is an arbitrary serial number, assigned in sequence. The last digit denotes a subclass of the explosive and provides for small changes in the specifications (different particle size materials used, for example).

The U.S. Air Force has adopted a nomenclature system that differs from the above systems for its recently developed explosives. This system consists of the capital letters AFX followed by a dash, then a three- or four-digit number. The number assigned to a given explosive depends upon the major ingredients of the composition. These major numerical subdivisions are listed in Table 5. For example, AFX-108 is a castable PBX consisting of RDX and an inert binder.

Table 5 Air Force explosives nomenclature

AFX-000[a] series	Explosive material
100	Castable composition whose main ingredient is RDX
200	Castable composition whose main ingredient is HMX
400	EAK compositions
500	Compositions containing TPM
700	Castable, aluminized RDX compositions
800	Castable, aluminized HMX compositions
900	Composition whose main ingredient is NQ
1000	Foamed explosives
1100	TNT-based, aluminized explosives

[a]AFX = Air Force explosive.

From Table 5, it can be seen that this nomenclature system also applies to TNT-based explosives (1100 series). From the nomenclature, there is no explicit way of determining whether the explosive has been approved for weapons use.

1.2.3 U.S. Military Explosive Compositions

A. Introduction. Most of the secondary high-explosive molecules listed in Table 1 cannot be used directly in warheads without some additional ingredients. A few of these molecules can be used in the pure state as booster materials. TNT and Explosive D can be used safely in the pure state as main charge explosives for certain warhead applications. The others, especially the most important and most powerful ones, HMX and RDX, must be diluted or desensitized to achieve a safe and reliable explosive material. Adding other materials to the high-explosive powders is called explosive formulating. The result of this is an explosive formulation or composition.

The minimum requirements placed upon secondary explosive compositions are listed in Table 6. These requirements significantly restrict the choice of component materials that can be used in explosives. An explosive must be insensitive over its total life cycle. If it is not insensitive as is, then it must be in a desensitized form throughout its life cycle from component ingredient manufacture, to explosive mixing and loading, and through its use in weapons, whether in storage or combat. Since almost all manufacturing processes involve the movement of component materials, explosive materials also must have the ability to flow during their manufacture. Except for the few injection moldable and extruded explosives, this usually means the explosive material must flow readily and reliably until it is loaded into a weapon. The two major manufacturing classes of explosives meet this requirement by being relatively viscous, but flowable, fluids, as in the case of castable explosives, or by being insensitive, flowing molding powders, as in the case of pressed explosives. With the exception of some specialty explosives, such as demolition explosives, explosives are loaded into some type of container. This may be a simple small cup for a booster case or a 1200-lb steel bomb case. After being loaded into its final container, the explosive fill is expected to maintain its physical and chemical form basically unchanged for the full life cycle of the item. Finally, since it is an explosive, by definition, it must detonate.

Table 6 Minimum requirements for secondary explosives

1. The explosive must be insensitive over its life cycle. Therefore, it must either:
    1. Be insensitive as is (for a specific application), or
    2. Be desensitized.
2. The explosive material must flow during manufacture.
3. In its end-use form, the explosive must maintain its physical integrity over its life cycle.
4. It must detonate.

These are minimum requirements for secondary explosives. The requirements beyond the minimum that have been placed on military explosives have evolved over the years and will continue to do so. Some of the current characteristics that must be possessed by new explosive compositions include 1) ease of manufacture, 2) chemical thermal stability and physical integrity of the explosive charge over a temperature range from −54°C to 71°C (the "military temperature range") for a minimum of 10 years, 3) the ability to function immediately (in some cases, within a few seconds or less) in a warhead, 4) insensitivity, with good vulnerability characteristics, and 5) high-performance characteristics.

Military explosives differ considerably from commercial explosives since they must meet much more stringent requirements on their properties and characteristics. These stringent requirements have precluded consideration of otherwise attractive low-cost commercial explosives. In a similar fashion, the huge list of potential explosive ingredients is reduced because of these restrictions to a relatively small number of materials that actually go into production.

Although a higher level of performance is quite often the goal of the developers of new explosive compositions, a review of the history of explosives and their use distinctly shows that almost all major explosive compositional developments were driven by the need for new materials that had the potential to avoid the repeated occurrences of prior accidents. One example is the development of the desensitized HBXs from the sensitive torpexes as a result of the major accidents involving the latter materials during World War II (Table 2).

The next two sections discuss the major explosive compositions being used or under consideration for use by the three services. The two sections reflect the significant differences between the older TNT-based and other "classical"-type explosive compositions and the more recently developed PBX technology. The initial acceptance of this newer and, in many cases, lower-performance PBX technology was again motivated by a history of accidents (within the U.S. Navy). Its continued acceptance is based on the desire to acquire insensitive explosives to reduce both accidents and ordnance vulnerability to enemy attack.

B. TNT-based and other classical explosives. The major TNT-based explosives and other "classical" explosives are listed in Table 7, along with their compositions by weight percent. They are grouped by their manufacturing technique. Part A lists the TNT-based melt-cast explosives, part B lists the pressed explosives. Table 8 gives the warhead applications of these materials, along with the specifications under which they are purchased, the type of explosive by warhead function and warhead effect, and the qualification status.

Until the development of the PBXs, TNT was the material of choice for all castable explosives. It melts at a reasonably low temperature (~ 80°C), producing a low-viscosity fluid that can be filled with other, higher-performing explosive powders, such as RDX or HMX. Because it is chemically compatible with a large number of materials, it can be filled with a variety of other solids, including aluminum, PETN, ammonium and potassium nitrates, and Explosive D, to provide a spectrum of compositions with a variety of properties. TNT can be filled to high levels with these solid powders resulting in a mixture with a viscosity low enough to be readily processible.

Table 7 Composition of TNT-based, pressed, and other classical explosives

A. Composition of TNT-based explosives

Explosive	% HMX	% RDX	% TNT	% Al	% D-2
Comp B	—	59.50	39.50	—	—
Cyclotol 75/25	—	75.00	25.00	—	—
Cyclotol Type I	—	25.00	75.00	—	—
Cyclotol Type II	—	29.00	71.00	—	—
Destex	—	—	74.70	18.70	4.70
H-6	—	45.10	29.20	21.00	4.70
HBX-1	—	40.40	37.80	17.10	4.70
HBX-3	—	31.30	29.00	34.80	4.90
HTA-3	49.00	—	29.00	22.00	—
Minol II	—	—	40.00	20.00	—
Minol IV(Q)	—	—	40.00	20.00	—
Octol 70/30	70.00	—	30.00	—	—
Octol 75/25	75.00	—	25.00	—	—
Pentolite	—	—	50.00	—	—
Picratol	—	—	48.00	—	—
TNT	—	—	100.00	—	—
Tritonal	—	—	80.00	20.00	—

B. Composition of pressed explosives

Explosive	% Pure molecule	% RDX	% Wax	% Al	% Calcium stearate
CH-6	—	97.50	—	—	1.50
Comp A-3	—	91.00	9.00	—	—
Comp A-3 (Aluminized)	—	63.70	6.30	30.00	—
Comp A-3 (Type I)(Q)	—	91.00	9.00	—	—
Comp A-4(Q)	—	97.00	3.00	—	—
Comp A-5	—	99.00	—	—	—
Comp A-5 (Type I)(Q)	—	98.50	—	—	—
Explosive D	100.00	—	—	—	—
Tetryl	100.00	—	—	—	—

PIB = polyisobutylene PE = polyethylene The compositions are listed as found in Ref. 33.

When loaded with RDX or HMX, these mixtures are the high-detonation-performance, brisant cyclotols or octols, respectively, that are used in metal-accelerating applications, such as rockets, missile warheads, and shaped charges. When aluminum powder is added to the cyclotols, the result is the HBX compositions, H-6, HBX-1, and HBX-3. The last two compositions are optimized to produce the maximum underwater damage and were used in Navy underwater mines and torpedoes. H-6 and tritonal, the aluminized TNT composition, are general purpose compositions with moderately good performance in a number of damage modes, including fragmentation and air-blast. Because of their use in Navy and Air Force bombs they are among the explosives with the largest production volume. Destex is a desensitized tritonal. HTA-3 is an aluminized

Table 7 *(continued)*

% Carbon black	% PETN	% AN	% Explosive D	% KN	% Wax
—	—	—	—	—	1.00
—	—	—	—	—	—
—	—	—	—	—	—
—	—	—	—	—	—
1.90	—	—	—	—	—
—	—	—	—	—	—
—	—	—	—	—	—
—	—	—	—	—	—
—	—	—	—	—	—
—	—	40.00	—	—	—
—	—	36.00	—	4.00	—
—	—	—	—	—	—
—	—	—	—	—	—
—	50.00	—	—	—	—
—	—	—	52.00	—	—
—	—	—	—	—	—
—	—	—	—	—	—

% Graphite	% PIB	% Stearic acid	% Barium stearate
0.50	0.50	—	—
—	—	—	—
—	—	—	—
—	—	—	0.50
—	—	—	—
—	—	1.00	—
—	—	1.50	—
—	—	—	—
—	—	—	—

octol that is used in applications where a blast capability is required along with some brisance. Picratol is an explosive used in applications that primarily require insensitivity, such as penetrating warheads, and can accept low levels of brisance. All the above explosives are employed as main charge fills. Pentolite, on the other hand, is considered a booster explosive by the Navy because of its sensitivity, which is due to use of the very sensitive PETN. It is considered too sensitive for use as a booster in Navy weapons, but it is safe enough to be used by explosives experts in experimental test charges. For this reason, it is quite often used, even in relatively large charges (> 50 lbs), in laboratory research programs as a performance standard (e.g., in underwater performance testing).

The compositions of pressed explosives are listed in Part B of Table 7, and their applications in Part B of Table 8. Explosive D and tetryl are insensitive

Table 8 Applications of TNT-based and pressed explosives

Explosive	Specification	Explosive type: By warhead function	Explosive type: By warhead effect
A. TNT-based explosives			
Comp B	MIL-E-401B	Main charge	Brisant
Cyclotol 75/25	MIL-C-13477B	Main charge	Brisant
Cyclotol Type I	WS 1724A	Main charge	Brisant
Cyclotol Type II	WS 1724A	Main charge	Brisant
Destex	WS 16221	Main charge	Blast
H-6	MIL-E-22267A	Main charge	GP; U/W
HBX-1	MIL-E-22267A	Main charge	U/W
HBX-3	MIL-E-22267A	Main charge	U/W
HTA-3	MIL-E-46495A	Main charge	Moderate brisant; blast
Minol II	MIL-M-14745	Main charge	GP
Minol IV(Q)		Main charge	GP
Octol 70/30	MIL-O-45445A	Main charge	Brisant
Octol 75/25	MIL-O-45445A	Main charge	Brisant
Pentolite		Booster	Brisant; booster
Picratol	JAN-A-166	Main charge	Low brisant, insensitive
TNT	MIL-T-248A	Main charge	Low brisant, insensitive
Tritonal		Main charge	GP
B. Pressed Explosives			
CH-6	MIL-R-21723	Booster	
Comp A-3	MIL-C-440B	Main charge	Brisant
Comp A-3 (Aluminized)		Main charge	Blast
Comp A-3 (Type I)(Q)		Booster	
Comp A-4(Q)		Booster	
Comp A-5	MIL-E-14970A	A—Main charge/ booster N—Booster	Brisant
Comp A-5 (Class I)(Q)	MIL-E-14970A	Booster	
Explosive D	JAN-A-166A	Main charge	Low brisant
Tetryl	MIL-T-00339B	Booster	

A—Army applications U/W—Underwater damage N—Navy applications GP—general purpose; blast and fragmentation Brisant—metal acceleration (shaped charge/fragmentation)

Table 8 (continued)

Status	Warhead applications
A. TNT-based explosives	
Service approved	Rocket, missiles
Service approved	Bombs, SUS
Service approved	Stabilized form in missile (Talos)
Service approved	Stabilized form in missile (Tartar)
Service approved	Harpoon missile
Service approved	Navy Mk 80 series bombs
Service approved	Mines
Service approved	Mines, torpedoes
Approved with waiver for Navy use	Navy Redeye missile
Approved with waiver for Navy bombs	Bombs (land-based operations)
Qualified	Bombs (land-based operations)
Service approved	Small bomblets
Service approved	Small bomblets
Experimental/laboratory; not approved for Navy weapons	None
Service approved	Missile (Bullpup)
Service approved	Demolition, depth charges
Service approved	Air Force bombs
B. Pressed explosives	
Service approved	Booster, leads
Service approved	Projectiles: 5 in./54 cal
Service approved	Demolition hose
Qualified	Experimental
Qualified	Experimental
Approved with waiver	Grenade
Qualified	Experimental
Service approved	Projectiles
Service approved; obsolete	Booster, leads

enough to be used without a desensitizer in their intended applications, that is, main charge and booster, respectively. The other pressed explosives are desensitized with a coating material, such as wax or stearic acid, applied to the RDX powders. Small amounts of other ingredients, such as graphite and barium stearate, are added as lubricants. Those coated RDX mixtures are also in powdered form and are called molding powders. These molding powders can be either pressed directly into a warhead cavity or pressed into a block, which is subsequently machined to the final charge shape and inserted into a warhead. The pressed compositions all contain high amounts ($> 90\%$) of RDX powders (or, in the case of aluminized Comp A-3, RDX and aluminum). Because of inherent, practical limitations of the packing of particulate matter in fluids, castable systems are limited to solid fills less than 90% by weight. If performance criteria require RDX or HMX levels above 90%, then pressed explosives or pressed PBXs must be used. However, at these levels, quite often the sensitivity is such that the composition is considered a booster material. This is especially true with the U.S. Navy.

Until the early 1970s, CH-6 and tetryl were the most widely used booster materials. As noted before, tetryl is no longer manufactured. CH-6 usage in new weapons is being reduced with the use of pressed PBX booster materials. Composition A-3 is a main charge fill used in projectiles and other weapons requiring high levels of brisance and is the second-highest Navy production volume explosive after H-6. The other Comp A–type materials, such as A-4 and A-5, have larger percentages of RDX and/or utilize stearic acid as the desensitizing binder, replacing the wax in Comp A-3.

Table 9 lists the chemical formulas, calculated assuming a molecular weight of 100 g for the mixtures, the theoretical maximum density (TMD), the nominal density usually found in production, and the calculated heats of formation of the TNT-based and pressed explosives. These data can be used to calculate estimates of the performance levels of these materials.

C. Plastic-bonded explosives. By the late 1950s, sufficient evidence had been accumulated indicating that possible improvements in explosive sensitivity could be achieved with the use of plastic binders. In the early 1960s, the first plastic-bonded explosives (PBXs) were introduced (Table 10). The first PBXs (PBXN-1, PBXN-3, and PBXN-4) were pressed compositions with moderately high-explosive powder contents bonded together with an inert nylon binder. Later, pressed PBXs were developed, using inert Viton as the binder, containing high-explosive powders (usually either HMX or RDX) at the 95% level. Castable PBXs were being introduced to military use at approximately the same time as the introduction of pressed PBXs. The first castable PBXs (PBXN-101 and PBXN-102) utilized an inert laminac/styrene binder. Although developed in the late 1950s, PBXN-103, the first energetic binder PBX, was not accepted for military use until 1970. The laminac/styrene binder used in the first castable PBXs resulted in explosive charges that were physically hard. The explosives using this binder (PBXN-101, PBXN-102, and later PBXN-104) eventually proved to be relatively sensitive in a number of large-scale sensitivity tests, as compared to PBXs developed later with rubbery, elastomeric

Table 9 Chemical formulas of TNT-based and pressed explosives

Explosive	TMD g/cc	Nominal density	Composition, moles C	H	N	O	Al	ΔHf, kcal/mole	Ref.
A. TNT-based explosives									
Comp B	1.727	1.68–1.72	2.093	2.62	2.129	2.651	0	0.94	34, 8
Cyclotol 75/25	1.765		1.784	2.577	2.356	2.687	0	3.32	34
Cyclotol Type I	1.690		2.649	2.327	1.666	2.657	0	−3.30	34
Cyclotol Type II	1.695		2.58	2.347	1.721	2.659	0	−2.77	34
Destex	1.730	1.694	2.763	2.237	0.992	1.998	0.693	−7.08	34, 8
H-6	1.814	1.75–1.76	1.812	2.453	1.61	2.014	0.778	−1.09	34, 8
HBX-1	1.777	1.72–1.75	2.013	2.516	1.596	2.115	0.634	−1.97	34, 8
HBX-3	1.899	1.84–1.85	1.632	2.101	1.235	1.637	1.29	−2.08	34, 8
HTA-3	1.946		1.556	1.962	1.707	2.09	0.815	1.05	34
Minol II	1.820	1.62–1.70	1.23	2.88	1.53	2.56	0.74	–	8
Minol IV(Q)									
Octol 70/30	1.822		1.87	2.551	2.287	2.683	0	2.26	34
Octol 75/25	1.835	1.80–1.82	1.783	2.576	2.356	2.686	0	2.89	34, 8
Pentolite	1.715		2.332	2.366	1.293	3.219	0	−23.65	34
Picratol	1.686		2.748	2.325	1.48	2.748	0	−23.04	34
TNT	1.654	1.56–1.59	3.082	2.202	1.321	2.642	0	−6.61	34, 8
Tritonal	1.793		2.466	1.761	1.057	2.114	0.741	−5.28	34
B. Pressed explosives									
CH-6	1.780	1.60–1.65	1.483	2.878	2.634	2.644	0.000	5.09	34
Comp A-3	1.667		1.872	3.743	2.458	2.458	0.000	2.49	34
Comp A-3 (Aluminized)	1.883		1.310	2.620	1.721	1.721	1.112	1.75	34
Comp A-4 (Q)	1.757		1.524	3.049	2.620	2.620	0.000	5.25	34
Comp A-5	1.790		1.401	2.801	2.674	2.682	0.000	5.76	34
Comp A-5 (Type I)(Q)	1.782		1.425	2.851	2.661	2.672	0.000	5.33	34
Tetryl	1.730	1.65	2.439	1.742	1.742	2.787	0	1.63	1, 5

Note: Chemical formulas calculated assuming a molecular weight of 100 g.

Table 10 Milestones in PBX development

Date	Milestone
1960	PBXN-3 approved for fleet use
ca 1962	PBXN-101/102 development completed PBXN-103 development completed
ca 1963	PBXN-4 approved for fleet use
ca 1963	PBXN-5 approved for fleet use
1970	PBXN-103 approved for use in Mk 46 torpedo
1972	PBXN-6 development completed
ca 1974	PBX(AF)-108 development completed
1974	PBXN-105 approved for use in the Mk 48 torpedo
1975	PBXN-106 approved for use in projectiles
ca 1975	PBXN-301 approved for fleet use
1989	PBXN-110 approved for use in Navy warhead
1989	PBXN-7 approved for fleet use

binders. Although never proven in any broad-based scientific sensitivity testing, this sensitivity testing led to the often-heard conclusion that rubbery, elastomeric binders provide sensitivity improvements over hard binders. All modern PBXs, such as PBXN-106, PBXN-107, and PBXN-110, utilize these soft, rubbery binders.

The properties of the most important, well-developed PBXs are given in Tables 11 through 13 where they are grouped according to their manufacturing process. Table 11 gives the compositions of these PBXs in terms of their major component ingredients. Although there are a large number of commercial products used as binder ingredients, there are a limited number of materials used as energetic powders. The only high-explosive powders used are RDX and HMX, along with some limited use of TATB, DATB, and PETN. The only oxidizer and metal fuel used are ammonium perchlorate (AP) and aluminum powder, respectively. Table 12 lists the applications of these PBXs and identifies the material specification used for manufacture, the type of explosive by warhead function and warhead damage effect, its current approval status, and some weapon applications of the material. Since each of the service-approved PBXs (i.e., those with N-numbers) had a different experimental designation, these prior designations are identified under "Other Names or Related PBXs." Table 13 lists the chemical formulas, calculated assuming a molecular weight of 100 g for the mixtures, the TMD, the nominal density usually found in production, and the calculated heats of formation.

Although at first glance it appears that there are many different castable and pressed PBXs, in actuality, many of these explosives are closely related and can be considered as part of a family of materials based on the similarity of the binders used. Typically, when a particular binder has been found to be successful in one explosive application, it is then applied in other related materials. From the explosive materials point of view, there are a limited number of damage mechanisms that can be achieved through the proper design of explosives. These damage mechanisms can be achieved with the use of one of the following three categories of explosives: 1) brisant (nonaluminized), 2) aluminized, and 3) underwater (aluminized; oxidizer). When a particular explosive, say a brisant PBX, has been successfully developed, then, quite often, aluminized and underwater versions are also developed using the same binder. These interrelated families of explosive materials are shown in Table 14. They are grouped according to the eight major polymer types used in the binders. Two of the binders (PNC and PEG) are energetic/detonable and six are inert. Two of the binders (nylon and Viton) are used in pressed compositions; the others are used in castable PBXs. These eight groupings encompass most of the PBX compositions given in Table 11, plus some additional experimental PBXs that are also listed in the table. Three PBXs (PBXN-101, PBXN-102, and PBXN-104) utilizing the laminac/styrene binder system have been approved for service use. Because of the above noted sensitivity deficiencies, these three PBXs and their binder are considered obsolete and have not found any applications in modern weapon systems. The nylon-based binder system has been employed in three pressed compositions (PBXN-1, PBXN-3, and PBXN-4). These materials have been essentially replaced in future explosive warhead applications by the Viton-binder-pressed PBXs and/or easier-to-manufacture castable PBXs with similar performance levels. PBXN-3 has been replaced by PBXN-5. PBXN-1, the only aluminized pressed PBX to be developed, has been replaced by the aluminized castable PBXs, such as PBXN-109, PBXC-117, or PBXW-114. The plastisol-grade nitrocellulose (PNC) binder system has been approved for service use in only one explosive, the underwater composition PBXN-103. The sensitivity of this binder to thermal inputs in any form, such as impact, friction, or electrostatic stimulation, has precluded its acceptability for fleet use in other compositions. Although PBXN-103 has in the past been the explosive of choice for the main charge explosive fill of many underwater weapons, it is expected to see fewer applications in the future as other more insensitive explosives, such as PBXW-115, are developed. The PEG binder, originally developed as a replacement for the PBXN-103 PNC binder, has found application in two PBX compositions approved for fleet use, PBXN-105 (developed as a replacement for PBXN-103) and PBXN-106. The PEG binder uses the mixed energetic plasticizer BDNP A/F = 50/50. This plasticizer has not been produced for many years, thus resulting in availability problems for the material and any system that utilizes it, such as the PEG binder or the PBXN-105 and PBXN-106 compositions. In addition to the past availability problems, the recent price (> $70.00/kg) of the mixed plasticizer has led to the consideration of other less costly PBXs, such as PBXN-110 and PBXW-115, as replacements for the PBXN-105 and PBXN-106 compositions, respectively. Thus, the compositions using the two energetic binders listed in Table 14 can be expected to see decreasing usage in future warheads.

The most viable candidates for weapon applications in the future are those PBXs utilizing the hydroxy-terminated polybutadiene (HTPB), acrylate, polypropylene glycol (PPG), or Viton inert binders. The HTPB binder has found application in a number of castable PBXs including PBXW-108, PBXN-109,

Table 11 Compositions of plastic-bonded explosives

A. Composition of castable PBXs

Explosive	% HMX	% RDX	% AP	% Al
PBX(AF)-108	—	82.00	—	—
PBXC-117(Q)	—	71.00	—	17.00
PBXN-101	82.00	—	—	—
PBXN-102	59.00	—	—	23.00
PBXN-103	—	—	40.00	27.00
PBXN-104	70.00	—	—	—
PBXN-105	—	7.00	49.80	25.80
PBXN-106	—	75.00	—	—
PBXN-107	—	86.00	—	—
PBXN-109	—	64.00	—	20.00
PBXN-110	88.00	—	—	—
PBXW-107(Q)	—	55.00	—	20.00
PBXW-108(Q)	—	85.00	—	—
PBXW-114(Q)	78.00	—	—	10.00
PBXW-115(Q)	—	20.00	43.00	25.00

B. Pressed PBXs

Explosive	% HMX	% RDX	% DATB	% TATB
Comp A-3 (Type II)(Q)	—	90.8	—	—
LX-14	95.5			
PBXN-1	—	68.00	—	—
PBXN-3	86.00	—	—	—
PBXN-4	—	—	94.00	—
PBXN-5	95.00	—	—	—
PBXN-6	—	95.00	—	—
PBXN-7	—	35.00	—	60.00
PBXW-7 (TYPE I)	—	35.00	—	60.00

C. Extrudable and injection moldable PBXs

Explosive	% RDX	% PETN	% Viton A	% Teflon
PBXN-201	83.00	—	12.00	5.00
PBXN-301	—	80.00	—	—

Table 11 (continued)

A. Composition of castable PBXs

Binder polymer/plasticizer	% Polymer	% Plasticizer	% Other binder components
PPG/IDP	10.83	5.30	1.87
Acrylate/-		–	
(Laminac/styrene)/-	18.00/-		
(Laminac/styrene)/-	18.00/-		
NC/(TMETN/TEGDN)	6.00	25.50	1.50
(Laminac/styrene)/-	30.00/-		
PEG/(BDNPA/F)	3.13	12.92	1.35
PEG/(BDNPA/F)	4.50	18.55	1.95
Acrylate/-		–	
R45/DOA	7.35	7.35	1.30
R45/IDP	5.37	5.37	1.27
PEG/(BDNPA/F)	4.50	18.55	1.95
R45/DOA	7.00	7.00	1.00
R45/IDP	5.37	5.37	1.27
R45/IDP	5.70	5.70	0.60

B. Pressed PBXs

% Al	% Nylon	% Viton	% Teflon	% Estane	% PE
—	—	—	—		9.2
				4.5	
20.00	12.00	—	—	—	
—	14.00	—	—	—	
—	6.00	—	—	—	
—	—	5.00	—	—	
	—	5.00		—	—
—	—	5.00	—	—	
—	—	—	5.00	—	

C. Extrudable and injection Moldable PBXs

% Sylgard 182
20.00

Table 12 Applications of plastic-bonded explosives

Explosive	Specification	Other names or related PBXs
A. Castable PBXs		
PBX(AF)-108		AFX-108
PBXC-117(Q)		–
PBXN-101	WS-3829	PBXC-104
PBXN-102	WS-3823	PBXC-105
PBXN-103	WS-12800	PBXW-100
PBXN-104	WS-11511	PBXC-115
PBXN-105	WS-13112	PBXW-104
PBXN-106	WS-13522	PBXW-106
PBXN-107		PBXC-116(Q)
PBXN-109	WS-23147	PBXW-109 (Type E)
PBXN-110	WS-22347	PBXW-113(Q)
PBXW-107(Q)	WS-14158	–
PBXW-108(Q)		–
PBXW-114(Q)	WS-23159	–
PBXW-115(Q)	WS-26580	–
B. Pressed PBXs		
Comp A-3 (Type II)(Q)		PBXW-6
LX-14		
PBXN-1	OS-11632	–
PBXN-3	OS-11641	–
PBXN-4	MIL-P-23625	PBXC-7
PBXN-5	MIL-E-81111	PBXC-6
PBXN-6	WS-12604	PBXC-12
PBXN-7	WS-23160	PBXW-7 (Type II)
PBXW-7(Type I)		
C. Extrudable and injection moldable PBXs		
PBXN-201	WS-11498	PBXC-202
PBXN-301	WS-12612A	PBXC-303, XTX-8003

GP–general purpose; blast and fragmentation.

Brisant—metal acceleration (shaped charge/fragmentation).

Table 12 (continued)

Explosive type			
By warhead function	By warhead effect	Status	Warhead applications
A. Castable PBXs			
Main charge	Low brisant, insensitive	Service approved	
Main charge	Brisant	Qualified	None
Main charge	Brisant	Service approved	Shrike missile
Main charge	Brisant	Service approved	Sidewinder—not in production
Main charge	Underwater	Service approved	Mk 46 torpedo, Quickstrike
Main charge	Low brisant	Service approved	Phoenix missile
Main charge	Underwater	Service approved	Mk 48 torpedo
Main charge	Brisant	Service approved	SM-II missile, Hi-Frag Projectile
Main charge	Brisant	Service approved	
Main charge	GP	Service approved	
Main/transfer charges	Brisant	Service approved	Navy warheads
Main charge	GP	Qualified	None
Main charge	Brisant	Qualified	None
Main charge	Brisant; U/W	Qualified	None
Main charge	Underwater	Qualified	Anti-U/W mines
B. Pressed PBXs			
Main charge; booster	Brisant		
Army—main charge Navy—booster	Brisant	Army—Service approved Navy—booster qualified	Hellfire missile
Main charge	Brisant	Service approved	Missiles
Main charge	Brisant	Service approved	Missiles
Main charge	Brisant	Service approved	Missiles
Booster	–	Service approved	Booster
Booster	–	Service approved	Booster
Booster	–	Qualified	Lead (Mk 50 torpedo WH)
Booster	–	Qualified	Developmental warheads
C. Extrudable and injection moldable PBXs			
Burster charge		Service approved	Bomb BLU-73/B
Transfer link		Service approved	Condor, Sidewinder

Table 13 Chemical formulas of PBXs

Explosive	TMD (g/cc)	Nominal density	Composition (moles) C	H	N	O
A. Castable PBXs						
PBXC-117	1.750		1.650	3.138	1.946	2.048
PBXN-103	1.887		0.739	2.572	0.695	2.483
PBXN-105	1.926		0.618	2.783	0.860	2.307
PBXN-106	1.663		1.765	3.316	2.381	2.632
PBXN-107	1.650		1.968	3.746	2.356	2.474
PBXN-109	1.663		1.904	3.476	1.740	1.829
PBXN-110	1.679		1.989	3.773	2.382	2.430
PBXW-107	1.771		1.495	2.776	1.840	2.092
PBXW-108	1.581		2.134	3.906	2.303	2.386
PBXW-109	1.679		1.864	3.365	1.763	1.846
PBXW-114	1.723		1.854	3.503	2.112	2.160
PBXW-115	1.790		1.080	3.407	0.911	2.053
B. Pressed PBXs						
Comp A-3 (Type II)(Q)						
LX-14	1.849		1.520	2.920	2.590	2.660
PBXN-1	1.796		1.566	3.031	1.938	1.938
PBXN-3	1.738		1.916	3.715	2.440	2.440
PBXN-4	1.771		3.086	5.951	4.528	4.598
PBXN-5	1.902		1.417	2.659	2.566	2.566
PBXN-6	1.808		5.002	9.666	6.968	7.038
C. Other PBXs						
PBXN-301	1.556	1.53	1.80	3.64	1.01	3.31

PBXN-110, PBXW-114, and PBXW-115. The acrylate and PPG binders are used in the RDX-based aluminized and nonaluminized castable PBXs [PBXN-107, PBXC-117, and PBX(AF)-108]. The Viton binder is used in a number of pressed compositions (PBXN-5, PBXN-6, and PBXN-7) that are used as booster explosives.

Some general observations can be made regarding the list of PBXs in these tables. All booster and lead materials are either pressed, extruded, or injection moldable; there are no booster materials available that are castable. Because of the size differences in warhead applications, it can be expected that the volume of main charge explosive production would far exceed that of booster and primary materials. Thus, with the exception of LX-14, used in large production Army antitank weapons, the majority of the PBXs used, by number and production volume, are castable materials. Because of processing limitations, the castable PBXs have solids (the high-explosive AP and Al powders) contents less than 90% by mass. The more recently developed pressed PBXs have solids contents equal to 95% by mass. Because of differences in explosives risk assessment, the

Table 13 (continued)

Al	Cl	F	Other	Δ hf, (kcal/mole)	Ref.
A. Castable PBXs					
0.630	0.000	0.000			3
1.001	0.340	0.000		−39.63	34
0.956	0.424	0.000	Sn	−39.42	34
0.000	0.000	0.000	Fe	−9.30	34
0.000	0.000	0.000			3
0.741	0.000	0.000	Fe	−2.40	34
0.000	0.000	0.000	P, Sn	0.95	34
0.741	0.000	0.000	Fe	−10.63	34
0.000	0.000	0.000		−0.49	34
0.741	0.000	0.000		−1.81	34
0.371	0.000	0.000	P, Sn	0.34	34
0.927	0.366	0.000	Sn	−28.78	34
B. Pressed PBXs					
0.000	0.000	0.000		1.50	1
0.741	0.000	0.000		−3.98	34
0.000	0.000	0.000		−4.69	34
0.741	0.000	0.000		−13.37	34
0.000	0.000	0.174		−3.14	34
0.741	0.000	0.000		−2.60	34
C. Other PBXs					
–	–	–	Si = 0.27	−39.00	1

Army and Navy have slightly different assessments of the dividing line between booster and main charge explosives. Compositions with RDX or HMX solids content greater than 91% to 92% are generally considered booster materials within the Navy but may be considered acceptable main charge fills within the Army. This point is illustrated by LX-14, which has been qualified as a booster material by the Navy (not as a main charge explosive) but is acceptable within the Army as a main charge explosive. This 91% to 92% solids dividing line has resulted, within the Navy, in having all current castable PBXs approved as main charge explosive fills and modern pressed PBXs approved as boosters.

D. Comparison of the characteristics of PBXs, TNT-based compositions, and other classical explosives. Both TNT-based and PBX compositions can be designed to cause the same type of damage; thus, either can be employed in a given warhead causing this type of damage. Table 15 compares equivalent PBX and non-PBX compositions in terms of their warhead application. For each category of damage or warhead type, there are currently available PBX compositions that provide almost equivalent performance levels to that possessed by the TNT-

Table 14 Families of PBX materials based upon binder type

General binder type*		Brisant HMX-based	Brisant RDX-based	Brisant Other	Aluminized	Underwater
Laminac/styrene	Inert castable	PBXN-101 PBXN-104			PBXN-102	
HTPB	Inert castable		PBXW-108		PBXN-109	PBXW-115
		PBXN-110			PBXW-114	
Acrylate	Inert castable		PBXN-107		PBXC-117	
PPG	Inert castable		PBX(AF)-108		AFX-720	
Nylon	Inert pressed	PBXN-3		PBXN-4	PBXN-1	
Viton	Inert pressed	PBXN-5	PBXN-6	PBXN-7		
PNC	Energetic castable	PBXW-116			PBXW-117	PBXN-103
PEG	Energetic castable		PBXN-106		PBXW-107	PBXN-105

*Grouped by the major polymer used in the binder.

Table 15 Performance type/applications equivalency of PBXs and non-PBX compositions

Type of performance or application	Warhead applications	Equivalent compositions Non-PBX	PBXs
GP	Bomb fill	H-6 Tritonal	PBXN-109 PBXC-117
High brisance	Shaped charges, fragmentation	Octol	PBXN-110
Brisance	Shaped charges, fragmentation	Comp B Cyclotols	PBXN-106 PBXN-107 PBXW-108 AFX-108
Underwater			
1. Shock & bubble	Torpedoes, mines; anti-devices	H-6 HBX-1	PBXN-103 PBXN-105 PBXW-115
2. Bubble	Torpedoes, mines	HBX-3	
Booster	Booster, lead	Tetryl CH-6	PBXN-5 PBXN-6 PBXN-7

based and other classical explosives. General purpose explosives must possess adequate performance levels in air-blast, fragmentation, and, in some cases, underwater performance. The aluminized explosives containing 20% aluminum, whether TNT-based (H-6 or tritonal) or PBXs (PBXN-109 or PBXC-117), have approximately the same overall general purpose performance levels, although there may be some differences in the air-blast and fragmentation performance. For metal-accelerating, brisant applications such as shaped charges and fragmenting warheads, PBX compositions usually provide slightly less performance when compared to an equivalent TNT-based explosive. For example, the HMX-containing explosives PBXN- 110 and 75/25 octol could be considered for the same warhead application. The detonation velocity and pressure of PBXN-110 will be lower than that of the octol. The RDX-containing explosives Comp B, cyclotols, PBXN-106, PBXN-107, PBXW-108, and PBXN-106 could be considered for the same warhead application. Again, the PBXs will possess slightly lower detonation performance than their equivalent TNT-based materials. The consideration of PBXs vs TNT-based explosives for metal-accelerating applications will entail a tradeoff between the slightly lower detonation performance of the PBXs vs possible improvements in other explosives characteristics, such as sensitivity or thermal stability.

Underwater explosives can be divided into two subcategories: 1) underwater explosives optimized for maximum underwater shockwave and bubble energy and 2) underwater explosives optimized for maximum bubble energy only. For the first category, the underwater PBXs, PBXN-103, PBXN-105, and PBXW-115, provide significant improvements in both shockwave and bubble energy when compared to H-6 and HBX-1. This improvement was achieved through the use of the ammonium perchlorate oxidizer. The use of ammonium perchlorate in TNT-based explosives had been examined prior to the development of PBXs but was discontinued because the use of AP made the compositions significantly more impact sensitive. The PBX binders, on the other hand, desensitize the AP mixtures to the point where these mixtures can safely be processed and used in warheads. For the second subcategory of underwater explosives, only the TNT-based HBX-3 has been optimized to produce the maximum bubble energy. With almost 35% aluminum, its bubble energy is significantly greater than the other aluminized TNT-based explosives. To date, no PBX that has reached an advanced state of development is optimized for maximum bubble energy only. With the use of AP, the current underwater PBXs (PBXN-103, PBXN-105, and PBXW-115) possess bubble energies comparable to or greater than that possessed by HBX-3. If there is a need, the current compositions can be optimized for maximum bubble energy by increasing their aluminum content. The PBX booster compositions, PBXN-5, PBXN-6, and PBXN-7, have effectively replaced tetryl and CH-6 in most modern warhead applications. PBXN-5 and PBXN-6 provide equal or greater detonation pressure output than either CH-6 or tetryl. The PBXN-7 composition was designed to provide comparable performance to that of CH-6 or tetryl, but with significantly less severe reactions in fast and slow cook-off environments.

Since the early to mid-1970s, PBXs have been increasingly selected as the explosive fill for new weapon systems. The extent to which PBXs have been introduced differs in each military service. Each service has its own safety and risk assessments and rules and regulations. Warhead production numbers or volumes differ according to service. With the current limited production base for PBXs, they can be expected to be selected for specialty weapons with limited production volumes. PBXs have seen the greatest usage within the Navy and to a lesser extent within the Air Force. The Army relies primarily on TNT-based technology because of its need for large-production-volume weapons. The increasing use of PBXs has been justified on the basis of a number of technical improvements achieved with their use. There are some less desirable characteristics associated with the use of these materials; most of these characteristics are not related directly to their use in warheads, but to their manufacture. Comparisons of the advantages and deficiencies of TNT-based explosives and castable PBXs are listed in Tables 16 and 17. These are very general, broad comparative statements on the two classes of materials based upon the accumulated experience of the explosives community. There is considerable variation in the properties of individual explosives, and not all members of each class will exhibit precisely the general characteristics listed.

The advantages and deficiencies of TNT-based explosives are listed in Table 16. When compared to the characteristics of PBXs, the major advantages of the use of TNT-based explosives are related to their manufacturing and cost. The component materials (TNT, RDX, aluminum) are readily available and

Table 16 Advantages and deficiencies of TNT-based explosives

A. Advantages of TNT-based explosives

1. Component materials (e.g., TNT) and compositions relatively inexpensive (many explosives < $4.50/kg)
2. Component materials readily available
3. Mixing and loading facilities in place such that they can be manufactured on a large scale
4. Can be loaded on a semicontinuous basis
5. Easy to demilitarize: remelt explosive charge

B. Deficiencies of TNT-based explosives

1. Susceptible to sympathetic detonation in magazines and storage areas
2. Susceptible to premature initiation upon impact with hard targets
3. Insufficient physical stability (to approx 85°C)
4. Increased performance results due to increased sensitivity
5. Higher probability of producing violent reactions in fuel fires
6. High shrinkage after casting with resulting large internal voids
7. Subject to cracking and liquid exudation during extended storage (due to temperature cycling)

Table 17 Advantages and deficiencies of PBXs

A. Advantages of PBXs

1. Reduced susceptibility to sympathetic detonation
2. Possible decrease in mechanical impact sensitivity because of elastomeric mechanical properties
3. Physical thermal stability to 150°C
4. Many exhibit minimal or negligible shrinkage after curing
5. Improved mechanical properties result in resistance to cracking
6. Improved cook-off behavior when subjected to fuel fires with reduced probability of violent reactions

B. Deficiencies of PBXs

1. In general, they are more expensive than TNT-based materials
2. For the same level of performance, PBXs require higher RDX or HMX content
3. Some PBXs utilize component materials of limited availablity or are very expensive or produced by a limited number of manufacturers (i.e., they are speciality chemicals)
4. They are produced on a batch basis—slower rate of production
5. Mixing facilities now in place are limited; very large production currently not possible
6. Their inherent physical stability makes them more difficult to demilitarize

produced in the United States on the scale of tens of millions of pounds per year. Being produced on this scale, the component raw material cost is relatively low. Since TNT-based technology dates to the pre–World War II period, there is considerable production plant capacity for mixing and loading these materials. Furthermore, TNT-based explosives can be loaded on a semicontinuous basis. Since TNT-based explosives melt at temperatures between 80° and 85°C, the TNT-based materials can be readily demilitarized at the end of their military service life by heating the warheads to within this temperature range and pouring out the explosive fill.

The majority of the deficiencies of these materials are related to their sensitivity and thermal stability. The significance of these deficiencies is viewed differently by each service. The more widespread use of PBXs as replacements for TNT-based explosives within the Navy is primarily due to the significance attached to deficiencies. Large-scale testing and fleet accidents have shown that TNT-based explosives have a higher probability than PBXs of producing violent reactions, either detonations or explosions, when subjected to heating in fuel fires.[35,36] Furthermore, these materials are susceptible to sympathetic detonation. Thus, any initial violent reaction in one warhead can be transmitted to a nearby warhead causing it also to be initiated violently. In general, increased performance with these systems of materials can be achieved only with some sacrifice in sensitivity or safety. Their sensitivity to impact and shock make them susceptible to unintentional or premature initiation during some operational deployments of TNT-based explosive–filled warheads, such as attempts to penetrate hard targets. Because of the relatively low melting point of TNT-based materials, which is an advantage in processing, loading, and demilitarizing, these materials may be unsuitable for certain warheads that are subjected to high-temperature environments. For example, long-range supersonic missiles may produce extremely high rates of aerodynamic heating. Under these conditions, TNT-based materials may undergo melting, the consequences of which are dependent upon the design and use of the warhead. Melting problems can be mitigated with the use of thermal protection, but at the cost of incorporating a volume of material that in most cases will reduce the volume of the damage-causing elements of the warhead. The difference in density of solid (1.654 g/cm^3) and liquid (1.48 g/cm^3) TNT will cause shrinkage in all TNT-based explosives,[12] with the probable creation of large internal voids found typically along the central warhead axis. With effort, this shrinkage can be minimized in warheads by carefully cooling the warhead and by refilling the void space with additional explosive. TNT-based materials respond poorly to temperature cycling. Being crystalline materials with low physical strength, they can be readily cracked when subjected to expansion and contraction forces during cycling. Also, high-temperature phase changes cause irreversible growth: when the temperature of a TNT sample is raised from room temperature to a high temperature (> 60°C) and then returned to room temperature, the volume of the sample is greater at the end of the cycle than its initial volume. This will be repeated with each temperature cycle until the sample volume has grown to greater than 5%. TNT-based explosives also have a tendency to exude at high temperatures. These growth and exudation problems can be mitigated by the use of relatively heavy and complete confinement of the explosive fill.

The advantages and deficiencies of PBXs are listed in Table 17 where the characteristics are compared with those of TNT-based explosives. The major advantages associated with the use of PBXs relate to their behavior after they have been loaded into warheads, whereas the major disadvantages relate to the process of mixing and loading them into the warheads. They exhibit improved fast cook-off behavior as compared to TNT-based materials; many exhibit improved slow cook-off behavior. When subjected to these thermal environments they decompose more mildly, resulting in a reduced probability of violent reactions. They are also less likely to be susceptible to sympathetic detonation. With their greater strength and other mechanical properties that are due to the elastomeric nature of the materials, they are able to better resist cracking under thermal cycling and to better resist severe impact environments. Many of these materials are both physically and chemically stable to above 150°C. When temperature cycled, most PBXs will exude none of their component materials. When loaded and cured, these materials do not undergo the large shrinkage observed with TNT-based materials; most will exhibit shrinkage of much less than 1%. Some will undergo no measurable shrinkage. The most important advantage acquired with the use of PBXs is the significant reduction in the probability of violent reactions when the materials are subjected to hazardous conditions, such as fast and slow cook-off, sympathetic detonation, and bullet or fragment attack.

The major deficiencies in the use of PBXs relate to their manufacture. The production facilities now in place cannot produce PBXs in quantities similar to the production levels of the TNT loading facilities. Although continuous processing is being evaluated at the research and development level, currently all castable PBXs are mixed and loaded on a batch basis. Most PBX production and loadings in the past were for low-volume weapons with less than 1000 to 2000 warheads loaded per year and with explosive fill weights under 80 kg. The PBX production base is continuously expanding, both at government facilities and in private industry; however, for the near future, PBX facilities will not be able to meet wartime or mobilization production requirements for mass-produced weapons. Many of the PBXs, especially those with inert binders, require higher percentages of the RDX or HMX powders within the composition to achieve performance levels comparable to equivalent TNT-based materials. In wartime, when RDX and HMX production quite often is insufficient, this need for greater amounts of these powders will put an additional strain on production facilities. Some PBXs, usually those utilizing energetic binder polymers or plasticizers, consist of component materials that are produced on a very limited scale or produced by a limited number of manufacturers, in some cases only one or two. These component materials are essentially specialty chemicals and, as such, are usually relatively expensive. In general, PBXs are more expensive than equivalent TNT-based materials, in both their raw component material cost and their loading cost. Because of their inherent physical stability they are much more difficult to demilitarize (demil) after their service life is completed.

In summary, the major advantages and disadvantages in the use of TNT-based explosives or PBXs are related to the economics of production, on one hand, and the quality of the produced product, on the other. When explosive product quality is measured in terms of its overall behavior in warheads, over the service life of the unit, in storage, in shipment, and in response to normal and hostile wartime environments, PBXs offer an improved-quality product. The price that has to be paid to acquire this quality product is a higher initial unit cost. The PBXs listed in Table 15, along with LX-14, can also be considered a list of the currently available, mature, viable explosive materials. When selected for use in an appropriately chosen warhead, these materials will provide a superior end product.

2. Explosive Performance, Effects, and Prediction Techniques

The focus of this section will be the factors influencing the behavior of explosive charges, the effects of the detonation on the nearby environments, and techniques developed to predict these behaviors and effects. The phenomena associated with the detonation of an explosive charge can be discussed on three levels: 1) charge detonation, the detonation of the charge itself and its influencing factors; 2) explosive effects, the effect of the detonation shockwave and expanding decomposition gases on the nearby environment without consideration of target interaction or damage (also called free-field effects); and 3) the damage effects of the detonation shockwave, expanding gases, and, if confined within a warhead, the metal fragments on any nearby structures. The behavior of the charge as it detonates (level 1) affects the shockwave and expanding decomposition gases (level 2); thus, charge detonation can be related to free-field effects. In a similar fashion, free-field effects can be related to the damage caused to any structures, such as armored vehicles, submarines, or buildings, near the detonating charge. One of the major thrusts of explosives research and development is directed toward understanding the relationship of the behavior at the lower, more fundamental levels (level 1 or 2) on the more complex behavior at the higher levels (level 2 or 3). Table 18 lists some of the factors affecting these relationships for the illustrative case of the effect of explosive performance on shaped charge performance. In the discussion below, emphasis will be placed on the behavior of the detonating charge. The subject of free-field explosive effects will be illustrated by a few examples. The complex and extensive subject of target damage will not be discussed; other sections of this book will address certain aspects of this subject.

Table 18 Explosives and shaped charges

Level	Factors
Charge detonation	Factors affecting detonation velocity and pressure and charge quality
Free-field effects	Effect of detonation pressure and velocity and charge quality on metal acceleration, shaped charge liner collapse, and jet tip velocity
Target damage	Interaction of the shaped change jet with target (jet penetration depth and hole volume)

2.1 Explosive Charge Performance

The general characteristics of a detonating charge are schematically shown in Fig. 1. Detonation of the cylindrical charge of diameter *d* has been initiated on the left-hand side. Sufficient time has elapsed such that all initiation transients have died out so that the detonation characteristics are those due to the properties of the explosive itself. The detonation front propagates with a velocity *D*, called the detonation velocity, from left to right. Behind the detonation or shock front there is a narrow region called the reaction zone, followed by the expanding gases released from the decomposing explosive. Figure 2 shows an enlargement of this region; schematically shown is the pressure as a function of distance behind the front. As the shock progresses supersonically through the explosive, it causes the rapid decomposition (on the microsecond scale) of the high-explosive molecules. Upon decomposition, these molecules release their energy exothermically. This energy is fed back into the propagating shock in such a way that there is a steady-state balance between the energy generated, the energy lost, and the energy supplying the shock propagation. This state is called a steady-state detonation. Not all of the total energy released from the decomposing high-energy molecules is released on a sufficient time scale (or equivalently within some distance behind the shock front) such that it has the opportunity to feed back into the propagating shock. All energy released in sufficient time to contribute to driving the detonation is released between the shock front and before what is called the Chapman-Jouget plane[38,39] or C-J plane (Fig. 2). This region between the shock front and the C-J plane is called the reaction zone (the region X_0 in Fig. 2). An ideal explosive, undergoing an ideal detonation, can be defined as one that completely releases its chemical energy within this zone. Only the pure high-explosive powders by themselves behave approximately like an ideal explosive. Most practical military explosives are somewhat nonideal in that they release a portion of their energy beyond the C-J

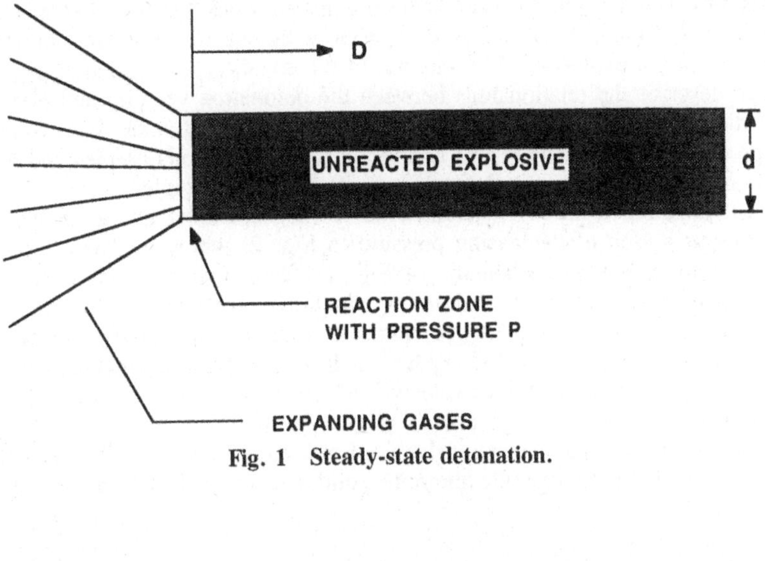

Fig. 1 Steady-state detonation.

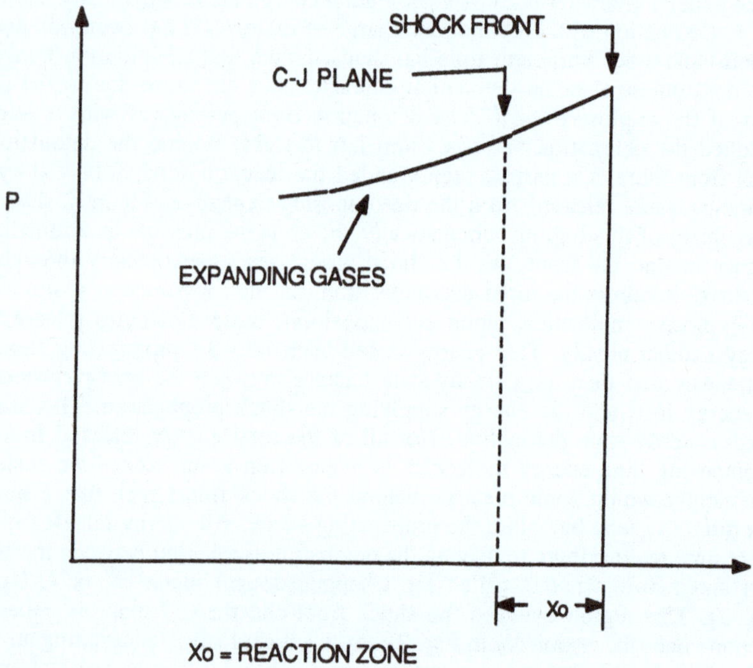

Fig. 2 Detonation front (from Ref. 37)

plane. The pressure within the reaction zone at the C-J plane is called the detonation pressure (P) or C-J pressure (P_{cj}) of the explosive. For a given explosive composition of a given charge diameter undergoing a steady-state detonation, the detonation velocity and pressure will be constant along the full charge length, provided the diameter is greater than what is known as the critical or failure diameter of the explosive. A discussion of the equations of detonation physics, which describe the relationships between the detonation velocity and pressure, densities, sound speed, particle velocities, and other variables, is beyond the scope of this article. The reader is referred to the numerous articles and books written on this subject (see Refs. 25 and 40–45, for example).

The transition from the reaction zone to the expanding gas is shown as a continuous region of decreasing pressure in Fig. 2, which is a more accurate representation than the schematic of Fig. 1. These expanding decomposition gases provide the necessary medium by which the explosive can do work, such as by accelerating metal in shaped charge or fragmenting warheads or accelerating water via gaseous bubble expansion. In contrast to the damage caused by the decomposition gases from explosives, thermites provide little if any damage capability because their high-energy output, greater than most explosives, is derived from a chemical reaction between two solids whose final products are also solids. Even though the thermite solid reaction products are extremely hot, they are gasless, and thus they provide no mechanism to do work (other than gravitationally). The expansion of the decomposition gases is a complex function of pressure, density, temperature, and energy. This gas expansion can be described or modeled by what is called the equation of state (EOS) of the detonation products (see Ref. 41, Sec. 2B). A number of different equations of state have been developed to describe the expansion process. The Jones-Wilkins-Lee (JWL) EOS[47] is accepted as one that accurately describes the expansion process for nearly ideal explosives[1,5]:

$$P = A[1 - (\omega/R_1 V)]e^{-R_1 V} + B[1 - (\omega/R_2 V)]e^{-R_2 V} + \omega E/V \qquad (1)$$

$$P_s = Ae^{-R_1 V} + Be^{-R_2 V} + CV^{-(\omega+1)} \qquad (2)$$

where

P = pressure
P_s = pressure along the expansion isentrope, i.e., pressure as a function of volume at constant entropy
A, B, C = linear coefficients in units of P and P_s
R_1, R_2, ω = nonlinear, unitless coefficients
$V = v/v_0$ = detonation product volume (v)/initial high-explosive volume (v_0)
E = detonation energy per unit volume

The coefficients are obtained by comparing EOS calculations with experimental expansion data. Almost all recently obtained EOS parameters are obtained using cylinder expansion (Cylex) data. When incorporated into computer codes, such as TIGER,[48] the JWL EOS will provide reasonable estimates of the detonation performance of large-diameter, nearly ideal explosives. Since the JWL parameters are obtained using Cylex data, the parameters are strictly valid to volume expansions similar to that found in the Cylex test, which is usually limited to volume expansions slightly greater than seven times the original explosive volume. Research is currently moving toward extending the JWL equation, in modified form, to greater expansions.[48] Table 19 lists the JWL parameters for a number of military explosives, along with the C-J detonation parameters used in obtaining this information.

2.2 Detonation Velocity, Pressure, and Energy

Three important parameters are considered as essential in the characterization of the charge performance of a explosive:

1) Detonation velocity [units: meters/second (m/s) or millimeters/microsecond (mm/μs)]
2) Detonation pressure [units: kilobars (kb) or gigapascals (GPa)]
3) Detonation energy [units: calories/gram (cal/g)]

These parameters are used in a number of models and predictive techniques to assess the ability of explosives to cause damage. The detonation velocity of an explosive is determined experimentally using either of the following techniques:

1) *The pin technique* uses a series of ionization probes that are embedded in the explosive charge in equal increments (δx) along the charge length. As the

Table 19 JWL equation-of-state parameters

Explosive	C-J parameters				
	Density, g/cc	P, GPa	D, mm/μs	E_0, GPa-m^3/m^3	γ
Molecules					
HMX	1.891	42.0	9.11	10.50	2.74
HNS	1.000	7.5	5.10	4.10	2.468
	1.400	14.4	6.34	6.00	2.881
	1.650	21.5	7.03	7.45	2.804
PETN	1.260	14.0	6.54	7.19	2.831
	1.500	22.0	7.45	8.56	2.788
	1.770	33.5	8.30	10.10	2.64
Tetryl	1.730	28.5	7.91	8.20	2.798
TNT	1.630	21.0	6.93	7.00	2.727
TNT-based					
Comp B[a]	1.717	29.5	7.98	8.50	2.706
Cyclotol 77/23	1.754	32.0	8.25	9.20	2.731
H-6	1.760	24.0	7.47	10.30	3.092
Octol 78/22	1.821	34.2	8.48	9.60	2.83
Pentolite	1.700	25.5	7.53	8.10	2.78
	1.650	23.5	7.36	8.00	2.78
Pressed explosives					
Comp A-3	1.650	30.0	8.30	8.90	2.79
Other classical explosives					
Comp C-4	1.601	28.0	8.19	9.00	2.838
Castable PBXs					
PBXN-106	1.634	26.0	7.84	8.20	–
PBXN-109	1.660	22.0	7.60	10.20	–
PBXN-110	1.672	27.5	8.33	8.70	3.2
PBXN-114	1.711	26.0	8.15	9.50	3.371
PBXN-115	1.782	12.0	5.70	6.00	3.819
Pressed PBXs					
LX-10-1	1.865	37.5	8.84	10.50	2.868
LX-14-0	1.835	37.0	8.80	10.20	2.841
PBXN-5[b]					

[a]Modified form of Composition B with RDX/TNT/wax = 64/36/1.

[b]See the values for LX-10-1, a similar composition.

Table 19 (continued)

			JWL EOS parameters			
A, GPa	*B*, GPa	*C*, GPa	*R*1	*R*2	*ω*	Ref.
Molecules						
778.3	7.071	0.643	4.2	1	0.3	50
162.7	10.82	0.658	5.4	1.8	0.25	51
366.5	6.75	1.163	4.8	1.4	0.32	51
463.1	8.873	1.349	4.55	1.35	0.35	51
573.1	20.16	1.267	6	1.8	0.28	50
625.3	23.29	1.152	5.25	1.6	0.28	50
617	16.926	0.699	4.4	1.2	0.25	50
586.8	10.671	0.774	4.4	1.2	0.28	49
371.2	3.231	1.045	4.15	0.95	0.3	49
TNT-based						
524.2	7.678	1.082	4.2	1.1	0.34	49
603.4	9.924	1.075	4.3	1.1	0.35	49
758.07	8.513	1.143	4.9	1.1	0.2	52
748.6	13.38	1.167	4.5	1.2	0.38	49
540.94	9.3726	1.033	4.5	1.1	0.35	53
531.77	8.933	0.976	4.6	1.05	0.33	5
Pressed explosives						
611.3	10.65	1.08	4.4	1.2	0.32	50
Other classical explosives						
609.77	12.95	1.043	4.5	1.4	0.25	1
Castable PBXs						
570.2	6.13	1.325	4.45	1.0	0.38	54
1341.3	32.7	1.334	6.00	2.0	0.2	5
950.4	10.98	1.816	5.00	1.4	0.4	55
1122.2	8.684	15.8	5.20	1.2	0.3	55
812.99	−134.98	3.424	5.00	3.5	0.6	55
Pressed PBXs						
880.7	18.36	1.296	4.62	1.32	0.38	1
826.1	17.24	1.296	4.55	1.32	0.38	1

detonation progresses along the charge, its pressure and/or ionized detonation gases electrically short the pin, closing an electronic circuit that measures the time (δt) from the shorting of the previous pin. The detonation velocity between the two pins is $D = \delta x / \delta t$. An average value is calculated from the values obtained for each of the pin increments.

2) *The streak camera technique* yields a photographic record of the detonation rate that is obtained on a single strip of film through a slit along the charge length.

The detonation pressure of an explosive is determined experimentally using one of the following techniques:

1) *The Manganin or Lagrange gauge technique*[56] uses gauges to directly measure the pressure behind the shock front.

2) *The Aquarium technique*[3,57,58] involves one end of a cylindrical explosive charge, usually with a 2-in. diameter, being placed in a water-filled aquarium. The change is initiated from the other end. When the detonation reaches the (water) end of the charge, a shock is propagated into the water. Cameras record the shock velocity in water. By using this shock velocity and the detonation velocity of the explosive, the detonation pressure can be calculated using standard hydrodynamic detonation equations.

3) Although not used widely now, an estimate of the detonation pressure can be obtained using the *witness plate dent test*[59] in which a thick steel witness plate is placed in contact with the (noninitiated) end of an explosive charge. When the charge is detonated, it leaves a dent in the witness plate. Depending upon the hardness of the steel, the depth of the dent can be approximately correlated to the detonation pressure of known explosives of the same diameter.

Table 20 lists experimentally determined detonation velocities and pressures for the major military explosives. Both detonation velocity and pressure measurements are influenced by charge density, diameter, confinement, and the specific experimental procedure used; some of these factors will be discussed below. The values listed are those from the literature that report the highest density and/or the largest diameter for the given explosive. The cited references should be reviewed to determine the precise experimental conditions involved for each determination.

The detonation energy or the heat of detonation has been experimentally determined for only a limited number of explosives, usually the high-explosive molecules such as HMX, RDX, tetryl, and so on. The chemical energy Q_c released when an explosive undergoes detonation can be estimated using the equation

$$Q_c = \frac{-\left\{\sum [\Delta H_f(\text{detonation products})] - \Delta H_f(\text{explosive})\right\}}{\text{Formula weight of the explosive}} \tag{3}$$

where ΔH_f is the heat of formation and the summation term ($\sum$) is calculated by

$$\sum[\Delta H_f(\text{detonation products})] = \sum(\text{moles of product})(\Delta H_f \text{ of that product}) \tag{4}$$

Table 20 **Detonation velocities and pressures of explosives**

Explosive	Detonation velocity			Detonation pressure		
	Charge density, g/cc	Velocity, m/sec	Ref.	Charge density, g/cc	Pressure, kbars	Ref.
Molecules						
DATB	1.80	7600	5	1.80	251	5
	1.78	7600	2	1.78	251	2
Explosive D	1.48	6700	5			
HMX	1.89	9110	1	1.90	387	1
				1.90	395	2
HNS	1.70	7000	70	1.69	208	70
NQ	1.78	8592	5	1.72	245	5
NTO	1.871	8120	71	1.853	260	6
	1.855	8200	71			
PETN	1.77	8310	72	1.77	320	72
				1.77	340	1
RDX	1.806	8950	5	1.767	337.9	61
				1.80	341	60
TATB	1.847	7660	65	1.847	259	65
Tetryl	1.73	7720	5	1.61	226.4	73
				1.68	239	69
TNETB	1.78	8460	5	1.69	265	5
TNT	1.64	6950	60	1.637	189.1	61
				1.64	177	60
lead azide	4.00	5180	4	3.7	158	74
	3.80	4500	75			
	4.60	5300	75			
lead styphnate	2.90	5200	74			
TNT-based						
Comp B	1.713	8018	61	1.67	264	
	1.717	7990	52	1.713	292.2	
Cyclotol 75/25	1.62	7950	5	1.62	265	
Destex	1.69	6745	5	1.695	179	5
H-6	1.76	7490	62	1.76	245	
HBX-1	1.712	7307	64	1.712	220.4	63, 64
HBX-3	1.85	7530	65	1.85	206	65
HTA-3	1.90	7866	4			
Minol II	1.82	6925	5			
Minol IV(Q)	–	5900	66			
Octol 75/25	1.81	8364	67	1.821	343.7	68
Pentolite	1.64	7530	5	1.66	280	1
	1.68	7650	69	1.68	246	5
				1.68	251	69
Picratol	1.63	6970	4			
Tritonal	1.69	6520	52			
Pressed explosives						
CH-6	1.72	8550	5			
Comp A-3	1.67	8520	5		260	5
Comp A-4	1.65	8956	5			
Comp A-4 Alum.	1.79	7140	5			
Comp A-5—I	1.65	8931	5		298	5

where the right-hand side is summed over all the detonation product molecular species. A knowledge of the number of moles of each species in the products is required to complete the calculation. If this is not known, as is usually the case, then estimates of the decomposition products can be obtained using arbitrary decomposition schemes in which the type of species present in the products is selected arbitrarily. Two decomposition schemes used quite frequently[60] are 1) the H_2O-CO_2 arbitrary decomposition scheme and 2) the $H_2O-CO-CO_2$ arbitrary decomposition scheme. The names of these arbitrary schemes reflect the priority in which the H_2 and C are oxidized. For example, in the first scheme all of the H_2 is oxidized by the oxygen in the decomposition products before any of the carbon is oxidized, all of which is oxidized to CO_2. In the second scheme, H_2 is given priority and the carbon is oxidized by the remaining oxygen, first to CO, then to CO_2, if sufficient quantities are available. As an example of the use of these schemes, consider an explosive whose elemental composition consists of only carbon, hydrogen, nitrogen, and oxygen and whose chemical formula is $C_aH_bN_cO_d$. Under the first scheme, the decomposition products are

$$C_aH_bN_cO_d \rightarrow (c/2)N_2 + (b/2)H_2O + [(d/2) - (b/4)]CO_2 + [a - (d/2) + (b/4)]C \tag{5}$$

With the decomposition products from Eq. (5), the chemical energy released can be calculated from Eqs. (3) and (4):

$$Q_c = \frac{28.9b + 47.0[d - (b/2)] + \Delta H_f(\text{explosive})}{12a + b + 14c + 16d} \tag{6}$$

Similar results can be obtained with the second decomposition scheme. The concept of an arbitrary decomposition scheme has been extended to non-CHNO explosives, such as those containing aluminum and ammonium perchlorate. In the former case, the aluminum is assumed to be oxidized completely by any available oxygen to form Al_2O_3. In the latter case, the chlorine in the AP is assumed to react with any hydrogen available to form HCl. This process can be further extended to other elemental species, such as fluorine, by the proper selection of reasonable decomposition product(s).

These decomposition schemes should be considered only as rough engineering approximations to the actual decomposition products and their energy release. The actual decomposition product and its energy are complex functions of chemical reaction kinetics, temperature, and expansion volume. Using the TIGER code, Renner and Short[83] have calculated the decomposition products of a large number of explosives as a function of expansion volume under conditions representing those found in underwater detonations. These calculations should be reviewed to develop an appreciation for the amount of detail lost with the use of arbitrary schemes. Table 21 lists the chemical decomposition energies of a number of explosive compositions calculated using the two decomposition schemes. These value should be considered as the maximum possible energy released upon detonation. In all practical applications, significantly less energy is released during the detonation process[84] than the calculated values of Table 21.

Figure 3 illustrates the approximate times available for the release of explosive energy in order for the energy to make a contribution to a number of detonation phenomena or free-field effects.[85] The upper block of the figure shows the approximate times for the given materials to release their energy. The single organic molecule HMX can decompose, releasing its energy, within 10^{-4} ms. When mixed with a binder to produce a practical explosive, such as LX-04 (HMX/Viton = 85/15), energy release may take slightly longer than 10^{-4} ms. The use of inorganic oxidizers and/or aluminum fuels to increase the total energy of the explosive will require 0.1 to 100 ms for the energy to be released. The lower block of the figure shows the approximate maximum times available for the explosive's chemical reactions to release their energy and still make a contribution to the effects given. Any energy released before 10^{-4} ms will make a contribution to the detonation velocity or pressure. To make a contribution to metal acceleration (shaped charges, fragmenting warheads), energy must be released within 0.1 ms, whereas energy contributions to the underwater shockwave must be made within one ms.[81b] The effects that allow the longest time, as much as 100 ms, for energy to be released are underwater bubble energy, airblast, and cratering.

A comparison of the times involved in the two blocks provides a fundamental understanding of the behavior of ideal and nonideal explosives and an understanding of the design philosophy of the various types of explosives in existence. Even simple formulations, such as LX-04, release some of their energy on time scales longer than that which permits a full contribution to the velocity or pressure (or, from another perspective, release some energy beyond the C-J plane), but they release their energy quickly enough to contribute to longer-time-scale phenomena, such as metal acceleration. The time scales involved in the use of oxidizers or metal fuels are far greater than the velocity time scale but are of the same order of magnitude or less than those for underwater shockwave and bubble, airblast, and cratering. Although shown to be on approximately the same time scales, the use of oxidizers or metal fuels in metal-accelerating applications has not been shown to be effective, even when efforts are made to use fine-particle-size materials to enhance the reaction times.[85] The best way to accelerate metal is to use a pure organic explosive. The use of oxidizers or metal fuels will always increase the total energy of the composition, but their incorporation into an explosive is made at the expense of a fast-reacting organic molecule. Thus, the detonation velocity and pressure and metal-accelerating capability will decrease with their use, as indicated by Fig. 3, but the additional total energy will be able to enhance the underwater explosive, air-blast, and cratering performance. This is the explanation for the selection of specific explosives for certain types of warheads and the explanation of classification of explosives by warhead effect (Tables 8 and 12). To drive metal, organic explosives, such as octol, PBXN-110, or PBXN-107, should be used. For air-blast and general purpose applications, aluminized explosives, such as H-6 or PBXN-109, should be used. For underwater and cratering applications, the nonideal

Table 21 Decomposition energies of explosives

Explosive	H_2O–CO–CO_2 arbitrary decomposition energy cal/g	cal/cc	H_2O–CO_2 arbitrary decomposition energy cal/g	cal/cc
Molecules				
Ammonium nitrate	353	609	353	609
Ammonium perchlorate	324	632	324	632
DATB	877	1612	1174	2157
Explosive D	752	1291	1087	1866
HMX	1198	2281	1476	2811
HNS	954	1660	1366	2377
NQ	884	1600	884	1600
NTO	740	1429	1057	2040
PETN	1383	2463	1513	2694
RDX	1204	2174	1482	2676
TATB	836	1619	1075	2083
Tetryl	1026	1774	1420	2457
TNT	977	1616	1295	2141
TNT-based				
Comp B	1120	1935		
Cyclotol 75/25	1147	2024		
Destex	1866	3228		
H-6	2031	3684		
HBX-1	1918	3408		
HBX-3	2553	4848		
HTA-3	2138	4161		
Octol 70/30	1132	2062		
Octol 75/25	1143	2097		
Pentolite	985	1689		
Picratol	860	1449		
TNT	977	1616		
Tritonal	1967	3526		
Cast PBXs				
PBXN-103	2243	4233	2243	4233
PBXN-105	2112	4067	2112	4067
PBXN-106	1123	1866	1323	2200
PBXN-109	1870	3109	1870	3109
PBXN-110	1244	2087	1355	2275
PBXW-107	1939	3434	1939	3434
PBXW-108	1238	1958	1327	2099
PBXW-109	1885	3165	1885	3165
PBXW-114	1670	2879	1670	2879
PBXW-115	2025	3625	2025	3625

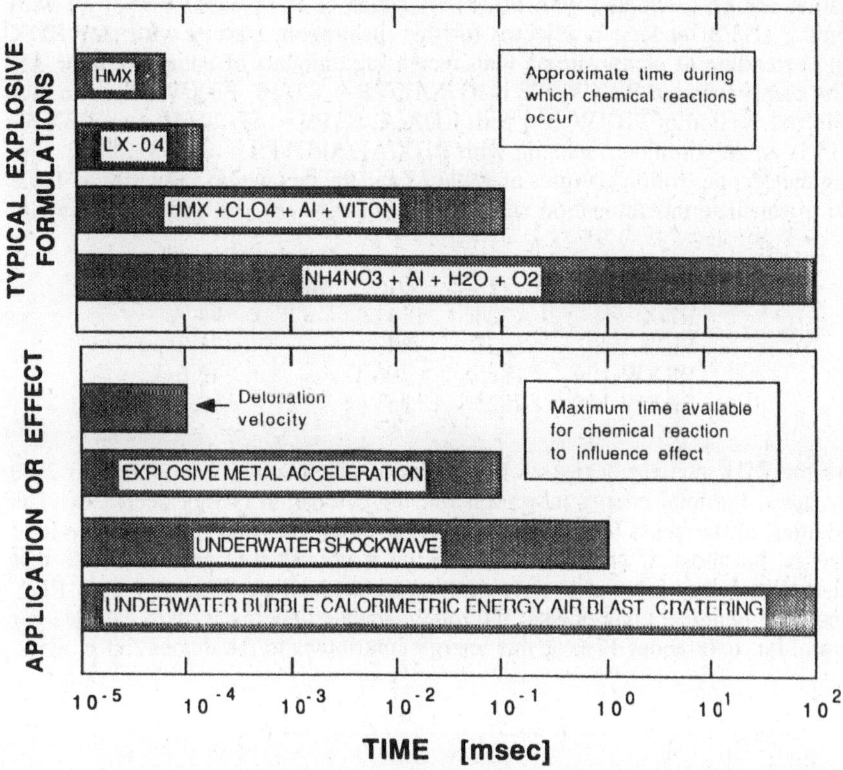

Fig. 3 Approximate variation of reaction times for typical explosive formulations.

AP and aluminized explosives, such as PBXN-103, PBXN-105, or PBXW-115, should be used.

The three quantities, pressure (P), velocity (D), and energy (Q) are related:

$$P = \rho D^2/(1 + \gamma) \tag{7}$$

$$Q_D = D^2/[2(\gamma^2 - 1)] \tag{8}$$

$$Q_D = P/2(\gamma - 1)\rho \tag{9}$$

where γ, the adiabatic exponent, defines the pressure-volume slope in the adiabatic expansion of the detonation gases, ρ is the explosive charge density, and Q_D is the energy released that drives the detonation. The units used for the variables in these equations are as follows: P, dynes/cm^2; D, cm/s; ρ, g/cm^3; and γ, unitless. If D is expressed in m/s, then Eq. (7), for example, becomes

$$P = 10^{-5} \rho D^2/(1 + \gamma)$$

Figure 4a illustrates the difference in an explosive's total energy calculated using the $H_2O-CO-CO_2$ decomposition scheme and Eq. (3) and the detonation energy calculated with Eq.(8). A series of RDX-based PBXs, all with similar HTPB binders, is selected for this illustration, starting with pure RDX and extending to compositions with increasing amounts of either inerts or AP. The composition of PBXW-108 is RDX/HTPB = 85/15. PBXW-109 is an aluminized version of PBXW-108 with RDX/Al/HTPB = 65/20/15, and PBXW-115 is an AP/aluminum version with RDX/AP/Al/HTPB = 20/43/25/12. Using the decomposition energies of Table 21 and the detonation velocities of Table 20 to calculate the detonation energy with Equation (8), we find the following:

Explosive	Q	Q_D	Detonation energy as percent of total energy
RDX	1204	1141	94.7
PBW-108	1238	883	71.3
PBXW-109	1885	796	42.6
PBXW-115	2025	351	17.3

As the RDX content decreases in this series of materials from 100% to 20% by mass, the total energy increases, but the detonation energy decreases. The addition of the inerts (aluminum and HTPB binder) and AP increases the total energy, but most, if not all, of this energy is not released on time scales that allow the energy to contribute to driving the detonation. With only 20% RDX and 80% nonideal components, the total calculated energy is 2025 calories per gram, but only about 17% of this energy contributes to the detonation process.

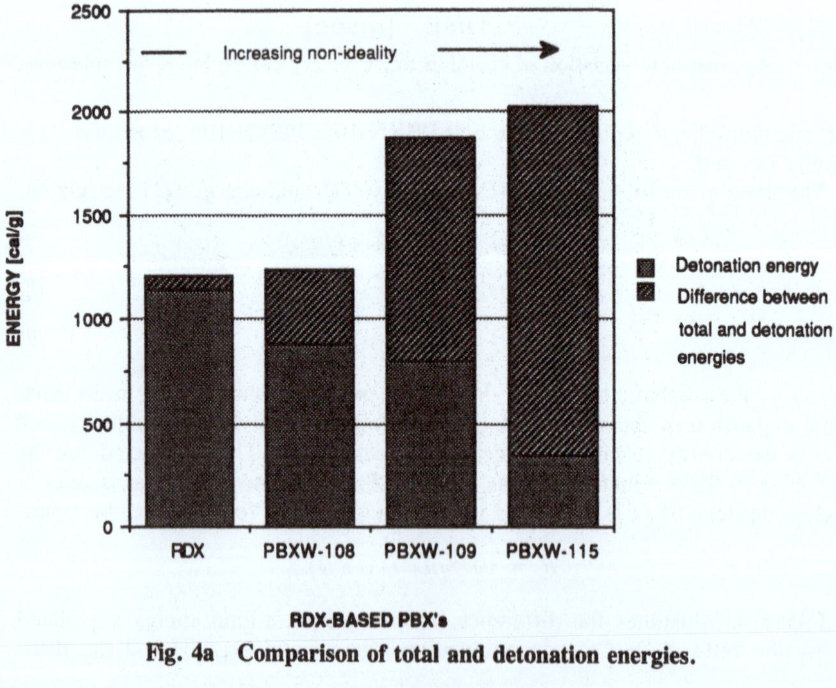

Fig. 4a Comparison of total and detonation energies.

This reduction in detonation manifests itself in a reduction in the detonation performance (Fig. 4b). The excess energy that is not applied to the detonation can be used in the other longer-time-scale applications indicated in Fig. 3. For example, the increased total energy obtained with the use of aluminum and AP in PBXW-109 and PBXW-115 can be exploited in general purpose/air-blast and underwater applications, respectively.

2.3 Predictive Techniques

A variety of predictive techniques have been developed that relate the chemical and physical properties of an explosive to its charge detonation behavior, free-field effects, and target damage capability (as illustrated in Table 18). Some of the fundamental properties of explosives used in the techniques or models are 1) density, 2) chemical composition, 3) chemical formula, 4) heat of formation, and 5) energy. These fundamental properties have been correlated using empirical techniques, usually simple curve fitting, to explosives detonation behavior (detonation velocity, pressure, and energy), to free-field effects (air shock peak pressure and impulse; underwater shockwave and bubble energies; fragment velocities), and to damage capability (shaped charge penetration; submarine hull

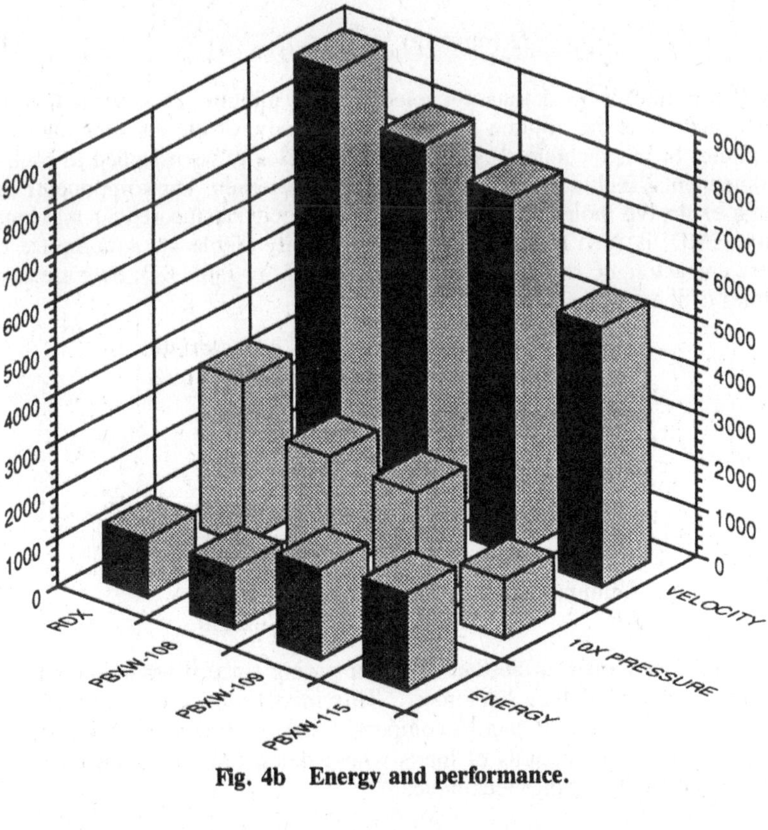

Fig. 4b Energy and performance.

rupture). These techniques are used quite often to make engineering assessments of explosive performance in cases where there is a lack of a complete or uniform database. They are particularly useful in early stages of explosive development work to predict the effect of compositional variations on detonation performance, free-field effect, or damage, especially when chemical and physical data are lacking, thus precluding a more sophisticated analysis being made by computer codes, such as TIGER.

The earliest work attempted to relate the explosive's energy, calculated by arbitrary decomposition schemes [Eqs. (5) and (6)], to detonation performance and free-field effects.[60,84,86] For mixtures, such as those found in typical military explosives, additive equations were used to predict the behavior of the mixture from the behavior of the components[86]:

$$A = \sum x_i A_i$$

where A is the property of the mixture in question, such as detonation velocity, A_i is the same property of the ith component of the mixture, and x_i is the percentage or fraction of the ith component in the mixture. These equations can be expressed on either a mass or a volume basis.

The additive principle was applied in the late 1940s by Urizar of LANL to predicting detonation velocities of mixtures[1,87]:

$$D\ (\text{mixture}) = \sum \phi_i D_i \tag{10}$$

where D (mixture) is the detonation velocity of the mixture, D_i is the detonation velocity, and ϕ_i is the volume fraction, respectively, of the ith component of the mixture. In this technique, a "characteristic velocity" is assigned to each of the components, including air, voids, and inert materials. For components that are high-explosive molecules, the detonation velocity at theoretical maximum density (TMD) is used as the characteristic velocity (Table 1). A complete list of these characteristic velocities is provided in Ref. 1, Table 8-3; here are some examples of these velocities:

Material	**Density (g/cm^3)**	**Characteristic Velocity (m/s)**
Beeswax	0.92	6500
Estane	1.20	5520
Polyethylene	0.93	5550
Teflon	2.15	5330
Viton A	1.82	5390
Air or void	—	1500
Aluminum	2.70	6850
AP	1.95	6250

This model is somewhat aesthetically unpleasing since it assumes that inert materials behave as if they "detonate." This model works well in curve fitting of detonation velocity data to component volume fractions, especially for explosives with large amounts of inerts whose detonation velocities cannot be accurately modeled by other techniques.[87]

Figure 5 illustrates a calculation of the detonation velocities at theoretical maximum density conditions of the cyclotol family of materials (RDX+TNT = 100%) using Eq. (10) in the form

$$D(\text{mixture}) = \sum \phi_i D_i = D(\text{RDX})\ \phi(\text{RDX}) + D(\text{TNT})\ \phi(\text{TNT})$$

with values for the detonation velocity at TMD for RDX and TNT from Table 22

$$D(\text{RDX}) = 8797 \text{ m/s}$$
$$D(\text{TNT}) = 7020 \text{ m/s}$$

Also shown for comparison are experimental values for specific cyclotols, from the cited references, that have been corrected to TMD conditions.

A number of empirical curve-fitting techniques have been developed that relate the detonation velocity and/or pressure to the more fundamental physical and chemical properties, such as the heat of detonation and moles of gaseous

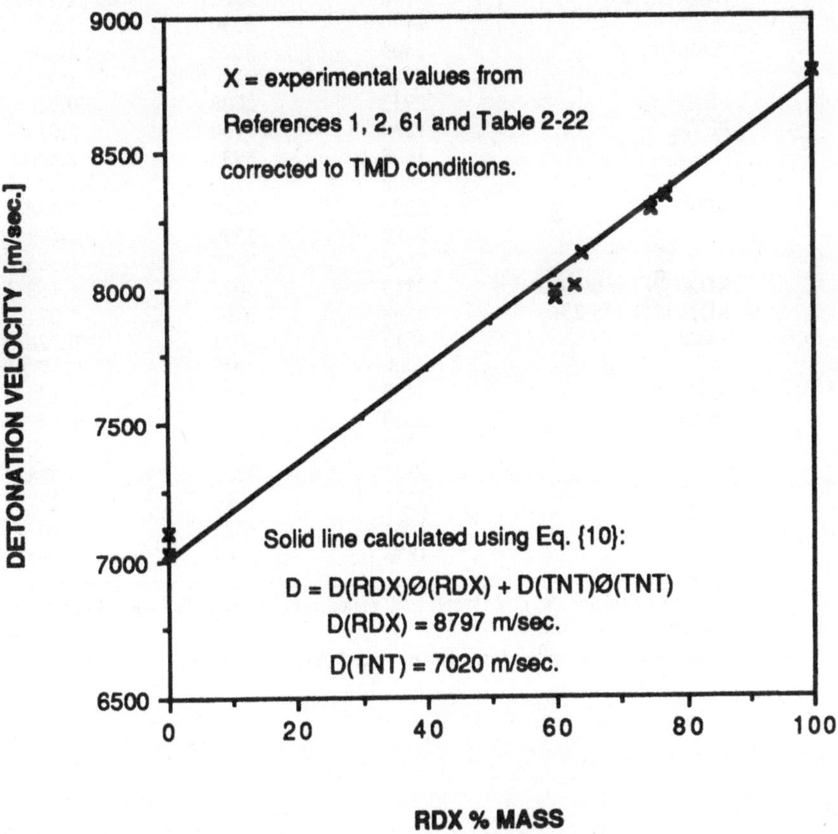

Fig. 5 Detonation velocity of RDX/TNT mixtures.

Table 22 Detonation velocity-density equations for explosives $D = A + B(density)$; $D = A[1 + C(density)]$

Explosive	Density equation parameters		
	A	B	C
CH-6	1533	3988	2.601
Comp A-3	1618	4118	2.545
DATB	2512	2834	1.128
	2495	2834	1.136
Explosive D	2323	2921	1.257
HBX-1	540	3960	7.333
	−63	4305	
HMX	621	4506	7.256
Octol (75/25)	2525	3265	1.293
	2120	3330	1.571
Pentolite	2506	3022	1.206
	2330	3170	1.361
PBXN-7	2932	2668	0.910
PETN	1839	3700	2.012
	1608	3933	2.446
	1820	3700	2.033
RDX	2509	3482	1.388
	2490	3590	1.442
	2560	3470	1.355
RDX/TNT (60/40)	2713	3073	1.133
RDX/TNT (75/25)	2568	3264	1.271
TATB	4813	1571	0.326
	343	3940	11.487
	2480	2852	1.150
Tetryl	2610	3020	1.157
	2742	2935	1.070
	2375	3225	1.358
TNT	2256	2880	1.277
	1785	3225	1.807
	1873	3187	1.702

[a]Ref. 88 calculated the given result from experimental or compiled data reported in the other cited references.

[b]Calculated by the author from data presented in cited reference.

Table 22 (continued)

Density range of validity			TMD,	D(TMD),
From	To	Ref	g/cc	m/s
1.450	1.720	88[a], 8	1.774	8608
1.000	1.670	88, 3, 5, 8, 102	1.672	8503
1.210	1.790	88	1.837	7718
–	–	1	1.837	7701
1.490	1.550	88, 94	1.717	7338
		102	1.717	7339
		64	1.717	7329
1.770	1.900	88, 94	1.905	9205
1.210	1.810	88, 1, 3, 94	1.835	8516
		102	1.835	8231
0.998	1.780	88, 1, 40, 94	1.715	7689
		102	1.715	7767
1.560	1.830	109[b]	1.885	7961
0.250	1.780	88, 40, 94	1.78	8425
0.570	1.590	5	1.78	8609
		103, 104	1.78	8406
0.900	1.800	88, 3, 94	1.806	8797
		102	1.806	8974
>1.00		1	1.806	8827
1.020	1.720	88, 1, 94, 102	1.742	8066
1.000	1.760	88, 1, 2, 94	1.765	8329
1.800	1.895	88, 3, 94	1.938	7858
>1.2		1	1.938	7979
		2	1.938	8007
1.000	1.730	88, 3, 8, 40, 94	1.73	7835
		2	1.73	7820
		103, 104	1.73	7954
1.000	1.650	88, 94, 3	1.654	7020
		103, 104	1.654	7119
		2	1.654	7144

detonation products[89] or to the chemical composition and structure of the high explosive.[90] One of the most widely used techniques involves the Kamlet-Jacobs equations,[91–94] which relate the detonation velocity and pressure of CHNO explosives to their chemical composition, density, and heat of formation:

$$P[\text{kb}] = 15.58\rho^2\phi \tag{11}$$

$$D[\text{mm}/\mu\text{s}] = 1.01(1 + 1.3\rho)\phi^{1/2} \tag{12}$$

$$\phi = NM^{1/2}Q^{1/2} \tag{13}$$

where
N = moles of gaseous detonation product per gram of high explosive
M = average molecular weight of the gaseous detonation products
Q = chemical energy of the detonation reaction

The CHNO explosive is assumed to decompose according to the H_2O–CO_2 arbitrary decomposition scheme with decomposition products given by Eq. (5). Q is calculated using Eq. (6). From the definitions of N and M, these parameters, for an explosive with the chemical formula $C_aH_bN_cO_d$, are equal to

$$N = \frac{2c + 2d + b}{48a + 4b + 56c + 64d}$$

$$M = \frac{56c + 88d - 8d}{2c + 2d + b}$$

The Kamlet-Jacobs equations are particularly useful and accurate for predicting the detonation performance of new, single-molecule explosives if the chemical formula, density, and heat of formation are known. They also accurately predict the performance of mixtures of high explosives and of mixtures of high-explosive molecules in energetic binders. These equations significantly underpredict the performance of many PBXs consisting of high-explosive molecules in inert binders, such as PBXN-110. The inability to predict performance of mixtures with inerts becomes significant at binder contents of at least 9% by mass; this includes all inert binder castable PBXs. Since the equations were developed for CHNO explosives, the equations do not, of course, apply to any explosive containing AP or aluminum. For explosives containing AP, aluminum, or appreciable amounts of inert materials, the additivity equation [Eq. (10)] must be used. If the additivity equation is used to calculate a detonation velocity for a material, or if an experimental detonation velocity is available and an estimate of the detonation pressure is desired, then the pressure can be calculated by using the velocity (designated here as D_e) value in Eq. (12) and solving for ϕ:

$$\phi_e = \{D_e/[1.01(1 + 1.3\rho)]\}^2 \tag{12a}$$

$$P_e = 15.58\rho^2\phi_e \tag{11a}$$

where the subscript e refers to the values obtained with the use of the experimental or additivity velocity.

For the Kamlet-Jacobs equations, an artificial molecule of the form $C_AH_BN_CO_D$ can be calculated from two or more component molecules with chemical formulas $C_{a1}H_{b1}N_{c1}O_{d1}$, $C_{a2}H_{b2}N_{c2}O_{d2}$, and so on, by setting

$$A = \sum (m_i/MW_i)a_i$$

$$B = \sum (m_i/MW_i)b_i$$

$$\vdots$$

where

m_i = mass of the ith component of the mixture
MW_i = molecular weight of the ith component of the mixture
m_i/MW_i = n_i, the number of moles of the ith component of the mixture

The heat of formation of the physical mixture, ΔH_f (mixture), can be calculated by assuming that each component contributes to the heat of formation of the physical mixture in proportion to the number of its moles present in the mixture:

$$\Delta H_f(\text{mixture}) = \sum (m_i/MW_i)\Delta H_{f_i} = \sum n_i \Delta H_{f_i}$$

The density of the mixture is calculated by Eq. (16). With this calculated information, the detonation velocity, pressure, gamma, and the Gurney constant (see below) can be calculated with the use of Kamlet-Jacob equations and other equations derived from them.

A number of empirical equations have been developed to relate the adiabatic exponent, γ, to the charge density, ρ, for organic explosives:

1) Defourneaux[95]: $\gamma = 1.90 + 0.6\rho$
2) Johansson and Persson[96]: $\gamma = \rho/(0.140 + 0.260\rho)$
3) Hardesty and Kennedy[97]: $\gamma = \rho/(0.136 + 0.238\rho)$
4) Kamlet and Short[98]: $\gamma = (0.655/\rho) + 0.702 + 1.107\rho$

The latter relationship can be obtained by the simple substitution of the Kamlet-Jacobs equations for P and D into Eq. (7) and solving for γ. These empirical relationships were developed from data obtained on CHNO-type explosives and caution should be exercised when extending the equations beyond their database. For explosives containing aluminum and/or ammonium perchlorate, γ can be calculated by solving for γ in Eq. (7), if the detonation velocity and pressure are known:

$$\gamma = (P/\rho D^2) - 1 \tag{14}$$

Values of γ obtained on military explosives range in approximate values from 2.50 to 3.30 (see Table 19 for some representative values).

2.4 Factors Affecting Charge Detonation

A number of factors affect the detonation velocity, pressure, and energy of an explosive charge. In the discussion of Figure 1, it was noted that, for a given explosive and a fixed diameter d, the detonation velocity and pressure would be constant (steady state) over the length of the charge provided the diameter was greater than the failure diameter of the explosive. Within experimental error, repeated measurements on the same explosive at the same diameter will yield the same value for the velocity and pressure. The focus of this section will be on the deviations from these conditions. What are the deviations such that it is no longer the "same explosive"? What is the effect of not having the same diameter?

The most important factors that affect the detonation velocity, pressure, and energy of an explosive charge are 1) charge density, 2) temperature, 3) explosive composition, 4) charge diameter effects (including general diameter effect and failure diameter effect), and 5) confinement. In addition to affecting charge performance, these factors also have a role in the sensitivity and safety characteristics of explosives.

2.4.1 *Effect of Explosive Charge Density*

The maximum density that can be possessed by a particular explosive is called its theoretical maximum density (TMD). For high-explosive crystalline powders, the TMD is obtained experimentally by x-ray crystallographic techniques for standard materials, such as those listed in Table 1, or it can be estimated using additivity methods based upon atomic or constituent volumes[99–101] for new materials. For a mixture of materials of total mass m_T the density ρ can be calculated by

$$\begin{aligned}\rho &= m_T/V_T \\ &= m_T/\sum[(m_i/\rho_i) + V_v]\end{aligned} \tag{15}$$

where

V_T = total volume
m_i = mass of ith component
ρ_i = density of the ith component
V_v = volume of voids or air in sample
$m_T = \sum m_i$

and where the summation is over all the components of the composition. If there are no voids or air in the sample ($V_v = 0$), then the sample is at its theoretical maximum density (ρ_{TMD} = TMD) and Eq. (15) reduces to

$$\mathrm{TMD} = \rho_{\mathrm{TMD}} = m_T/\sum(m_i/\rho_i) \tag{16}$$

The densities of the high-explosive powders in Table 1 are experimentally determined, as noted, whereas almost all reported TMDs for explosive compositions, including those listed in Tables 9 and 13, are calculated values using Eq. (16). Equations (15) and (16) assume there is no interaction between the component ingredients, such as chemical reactions or one component being soluble in another, that would reduce the components' volume in the mixture. This is the case for almost all TNT-based materials; there is little or no interaction between materials such as TNT, RDX, HMX, and aluminum. The TMD of many PBXs, but not all, can be calculated with the use of Eq. (16). Care should be taken in considering the possible interactions between the complex components of a typical PBX. The TMD values cited from Ref. 34 in Table 13 were calculated using Eq. (16), for example.

The density of a material can be expressed as a percent of the TMD:

$$\%\ \text{TMD} = 100\rho/\rho_{\text{TMD}}$$

When an explosive has a density less than the TMD, it means that there is entrapped air present in the explosive, either in the form of distinct voids or in the form of air dispersed throughout the sample.

The effect of explosive density on charge detonation velocity D is usually represented by a linear equation in ρ:

$$D(\rho) = A + B\rho \tag{17a}$$

$$= A(1 + C\rho) \tag{17b}$$

where the parameters A, B, and C $(= B/A)$ are experimentally determined by measuring the detonation velocity using large-diameter charges of different densities. A linear least-squares fit to the $D(\rho) - \rho$ data will yield the parameters A and B. Higher order polynomials are occasionally used to represent the $D(\rho)-\rho$ data.[1,5]

Table 22 lists the velocity-density equation parameters A, B, and C used in Eqs. (17a) and (17b) for a number of explosives. Data of this type have been acquired primarily on pressed compositions, such as CH-6 or Comp A-3, or the high-explosive powders that can be compressed to a high percentage of the TMD. Some data exists on castable TNT-based materials, such as HBX-1, octol, and the cyclotols, which were manufactured as cast materials. Since they are crystalline materials, TNT-based explosives can be readily ground to a fine powder and pressed into charges of the appropriate density for the detonation testing. Because they are rubbery, elastomeric materials, it is much more difficult to grind castable PBXs into fine powders that represent a sample of the composition. Thus, no systematic studies have been published on the effect of density on performance of castable PBXs. In warhead applications, pressed compositions are pressed quite often to densities well below the TMDs for sensitivity reasons, whereas TNT-based castable explosives are quite often manufactured at densities below their TMDs because of void formation upon shrinkage and the lack of the use of vacuum casting to remove entrapped air. Castable PBXs, which almost universally use vacuum casting to remove entrapped air, are typically manufactured at high percentages of the TMD.

Using the TMD of a composition as the density in Eq. (16) allows the calculation of the detonation velocity of the material of TMD, D(TMD), which is an intrinsic property of the material. Table 22 lists these detonation velocities extrapolated to TMD conditions for explosives listed. Figure 6 illustrates the $D(\rho) - \rho$ behavior for the explosive powders RDX and TNT (Fig. 6a) and for pressed compositions CH-6 and Comp A-3 (Fig. 6b). The straight lines are linear fits of the data to Eq. (17a).

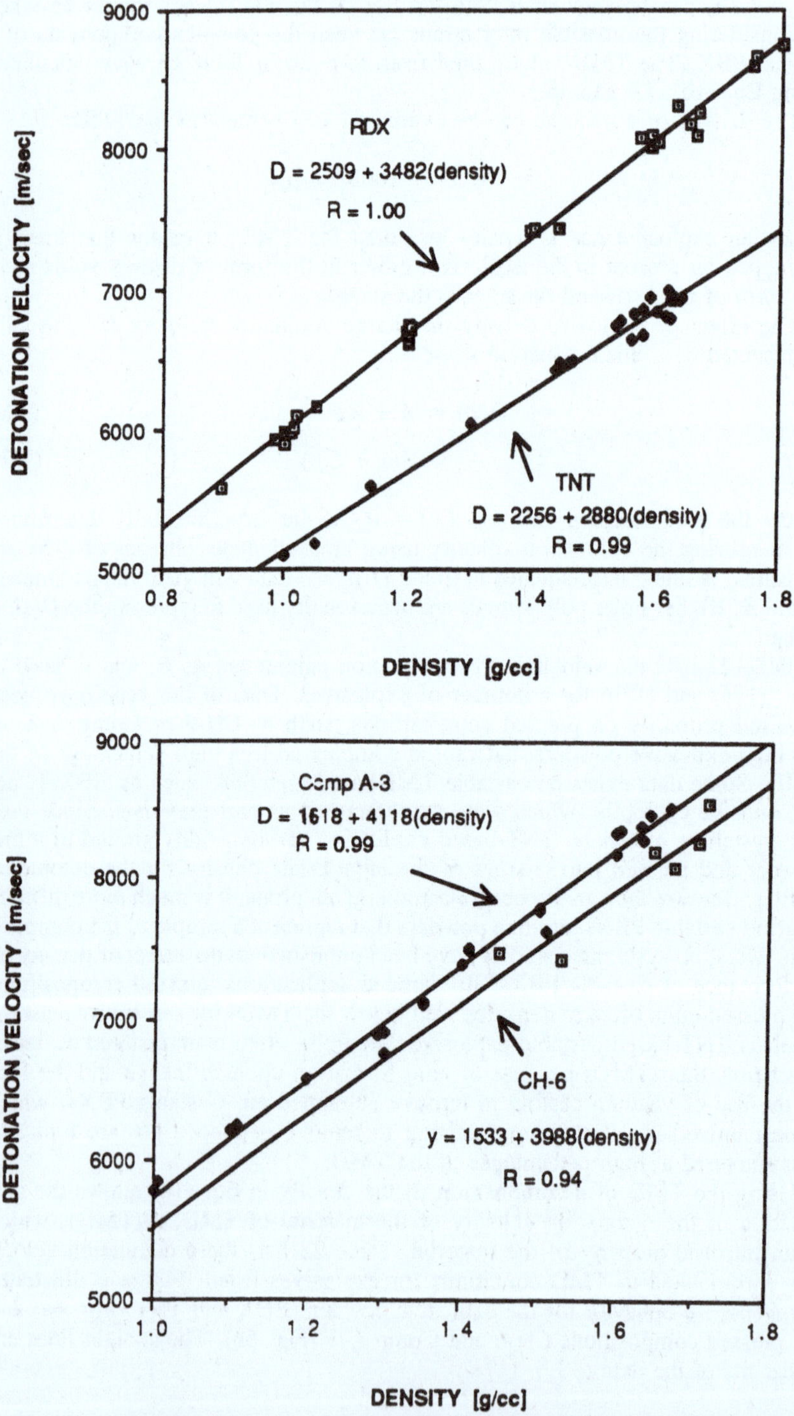

Fig. 6 Detonation velocities of RDX and TNT, CompA − 3 and CH − 6.

The behavior of detonation pressure as a function of density is usually represented by the equation

$$P(\rho) = A\rho^n \quad (18)$$

where A and n are experimentally determined parameters obtained from a least-squares fit of $P(\rho) - \rho$ data in the form $\ln P(\rho) = \ln A + n \ln \rho$. There are, as usual, considerably less $P(\rho) - \rho$ data than there are $D(\rho) - \rho$ data. Table 23 presents $P(\rho) - \rho$ data for five explosive powders. Using the TMD as the density in Eq. (18), the detonation pressure can be extrapolated to TMD conditions, as shown in the table. Figure 7 illustrates the $P(\rho) - \rho$ behavior of PETN. The

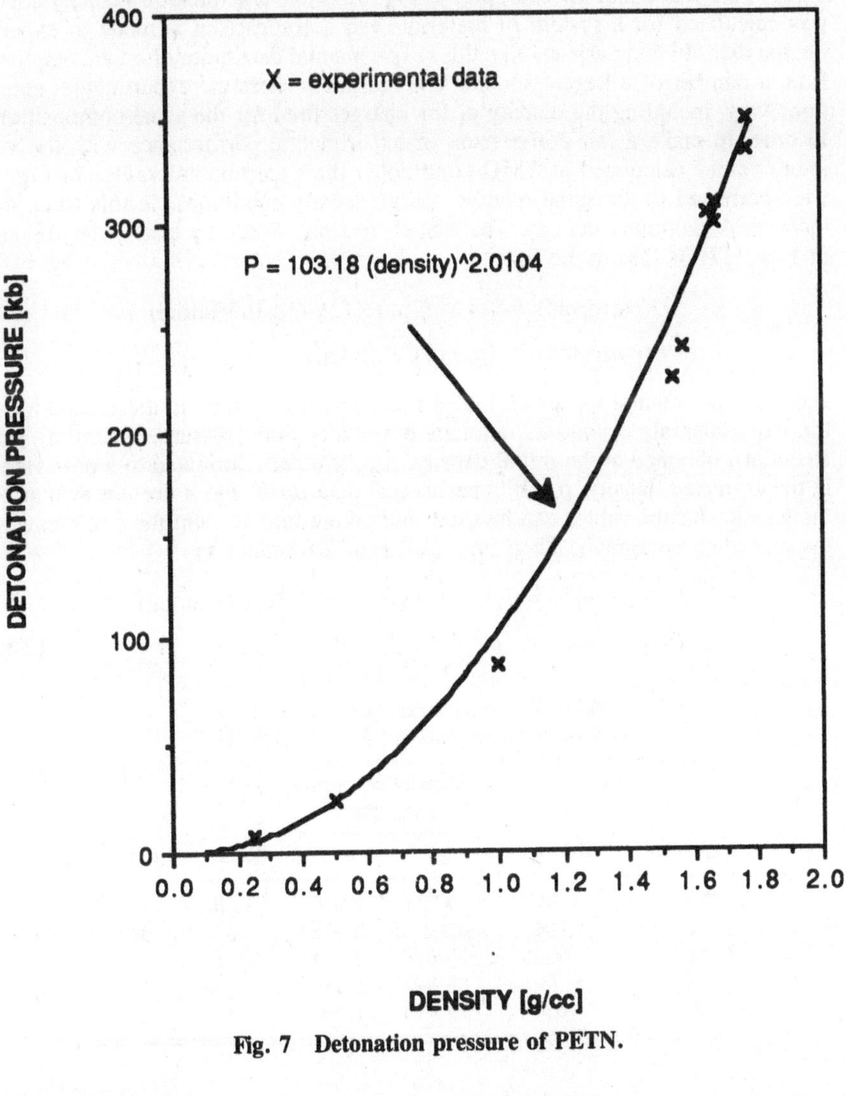

Fig. 7 Detonation pressure of PETN.

solid line represents calculated values using Eq. (18) with PETN values from Table 23 for the parameters A and n and the experimental data used to calculate the parameters represented by x's.

In their analysis, Kamlet-Jacobs[93,94] found that the best fit of the $D(\rho) - \rho$ and $P(\rho) - \rho$ data for a large number of different explosives yielded values of $A = 1.01\phi^{1/2}$ and $C = 1.3$, resulting in Eq. (12), and $A = 15.58\phi$ and $n = 2.00$, resulting in Eq. (11). A review of Tables 22 and 23 shows that considerable deviations from the $C = 1.3$ and $n = 2.00$ values can occur with individual explosives.

When comparing velocity or pressure data obtained at two different densities, it is necessary to correct the density of one value to the density of the other value. This was done, for example, in Fig. 5, where a detonation velocity curve was calculated for a system of materials and a comparison is made to experimental data. In comparisons like this, experimental data quite often are acquired from a number of different sources and facilities where the experimental conditions vary, including the density of the charges fired for the same composition. In order to make a fair comparison of experimental performance with the velocity values calculated at TMD conditions, the experimental values in Fig. 5 were corrected to the same relative charge density conditions, in this case, the theoretical maximum density. The best corrections would be made with the use of Eqs. (17) or (18) in the form

$$D(\text{corrected}) = [(1 + C\rho_c)/(1 + C\rho_i)]D(\text{initial}) \tag{19}$$

$$P(\text{corrected}) = (\rho_c/\rho_i)^n P(\text{initial}) \tag{20}$$

and the experimental values of Tables 22 or 23 for C and n. In these equations, the experimental, or initial, detonation velocity and pressure, D(initial) and P(initial), obtained at the initial density, ρ_i, have been corrected to a new value at the corrected density, ρ_c. If experimental data for C and n are not available, the Kamlet-Jacobs values can be used, but taking into account the comments in the preceding paragraph. Then Eqs. (19) and (20) reduce to

$$D(\text{corrected}) = [(1 + 1.3\rho_c)/(1 + 1.3\rho_i)]D(\text{initial}) \tag{19a}$$

$$P(\text{corrected}) = (\rho_c/\rho_i)^2 P(\text{initial}) \tag{20a}$$

Table 23 Detonation pressure-density equations for explosives: $P = A\ (\text{Density})^n$

		Density equation parameters			
Explosive	TMD, g/cc	A	n	Refs.	P(TMD)
PETN	1.78	103.18	2.0104	93, 40	329
RDX	1.806	103.152	2.1185	93	361
TATB	1.938	65.634	2.3543	40, 5	312
Tetryl	1.73	56.843	2.8846		2.76
TNT	1.654	75.975	1.9663	93	204

2.4.2 *Effect of Explosive Charge Temperature*

There are limited data on the effect of temperature on detonation velocity. Dobratz[1] has compiled some temperature effect data in the form

$$\Delta D/\Delta T = a \times 10^{-3}$$

where ΔD is in mm/μs and ΔT is in °C. Some of the reported values of a are

Explosive	a
Comp B (RDX/TNT=64/36)	−0.5
LX-04-1	−1.55
Pentolite	−0.4
PBXN-301(XTX-8003)	−2.34

2.4.3 *Effect of Explosive Composition Ingredients*

The most important factor affecting the explosive charge performance is the composition of the explosive itself. For example, whether the charge consists of a single explosive or a mixture of explosives or whether it consists of a mixture of a high-explosive powder and an inert has a profound effect on the charge performance and all other characteristics of the charge, including sensitivity, physical properties, and so on. The choice of a specific component or ingredient and its amount in an explosive will affect the detonation velocity, pressure, and energy. Most ingredient effect studies have evaluated the effect of one or more ingredients and their compositional percentage on only the detonation velocity. The most important components studied are 1) single high-explosive molecules, 2) mixtures of explosives, 3) inerts, and 4) metal fuels, especially aluminum. From the large number of possible combinations of these materials, a limited number of examples will be discussed below to illustrate the effect of these materials on detonation performance.

The detonation performance of single high-explosive molecules has been discussed throughout this chapter. The properties of the explosive molecules listed in Table 1 can be found in Tables 19 through 23, for example.

The effect of physically mixing two or more high-explosive molecules can be predicted by the use of computer codes or the use of the Kamlet-Jacob equations,[11,12] if the physical and chemical properties of the components are known, or by the use of the additivity equations [Eq. (10)] if only the detonation velocity is known. Some of the major characteristics exhibited by mixtures of explosive materials are illustrated in Fig. 8. In this figure, each curve represents the calculated detonation velocity (top) and pressure (bottom) for mixtures, at TMD conditions, consisting of TNT and one of following high-explosive (HE) powders: HMX, RDX, PETN, TATB, tetryl, DATB, or Explosive D. The detonation velocity of a given mixture was calculated using Eq. (10):

$$D(\text{mixture}) = \sum \phi_i D_i \tag{10}$$

$$= \phi(\text{TNT})D(\text{TNT}) + \phi(\text{HE powder})D(\text{HE powder})$$

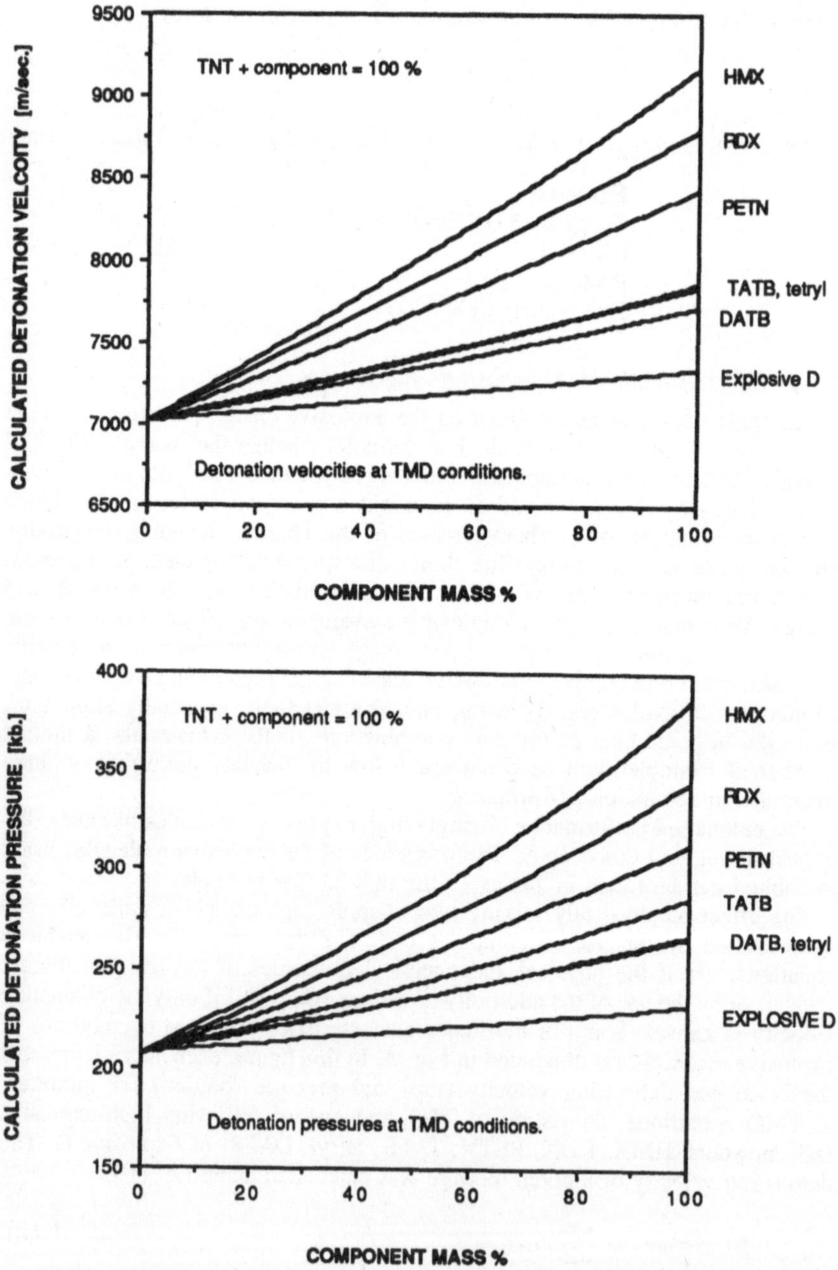

Fig. 8 Calculated detonation velocities and pressures of TNT mixtures.

Or, since $\phi(\text{TNT})+\phi(\text{HE powder})= 1.00$,

$$D(\text{mixture}) = D(\text{TNT}) + [D(\text{HE powder}) - D(\text{TNT})]\phi(\text{HE powder}) \quad (21)$$

The values for D(HE powder) and D(TNT), at TMD conditions, used in the calculations were those from Table 22. The pressures in Fig. 8b were calculated using Eqs. (11a) and (12a). The velocity curves [D (mixture) vs ϕ(HE powder)] are straight lines whose slopes are given by

$$\text{slope} = \partial D/\partial \phi = D(\text{HE powder}) - D(\text{TNT})$$

That is, the slopes are equal to the difference in the velocities of TNT and the HE powder. Although the velocities in Fig. 8a are plotted as a function of the percent mass (and not the volume fraction) of the HE powders, an almost linear slope is still observed. The greater the detonation velocity of the HE powder used, the greater the detonation velocity of the mixture for a given amount of powder. There is an upper limit on the amount of powder that can be loaded into a fluid, such as TNT in this case; practical castable compositions will be limited to less than 80% solid powder by mass. Figure 8b shows a nonlinear relationship between the detonation pressures and the amount of HE powder in TNT, reflecting the effect of the HE powders' density and energy differences.

The use of inert materials reduces an explosive's charge performance since they do not release energy upon decomposition in the detonation. Inerts are necessary components of many explosives and are used either as a desensitizer, as is the case with Comp D-2 in the HBXs and Comp B, or as desensitizing binders in PBXs. Most inert PBX binders absorb energy in decomposition. In addition, many inerts, such as inert PBX binders and Comp D-2, are low-density materials that reduce the overall density of the explosive composition. The deleterious effect of inerts on charge performance is illustrated in Fig. 9, which is a plot of the detonation velocity data as a function of the volume fraction occupied by the inert, ϕ(binder). Data obtained on two families of materials are shown: 1) castable compositions of HMX (Class III/Class II ratio = 3/1) in an HTPB binder and 2) pressed compositions of HMX coated with Viton. For each family of materials, an equation of the form of Eq. (21)

$$D = D(\text{HMX}) - [D(\text{HMX}) - D(\text{binder})]\phi\ (\text{binder})$$

is fit to the data. The D(binder) is the "characteristic velocity" of the inert: 5595 m/s for Viton and 6023 m/s for the HTPB binder.

Examples of the effect of aluminum content on detonation velocity and pressure are shown in Fig. 10 with pressed compositions of TNETB or RDX with aluminum powder. The data in the top graph were obtained on TNETB/Al charges pressed to densities at 94% to 95% of their respective TMDs. The detonation velocity and pressure data are presented as values relative to TNETB values at 94.94% TMD and as a function of the aluminum volume fraction. The data were obtained on five compositions consisting of TNETB filled with

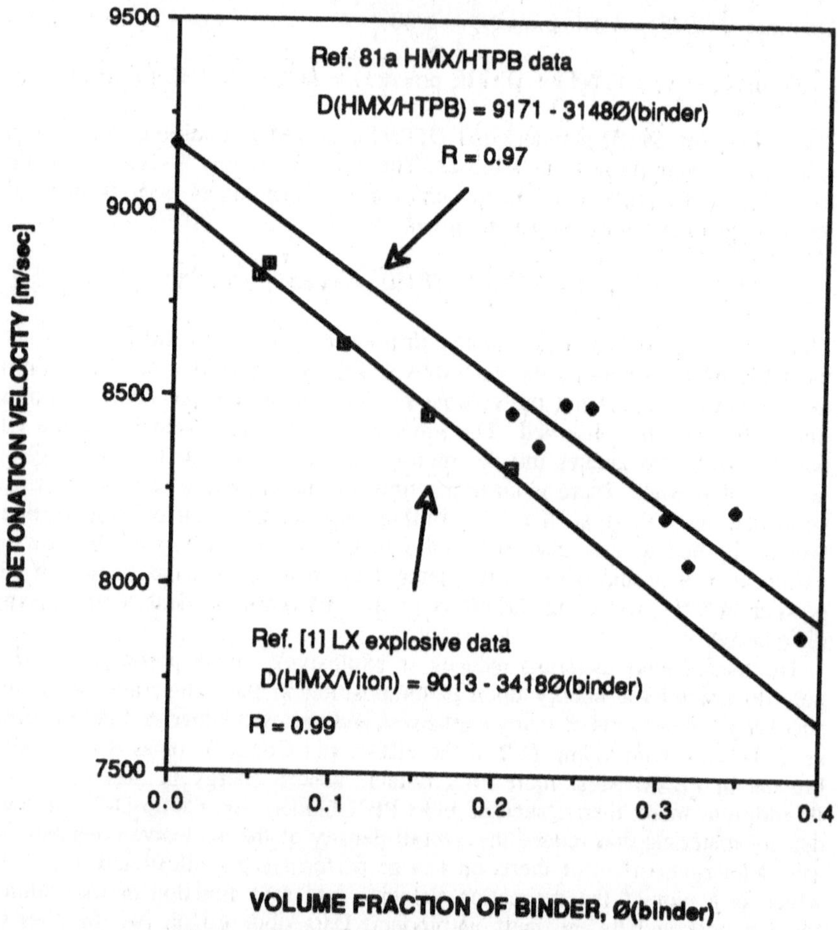

Fig. 9 Effect of inerts on detonation velocity.

0%, 10%, 20%, 30%, and 40% aluminum, respectively. The data in the lower graph were obtained on RDX/Al charges pressed to densities at 87.3% to 89.5% of their respective TMDs and are presented as values relative to RDX values at 88.8% TMD. The data were obtained on six compositions consisting of RDX-filled explosives with 0%, 10%, 20%, 30%, 40%, and 50% aluminum, respectively. Being basically inert within the reaction zone, aluminum provides little, if any, energy to drive the detonation, and thus its incorporation into an explosive composition will reduce both the velocity and pressure of the material, as indicated with the TNETB and RDX examples of Fig. 10. The presence of aluminum has a more pronounced effect on pressure than it does on velocity. Data obtained with the use of other metals in explosives, such as lead, show similar results.[40]

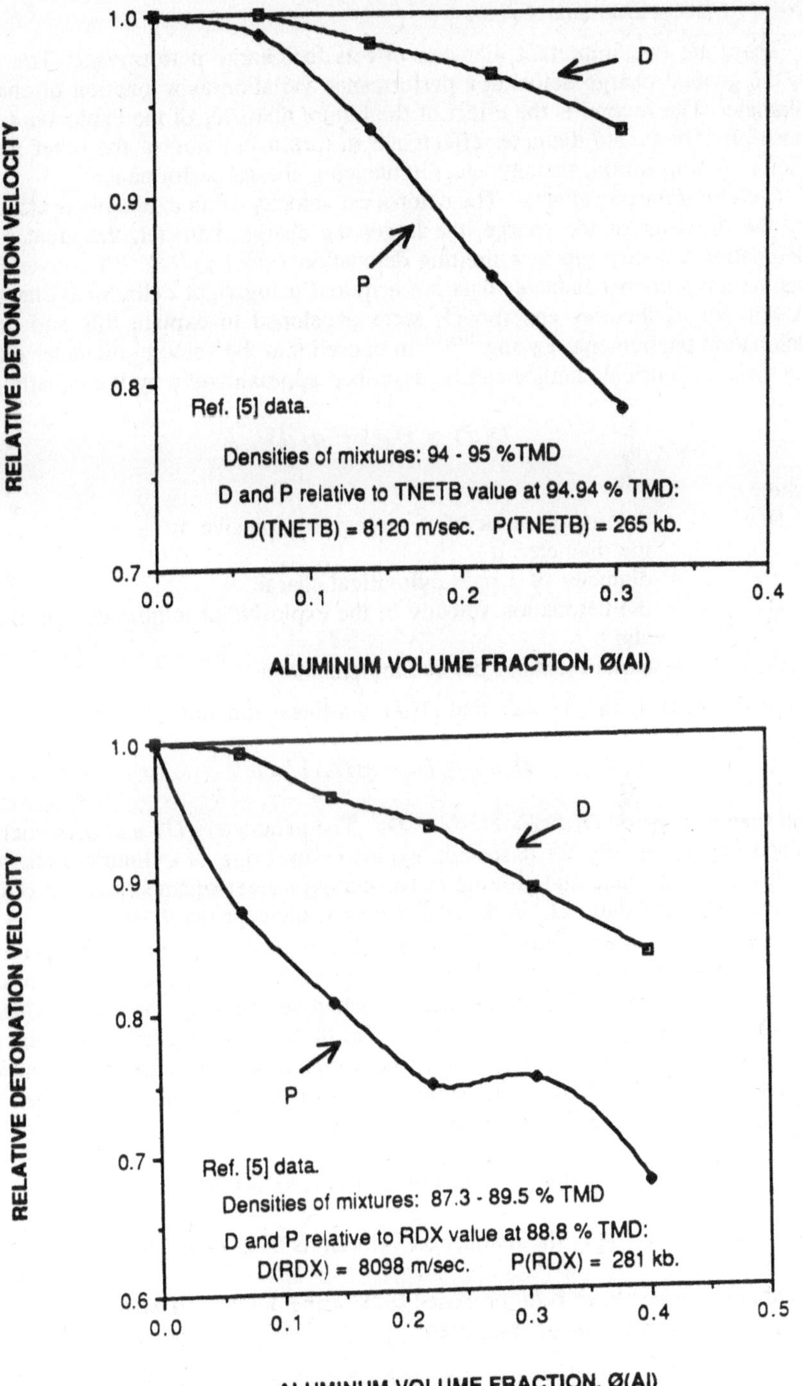

Fig. 10 Relative detonation velocities and pressures of TNETB/Al and RDX/Al.

2.4.4 *Charge Diameter Effects*

There are two important diameter effects for charge performance. The first is the general charge detonation performance variation as a function of charge diameter. The second is the effect of the failure diameter of the explosive composition. These two diameter effects are in turn a function of the other major factors (composition, density, etc.) influencing charge performance.

General diameter effect. The detonation velocity of an explosive is affected by the diameter of the charge: the larger the charge diameter, the greater the detonation velocity (up to a limiting detonation velocity, D_∞). By convention, detonation velocity-diameter data are acquired using right cylindrical charges. A number of theories and models were developed to explain this and other detonation phenomena. Eyring [105,106] proposed that the velocity-diameter effect for right cylindrical charges can be described approximately by the equation

$$D(d) = D_\infty(1 - a/d) \tag{22}$$

where

$D(d)$	=	detonation velocity of a specific explosive at the diameter, d
d	=	diameter of a right cylindrical charge
D_∞	=	the detonation velocity of the explosive at infinite charge diameter
a	=	experimentally determined parameter

From Eq. (22), it can be seen that $D(d)$ is a linear function of d^{-1}:

$$D(d) = D_\infty - aD_\infty(1/d)$$

and that as $d \rightarrow \infty$, $D(d) \rightarrow D(\infty) = D_\infty$. The parameters D_∞ and a are determined experimentally for a specific explosive by firing of cylindrical charges of different diameters and plotting or calculating a least-squares linear fit of the velocity vs. d^{-1} data. D_∞ is the velocity axis intercept (as $d \rightarrow \infty$, $d^{-1} \rightarrow 0$) and aD_∞ is the slope of the $D(d)$ vs d^{-1} curve. The infinite diameter detonation velocity (D_∞) is an intrinsic property of each explosive composition and is the velocity calculated by predictive techniques such as the Kamlet-Jacob equations. Because of practical limitations in firing extremely large charges, all reported infinite diameter detonation velocity data are linear extrapolations to infinite diameter conditions. The significance of the parameter a can be seen by setting the diameter in Eq. (22) equal to $100a$:

$$D(100a) = D_\infty(1 - a/100a) = 0.99D_\infty$$

At a diameter of $100a$, the detonation velocity is 99% of the infinite diameter velocity.

Other equations have been proposed to describe detonation velocity-diameter effect. Gibbs and Popalato[2] proposed

$$D(d) = D_\infty[1 - A/r - B/r(r - C)]$$

whereas Campbell and Engelke[108] use the relationship

$$D(d) = D_\infty[1 - E/(r - r_c)]$$
r = charge radius
A, B, C, E, r_c = experimentally determined parameters

When $B \to 0$ or $r_c \to 0$, these two equations reduce to the Eyring form, Eq. (22).

Table 24 lists the infinite diameter detonation velocity and the parameter a for a number of explosive compositions. Illustrative examples of the $D(d)$ vs diameter behavior are shown in Fig. 11 where the curves for the explosives PBXN-110, PBXW-114, TATB, PBXN-301, PBXN-103, and PBXW-115 were calculated using Table 24 parameters. The more ideal explosive materials, such as TATB or PBXN-301, show little variation in velocity as a function of diameter on thc scale shown in the figure; most of the variation occurs below 1 cm for these types of materials. These two materials reach 99% of their infinite diameter velocity at a diameter equal to $100a$, or 0.43 cm and 0.08 cm, respectively. With the use of an inert binder (PBXN-110) or inert binder with aluminum (PBXW-114) and with the corresponding increase in nonideality, the diameter effect becomes more pronounced. PBXN-110 and PBXW-114 reach 99% of their D_∞ at a diameter equal to 3.9 cm and 5.5 cm, respectively. The very nonideal explosives PBXN-103 and PBXW-115 exhibit significant diameter effects even above charge diameters of 5 cm (2 in.). They reach 99% of their respective D_∞ at diameters of 18.24 cm and 53.82 cm (= 21.2 in.), respectively.

Failure diameter effect. For a cylindrical explosive charge composed of a specific explosive, the failure diameter, d_f, is the maximum diameter at which a steady-state detonation will fail to propagate. If an explosive charge with a diameter less than the failure diameter for that specific explosive is

Table 24 Detonation velocity-diameter equations: $D = D(\infty)[1 - (a/d)]$

Explosive	$D(\infty)$, m/s	a, cm	Comment	Ref.
A. Molecules				
DATB	7520	0.0140	density = 1.788	1
TATB	7790	0.0043	density = 1.876	1
B. Compositions				
Comp B[a]	7990	0.0189	density = 1.715	1
75/25 Cyclotol	8298	0.0139	density = 1.755; RDX = 77%	1
Destex	6816	0.1185	density = 1.696	2
75/25 Octol	8480	0.0153	density = 1.814; HMX = 77%	1, 2
PBXN-103	6305	0.1824	density = 1.88	5
PBXN-110	8418	0.0390	density = 1.68	88
PBXN-301	7260	0.0008	density = 1.53; XTX-8003	1
PBXW-114	8234	0.0550	density = 1.72	88
PBXW-115	6195	0.6870	density = 1.786–1.796	80
PBXW-115	5969	0.5382		88

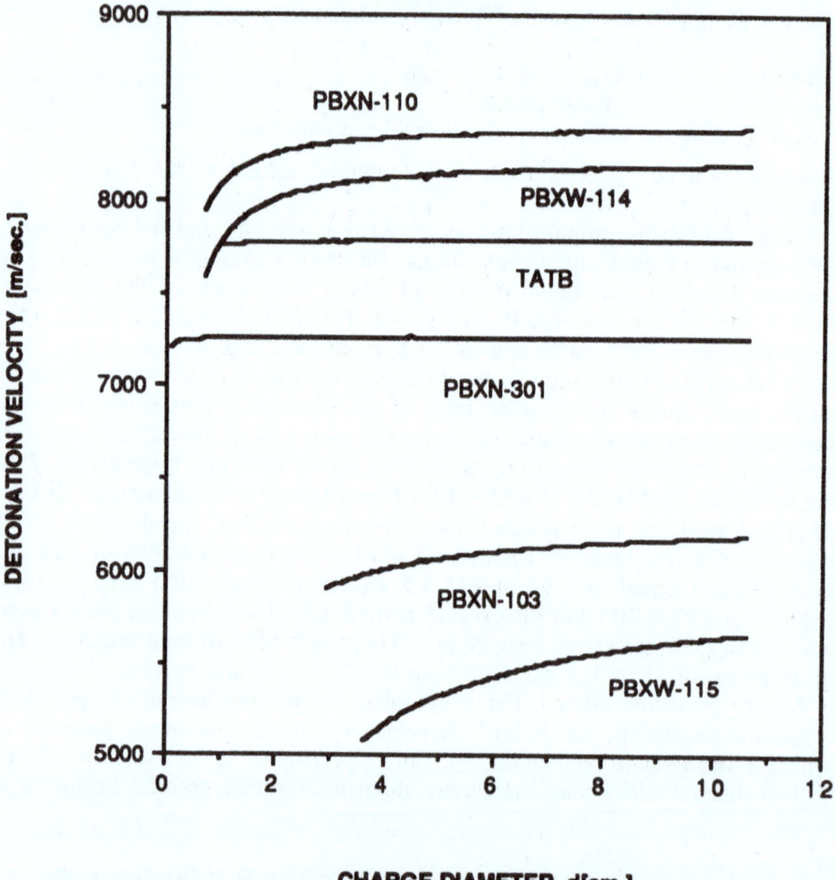

Fig. 11 Detonation velocity as a function of diameter.

initiated by a booster explosive, the detonation either will immediately fail or will propagate with a decreasing velocity along the charge length until it fails. With an appropriately strong booster explosive as a initiation source, an ideal explosive either will be initiated into a steady-state detonation, if the charge diameter is above the failure diameter, or will fail to initiate, if below the failure diameter. On the other hand, with an appropriately strong booster explosive as a initiation source, a nonideal explosive will be similarly initiated into a steady-state detonation, if the charge diameter is above the failure diameter, but if the charge diameter is below the failure diameter, the detonation may propagate a considerable distance with a decreasing velocity before dying out.

This nonideal explosive behavior is illustrated in Figure 12. Detonation velocity data were acquired using the ionization pin technique on cylindrical charges of two nonideal explosives: 1) PBXW-115 using charge diameters of 1.50 in.

and 1.75 in. and 2) a similar RDX/AP/Al/HTPB binder composition at a diameter of 2.00 in., in which the RDX is all Class I as compared with Class I/Class V = 60/40 RDX of PBXW-115. The velocity data are the average velocities between the distances shown and the position of the preceding ionization pin; the distances L are measured from the booster end of the charge. For example, the data at $L = 2.0$ in. are the average velocities from $L = 0$ to $L = 2.0$ in. PBXW-115 has a failure diameter between 1.50 and 1.625 in. After the first 2-in. charge increment, where the effect of the booster explosive is still evident, the PBXW-115 1.75-in.-diameter charge ($d > d_f$) settles down to a steady-state detonation with a velocity of 5198 m/s. With the 1.50-in.-diameter charge ($d < d_f$), a steady-state detonation is not established; a decaying reaction does progress through the charge at a decreasing velocity until it dies out at between 6 and 8 in. The same type of failure occurs with a 2-in.-diameter charge composed of a RDX/AP/Al/HTPB binder explosive with the same composition as PBXW-115, except the RDX is a more coarse particle size than found in PBXW-115. In this case, the reaction progresses out to 10 to 12 in. before dying out.

The critical diameter of an explosive is a related concept. For a cylindrical explosive charge composed of a specific explosive, the critical diameter, d_c, is the minimum diameter at which a steady-state detonation will propagate. If the critical diameter of an explosive has been determined, then reducing the diameter by a extremely small amount will result in a charge with a diameter equal to the failure diameter. Thus,

$$d_f < d_c$$

$$d_f = d_c + \delta$$

where δ is a very small value. For most practical purposes, $d_f \approx d_c$.

The critical, or failure, diameter of an explosive can be determined by attempting to detonate the test explosive at different charge diameters and determining the minimum diameter at which a steady-state detonation occurs and the maximum diameter at which detonation fails to occur. These two values are usually determined to within 0.318 cm ($\frac{1}{8}$ in.) to 0.635 cm ($\frac{1}{4}$ in.) of each other. The critical or failure diameter is usually reported as the midpoint between these two values plus or minus half the distance between these two values. Experimental arrangements have been developed to reduce the number of charge firings needed to pinpoint the failure diameter. One technique is to use stacked charges of different diameters with the large-diameter charge initiating a smaller-diameter charge, which in turn initiates a smaller-diameter charge. This continues until one of the smaller-diameter charges fails to propagate a detonation because it is below the critical diameter. Also, conical charges initiated at the base of the cone are used to obtain the approximate failure diameter. As the detonation travels down the axis of the cone from the initiated base end, the detonation progresses through the explosive material with a decreasing effective diameter. At some point the cross-sectional diameter will be less than the critical diameter of the explosive, and failure will occur. Depending upon the angle of the cone, there is the possibility that the detonation will be overdriven and failure will occur at a diameter somewhat less than the critical diameter of the explosive as measured with cylindrical charges. The conical results are confirmed by firing cylindrical charges with diameters above and below the conical failure diameter in order to verify the measurement. This is especially important for nonideal explosives. For example, the information provided in Fig. 12 would indicate that PBXW-115-type materials could be significantly overdriven in a conical configuration.

The factors discussed above that affect the detonation performance of explosives also affect the failure or critical diameter. Charge density or porosity, composition, and temperature have an influence on the failure diameter of an explosive charge.[37,42] One of the most important factors, charge confinement, will be discussed in the next section. Price[37] provides a thorough review of many of the factors affecting failure or critical diameter and briefly discusses the limited data available on noncylindrical charges.

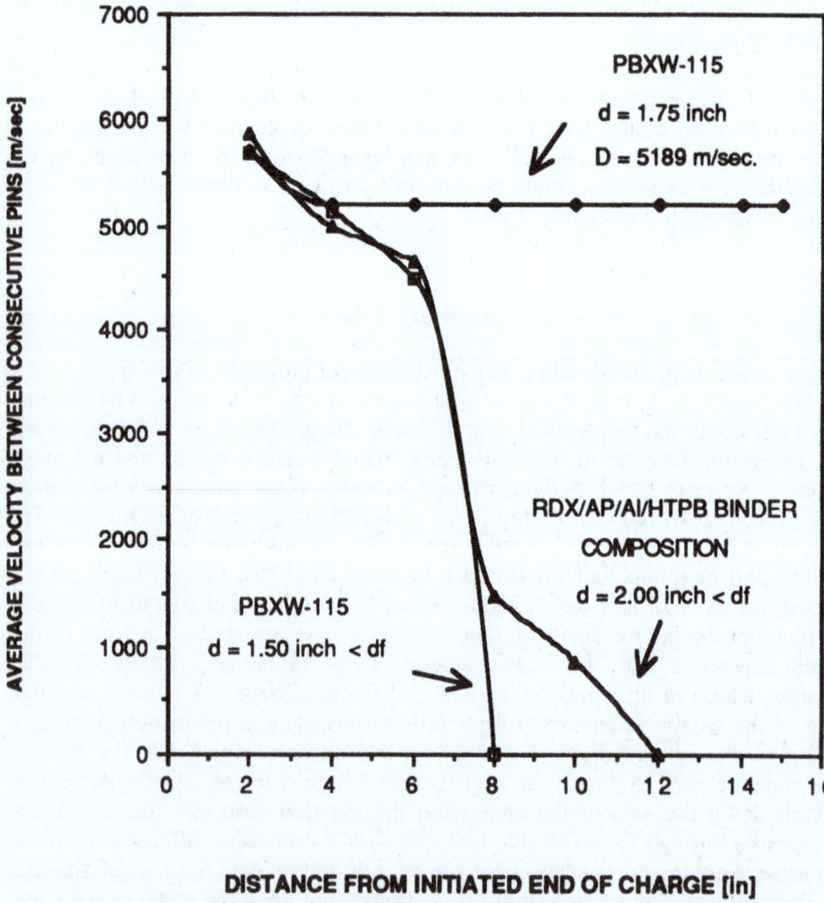

Fig. 12 Steady-state and non-steady-state detonations in nonideal explosives.

The failure diameters of standard explosives are listed in Table 25. Unless otherwise specified, the listed failure diameters were obtained on bare cylindrical charges. From the data given in this table, some very general, qualitative observations can be made regarding the effect of composition. The failure diameters of primary explosives are less than 1 mm. The high-explosive powders possess failure diameters that range in value from less than 1 mm for materials used in boosters such as HNS, PETN, and tetryl to as large as 40 mm for insensitive materials such as TNT. Figure 13 illustrates the effect of compositional changes on the failure diameter of RDX-based PBXs. The ~ 2-mm failure diameter of RDX is increased to 10 mm within an inert binder (PBXW-108). The addition of 20% aluminum (PBXW-109) slightly increases the failure diameter to ~ 13 mm, whereas the presence of 68% aluminum and AP increase the diameter to 40 mm.

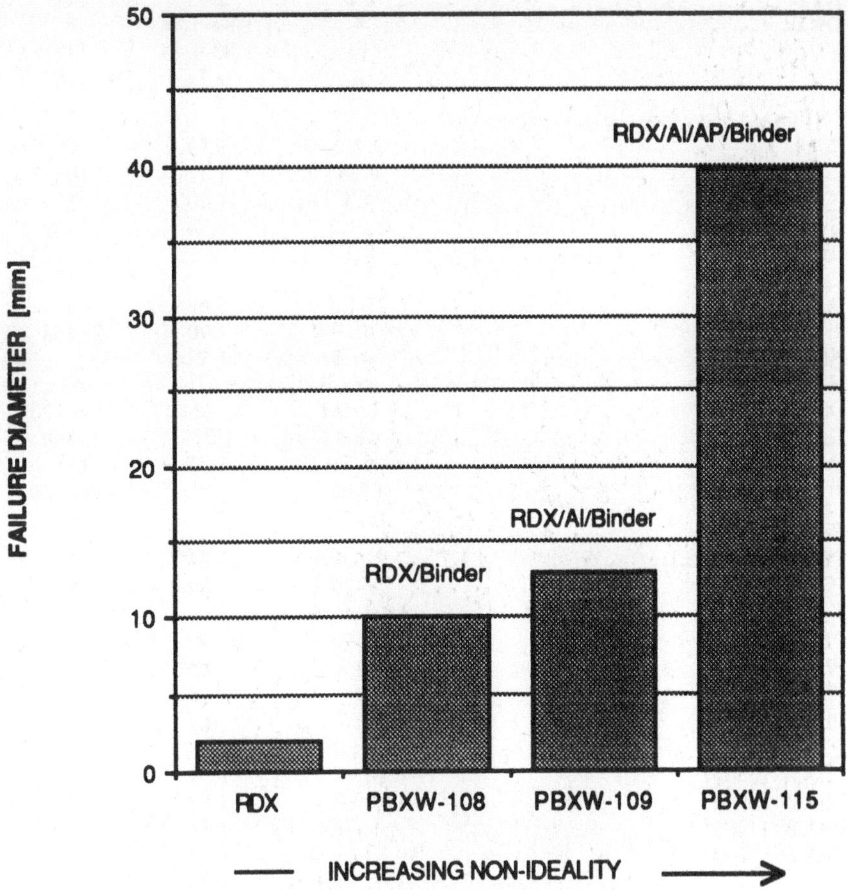

Fig. 13 Failure diameters of RDX-based PBXs.

Table 25 Failure diameters of explosives

Explosive	df, mm	Density, g/cc	Ref.
Primary explosives			
DXW-1	≪1.27[a]		81c
Lead azide	0.06	3.50	110
Lead azide, dextrinated	0.4–0.6	3.14	8, 111
Molecules			
DATB	5.3	1.78–1.80	112
Explosive D	<25.4	1.65	121
HNS	0.5		5
NQ	1.27–1.43	1.52	113
PETN	≈0.2	1.68	114
RDX	3	1.00	115
	5.2	0.90	116
	2	1.20	25
TNT	13–40	1.62±0.1	8
Tetryl	0.6	1.65	117
TATB	4	1.86–1.88	2
	13	1.80–1.84	112
	8	1.71	118
TNT-based and other classical explosives			
Comp A-3	<2.2	1.63	37
Comp B	4.28	1.71	108
	4.3	1.70	2
Cyclotol 77/23	6±1.2	1.74	2, 108
Cyclotol 75/25	8.1		112, 124
Cyclotol 60/40	6.2		112
Destex	21–23	1.69	8
	28.6±3.2	1.70	2, 108
HBX-1	>6.4	1.72	
H-6	6.4±1.3		57
Octol 75/25	6.1±0.1	1.814	119, 122
Octol 77/23	<6.4	1.814	2, 108
Pentolite	6.7	1.70	119
Tritonal	18.3		119, 124
PBXs			
PBXW-7 (TYPE I)	<6.4	1.84	123
PBXN-103	27–29	1.89	5
	14.0–16.5		5
PBXN-105	60–90	1.90	5
PBXN-106	3	1.65	120
PBXN-109	7	1.66	5
PBXN-110	6.1±0.9	1.67	57
PBXN-301	0.36±0.1[b]	1.53	2
PBXW-108(Q)	10±2	1.59	57
PBXW-109(I)	12.9±0.7	1.69	57
PBXW-114(Q)	6.1±0.8	1.74	57
PBXW-115(Q)	39.69±1.59	1.78	81a
	37.1±1.6	1.79	80

[a]Confined.

[b]Confined; $\frac{1}{2}$ cylinder.

For most practical warhead applications, the failure diameter of most main charge explosives is not of significance since most warhead diameters far exceed the failure diameter. For the very nonideal explosives, such as PBXN-103 and PBXW-115, the failure diameter should be considered in applications where the warhead diameters are on the order of 2.5 cm to 5.0 cm. In most cases, the diameter effect [Eq. (22)] has a more significant role to play than the failure diameter in determining the performance of the warhead. Although the failure diameter of an explosive should be considered when designing small-diameter boosters or leads, most materials used in these applications have failure diameters well under 6 mm. If failure-diameter problems arise, they can in many cases be mitigated with the proper choice of confinement.

2.4.5 Confinement Effects

Confinement refers to the use of any nonexplosive material in intimate contact with the outer surface of an explosive charge. To provide confinement, the material must be at approximately right angles to the direction of the detonation wave progression as shown in Fig. 14. Here a slab of explosive is sandwiched between two metal slabs. If the explosive is initiated from the right (Fig. 14a) with the detonation front traveling from right to left, confinement is provided by the upper and lower slabs of metal. If, on the other hand, the explosive is initiated through the upper metal slab with the resulting detonation front traveling from the top to the bottom of the explosive (Fig. 14b), then the explosive is unconfined. Confinement is often described as being either light, moderate, or heavy. Light confinement refers to any low-mass or low-density material surrounding the charge, such as paper or low-density plastics. Moderate confinement is provided by the low-density metals, such as aluminum, whereas heavy confinement is provided by the dense metals and alloys, such as steel, brass, and copper.

Confinement slows down the rate of expansion of the explosive's decomposition product gases. This allows more product gas reaction time at high temperature and pressure. With confinement, the explosive behaves as if it had a larger "effective diameter."[37,80] For small-diameter charges, below the infinite diameter, confinement will provide an effective diameter closer to infinite-diameter conditions. For ideal explosives, which release most of their total energy within the reaction zone, the effect of confinement should diminish as the charge diameter increases; both the confined and unconfined charges should approach the same infinite-diameter velocity. For very nonideal explosives, one could speculate that the heavy confinement would allow a larger fraction of the total energy to be released with the reaction zone resulting in a larger infinite diameter detonation velocity; currently there are no experimental data to support or disprove this speculation.

Confinement also affects the failure diameter of the explosive by reducing this diameter below that observed with bare charges. This effect is most pronounced and more easily observed with nonideal explosives. For example, the nonideal underwater explosives have bare charge failure diameters above 2.54 cm, but

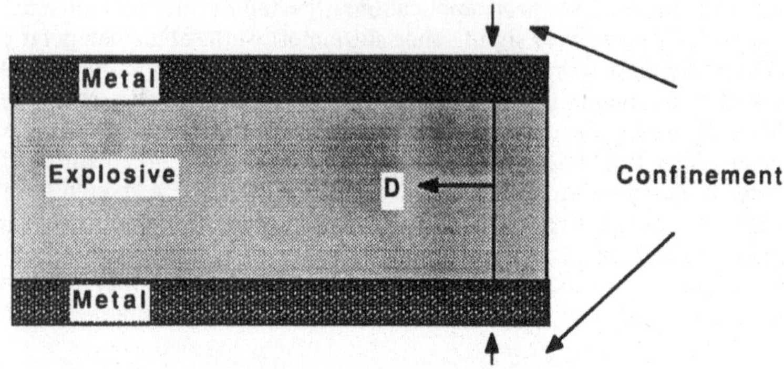

A. Confinement by side walls at right angles to the propagation of the detonation wave D.

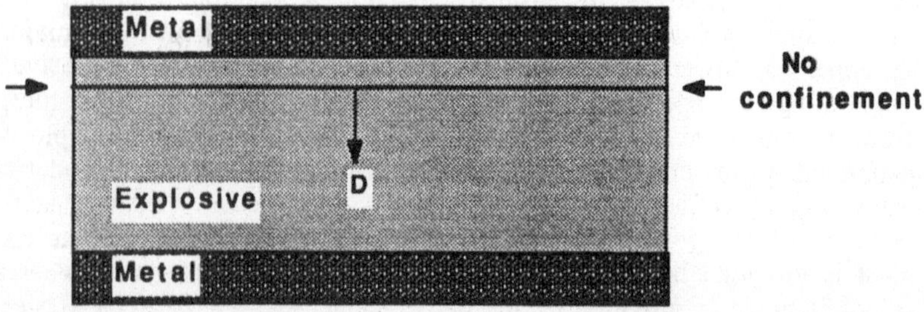

B. No confinement at right angles to propagation of the detonation wave D.

Fig. 14 Confinement in a sandwich.

all readily detonate at a 2.54-cm diameter when confined in heavy steel (wall thickness = t = 0.22 in. = 0.55 cm):

Explosive	d_f, cm	Detonation velocity, m/s (d = 2.54 cm; confined by steel)
PBXN-103	2.8 ± 0.1	5400–5500
PBXW-115	3.97 ± 0.16	5359
PBXN-105	7.5 ± 1.5	5180

Fig. 15 depicts the detonation velocity of PBXW-115 as a function of diameter and as a function of the type of confinement. The bare charge velocity vs diameter data are compared with data obtained on charges lightly confined with polyurethane (PU) of t = 0.635 cm and moderately confined with aluminum of t = 0.21 cm. Individual experimental data points show the effect of heavy steel

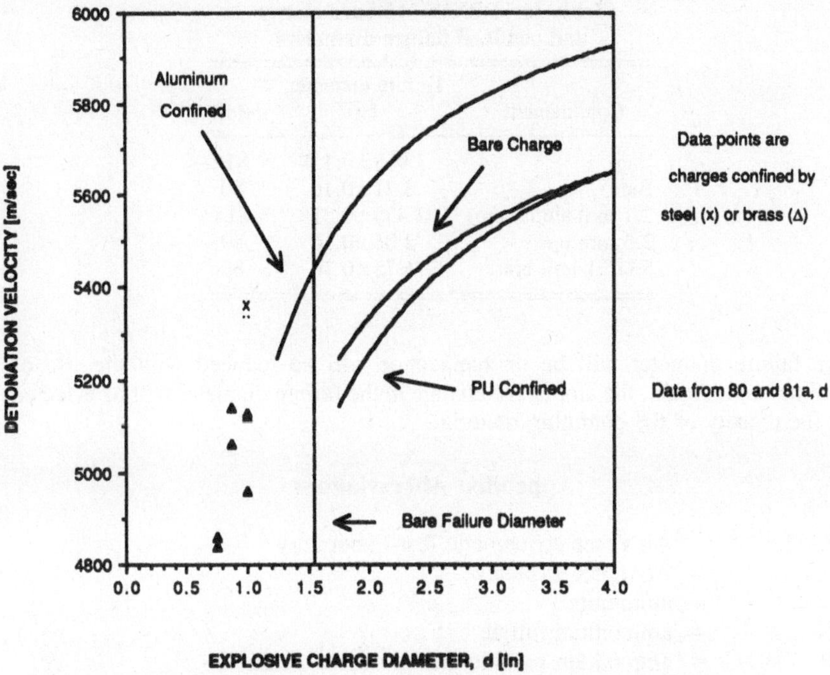

Fig. 15 Effect of confinement on the detonation velocity of PBXW-115.

confinement of $t = 0.55$ cm and brass confinement of $t = 2.5$ and 5.0 mm. The vertical line indicates the bare charge failure diameter of the material. The light confinement of the polyurethane does not change the detonation characteristics of the explosive, whereas the moderate confinement of aluminum increases the detonation velocity over that of the bare charge for a given charge diameter and allows detonation below that of the bare charge. The heavy confinement of steel or brass allows steady-state detonation to occur at charge diameters as low as one-half the bare charge failure diameter. The effects of various confinement levels on the failure diameter of PBXW = 115 are summarized in Table 26.

The effect of confinement on the detonation properties of an explosive can be summarized as follows:

$$D\{\text{confined}, d_s\} \geq D\{\text{unconfined}, d_s\}$$

$$d_f\{\text{confined}\} \leq d_f\{\text{unconfined}\}$$

The addition of confinement around a charge either will have little or no effect or will increase the detonation velocity at a specific charge diameter d_s; confinement will not reduce the detonation velocity. Dense materials, such as metals, provide better confinement than low-density materials, such as plastics.

Table 26 PBXW-115 bare charge and confined failure diameters

Confinement	Failure diameter, cm	Ref.
Bare	3.969±0.159	81a
Bare	3.71±0.16	80
2.1 mm aluminum	2.435±0.315	81d
2.5 mm brass	2.06±0.16	80
5.0/5.1 mm brass	1.75±0.16	80

The failure diameter will be unchanged or will be reduced with the use of confinement. Again, the amount of change in the failure diameter will be affected by the density of the confining material.

Appendix: Abbreviations

AFATL = Air Force Armament Test Laboratory
AFX = Air Force explosive
Al = aluminum
AN = ammonium nitrate
AP = ammonium perchlorate
APL = Applied Physics Laboratory, Johns Hopkins University
BDNP A/F = bis (2,2-dinitropropyl) acetal/formal 50/50 mixture; energetic plasticizer used in PBXN-105 and PBXN-106
CPIA = Chemical Propulsion Information Agency
cyclotols = explosives consisting of mixtures of RDX and TNT
Cylex = cylinder expansion (test)
DATB = 1,3-diamino-2,4,6-trinitrobenzene
DOP = dioctyl phthalate; inert plasticizer
DXW = detonation explosive, White Oak; nomenclature for primary explosives
D-2 = Composition D-2; desensitizer for aluminized TNT-based explosive; wax/NC/lecithin = 84/14/2
EOS = equation of state
EAK = ethylenediamine dintrate/ammonium nitrate/potassium nitrate explosive compositions; Air Force AFX-400 series
Explosive D = ammonium picrate
GP = general purpose
HBXs = explosive consisting of a mixture of RDX, aluminum, TNT, and a desensitizer; aluminized and desensitized cyclotols; high blast explosive
HMX = octahydro-1,3,5,7-tetranitro-1,3,5,7-tetrazocine; octogen (European); high melting explosive
HNS = 2,2′,4,4′,6,6′-hexanitrostilbene

HTPB = hydroxyl terminated polybutadiene; inert polymer resin
IDP = isodecyl pelargonate; inert plasticizer
LANL = Los Alamos National Laboratory
LANSL = Los Alamos National Scientific Laboratory, Los Alamos, NM
LASL = Los Alamos Scientific Laboratory, now called LANSL
LLL = Lawrence Livermore Laboratory, now called LLNL
LLNL = Lawrence Livermore National Laboratory, Livermore, CA
LSGT = large-scale gap test
LX = Livermore explosive
minols = explosives consisting of mixtures of RDX, TNT, Al, and AN
MW = molecular weight (g)
NAVORD = Bureau of Naval Ordnance; Naval Ordnance Systems Command
NC = nitrocellulose
NOL = Naval Ordnance Laboratory (now NSWC)
NTO = 3-Nitro-1,2,4-Triazole-5-One; also ONTA (French)
NQ = nitroguanidine
NSWC = Naval Surface Warfare Center (formerly Naval Surface Weapons Center)
NSWC/WO = Naval Surface Warfare Center, White Oak Laboratory, Silver Spring, MD
octols = explosives consisting of mixtures of HMX and TNT
ONR = Office of Naval Research
PBX = plastic bonded explosive; also used in LANL explosive nomenclature
PE = polyethylene
PEG = polyethylene glycol; inert polymer resin; used in PBXN-105 and PBXN-106 binders
PETN = 2,2-bis[(nitroxy) methyl]-1,3-propanediol dinitrate or pentaerythritol tetranitrate
PIB = polyisobutylene
PPG = polypropylene glycol; inert polymer resin; used in AFX-108
RDX = hexahydro-1,3,5-trinitro-1,3,5-triazine; hexogen (European); research department explosive
R-45 = Atochem Corp. trade name for its HTPB product
TATB = 1,3,5-Triamino-2,4,6-trinitrobenzene
TEGDN = triethylene glycol dinitrate; one of the energetic plasticizers used in PBXN-103 binder
TMD = theoretical maximum density
TMETN = trimethylolethane trinitrate or metriol trinitrate; one of the energetic plasticizers used in PBXN-103 binder
TNT = trinitrotoluene; 2-methyl-1,3,5-trinitrobenzene
TPM = tripicryl melamine
torpex = explosives consisting of mixtures of RDX, aluminum, and TNT (UK usage of this terminology includes the desensitized torpexes, i.e., the HBXs)
U/W = underwater

References

[1]Dobratz, B. M., and Crawford, P. C., *LLNL Explosives Handbook—Properties of Chemical Explosives and Explosive Simulants*, Lawrence Livermore National Laboratory, UCRL-52997, Jan. 31, 1985.

[2]Gibbs, T. R., and Popolato, A., *LASL Explosive Property Data*, University of California Press, Berkeley, CA, 1980.

[3]Mader, C., et al., *Los Alamos Explosives Property Data*, University of California Press, Berkeley, CA, 1982.

[4]*Engineering Design Handbook: Properties of Explosives of Military Interest*, Army Materiel Command, AMC Pamphlet AMCP 706-177, Jan. 1971.

[5]Hall, T. N., and Holden, J. R., *Navy Explosives Handbook Explosion Effects and Properties, Part III, Properties of Explosives and Explosive Compositions*, NSWC MP 88-116, Oct. 1988.

[6]Lee, K. Y., and Coburn, M. D., "3-Nitro-1.2.4-Triazol-5-One, A Less Sensitive Explosive," Los Alamos National Laboratory, LA10302-MS, Feb. 1985; also *Journal of Energetic Materials*, Vol. 5, No. 1, 1987, pp. 27–33.

[7]"Properties of ONTA," SNPE Corp.,

[8]Drimmer, B. E., *Navy Bank of Explosives Data (NAVBED)*, NSWC MP 83-230, June 30, 1983.

[9]*Solid Propellant Ingredients Manual*, CPIA/APL, CPIA/M3, Aug. 1979.

[10]Shipp, K. G., and Hill, M. E., *Heat Resistant Explosives 11.1.3-Diamino-2.4.6-trinitrobenzene, DATB*, NOL, NAVORD Rept. 6016, March 1958.

[11]Beauregard, R. L., *History of Navy Use of Composition A-3 and Explosive D in Projectiles*, Naval Ordnance Systems Command, NAVORD TR 71 -1, Jan. 1, 1971.

[12]Heller, H., et al., *Feasibility of Development of an Insensitive, High Detonation Velocity Explosive Based Upon Nitroguanidine (NQ)*, Naval Ordnance Laboratory, NAVORD Rept. 6688, July 15, 1959.

[13]Kaplan, L. A., and Taylor, F., *High Temperature Stable Explosives I. Process Development Study of 1.3.5-Triamino-2.4.6-trinitrobenzene*, NOL, NAVORD Rept. 6017, March 1958.

[14]Chipen, G. I., et al., "1,2,4-Triazol-3-One and Its Nitro and Amino Derivatives," *Khimiia Geterotsiklicheskikh Soedinenii*, Vol. 2, No. 1, 1966, pp. 110–116.

[15]"Summary Report on Elimination of Tetryl from Munitions," JTCG/ALNNO/WPE, March 20, 1973.

[16]Shipp, K. G., *Heat Resistant Explosives, XIV, Synthesis and Properties of 2.2'.4.4'.6.6'-Hexanitrostilbene*, NOLTR 62-107, June 1962.

[17]Shipp, K. G., and Kaplan, L. A., "Reaction of a Substituted Polynitrotoluene. II. The Generation and Reactions of 2,4,6-trinitrobenzyl Amion," *Journal of Organic Chemistry*, Vol. 31, No. 3, 1966, p. 857.

[18]Kilmer, E. E., *Development of a High Temperature Resistant Mild Detonating Fuze (MDF)*, NOLTR 63-258, April 1964.

[19]Kilmer, E. E., "Heat-Resistant Explosives for Space Applications," *Journal of Spacecraft and Rockets*, Vol. 5, No. 10, 1968, pp. 1216–1219.

[20]Urman, S., and Kabik, I., *Hazards of Copper Azide*, NOLTR 72-113, Sept. 8, 1972.

[21]Nelson, H. R., *Explosive Accidents Involving Naval Munitions*, Naval Weapons Station, Yorktown, NWSY TR 85-5, July 1985.

[22]*The Use of Lead Azide in Explosive Component Design*, Naval Sea Systems Command Instruction 8020.3, November 8, 1974.

[23]Gilligan, W. H., and Kamlet, M. J., *Synthesis of Mercuric 5-Nitrotetrazole*, NSWC/WOL TR 76-146, Dec. 1976.

[24]Swisdak, M. M., *A Survey of Tritonal and H-6 as Explosive Bomb Fills*, NOLTR 73-92, June 18, 1973.

[25]Cook, M. C., *The Science of High Explosives*, Robert E. Krieger Publishing Co., Huntington, NY, 1971.

[26]Bebie, J., *Manual of Explosives, Military Pyrotechnics and Chemical Warfare Agents*, Macmillan, New York, 1943.

[27]Ablard, J. E., *HBX-1: Its History and Properties*, Naval Sea Systems Command, NAVSEA-03-TR-021, May 1, 1975.

[28]Strehlow, R. A., "Accidental Explosions," *American Scientist*, Vol. 68, 1980, pp. 420–428.

[29]Baker, W. E., et al., *Explosion Hazards and Evaluation*, Elsevier, Amsterdam, 1983.

[30]*Safety and Performance Tests for Qualification of Explosives*, Naval Ordnance Systems Command, NAVORD OD 4481 1, Vol. 1, Jan. 1, 1972.

[31]Price, D., "Contrasting Patterns in the Behavior of High Explosives," *Eleventh Symposium (International) on Combustion*, Combustion Institute, Pittsburgh, PA, 1967, pp. 693–702.

[32]Ablard, J. E., *H-6 Explosive History and Properties*, Naval Sea Systems Command, NAVSEA-03-TR-044, Oct. 1, 1977.

[33]*List of Explosives for U. S. Naval Weapons*, Naval Sea Systems Command, NAVSEA OP 3613, rev. 2, Aug. 30, 1979.

[34]Anderson, E. W., *Chemical Formulas for Selected Explosives*, NSWC TR 88-56 (in preparation).

[35]McQuaide, P. *Fast Cook-off Characteristics of Air-launch Inservice Weapons*, rev. 2, Pacific Missile Test Center, Pt. Mugu, TP 75-22, June 15, 1985.

[36]Dettling, C. A., et al., *Summary Report of Fast Cookoff Times, Reactions, and Hazards of Bombs, Rockets, Aircraft Guns, Air-Launched Missiles, Mines and Torpedoes*, NWC/CL TP 6914, Sept. 1989.

[37]Price, D., *Critical Parameters for Detonation Propagation and Initiation of Solid Explosives*, NSWC TR 80-339, Sept. 10, 1981.

[38]Chapman, D. L., "On the Rate of Explosion in Gases," *Philosophical Magazine*, Vol. 47, 1899, pp. 90–104.

[39]Jouguet, E., "On the Propogation of Chemical Reactions in Gases," *Journal de Mathematiques Pures et Appliquées*, Vol. 1, 1905, pp. 347–425; and Vol. 2, 1906, pp. 5–85.

[40]Mader, C. L., *Numerical Modeling of Detonations*, University of California Press, Berkeley, CA, 1979.

[41]Fickett, W., and Davis, W. C., *Detonation*, University of California Press, Berkeley, CA, 1979.

[42]Johansson, C. H., and Persson, P. A., *Detonics of High Explosives*, Academic Press, New York, 1970.

[43]Zerilli, F. J., *Notes from Lectures on Detonation Physics*, NSWC MP 81-399, Oct. 1981.

[44]Gruschka, H. D., and Wecken, F., *Gasdynamic Theory of Detonation*, Gordon & Breach, New York, 1971.

[45]First through Ninth Symposia (International) on Detonation, Office of Naval Research.

[46]Lee, E. L., Hornig, H. C., and Kury, J. W., *Adiabatic Expansion of High Explosive Detonation Products,* LLNL Rept. UCRL-50422, 1968.

[47]Cowperthwaite, M., and Zwisler, W. H., *TIGER Computer Program Documentation,* Stanford Research Institute, Publication No. Z1 06, 1973.

[48]Sternberg, H. M., and Hudson, L., "Equations of State for Underwater Explosives," *Proceedings of the International Symposium on Pyrotechnics and Explosives,* Beijing, China, Oct. 15, 1987, p. 5.

[49]Lee, E., et al., *JWL Equation of State Coefficients for High Explosives,* LLNL Rept. UCID-16189, 1973.

[50]Lee, E. L., et al., *Equations of State for Detonation Products of High Energy PBX Explosives,* LLNL Rept. UCID-1 7540, 1977.

[51]Lee, E., et al., *Equation of State for the Detonation Products of Hexanitrostilbene at Various Charge Densities,* LLNL Rept. UCID-17134, 1976.

[52]Slape, R. J., et al., *Some Sensitivity and Performance Characteristics of the Explosives H-6 and Tritonal,* Mason & Hanger—Silas Mason Co., Inc., AFATL-TR-74-104, June 1974.

[53]Lee, E. L., and Walton, J. R., *Equation of State for Pentolite,* LLNL Rept. UCID-16953, 1975.

[54]Hudson, L., and Sternberg, H. M., *JWL Equation of State Constants for PBXN-106,* Advanced Technology and Research, Inc., May 14, 1986.

[55]Breithaupt, D., et al., *Cylinder Test Performance of PBXW-113, 114, and 115,* LLNL Rept. UCID-20159, Aug. 31, 1984.

[56]Cowperthwaite, M., and Rosenberg, J. T., *Lagrange Gage Studies of Composition B and H-6,* SRI Rept., Project PYU, May 1982.

[57]Forbes, J. W., et al., *Sensitivity and Performance of Several Select and Insensitive Plastic Bonded Explosives,* NSWC TR 83-74, March 12, 1984.

[58]Kramer, P. E., and Kramer, A., *Reduction of C-J Pressure Data,* Mason & Hanger—Silas Mason Co., TR MHSMP 73-6, Dec. 1972.

[59]Smith, L. C., "On Brisance, and a Plate-Denting Test for the Estimation of Detonation Pressure," *Explosivstoffe,* Nr. 5, 1967, pp. 106–110, and Nr. 6, 1967, pp. 130–134.

[60]Price, D., "Dependence of Damage Effects upon Detonation Parameters of Organic High Explosives," *Chemical Reviews,* Vol. 59, No. 5, Oct. 1959, pp. 801–825.

[61]Deal, W. E., "Measurement of Chapman-Jouguet Pressure for Explosives," *Journal of Chemical Physics,* Vol. 27, 1957, pp. 796–800.

[62]Clairmont, A. R., *PBX's in Large Navy Munitions Program, Vol. 7. Cylinder Expansion Testing,* NSWC TR 80-43, Nov. 15, 1988.

[63]Roslund, L. A., and Coleburn, N. L., *Hydrodynamic Behavior of HBX-1 and Equation of State of the Detonation Products Below the Chapman-Jouguet State,* NOLTR 70-133, Nov. 6, 1970.

[64]Roslund, L. A., and Coleburn, N. L., "Hydrodynamic Behavior and Equation of State of the Detonation Products Below the Chapman-Jouguet State," *Proceedings of the Fifth Symposium (International) on Detonation,* ACR-1 84, Aug. 18–21, 1970, pp. 523–532.

[65]Coleburn, N. L., and Liddiard, T., "Hugoniot Equations of State of Several Unreacted Explosives," *Journal of Chemical Physics,* Vol. 44, No. 5, 1966, pp. 1929–1936.

[66]Ayres, J. N., et al., "Qualification of Minol IV in the Mk 82 Bomb," unpublished data, NSWC Rept., 1975.

[67]Costain, T. S., and Motto, R. V., *The Sensitivity, Performance and Material Properties of Some High Explosive Formulations,* Picatinny Arsenal TR 4587, Sept. 1973.

[68]Deal, W. E., "Low Pressure Points on the Isentropes of Several High Explosives," *Third Symposium on Detonation,* ONR, ACR-52, 1960, p. 386.

[69]Dremin, A. N., et al., *Detonation Waves in Condensed Media* lzd-vo Nauka, Moscow, 1970. (Also available in translation by John Sechovicz, Foreign Technology Division, Wright Patterson AFB, Rept. FTD-H-T-23-1889-71.)

[70]Kilmer, E., *Annual Report on Investigation of High and Low Temperature Resistant Explosive Devices,* NOLTR 66-94, 1969.

[71]Chapman, L. B., "NTO Development at Los Alamos," *The Ninth Symposium (International) on Detonation,* Aug. 28–Sept. 1, 1989, Preprints Vol. 1, Paper 160.

[72]Lee, E. L., and Hornig, H. C., "Equation of State of Detonation Product Gases," *Proceedings of the Twelfth Symposium (International) on Combustion,* Poitiers, France, July 14–20, 1968, pp. 493–499.

[73]Coleburn, N. L., *Chapman-Jouguet Pressure of Several Pure and Mixed Explosives,* NOLTR 64-58, June 25, 1964.

[74]Chaudhri, M., and Field, J., "Fast Decompositions in the Inorganic Azides," *Energetic Materials,* edited by H. D. Fair and R. F. Walker, Plenum Press, New York, 1977.

[75]Meyer, R., *Explosives,* Verlag Chemie, Weinheim/New York, 1977.

[76]West, G. T., et al., *Some Sensitivity & Performance Characteristics of the Plastic Bonded Explosive PBXW-106,* Mason & Hanger—Silas Mason Co., AFATL-TR-70-27, March 1970.

[77]Harp, D. L., et al., *Cylinder Expansion Tests of Six Plastic-Bonded Explosives,* NWC TP 5899, Sept. 1976.

[78]Anderson, E. W., *Explosive Performance of PBXW-113 and PBXW-114,* NSWC TR 86-518, July 1, 1988.

[79]Slape, R. J., *Additional Sensitivity and Performance Characteristics of the Plastic Bonded Explosives PBXW-106 and PBXW-107,* Mason & Hanger—Silas Mason Co., AFATL-TR-74-184, Nov. 1974.

[80]Forbes, J. W., et al., "Detonation Wave Propagation in PBXW-115," *The Ninth Symposium (International) on Detonation,* Aug.–Sept. 1, 1989, Preprints Vol. 111, Paper 59.

[81]Unpublished data obtained from a) E. Anderson, NSWC; b) H. Sternberg, ATR Inc.; c) G. Laib, NSWC; d) M. Lerardi, NSWC.

[82]Baudler, B., and Montesi, L. J., *Qualification Test Results of PBXW-7 Mod 1,* NSWC TR 86-22, May 3, 1986.

[83]Renner, R. H., and Short, J. M., *Chemical Products of Underwater Explosions,* NSWC/WOL TR 78-87, Feb. 20, 1980.

[84]Price, D., *Explosive Effects and Properties,* Naval Ordnance Laboratory, NAVORD Rept. 5678, Sept. 24, 1957.

[85]Finger, M., et al., "Metal Acceleration by Composite Explosives," *Proceedings of the Fifth Symposium (International) on Detonation,* Office of Naval Research ACR-184, Aug. 18–21, 1970, pp. 137–149.

[86]Price, D., *Inter-Relationships of Explosive Characteristics II,* Naval Ordnance Laboratory, NAVORD Rept. 4230, April 10, 1956.

[87]McMurphy, F. E., *Cylinder Tests of HE Materials,* Lawrence Livermore National Lab., M-118, 1980.

[88]Anderson, E. W., *Elements of Explosives Technology - Course, Lecture Notes*, NSWC MP 92/22 (in preparation).

[89]Pepekin, V. I., and Lebedev, Y. A., Combust. Expl. Shock Waves, *Academy of Sciences USSR Doklady,* Vol. 14, 1977, pp. 630–633.

[90]Rothstein, L. R., and Peterson, R., "Predicting High Explosive Detonation Velocities from their Composition and Structure," *Propellants and Explosives,* Vol. 4, 1979, pp. 56–60.

[91]Kamlet, M. J., and Jacobs, S. J., "Chemistry of Detonations. I. A Simple Method of Calculating Detonation Properties of C-H-N-O Explosives," *Journal of Chemical Physics,* Vol. 48, No. 1, 1968, pp. 23–35.

[92]Kamlet, M. J., and Dickinson, C., "Chemistry of Detonations. II. Buffered Equilibria," *Journal of Chemical Physics,* Vol. 48, No. 1, 1968, pp. 36–42.

[93]Kamlet, M. J., and Dickinson, C., "Chemistry of Detonations. III. Evaluation of the Simplified Calculational Method for Chapman-Jouguet Detonation Pressures on the Basis of Available Experimental Information," *Journal of Chemical Physics,* Vol. 48, No. 1, 1968, pp. 43–50.

[94]Kamlet, J., and Hurwitz, H., "Chemistry of Detonations. IV. Evaluation of Simple Predictional Method for Detonation Velocities of C-H-N-O Explosives," *Journal of Chemical Physics,* Vol. 48, No. 8, April 15, 1968, pp. 3685–3692.

[95]Defourneaux, M., *Science et Techniques de l'Armement,* Vol. 47, p. 723, 1973; English translation in UCRL-Trans-1 0788, 1974.

[96]Johansson, C. H., and Persson, P., "Density and Pressure in the Chapman Jouguet Plane as Functions of Initial Density of Explosives," *Nature,* Vol. 212, 1966, p. 1230.

[97]Hardesty, D. R., and Kennedy, J. E., "Thermo Chemical Estimation of Explosive Energy Output," *Combustion and Flame,* Vol. 28, 1977, p. 45.

[98]Kamlet, J., and Short, M., "Chemistry of Detonations. VI. A Rule for Gamma as a Criterion for Choice Among Conflicting Detonation Pressure Measurements," *Combustion and Flame,* Vol. 38, 1980, pp. 221–230.

[99]Cichra, D. A., et al., *Estimation of Normal Densities of Explosive Compounds from Empirical Atomic Volumes*, NSWC TR 79-273, Feb. 19, 1980.

[100]Stine, J. R., *Prediction of Crystal Densities of Organic Explosives by Group Additivity,* LANL Rept. LA-8920 UC-45, Aug. 1981.

[101]Cady, H. H., *Estimation of the Density of Organic Explosives from Their Structural Formulas,* LANL Rept., LA-7760-MS, Aug. 1979.

[102]Coleburn, N. L., and Liddiard, T. P., *The Rates of Detonation of Several Pure and Mixed Explosives,* Naval Ordnance Laboratory, NAVORD Rept. 2611, Sept. 22, 1952.

[103]Price, D., *The Detonation Velocity-Loading Density Relation for Selected Explosives and Mixtures of Explosives,* NSWC TR 82-298, Aug. 26, 1982.

[104]Price, D., "The Detonation Velocity Loading Density Relation for Selected Explosives and Mixtures of Explosives," *Journal of Energetic Materials,* Vol. 1, No. 1, 1983, pp. 55–82.

[105]Eyring, H., et al., "The Stability of Detonation," Princeton University, NAVORD Rept. 70-46, June 1946.

[106]Eyring, H., et al., "The Stability of Detonation," *Chemical Reviews,* Vol. 45, 1949, pp. 69–181.

[107]Jones, H., "The Theory of the Dependence of the Rate of Detonation of Solid Explosives on the Diameter of the Charge," *Proceedings of the Royal Society of London,* Vol. A189, 1947, p. 415.

[108]Campbell, A. W., and Engelke, R., "The Diameter Effect in High Density Heterogenous Explosives," *Sixth Symposium (International) on Detonation*, ACR, ONR, U.S. Government Printing Office, Washington, DC, 1978, pp. 642–663.

[109]Baudler, B., and Montesi, L., "Variations in Sensitivity and Performance of PBXW-7 Type II," NSWC TR 87-76, Sept. 1, 1988.

[110]Voreck, W., et al., "The Role of Azides in Explosive Trains," *Energetic Materials*, edited by H. D. Fair and R. F. Walker, Plenum Press, New York, 1977.

[111]Bowden, F. P., and McLaren, A. C., "Detonation of Azides When the Dimensions Are Comparable with the Length of the Reaction Zone," *Second ONR Symposium on Detonation*, ONR, 1955, p. 443.

[112]Coleburn, N. L., and Drimmer, B. E., *The Explosive Properties of the Amino-Substituted Symmetrical Trinitrobenzenes*, NOLTR 63-81, May 1963.

[113]Price, D., and Clairmont, A. R., "Explosive Behavior of Nitroguanidine," *Proceedings of the 12th Symposium (International) on Combustion*, Combustion Institute, Pittsburgh, 1969, pp. 761–770.

[114]Price, D., et al., *The Large Scale Gap Test III. Compilation of Unclassified Data and Supplementary Information on Interpretation of Results*, NOL TR 74-40, 1974.

[115]Evans, M. W., "Detonation Sensitivity and Failure Diameter in Homogenous Condensed Materials," *Journal of Chemcial Physics*, Vol. 36, No. 1, 1962, p. 193.

[116]Atkimova, L. N., et al., "Parameters of the Detonation of Low Density Condensed Explosion," *Combustion, Explosion, and Shockwaves*, Vol. 14, No. 2, 1978, pp. 248–251.

[117]Belyaev, F., and Sukoyan, M. K., "Detonability of Some Explosives With Increase in External Pressure," *Combustion, Explosion and Shockwaves*, Vol. 3, No. 1, 1967, pp. 11–13.

[118]Dobratz, B. M., et al., *The Sensitivity of Triaminotrinitro-benzen (TATB) and TATB Formulations: A Summary Report*, LLNL Rept. UCID-17808, Sept. 1978.

[119]Jaffe, I., and Price, D., *Determination of the Critical Diameter of Explosive Materials*, *ARS Journal*, Vol. 32, No. 7, 1962, p. 1060.

[120]Heller, H., et al., *PBXW-106, A New Impact Insensitive, Heat Resistant Castable Explosive*, NOLTR 72-86, July 11, 1972.

[121]Price, D., *Burning Rates and Sensitivity Data for Explosive D, Comp A-3 and PBXW-106*, NOLTR 70-137, July 28, 1970.

[122]August, W. T., and Solem, A. D., *Shaped Charge Performance with Various Explosive Loadings*, Naval Ordnance Laboratory, NAVORD Rept. 2767, Feb. 16, 1953.

[123]Lee, J. D., and Montesi, L. J., *Characterization and Performance of PBXW-7 and Its Applications in the FMU-1 17 Bomb Fuze*, NSWC TR 80-187, Oct. 20, 1981.

[124]Jaffe, I., *A Method for the Determination of the Critical Diameters of Explosives*, NOL NAVWEPS Rept. 7360, Dec. 20, 1960.

Chapter 3

Material Characterization for Warhead Computations

Gordon R. Johnson*

Alliant Techsystems Inc., Hopkins, Minnesota 55343

1. Introduction

COMPUTER simulations are now viewed as a necessary step for the design and analysis of warheads. During recent years there have been significant advances both in computers and in the computer codes used for warhead simulations. As more and more warhead computations have been performed, it has become increasingly apparent that better material models and data are necessary to increase the accuracy of these computations. As a result, there has been an increased emphasis on obtaining improved computational *material* models and data.

The purpose of this chapter is to discuss various aspects of computational material models. It will be limited to metals, as explosive materials are discussed in Chapter 2. It will begin by briefly summarizing the basic computational algorithms for material response that are used in most of the applicable computer codes. Then it will proceed by considering two example materials (OFHC copper and Armco iron) in detail. Laboratory test data will be presented, followed by analyses of these data for both strength and fracture models. This chapter is not intended to be a comprehensive review of all related work but rather an example of how one might proceed to obtain adequate material models and data for warhead computations.

2. Computational Algorithms for Dynamic Material Response

Most computer codes use a computational procedure similar to that initially presented by Wilkins[1] for use in the HEMP code. For axisymmetric geometry the three normal stresses are expressed as

$$\sigma_r = s_r - (P + Q) \tag{1}$$

$$\sigma_z = s_z - (P + Q) \tag{2}$$

*Chief Engineering Fellow.

$$\sigma_\theta = s_\theta - (P + Q) \tag{3}$$

where

s_r, s_z, s_θ = deviator stresses
P = hydrostatic pressure
Q = artificial viscosity

The shear stress is represented by τ_{rz}. The deviator and shear stresses are updated in an incremental manner for each time increment. Trial values of these stresses at time $= t + \Delta t$ are

$$s_r^{t+\Delta t} = s_r^t + 2G\dot{e}_r\Delta t - 2\tau_{rz}^t\omega_{rz}\Delta t \tag{4}$$

$$s_z^{t+\Delta t} = s_z^t + 2G\dot{e}_z\Delta t + 2\tau_{rz}^t\omega_{rz}\Delta t \tag{5}$$

$$s_\theta^{t+\Delta t} = s_\theta^t + 2G\dot{e}_\theta\Delta t \tag{6}$$

$$\tau_{rz}^{t+\Delta t} = \tau_{rz}^t + G\dot{\gamma}_{rz}\Delta t + (\sigma_r^t - \sigma_z^t)\omega_{rz}\Delta t \tag{7}$$

In Eq. (4), $s_r^{t+\Delta t}$ is the trial radial deviator stress at time $= t+\Delta t$, s_r^t is the radial deviator stress at time $= t$, G is the elastic shear modulus, $\dot{e}_r$ is the deviator strain rate in the radial direction, Δt is the integration time increment, τ_{rz}^t is the shear stress at time $= t$, and ω_{rz} is the rate of rotation in the r–z plane. The second term $(2G\dot{e}_r\Delta t)$ is the incremental stress due to the incremental strain $(\dot{e}_r\Delta t)$ during the time increment. The third term $(2\tau_{rz}^t\omega_{rz}\Delta t)$ is due to shear stresses from the previous time increment, which now act as normal stresses due to the new orientation of the element, caused by an incremental rotation $(\omega_{rz}\Delta t)$ during the time increment.

The axial, hoop, and shear stresses $(s_z, s_\theta, \tau_{rz})$ have a similar form except that the hoop stress does not have a contribution from rotated shear stresses.

The updated stresses in Eqs. (4–7) are trial stresses only, and they may need to be reduced if they are greater than the strength of the material. This determination can be made by computing the von Mises equivalent stress,

$$\overline{\sigma} = \{\tfrac{1}{2}[(s_r - s_z)^2 + (s_z - s_\theta)^2 + (s_\theta - s_r)^2] + 3\tau_{rz}^2\}^{1/2} \tag{8}$$

If $\overline{\sigma}$ is not greater than the equivalent tensile strength of the material $\sigma_{\max}$, the final deviator and shear stresses are in the elastic range and are as given in Eqs. (4–7). If $\overline{\sigma}$ is greater than $\sigma_{\max}$, then the material is in the plastic range and the stresses in Eqs. (4–7) must be multiplied by the factor $(\sigma_{\max}/\overline{\sigma})$. When the reduced deviator and shear stresses are put into Eq. (8) the result is always $\overline{\sigma} = \sigma_{\max}$. The purpose of the strength model is to determine $\sigma_{\max}$, and this will be described in more detail later.

The hydrostatic pressure is dependent on the volumetric strain and the internal energy in the element. The Mie-Gruneisen Equation of State is commonly used for metals,[2] and it can be expressed in the following form:

$$P = (K_1\mu + K_2\mu^2 + K_3\mu^3)(1 - \Gamma\mu/2) + \Gamma E_s(1 + \mu) \tag{9}$$

where $\mu = \rho/\rho_0 - 1 = V_0/V - 1$ and E_s is internal energy per initial volume. The initial and current densities are ρ_0 and ρ, and the initial and current volumes are V_0 and V. The four material constants are K_1, K_2, K_3, and the Gruneisen coefficient Γ. For small spherical volumetric strains, Eq. (9) reduces to $P = K_1\mu$, where K_1 is analogous to the elastic bulk modulus. For high pressures the effect of the internal energy E_s can be significant, and it is therefore necessary to solve the pressure and internal energy equations simultaneously to ensure that the pressure and energy are consistent with one another.

The artificial viscosity Q tends to damp out localized oscillations near the leading edge of wave fronts. It is called artificial because it is a numerically induced viscosity rather than a real material viscosity. More details are presented in Chapter 5.

There appears to be a general consensus that the equation of state models (pressure as a function of volume and energy) and data are adequate for most warhead computations. As a result, this chapter will primarily focus on strength and fracture models, which are not so well developed as the corresponding pressure models. More information on equation of state models and data is given in Refs. 2 and 3.

3. Test Data

Physical properties of OFHC copper and Armco iron are given in Table 1. These two materials will be used to demonstrate how constants can be obtained for computational strength and fracture models. The laboratory test data are obtained from torsion tests at various strain rates, Hopkinson bar tests at various

Table 1 Physical properties of OFHC copper and Armco iron

Property	OFHC copper	Armco iron
Heat Treatment		
Temperature, K	700	1200
Time at temperature, min	60	60
Hardness (Rockwell)	F-30	F-72
Grain size, mm	0.060–0.090	0.090–0.150
Elastic properties		
Elastic modulus, E, GPa	124	207
Poisson's ratio, ν	0.34	0.29
Shear modulus, $E/2(1+\nu)$, GPa	46	80
Bulk modulus, $E/3(1-2\nu)$, GPa	129	164
Thermal properties		
Density, kg/m^3	8960	7890
Conductivity, k, W/mk	386	73
Specific heat, c_p, J/kgK	383	452
Diffusivity, $k/\rho c_p$, m^2/s	0.000113	0.000020
Expansion coefficient (vol.), α, K^{-1}	0.000050	0.000032
Melting temperature, T_{melt}, K	1356	1811

temperatures, quasistatic tensile tests with various notch geometries (notched and unnotched), and cylinder impact tests.

Torsion data are shown in Fig. 1.[4] Some desirable features of this testing technique are that the state of stress in the specimen is well defined, large strains can be achieved without forming geometric instabilities, and a wide range of strain rates can be obtained with the same testing technique. A description of the torsion testing apparatus is given in Ref. 5.

Both of the materials in Fig. 1 exhibit strain rate sensitivity and some thermal softening at the higher strain rates. The OFHC copper is very ductile and does not fracture at the maximum attainable strain of 7.0.

Hopkinson bar test data at various temperatures are shown in Fig. 2.[4] This test method allows for higher strain rates through the use of wave propagation techniques. A description of the Hopkinson bar test apparatus is given in Ref. 6. The elevated temperatures are obtained by surrounding the in-place test specimen by an oven such that the temperatures are applied for several minutes prior to testing. Although it is possible to test materials to greater strains than those shown in Fig. 2, these Hopkinson bar data could not be accurately evaluated after necking began to occur in the tensile specimen. Furthermore, at larger strains the effect of adiabatic heating can also complicate the results. The elevated temperatures show a distinct softening effect on the strength of the materials.

Photographs of the fracture surfaces of the Hopkinson bar specimens are shown in Fig. 3.[7] The cross-sectional areas of these surfaces can be used to obtain fracture characteristics.

In Fig. 4 are shown quasistatic stress-strain data for various tension tests.[7] The stress is based on the current area of the neck, and the strain is defined as $\ell n(A_o/A)$, where A_o and A represent the initial and current areas of the neck.

The notched tension specimens give a concentration of hydrostatic tension in the test specimen,[8] as one clearly sees in the data of Fig. 4. For a specified true strain, the stress (including hydrostatic tension) increases for the notched geometry. It can also be seen that the presence of the hydrostatic tension significantly decreases the strain at which the material fractures.

The test data of Figs. 1–4 allow for variations in strain, strain rate, temperature, and pressure. Unfortunately, these variations do not cover the extreme conditions encountered in many warhead computations. Although new test techniques are being developed, it is still very difficult to obtain accurate stress-strain data at large strains ($\varepsilon > 1.0$) and high strain rates ($\dot{\varepsilon} > 10^5 s^{-1}$).

Another test that has been used extensively is the cylinder impact test. This test was developed by Taylor in 1948,[9] and it has been used by many others to obtain dynamic flow stresses of various materials.[10–12] This test consists of impacting a cylinder against a relatively rigid surface and then measuring the deformed geometry. Photographs of deformed test specimens are shown in Fig. 5.[4]

The merits of this test have often been debated. Some would argue that this test should not be used to obtain constitutive data because a wide range of unknown strains, strain rates, temperatures, and stresses vary throughout the test specimen and throughout the duration of the test. It is better, they would say, to obtain data from dynamic tension, compression, and/or torsion tests, in which the variables are better defined, and then use the cylinder impact test to provide an independent check of the data.

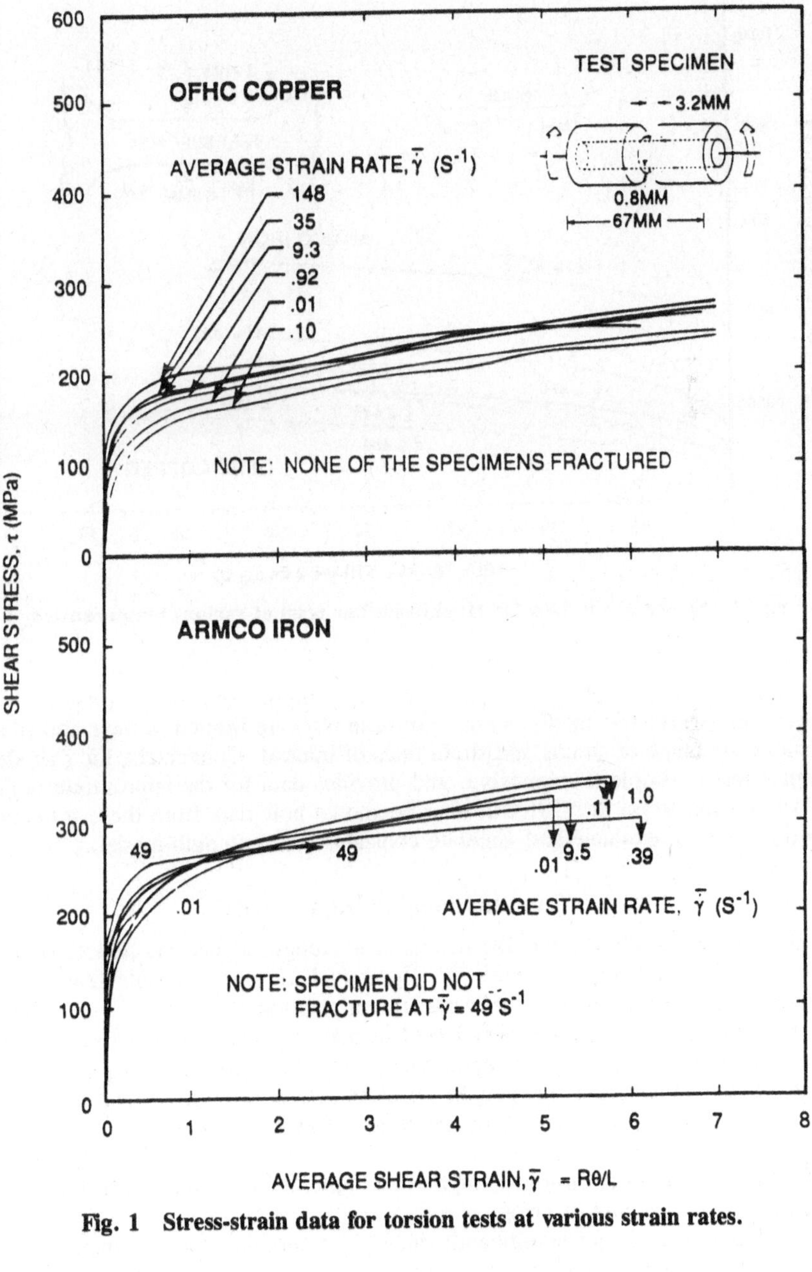

Fig. 1 Stress-strain data for torsion tests at various strain rates.

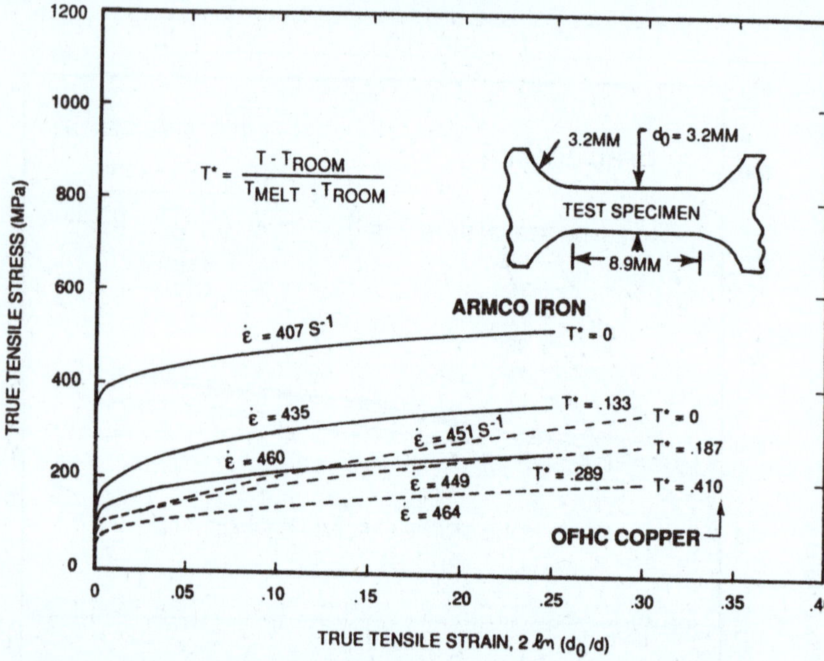

Fig. 2 Stress-strain data for Hopkinson bar tests at various temperatures.

Others would argue that dynamic tension, compression, and/or torsion tests require expensive test facilities and that these tests are limited in their ability to achieve the range of strains and strain rates of interest. Conversely, the cylinder impact test is simple, inexpensive, and provides data for the appropriate range of strains and strain rates. It will later be shown how data from these tests can be used both to evaluate and generate constants for a strength model.

4. Computational Strength Models

Before getting into the specific details of a strength model and how to obtain constants for that model, it will be helpful to list some desirable features of strength and fracture models for warhead computations:

- Models should be physically based so that they may be used over a wide range of strains, strain rates, temperatures, and stress states.
- Models should use the von Mises flow rule and express the stress as a function of strain, strain rate, temperature, pressure, and other macroscopic variables if necessary.
- Models should be simple enough to be implemented into existing computer codes with a minimum of effort.
- Models should not significantly increase computing time or memory requirements of existing codes.
- Models should be usable for a variety of metals.

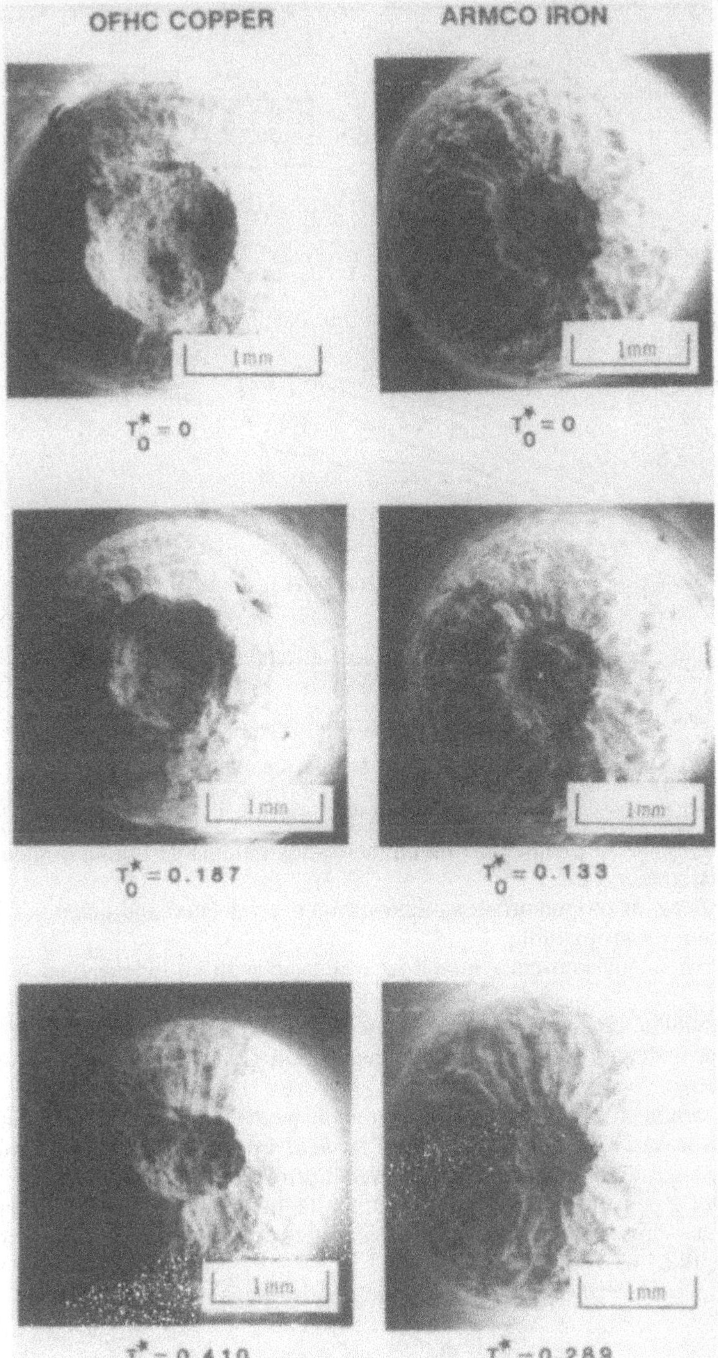

Fig. 3 Photographs of the fracture surfaces of the Hopkinson bar test specimens.

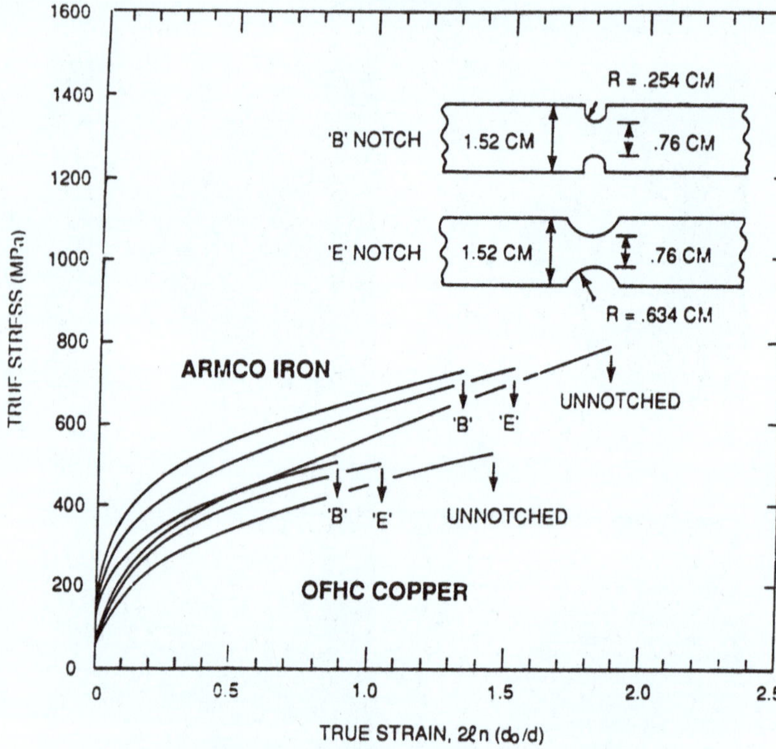

Fig. 4 Stress-strain data for quasi-static tension tests.

- Constants for the models should be readily obtainable from a limited number of laboratory tests.
- Effects of the important variables should be isolated and quantified to the maximum extent possible.
- Models and constants should be evaluated with an independent series of tests.

It should be emphasized that these are desirable (not necessary) features only, and that it may not be possible to adhere to all of these features for a variety of reasons.

The origin of the Johnson-Cook strength model, which will be considered here, is given in Ref. 4. It should be recalled from the discussion on computational algorithms that the strength model provides the maximum equivalent (von Mises) tensile, plastic flow stress that the material can develop in its current state. This stress was previously designated as σ_{max} for clarity, but will now be simplified to $\sigma = \sigma_{max}$.

The model for this equivalent tensile flow stress is

$$\sigma = [A + B\varepsilon^n][1 + C \ln \dot{\varepsilon}*][1 - T*^m] \qquad (10)$$

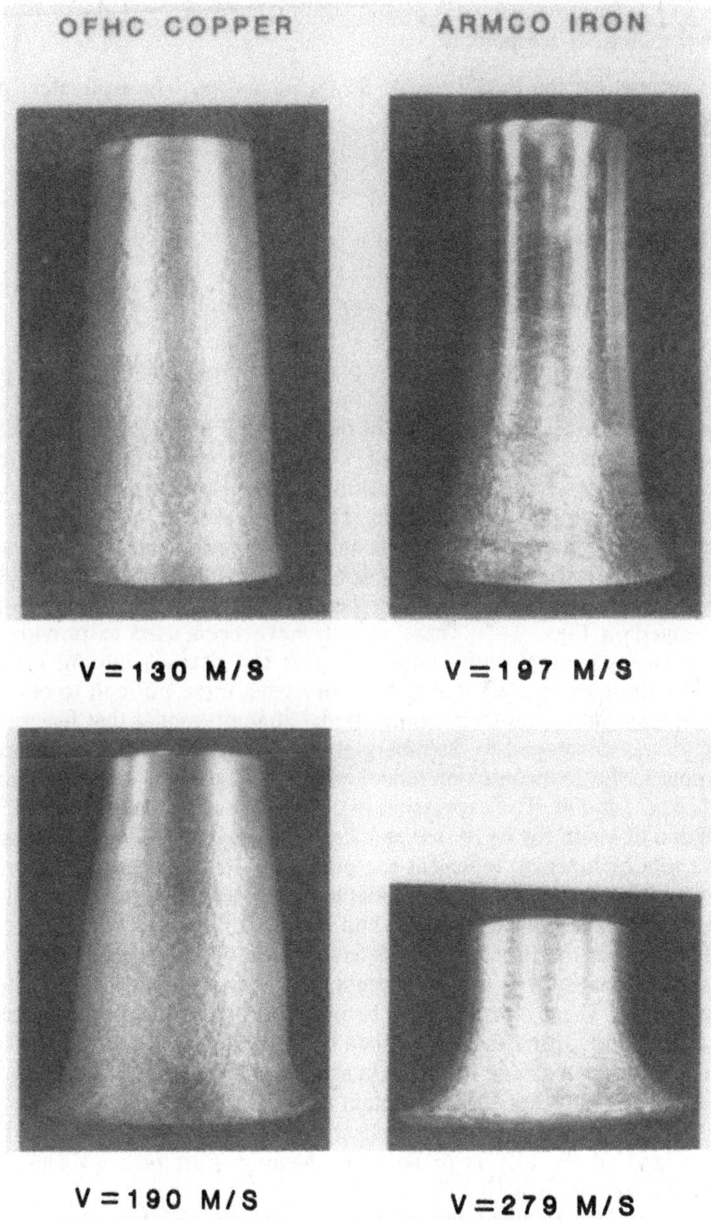

Fig. 5 Deformed shapes of cylinder impact test specimens.

where

ε = equivalent plastic strain

$\dot{\varepsilon}* = \dot{\varepsilon}/\dot{\varepsilon}_o$, the dimensionless plastic strain rate for $\dot{\varepsilon}_o = 1.0\ s^{-1}$

$T*$ = homologous temperature

The constants for the model are A, B, C, n, and m. The equivalent plastic strain is obtained from

$$\varepsilon = \sum \dot{\varepsilon}\Delta t \tag{11}$$

where the equivalent plastic strain rate is given by

$$\dot{\varepsilon} = \{\tfrac{2}{9}[(\dot{e}_r - \dot{e}_z)^2 + (\dot{e}_z - \dot{e}_\theta)^2 + (\dot{e}_\theta - \dot{e}_r)^2] + \tfrac{1}{3}\dot{\gamma}_{rz}\}^{1/2} \tag{12}$$

where $\dot{e}_r$, $\dot{e}_z$, $\dot{e}_\theta$, are the three deviator plastic strain rates and $\dot{\gamma}_{rz}$ is the plastic shear strain rate.

Although this model exhibits many of the desirable features noted previously, its major shortcoming is that it is not a physically based model. There are, however, several researchers now working on the development of improved physically based computational models. The Zerilli-Armstrong[13] strength models (one for body-centered cubic metals and one for face-centered cubic metals) are more physically based and are of special interest here because they were developed and evaluated with the specific OFHC copper and Armco iron test data presented in Figs. 1–5. These models have been used to provide good computational agreement with cylinder impact test data. A current drawback associated with these models is that it is somewhat more difficult to obtain the appropriate constants. Another computational strength model that has received wide usage was developed by Steinberg et al.[14]

Returning to the Johnson-Cook model of Eq. (10), the five material constants are A, B, n, C, and m. The expression in the first set of brackets gives the stress as a function of strain for $\dot{\varepsilon}* = 1.0$ and $T* = 0$. The expressions in the second and third sets of brackets represent the effects of strain rate and temperature, respectively. The basic form of the model is readily adaptable to most computer codes since it uses the variables ε, $\dot{\varepsilon}*$, and $T*$, which are available in the codes.

The first step in the process is to determine the constants in the first set of brackets. The quantity A is the yield stress, and B and n represent the effect of strain hardening. These constants can be obtained independently from both the torsion data of Fig. 1 and the tensile data of Figs. 2 and 4.

Because the torsion data in Fig. 1 include the strain rate of interest ($\dot{\varepsilon}* = 1.0$), it is a straightforward procedure to obtain the strain hardening constants A, B, and n for this strain rate. However, these constants are for isothermal conditions ($T* = 0$), and so the effects of adiabatic heating must be considered when analyzing the torsion data.[15]

This process also requires the conversion of torsional stress-strain data to equivalent tensile stress-strain data. From the von Mises flow rule, the equivalent tensile stress is $\overline{\sigma} = \sqrt{3}\tau$, the equivalent tensile strain is $\varepsilon = \gamma/\sqrt{3}$, and the equivalent tensile strain rate is $\dot{\varepsilon} = \dot{\gamma}\sqrt{3}$.

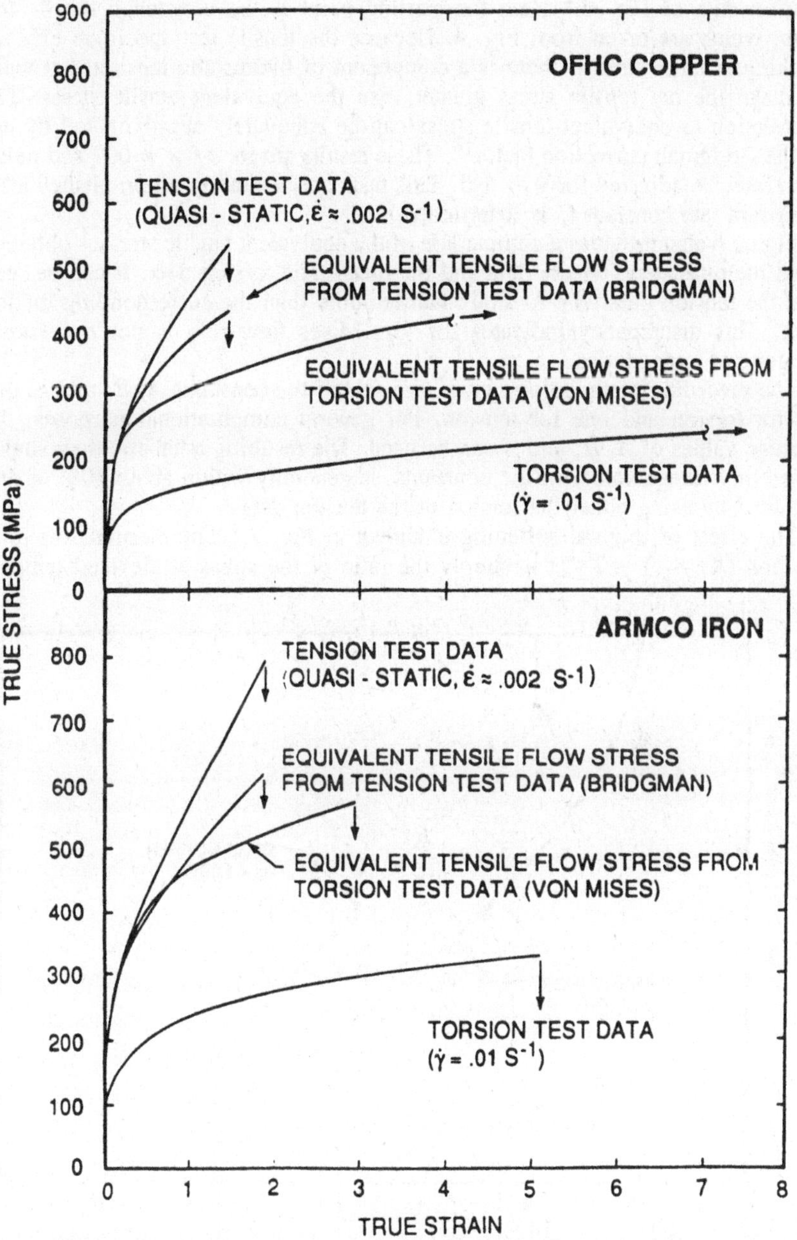

Fig. 6 Equivalent tensile flow stress as determined from quasistatic tension and torsion tests.

The three constants A, B, and n can also be derived from the quasistatic tension data of Fig. 6.[4] Here the starting point is the unnotched tensile test data, which are taken from Fig. 4. Because the tensile test specimen exhibits necking at large strains, there is a component of hydrostatic tension that tends to make the net tensile stress greater than the equivalent tensile stress. The conversion to equivalent tensile stress can be adequately approximated by use of the Bridgman correction factor[16]. These results are for $\dot{\varepsilon}* = 0.002$ and must, therefore, be adjusted for $\dot{\varepsilon} = 1.0$. This task can be readily accomplished after the strain rate constant C is determined.

Figure 6 also provides a comparison of the equivalent tensile stresses obtained from the quasistatic tensile data and the quasistatic torsion data. It can be seen that the tension data tend to strain-harden more than the corresponding torsion data. This discrepancy indicates the von Mises flow rule is not necessarily accurate at large strains.

The preceding methodology gives two sets of the constants A, B, and n: one set for torsion and one for tension. For general computational purposes, the average values of A, B, and n can be used. The resulting adiabatic flow stress, based on average values of the constants, is generally within about 10% of that obtained by using either the torsion or the tension data.

The effect of thermal softening is shown in Fig. 7.[4] The thermal softening fraction ($K_T = 1 - T*^m$) is simply the ratio of the stress at elevated temper-

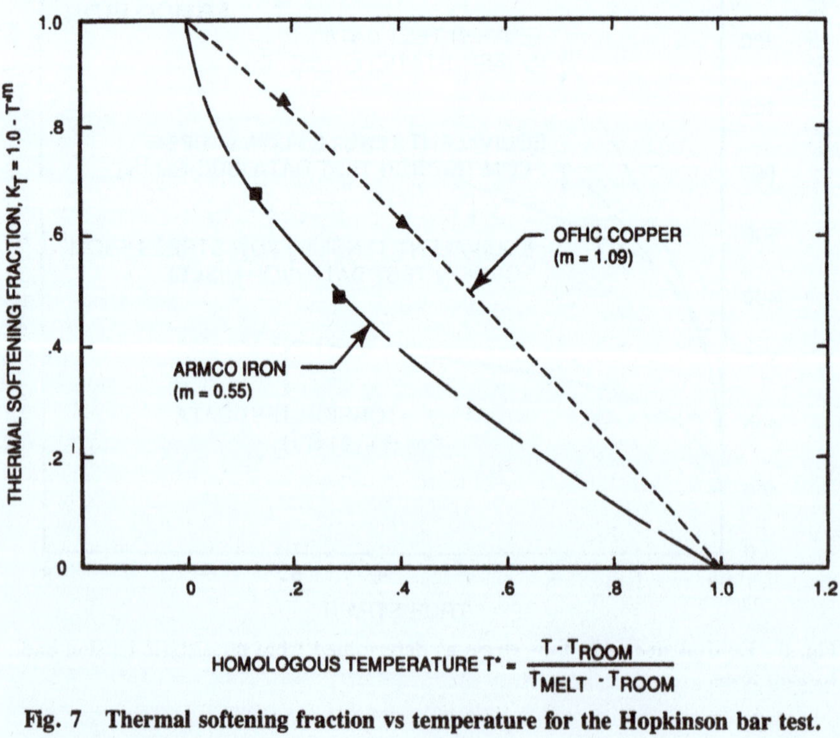

Fig. 7 Thermal softening fraction vs temperature for the Hopkinson bar test.

ature to that at room temperature. The data are taken from the Hopkinson bar tests of Fig. 2. The OFHC copper shows an approximately linear dependence on $T*$, whereas the Armco iron tends to soften more rapidly with increasing temperature.

The strain rate constant C is determined next. For the tensile mode of deformation it can be obtained from the tensile data of Fig. 8.[4] The strain rate constant for torsion can be obtained from Fig. 8 in a similar manner.[4] The torsion data also allow the combination of strain rate hardening and thermal softening to be evaluated at large strains. This can be done by numerically simulating the torsion test at the higher strain rates. Strain hardening, strain rate hardening, thermal softening, and heat conduction effects are considered.[15] In some instances the torsional strain rate constant is adjusted to give better agreement with the torsion test data at the higher strain rates. In Fig. 9 computed and test results are compared using the torsional constants.[4] The final strain rate constant is again taken as an average of those constants obtained from the tensile data and the torsion data.

The resulting strength model constants for OFHC copper and Armco iron are shown in Table 2. Similar data for other materials[4] are also provided in this table. The corresponding stress-strain relationships for OFHC copper and Armco iron are shown in Fig. 10.[4] The temperature for the adiabatic relationships is due to the plastic work of deformation. In both cases, the adiabatic stresses reach a maximum and then decrease with increasing strain. This point of maximum stress is important inasmuch as it represents the strain at which localized instabilities may begin to occur.

An evaluation of the model and data can be made by comparing computed results with data from the cylinder impact results in Fig. 5. These tests provide strain rates in excess of $10^5 s^{-1}$ and strains in excess of 1.5. Also, a range of tensile and shear deformation is experienced. The results of these comparisons are shown in Fig. 11.[4] The OFHC copper strength model is too strong and the Armco iron model appears to be too weak. Nevertheless, there is good general agreement in both cases. It should be emphasized that the results of the cylinder impact tests have not been incorporated into the data, and they represent a totally independent check.

Another independent check case is the explosively formed penetrator (EFP) computation shown in Fig. 12. This computation was also performed with the EPIC-2 code,[17] using the Armco iron constants in Table 2. It can be seen that the computed shape is in good general agreement with the test results and that the majority of the EFP is highly strained in the range of $1.0 < \varepsilon < 3.0$.

It was previously noted that cylinder impact test results can be used to evaluate models derived from other methods (as shown in Fig. 11), or they can also be used directly to develop models and/or obtain constants for existing models.

This procedure can be illustrated by considering the four EPIC-2 computations in Fig. 13.[12] The experimental data for the OFHC copper cylinder are identical to those shown in Figs. 5 and 11. The adiabatic stress-strain relationships are given in Fig. 14, and the strength model constants used for the four cases are given in Table 3.[12] The adiabatic stresses are shown only to the maximum strains attained in the computed results.

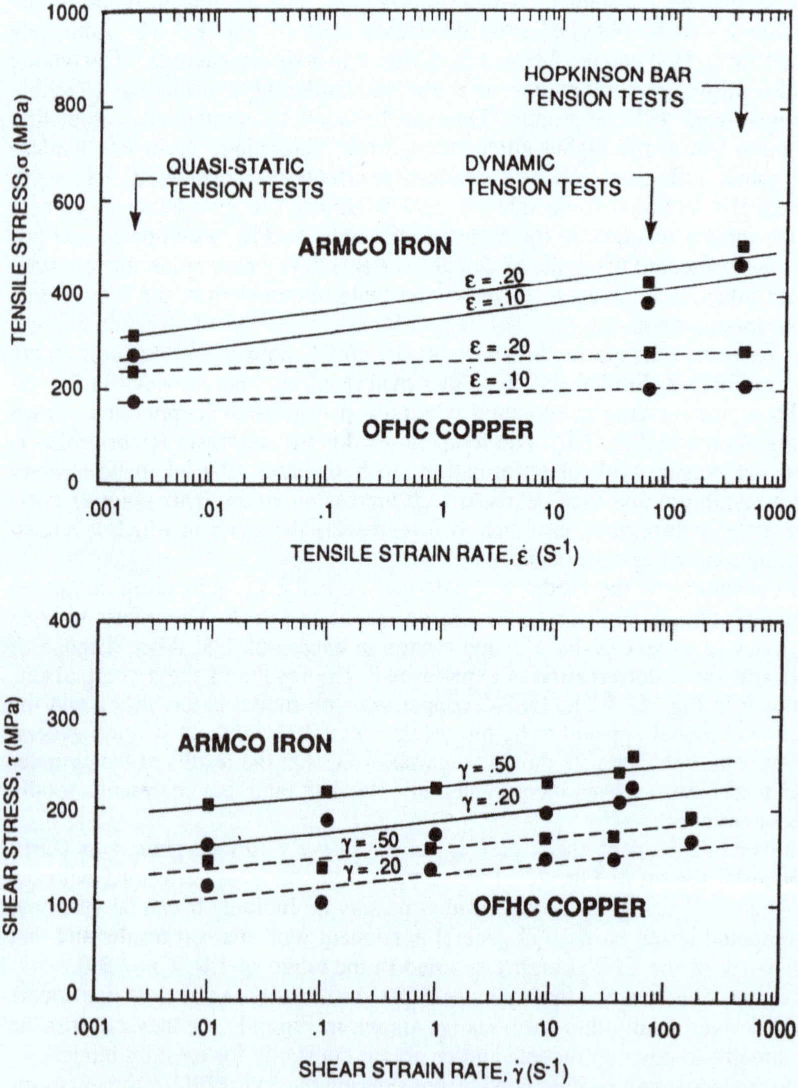

Fig. 8 Stress vs strain rate for tension and torsion tests.

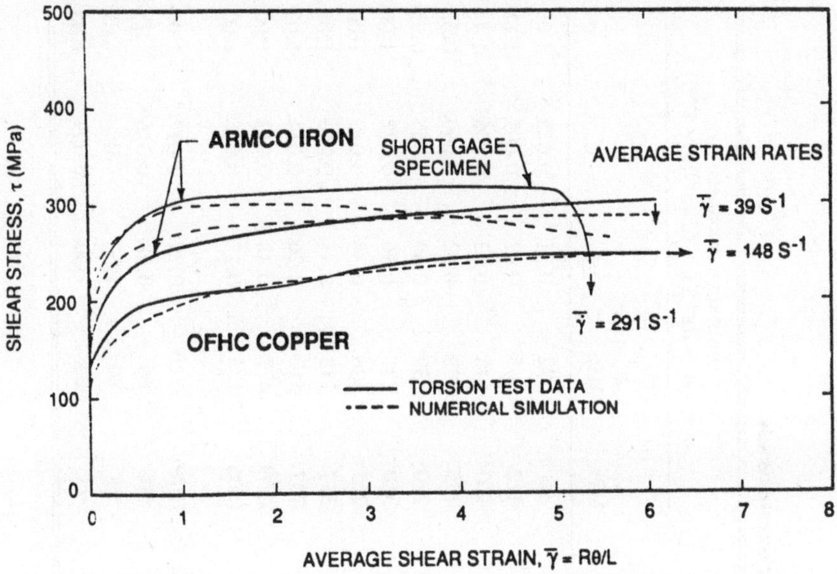

Fig. 9 Comparison of computed and test results for stress vs strain rate for tension and torsion.

For Case A the length L of the deformed cylinder is matched with the computational result. Because only one deformed dimension is matched, only one independent strength constant (representing a constant flow stress) can be obtained. One can see significant discrepancies between the test and computational results at the deformed end of the cylinder.

Case B allows for linear strain hardening. By matching both the deformed length L and the maximum diameter D, it is possible to obtain the two constants for linear hardening. Here the computed result is in good general agreement with the test result. If this model were to be applied to larger strains than those experienced in the computed results of Case B, however, it would probably lead to excessively high stresses. It is well known that thermal softening tends to decrease the rate of strain hardening at large strains.

Cases C and D are based on the Johnson-Cook model of Eq. (10). Although there are five constants in this model (A, B, n, C, m), only the three strain-hardening constants A, B, and n are determined from the test data. The strain rate constant C and the thermal softening constant m must be approximated or obtained from other sources. Because there are now three constants, it is possible to match three deformed dimensions—the length L, the maximum diameter D, and an intermediate diameter W. A direct method for obtaining these constants is given in Refs. 10–12.

Case C is the result obtained if nothing is known about the strain rate or thermal softening characteristics of the material. Here, the strain rate constant is set to $C = 0$ and the thermal softening is assumed to be linear, or $m = 1.0$.

Table 2 Strength model constants for various materials

Material	Description				Constitutive constants for $\sigma = [A + B\epsilon^n][1 + C\ln\dot{\epsilon}^*][1 - T^{*m}]$				
	Hardness (Rockwell)	Density, kg/m^3	Specific heat, J/kgK	Melting temp, K	A, MPa	B, MPa	n	C	m
OFHC copper	F-30	8960	383	1356	90	292	0.31	0.025	1.09
Cartridge brass	F-67	8520	385	1189	112	505	0.42	0.009	1.68
Nickel 200	F-79	8900	446	1726	163	648	0.33	0.006	1.44
Armco iron	F-72	7890	452	1811	175	380	0.32	0.060	0.55
Carpenter Electrical iron	F-83	7890	452	1811	290	339	0.40	0.055	0.55
1006 steel	F-94	7890	452	1811	350	275	0.36	0.022	1.00
2042-T351 aluminum	B-75	2770	875	775	265	426	0.34	0.015	1.00
7039 aluminum	B-76	2770	875	877	337	343	0.41	0.010	1.00
4340 steel	C-30	7830	477	1793	792	510	0.26	0.014	1.03
S-7 tool steel	C-50	7750	477	1763	1539	477	0.18	0.012	1.00
Tungsten alloy (.07Ni, .03Fe)	C-47	17000	134	1723	1506	177	0.12	0.016	1.00
DU-.75Ti	C-45	18600	117	1473	1079	1120	0.25	0.007	1.00
1100 aluminum	—	2700	905	923	116	69	0.58	0.016	1.13
Polycarbonate	—	1200	1340	533	76	69	1.00	0	1.85
Polyethersulfone	—	1370	1120	617	97	69	1.00	0.009	1.85

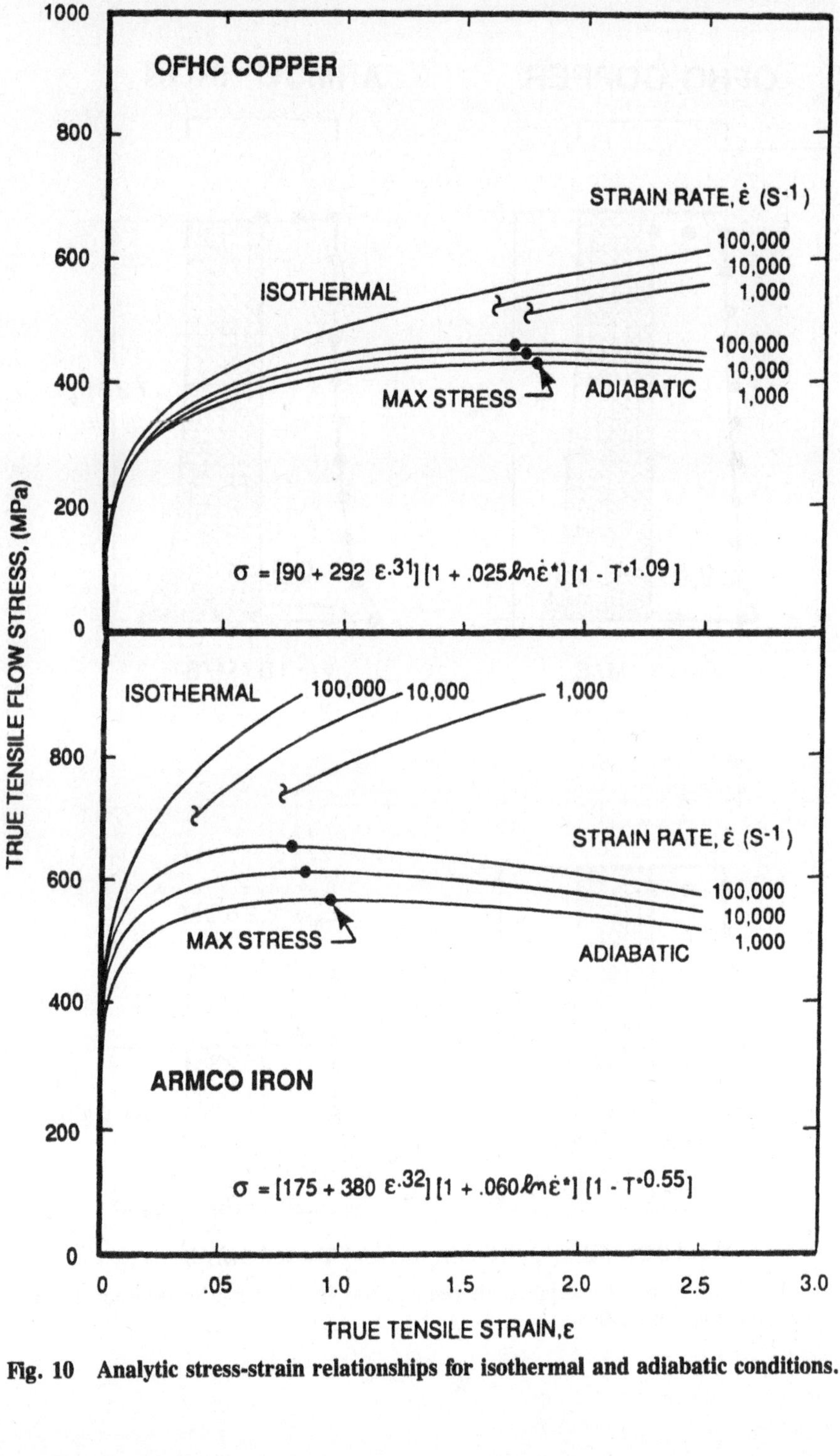

Fig. 10 Analytic stress-strain relationships for isothermal and adiabatic conditions.

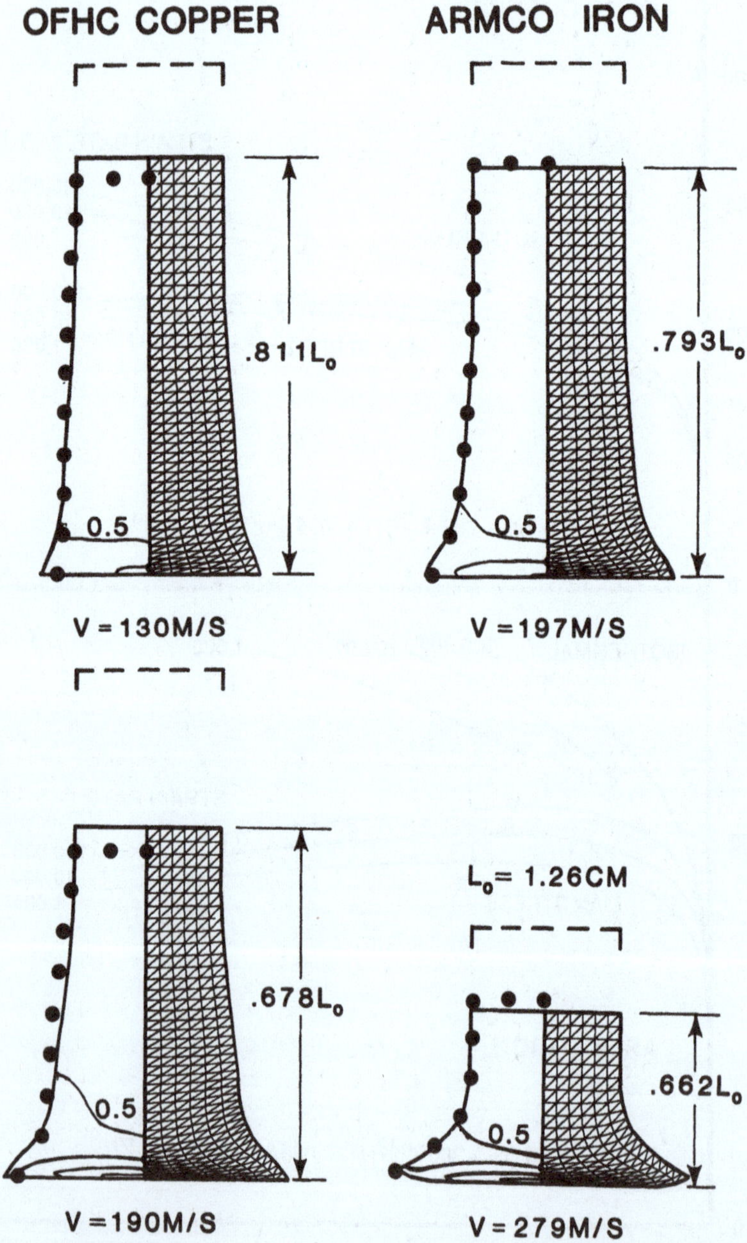

Fig. 11 Comparison of computed shapes and test results for cylinder impact tests at various impact velocities.

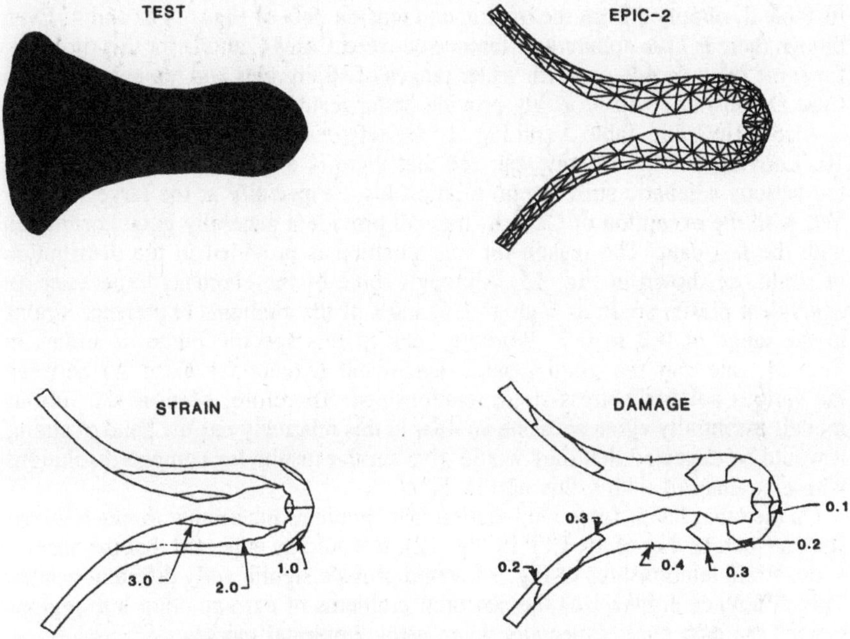

Fig. 12 Comparison of computational and test results for an Armco iron, explosively formed penetrator.

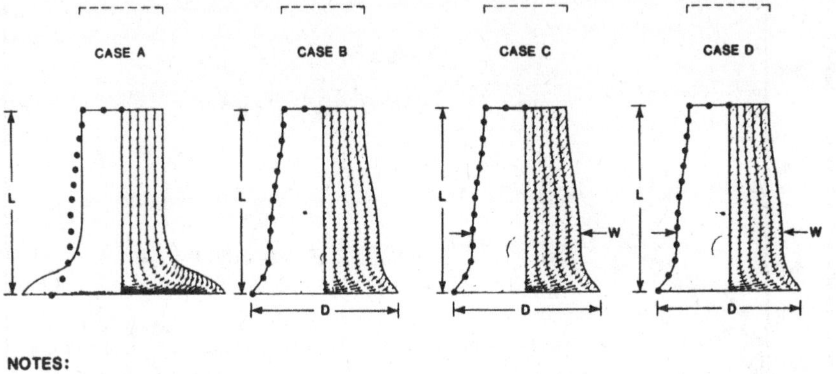

NOTES:

- OFHC COPPER CYLINDER (L_o = 25.4mm, D_o = 7.6mm, L = 16.2mm, D = 13.5mm, W = 10.1mm)
- IMPACT VELOCITY = 190 m/s
- TEST RESULTS INDICATED BY DOTS ••••
- EQUIVALENT PLASTIC STRAIN CONTOURS SHOWN AT INTERVALS OF 0.5

Fig. 13 Comparison of test results with computed results using various strength model constants.

Case D uses the previously determined strain rate and thermal softening constants in Table 2, obtained from the torsion and tension data of Figs. 1, 2, and 4. Even though there is little apparent difference between Cases C and D for this problem, for some other problems (with wider ranges of strain rates and temperatures) the Case D constants will probably provide better results.

Also included in Table 3 and Fig. 14 are reference data from Table 2 and Fig. 10. Looking at Fig. 14, one can see that there is a significant discrepancy in the various adiabatic stress-strain relationships, especially at the larger strains. Yet, with the exception of Case A, they all provide a generally good correlation with the test data. The reason for this situation is provided in the distribution of strain, as shown in Fig. 15. Although some of the elements experience an equivalent plastic strain as high as 1.6, most of the elements experience strains in the range of 0.3 to 0.7. Working back to this specific range of strains in Fig. 14, one can see good general agreement (except for Case A) between the various adiabatic stress-strain relationships. Therefore, because the various models essentially agree with one another in this relatively narrow band of strain, it would be expected that they would give similar results for computed solutions whose strains fall within this narrow band.

On the other hand, for a computation that simulates material with much higher strains (such as that of the EFP in Fig. 12), it would be expected that the various stress-strain relationships of Fig. 14 would provide significantly different results. This difference emphasizes the potential problems of extrapolating into regions beyond the data base, especially when using empirical models.

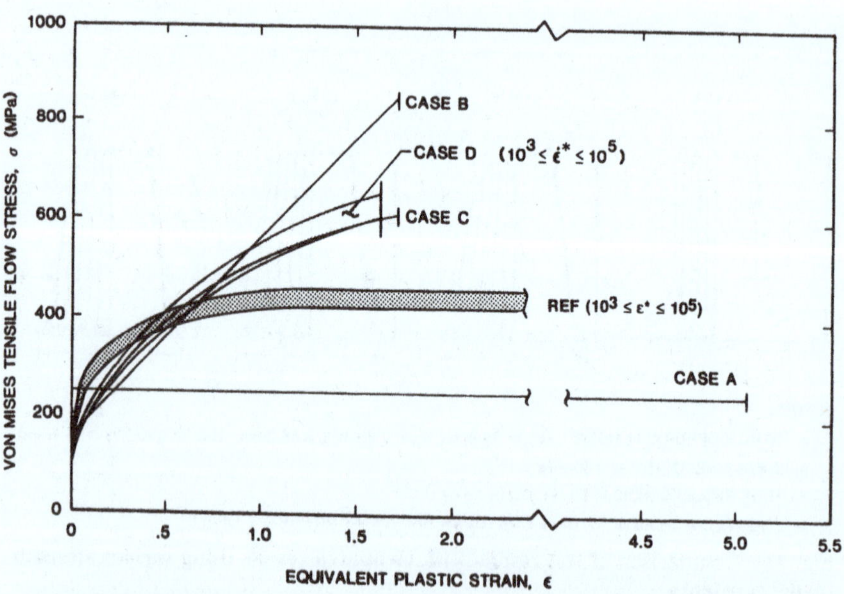

Fig. 14 Adiabatic stress-strain relationships for the various strength model constants.

Table 3 Strength model constants for OFHC copper as determined from cylinder impact test results

	Case A	Case B	Case C	Case D	Ref. data from Table 2
Strength constants for $\sigma = [A + B\epsilon^n][1 + C\, \ell n\, \dot{\epsilon}^*][1 - T^{*m}]$					
A, MPa	250	160	103	98	90
B, MPa	0	400	469	368	292
n	—	1.00	.65	.70	.31
C	0	0	0	.025	.025
m	—	—	1.00	1.09	1.09

$\dot{\epsilon}^* = \dot{\epsilon}/\dot{\epsilon}_0$ for $\dot{\epsilon}_0 = 1.0\ \mathrm{s}^{-1}$
$T^* = (T - T_{\mathrm{room}})/(T_{\mathrm{melt}} - T_{\mathrm{room}})$

To summarize the status of strength modeling for warhead computations, two points should be made. The first point is that there is much uncertainty about how materials behave under the extreme strains, strain rates, temperatures, and pressures that are experienced by warheads. Proven physically based models are not generally available and it is difficult to obtain well-defined material test data at the appropriate strains, strain rates, temperatures, and pressures.

On the other hand, the second point is that existing strength models and data are currently being used to perform very meaningful warhead computations, such as shown in Fig. 12. Even if there is not perfect agreement with experiment,

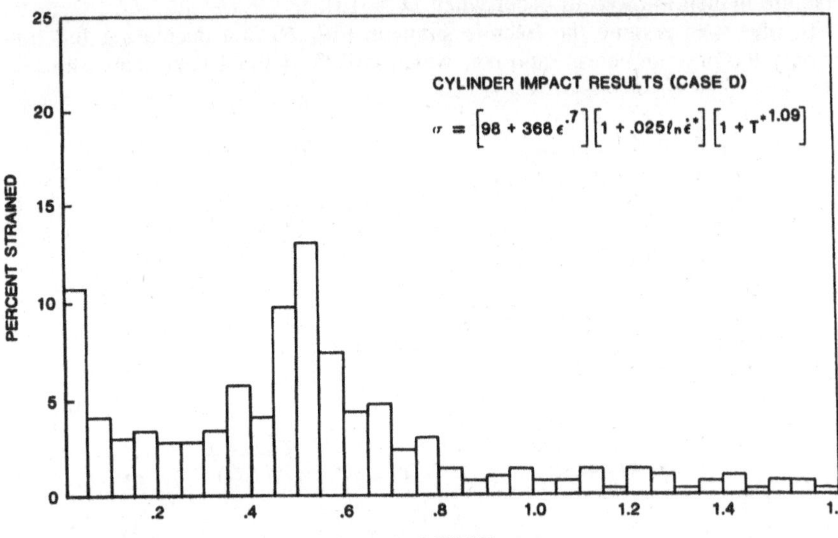

Fig. 15 Distribution of element strain for the Case D cylinder impact computation.

computations have been very helpful in analyzing and designing various types of warheads.

5. Computational Fracture Models

Throughout the past years there has been a great deal of research directed at fracture. Although much is understood about brittle fracture and crack propagation, there has been relatively little effort directed at computational models for ductile fracture under conditions of large strains, high strain rates, high temperatures, and high pressures. As a result, current models generally cannot make accurate predictions of fracture in warhead computations.

A notable exception is the work performed by Seaman, Curran, and Shockey during the past decade[18,19]. Their computational fracture models have contributed significantly to the understanding of fracture. These models are rather complex, however, and require extensive effort to obtain the appropriate constants. As a result, they have not been widely used for warhead computations.

The origin of the Johnson-Cook fracture model, which will be considered here, is given in Ref. 7. The basic form of the model is

$$D = \sum \frac{\Delta\epsilon}{\epsilon^f} \tag{13}$$

where

D = damage to an element

$\Delta\epsilon$ = increment of equivalent plastic strain that occurs during an integration cycle

ϵ^f = equivalent strain to fracture under the current conditions of strain rate, temperature, pressure, and stress

Fracture is then allowed to occur when $D = 1.0$.

To illustrate, assume the fracture strain in Fig. 16 is a decreasing function of only the pressure-stress ratio $\sigma*$, which will be defined later. This assumed

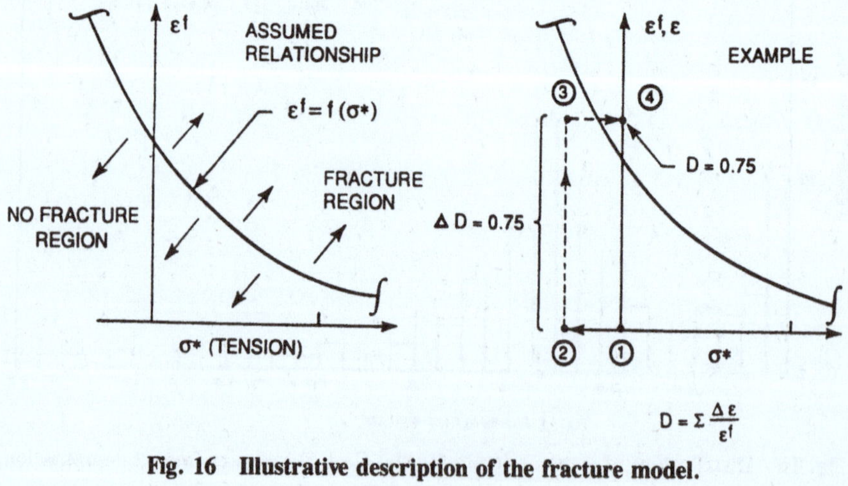

Fig. 16 Illustrative description of the fracture model.

relationship indicates the material can be more severely strained when it is under hydrostatic compression. The example in Fig. 16 shows how damage is accumulated under compression. From the initial conditions at point 1, the material is put under compression at point 2. It is strained to point 3 under constant compression, and the compression is then released as the material goes to point 4. When the pressure is released to $\sigma* = 0$, $D = 0.75$ even though the plastic strain ϵ is greater than ϵ^f at $\sigma* = 0$. The importance of path-dependency is evident, since a model based only on current conditions would incorrectly predict fracture for these conditions.

The general expression for the strain at fracture is given by

$$\epsilon^f = [D_1 + D_2^{D_3\sigma*}][1 + D_4 \ln \dot{\epsilon}*][1 + D_5 T*] \tag{14}$$

for constant values of the variables $\sigma*$, $\dot{\epsilon}*$, and $T*$ and for $\sigma* \leq 1.5$. The dimensionless pressure-stress ratio is defined as $\sigma* = \sigma_m/\overline{\sigma}$, where σ_m is the average of the three normal stresses and $\overline{\sigma}$ is the von Mises equivalent stress. The dimensionless strain rate $\dot{\epsilon}*$ and homologous temperature $T*$ are identical to those used in the strength model of Eq. (12).

The five constants are $D_1, \ldots, D_5$. The expression in the first set of brackets follows the form presented by Hancock and Mackenzie.[20] Because D_3 is generally negative, it essentially says that the strain to fracture decreases as the hydrostatic tension σ_m increases. The expression in the second set of brackets represents the effect of strain rate, and that in the third set of brackets represents the effect of temperature. For high values of hydrostatic tension ($\sigma* > 1.5$) a different relationship is used; it will be presented later.

The first step required to obtain the material constants is to determine the effect of the dimensionless pressure-stress ratio $\sigma*$. This can be done by considering the quasistatic tensile data of Fig. 4 and the torsion fracture data of Fig. 17.[7] Numerical simulations, using the EPIC-2 code, are performed as shown in Fig. 18.[7]

In Fig. 19 is shown the computed pressure-stress ratio as a function of the equivalent strain in the center of the specimen.[7] The results for the unnotched specimens show good agreement with the Bridgman data,[16] which are for a range of materials. Although the computed results for both the OFHC copper and the Armco iron fall below the Bridgman prediction, materials with less strain hardening, such as 4340 steel, give computed results that lie above the Bridgman prediction.[7] For the 'E' notch specimens, $\sigma*$ tends to increase as ϵ increases. For the 'B' notched specimens, however, $\sigma*$ decreases as ϵ increases.

The strain to fracture as a function of the pressure-stress ratio for quasistatic conditions is presented in Fig. 20.[7] For a given pressure-stress ratio, strains equal to or greater than those indicated by the curves will give fracture. The data at $\sigma* = 0$ are from the torsion tests. For both materials the torsion tests provide fracture strains greater than those of the tension tests. This result is as expected based on the trend of the tension data in Fig. 20 and similar data reported elsewhere.[8,20,21]

The analytic representations through the test data (dashed lines) are obtained by varying the appropriate constants until $D \approx 1.0$ at fracture for the three

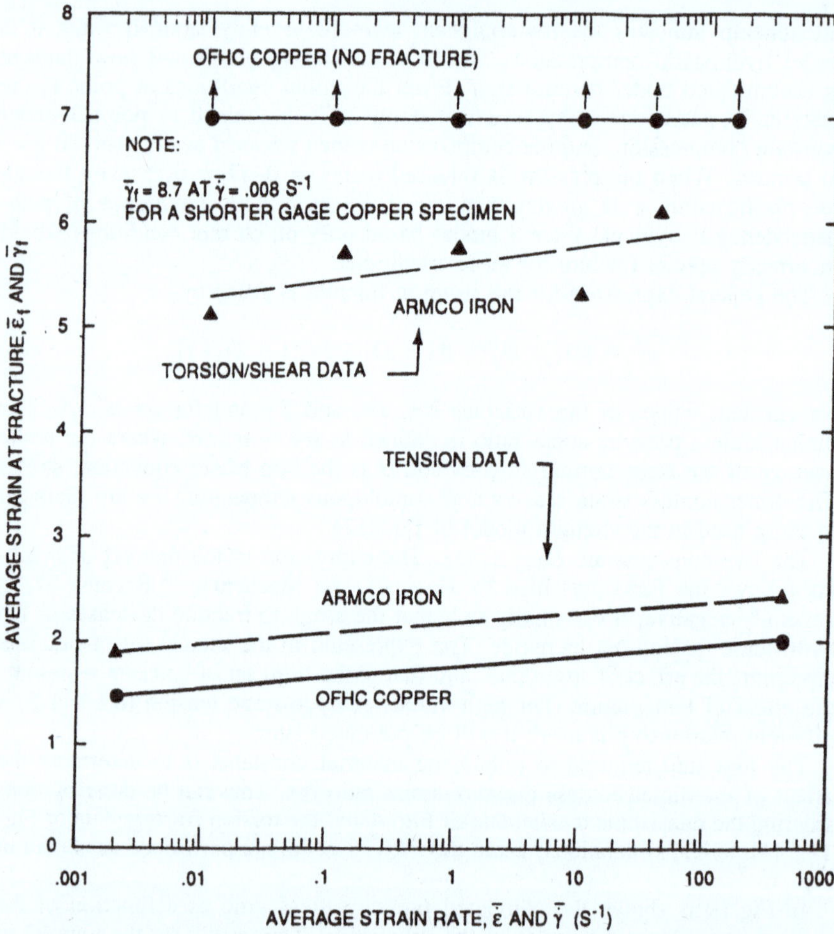

Fig. 17 Average fracture strain vs strain rate for tension and torsion tests.

tension tests and torsion test. The three constants (for the analytic representation) for each material in Fig. 20 are proportional to D_1, D_2, and D_3 and need only be adjusted to correspond to a strain rate of $\dot{\epsilon}* = 1.0$.

The effect of strain rate and temperature on the strain to fracture is shown in Fig. 21.[7] The fracture strains are expressed as the Hopkinson bar fracture strains divided by the quasistatic tensile fracture strains. The Hopkinson bar fracture strains are approximately determined by measuring the cross-sectional area of the post-tested specimens, as shown in Fig. 3. The temperatures for each test range from the initial temperature at the beginning of the test to the computed temperature at the strain where fracture occurs.

The straight lines drawn through the data are least squares fits to the midpoints of the range of temperatures. Because of the approximate nature of the data and

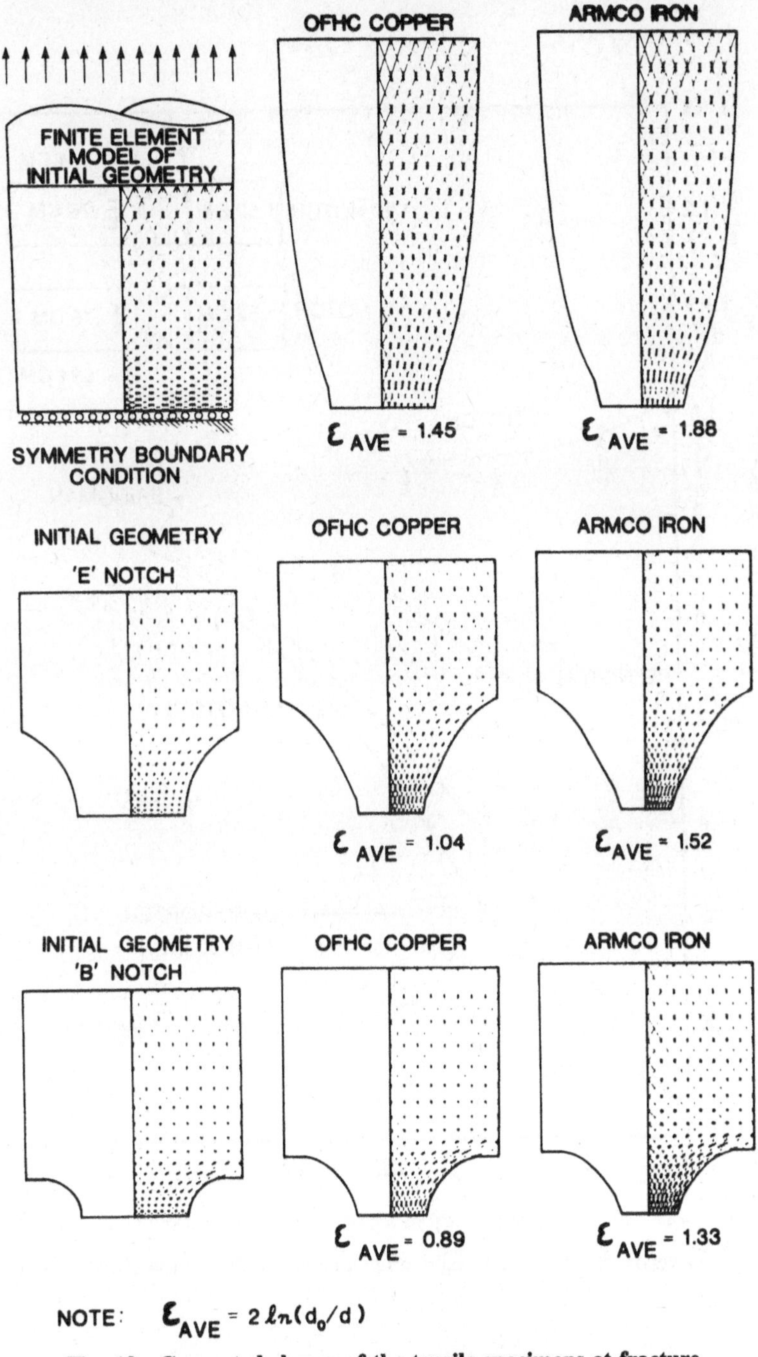

Fig. 18 Computed shapes of the tensile specimens at fracture.

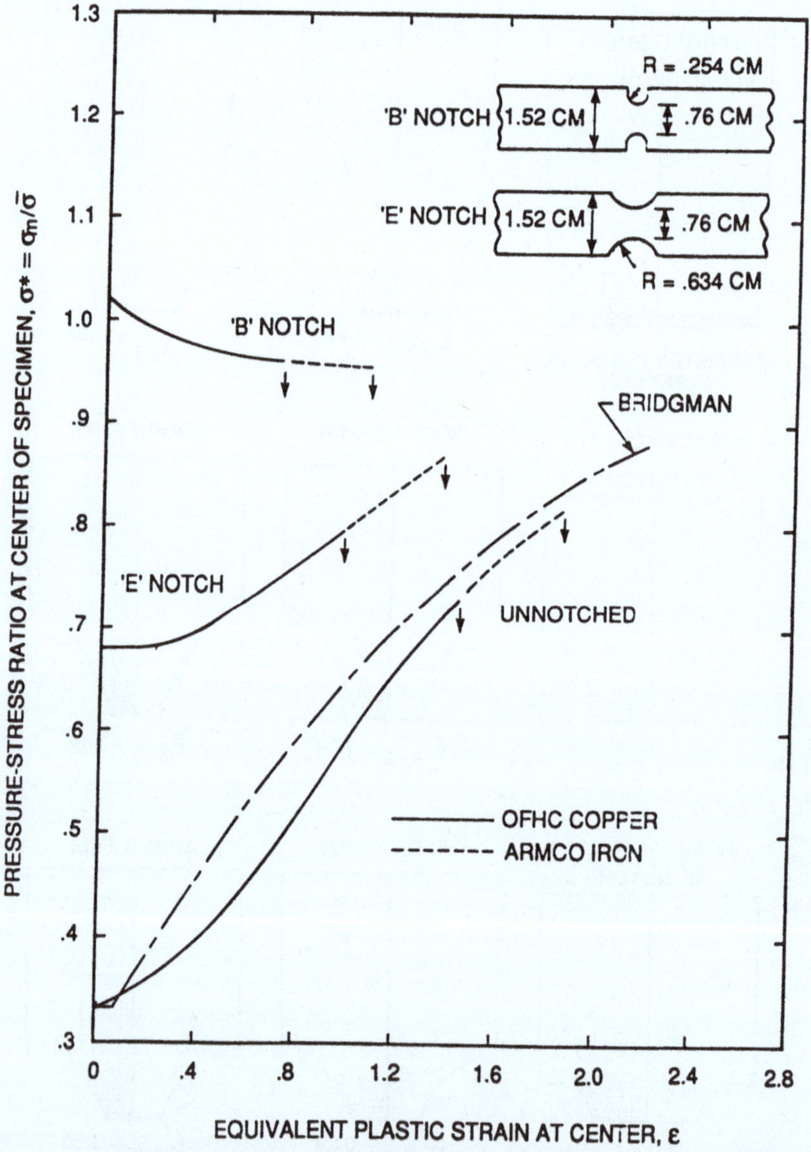

Fig. 19 Pressure-stress ratio vs equivalent plastic strain for the tensile specimens.

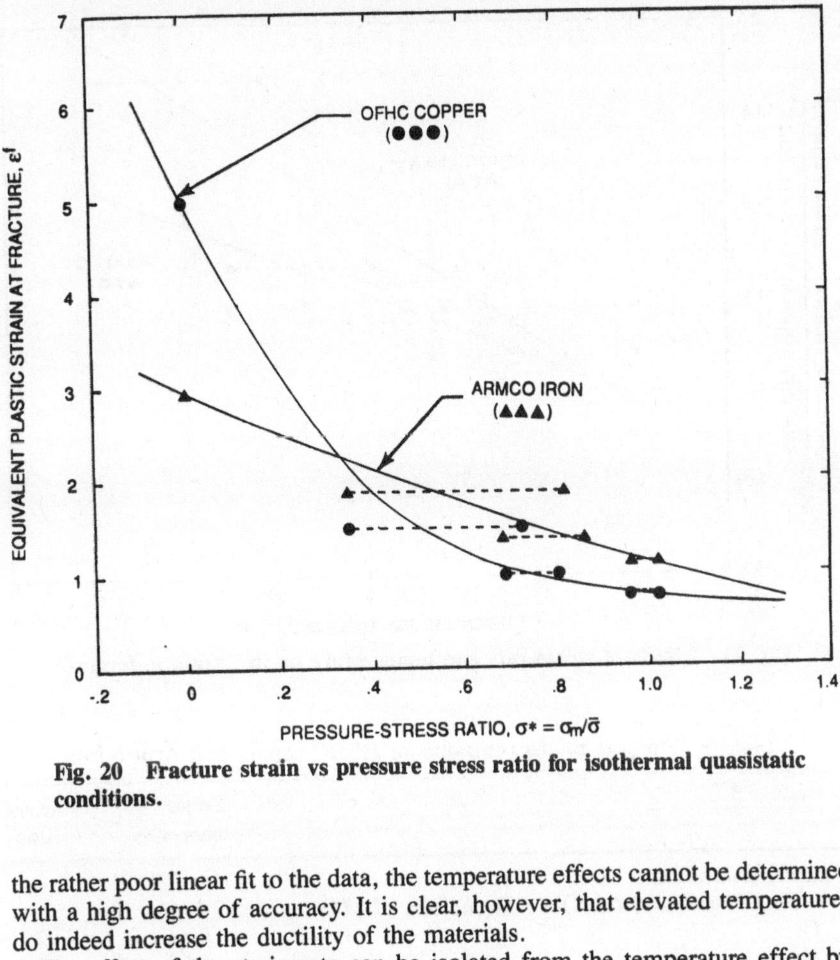

Fig. 20 Fracture strain vs pressure stress ratio for isothermal quasistatic conditions.

the rather poor linear fit to the data, the temperature effects cannot be determined with a high degree of accuracy. It is clear, however, that elevated temperatures do indeed increase the ductility of the materials.

The effect of the strain rate can be isolated from the temperature effect by considering the ratio of the fracture strains at $T* = 0$. For both materials, the ratio is greater than 1.0 at $T* = 0$, indicating that the strain to fracture increases slightly as the strain rate increases. This same trend is verified by the fracture data of Fig. 17 and by those of other materials.

Now the fracture constants can be determined. The strain rate constant D_4 is obtained from the data of Fig. 21 at $T* = 0$. The constants related to the pressure-stress ratio (D_1, D_2, D_3) are those of Fig. 20, adjusted from quasistatic conditions ($\dot{\epsilon} \approx 0.002s^{-1}$) to $\dot{\epsilon}* = 1.0$. Finally, the temperature constant D_5 is obtained from the data of Fig. 21.

The resulting relationships are shown in Fig. 22.[7] It would appear that the pressure-stress ratio is of primary importance. As the hydrostatic tension is increased, the strain to fracture decreases rapidly. The strain rate and temperature effects appear to be less important. A summary of the fracture constants is given in Table 4.[7]

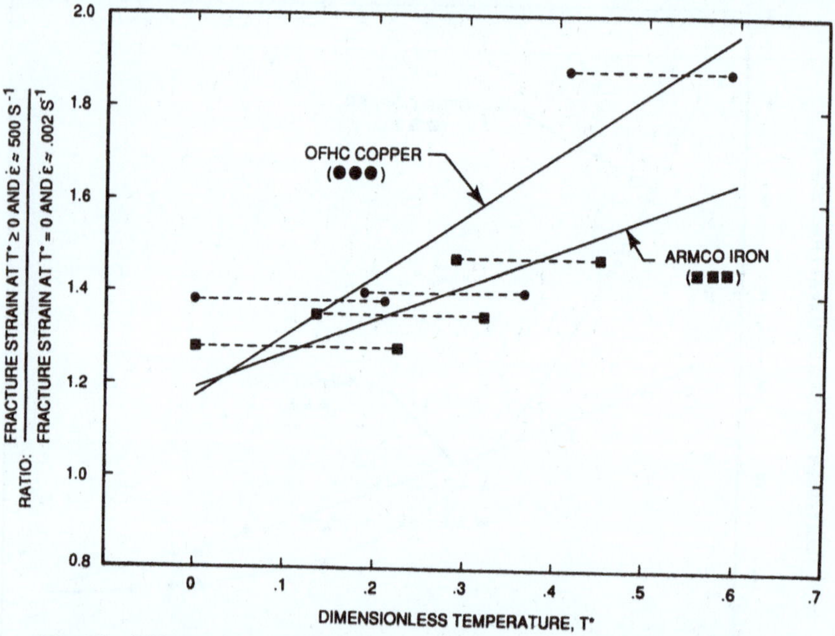

Fig. 21 Effects of strain rate and temperature on the strain to fracture.

Table 4 Fracture model constants for OFHC copper and Armco iron

	OFHC copper	Armco iron
Fracture constants for		
$\epsilon^f = (D_1 + D_2 \exp D_3\sigma^*)(1 + D_4 ln \dot{\epsilon}^*)(1 + D_5 T^*)$		
D_1	0.54	−2.20
D_2	4.89	5.43
D_3	−3.03	−0.47
D_4	0.014	0.16
D_5	1.12	0.63
Spall stress σ_{spall}, MPa	—	—
ϵ^f_{min}	—	—

$\sigma^* = \sigma_m/\bar{\sigma}$
$\dot{\epsilon}^* = \dot{\epsilon}/\dot{\epsilon}_0$ for $\dot{\epsilon}_0 = 1.0\ \text{s}^{-1}$
$T^* = (T - T_{\text{room}})/(T_{\text{melt}} - T_{\text{room}})$

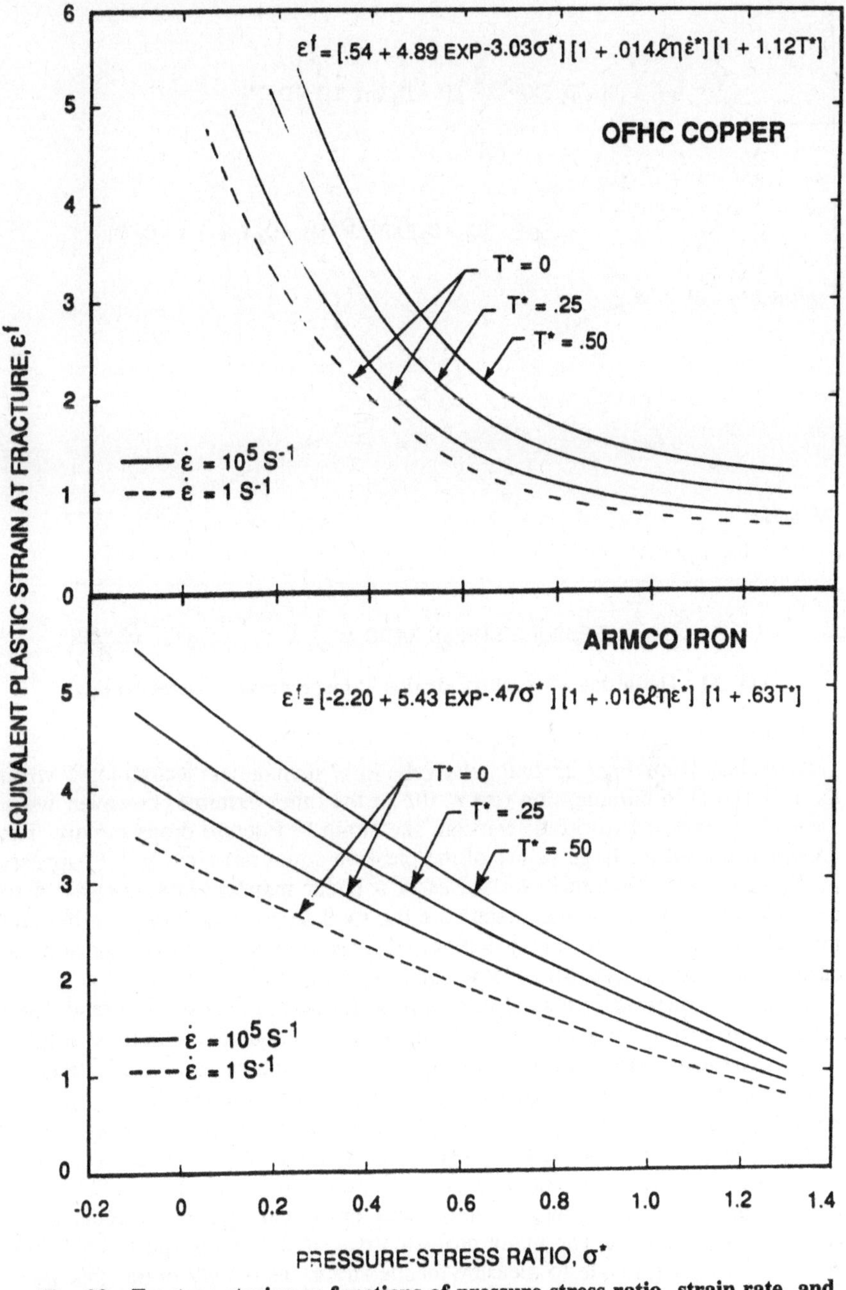

Fig. 22 Fracture strains as functions of pressure stress ratio, strain rate, and temperature.

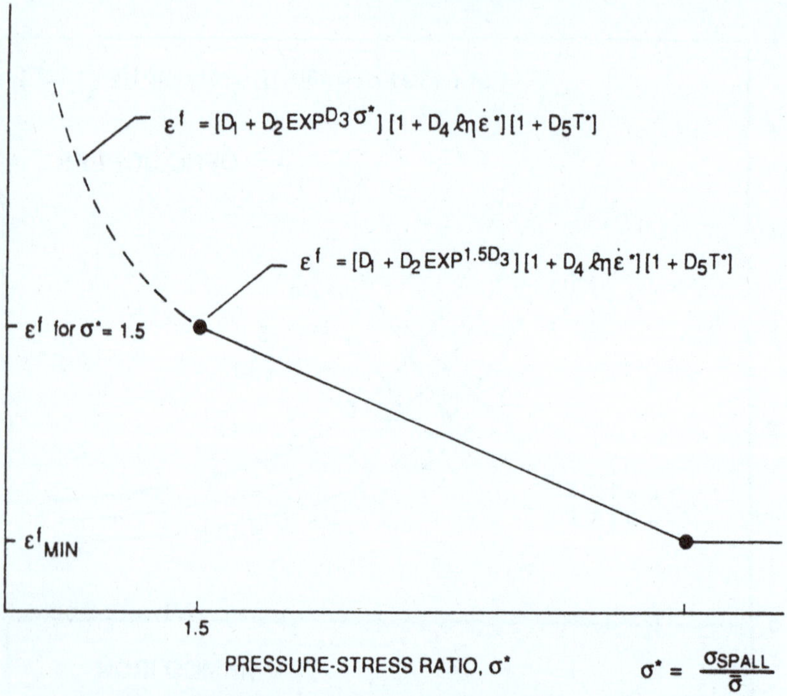

Fig. 23 Definition of fracture strains at large pressure-stress ratios.

It is clear from Fig. 22 that relatively little damage is accumulated when the material is in compression ($\sigma* < 0$). At the other extreme, however, when there is significant hydrostatic tension, the strain to fracture drops rapidly. The relationship used for large values of the pressure-stress ratio ($\sigma* > 1.5$) appears in Fig. 23.[7] The fracture strain varies in a linear manner, from $\sigma* = 1.5$ to $\sigma*_{spall}$ at ϵ^f_{min}. The input parameters are the spall stress σ_{spall} and the minimum fracture strain ϵ^f_{min}. The dimensionless σ^*_{spall} is computed from σ_{spall} and the current value of the von Mises flow stress $\bar{\sigma}$.

A series of quasistatic biaxial tests was performed to evaluate the model and data. Unfortunately there is a degree of uncertainty associated with these results. The biaxial test consists of a torsion specimen, as shown in Fig. 1, subjected to a torsional strain, followed by tensile strain until fracture occurs. A thin-walled specimen, as opposed to a solid specimen, was selected so the maximum damage due to both torsion and tension would occur in the same location. A solid specimen was not selected because the maximum torsional damage would occur at the outer radius, and the maximum tensile damage would occur at the center of the specimen. The major problem associated with the specimen of Fig. 9 is that it is not possible to measure tensile strains accurately in the specimen.

The results of biaxial tests for OFHC copper are summarized in Fig. 24.[7] The tensile strains are estimated from the post-tested cross-sectional areas of

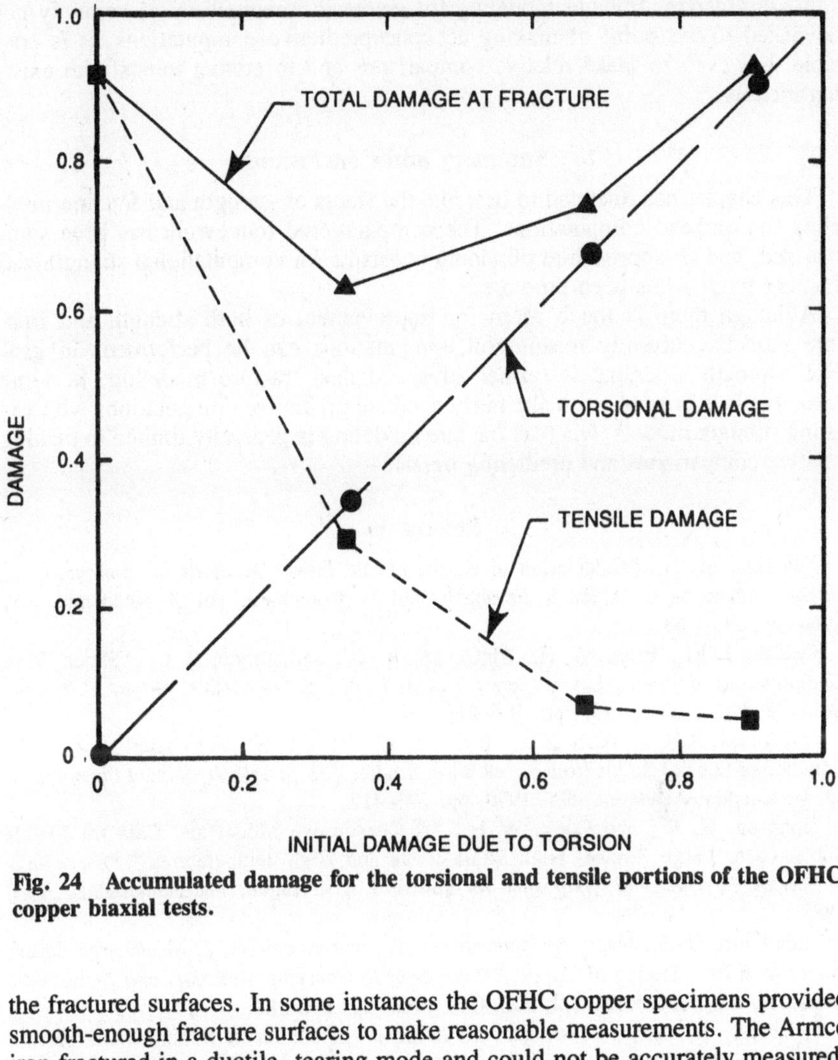

Fig. 24 Accumulated damage for the torsional and tensile portions of the OFHC copper biaxial tests.

the fractured surfaces. In some instances the OFHC copper specimens provided smooth-enough fracture surfaces to make reasonable measurements. The Armco iron fractured in a ductile, tearing mode and could not be accurately measured.

One can see from Fig. 24 that the model tends to underpredict the damage for fracture when there are combinations of tensile and torsional deformations. In other words, the material fractures before the damage approaches $D = 1.0$.

This same trend is exhibited in the EFP computation of Fig. 12. Here the maximum damage is $D = 0.40$, and although the EFP did not fracture, a minor variation in this design led to fracture in the vicinity of the maximum damage. This would appear to indicate that the point of maximum damage is properly located, but that the magnitude of the damage is again under-predicted. It should be noted that the location of the maximum damage does not coincide with that of the maximum strain.

To summarize, fracture modeling for warhead computations is certainly not advanced to the point of making accurate predictive computations. It is possible, however, to make relative comparisons and to predict trends with existing models.

6. Summary and Conclusions

This chapter has intended to describe the status of strength and fracture modeling for warhead computations. The computational framework has been summarized, and an approach to obtaining constants for computational strength and fracture models has been presented.

Although there is much room for improvement of both strength and fracture models, currently meaningful computations can be performed. In general, strength modeling is further advanced than fracture modeling. In some instances it is possible to make fairly accurate predictive computations with existing strength models, whereas fracture modeling is generally limited to making relative comparisons and predicting trends.

References

[1]Wilkins, M. L., "Calculation of Elastic-Plastic Flow," *Methods in Computational Physics,* edited by B. Alder, S. Fernbach, and M. Rotenberg, Vol. 3, Academic, NY, 1964, pp. 211–263.

[2]Walsh, J. M., Rice, M. H., McQueen, R. G., and Yarger, F. L., "Shock-Wave Compressions of Twenty-Seven Metals. Equations of State of Metals," *Physical Review,* Vol. 108, No. 2, Oct. 1957, pp. 196–216.

[3]McQueen, R. G., Walsh, S. P., Taylor, J. W., Fritz, J. N., and Carter, W. J., "The Equation of State of Solids from Shock Wave Studies," *High-Velocity Impact Phenomena,* ed. R. Kinslow, Academic, NY, 1970, pp. 293–417.

[4]Johnson, G. R., and Cook, W. H., "A Constitutive Model and Data for Metals Subjected to Large Strains, High Strain Rates and High Temperatures," *Proceedings of Seventh International Symposium on Ballistics,* The Hague, The Netherlands, April 1983.

[5]Lindholm, U. S., Nagy, A., Johnson, G. R., and Hoegfeldt, J. M., "Large Strain, High Strain Rate Testing of Copper," *Journal of Engineering Materials and Technology, Transactions of the ASME*, Vol. 102, Oct. 1980, pp. 376–381.

[6]Nicholas, T., "Tensile Testing of Materials at High Rates of Strain," *Experimental Mechanics,* May 1981, pp. 177–185.

[7]Johnson, G. R., and Cook, W. H., "Fracture Characteristics of Three Metals Subjected to Various Strains, Strain Rates, Temperatures, and Pressures," *Engineering Fracture Mechanics,* Vol. 21, No. 1, 1985, pp. 31–48.

[8]Mackenzie, A. C., Hancock, J. E., and Brown, D. K., "On the Influence of State of Stress on Ductile Failure Initiation in High Strength Steels," *Engineering Fracture Mechanics,* 1977, pp. 167–188.

[9]Taylor, G. I., "The Use of Flat-Ended Projectiles for Determining the Dynamic Yield Stress," *Proceedings of The Royal Society,* 1948, pp. 289–299.

[10]Wilkins, M. L., and Guinan, M. W., "Impact of Cylinders on a Rigid Body," *Journal of Applied Physics,* Vol. 44, No. 3, 1973, pp. 1200–1206.

[11]Erlich, D. C., and Chartegnac, P., "Determination of Dynamic Flow Curve of Metals at Ambient and Elevated Temperatures by Rod Impact Techniques," *Proceedings of International Conference on Mechanical and Physical Behavior of Materials Under Dynamic Loading*, Paris, France, Sept. 1985, pp. 455–462.

[12]Holmquist, T. J., and Johnson, G. R., "A Direct Method to Obtain Constitutive Model Constants from Cylinder Impact Test Results," *Work-in-Progress at Army Symposium on Solid Mechanics*, MTL MS 86-3, West Point, NY, Oct. 1986.

[13]Zerilli, F. J., and Armstrong, R. W., "Dislocation Mechanics Based Constitutive Relations for Material Dynamics Calculations," *Journal of Applied Physics*, Vol. 61, No. 5, March 1987, pp. 1816–1825.

[14]Steinberg, D. J., Cochran, S. G., and Guinan, M.W., "A Constitutive Model for Metals Applicable at High-Strain Rate," *Journal of Applied Physics*, Vol. 51, No. 3, March 1980, pp. 1498–1504.

[15]Johnson, G. R., "Dynamic Analysis of a Torsion Test Specimen Including Heat Conduction and Plastic Flow," *Journal of Engineering Materials and Technology, Transactions of the ASME*, Vol. 103, No. 1, March 1981, pp. 201–206.

[16]Bridgman, P. W., *Studies in Large Plastic Flow and Fracture*, McGraw-Hill, New York, 1952.

[17]Johnson, G. R., and Stryk, R. A., "User Instructions for the EPIC-2 Code," AFATL-TR-86-51, Air Force Armament Laboratory, Eglin Air Force Base, FL, Sept. 1986.

[18]Seaman, L., Curran, D. R., and Shockey, D. A., "Computational Models for Ductile and Brittle Fracture," *Journal of Applied Physics*, Vol. 47, No. 11, Nov. 1976, pp. 4814–4826.

[19]Curran, D. R., Seaman, L., and Shockey, D. A., "Dynamic Failure of Solids," *Physics Reports*, Vol. 147, Nos. 5 and 6, Mar. 1987, pp. 253–388.

[20]Hancock, J. W., and Mackenzie, A. C., "On the Mechanisms of Ductile Failure in High-Strength Steels Subjected to Multi-Axial Stress States," *Journal of the Mechanics and Physics of Solids*, June 1976, pp. 147–169.

[21]Johnson, G. R., Hoegfeldt, J. M., Lindholm, U. S., and Nagy, A., "Response of Various Metals Subjected to Large Torsional Strains over a Large Range of Strain Rates—Part 1: Ductile Metals," *Journal of Engineering Material and Technology, Transactions of the ASME*, Vol. 105, No. 1, Jan. 1983, pp. 42–47.

Chapter 4

Velocity of Explosively Driven Liners

J. M. Conner* and A. A. Quong†
Aerojet General Corporation, Azusa, California 91702

1. Introduction

IN this chapter we review the development of certain formulae that describe the velocity of explosively driven liners in one-dimensional geometries. These formulae are generally referred to as *Gurney relations*[1–3] as a tribute to the British physicist Ronald W. Gurney, who derived the relations for the cylinder and sphere while assigned to the Ballistics Research Laboratory at the Aberdeen Proving Ground in Maryland during World War II. Although one-dimensional in nature, these relations are commonly used to obtain engineering estimates of the behavior of more complex explosive mechanisms, such as in shaped charge design and analysis.[4]

The one-dimensional geometries considered are planar, cylindrical, and spherical. By applying the principles of conservation of energy and conservation of linear momentum to selected control masses, the equations for the asymptotic velocity of the liners may be derived. In the cases where a symmetry boundary falls within the definition of the control mass, as in the symmetric sandwich and exploding cylinder and sphere, only the equation for conservation of energy is needed. When no symmetry boundary is defined, or when the boundary falls outside the control mass, an equation for conservation of linear momentum must be developed. As has been traditional in the development of Gurney relations, we refer to the mass of the explosive charge as C, and the mass of the liner or liners as M or N, where M is the liner mass whose velocity we are concerned with, and N is tamping mass. Naturally, in the case of planar geometry we speak of a mass per unit area, whereas in the case of cylindrical geometry we are referring to a mass per unit length.

The principal assumption underlying these formulae is a linear velocity-material position profile in the explosive detonation products at their maximum expansion. This assumption is applicable for a wide range of liner mass-to-explosive mass ratios, or M/C. Because the assumption has the effect of averaging shock

*Staff Engineer; currently, Martin Marietta Corporation, Orlando, Florida 32855.

†Research Associate; currently, Naval Research Laboratory, Washington, DC 20375.

wave reverberations in both the explosive and liner materials, we note that for M/C less than about 0.1 significant acceleration of the liner occurs before several shock reflections across the explosive detonation products are able to deliver all of the available impulse. The actual value for M/C where the underlying assumption does apply will depend primarily on the shock transmission-related material characteristics of the explosive detonation products and the liner. For M/C values greater than 0.1, however, these formulae have found wide application. Clearly, for those cases where shock wave interactions dominate the behavior of interest, the underlying assumption of the Gurney formulae does not apply and resort must be made either to analysis that uses the method of characteristics[5–7] or to detailed hydrodynamic computer code analysis.

The unsteady velocity-time histories of liners in the various geometries were also studied with the one-dimensional finite difference program SIN.[8] These computer calculations used a simple constant gamma-law equation of state[9] for the explosive detonation products. The resulting velocity-time profiles of the liner center of mass were then fitted to an exponential function.

In Section 2 of this chapter, we review the derivation of the Gurney formula for the symmetric sandwich, cylinder, and sphere. In Section 3, the Gurney formulae for the asymmetric sandwich and for concentric cylindrical and spherical shells are derived. Finally, in Section 4 we present the results of our one-dimensional computer calculations to obtain the characteristic time constant of these systems. Previous reviews on this subject that are of particular note have been published by Kennedy[10] and Henry.[11] These are recommended for further study of some of the topics covered in this chapter.

2. Symmetric Geometries

In Fig. 1 is a diagram of the control mass used for the symmetric sandwich, or plane parallel plates, where X refers to the Lagrangian or material coordinate position. If we assume that the explosive material velocity V_c is proportional to the distance from the symmetry boundary $X = 0$, we have

$$V_c(X) = \left(\frac{X}{X_m}\right)V_m \tag{1}$$

where V_m is the velocity of the plates. The plate and explosive mass per unit area are defined as

$$M = \rho_m t_m \tag{2}$$

and

$$C = \rho_c X_m \tag{3}$$

respectively. The equation for the total kinetic energy per unit area of the control mass E_{ke} may be written as

$$E_{ke} = \int_0^{X_m} \frac{1}{2}\rho_c V_c^2(X)\,\mathrm{d}X + \int_{X_m}^{X_m+t_m} \frac{1}{2}\rho_m V_m^2\,\mathrm{d}X \tag{4}$$

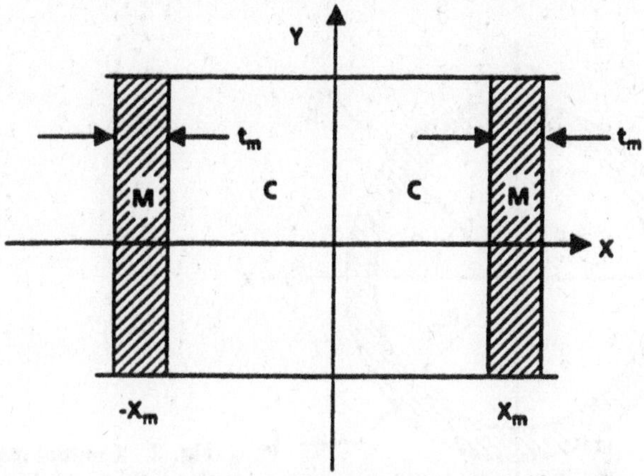

Fig. 1 Control mass used for the symmetric sandwich.

Equation (4) may be integrated using the definitions in Eqs. (1– 3). We have

$$2E_{ke} = \left(\frac{1}{3}C + M\right)V_m^2 \tag{5}$$

If we define E as the kinetic energy per unit mass of explosive, or E_{ke}/C, Eq. (5) simplifies to

$$\frac{V_m}{\sqrt{2E}} = \left(\frac{M}{C} + \frac{1}{3}\right)^{-1/2} \tag{6}$$

where the term $\sqrt{2E}$ is an explosive material constant with units of velocity, or the *Gurney velocity* of the explosive. The term $\sqrt{2E}$ is also called the *Gurney constant* or the *Gurney energy*. It is the single material constant needed to estimate the ability of an explosive to launch a mass of material. The Gurney constant is most often obtained experimentally for the particular explosive system of interest because this allows for any deviation of the system geometry from the one-dimensional ideal case and for tighter correlation over the range of M/C of interest. We will later present for reference, in Section 4, tabulated values of the Gurney constant for a variety of explosives developed from one-dimensional hydrodynamic computer calculations.

In Fig. 2 is a diagram of the control mass used for the cylinder, where R refers to the Lagrangian radial position coordinate. Again we approximate the velocity profile in the explosive as linear, or

$$V_c(R) = \left(\frac{R}{R_m}\right)V_m \tag{7}$$

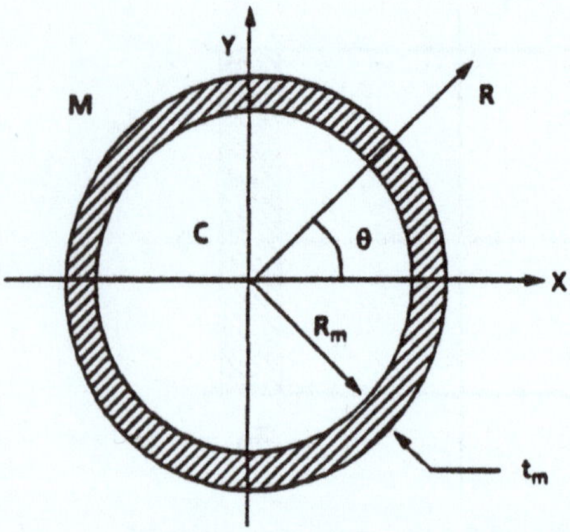

Fig. 2 Control mass used for the cylinder.

where V_m is the radial velocity of the expanding liner. For the cylinder, the definitions of liner and explosive mass per unit length are

$$M = \pi\rho_m[(R_m + t_m)^2 - R_m^2] \tag{8}$$

and

$$C = \pi\rho_c R_m^2 \tag{9}$$

respectively. The total kinetic energy per unit length E_{ke} may be written for the cylinder as

$$E_{ke} = \int_0^{2\pi}\int_0^{R_m} \frac{1}{2}\rho_c V_c^2(R) R \, \mathrm{d}R \, \mathrm{d}\theta + \int_0^{2\pi}\int_{R_m}^{R_m+t_m} \frac{1}{2}\rho_m V_m^2 R \, \mathrm{d}R \, \mathrm{d}\theta \tag{10}$$

Integration of Eq. (10) gives

$$2E_{ke} = \left(\frac{1}{2}C + M\right)V_m^2 \tag{11}$$

With the definition of E given before, Eq. (11) simplifies to

$$\frac{V_m}{\sqrt{2E}} = \left(\frac{M}{C} + \frac{1}{2}\right)^{-1/2} \tag{12}$$

which is the Gurney equation for the cylinder.

For the sphere, M and C are defined as

$$M = \frac{4}{3}\pi\rho_m[(R_m + t_m)^3 - R_m^3] \tag{13}$$

and

$$C = \frac{4}{3}\pi\rho_c R_m^3 \tag{14}$$

respectively. The total kinetic energy for the sphere is then

$$E_{ke} = \int_0^{2\pi}\int_0^{\pi}\int_0^{R_m} \frac{1}{2}\rho_c V_c^2(R)R^2 \sin\phi \; dR\; d\phi\; d\theta$$

$$+ \int_0^{2\pi}\int_0^{\pi}\int_{R_m}^{R_m+t_m} \frac{1}{2}\rho_m V_m^2 R^2 \sin\phi \; dR\; d\phi\; d\theta \tag{15}$$

If we substitute for the velocity profile Eq. (7) and integrate, we have

$$2E_{ke} = \left(\frac{3}{5}C + M\right)V_m^2 \tag{16}$$

The Gurney equation for the sphere is then

$$\frac{V_m}{\sqrt{2E}} = \left(\frac{M}{C} + \frac{3}{5}\right)^{-1/2} \tag{17}$$

The variation of normalized liner velocity with M/C is shown in Fig. 3 for the symmetric sandwich, cylinder, and sphere. With a given M/C ratio and the Gurney constant for an explosive, one may then determine the liner velocity that may be achieved using Eqs. (6), (12), and (17), appropriately.

3. Asymmetric Geometries

For one-dimensional geometries that do not contain a symmetry boundary within the control mass, the equation for conservation of linear momentum is used to determine the material location in the explosive that experiences, on average, no change in position. The approximation of a linear velocity distribution in the explosive is then used to integrate the energy equation, as it was in the case of the symmetric geometries discussed in Section 2.

For the asymmetric sandwich, or plane parallel plates, a diagram of the control mass is shown in Fig. 4. The velocity in the explosive is expressed as

$$V_c(X) = V_m\frac{(X - X_s)}{(X_s - X_m)} \tag{18}$$

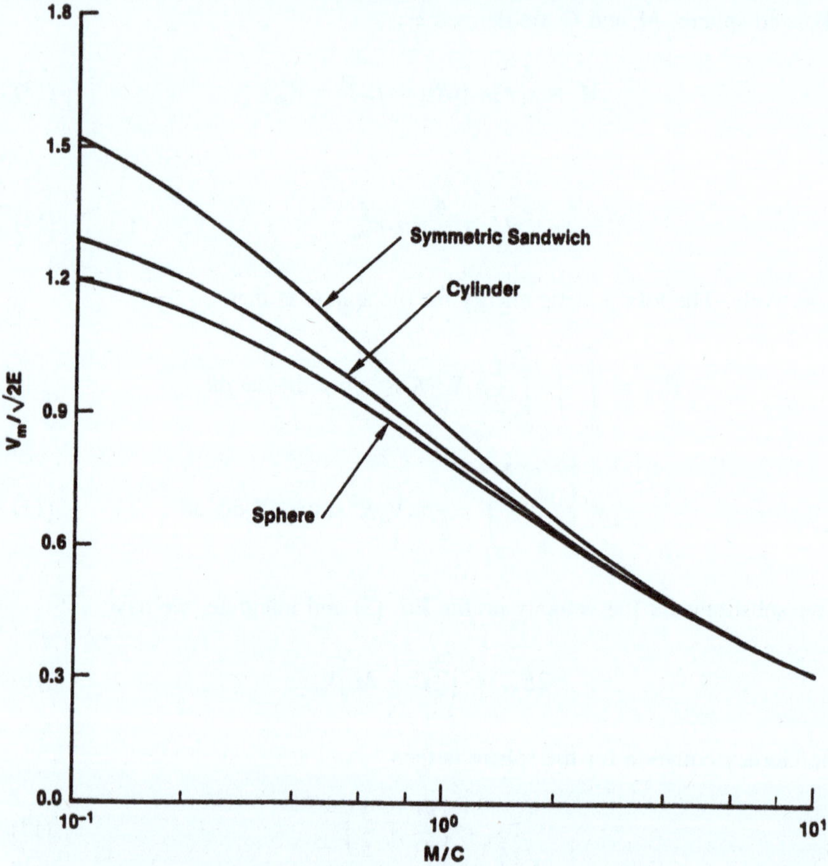

Fig. 3 Variation of the normalized liner velocity with liner-to-charge mass ratio for the symmetric sandwich, cylinder, and sphere.

for the left-moving portion, and

$$V_c(X) = V_n \frac{(X - X_s)}{(X_n - X_s)} \tag{19}$$

for the right-moving portion. Or, combining Eqs. (18) and (19), we have

$$\frac{V_m}{V_n} = \frac{(X_s - X_m)}{(X_n - X_s)} \tag{20}$$

The liner and explosive mass per unit areas M, N, and C are defined as

$$M = \rho_m t_m \tag{21}$$

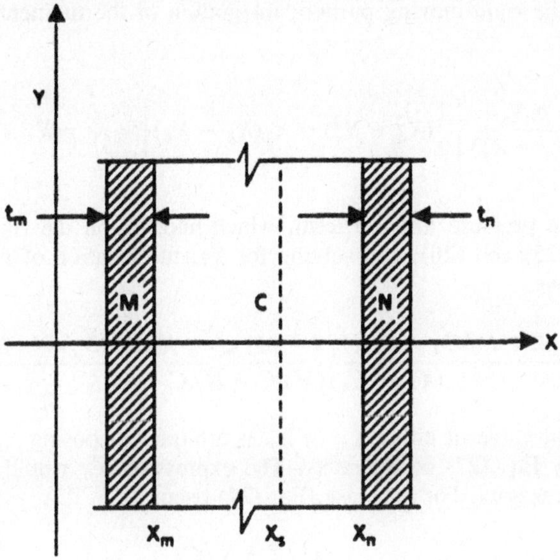

Fig. 4 Control mass used for the asymmetric sandwich

$$N = \rho_n t_n \tag{22}$$

and

$$C = \rho_c (X_n - X_m) \tag{23}$$

respectively.

Conservation of linear momentum for the leftward-moving portion of the explosive and the M plate may be written as

$$-\int_{X_m - t_m}^{X_m} \rho_m V_m \, \mathrm{d}X + \int_{X_m}^{X_s} \rho_c V_c(X) \, \mathrm{d}X = -\int_0^T p(X_s, t) \, \mathrm{d}t \tag{24}$$

where $p(X_s, t)$ is the pressure at the hypothetical stationary surface and T is the time required for the explosive to deliver its impulse to the liners. Substituting Eq. (18) for $V_c(X)$ into Eq. (24) and integrating, we have

$$-MV_m + \frac{\rho_c V_m}{(X_s - X_m)} \left[\frac{1}{2}(X_s^2 - X_m^2) - X_s(X_s - X_m) \right] = -\int_0^T p(X_s, t) \, \mathrm{d}t \tag{25}$$

Similarly for the right-moving portion, integration of the momentum equation gives

$$NV_n + \frac{\rho_c V_n}{(X_n - X_s)}\left[\frac{1}{2}(X_n^2 - X_s^2) - X_s(X_n - X_s)\right] = \int_0^T p(X_s, t)\,\mathrm{d}t \qquad (26)$$

Eliminating the pressure impulse term, which appears on the right-hand side of both Eqs. (25) and (26), and solving for X_s, the location of the stationary surface, we have

$$X_s = \frac{X_n^2(1/2 + N/C) - X_m^2(1/2 + M/C) - X_m X_n(N/C - M/C)}{(X_n - X_m)(M/C + N/C + 1)} \qquad (27)$$

Since the actual value of either X_m or X_n is arbitrary, choosing X_m to be zero would simplify Eq. (27) considerably. The expression X_n would then be the explosive thickness t_c. For this case, Eq. (27) becomes

$$X_s = t_c \frac{(1/2 + N/C)}{(M/C + N/C + 1)} \qquad (28)$$

The equation for conservation of kinetic energy E_{ke} may now be written for the control mass. We have

$$E_{ke} = \int_{X_m - t_m}^{X_m} \frac{1}{2}\rho_m V_m^2\,\mathrm{d}X + \int_{X_m}^{X_n} \frac{1}{2}\rho_c V_c^2(X)\,\mathrm{d}X + \int_{X_n}^{X_n + t_n} \frac{1}{2}\rho_n V_n^2\,\mathrm{d}X \qquad (29)$$

Substituting Eqs. (18) and (19) for $V_c(X)$ and integrating, we arrive at the Gurney equation for the left-moving plate of the asymmetric sandwich, or

$$\frac{V_m}{\sqrt{2E}} = \left\{\frac{M}{C} + \frac{N}{C}\frac{(X_n - X_s)^2}{X_s^2} + \frac{1}{3}\frac{X_s}{X_n}\left[1 + \frac{(X_n - X_s)^3}{X_s^3}\right]\right\}^{-1/2} \qquad (30)$$

The velocity of the right-moving plate may then be found using Eq. (20). Note that as N becomes large compared with M and C, the stationary surface approaches X_n, and Eq. (30) becomes the Gurney equation for the symmetric sandwich, or Eq. (6). In Fig. 5 we have plotted the variation of normalized liner velocity with M/C for the asymmetric sandwich. To illustrate this, it was useful to define a new parameter η, which is equal to N/M. As η increases, so does the velocity of the M plate until an almost rigid boundary is created at X_n, as shown in the figure.

Concentric cylindrical and spherical shells differ significantly from the planar sandwiches because additional terms must be considered in the momentum equation owing to a curvature effect. This effect is discussed in more detail by Carleone and Flis,[12] for whom it presented something of an obstacle to

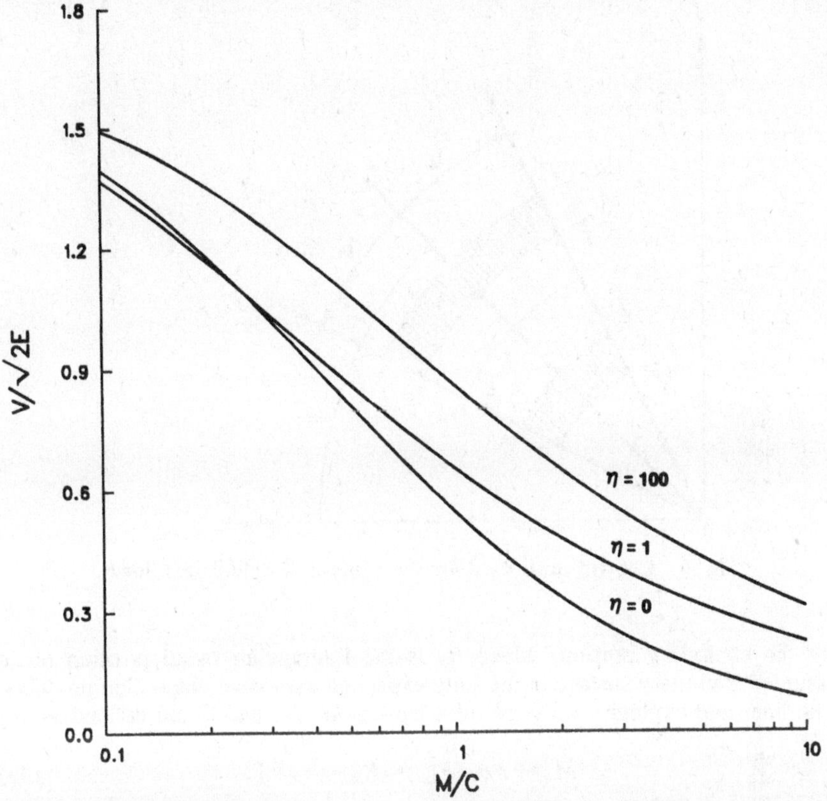

Fig. 5 Variation of normalized liner velocity with liner-to-charge mass ratio for the asymmetric sandwich.

derivation of a complete Gurney formula for the imploding cylinder. Here we make some additional approximations that allow the derivation to be completed.

The control mass used for the cylindrical shells is shown in Fig. 6, where the addition of a net outward impulse delivered to the sides of the control mass by the surrounding explosive must be considered. To begin, the velocity profile in the explosive is approximated as

$$V_c(R) = V_m \frac{(R - R_s)}{(R_s - R_m)} \tag{31}$$

for the imploding portion and

$$V_c(R) = V_n \frac{(R - R_s)}{(R_n - R_s)} \tag{32}$$

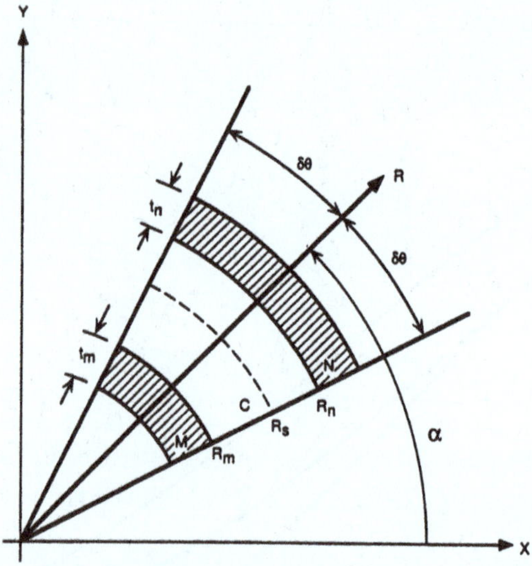

Fig. 6 Control mass used for the concentric cylindrical liners.

for the exploding portion, where R_s is the Lagrangian radial position of an assumed stationary surface in the fully expanded explosive detonation products. The liner and explosive mass per unit lengths M, N, and C are defined as

$$M = \pi\rho_m[R_m^2 - (R_m - t_m)^2] \tag{33}$$

$$N = \pi\rho_n[(R_n + t_n)^2 - R_n^2] \tag{34}$$

and

$$C = \pi\rho_c(R_n^2 - R_m^2) \tag{35}$$

respectively, for the cylinder.

Again, we assume a stationary surface at R_s, which, rigorously, does not exist, but serves as an adequate approximation, and write a conservation of linear momentum expression for the imploding portion of the control mass as

$$-\int_{\alpha-\delta\theta}^{\alpha+\delta\theta}\int_{R_m-t_m}^{R_m} \rho_m V_m R\cos(\theta-\alpha)\,\mathrm{d}R\,\mathrm{d}\theta + \int_{\alpha-\delta\theta}^{\alpha+\delta\theta}\int_{R_m}^{R_s} \rho_c V_c(R) R\cos(\theta-\alpha)\,\mathrm{d}R\,\mathrm{d}\theta$$

$$= 2\int_0^T\int_{r_m(t)}^{R_s} p(r,t)\sin\delta\theta\,\mathrm{d}r\,\mathrm{d}t - \int_0^T\int_{\alpha-\delta\theta}^{\alpha+\delta\theta} p(R_s,t)R_s\cos(\theta-\alpha)\,\mathrm{d}\theta\,\mathrm{d}t \tag{36}$$

where $r_m(t)$ is the Eulerian coordinate of the imploding explosive liner surface, $p(r, t)$ is the pressure in the explosive along the side boundaries of the control mass, and $p(R_s, t)$ is the pressure in the explosive at the stationary surface. The velocity profile for the imploding portion is approximated using Eq. (31). However, Eq. (36) may now be integrated with the following additional approximations:

$$p(r, t) = p(t) \tag{37}$$

$$\int_0^T p(t)dt = PT \tag{38}$$

and

$$\int_0^T \int_{r_m(t)}^{R_s} p(t)\, dr\, dt = PT(R_s - R_m) \tag{39}$$

where P is the average pressure. Substituting Eq. (31) into the momentum equation and performing the integration using Eqs. (37–39) gives

$$-\frac{1}{2\pi} M V_m + \frac{\rho_c V_m}{(R_s - R_m)}\left[\frac{1}{3}(R_s^3 - R_m^3) - \frac{1}{2}R_s(R_s^2 - R_m^2)\right] = -PTR_m \tag{40}$$

Similarly, for the exploding portion the equation for conservation of linear momentum is

$$\int_{\alpha-\delta\theta}^{\alpha+\delta\theta} \int_{R_n}^{R_n+t_n} \rho_n V_n R \cos(\theta - \alpha)\, dR\, d\theta + \int_{\alpha-\delta\theta}^{\alpha+\delta\theta} \int_{R_s}^{R_n} \rho_c V_c(R) R \cos(\theta - \alpha)\, dR\, d\theta$$

$$= 2\int_0^T \int_{R_s}^{r_n(t)} p(r, t) \sin\delta\theta\, dr\, dt + \int_0^T \int_{\alpha-\delta\theta}^{\alpha+\delta\theta} p(R_s, t) R_s \cos(\theta - \alpha)\, d\theta\, dt \tag{41}$$

where $r_n(t)$ is the Eulerian coordinate of the outward-moving explosive liner interface. The linear velocity profile for the exploding portion is approximated using Eq. (32). Assuming Eqs. (37) and (38) hold, and

$$\int_0^T \int_{R_s}^{r_n(t)} p(t)\, dr\, dt = PT(R_n - R_s) \tag{42}$$

we may integrate Eq. (41) to yield

$$\frac{1}{2\pi}NV_n + \frac{\rho_c V_n}{(R_n - R_s)}\left[\frac{1}{3}(R_n^3 - R_s^3) - \frac{1}{2}R_s(R_n^2 - R_s^2)\right] = PTR_n \tag{43}$$

Eliminating the term PT between Eqs. (40) and (43) for the imploding and exploding portions, respectively, and solving for R_s, the position of the stationary surface, we have

$$R_s^3 + 3R_s\left[(R_n + R_m)\left(\frac{M}{C}R_n + \frac{N}{C}R_m\right) + R_mR_n\right]$$
$$- 3R_mR_n(R_n + R_m)\left[\frac{2}{3} + \left(\frac{M}{C} + \frac{N}{C}\right)\right] = 0 \tag{44}$$

which is a third-order polynomial in R_s. This may be solved analytically or numerically using Newton's method. This relation was first derived by Chanteret, whose application included the analysis of shaped charge liner collapse.[13]

Given the location of the stationary surface R_s from the solution of Eq. (44), the asymptotic velocity of the imploding liner can be derived from conservation of kinetic energy for the control mass, or

$$E_{ke} = \int_0^{2\pi}\int_{R_m - t_m}^{R_m} \frac{1}{2}\rho_m V_m^2 R \, dR \, d\theta + \int_0^{2\pi}\int_{R_m}^{R_n} \frac{1}{2}\rho_c V_c^2(R) R \, dR \, d\theta$$
$$+ \int_0^{2\pi}\int_{R_n}^{R_n + t_n} \frac{1}{2}\rho_n V_n^2 R \, dR \, d\theta \tag{45}$$

Integration of Eq. (45) using the velocity profiles in the explosive detonation products expressed in Eqs. (31) and (32) gives

$$\frac{V_m}{\sqrt{2E}} = \left\{\frac{M}{C} + \frac{N}{C}\frac{(R_n - R_s)^2}{(R_s - R_m)^2} + \frac{1}{6}\frac{(R_s - R_m)(3R_m + R_s)}{(R_n^2 - R_m^2)}\right.$$
$$\left. + \frac{1}{6}\frac{(R_n - R_s)^3(3R_n + R_s)}{(R_n^2 - R_m^2)(R_s - R_m)^2}\right\}^{-1/2} \tag{46}$$

which is the Gurney equation for the imploding cylindrical liner. The velocity for the exploding liner may then be determined by taking the ratio of Eqs. (31) and (32). To illustrate the variation of normalized liner velocity with M/C for the imploding portion, let us define a curvature parameter ϵ as

$$\epsilon = \frac{R_m}{(R_n - R_m)} \tag{47}$$

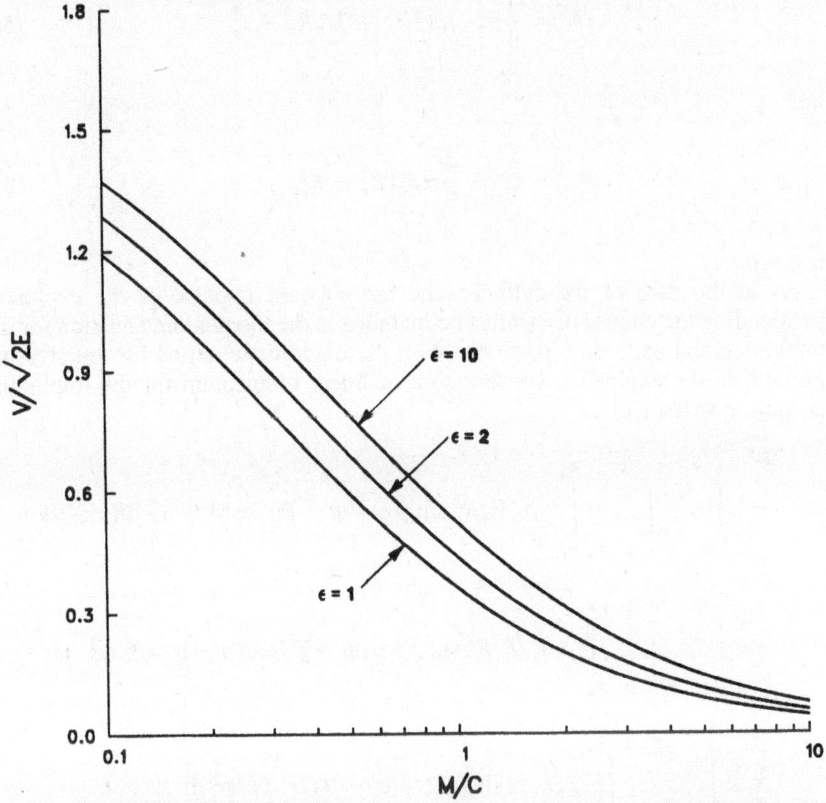

Fig. 7 Variation of the normalized liner velocity with liner-to-charge mass ratio for the imploding cylinder, η = 0.

Clearly, for large ϵ, effects of curvature are small and vice versa. In Fig. 7 we have plotted the normalized velocity of the imploding liner vs M/C for increasing values of ϵ, and for $\eta = 0$. Note that as ϵ increases, the curve approaches that for the open-faced sandwich shown in Fig. 5.

The control mass for the concentric spherical liners, in spherical coordinates (R, ϕ, θ) is defined as

$$
\begin{aligned}
R_m - t_m \leq R \leq R_n + t_n \\
\beta - \delta\phi \leq \phi \leq \beta + \delta\phi \\
\alpha - \delta\theta \leq \theta \leq \alpha + \delta\theta
\end{aligned}
\tag{48}
$$

The equations for the masses of the spherical liners and explosive are

$$
M = \frac{4}{3}\pi \rho_m t_m (3R_m^2 - 3t_m R_m + t_m^2) \tag{49}
$$

$$N = \frac{4}{3}\pi\rho_n t_n(3R_n^2 + 3t_n R_n + t_n^2) \tag{50}$$

and

$$C = \frac{4}{3}\pi\rho_c(R_n^3 - R_m^3) \tag{51}$$

respectively.

As in the case of the cylinder, the net outward impulse of the explosive surrounding the control mass must be included in the momentum equation for the imploding and exploding portions. With these additional terms for the pressure impulse in the explosive, conservation of linear momentum for the imploding portion is written as

$$-\int_{\alpha-\delta\theta}^{\alpha+\delta\theta}\int_{\beta-\delta\phi}^{\beta+\delta\phi}\int_{R_m-t_m}^{R_m} \rho_m V_m R^2 \sin\phi\cos(\phi-\beta)\cos(\theta-\alpha)\,dR\,d\phi\,d\theta$$

$$-\int_{\alpha-\delta\theta}^{\alpha+\delta\theta}\int_{\beta-\delta\phi}^{\beta+\delta\phi}\int_{R_m}^{R_s} \rho_c V_c(R) R^2 \sin\phi\cos(\phi-\beta)\cos(\theta-\alpha)\,dR\,d\phi\,d\theta$$

$$= 2\int_0^T\int_{r_m(t)}^{R_s}\int_{\beta-\delta\phi}^{\beta+\delta\phi} p(r,t)\sin\delta\theta\cos(\phi-\beta)2r\,d\phi\,dr\,dt$$

$$+\int_0^T\int_{r_m(t)}^{R_s}\int_{\alpha-\delta\theta}^{\alpha+\delta\theta} p(r,t)\sin\delta\phi\cos(\theta-\alpha)\sin(\beta-\delta\phi)r\,d\theta\,dr\,dt$$

$$+\int_0^T\int_{r_m(t)}^{R_s}\int_{\alpha-\delta\theta}^{\alpha+\delta\theta} p(r,t)\sin\delta\phi\cos(\theta-\alpha)\sin(\beta+\delta\phi)r\,d\theta\,dr\,dt$$

$$-\int_0^T\int_{\alpha-\delta\theta}^{\alpha+\delta\theta}\int_{\beta-\delta\phi}^{\beta+\delta\phi} p(R_s,t)\cos(\phi-\beta)\cos(\theta-\alpha)R_s^2\sin\phi\,d\phi\,d\theta\,dt \tag{52}$$

Approximations regarding the pressure integrals must also be made before Eq. (52) may be integrated and an equation explicit in R_s found. The approximations used here are Eqs. (37) and (38) in addition to the following:

$$\int_0^T \int_{r_m(t)}^{R_s} p(t) r \, dr \, dt = \frac{1}{2} PT(R_s^2 - R_m^2) \tag{53}$$

and

$$\int_0^T \int_{R_s}^{r_n(t)} p(t) r \, dr \, dt = \frac{1}{2} PT(R_n^2 - R_s^2) \tag{54}$$

Using the preceding approximations and the explosive velocity profile of Eq. (31), Eq. (52) then simplifies to

$$-\frac{1}{4\pi} M V_m + \frac{\rho_C V_m}{(R_s - R_m)} \left[\frac{1}{4}(R_s^4 - R_m^4) - \frac{1}{3} R_s (R_s^3 - R_m^3) \right] = -\frac{1}{2} PT(3R_m^2 - R_s^2) \tag{55}$$

Similarly for the exploding portion, the equation for conservation of linear momentum simplifies to

$$\frac{1}{4\pi} N V_n + \frac{\rho_C V_n}{(R_n - R_s)} \left[\frac{1}{4}(R_n^4 - R_s^4) - \frac{1}{3} R_s (R_n^3 - R_s^3) \right] = \frac{1}{2} PT(3R_n^2 - R_s^2) \tag{56}$$

Combining Eqs. (55) and (56) gives the equation for the hypothetical stationary surface R_s as

$$\begin{aligned}
&R_s^4 \left[\frac{3}{4}(R_n^2 - R_m^2) \right] - R_s^3 \left[(R_n^3 - R_m^3) \left(1 + \frac{M}{C} + \frac{N}{C} \right) \right] \\
&+ R_s^2 \left[\frac{3}{4}(R_n^4 - R_m^4) + \left(\frac{N}{C} R_n^4 - \frac{M}{C} R_m^4 \right) + R_m R_n \left(\frac{M}{C} R_n^2 - \frac{N}{C} R_m^2 \right) \right] \\
&+ 3R_s \left[\left(\frac{M}{C} R_n^5 - \frac{N}{C} R_m^5 \right) + R_m^2 R_n^2 \left(\frac{N}{C} R_n - \frac{M}{C} R_m + R_n - R_m \right) \right] \\
&- 3 \left[R_m R_n \left(\frac{M}{C} R_n^4 - \frac{N}{C} R_m^4 \right) + R_m^2 R_n^2 \left(\frac{N}{C} R_n^2 - \frac{M}{C} R_m^2 + \frac{3}{4}(R_n^2 - R_m^2) \right) \right] = 0
\end{aligned} \tag{57}$$

a fourth-order polynomial in R_s, which may be solved numerically using Newton's method.

The asymptotic velocity of the imploding liner may now be derived from conservation of kinetic energy for the imploding portion of the control mass, or

$$E_{ke} = \int_0^{2\pi}\int_0^{\pi}\int_{R_m-t_m}^{R_m} \frac{1}{2}\rho_m V_m^2 R^2 \sin\phi \, dR \, d\phi \, d\theta$$
$$+ \int_0^{2\pi}\int_0^{\pi}\int_{R_m}^{R_n} \frac{1}{2}\rho_c V_c^2(R) R^2 \sin\phi \, dR \, d\phi \, d\theta$$
$$+ \int_0^{2\pi}\int_0^{\pi}\int_{R_n}^{R_n+t_n} \frac{1}{2}\rho_n V_n^2 R^2 \sin\phi \, dR \, d\phi \, d\theta \qquad (58)$$

Integration of Eq. (58) using the velocity profile of Eqs. (31) and (32) gives

$$\frac{V_m}{\sqrt{2E}} = \left\{\frac{M}{C} + \frac{N}{C}\frac{(R_n - R_s)^2}{(R_s - R_m)^2}\right.$$
$$+ \frac{1}{10}\frac{(R_s - R_m)}{(R_n^3 - R_m^3)}[6(R_s - R_m)^2 - 15R_s(R_s - R_m) + 10R_s^2]$$
$$\left. + \frac{1}{10}\frac{1}{(R_s - R_m)^2}\frac{(R_n - R_s)^3}{(R_n^3 - R_m^3)}[6(R_n - R_s)^2 + 15R_s(R_n - R_s) + 10R_s^2]\right\}^{-1/2} \qquad (59)$$

the Gurney equation for the imploding spherical liner. The velocity for the exploding liner may then be found using the same relation developed before for the cylinder. This equation is illustrated in Fig. 8 where, for $\eta = 0$, we have plotted the normalized liner velocity vs M/C for increasing values of ϵ. As was the case for the imploding cylinder, the curve approaches that for the open-faced sandwich as ϵ is increased.

4. Hydrodynamic Calculations

The preceding equations define the asymptotic velocity of the various liners but give no information about the unsteady velocity-time relationship. To provide some insight into the unsteady behavior of these systems, we have performed a number of computer calculations for the different geometries using the finite difference computer program SIN.[8] The constant gamma-law equation of state was used for the explosive detonation products,[9] and the various liners were modeled as either aluminum or copper plates and shells using the Mie-Gruneisen equation of state. The metals were treated without strength, that is, hydrodynamically. We varied the liner-to-explosive mass ratios from 0.1 to 10 in order to explore fully the behavior of these systems over the range in which the Gurney model for the asymptotic velocity is valid.

The selection of the gamma-law equation of state for the explosives was made because of its simplicity. Only three coefficients are required to define

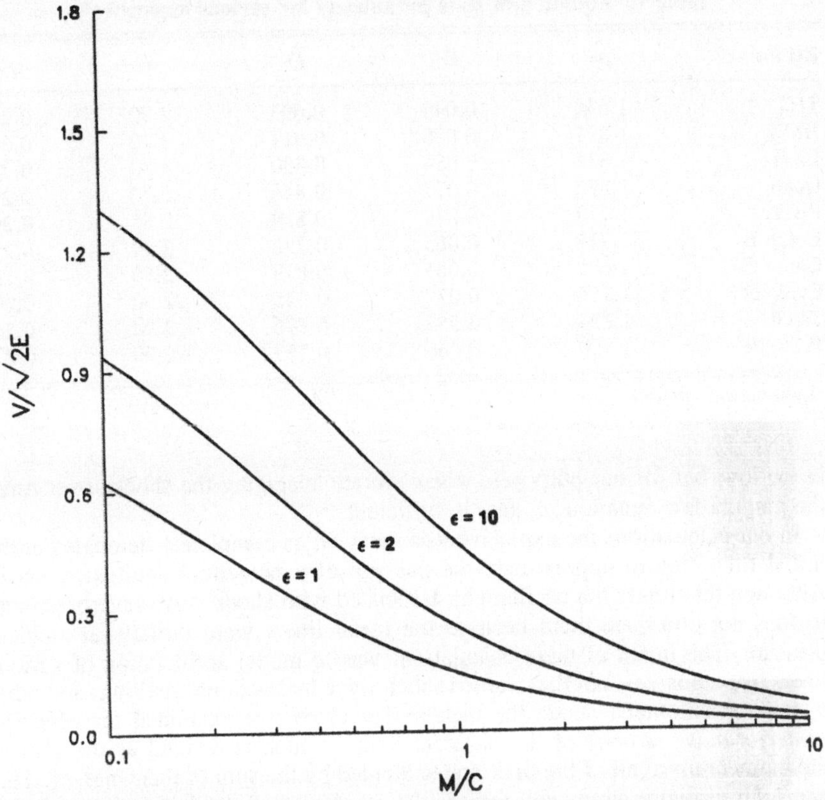

Fig. 8 Variation of the normalized liner velocity with liner-to-charge mass ratio for the imploding sphere, $\eta = 0$.

an explosive: the initial density ρ_0, the initial internal energy per unit volume E_0, and the gamma-law coefficient γ. These determine the pressure-density relationship for the material as

$$P(\rho) = (\rho/\rho_0)(\gamma - 1)E_0 \tag{60}$$

Calibration of the gamma-law equation of state for an explosive is accomplished by adjusting E_0 and γ until the computer calculations fit experimental data for the velocity-time history of explosive-metal systems. For a description of this procedure we refer the reader to the literature.[6] In performing our fits to the experimental data,[14] we included data from only these experiments. In many other works, the initial internal energy is fit to other experiments—for example, the internal energy may be determined by measuring the heat formed during the burning of the explosive. Therefore, the values we obtain are very different from other tabulated values. We have included our numerical fits in Table 1. The drawback to this approach is that the initial shock transferred from the explosive

Table 1 Equation-of-state parameters for various explosives[a]

Explosive	ρ_0	E_0	D	γ	$\sqrt{2E}$
TNT	1.630	0.048	0.693	2.90	0.240
HMX	1.891	0.086	0.911	3.10	0.301
LX-14	1.835	0.084	0.880	3.30	0.300
Octol	1.822	0.077	0.848	2.70	0.291
PETN	1.770	0.078	0.830	2.85	0.297
Comp B	1.717	0.065	0.798	2.90	0.275
Comp C4	1.601	0.065	0.819	2.50	0.282
Cyclotol	1.754	0.077	0.825	2.50	0.296
TATB	1.854	0.055	0.776	2.90	0.244
Pentolite	1.700	0.066	0.753	2.90	0.277

[a]Units are cm, g, μs.

is too low, but for our purposes, where we are averaging the shocks over time, the gamma-law equation of state is sufficient.

In our calculations the explosive was modeled as completely detonated at the initial time step to approximate the passage of a tangential detonation wave. Although this limits the oscillations associated with shock wave reverberations, it does not eliminate them because the metal liners were initially at ambient pressure. The intent of these calculations was to model acceleration of various liners by explosives, not the detailed shock wave interactions, spalling, and other dynamic phenomena inside the metals. Therefore, we examined the effective *center-of-mass velocity* of the different liners, which is defined as the sum of the momentum of all of the liner nodes divided by the sum of their masses. This approach virtually eliminated the oscillatory effects of shock reverberations on the velocity-time histories of the liners.

An example is shown in Fig. 9 where, for exploding cylinders of M/C = 0.1, 1.0, and 10.0, we have plotted the computed liner velocities vs time as the solid curves. The dotted curves represent best fit approximations to the computed values using the exponential function

$$V(t) = V_m(1 - e^{-t/\tau}) \tag{61}$$

where V_m is the asymptotic velocity of the liners as discussed before and τ is the system time constant. Therefore, to describe the unsteady behavior of a one-dimensional explosive-liner system we require two parameters: V_m, the asymptotic velocity of the liner, and τ, the system time constant.

The results of our computer calculations have shown that the explosive equation-of-state parameters may be directly related to these dynamic system parameters. The effect of varying γ is shown in Fig. 10, where we have plotted the velocity history of an exploding cylindrical liner M/C = 1.0 for various values of γ ranging from 2 to 3. The explosive initial energy per unit volume E_0 and density ρ_0 were held constant. Changing γ does not affect the terminal velocity of the shell but does change the time-dependence of the velocity. In-

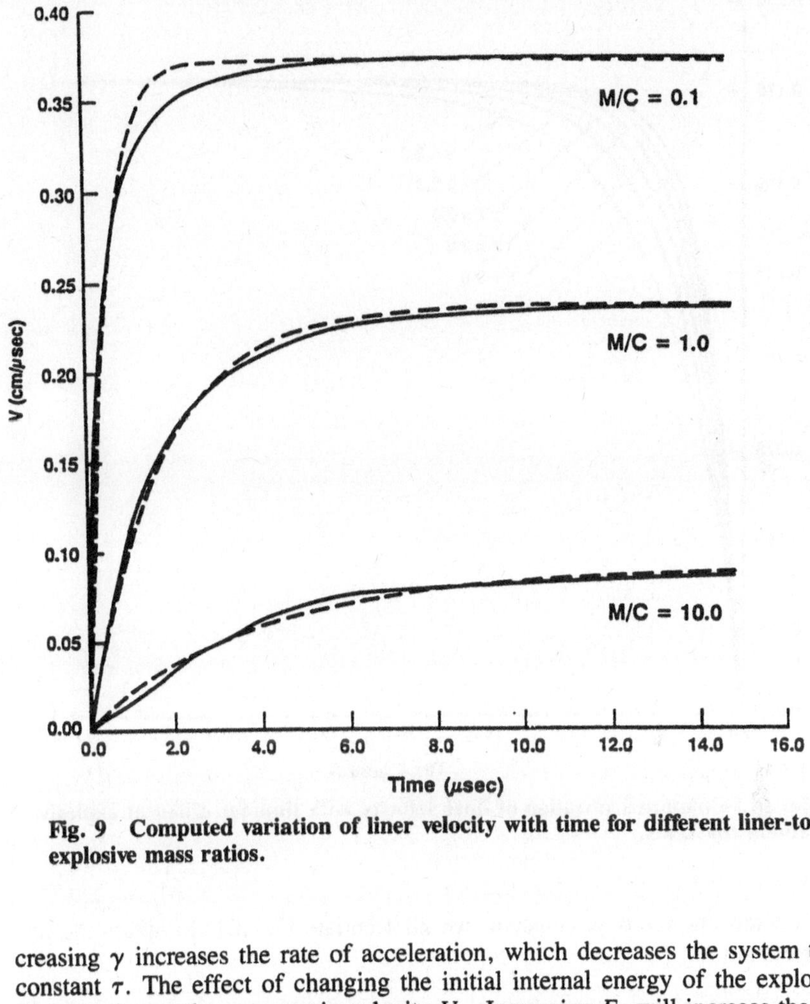

Fig. 9 Computed variation of liner velocity with time for different liner-to-explosive mass ratios.

creasing γ increases the rate of acceleration, which decreases the system time constant τ. The effect of changing the initial internal energy of the explosive E_0 is to change the asymptotic velocity V_m. Increasing E_0 will increase the V_m. The following relationship was established as a result of these calculations:

$$\sqrt{2E} = \left(\frac{2E_0}{\rho_0}\right)^{1/2} \tag{62}$$

where $\sqrt{2E}$ is the Gurney constant for the explosive. We note that the calculated values of the Gurney constant agree very well with the tabulated values. By using this relation, the initial internal energy of the explosive for the gamma-law equation of state may be determined directly from the experiments that measure asymptotic velocity V_m. Then γ can be determined from the velocity-time relationship.

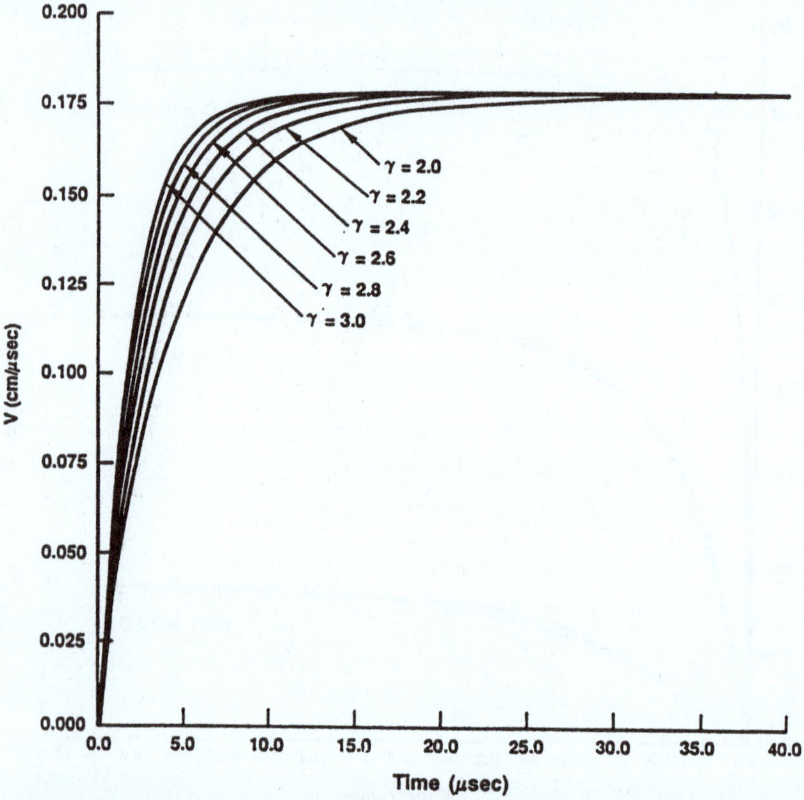

Fig. 10 Computed variation of liner velocity with time for different explosive gamma constants.

To determine the time constant, we differentiate Eq. (61) to obtain the liner acceleration

$$\frac{dV}{dt} = \left(\frac{V_m}{\tau}\right)e^{-t/\tau} \tag{63}$$

If we assume the average pressure on the liner is proportional to the initial pressure of the detonated explosive that is given by the gamma-law term $(\gamma - 1)E_0$ and refer to m as the liner mass per unit area, Newton's second law gives the following relation for the specific impulse delivered by the explosive to the liner I:

$$I = E_0(\gamma - 1)\tau/m \approx f(V_m) \tag{64}$$

where f is some function of V_m. Using this relation between the specific impulse I and the liner asymptotic velocity V_m, we have plotted all of the results from the calculations performed for the different geometries in Fig. 11. Both the symmetric and asymmetric geometries are included, which suggest that the time constant depends on the tamping only through geometry and velocity. The points plotted were obtained by fitting the computed velocity-time histories of the liners to Eq. (61). This step established the time constant τ. The curves drawn represent third-order polynomial fits to the points shown of the form

$$I = f(V_m) = a_1 V_m + a_2 V_m^2 + a_3 V_m^3 \tag{65}$$

The solid dots are the values obtained from the hydrodynamic calculations and the solid curves are the best fits to Eq. (65). The fitted coefficients obtained for the planar, cylindrical, and spherical geometries are listed in Table 2, where the units of velocity are cm/μs.

5. Conclusions

We have presented a set of relations that allow one to predict, with reasonable accuracy, the complete velocity-time history of an explosively driven liner in various one-dimensional geometries. The information required is essentially the

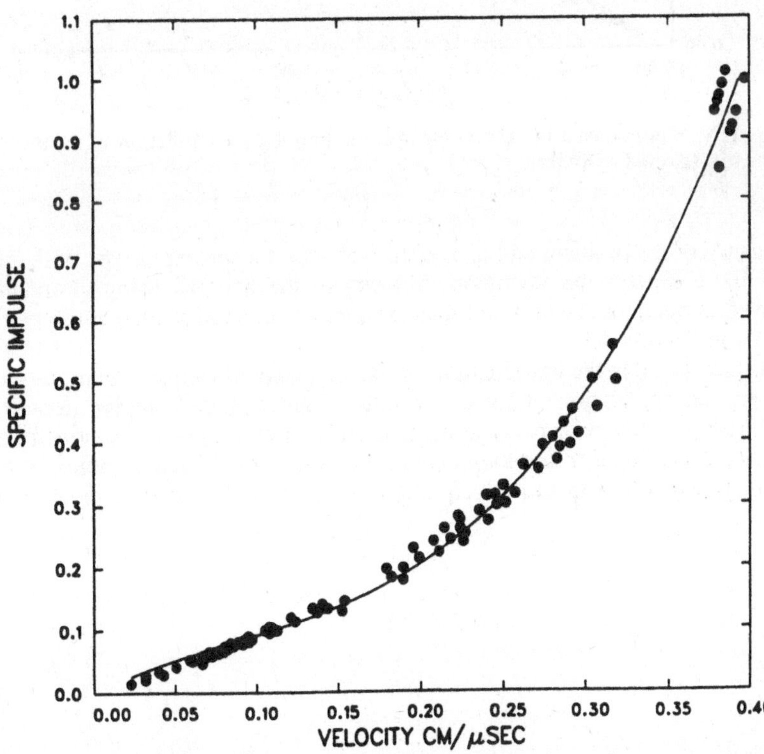

Fig. 11a Comparison of results for specific impulse as a function of asymptotic velocity: asymmetric sandwich.

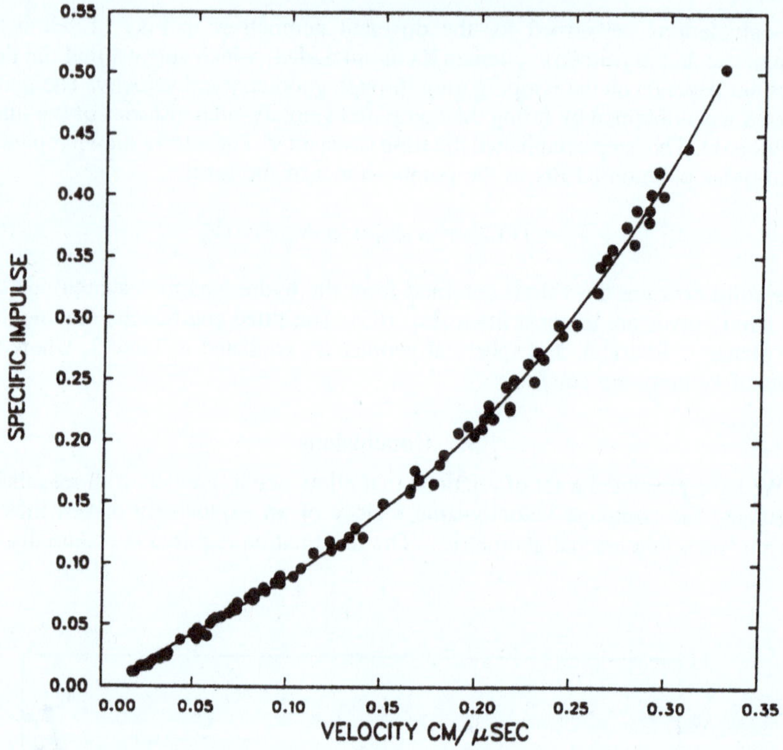

Fig. 11b Comparison of results for specific impulse as a function of asymptotic velocity: tamped cylinder.

geometry of the problem and an equation of state for the explosive. With these, one first computes the asymptotic velocity of the liner V_m using the relations derived in Sections 2 and 3 and then determines the system time constant τ, as shown in Section 4.

As mentioned in the introduction, the development of tactical munitions relies, in part, on application of these relations to much more complex geometries and systems. However, because of their strong foundation in the principles of conservation of energy and momentum, the behaviors of these complex systems often closely follow quantitatively the trends predicted by these simple Gurney models.

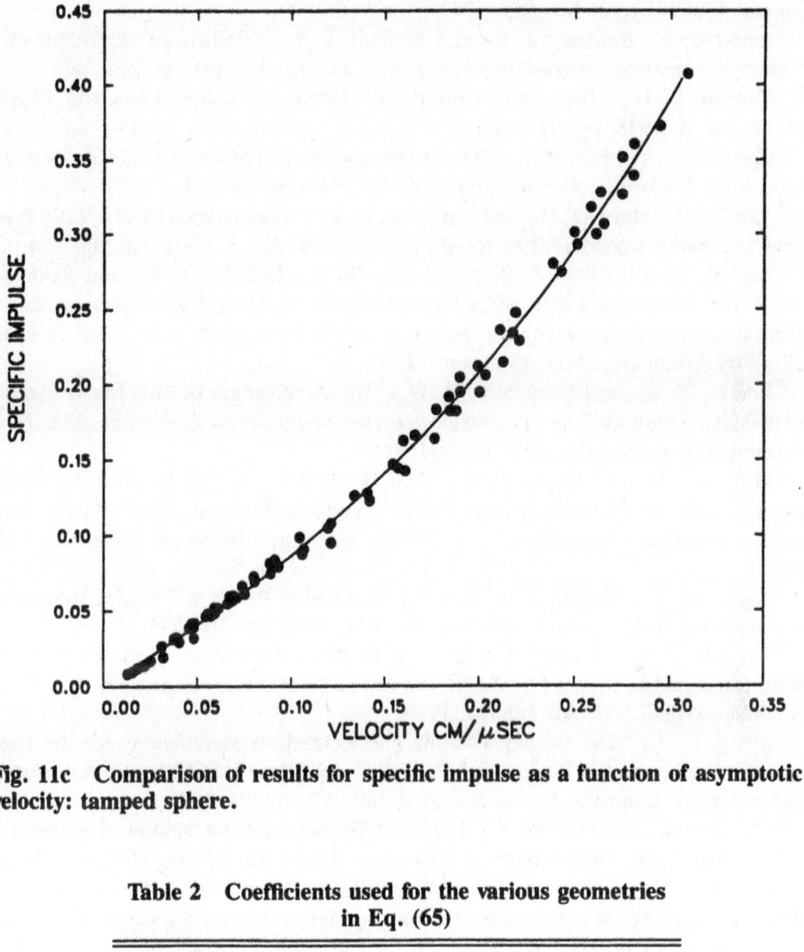

Fig. 11c Comparison of results for specific impulse as a function of asymptotic velocity: tamped sphere.

Table 2 Coefficients used for the various geometries in Eq. (65)

Geometry	a_1	a_2	a_3
Planar	1.275	−5.871	23.05
Cylindrical	0.856	−0.578	7.94
Spherical	0.782	0.381	4.49

References

[1]Gurney, R. W., "The Initial Velocity of Fragments from Bombs, Shells, and Grenades," BRL Report No. 405, 1943.

[2]Jones, G. E., Kennedy, J. E., and Bertholf, L. D., "Ballistics Calculations of R. W. Gurney," *American Journal of Physics*, Vol. 48, No. 4, 1980, pp. 264–269.

[3]Condon, E. U., "Tunneling—How It All Started," *American Journal of Physics*, Vol. 46, No. 4, 1978, pp. 319–323.

[4]Chou, P. C., and Flis, W. J., "Recent Developments in Shaped Charge Technology," *Propellants, Explosives, Pyrotechnics*, Vol. 11, 1986, pp. 99–114.

[5]Aziz, A. K., Hurwitz, H., and Sternberg, H. M., "Energy Transfer to a Rigid Piston Under Detonation Loading," *Physics of Fluids*, Vol. 4, No. 3, 1961, pp. 380–384.

[6]Hoskin, N. E., Allan, J. W. S., Bailey, W. A., Lethaby, J. W., and Skidmore, I. C., "The Motion of Plates and Cylinders Driven by Detonation Waves at Tangential Incidence," *Fourth Symposium (International) on Detonation*, U.S. Naval Ordnance Laboratory, White Oak, MD, 1965, pp. 14–26.

[7]Hoskin, N. E., and Lambourn, B. D., "The Acceleration of Two Metal Plates in an HE-Metal Sandwich," *Seventh Symposium (International) on Detonation*, U.S. Naval Academy, Annapolis, MD, 1981, pp. 811–825.

[8]Mader, C. L., and Gage, W. R., "FORTRAN SIN—A One-Dimensional Hydrodynamic Code for Problems Which Include Chemical Reaction, Elastic-Plastic Flow, Spalling, and Phase Transitions," Los Alamos Scientific Laboratory Report LA-3720, 1967.

[9]Davis, W. C., "Equation of State for Detonation Products," *Eighth Symposium (International) on Detonation*, Albuquerque, NM, 1985, pp. 785–795.

[10]Kennedy, J. E., "Gurney Energy of Explosives: Estimation of the Velocity and Impulse Imparted to Driven Metal," Sandia National Laboratory Report No. SC-RR-70-790, Albuquerque, NM, AD 886093, 1970.

[11]Henry, I. G., "The Gurney Formula and Related Approximations for the High-Explosive Deployment of Fragments," Presented to the American Ordnance Association, Hughes Aircraft Company, Report No. PUB-189, AD 813398, 1967.

[12]Carleone, J., and Flis, W. J., "Improved Formulas for the Motion of Explosively Driven Liners," Air Force Armament Laboratory Report No. AFATL-TR-82-01, 1982.

[13]Chanteret, P. Y., "An Analytical Model for Metal Acceleration by Grazing Detonation," *Proceedings of the Seventh International Symposium on Ballistics*, The Hague, The Netherlands, 1983, pp. 515–524.

[14]Lawrence Livermore National Laboratory Cylinder Test Data, private communication.

Chapter 5

Warhead Simulation Techniques: Hydrocodes

Robert R. Karpp*
Los Alamos National Laboratory, Los Alamos, New Mexico 87545

Nomenclature

A = area
a_i = material coordinates
c = sound speed (bulk)
c_L = longitudinal sound speed
d_{ij} = deformation rate
e = specific internal energy
e_t = specific total energy
e^p = equivalent plastic strain
$\overline{e}_i$ = unit vector
e_{ij} = strain
F = burn fraction
$\mathbf{F}$ = nodal force
f = yield function
f = general variable
G = amplification factor
G = shear modulus
J_2 = second invariant of stress deviator
K = bulk modulus
k = yield stress in simple shear
k = wave number
M = nodal mass
n_i = component of unit outward normal vector
P = pressure
q = artificial viscous pressure
r = radial cylindrical coordinate (spatial)
$\mathbf{r}$ = position vector
S = surface area
s = arc length

*Staff Member.

s = entropy
$\boldsymbol{T}$ = stress vector
t = time
u = general variable
u_i = displacement component
V = specific volume
V = volume
v_i = velocity component
$\dot{v}_i$ = acceleration component
x_i = spatial coordinates
x, y = two-dimensional Cartesian coordinates (spatial)
z = axial cylindrical coordinate (spatial)
α_{ij} = direction cosines
γ = ratio for specific heats for an ideal gas
Γ = Gruneisen parameter
δ_{ij} = Kronecker delta
μ = $\rho/\rho_o - 1$
η = $1 - \rho_o/\rho$
λ = proportionality factor
λ = Lamé constant
ρ = density
σ = constant
$\overline{\sigma}$ = equivalent stress
σ_{ij} = stress
$\overset{\nabla}{\sigma}_{ij}$ = Jaumann stress rate
θ = angular cylindrical coordinate (spatial)
$\dot{\theta}$ = angular velocity
ω_{ij} = spin
$\mathrm{D}/\mathrm{D}t$ = material derivative
$(\,)^E$ = elastic part
$(\,)^P$ = plastic part
$(\,)'_{ij}$ = deviator component
$T_{(\,)}$ = a tentative value that is correct only if the deformation is elastic

1. Introduction

AN introduction to the computational methods that are currently being employed to simulate the dynamics of warheads is presented in this chapter. These terminal ballistic problems are characterized by intense dynamic loads that are generated by impact or by explosive detonation and that act over time periods measured in microseconds. Therefore, inertial effects are dominant in the initial stages, and wave propagation must be adequately simulated.

The general continuum mechanics problem used to simulate these phenomena is based upon the conservation or balance equations of mass, momentum, and energy along with response functions describing the behavior of materials. The formulation is completed by specifying initial and boundary conditions appropriate to the problem of interest. The resulting system of partial differential equations is nonlinear, and analytic solutions are not usually possible because of the complexity of the initial value problem. Approximate numerical techniques are the only methods currently available for obtaining complete solutions. Computer programs constructed to produce numerical solutions to these unsteady, continuum mechanics problems are known variously as continuum mechanics codes, wave propagation codes, hydrodynamic codes, or hydrocodes.

Currently, about a dozen hydrocodes are routinely used for these simulations. Each code has some features that make it unique, but no single code or numerical method has emerged as clearly superior. Considerable effort is being made to increase the accuracy of numerical solutions and to decrease the required computing time. A similar experimental and theoretical effort is also being made to characterize materials better under conditions of impact and explosive attack.

The usefulness of hydrocodes is, of course, linked to the availability of high-speed computers. Most practical problems involve simulations in two and three spatial dimensions and, therefore, require large modern computing facilities. Two-dimensional simulations have been practical for a number of years, but only recently has the hardware been available for full three-dimensional simulations.

The results of hydrocode calculations depend upon the models used to describe the materials. Since the physical phenomena range from the detonation of explosives to the flow and fracture of solids, such complex behavior can at best be modeled only approximately, and the simulations can be expected to be no more accurate than the material models. However, in many cases, simulations based upon currently available material models are in good agreement with experiments. In these cases, the calculations can provide detailed information that is not otherwise obtainable. These data can be used to supplement experimental data, and the use of both may lead to a more comprehensive understanding of the physical process.

Hydrocodes have emerged as important design tools in the field of warhead engineering. Their use is most effective when integrated directly into the design iteration cycle. For example, the preliminary design of an explosively formed projectile (EFP) can be developed from hydrocode calculations that predict the shape and velocity of the projectile. This preliminary design can be built and tested, and flash x-ray pictures can reveal the final shape and velocity of the projectile. The test results may differ somewhat from the simulation, and the desired performance probably will not be achieved by the initial design. At this point, the analyst may choose to modify the computational model to obtain better agreement with test results. Modifications are usually made by adjusting constants in the material models or by changing the initial setup of the problem, for example, to resolve certain regions better. This modified computational model can then be used as the design tool for the remainder of the study. In many cases, code calculations have significantly reduced the number of design and test iterations necessary to meet the required performance.

In many applications, the use of hydrocode simulations requires a great deal of care since inadequate material models or inaccurate numerical treatments can produce results that are not even qualitatively correct. The experience of the analyst with a particular code is important since questionable results can usually be recognized and eliminated from consideration.

In this chapter, we first review the basic principles and associated equations of continuum mechanics that are solved by hydrocodes. Then, a brief review of numerical solution techniques for partial differential equations is presented. The specific numerical techniques used in many hydrocodes to solve the applicable equations of continuum mechanics are then discussed. The chapter is concluded by illustrating the results of typical hydrocode simulations. Several hydrocodes that have been used on a production basis for warhead simulations are listed in Table 7.

2. Governing Equations

In this section, we review the basic principles of mechanics and fundamental equations that are incorporated into hydrocodes that simulate warhead and target dynamics. A more thorough discussion of these equations can be found in textbooks on continuum mechanics (see, for example, Refs. 1–4).

2.1 Kinematics

To describe the motion of a body, we refer its configuration to a fixed, three-dimensional, Cartesian coordinate system as indicated in Fig. 1. The material particles of the body in its initial configuration (time $t = 0$) can be identified by their coordinate values (a_1, a_2, a_3) along the 1-, 2-, and 3-axes. At a later time, the body moves to a new configuration, and a typical material particle with initial coordinates (a_1, a_2, a_3) moves to a new position with coordinates (x_1, x_2, x_3). The motion of the body is described by functions that give the current position of its particles in terms of their initial position and time; that is,

$$
\begin{aligned}
x_1 &= x_1(a_1, a_2, a_3, t) \\
x_2 &= x_2(a_1, a_2, a_3, t) \\
x_3 &= x_3(a_1, a_2, a_3, t)
\end{aligned}
$$

By using standard indicial notation, these three functions can be written more compactly as

$$x_i = x_i(a_j, t) \tag{2.1}$$

where $i = 1, 2,$ or 3 and $j = 1, 2,$ and 3. These mapping functions are numerically determined in Lagrangian hydrocode calculations. Equation (2.1) can be inverted to give the initial position of a particle in terms of its current position

$$a_i = a_i(x_j, t) \tag{2.2}$$

When the dependent variables of a problem are expressed as functions of the independent variables a_1, a_2, a_3, and t, the description is called material or Lagrangian, and when the independent variables x_1, x_2, x_3 and t are used, the description is called spatial or Eulerian.

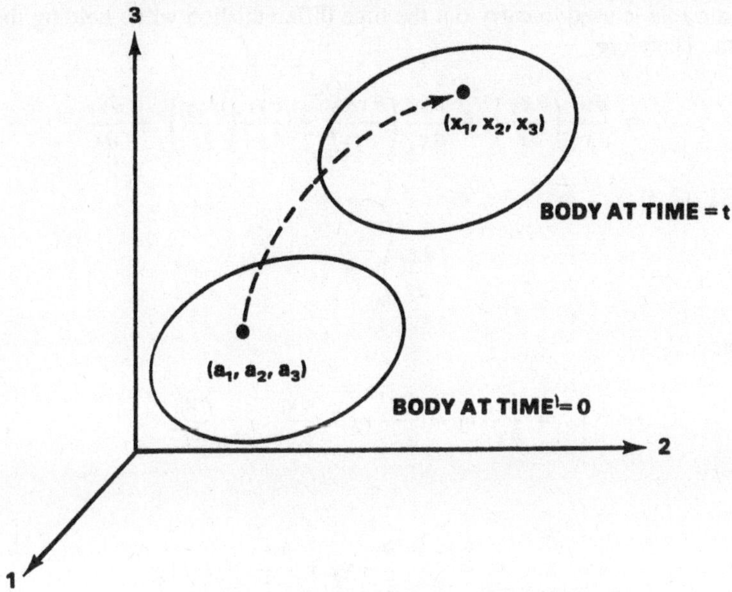

Fig. 1 Body in initial and current configurations.

The velocity components v_i of a particle (a_1, a_2, a_3) are defined by

$$v_i \equiv \left(\frac{\partial}{\partial t}\right)_a x_i(a_j, t) \tag{2.3}$$

where the partial time derivative is taken with the a_j held constant. The acceleration $\dot{v}_i$ is defined by

$$\dot{v}_i \equiv \left(\frac{\partial}{\partial t}\right)_a v_i(a_j, t) \tag{2.4}$$

In Eq. (2.4), the velocity is expressed in terms of the Lagrangian description. If the velocity is expressed as a function of the current position, $v_i(x_j, t)$, Eq. (2.4) still defines the acceleration; however, to apply Eq. (2.4), the velocity must first be transformed to Lagrangian coordinates,

$$v_i(x_j, t) = v_i[x_j(a_k, t), t]$$

before the time derivative is taken. The acceleration is calculated from

$$\dot{v}_i = \left(\frac{\partial}{\partial t}\right)_a v_i[x_j(a_k, t), t] = \left(\frac{\partial}{\partial t}\right)_a v_i[x_1(a_k, t), x_2(a_k, t), x_3(a_k, t), t]$$

The chain rule is used to carry out the time differentiation while holding the a_j constant. Therefore,

$$\dot{v}_i = \frac{\partial v_i}{\partial x_1}\left(\frac{\partial x_1}{\partial t}\right)_a + \frac{\partial v_i}{\partial x_2}\left(\frac{\partial x_2}{\partial t}\right)_a + \frac{\partial v_i}{\partial x_3}\left(\frac{\partial x_3}{\partial t}\right)_a + \frac{\partial v_i}{\partial t}$$

From Eq. (2.3),

$$\left(\frac{\partial x_i}{\partial t}\right)_a = v_i$$

therefore,

$$\dot{v}_i = \frac{\partial v_i}{\partial x_1}v_1 + \frac{\partial v_i}{\partial x_2}v_2 + \frac{\partial v_i}{\partial x_3}v_3 + \frac{\partial v_i}{\partial t}$$

or

$$\dot{v}_i = \sum_{j=1}^{3} \frac{\partial v_i}{\partial x_j}v_j + \frac{\partial v_i}{\partial t}$$

If we use the standard summation convention that implies a sum be taken over any repeated index occurring in the same term, the summation sign can be suppressed, and the preceding equation takes the form

$$\dot{v}_i = v_j \frac{\partial v_i}{\partial x_j} + \frac{\partial v_i}{\partial t} \tag{2.5}$$

Equation (2.5) is the rule for determining the acceleration when the velocity is expressed in the spatial description. The derivative operator on the right side of Eq. (2.5), which can be applied to any quantity expressed in Eulerian coordinates, is written as

$$\frac{\mathrm{D}}{\mathrm{D}t} = \frac{\partial}{\partial t} + v_j \frac{\partial}{\partial x_j} \tag{2.6}$$

and is known as the material derivative. It gives the rate of change experienced at a material particle as the particle moves through space. Note that the material derivative of the particle position x_i is

$$\frac{\mathrm{D}x_i}{\mathrm{D}t} = \frac{\partial x_i}{\partial t} + v_j \frac{\partial x_i}{\partial x_j} = v_i$$

because x_1, x_2, x_3, and t are independent.

A measure of the rate at which a material particle is deforming is given by the Eulerian deformation rate tensor. In Fig. 2 the configuration of a body at

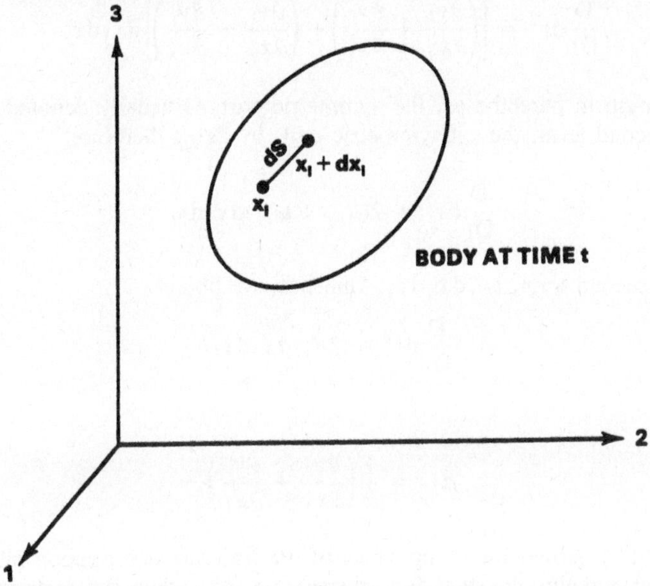

Fig. 2 Distance between two particles in the current configuration.

time t is indicated. Consider an arbitrary point in the body x_i and a second arbitrary point $(x_i + \mathrm{d}x_i)$ in the neighborhood of x_i. The deformation rate is based upon the rate at which the distance between these two points is changing.

The material derivative of the distance squared $\mathrm{d}s^2$ between these points is

$$\frac{\mathrm{D}}{\mathrm{D}t}\mathrm{d}s^2 = \frac{\mathrm{D}}{\mathrm{D}t}\mathrm{d}x_i\,\mathrm{d}x_i$$

It can be shown[1] that

$$\frac{\mathrm{D}}{\mathrm{D}t}\mathrm{d}x_i = \frac{\partial v_i}{\partial x_j}\mathrm{d}x_j$$

therefore,

$$\frac{\mathrm{D}}{\mathrm{D}t}\mathrm{d}s^2 = 2\frac{\partial v_i}{\partial x_j}\mathrm{d}x_j\,\mathrm{d}x_i$$

The velocity gradient $\partial v_i/\partial x_j$ can be written as the sum of a symmetric and an antisymmetric part. (A tensor T_{ij} is symmetric if $T_{ij} = T_{ji}$ and antisymmetric if $T_{ij} = -T_{ji}$.) Therefore,

$$\frac{D}{Dt}ds^2 = \left[\left(\frac{\partial v_i}{\partial x_j} + \frac{\partial v_i}{\partial x_i}\right) + \left(\frac{\partial v_i}{\partial x_j} - \frac{\partial v_i}{\partial x_i}\right)\right] dx_j dx_i$$

The first term in parentheses, the symmetric part, is usually denoted by $2d_{ij}$ and the second term, the antisymmetric part, by $2\omega_{ij}$; therefore,

$$\frac{D}{Dt}ds^2 = 2(d_{ij} + \omega_{ij}) dx_j dx_i$$

Since the second term, $\omega_{ij} dx_j dx_i$, vanishes, we obtain

$$\frac{D}{Dt}ds^2 = 2d_{ij} dx_j dx_i$$

where

$$d_{ij} = \frac{1}{2}\left(\frac{\partial v_i}{\partial x_j} + \frac{\partial v_i}{\partial x_i}\right) \tag{2.7}$$

Equation (2.7) defines the components of the Eulerian deformation rate tensor d_{ij}. If ds^2 remains constant for arbitrary $dx_j dx_i$, then the material is not deforming (rigid body motion) and the deformation rate will vanish.

The antisymmetric part of the velocity gradient ω_{ij} is a measure of the rotation rate of a material particle. The meaning of ω_{ij} is illustrated in Fig. 3, which shows, in two dimensions, the current position of a material particle located at

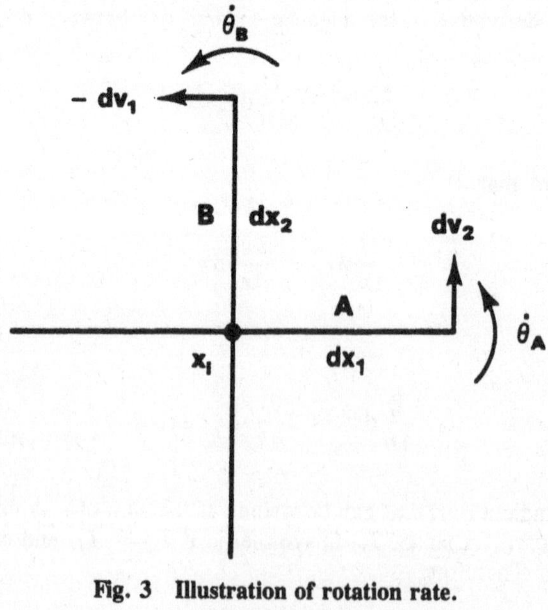

Fig. 3 Illustration of rotation rate.

x_i. Line A, of length dx_1, is a material line that is horizontal at time t, and line B, of length dx_2, is vertical at time t. The relative transverse velocity of the end of line A with respect to point x_i is dv_2, where

$$dv_2 = \frac{\partial v_2}{\partial x_1} dx_1$$

Also, for line B,

$$dv_1 = \frac{\partial v_1}{\partial x_2} dx_2$$

The rotation rate of line A, $\dot{\theta}_A$, is related to dv_2 by

$$dv_2 = dx_1 \dot{\theta}_A; \qquad \text{therefore,} \qquad \dot{\theta}_A = \frac{\partial v_2}{\partial x_1}$$

Similarly,

$$-dv_1 = dx_2 \dot{\theta}_B; \qquad \text{therefore,} \qquad \dot{\theta}_B = -\frac{\partial v_1}{\partial x_2}$$

The average rotation rate of these two orthogonal lines is

$$\dot{\theta}_{\text{average}} = \frac{1}{2}\left(\dot{\theta}_A + \dot{\theta}_B\right) = \frac{1}{2}\left(\frac{\partial v_2}{\partial x_1} - \frac{\partial v_1}{\partial x_2}\right) = \omega_{21}$$

Thus, ω_{21} is the average rotation rate about the 3-axis of two material lines parallel to the 1- and 2-axes. The quantity ω_{ij} is usually called the spin tensor.

2.2 Stress

The stress tensor of Cauchy is used in most hydrocodes. The components of this tensor are illustrated in Fig. 4. Unit vectors $\boldsymbol{e}_i$ are defined parallel to the three coordinate axes as shown. At time t, an infinitesimal parallelpiped with faces perpendicular to the axes is isolated from the body. In general, contact forces will act on these faces. The stress vector acting on any face i of the element is defined as

$$\overset{i}{\boldsymbol{T}} = \frac{\text{force vector acting on face } i}{\text{current area of face } i}$$

where face i is the face normal to $\boldsymbol{e}_i$. The stress vector $\overset{i}{\boldsymbol{T}}$ can be resolved into components parallel to the three coordinate axes,

$$\overset{i}{\boldsymbol{T}} = \sigma_{i1}\boldsymbol{e}_1 + \sigma_{i2}\boldsymbol{e}_2 + \sigma_{i3}\boldsymbol{e}_3 = \sigma_{ij}\boldsymbol{e}_j \tag{2.8}$$

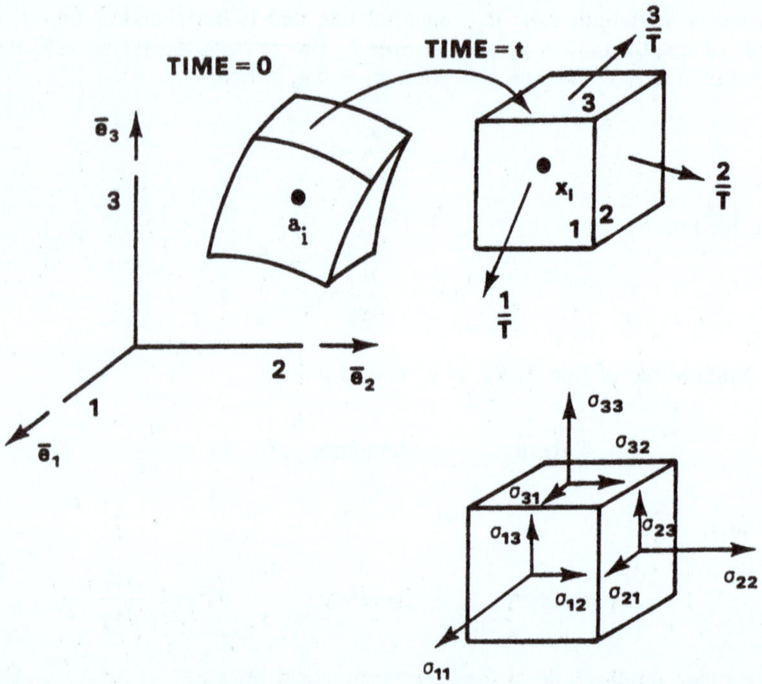

Fig. 4 Illustration of the stress vector and components of the Cauchy stress tensor.

Equation (2.8) defines the nine components σ_{ij} of the Cauchy stress tensor. The stress components acting on a two-dimensional element are shown in Fig. 5. The sign convention for stress is that a stress component σ_{ij} acting on the positive i-face (face with unit outward normal $+ \boldsymbol{e}_i$) is positive if it acts in the positive $\boldsymbol{e}_j$-direction. The stress component σ_{ij} acting on the negative i-face (face with outward normal $-\boldsymbol{e}_i$) is positive if it acts in the negative $\boldsymbol{e}_j$-direction. All stress components in Fig. 5 are shown positive.

By applying the angular momentum principle about the 3-axis through the center of the two-dimensional element of Fig. 5 and letting the size of the element approach zero, we find that $\sigma_{12} = \sigma_{21}$. In general, for the force system considered, the stress tensor is symmetric; that is,

$$\sigma_{ij} = \sigma_{ji} \tag{2.9}$$

When the coordinate system is changed, the components of stress change. Consider two Cartesian coordinate systems with a common origin, but one rotated with respect to the other. The stress components referred to each coordinate system are related by the stress transformation law, a law that applies to any tensor of rank 2. We indicate the derivation of the transformation law in two dimensions with the aid of Fig. 6. If the stress components σ_{ij} referred to the x_1, x_2-coordinate system are known, the stress components σ'_{ij}, referred to

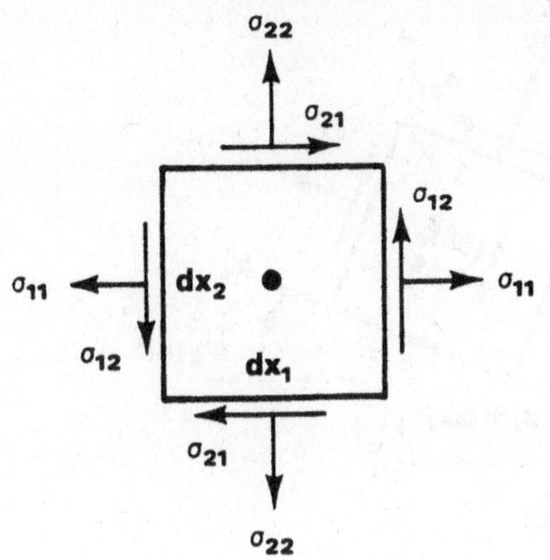

Fig. 5 Stress components acting on a two-dimensional element.

the x'_1, x'_2-coordinate system can be calculated in terms of the rotation between the axes. The calculation is performed by applying the equation of motion in the x'_1- and x'_2-directions and letting the size of the element approach zero. The two equations that result from this calculation can be summarized as

$$\sigma'_{kr} = \alpha_{ki}\alpha_{rj}\sigma_{ij} \tag{2.10}$$

where α_{ki} is the cosine of the angle between the x'_k-direction and the x_i-direction. Equation (2.10) is the stress transformation law, which, although indicated for two dimensions, also applies in three dimensions.

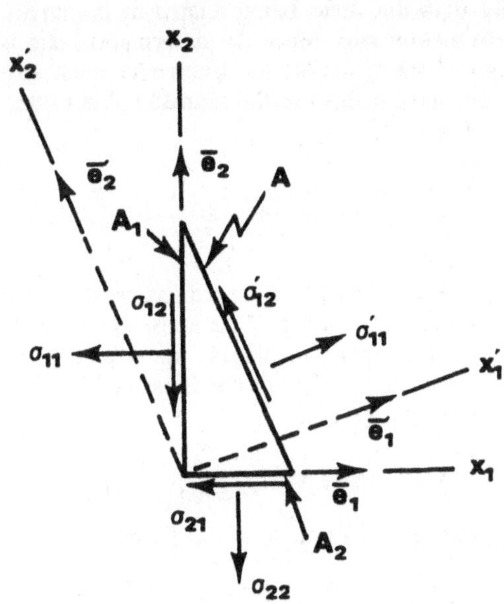

Fig. 6 Stress components referred to a rotated coordinate system.

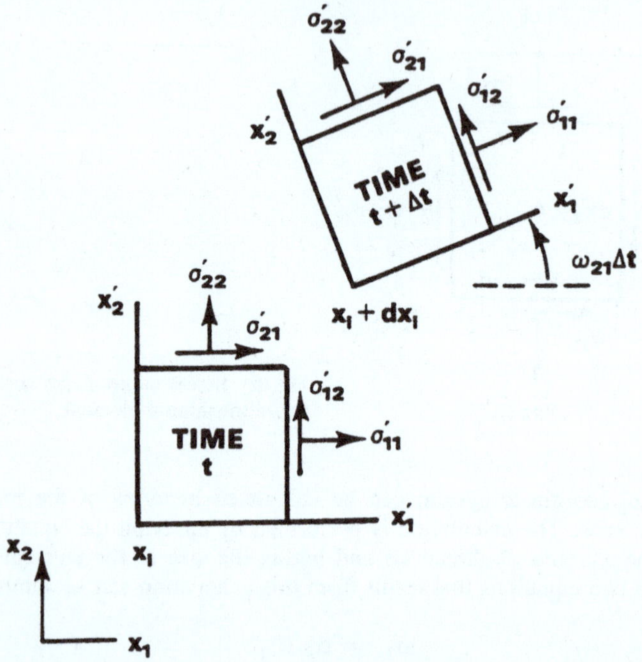

Fig. 7 Translation and rotation of a material element over a small time interval.

In most hydrocodes, the constitutive relations are formulated as rate laws or incremental laws; that is, rather than the stress being related to the strain, the stress rate is related to the deformation rate. Since the deformation rate is unaffected by a rigid body rotation of the material, the stress rate must also possess this property. For this reason, most codes use the Jaumann stress rate.

The Jaumann stress rate $\overset{\nabla}{\sigma}_{ij}$ is defined[2] by

$$\overset{\nabla}{\sigma}_{ij} = \lim_{\Delta t \to 0} \frac{\sigma'_{ij}(t + \Delta t) - \sigma'_{ij}(t)}{\Delta t} \tag{2.11}$$

where σ'_{ij} are the stress components referred to the x'_i coordinate system that translates and rotates with the material particle. In Fig. 7 the stress components at time t and at time $t + \Delta t$ for the two-dimensional case are depicted. To evaluate the right side of Eq. (2.11), we first transform the stress components to the fixed x_i-coordinate system. From Eq. (2.10),

$$\sigma'_{kr}(t + \Delta t) = \alpha_{ki}\alpha_{rj}\sigma_{ij}(t + \Delta t)$$

Since the x'_i-axes have rotated through the small angle $\omega_{21}\Delta t$ in time Δt, the cosines α_{ij} become

$$\alpha_{11} = \cos(\omega_{21}\Delta t) \simeq 1, \qquad \alpha_{22} \simeq 1$$

$$\alpha_{12} = \cos\left(\frac{\pi}{2} - \omega_{21}\Delta t\right) = \sin(\omega_{21}\Delta t) \simeq \omega_{21}\Delta t$$

and

$$\alpha_{21} = \alpha_{12}$$

These relations can be summarized as

$$\alpha_{ij} = \delta_{ij} + \omega_{ij}\Delta t$$

where $\delta_{ij} = 1$ if $i = j$, and $\delta_{ij} = 0$ if $i \neq j$. Also, $\sigma'_{ij}(t) = \sigma_{ij}(t)$. When these relations are substituted into Eq. (2.11), we obtain

$$\overset{\nabla}{\sigma}_{ij} = \lim_{\Delta t \to 0}\left[\frac{\sigma_{ij}(t + \Delta t) - \sigma_{ij}(t)}{\Delta t}\right] + \omega_{ki}\sigma_{kj} + \omega_{rj}\sigma_{ir}$$

The first term on the right is the material derivative of the stress, since it gives the rate of change as observed from a coordinate system moving with the particle but not rotating with it. The second and third terms account for the rotation of the material. The Jaumann rate then takes the form

$$\overset{\nabla}{\sigma}_{ij} = \frac{\mathrm{D}}{\mathrm{D}t}\sigma_{ij} + \omega_{ki}\sigma_{kj} + \omega_{rj}\sigma_{ir} \tag{2.12}$$

If the stress is not changing in a coordinate system that translates and rotates with the material, then the Jaumann rate will vanish. Although derived in two dimensions, Eq. (2.12) also holds in three dimensions.

2.3 Conservation Laws

The fundamental principles that are assumed to govern the motion of any continuum are the conservation of mass, the conservation of linear and angular momentum, and the conservation of energy. The equations expressing these principles are described in this section.

Reynold's transport theorem, derived in Ref. 4, is used to evaluate the material derivative of a volume integral. The theorem can be stated as

$$\frac{\mathrm{D}}{\mathrm{D}t}\int_{V(t)} f(x_i, t)\mathrm{d}V = \int_{V(t)}\left(\frac{\mathrm{D}f}{\mathrm{D}t} + f\frac{\partial v_i}{\partial x_i}\right)\mathrm{d}V \tag{2.13}$$

where the integrals are evaluated over the volume $V(t)$ and $V(t)$ is a varying volume that moves with and encloses the same material particles, and f is any function in the spatial description.

The mass of a body enclosed by $V(t)$ is given by

$$m = \int_{V(t)} \rho \, \mathrm{d}V$$

Conservation of mass implies

$$\frac{\mathrm{D}m}{\mathrm{D}t} = 0$$

Using Eq. (2.13) to evaluate this equation, we obtain

$$\int_{V(t)} \left(\frac{\mathrm{D}\rho}{\mathrm{D}t} + \rho \frac{\partial v_i}{\partial x_i} \right) \mathrm{d}V = 0$$

Since the preceding equation is true for an arbitrary volume $V(t)$,

$$\frac{\mathrm{D}\rho}{\mathrm{D}t} + \rho \frac{\partial v_i}{\partial x_i} = 0 \tag{2.14}$$

Equation (2.14) is the conservation of mass equation, or continuity equation.

The linear momentum principle states that the rate of change of linear momentum of a body is equal to the resultant of the applied forces. Therefore, at any instant

$$\frac{\mathrm{D}}{\mathrm{D}t} \int_{V(t)} \rho v_i \, \mathrm{d}V = \int_{S(t)} \overset{n}{T}_i \, \mathrm{d}S \tag{2.15}$$

where $S(t)$ is the surface of $V(t)$, as shown in Fig. 8, and $\overset{n}{T}_i$ is a component of the stress vector, force per unit current area, acting on $S(t)$. The stress vector is expressed in terms of the stress tensor by Cauchy's formula[4]

$$\overset{n}{T}_i = \sigma_{ji} n_j \tag{2.16}$$

where n_j is a component of the unit outward normal vector to the surface $S(t)$.

Equation (2.15) can be rearranged by substituting Eq. (2.16) into Eq. (2.15) and converting the surface integral to a volume integral by use of the divergence theorem. An application of Reynold's transport theorem and use of Eq. (2.14) then yield the linear momentum equation, or equation of motion

$$\rho \frac{\mathrm{D}v_i}{\mathrm{D}t} = \frac{\partial \sigma_{ji}}{\partial x_j} \tag{2.17}$$

Note that gravity, a body force, has not been included because it is negligible for the problems of current interest. The vector equation (2.17) represents three scalar equations in three-dimensional Cartesian coordinates.

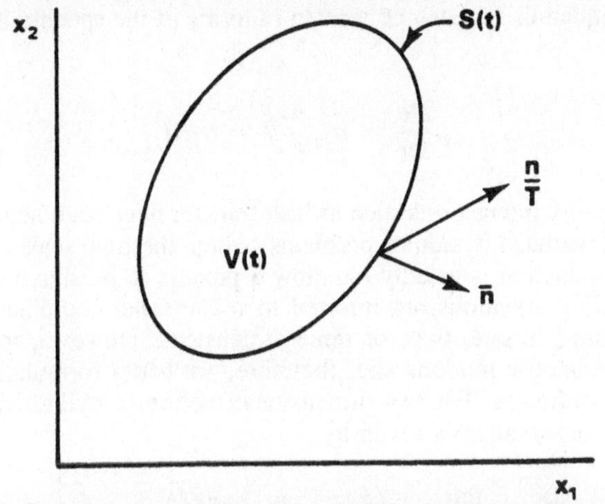

Fig. 8 Stress vector acting on the boundary of a material body.

The angular momentum principle states that the rate of change of angular momentum of a body about any point equals the resultant momentum of applied forces about the same point. This principle is stated by the vector equation

$$\frac{D}{Dt}\int_{V(t)} \rho(\boldsymbol{r} \times \boldsymbol{v})dV = \int_{S(t)} (\boldsymbol{r} \times \overset{n}{\boldsymbol{T}})dS \tag{2.18}$$

where $\boldsymbol{r}$ is the position vector from the origin to a generic point in the body and $\times$ indicates the vector cross product. By using the preceding relations to rearrange Eq. (2.18), we can show that the angular momentum principle implies

$$\sigma_{ij} = \sigma_{ji} \tag{2.19}$$

which states that the stress tensor is symmetric.

The energy conservation principle states that the rate of increase of energy of a body is equal to the rate at which the applied forces do work on the body. The principle can be stated as

$$\frac{D}{Dt}\int_{V(t)} \rho e_t dV = \int_{S(t)} \overset{n}{T}_i v_i \, dS \tag{2.20}$$

where $e_t = e + \frac{1}{2}v_i v_i$ is the specific total energy of the material, equal to the sum of the specific internal energy and the specific kinetic energy. Using manipulations similar to those indicated before, we can write the conservation of energy equation as

$$\rho\frac{De_t}{Dt} = \frac{\partial}{\partial x_i}(\sigma_{ij}v_j) \tag{2.21}$$

The energy equation can also be written in terms of the specific internal energy e as

$$\rho \frac{\mathrm{D}e}{\mathrm{D}t} = \sigma_{ij} \frac{\partial v_j}{\partial x_i} = \sigma_{ij} d_{ij} \tag{2.22}$$

Additional energy interactions such as heat transfer have been neglected in these equations. In warhead dynamics problems, where the time scale is in microseconds, heat conduction is usually too slow a process to be significant.

The preceding equations are referred to a Cartesian coordinate system and apply to motions in one, two, or three dimensions. However, many problems involve axisymmetric motions and, therefore, are better formulated in terms of cylindrical coordinates. For two-dimensional motion in cylindrical coordinates (r, z), mass conservation is given by

$$\frac{\mathrm{D}\rho}{\mathrm{D}t} + \rho\left(\frac{\partial v_r}{\partial r} + \frac{v_r}{r} + \frac{\partial v_z}{\partial z}\right) = 0 \tag{2.23}$$

The equations of motion are

$$\rho \frac{\mathrm{D}v_r}{\mathrm{D}t} = \frac{\partial}{\partial r}\sigma_{rr} + \frac{\partial}{\partial z}\sigma_{rz} + \frac{1}{r}(\sigma_{rr} - \sigma_{\theta\theta}) \tag{2.24}$$

and

$$\rho \frac{\mathrm{D}v_z}{\mathrm{D}t} = \frac{\partial}{\partial z}\sigma_{zz} + \frac{\partial}{\partial r}\sigma_{rz} + \frac{1}{r}\sigma_{rz}$$

The energy equation is

$$\rho \frac{\mathrm{D}e}{\mathrm{D}t} = \sigma_{rr} d_{rr} + 2\sigma_{rz} d_{rz} + \sigma_{zz} d_{zz} + \sigma_{\theta\theta} d_{\theta\theta} \tag{2.25}$$

where the components of the deformation rate are given by

$$\begin{aligned} d_{rr} &= \frac{\partial v_r}{\partial r}, \qquad & d_{zz} &= \frac{\partial v_z}{\partial z} \\ d_{\theta\theta} &= \frac{1}{r} v_r, \qquad & d_{rz} &= \frac{1}{2}\left(\frac{\partial v_z}{\partial r} + \frac{\partial v_r}{\partial z}\right) \end{aligned} \tag{2.26}$$

The material derivative operator is given by

$$\frac{\mathrm{D}}{\mathrm{D}t} = \frac{\partial}{\partial t} + v_r \frac{\partial}{\partial r} + v_z \frac{\partial}{\partial z} \tag{2.27}$$

2.4 Constitutive Equations

The basic conservation laws of mass, Eq. (2.14), momentum, Eq. (2.17), and energy, Eq. (2.22), apply under the conditions of the present formulation to any material. However, additional equations that relate the stress developed in a material to the deformation of the material must be added in order to complete the system of equations. These relations, in general, will be different for different materials. For the problems considered, we need to describe compressible fluids and deformable solids. Solids are materials that can resist shear deformation. In this section, we illustrate only the simple elastic-plastic description of solids. Many hydrocodes, however, include additional effects such as rate-dependent deformation and fracture.

Elastic Relations

In linear elasticity, the generalized Hooke's law for an isotropic, elastic material is

$$\sigma_{ij} = \lambda\delta_{ij}e_{kk} + 2Ge_{ij} \tag{2.28}$$

where λ and G are the Lamé constants; G is also known as the shear modulus. In Eq. (2.28), e_{ij} is the linear strain tensor

$$e_{ij} = \frac{1}{2}\left(\frac{\partial u_i}{\partial x_j} + \frac{\partial u_j}{\partial x_i}\right)$$

where u_i is the displacement vector. Deviatoric stress and strain tensors σ'_{ij} and e'_{ij} are defined by

$$\sigma'_{ij} = \sigma_{ij} - \frac{1}{3}\sigma_{kk}\sigma_{ij} \tag{2.29}$$

and

$$e'_{ij} = e_{ij} - \frac{1}{3}e_{kk}\delta_{ij} \tag{2.30}$$

The term $-\frac{1}{3}\sigma_{kk}$ in Eq. (2.29) is usually defined as the pressure

$$P = -\tfrac{1}{3}\sigma_{kk} = -\tfrac{1}{3}(\sigma_{xx} + \sigma_{yy} + \sigma_{zz})$$

and the term e_{kk} in Eq. (2.30) is the relative volume change of an element

$$\frac{\Delta V}{V} = e_{kk} = e_{xx} + e_{yy} + e_{zz}$$

With this notation, the generalized Hooke's law, Eq. (2.28), may be written in two parts, a bulk response part

$$P = -(\lambda + \frac{2}{3}G)\frac{\Delta V}{V} = -K\frac{\Delta V}{V} \tag{2.31}$$

and a deviatoric response part

$$\sigma'_{ij} = 2Ge'_{ij} \tag{2.32}$$

In Eq. (2.31), K is the bulk modulus. The preceding two equations are equivalent to the generalized Hooke's law, Eq. (2.28).

Plastic Relations

When ductile materials are strained beyond their elastic limit, plastic deformation occurs, and Hooke's law is no longer applicable. In plasticity theory, the strain is usually expressed as the sum of an elastic strain, e^E_{ij}, plus a plastic strain, e^P_{ij}, or

$$e_{ij} = e^E_{ij} + e^P_{ij}$$

The plastic strain is assumed to be incompressible, that is,

$$e^P_{kk} = 0$$

Because of this relation, the plastic strain deviator tensor is equal to the plastic strain tensor.

To determine whether a material is in an elastic or plastic state, the existence of a yield function is postulated. This scalar function f is a function of the state of stress and possibly the state of plastic strain, that is,

$$f = f(\sigma_{ij}, e^P_{ij})$$

The yield function describes the state of the material in the following manner. If $f < 0$, the material is in an elastic state. If $f = 0$, plastic deformation may occur depending upon the time rate of change of f. If $f = 0$ and $\dot{f} < 0$, the material is also in an elastic state; however, if $f = 0$ and $\dot{f} > 0$, then plastic deformation occurs. The state corresponding to $f > 0$ cannot be achieved.

Many materials approximately satisfy the von Mises yield function, which is

$$f = J_2 - k^2 \tag{2.33}$$

where J_2 is the second invariant of the stress deviator,

$$J_2 = \frac{1}{2}\sigma'_{ij}\sigma'_{ij}$$

and k is the yield strength in simple shear. For strain-hardening materials, k is a function of the plastic strain. The von Mises yield surface $f = 0$ is shown in Fig. 9, where it is plotted in principal stress space. It is a circular cylinder with

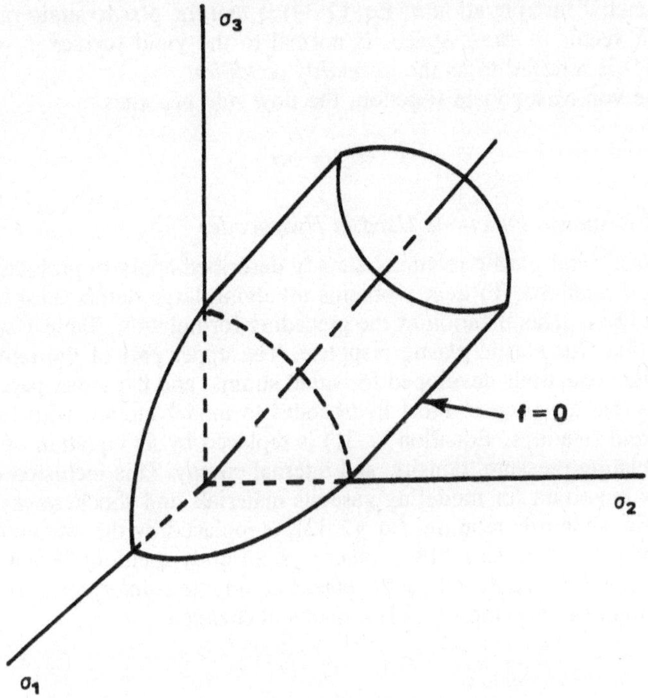

Fig. 9 Von Mises yield surface in principal stress space.

radius $\sqrt{2}k$ whose axis is equally inclined to the coordinate axes. Elastic states correspond to points inside the cylinder, and plastic states correspond to points on the surface. The von Mises yield function is an option available in nearly all hydrocodes.

If the material is in a plastic state as determined from the yield function ($f = 0, \dot{f} > 0$), an additional relation must be specified in order to calculate the resulting plastic deformation. This additional relation is the flow rule, which can be stated as

$$\dot{e}^P_{ij} = \lambda \frac{\partial f}{\partial \sigma_{ij}} \tag{2.34}$$

where λ is a proportionality factor. Since $f = 0$ defines a surface in nine-dimensional stress space, the outward normal vector to this surface has the direction given by the gradient

$$\frac{\partial f}{\partial \sigma_{ij}}$$

The geometric interpretation of Eq. (2.34) is that the plastic strain rate, considered as a vector in stress space, is normal to the yield surface $f = 0$. Thus, Eq. (2.34) is referred to as the *normality condition*.

For the von Mises yield function, the flow rule becomes

$$\dot{e}^P_{ij} = \lambda \sigma'_{ij} \tag{2.35}$$

Material Response Functions Used in Hydrocodes

The elastic and plastic relations already described apply to problems of small strains and rotations. To treat problems involving large deformations, most hydrocodes use a generalization of the preceding formulation. Table 1 summarizes the equations for elastic-plastic response. The upper part of the table presents the standard equations developed for small strains, and the lower part shows the equations that are used in most hydrocodes to model motion with large deformations and rotations. Equation (2.31) is replaced by an equation of state, Eq. (2.36), relating pressure, density, and internal energy. This inclusion of internal energy is important for modeling gaseous materials and shock waves in solids. The elastic deviatoric relation, Eq. (2.32), is replaced by the rate equation, Eq. (2.37), which is invariant with respect to rigid-body rotations. Equation (2.35) is replaced by Eq. (2.38), where the plastic strain rate is interpreted as the plastic deformation rate. Equation (2.33) remains unchanged.

Isotropic Strain Hardening

Strain hardening is incorporated into the formulation of many hydrocodes. To include this effect, two scalar variables, equivalent stress $\overline{\sigma}$ and equivalent plastic strain $\overline{e}^P$, are introduced. These quantities are defined by

$$\overline{\sigma} = \left(\frac{3}{2}\sigma'_{ij}\sigma'_{ij}\right)^{1/2} \tag{2.39}$$

Table 1 Material response functions used in hydrocodes

Equations used to describe small deformations and rotations			
Elastic		Plastic	
$P = -K\frac{\Delta V}{V} = K(\rho/\rho_o - 1)$	(2.31)	$f = J_2 - k^2$	(2.33)
$\sigma'_{ij} = 2Ge^{E\,\prime}_{ij}$	(2.32)	$\dot{e}^P_{ij} = \lambda\sigma'_{ij}$	(2.35)
Equations used in hydrocodes to describe finite deformations and rotations			
Elastic		Plastic	
$P = P(\rho, e)$	(2.36)	$f = J_2 - k^2$	(2.33)
$\overset{\nabla}{\sigma}{}'_{ij} = 2Gd^{E\,\prime}_{ij}$	(2.37)	$d^P_{ij} = \lambda\sigma'_{ij}$	(2.38)

and

$$\bar{e}^p = \int_0^t d\bar{e}^p \tag{2.40}$$

where

$$d\bar{e}^p = \left(\frac{2}{3} de_{ij}^p de_{ij}^p\right)^{1/2} \tag{2.41}$$

For the case of uniaxial stress as in a tension test,

$$\bar{\sigma} = \sigma_{xx}, \qquad \text{and} \qquad \bar{e}^p = e_{xx}^p$$

The stress-strain data from a tension test may be used to describe a functional relationship between stress and plastic strain,

$$\bar{\sigma} = F_p(\bar{e}^p) \tag{2.42}$$

The typical shape of such a function is shown in Fig. 10. To include the general case of multiaxial stress and strain, we assume a universal curve; that is, we assume Eq. (2.42) holds in general.

Equations of State

The form of the equation of state required by most hydrocodes is indicated by Eq. (2.36). Two main classes of materials are distinguished: gaseous products from the detonation of high explosives (HE) and inert solids.

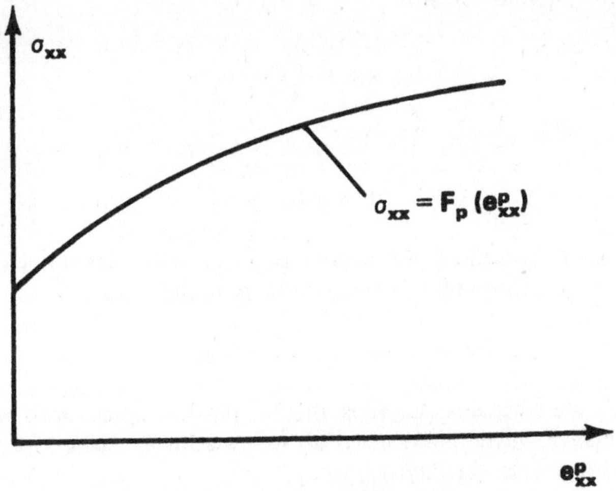

Fig. 10 Typical strain hardening function.

The equation of state most commonly used to describe the behavior of the detonation products of explosives in applications of metal acceleration is the Jones-Wilkins-Lee (JWL) equation,[5] given by

$$P = A\left(1 - \frac{\omega}{R_1 V'}\right)e^{-R_1 V'} + B\left(1 - \frac{\omega}{R_2 V'}\right)e^{-R_2 V'} + \frac{\omega}{V'}E \tag{2.43}$$

In Eq. (2.43), A, B, R_1, R_2, and ω are constants for a particular explosive, V' is the relative volume (volume of products/volume of undetonated HE), and E is the relative internal energy (energy/volume of undetonated HE). That is,

$$V' = \rho_0/\rho \qquad \text{and} \qquad E = \rho_0 e$$

where ρ_0 is the undetonated density. The constants for many explosives are listed in Table 2. The ideal gas equation of state, $P = (\gamma - 1)\rho e$, is used extensively in hydrocodes to describe inert gases.

The equation of state most commonly used to describe inert solids is the Mie-Gruneisen equation given by

$$P = P_H + \rho\Gamma(e - e_H) \tag{2.44}$$

In Eq. (2.44), P_H and e_H, which are functions of density only, represent the pressure and internal energy on the principal Hugoniot curve, and Γ is the Gruneisen parameter defined by

$$\Gamma = \frac{1}{\rho}\left(\frac{\partial P}{\partial e}\right)_\rho \tag{2.45}$$

From the shock jump relations (Hugoniot relation), the internal energy e_H is related to the pressure P_H by

$$e_H = \frac{1}{2}P_H\frac{\mu}{\rho}$$

where

$$\mu = \rho/\rho_0 - 1$$

The quantities $P_H(\rho)$ and Γ are usually experimentally determined. Some hydrocodes use P_H expressed as a function of μ in the form

$$P_H = C_1\mu + C_2\mu^2 + C_3\mu^3 + \cdots \tag{2.46}$$

where the C_k are constants chosen so that the pressure agrees with experimental measurements. However, many materials follow a linear relation for shock wave velocity U vs particle velocity u_p, or

$$U = c_0 + s u_p \tag{2.47}$$

Table 2 Parameters for the JWL equation of state*

Explosive	ρ_o^a, Mg/m^3	D^b, km/s	E_o^c, GPa	A, GPa	B, GPa	R_1	R_2	ω
Comp B (Grade A)	1.717	7.98	8.50	524.2	7.678	4.20	1.10	0.34
Comp C-4	1.601	8.19	9.00	609.8	12.95	4.50	1.40	0.25
Cyclotol 77/23	1.754	8.25	9.20	603.4	9.924	4.30	1.10	0.35
H-6	1.760	7.47	10.3	758.1	8.513	4.90	1.10	0.20
HMX	1.891	9.11	10.5	778.3	7.071	4.20	1.00	0.30
LX-04-1	1.865	8.47	9.5	849.8	15.277	4.65	1.30	0.35
LX-07	1.865	8.64	10.0	871.0	13.896	4.60	1.15	0.30
LX-09	1.838	8.84	10.5	868.4	18.711	4.60	1.25	0.25
LX-10	1.860	8.82	10.4	880.2	17.437	4.60	1.20	0.30
LX-11	1.875	8.32	9.0	779.1	10.668	4.50	1.15	0.30
LX-13	1.540	7.35	6.6	2714.0	17.930	7.00	1.60	0.35
LX-14-0	1.835	8.80	10.2	826.1	17.240	4.55	1.32	0.36
LX-17-0	1.900	7.60	6.90	446.0	13.399	3.85	1.30	0.46
Nitromethane	1.128	6.28	5.10	209.2	5.689	4.40	1.20	0.30
Octol	1.821	8.48	9.60	748.6	13.380	4.50	1.20	0.38
PBX-9010	1.787	8.39	9.00	581.4	6.801	4.10	1.00	0.35
PBX-9011	1.777	8.50	8.90	634.7	7.998	4.20	1.00	0.30
PBX-9404-3	1.840	8.80	10.20	854.5	20.493	4.60	1.35	0.25
PBX-9407	1.600	7.91	8.60	573.19	14.639	4.60	1.40	0.32
PBX-9501[d]	1.840	8.80	10.12	852.4	18.02	4.60	1.30	0.38
PBX-9502[d]	1.895	7.62	7.07	460.3	9.544	4.00	1.70	0.48
Pentolite	1.670	7.47	8.00	491.1	9.061	4.40	1.10	0.30
PETN	1.770	8.30	10.10	617.0	16.926	4.40	1.20	0.25
Tetryl	1.730	7.91	8.20	586.8	10.671	4.40	1.20	0.28
TNT	1.630	6.93	7.00	371.2	3.231	4.15	0.95	0.30
Deta Sheet C[d]	1.48	7.00	3.69	349.0	4.524	4.10	1.20	0.30

* Most data taken from Ref. 6. (Courtesy of University of California, Lawrence Livermore National Laboratory, and the Department of Energy).

[a] ρ_o = undetonated explosive density.

[b] D = detonation velocity.

[c] E_o = detonation energy.

[d] Data contributed by J. Jacobson, Group M-4, Los Alamos National Laboratory.

where c_0 and s are constants. For these materials, P_H can be expressed as

$$P_H = \frac{\rho_0 \eta c_0^2}{(1 - s\eta)^2} \tag{2.48}$$

where

$$\eta = 1 - \rho_0/\rho$$

Equation (2.48) can be derived from the shock jump equations for mass and momentum and Eq. (2.47).

Table 3 Parameters for the Mie-Gruneisen equation of state

Material	ρ_0, Mg/m^3	c_0, km/s	s	Γ_0
2024 aluminum	2.79	5.33	1.34	2.00
6061 aluminum[a]	2.70	5.29	1.38	2.14
Beryllium	1.85	8.00	1.12	1.16
Copper	8.93	3.94	1.49	1.96
Lexan[a]	1.20	2.33	1.57	1.00
Lucite	1.18	2.26	1.82	0.75
Magnesium[a]	1.78	4.52	1.26	1.43
Nickel[a]	8.86	4.80	1.25	1.50
Stainless steel	7.90	4.57	1.49	2.17
Steel	7.85	3.60	1.90	1.70
Tantalum	16.66	3.41	1.20	1.68
Titanium[a]	4.51	4.70	1.15	1.14
Tungsten	19.24	4.03	1.24	1.54
Uranium[a]	19.00	2.51	1.51	2.03

[a]These values were obtained from J. D. Jacobson, Los Alamos National Laboratory. The remaining values were obtained from Ref. 7. (Reprinted from *High-Velocity Impace Phenomena*, edited by Ray Kinslow. Copyright©1970 by Academic Press, Inc. reprinted with permission of Academic press, Inc.

When one uses the Mie-Gruneisen equation of state, the assumption of constant Γ or constant $\rho\Gamma$ is usually made. If Eq. (2.47) holds, we need values for the constants c_0, s, Γ, and ρ_0 in order to use Eq. (2.44). Values for these constants are given in Table 3 for some materials.

Additional equations of state such as the Osborne[7] and Tillotson[8] equations have also been used extensively in hydrocodes. Several codes also use tabulated data to describe the equation of state.

3. Numerical Solution Techniques

In this section numerical methods for obtaining approximate solutions to initial value problems governed by partial differential equations are discussed. The method most commonly employed by hydrocodes is the method of finite differences. In its simplest form, derivatives occurring in the differential equations are replaced by difference approximations. This replacement produces a system of algebraic equations that can be solved for the dependent variables at the node points, or lattice points, of a finite difference mesh. This technique is illustrated for a single, linear partial differential equation, the one-dimensional heat conduction equation. We will demonstrate that instabilities that completely destroy its accuracy can develop in the numerical solution. Conditions necessary for stable solutions are derived, and the Courant-Friedrichs-Lewy condition is stated. Lax's equivalence theorem, which links stability and convergence for linear systems, is also stated.

3.1 Finite Difference Approximations

The Taylor series expansion can be used to develop difference approximations to partial derivatives. Consider a dependent variable u that is a function of two independent variables x and t; that is,

$$u = u(x, t)$$

where x is the spatial variable and t is time. A mesh or grid of lines is established in the x,t-plane at finite intervals Δx and Δt, as shown in Fig. 11. In this section we assume, for simplicity, that the grid spacing is uniform. Values of u at the lattice points of the grid are denoted by

$$u_j^n = u(j\Delta x, n\Delta t)$$

where j and n are integers indicating line numbers. The value of u at point $(j, n+1)$ can be determined from a Taylor series expansion of u about the point (j, n). This series expansion in time is given by

$$u_j^{n+1} = u_j^n + \left(\frac{\partial u}{\partial t}\right)_j^n \Delta t + \frac{1}{2}\left(\frac{\partial^2 u}{\partial t^2}\right)_j^n \Delta t^2 + O(\Delta t^3) \tag{3.1}$$

where $O(\Delta t^3)$ indicates terms of the order of Δt^3 and higher. Similarly, series expansions about the point (j, n) in the spatial coordinate give

$$u_{j+1}^n = u_j^n + \left(\frac{\partial u}{\partial x}\right)_j^n \Delta x + \frac{1}{2}\left(\frac{\partial^2 u}{\partial x^2}\right)_j^n \Delta x^2 + \frac{1}{6}\left(\frac{\partial^3 u}{\partial x^3}\right)_j^n \Delta x^3 + O(\Delta x^4) \tag{3.2}$$

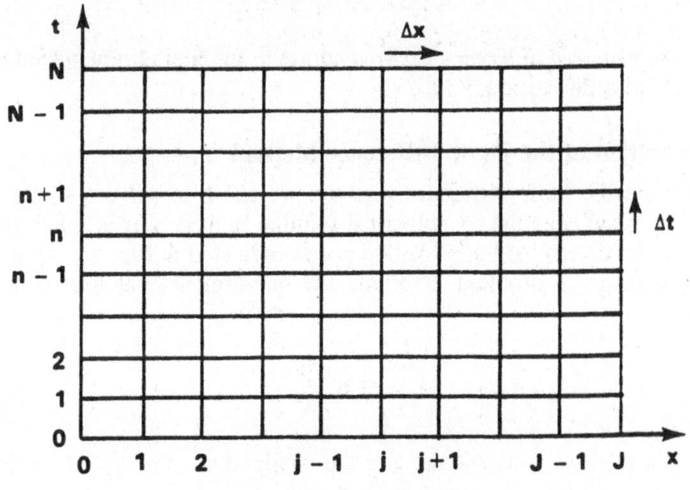

Fig. 11 Finite difference mesh in the x, t-plane.

and

$$u_{j-1}^{n} = u_j^n - \left(\frac{\partial u}{\partial x}\right)_j^n \Delta x + \frac{1}{2}\left(\frac{\partial^2 u}{\partial x^2}\right)_j^n \Delta x^2 - \frac{1}{6}\left(\frac{\partial^3 u}{\partial x^3}\right)_j^n \Delta x^3 + O(\Delta x^4) \quad (3.3)$$

From Eq. (3.1),

$$\frac{u_j^{n+1} - u_j^n}{\Delta t} = \left(\frac{\partial u}{\partial t}\right)_j^n + \frac{1}{2}\left(\frac{\partial^2 u}{\partial t^2}\right)_j^n \Delta t + O(\Delta t^2)$$

Neglecting terms of order Δt and higher, we can write

$$\frac{u_j^{n+1} - u_j^n}{\Delta t} \approx \left(\frac{\partial u}{\partial t}\right)_j^n \quad (3.4)$$

Equation (3.4) gives the *forward difference approximation* to the partial time derivative at point (j, n), which is accurate to first order in Δt. By adding Eqs. (3.2) and (3.3), we obtain

$$\frac{u_{j+1}^n - 2u_j^n + u_{j-1}^n}{\Delta x^2} \approx \left(\frac{\partial^2 u}{\partial x^2}\right)_j^n \quad (3.5)$$

which is the *centered difference approximation* to the second partial spatial derivative of second-order accuracy in Δx. By subtracting Eq. (3.3) from (3.2), we obtain

$$\frac{u_{j+1}^n - u_{j-1}^n}{2\Delta x} \approx \left(\frac{\partial u}{\partial x}\right)_j^n \quad (3.6)$$

which is the *centered difference approximation* to the first partial spatial derivative of second-order accuracy in Δx.

3.2 Illustration of the Finite Difference Method

To illustrate the finite difference method, we use it to solve a single, linear partial differential equation. A numerical solution to time $N\Delta t$ is achieved when values of u are determined at all lattice points indicated in Fig. 11. As a simple example, consider a problem involving the one-dimensional heat conduction equation

$$\frac{\partial u}{\partial t} = \sigma \frac{\partial^2 u}{\partial x^2} \quad (3.7)$$

where σ is a positive constant and u is the temperature. The initial condition is

$$u(x, 0) = f(x)$$

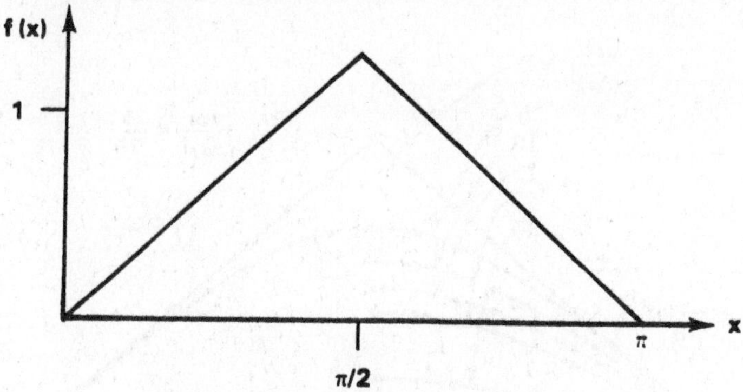

Fig. 12 Initial temperature distribution.

where $f(x)$ is the triangular function indicated in Fig. 12, and the boundary conditions are

$$u(0, t) = u(\pi, t) = 0$$

The finite difference method is based upon replacing the partial derivatives by finite difference approximations. The time derivative in Eq. (3.7) is replaced by the difference approximation Eq. (3.4), and the spatial derivative is replaced by Eq. (3.5). Equation (3.7) then takes the form

$$\frac{u_j^{n+1} - u_j^n}{\Delta t} = \sigma \frac{u_{j+1}^n - 2u_j^n + u_{j-1}^n}{\Delta x^2} \tag{3.8}$$

which is accurate to first order in Δt.

Solving for u_j^{n+1}, we obtain the system of difference equations

$$u_j^{n+1} = u_j^n + \frac{\sigma \Delta t}{\Delta x^2}\left(u_{j+1}^n - 2u_j^n - u_{j-1}^n\right) \tag{3.9}$$

which can be used to determine u_j^{n+1} at node locations j.

The system of Eqs. (3.9) is termed *explicit* since values of u at time $(n+1)\Delta t$ are explicitly given in terms of the values of u at time $n\Delta t$. Thus, Eqs. (3.9) can be used to determine u at time Δt from the known initial values of u. Values of u at the boundary points are already known from the specified boundary conditions. Thus, the solution can be advanced from time zero to time Δt. By a repetition of this process the numerical solution can be advanced in time until the desired time $N\Delta t$ is reached.

The results of this method[9] for several values of dimensionless time $\sigma t/(\Delta x)^2$ are shown in Fig. 13. This numerical solution, which agrees well with the exact solution, was developed with the factor in Eq. (3.9), $\sigma \Delta t/(\Delta x)^2$, equal to 5/11. The results of a second calculation, similar to the first but with the factor

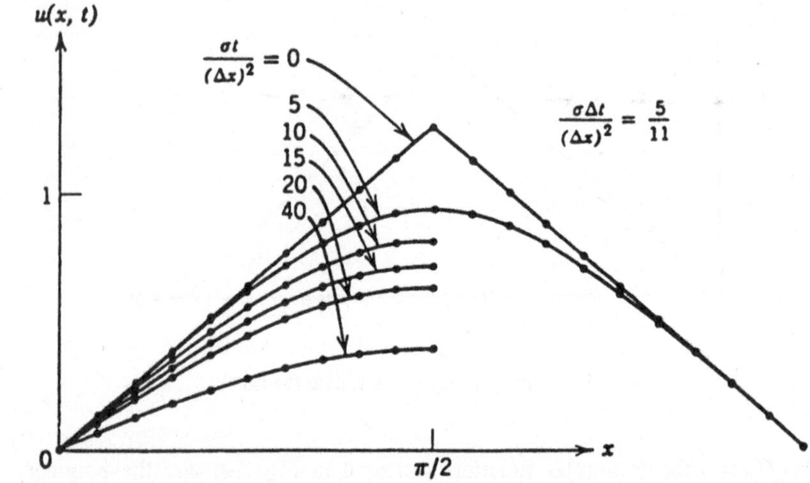

Fig. 13 Numerical solution of the heat conduction problem obtained with $\sigma\Delta t/(\Delta x)^2 = 5/11$ (from Ref. 9). The numerical solution is indicated by dots, and the exact solution is shown by solid lines. (Reprinted from Difference Methods for Initial-Value Problems, by Robert D. Richtmyer and K. W. Morton, 2nd edition. Interscience Publishers Copyright © 1967. Reprinted by John Wiley and Sons, Inc.)

equal to 5/9, are shown in Fig. 14. In this case, a substantial difference exists between the exact and numerical solutions. The numerical solution oscillates with increasing amplitude as the solution is advanced in time. This oscillation completely destroys the accuracy of the numerical solution. The numerical solution shown in Fig. 14 is unstable.

3.3 Stability of Finite Difference Solutions

The differential equation in Eq. (3.7) is mathematically stable. To demonstrate this fact, assume that $u_0(x, t)$ is the exact solution to our problem, and add a small perturbation to the exact solution. Then determine whether the perturbation grows (unstable), decays (stable), or maintains a constant amplitude (neutral stability). Assume that the perturbation is of the form

$$u' = A(t)\cos(kx)$$

where $A(t)$ is a time-dependent amplitude function and k is a constant, the wave number. We seek a solution of the form

$$u(x, t) = u_o(x, t) + A(t)\cos(kx)$$

By substituting this expression in Eq. (3.7), we obtain

$$\frac{\partial u_0}{\partial t} + \frac{dA}{dt}\cos(kx) = \sigma\left\{\frac{\partial^2 u_0}{\partial x^2} - Ak^2\cos(kx)\right\}$$

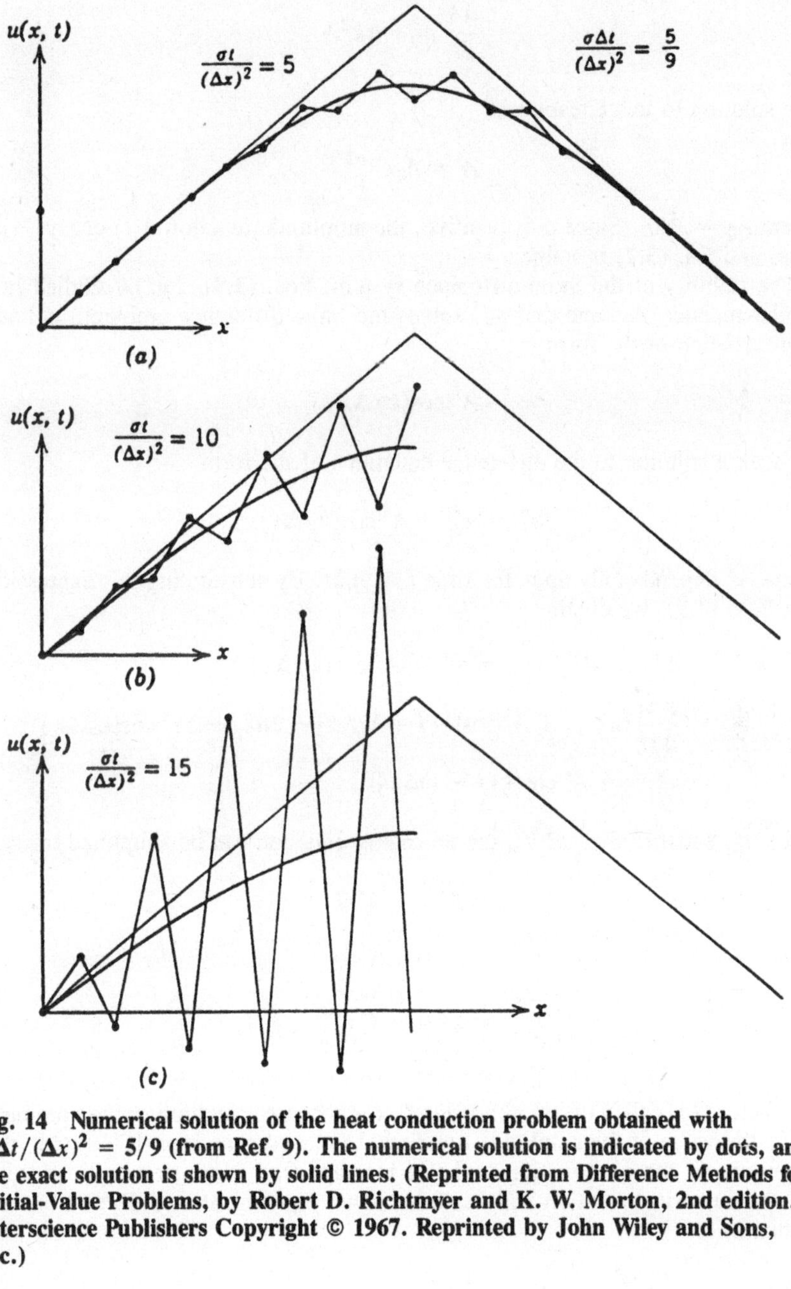

Fig. 14 Numerical solution of the heat conduction problem obtained with $\sigma\Delta t/(\Delta x)^2 = 5/9$ (from Ref. 9). The numerical solution is indicated by dots, and the exact solution is shown by solid lines. (Reprinted from Difference Methods for Initial-Value Problems, by Robert D. Richtmyer and K. W. Morton, 2nd edition. Interscience Publishers Copyright © 1967. Reprinted by John Wiley and Sons, Inc.)

or, since u_0 solves Eq. (3.7),

$$\frac{dA}{dt} = -\sigma k^2 A$$

The solution to this equation is

$$A = A_0 e^{-\sigma k^2 t}$$

where $A_0 = A(0)$. Since σ is positive, the amplitude function $A(t)$ decays with time, and Eq. (3.7) is stable.

The stability of the finite difference system, Eqs. (3.9), can be studied in a similar manner. Assume that u^n_{0j} solves the finite difference problem, and add a perturbation of the form

$$A^n \cos(kj\Delta x)$$

We seek a solution to the difference equations of the form

$$u^n_j = u^n_{0j} + A^n \cos(kj\Delta x)$$

where A^n depends only upon the time $t = n\Delta t$. By substituting this expression into Eqs. (3.9), we obtain

$$\begin{aligned} u^{n+1}_{0j} + A^{n+1}\cos(kj\Delta x) &= u^n_{0j} + A^n\cos(kj\Delta x) \\ &+ \frac{\sigma\Delta t}{\Delta x^2}\{u^n_{0j+1} + A^n\cos[k(j+1)\Delta x] - 2u^n_{0j} - 2A^n\cos(kj\Delta x) \\ &+ u^n_{0j-1} + A^n\cos[k(j-1)\Delta x]\} \end{aligned}$$

Since u^n_{0j} satisfies Eqs. (3.9), the preceding equation can be simplified to

$$A^{n+1} = A^n G$$

where

$$G = 1 + 2\frac{\sigma\Delta t}{\Delta x^2}[\cos(k\Delta x) - 1]$$

The factor G is called the amplification factor because it multiplies A^n to produce A^{n+1}, the value of the amplitude function at the next time step. If $|G| \le 1$, the amplitude of the perturbation will not increase from time step to time step, and the numerical solution will be stable. The requirement that $|G| \le 1$ can be written as

$$-2 \le 2\frac{\sigma\Delta t}{\Delta x^2}[\cos(k\Delta x) - 1] \le 0$$

The term in the brackets varies from 0 to -2 for all values of k. The most stringent requirement for the satisfaction of the inequality occurs when the term in brackets equals -2. For that case,

$$-2 \leq -4\frac{\sigma \Delta t}{\Delta x^2} \leq 0$$

This inequality is satisfied if

$$\frac{\sigma \Delta t}{\Delta x^2} \leq \frac{1}{2} \tag{3.10}$$

Therefore, for a stable numerical solution, inequality (3.10) must be satisfied. The numerical solution illustrated in Fig. 13 was obtained with this inequality satisfied, but the inequality was violated for the solution illustrated in Fig. 14. Inequality (3.10) is thought of as a restriction on the allowable size of the time step. For a fixed spatial mesh size Δx, the relation $\Delta t \leq \Delta x^2/2\sigma$ fixes the size of the maximum time step that can be used with Eq. (3.9).

3.4 Implicit and Explicit Difference Equations

An *explicit* system of difference equations, approximating the differential equation (3.7), was analyzed in the previous section. Instead of Eqs. (3.8), we could also write the difference equation approximating Eq. (3.7) as

$$\frac{u_j^{n+1} - u_j^n}{\Delta t} = \sigma \frac{u_{j+1}^{n+1} - 2u_j^{n+1} + u_{j-1}^{n+1}}{\Delta x^2} \tag{3.11}$$

where the spatial derivative has been approximated at time $(n+1)\Delta t$. Solving Eq. (3.11) for u_j^{n+1}, we obtain the system of difference equations

$$u_j^{n+1} = \frac{1}{1 + \frac{2\sigma\Delta t}{\Delta x^2}} \left[u_j^n + \frac{\sigma \Delta t}{\Delta x^2}(u_{j+1}^{n+1} + u_{j-1}^{n+1}) \right] \tag{3.12}$$

The system of Eqs. (3.12) is termed *implicit* since u_j^{n+1} is expressed in terms of the as yet undetermined values of u at time $(n+1)\Delta t$. This system must be solved by writing equations for each point j, and then simultaneously solving for the values u_j^{n+1}.

The stability of Eqs. (3.12) can be analyzed by the same method used in Sec. 3.3. The amplification factor is

$$G = \frac{1}{1 + (\frac{2\sigma\Delta t}{\Delta x^2})[1 - \cos(k\Delta x)]}$$

In this case, $|G| \leq 1$ is satisfied for all possible values of $\sigma\Delta t/\Delta x^2$, and the system of Eqs. (3.12) is unconditionally stable. Therefore, $\sigma\Delta t/\Delta x^2$ may be as large as desired, and the solution will be stable. However, for large values of Δt, the solution may not be very accurate.

As indicated, implicit methods possess the advantage of stability; therefore, large time steps can be used. However, implicit methods require the solution of a large system of coupled equations at each time step. For implicit methods to be competitive, the increased solution time at each time step must be offset by larger time steps. For problems involving wave propagation where variables change significantly in short periods of time, small time steps are required for accuracy. Therefore, most hydrocodes that are routinely applied to warhead problems use explicit methods.

3.5 Courant-Friedrichs-Lewy Condition

A simple system of equations that is representative of compressible fluid dynamics problems is the system describing the propagation of sound waves. The equations can be obtained by linearizing the momentum and mass equations about a state of static equilibrium. They can be written as

$$\frac{\partial v}{\partial t} + c\frac{\partial \rho}{\partial x} = 0 \qquad \text{and} \qquad \frac{\partial \rho}{\partial t} + c\frac{\partial v}{\partial x} = 0 \tag{3.13}$$

where, in these equations, $\rho = \rho'/\rho_0$ and $v = v'/c$. The primed quantities represent perturbations of density and velocity about the equilibrium state, and the sound speed c is evaluated at the equilibrium state. A difference system appropriate to these equations is given[9] by

$$v_j^{n+1} = v_j^n - \left(\frac{c\Delta t}{\Delta x}\right)(\rho_{j+1/2}^n - \rho_{j-1/2}^n)$$

and

$$\rho_{j-1/2}^{n+1} = \rho_{j-1/2}^n - \left(\frac{c\Delta t}{\Delta x}\right)(v_j^{n+1} - v_{j-1}^{n+1})$$

A stability analysis of this system shows that the solution will be stable if

$$\frac{c\Delta t}{\Delta x} < 1$$

This condition is the Courant-Friedrichs-Lewy (CFL) condition, and the quantity $c\Delta t/\Delta x$ is the Courant number. Since c gives the wave propagation speed, the CFL condition states that Δt should be chosen such that a wave cannot travel across a zone in one time step.

The CFL condition has been applied to more general systems of hyperbolic equations. For these equations, the time steps should be chosen such that the domain of dependence of any mesh point as determined by the difference equations is not less than the domain of dependence determined by the differential equations. Fig. 15 is a geometric illustration of this condition for one-dimensional problems. Two-dimensional problems are discussed in Ref. 10.

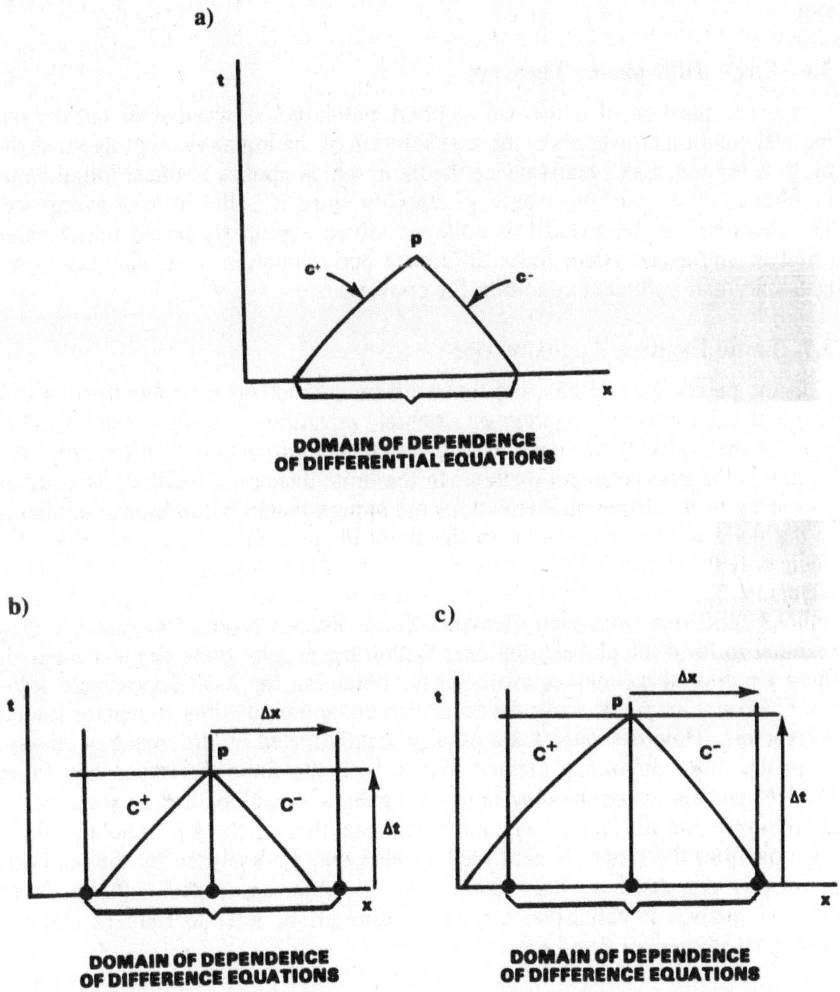

Fig. 15 Geometric illustration of the CFL stability condition: a) domain of dependence of point *P* as determined by the differential equations is illustrated, where C^+ and C^- represent the fastest running right and left characteristics; b) example of a stable calculation according to the CFL condition; c) example of an unstable calculation according to the CFL condition.

Although this section deals with linear equations, the general technique of stability analysis and stability conditions has been applied to the nonlinear case. In practice, the analysis of nonlinear systems is usually performed by first linearizing the equations at a certain value of time. The stability conditions that result are, therefore, approximate but serve as a useful guide for selecting a time step.

3.6 Lax's Equivalence Theorem

A basic question of numerical solution techniques is whether or not the numerical solution converges to the true solution of the initial value problem as the mesh is refined. Lax's equivalence theorem, which applies to linear initial value problems, shows the importance of stability since it is linked to convergence. The theorem can be stated[11] as follows: Given a properly posed initial value problem and a consistent finite difference approximation to it, stability is the necessary and sufficient condition for convergence.

3.7 Finite Element Appoximations

In the preceding sections, the finite difference method for appoximating differential equations by a system of algebraic equations was discussed. Another popular method that has been used in hydrocodes to accomplish this approximation is the finite element method. In the finite difference method, derivatives appearing in the differential equations are approximated by difference quotients at the node points of the grid; in the finite element method, the problem domain is first subdivided into a large number of nonoverlapping subregions or elements. Solutions to the differential equations are then approximated by continuous functions over each element. These local solutions are than patched together to form the global solutions. Within the general finite element method, there are several techniques available for obtaining the local appoximate solutions. Several hydrocodes use the finite element approximations to replace spatial derivatives. Time derivatives are usually approximated by difference quotients, as in the finite difference method. Since both the finite difference and finite element techniques produce systems of algebraic equations that in some sense approximate the differential equations, the stability of the algebraic equations resuting from the finite element method also may be analyzed by the methods previously discussed in this section. The literature associated with the finite element method is extensive; for more information, refer to Ref. 12 and the additional references listed there.

4. Numerical Solution Techniques Used in Hydrocodes

In this section, we review some of the standard numerical methods that have been successfully used in hydrocodes to solve the equations of continuum mechanics discussed in Sec. 2.

4.1 Lagrangian and Eulerian Calculation Grids

Numerical solution techniques have traditionally used either Lagrangian or Eulerian grids. These grids or meshes are illustrated in Fig. 16 for a simple two-

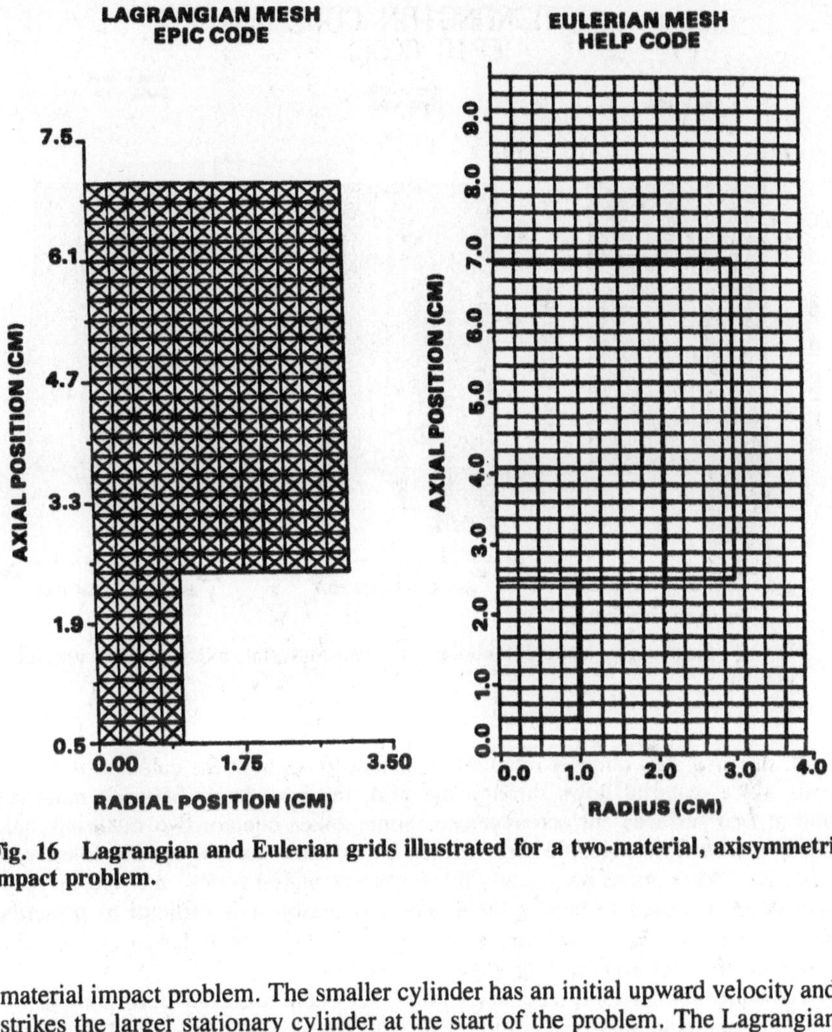

Fig. 16 Lagrangian and Eulerian grids illustrated for a two-material, axisymmetric impact problem.

material impact problem. The smaller cylinder has an initial upward velocity and strikes the larger stationary cylinder at the start of the problem. The Lagrangian grid, taken from an EPIC-2 calculation,[13] is made up of triangular zones or finite elements. The lattice points or nodes of this grid are moved with the local particle velocity. As the solution is advanced in time, the grid moves and deforms with the material. The solution, advanced to 20 μ, is shown in Fig. 17. Severe grid distortions have developed in the neighborhood of the interface between target and penetrator. These distortions decrease the zone width, and, because of the stability condition, the allowable time step may eventually become prohibitively small. In order to continue the solution, some rezoning technique must be used to improve the zone shape so that larger time steps can be used.

An Eulerian grid, taken from a HELP[14] code calculation, is also illustrated in Fig. 16. Since this grid remains fixed in space, the zone distortion problem is eliminated. Fig. 18 shows this solution advanced to 60 μ. With the Eulerian

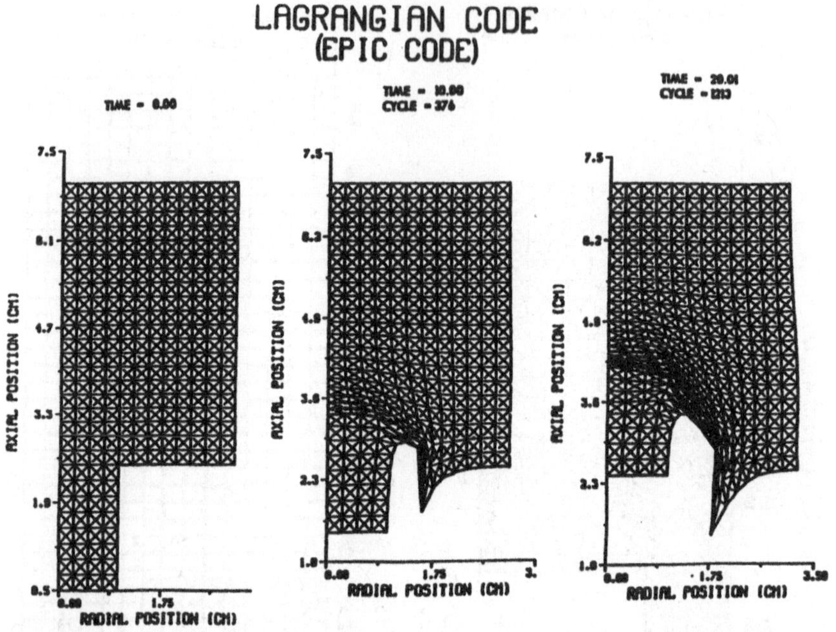

Fig. 17 Lagrangian code calculation of a two-material, axisymmetric impact problem.

grid, the time step remains relatively constant throughout the calculation. However, since material flows through the grid, the boundaries between materials and at free surfaces cut across zones. Some zones contain two materials, and some zones, which contain free surface boundaries, are only partly filled with material. These zones are usually referred to as mixed zones, and special techniques are required to handle them. For this reason it is difficult to prescribe the precise boundary condition at material interfaces in Eulerian codes. This problem does not arise in Lagrangian codes.

For conventional solid mechanics problems where moderate deformations occur, a Lagrangian approach is more efficient; however, for fluid dynamics problems where extreme deformations are common, an Eulerian approach appears more efficient. Warhead mechanics problems fall somewhere in between, and the two approaches are competitive.

Two additional zoning approaches are the Arbitrary-Lagrangian-Eulerian (ALE) technique[15] and the Free-Lagrangian method.[16] The ALE technique uses a mesh with lattice points that can be moved with the fluid (Lagrangian), held fixed (Eulerian), or moved in any other prescribed way. This mesh flexibility allows some problems to be completed where fully Lagrangian treatments are not possible. The advantage of the Free-Lagrangian method is that mesh points are not permanently linked by one mesh for the duration of the calculation, but as each cycle begins the mesh can be reestablished if necessary.

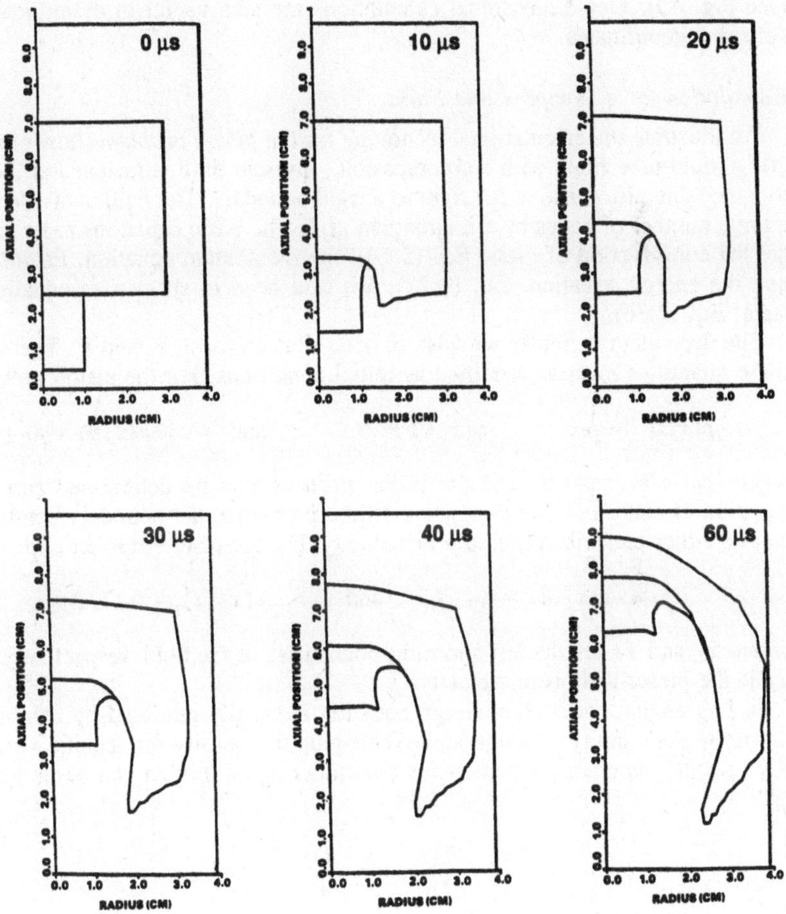

Fig. 18 Eulerian code calculation of a two-material, axisymmetric impact problem.

4.2 Lagrangian Finite Difference Method for One-Dimensional Motion

Although we are primarily interested in two- and three-dimensional problems, it is instructive to discuss the simpler one-dimensional problem to establish clearly the calculation method. One-dimensional calculations in Cartesian coordinates can be used to model problems in slab geometry where the transverse dimensions are much larger than the dimension in the direction of motion (see Fig. 19). One-dimensional calculations are also useful in cylindrical and spherical coordinates.

Calculation for a Compressible Fluid

We illustrate the calculational technique for the *piston problem* shown in Fig. 20. A rigid tube filled with a compressible, inviscid fluid is fastened at the left to a movable piston and at the right to a rigid boundary. The fluid is divided into a large number of zones by a Lagrangian grid. The basic equations to be solved are the conservation of mass, Eq. (2.14), the momentum equation, Eq. (2.17), and the energy equation, Eq. (2.22). We also need to specify an equation of state, Eq. (2.36).

The dependent variables we wish to determine are ρ, v, e, and P. Therefore, these quantities must be specified as initial conditions. For the piston problem,

$$\rho(x,0) = \rho_0, \qquad v(x,0) = 0, \qquad \text{and} \qquad e(x,0) = 0$$

where ρ_0 is a constant, and the initial pressure can be determined from the equation of state. For one-dimensional fluid problems, the boundary conditions can be either prescribed pressure or velocity. For the piston problem, we set

$$v(x_L, t) = v_p \qquad \text{and} \qquad v(x_R, t) = 0$$

where x_L and x_R are the left and right boundaries of the fluid, respectively, and v_p is the prescribed piston velocity.

In Lagrangian codes, the energy equation is usually modified by combining it with the continuity equation and exchanging the density for specific volume. As a result, the energy equation for one-dimensional motion can be expressed as

$$\frac{De}{Dt} = \sigma_{xx} \frac{DV}{Dt}$$

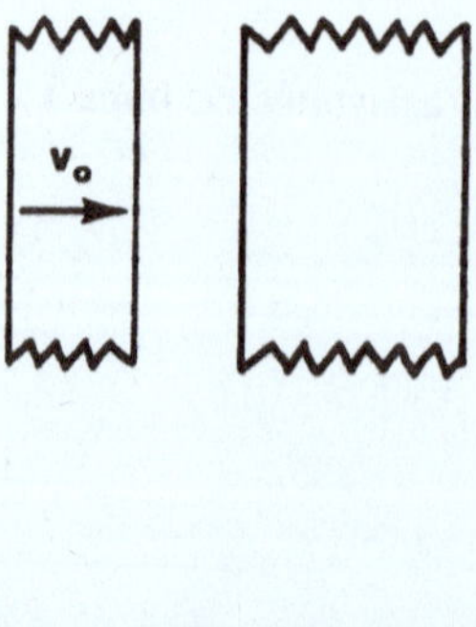

Fig. 19 One-dimensional impact in slab geometry.

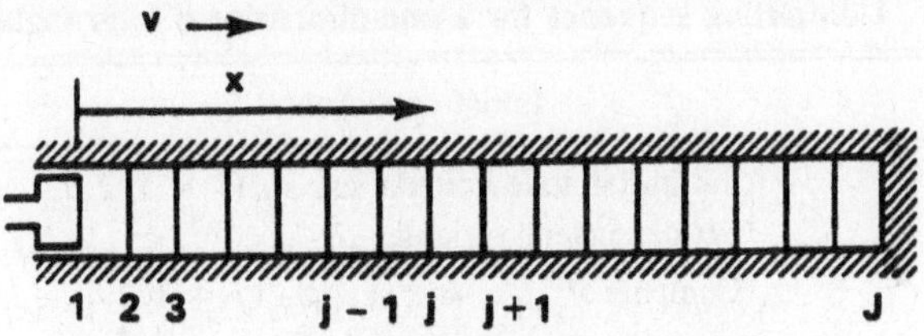

Fig. 20 Notation for the *piston problem.*

where $\sigma_{xx} = -P$ for the inviscid fluid case, and $V = 1/\rho$ is the specific volume.

The dependent variables, ρ, P, and e are associated with the zone centers $j + \frac{1}{2}$, and the variables x and v are associated with the zone boundaries j as illustrated in Fig. 21. The typical computing sequence is summarized in Table 4. The initial values of the variables are usually computed in a *setup* routine of the hydrocode. The mass of each zone $m_{j+1/2}$ is computed from

$$m_{j+1/2} = \rho^0_{j+1/2}(x^0_{j+1} - x^0_j)$$

where, in a one-dimensional calculation, $m_{j+1/2}$ is the mass per unit cross-sectional area. Since zone boundaries move with the material, conservation of mass requires that

$$m_{j+1/2} = \text{constant}$$

for each zone. After the setup of the problem is completed, the cycle calculation uses the values of the variables at time step n to produce values at the next time step, $n + 1$. The cycle calculation is, therefore, used repetitively to advance the solution in time until the final problem time is reached. The differential equations and the difference approximations to them that are used in the cycle calculation are given in Table 5. In computer code calculations, the values of the variables are usually output at preselected values of time.

In Table 5, central difference approximations have generally been used in order to obtain second-order accuracy. Because all variables are referred to the Lagrangian grid, the material derivative can be directly differenced as

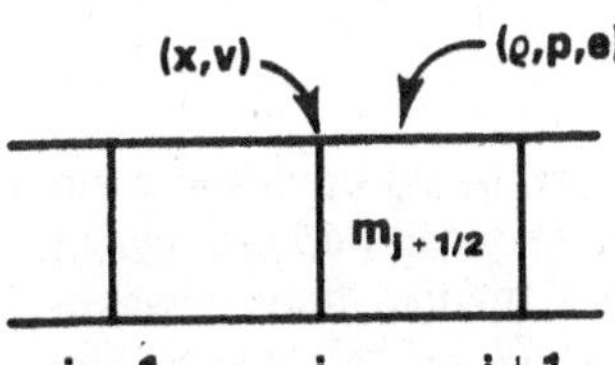

Fig. 21 Dependent variables referenced to the Lagrangian grid. The variables x and v are associated with the zone boundaries, and the remaining variables are associated with the zone centers.

Table 4 Computing sequence for a one-dimensional Lagrangian calculation

	Initial conditions
I.1	Set initial zone boundaries, x_j ($j = 1, 2, \ldots, J$)
I.2	Set dependent variables $\rho^0_{j+1/2}$, v^0_j, $e^0_{j+1/2}$ ($j = 1, 2, \ldots, J$-1)
I.3	Compute $P^0_{j+1/2}$ and $m_{j+1/2}$ ($j = 1, 2, \ldots, J$-1)
I.4	Choose initial time steps Δt^0 and $\Delta t^{1/2}$
	Cycle calculation
C.1	Compute new value of $v^{n+1/2}_j$ from momentum equation
C.2	Compute new value of x^{n+1}_j from definition of velocity
C.3	Compute new value of $\rho^{n+1}_{j+1/2}$ from mass conservation
C.4	Compute new value of $V^{n+1}_{j+1/2}$ from $V = 1/\rho$
C.5	Compute new value of $q^{n+1/2}_{j+1/2}$ from viscosity equation
C.6	Compute new value of $e^{n+1}_{j+1/2}$ from energy equation
C.7	Compute new value of $P^{n+1}_{j+1/2}$ from equation of state
C.8	Compute new value of $c^{n+1}_{j+1/2}$ from equation of state
C.9	Compute new value of $\Delta t^{n+3/2}$ from Courant condition

$$\frac{Df}{Dt} = \frac{f^{n+1} - f^n}{\Delta t^{n+1/2}}$$

where f is any variable and Df/Dt is evaluated at time $n+1/2$. Note that, in the finite difference equations of Table 5, v is evaluated at the half time step $n+1/2$. In Lagrangian calculations, two time steps are used to advance the dependent variables. The time step $\Delta t^n = t^{n+1/2} - t^{n-1/2}$ is used to advance the velocity from time $n - 1/2$ to time $n + 1/2$, and the time step $\Delta t^{n+1/2} = t^{n+1} - t^n$ is used to advance the position coordinate of zone boundaries from time n to time $n + 1$. In Eq. (C.1), the quantity $\rho^n_j(x^n_{j+1/2} - x^n_{j-1/2})$ represents the mass of material between the zone centers $x^n_{j+1/2}$ and $x^n_{j-1/2}$ as indicated in Fig. 22. This quantity m_j remains constant in time and is computed from

$$m_j = \frac{1}{2}(m_{j+1/2} + m_{j-1/2})$$

Artificial Viscosity

The differential equations presented in Sec. 2 are based upon the assumption that the variables are continuous. However, in most warhead dynamics problems, shock waves are present. Shock waves involve finite changes in the variables over small spatial dimensions. In analytical treatments, shocks are usually approximated by discontinuities in the flow field since their thickness

Table 5 Differential equations and finite difference approximations used in a one-dimensional Lagrangian calculation

Differential equations	Difference approximations	
Momentum equation		
$\frac{Dv}{Dt} = \frac{1}{\rho}\frac{\partial \sigma_x x}{\partial x}$	$\frac{v_j^{n+1/2} - v_j^{n-1/2}}{\Delta t^n} = \frac{(\sigma_{xx})_{j+1/2}^n - (\sigma_{xx})_{j-1/2}^n}{\rho_j^n\left(x_{j+1/2}^n - x_{j-1/2}^n\right)}$ $= \frac{(\sigma_{xx})_{j+1/2}^n - (\sigma_{xx})_{j-1/2}^n}{m_j}$	(C.1)
Velocity		
$\frac{Dx}{Dt} = v$	$\frac{x_j^{n+1} - x_j^n}{\Delta t^{n+1/2}} = v_j^{n+1/2}$	(C.2)
Conservation of mass		
$m_{j+1/2} = \text{constant}$	$\rho_{j+1/2}^{n+1} = \frac{m_{j+1/2}}{x_{j+1}^{n+1} - x_j^{n+1}}$	(C.3)
Specific volume		
$V = 1/\rho$	$V_{j+1/2}^{n+1} = 1/\rho_{j+1/2}^{n+1}$	(C.4)
Artificial viscosity		
$q = B_2^2\rho\left(\frac{\partial v}{\partial x}\right)^2(\Delta x)^2 - B_1 c\rho\left(\frac{\partial v}{\partial x}\right)\Delta x$	$q_{j+1/2}^{n+1/2} = B_2^2\rho_{j+1/2}^{n+1/2}(\Delta v)^2 - B_1 c_{j+1/2}^n \rho_{j+1/2}^{n+1/2}(\Delta v)$ if $\Delta v < 0$ $= 0$ if $\Delta v > 0$ where $\Delta v = v_{j+1}^{n+1/2} - v_j^{n+1/2}$	(C.5)
Conservation of energy		
$\frac{De}{Dt} = \sigma_{xx}\frac{DV}{Dt}$	$e_{j+1/2}^{n+1} - e_{j+1/2}^n = (\sigma_{xx})_{j+1/2}^n\left(V_{j+1/2}^{n+1} - V_{j+1/2}^n\right)$	(C.6)
Equation of state		
$P = P(\rho, e)$	$P_{j+1/2}^{n+1} = P\left(\rho_{j+1/2}^{n+1}, e_{j+1/2}^{n+1}\right)$	(C.7)
$c = c(\rho, e)$	$c_{j+1/2}^{n+1} = c\left(\rho_{j+1/2}^{n+1}, e_{j+1/2}^{n+1}\right)$	
Stress		
$\sigma_{xx} = -(P + q)$	$(\sigma_{xx})_{j+1/2}^{n+1} = -\left(P_{j+1/2}^{n+1} + q_{j+1/2}^{n+1/2}\right)$	(C.8)
Courant condition		
$\Delta t \leq \Delta x / c$	$\Delta t_{j+1/2}^{n+3/2} = K_0 \frac{x_{j+1}^{n+1} - x_j^{n+1}}{c_{j+1/2}^{n+1}}$	(C.9)

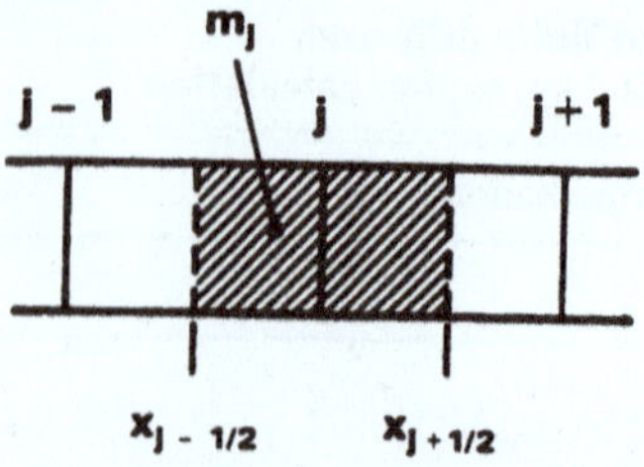

Fig. 22 Mass m_j associated with the zone boundary j.

is usually small compared to other dimensions of the problem. In some numerical methods,[10] shocks are modeled as discontinuities; however, this treatment increases the complexity of the computer program. Therefore, most production codes use the *artificial viscosity* method introduced by von Neumann and Richtmyer.[17] This technique uses an artificially large viscous pressure to spread the shock front over a few computational zones, typically four to five zones. If a physical value of viscosity were to be used in the numerical calculation, an inordinate number of zones would be required for adequate resolution of the shock. For methods that treat the shock as a continuous wave, some viscosity is required to model the irreversible shock process correctly.

In Eq. (C.5) of Table 5, the form of artificial viscosity used in many hydrocodes is presented. This form consists of a term that is quadratic in the velocity gradient and a term that is linear. The quantities B_1 and B_2 are constants, and q is the viscous pressure, which, as indicated by Eq. (C.8), is added to the thermodynamic pressure. Values of $B_1 = 1/2$ and $B_2 = 2$ have been recommended by Wilkins.[18] Also, q is set equal to zero in all zones that are in a state of expansion ($\Delta v > 0$).

Energy Equation

In Eq. (C.6), the stress σ_{xx}, which is the negative of the pressure for the fluid example, should be expressed at time $n + (1/2)$ rather than at time n. This centering, which is required for second-order accuracy, can be conveniently accomplished if the equation of state is of the form

$$P = A(\rho) + B(\rho) \cdot e$$

where $A(\rho)$ and $B(\rho)$ are functions of density. Following Wilkins,[18] the energy equation and the equation of state can be written as

$$e^{n+1}_{j+1/2} - e^{n}_{j+1/2} = \frac{1}{2}(-P^{n+1} - q^{n+1/2} - P^{n} - q^{n-1/2})_{j+1/2}\,\Delta V$$

$$P^{n+1}_{j+1/2} = A(\rho^{n+1}_{j+1/2}) + B(\rho^{n+1}_{j+1/2})e^{n+1}_{j+1/2}$$

where

$$\Delta V = V^{n+1}_{j+1/2} - V^{n}_{j+1/2}$$

The preceding two equations can be solved for the new value of internal energy. The result is

$$e^{n+1}_{j+1/2} = \frac{e^n_{j+1/2} - \frac{1}{2}[A(\rho^{n+1}_{j+1/2}) + P^n_{j+1/2} + q^{n+1/2}_{j+1/2} + q^{n-1/2}_{j+1/2}]\Delta V}{1 + \frac{1}{2}B(\rho^{n+1}_{j+1/2})\Delta V} \tag{4.1}$$

Most codes use this difference equation in place of Eq. (C.6) when the equation of state is linear in internal energy. To achieve second-order accuracy for more general equations of state, an iteration process can be used to solve for new values of internal energy and pressure.

The sound speed c in Eq. (C.7) can be evaluated from the equation of state by the formula

$$c^2 = \left(\frac{\partial P}{\partial \rho}\right)_s$$

where s is entropy. To compute the sound speed, the equation of state must first be expressed as a function of density and entropy in the form

$$P = P(\rho, e) = P[\rho, e(\rho, s)]$$

The sound speed squared is then computed from

$$c^2 = \left(\frac{\partial P}{\partial \rho}\right)_e + \left(\frac{\partial P}{\partial e}\right)_\rho \left(\frac{\partial e}{\partial \rho}\right)_s$$

From the first law of thermodynamics

$$\left(\frac{\partial e}{\partial \rho}\right)_s = P/\rho^2$$

therefore, the sound speed can be calculated from

$$c^2 = \left(\frac{\partial P}{\partial \rho}\right)_e + \frac{P}{\rho^2}\left(\frac{\partial P}{\partial e}\right)_\rho$$

for any equation of state. The sound speed c is sometimes called the bulk sound speed.

The time step, $\Delta t^{n+3/2}$, to be used as $\Delta t^{n+1/2}$ in the next calculational cycle is taken to be the minimum, over all zones, of $\Delta t^{n+3/2}_{j+1/2}$ computed from Eq. (C.9).

Also,

$$\Delta t^{n+1} = \frac{1}{2}(\Delta t^{n+3/2} + \Delta t^{n+1/2})$$

The constant K_o in Eq. (C.9) is chosen to be less than unity to satisfy the Courant condition. When artificial viscosity is active in the calculation (regions undergoing compression), the Courant condition must be modified. In Ref. 19, the time step limitation, which can be used in place of Eq. (C.9), is given by

$$\Delta t \le \frac{x_{j+1} - x_j}{B_1 c - B_2^2(\Delta v) + [(B_1 c - B_2^2(\Delta v))^2 + c^2]^{1/2}} \qquad \text{for} \qquad \Delta v < 0$$

where

$$\Delta v = v_{j+1}^{n+1/2} - v_j^{n+1/2}$$

The time step increase on one cycle is usually limited to about 10% of the preceding time step.

Numerical Results

To illustrate the effect of artificial viscosity, a numerical calculation has been performed using the procedure indicated in Tables 4 and 5, but the artificial viscous pressure q has been set to zero. The results are shown in Fig. 23 for both a rarefaction wave (piston moved to the left) and a shock wave (piston moved to the right). In Fig. 23 we see the rarefaction wave is adequately represented by the numerical solution, but, as shown in Fig. 23, the shock wave propagation is incorrectly simulated. The solution to the same two wave propagation problems, with the artificial pressure included, is shown in Fig. 24. In this case, both waves are correctly simulated.

Calculation of Elastic-Plastic Flow

For materials with strength, additional calculations for the deviatoric stress components must be inserted between steps C.4 and C.5 of Table 4. For simplicity, these calculations are illustrated for a material with a constant yield strength.

The deformation rate components are calculated from Eq. (2.7). For the one-dimensional case considered,

$$d_{xx} = \frac{\partial v}{\partial x}$$

and all other components are zero. The difference approximation is

$$(d_{xx})_{j+1/2}^{n+1/2} = \frac{v_{j+1}^{n+1/2} - v_j^{n+1/2}}{\frac{1}{2}(x_{j+1}^{n+1} + x_{j+1}^n) - \frac{1}{2}(x_j^{n+1} + x_j^n)}$$

The deviatoric components d'_{ij} are calculated from

$$d'_{ij} = d_{ij} - \frac{1}{3} d_{kk} \delta_{ij}$$

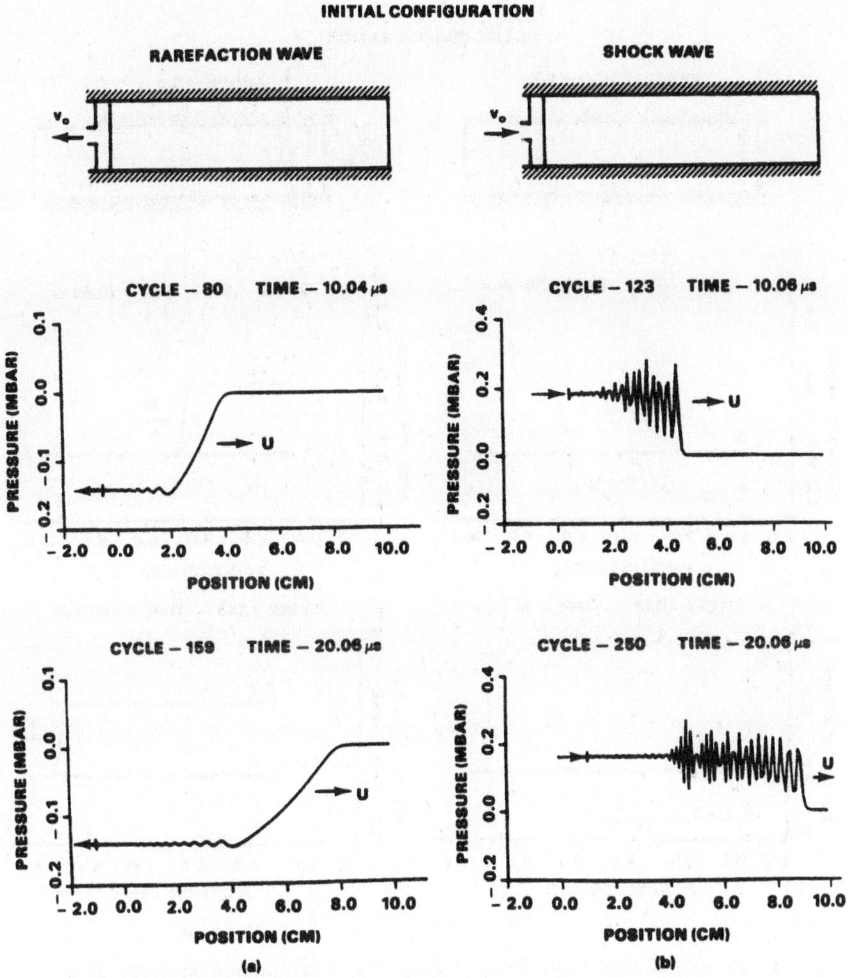

Fig. 23 Results of a numerical calculation of a) a rarefaction wave and b) a compression wave (shock) obtained from the Lagrangian technique without the use of artificial viscosity (ν_C = constant).

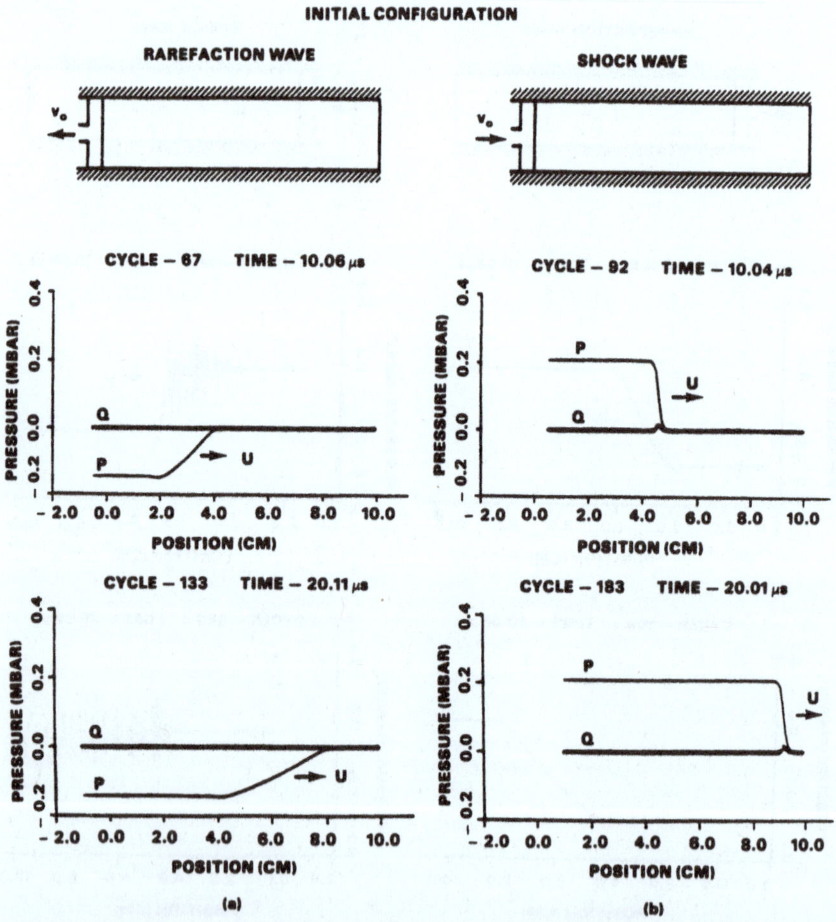

Fig. 24 Results of a numerical calculation of a) a rarefaction wave and b) a compression wave (shock) obtained from the Lagrangian technique using artificial viscosity (Q is the artificial viscous pressure, and v_o = constant).

For the one-dimensional case, $d_{kk} = d_{xx}$; therefore,

$$d'_{xx} = d_{xx} - \frac{1}{3}d_{xx} = \frac{2}{3}d_{xx}$$

$$d'_{yy} = -\frac{1}{3}d_{xx}$$

$$d'_{zz} = -\frac{1}{3}d_{xx}$$

and all other components are zero. It is first assumed that the deformation from time step n to $n+1$ is elastic. The stress deviator components can then be calculated from Eq. (2.37). Since rotation cannot occur in one-dimensional motion,

$$\overset{\nabla}{\sigma}{}'_{ij} = \frac{\mathrm{D}}{\mathrm{D}t}\sigma'_{ij}$$

and, therefore,

$$\frac{\mathrm{D}}{\mathrm{D}t}\sigma'_{ij} = 2G\mathrm{d}'_{ij}$$

The difference approximation to this equation is written as

$$\left({}^{T}\sigma'_{xx}\right)^{n+1}_{j+1/2} = \left(\sigma'_{xx}\right)^{n}_{j+1/2} + 2G\left(d'_{xx}\right)^{n+1/2}_{j+1/2}\Delta t^{n+1/2}$$

where the superscript T implies a tentative value based on the elastic assumption. The value of the yield function is computed from Eq. (2.33). For the one-dimensional case,

$$J_2 = \frac{1}{2}\left[\left(\sigma'_{xx}\right)^2 + \left(\sigma'_{yy}\right)^2 + \left(\sigma'_{zz}\right)^2\right]$$

however, by symmetry, $\sigma'_{yy} = \sigma'_{zz}$, and, from the definition of deviatoric components, $\sigma'_{xx} + \sigma'_{yy} + \sigma'_{zz} = 0$. Therefore,

$$J_2 = \frac{3}{4}(\sigma'_{xx})^2$$

and, at time $n+1$,

$$(J_2)^{n+1}_{j+1/2} = \frac{3}{4}\left({}^{T}\sigma'^{\,2}_{xx}\right)^{n+1}_{j+1/2}$$

If

$$f^{n+1}_{j+1/2} \leq 0$$

where

$$f_{j+1/2}^{n+1} = (J_2)_{j+1/2}^{n+1} - k^2$$

the material in zone $j + 1/2$ is elastic, and

$$\left(\sigma'_{xx}\right)_{j+1/2}^{n+1} = \left({}^{T}\sigma'_{xx}\right)_{j+1/2}^{n+1}$$

If

$$f_{j+1/2}^{n+1} > 0$$

the material in zone $j + 1/2$ is plastic, and the tentative value of stress is too large in magnitude and must be adjusted to lie on the yield surface $f = 0$. Therefore,

$$\sigma'_{xx} = \pm\sqrt{\frac{4}{3}}k,$$

and the new stress component is calculated from

$$(\sigma'_{xx})_{j+1/2}^{n+1} = \operatorname{sgn}\left({}^{T}\sigma'_{xx}\right)_{j+1/2}^{n+1}\sqrt{\frac{4}{3}}k$$

The energy equation Eq. (4.1) must be modified to include the work done by the deviator stress component, σ'_{xx}. The modified equation is

$$e_{j+1/2}^{n+1} = \frac{e_{j+1/2}^{n} - 1/2\left[A\left(\rho_{j+1/2}^{n+1}\right) + p_{j+1/2}^{n} + q_{j+1/2}^{n+1/2} + q_{j+1/2}^{n-1/2} - (\sigma'_{xx})_{j+1/2}^{n+1} - (\sigma'_{xx})_{j+1/2}^{n}\right]\Delta V}{[1 + \frac{1}{2}B(\rho_{j+1/2}^{n+1})\Delta V]}$$

Equation (C.8) in Table 5 must be modified to include the stress deviator; therefore, the equation is written as

$$(\sigma_{xx})_{j+1/2}^{n+1} = -(P_{j+1/2}^{n+1} + q_{j+1/2}^{n+1/2} - (\sigma'_{xx})_{j+1/2}^{n+1})$$

When strength is included in the calculation, the time step calculation should use the longitudinal sound speed c_L in place of the bulk sound speed c. The sound speed c_L is given by

$$c_L^2 = -\left(\frac{\partial \sigma_{xx}}{\partial \rho}\right)_s = c^2 + \frac{4}{3}\frac{G}{\rho}$$

In the preceding calculation for the deviatoric stress component, the flow rule, Eq. (2.34), did not have to be used. This simplification results in one-dimensional motion because the yield function can be expressed in terms of only one deviatoric stress component σ'_{xx}, since the transverse stresses are equal. However, for two- or three-dimensional motion the flow rule must be used to determine the state of stress. For one-dimensional motion in the radial direction in spherical coordinates, this simplification also occurs; however, in cylindrical coordinates, it does not.

The results of these calculations applied to the piston problem with an elastic-perfectly plastic material are illustrated in Fig. 25. In this case a two-wave structure consisting of an elastic wave propagating with velocity U_e followed by

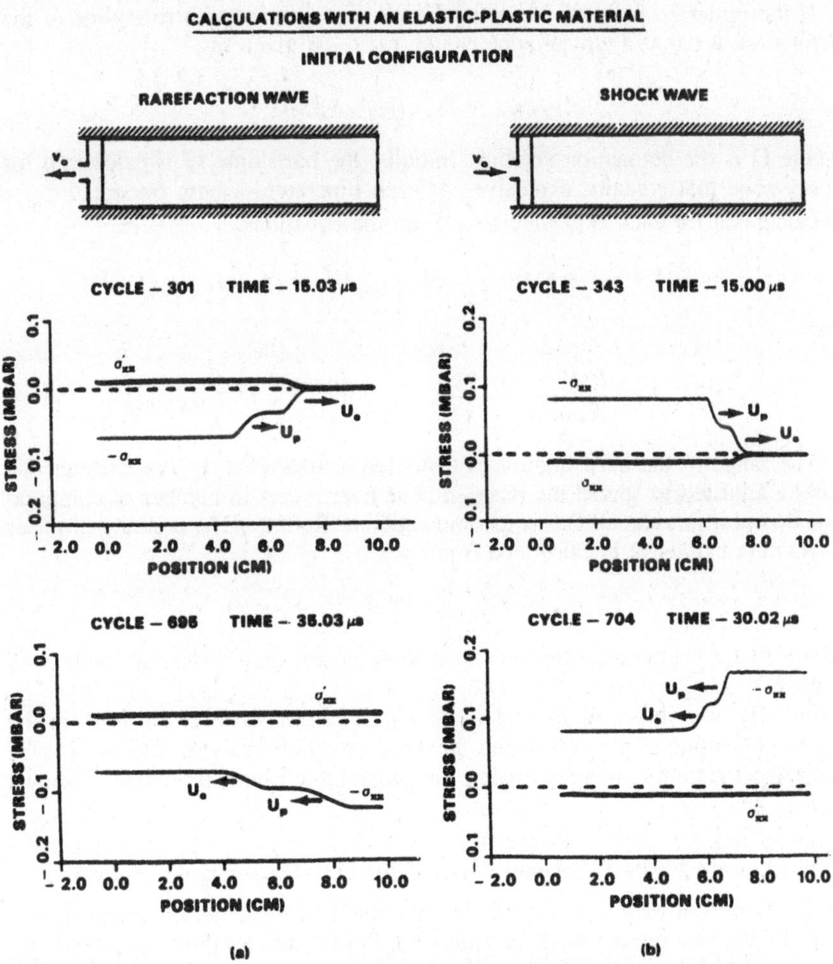

Fig. 25 Results of a numerical calculation of a) a rarefaction wave and b) a compression wave obtained from the Lagrangian technique for an elastic-perfectly plastic material.

a plastic wave propagating with velocity U_p develops. This numerical solution agrees well with the exact solution to the problem.

Detonation of High Explosives

Many warhead problems involve the detonation of high explosives. The *programmed burn* technique is a simple method that is frequently used in hydrocode calculations to simulate detonations. We assume that the pressure is zero in the undetonated explosive; therefore, an equation of state for the unreacted explosive is not required. The detonation wave is programmed to propagate at the Chapman-Jouguet detonation velocity.

If the initiation point of the detonation is at x_D, then the arrival time of the detonation wave at a typical zone center $x_{j+1/2}$ is given by

$$t_{Bj+1/2} = |x^0_{j+1/2} - x_D|/D$$

where D is the detonation velocity. Initially, the burn time t_B is calculated for every zone that contains explosive. At each time step, a burn fraction $F^n_{j+1/2}$ is calculated for each explosive zone from the equations

$$F^n_{j+1/2} = 0 \qquad \text{if} \qquad t^n < t_{Bj+1/2}$$

and

$$F^n_{j+1/2} = \frac{D(t^n - t_{Bj+1/2})}{K_B(x^n_{j+1} - x^n_j)} \qquad \text{if} \qquad t^n > t_{Bj+1/2}$$

The range of the burn fraction is restricted to $0 \leq F \leq 1$. The constant K_B can be adjusted to spread the reaction zone over a certain number of computational zones. A value of 2.5 is recommended in Ref. 19. The pressure in zones containing explosive is calculated from

$$P^n_{j+1/2} = F^n_{j+1/2} \cdot P(\rho^n_{j+1/2}, e^n_{j+1/2})$$

where $P(\rho, e)$ is the pressure calculated from the equation of state of the reacted explosive.

Initially, the values of ρ_0 and $e_0 = E_0/\rho_0$, from Table 2, are used in the zones containing explosives when the JWL equation is used. The result of a typical detonation calculation using the programmed burn technique is shown in Fig. 26.

4.3 Eulerian Finite Difference Method for One-Dimensional Motion

In this section we illustrate the Eulerian finite difference method by applying it to a simple, one-dimensional, compressible fluid dynamics problem. The donor cell version of this method will be illustrated since it was one of the first Eulerian techniques to be successfully applied to warhead dynamics problems.[14]

The equations expressing conservation of mass, Eq. (2.14), conservation of linear momentum, Eq. (2.17), and conservation of energy, Eq. (2.21), can be written for one-dimensional motion in Cartesian coordinates as

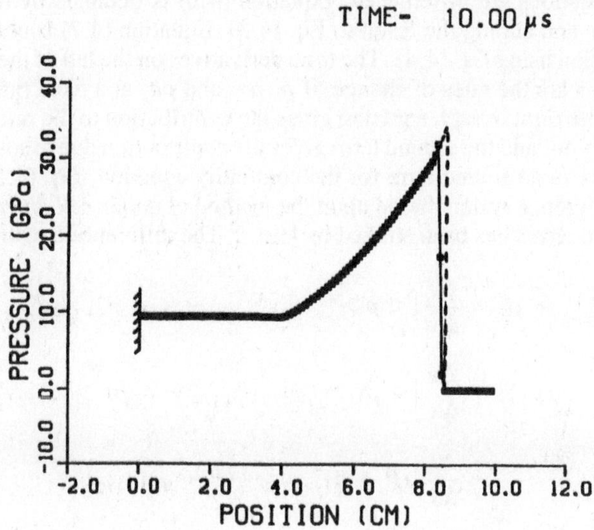

Fig. 26 Numerical simulation of a detonation wave obtained by the Lagrangian technique (100 zones). ———Numerical result,---------Exact result.

$$\frac{\partial \rho}{\partial t} + \frac{\partial}{\partial x}(\rho v) = 0 \tag{4.2}$$

$$\rho\left(\frac{\partial v}{\partial t} + v\frac{\partial v}{\partial x}\right) = -\frac{\partial P}{\partial x} \tag{4.3}$$

and

$$\rho\left(\frac{\partial e_t}{\partial t} + v\frac{\partial e_t}{\partial x}\right) = -\frac{\partial}{\partial x}(Pv) \tag{4.4}$$

where the material derivative has been expanded according to Eq. (2.6) and $\sigma_{xx} = -P$. For many Eulerian calculations, these equations are rearranged into the *conservative* form

$$\frac{\partial \rho}{\partial t} = -\frac{\partial}{\partial x}(\rho v) \tag{4.5}$$

$$\frac{\partial}{\partial t}(\rho v) = -\frac{\partial}{\partial x}(\rho v^2) - \frac{\partial P}{\partial x} \tag{4.6}$$

and

$$\frac{\partial}{\partial t}(\rho e_t) = -\frac{\partial}{\partial x}(\rho v e_t) - \frac{\partial}{\partial x}(Pv) \tag{4.7}$$

before the equations are differenced. Equation (4.6) is obtained by multiplying Eq. (4.2) by v and adding the result to Eq. (4.3). Equation (4.7) is obtained by a similar operation using Eq. (4.4). The time derivatives on the left of the preceding three equations are the rates of change of ρ, ρv, and ρe_t at a fixed position. The first term on the right in each equation gives the contribution to the rate of change due to convection, and the second term gives the contribution due to source terms. Note that there is no source term for the continuity equation, Eq. (4.5).

A finite difference system based upon the method of *donor cell differencing* for the convection terms has been studied by Hirt.[20] The difference equations are

$$\rho_j^{n+1} = \rho_j^n - \frac{\Delta t}{\Delta x}[<\rho>_{j+1/2}^n v_{j+1/2}^n - <\rho>_{j-1/2}^n v_{j-1/2}^n] \quad (4.8)$$

$$(\rho v)_j^{n+1} = (\rho v)_j^n - \frac{\Delta t}{\Delta x}[<\rho v>_{j+1/2}^n v_{j+1/2}^n - <\rho v>_{j-1/2}^n v_{j-1/2}^n]$$
$$- \frac{\Delta t}{\Delta x}[(P+q)_{j+1/2}^n - (P+q)_{j-1/2}^n] \quad (4.9)$$

$$(\rho e_t)_j^{n+1} = (\rho e_t)_j^n - \frac{\Delta t}{\Delta x}[<\rho e_t>_{j+1/2}^n v_{j+1/2}^n - <\rho e_t>_{j-1/2}^n v_{j-1/2}^n]$$
$$- \frac{\Delta t}{\Delta x}[(P+q)_{j+1/2}^n v_{j+1/2}^n - (P+q)_{j-1/2}^n v_{j-1/2}^n] \quad (4.10)$$

The labeling for zone centers and zone boundaries is illustrated in Fig. 27. All dependent variables are referred to zone centers, and variables at zone boundaries are obtained by simple averages, for example,

$$v_{j+1/2}^n = \frac{1}{2}(v_{j+1}^n + v_j^n)$$

The donor cell values for any variable Q are denoted by $<Q>$ and are defined by

$$<Q>_{j+1/2}^n = Q_j^n \quad \text{if} \quad v_{j+1/2}^n \geq 0$$
$$= Q_{j+1}^n \quad \text{if} \quad v_{j+1/2}^n < 0$$

The linear form of artificial viscosity,

$$q = -B_1 \rho c \frac{\partial v}{\partial x}\Delta x \quad \text{for} \quad \frac{\partial v}{\partial x} < 0$$

and

$$q = 0, \quad \text{for} \quad \frac{\partial v}{\partial x} \geq 0 \quad (4.11)$$

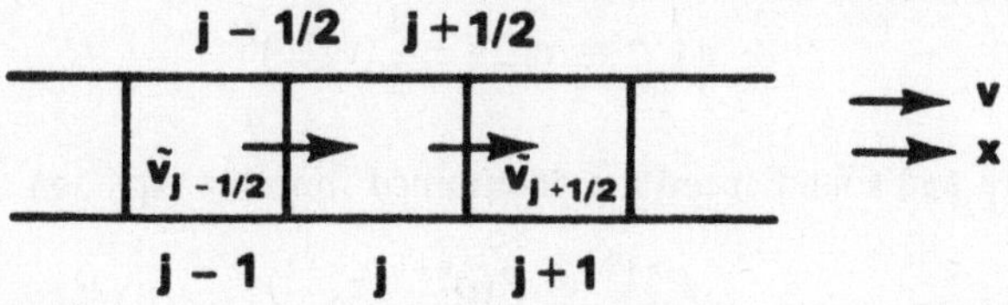

Fig. 27 Notation for zone centers and zone boundaries used with the Eulerian technique.

has been used with these difference equations. This calculational technique is conservative in the sense that, if mass does not cross and forces do not act upon the boundaries of the calculational grid, the total mass, momentum, and energy of the system are conserved in exact agreement with physical principles. This global conservation results from differencing the conservative form of the differential equations. Several hydrocodes have been based upon this donor cell technique.[14,21]

In the HELP code,[14] the calculations are conveniently divided into two steps or phases. In phase I of the calculation, new values of ρ, ρv, and ρe_t are obtained by considering changes due only to source terms. The difference equations for phase I are

$$\tilde{\rho}_j = \rho_j^n$$

$$\tilde{\rho}_j \tilde{v}_j = \rho_j^n v_j^n - \frac{\Delta t}{\Delta x}(P_{j+1/2}^n - P_{j-1/2}^n) \tag{4.12}$$

$$\tilde{\rho}_j \tilde{e}_{tj} = \rho_j^n e_{tj}^n - \frac{\Delta t}{\Delta x}\left[(Pv)_{j+1/2}^n - (Pv)_{j-1/2}^n\right]$$

Note that the artificial viscous pressure q has not been explicitly included.

Equations (4.12) determine the post-phase-I values $\tilde{\rho}_j$, $\tilde{v}_j$, and $\tilde{e}_{tj}$. With these intermediate values determined, the phase II calculation considers changes due to convection. The equations for phase II are

$$\rho_j^{n+1} = \tilde{\rho}_j - \frac{\Delta t}{\Delta x}\left[<\tilde{\rho}>_{j+1/2}\tilde{v}_{j+1/2} - <\tilde{\rho}>_{j-1/2}\tilde{v}_{j-1/2}\right]$$

$$(\rho v)_j^{n+1} = (\tilde{\rho}\tilde{v})_j - \frac{\Delta t}{\Delta x}\left[<\tilde{\rho}\tilde{v}>_{j+1/2}\tilde{v}_{j+1/2} - <\tilde{\rho}\tilde{v}>_{j-1/2}\tilde{v}_{j-1/2}\right] \tag{4.13}$$

$$(\rho e_t)_j^{n+1} = (\tilde{\rho}\tilde{e}_t)_j - \frac{\Delta t}{\Delta x}\left[<\tilde{\rho}\tilde{e}_t>_{j+1/2}\tilde{v}_{j+1/2} - <\tilde{\rho}\tilde{e}_t>_{j-1/2}\tilde{v}_{j-1/2}\right]$$

Equations (4.13) determine the new values ρ_j^{n+1}, v_j^{n+1}, and e_{tj}^{n+1}. The new specific internal energy is determined from

$$e_j^{n+1} = e_{tj}^{n+1} - \frac{1}{2}(v_j^{n+1})^2$$

The pressure and sound speed are determined from the equation of state by

$$P_j^{n+1} = P(\rho_j^{n+1}, e_j^{n+1})$$

and

$$c_j^{n+1} = c(\rho_j^{n+1}, e_j^{n+1})$$

The time step for the next cycle, chosen to satisfy the Courant condition, is calculated from

$$\Delta t = K_0 \operatorname*{MIN}_j \left(\frac{\Delta x_j}{|v_j^{n+1}| + c_j^{n+1}} \right)$$

where K_0 is a constant less than one, usually about 1/2.

The results of applying this calculational technique to a one-dimensional impact problem are shown in Fig. 28 for two different zonings. In this problem, both materials are the same compressible fluid. At impact, shock waves form at the interface and propagate in both directions. Significant oscillations occur behind the shocks; however, these oscillations are not so severe as those resulting from the Lagrangian calculation without artificial viscosity (see Fig. 23). This calculational technique introduces significant damping and can produce meaningful results without the addition of an explicit artificial viscosity; however, most Eulerian codes now use the artificial viscosity formula given by Eq. (C.5) in Table 5. If the artificial viscosity given by Eq. (4.11) is added to the pressure in the preceding Eulerian method, the results illustrated in Fig. 29 are obtained. These results are in close agreement with the exact solution.

Eulerian calculations of elastic-plastic flow generally follow the method given in Sec. 4.2. The deformation rate is calculated from

$$(d_{xx})_j^n = \frac{v_{j+1/2}^n - v_{j-1/2}^n}{\Delta x}$$

and the deviator component for the one-dimensional case is given by

$$(d'_{xx})_j^n = (d_{xx})_j^n - \frac{1}{3}(d_{xx})_j^n$$

The stress rate-deformation rate relation

$$\frac{D\sigma'_{xx}}{Dt} = 2G d'_{xx}$$

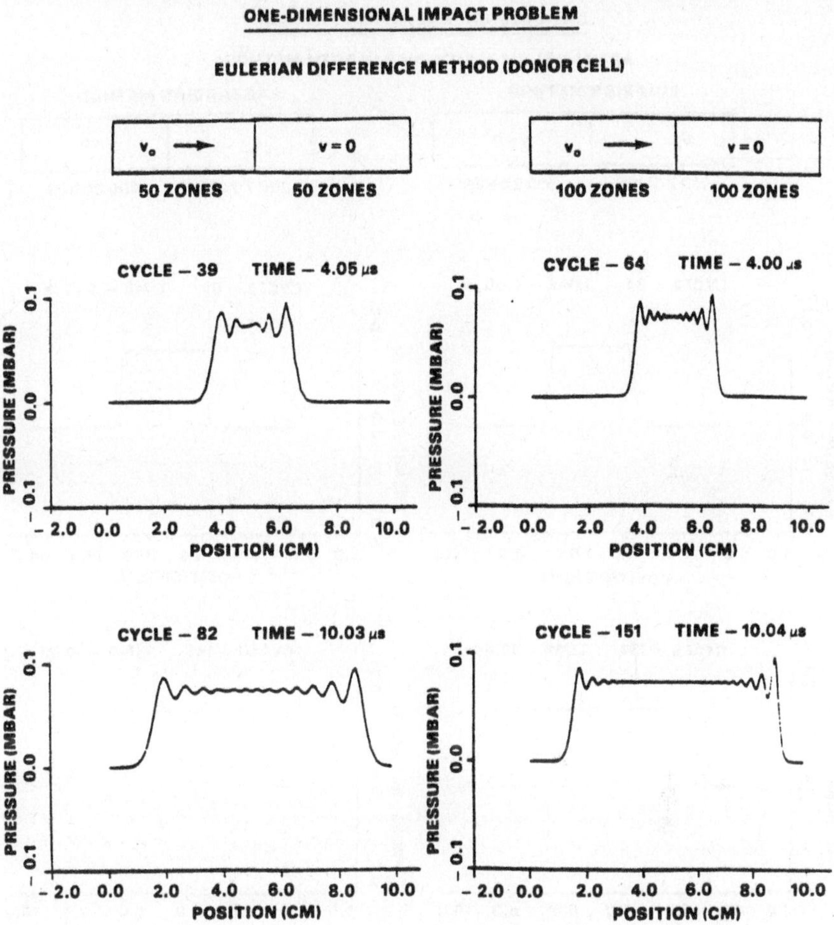

Fig. 28 Numerical results of a one-dimensional impact problem obtained from the Eulerian technique without the use of an explicit artificial viscosity.

is written in the form

$$\frac{\partial \sigma'_{xx}}{\partial t} = -v \frac{\partial \sigma'_{xx}}{\partial x} + 2Gd'_{xx}$$

This relation can be written in the difference form

$$\left({}^{T}\sigma'_{xx}\right)_j^{n+1} = \left(\sigma'_{xx}\right)_j^n - \frac{\Delta t}{\Delta x} v_j^n \left[\left(\sigma'_{xx}\right)_{j+1/2}^n - \left(\sigma'_{xx}\right)_{j-1/2}^n\right] + 2G\left(d'_{xx}\right)_j^n \Delta t$$

where the superscript T indicates a tentative value of stress that is correct if the deformation is elastic. If plastic flow occurs during the time step, the tentative value of stress must be modified as indicated in Sec. 4.2. The effects of

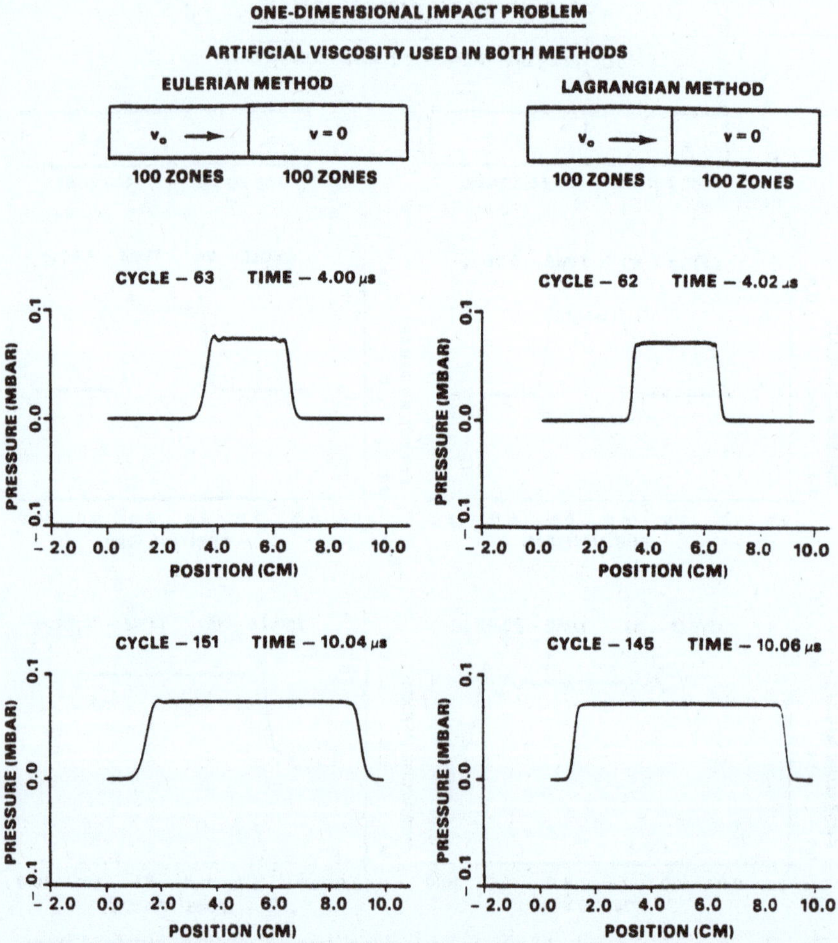

Fig. 29 Numerical results of a one-dimensional impact problem obtained from the Eulerian technique using an explicit artificial viscosity compared with results from a Lagrangian calculation.

the deviatoric stress and viscous pressure are incorporated into Eqs. (4.12) by replacing the pressure P by the stress $-\sigma_{xx}$, where

$$\sigma_{xx} = -P - q + \sigma'_{xx}$$

It is instructive to compare the results of the preceding Eulerian method with results from the Lagrangian method of Sec. 4.2. Fig. 30 shows the results of these two methods for the simple one-dimensional impact problem of an elastic-plastic material. In Table 6, the values of the variables behind the elastic and plastic waves obtained from the two numerical solutions are compared with

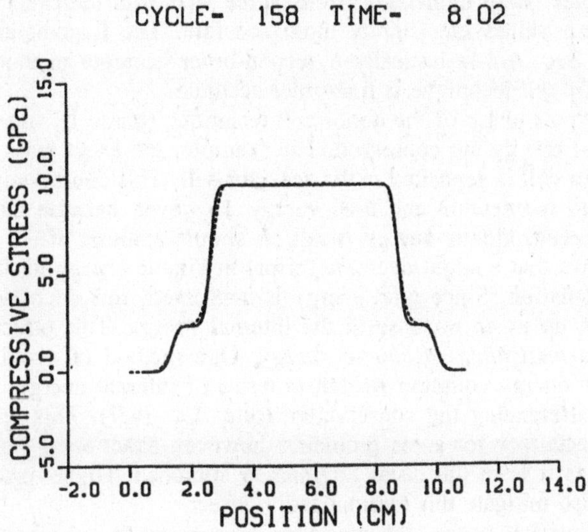

Fig. 30 Comparison of results from a Lagrangian (———) and an Eulerian (----------) calculation of a one-dimensional impact problem of an elastic-perfectly plastic material.

Table 6 Comparison of numerical results for the one-dimensional impact problem[a]

Variable	Lagrangian	Eulerian	Exact
	Values behind the elastic wave		
ρ (gm/cm^3)	8.978	8.979	8.977
P (GPa)	1.193	1.206	1.192
σ'_{xx} (GPa)	1.155	1.155	1.155
v (m/s)	47.83	48.54	47.68
e (m/s^2)	0.1150	0.1201	0.1137
	Values behind the plastic wave		
ρ (gm/cm^3)	9.452	9.454	9.453
P (GPa)	8.501	8.525	8.505
σ'_{xx} (GPa)	1.155	1.155	1.155
v (m/s)	250.0	249.5	250.0
e (m/s)^2	3.542	3.551	3.474

[a]An impact velocity of 500 m/s was used. The equation of state was $P = 137(\rho/\rho_o - 1)$ GPa with $\rho_o = 8.9$ gm/cm^3. Values of $G = 100.0$ GPa and $k = 1.0$ GPa were used. The values have been averaged over a few zones and rounded to four significant figures, and 200 zones were used. The values listed apply to the wave traveling to the right at 8 μs after impact.

the exact values. Both numerical results agree well with the exact results, but the Lagrangian values are slightly more accurate. The Lagrangian technique discussed in Sec. 4.2 is basically a second-order accurate method while the Eulerian donor cell technique is first-order accurate.

In the transport phase of the donor cell technique (phase II), mass, momentum, and total energy are conserved. For example, the exact mass transported from the donor cell is deposited in the acceptor cell. This conservative transport also applies to momentum and total energy. However, because momentum is exactly conserved, kinetic energy is not. A simple analysis of the momentum transport shows that a slight decrease (error) in kinetic energy results from the transport calculation. Since total energy is conserved, this decrease in kinetic energy shows up as an increase in the internal energy. This effect is usually known as *thermalization of kinetic energy*. One method of avoiding it is to difference the energy equation written in terms of internal energy, Eq. (2.22), rather than differencing the conservative form, Eq. (4.7). This approach can improve the accuracy for some problems; however, exact energy conservation is sacrificed as it is in the usual Lagrangian approach. Higher-order Eulerian approaches also mitigate this thermalization effect.

In many Eulerian codes, the pressures or stresses at zone boundaries are determined by inverse density weighting rather than by simple averaging. For example,

$$P^n_{j+1/2} = \frac{P^n_{j+1}\rho^n_j + P^n_j\rho^n_{j+1}}{\rho^n_j + \rho^n_{j+1}}$$

In many Eulerian codes, the velocities at zone boundaries are determined by direct density weighting, for example,

$$v^n_{j+1/2} = \frac{v^n_{j+1}\rho^n_{j+1} + v^n_j\rho^n_j}{\rho^n_j + \rho^n_{j+1}}$$

These values produce more realistic results when large density variations exist between zones.

Another two-step method has been successfully used in Eulerian hydrocodes.[22,23] In this method, the two phases are thought of as a Lagrangian phase followed by an Eulerian remap phase. In the Lagrangian phase, the dependent variables are updated according to the Lagrangian method. In the second phase, the dependent variables, which are now referred to the slightly displaced Lagrangian grid, are mapped back to the fixed Eulerian grid by a method that is consistent with the conservation laws.

Another technique that uses an Eulerian grid is the Particle-in-Cell (PIC) method.[24] In this method, the continuous materials are represented by a large number of particles (mass points) that are distributed over the regions occupied by the materials. In the transport phase of the calculation, the particles are moved according to the local velocity field. When a particle leaves one zone and enters another, it carries with it a certain amount of mass, momentum, and energy. The transport is, therefore, quantized. This method automatically determines material boundaries and was one of the first methods to treat multimaterial problems involving large deformations.

One problem that arises in Eulerian calculations but not in Lagrangian calculations is the diffusion of properties caused by the transport process. And extreme example of mass diffusion is illustrated in Fig. 31. In this problem, an initial density discontinuity is present in the flow field. The pressure is maintained at a value of zero, and all of the material is given a uniform velocity to the right. Changes in density result only from transport. Ideally, the high-density region should move to the right with unchanged shape. However, the solution at 18 μs, obtained from the first-order donor cell technique, gives a distorted shape. When a small amount of mass is transported from the high-density donor cell to the low-density acceptor cell, the density of the acceptor cell is immediately increased. On the following cycle, this acceptor cell will transport some of its slightly higher-density material to the next downstream acceptor cell. Repitition of the process causes a diffusion of mass that destroys the initial shape of the high-density region. Any property transported with the material will also be diffused. Higher-order transport techniques can substantially reduce this diffusion process.[25]

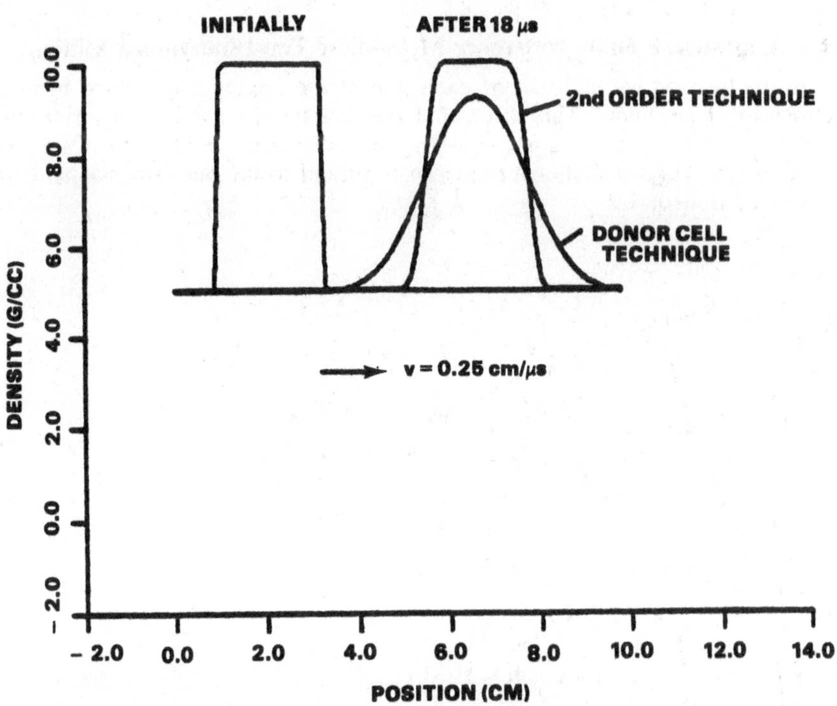

Fig. 31 An example of mass diffusion of an initially sharp density discontinuity computed by the Eulerian "donor cell" technique compared with a second order transport technique described in Ref. 25 (scheme 1).

The treatment of problems involving more than one material increases the complexity of Eulerian calculations. Some zones will then contain two or more materials, and special techniques must be used to treat these *mixed zones*. The calculation of pressure in a mixed zone is usually accomplished by an iteration process that adjusts the densities and possibly the internal energies until pressure equilibrium is achieved. However, some codes compute pressures for each material in a zone and then compute the cell-centered pressure as a volume weighted average. Both methods work, but the second method avoids the time-consuming iteration. Most Eulerian codes do not precisely define material interfaces, but rely upon the position of mixed cells to approximate material boundaries. Special transport rules are used to limit the diffusion of one material into another.[22,23] For example, if the donor cell contains two materials, material A and material B, and if the acceptor cell contains only material A, then transport of only material A would be calculated. Material B would not be transported until material A is depleted from the donor cell.

Some codes use a line of tracer particles to define the material interface.[14] These tracer particles are moved with the local flow velocity, and the position of the interface within the donor cell is used in the transport prescription. This method should produce increased accuracy, but difficulties can result when severe distortions occur at the interface.

4.4 Lagrangian Finite Difference Method for Two-Dimensional Motion

In this section we indicate the extension of the Lagrangian method to two-dimensional problems. This discussion summarizes the methods described by Wilkins[18] and Swegle.[26] For simplicity, we indicate the method for Cartesian coordinates. The calculational sequence is similar to the one-dimensional case discussed in Sec. 4.2.

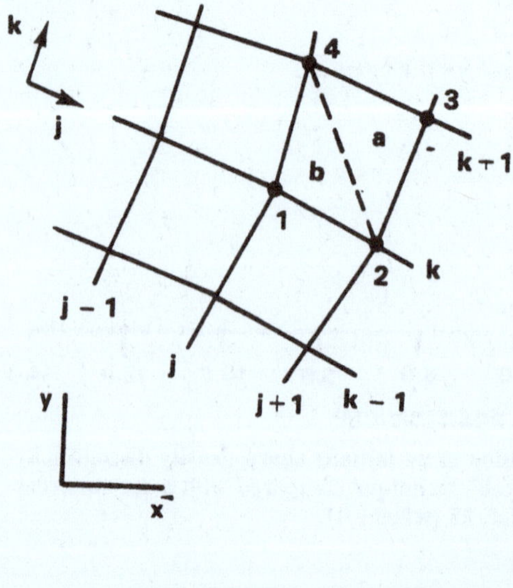

Fig. 32 Notation used for the two-dimensional Lagrangian mesh.

Conservation of Mass

The notation used to describe nodes and zones in the Lagrangian mesh is shown in Fig. 32. The mass per unit depth in a typical quadrilateral zone with nodes 1–4 is calculated from

$$m_{j+1/2,k+1/2} = \rho^n_{j+1/2,k+1/2}(A^n_a + A^n_b)$$

where the areas of the two triangles are given by

$$A^n_a = \frac{1}{2}[x_2(y_3 - y_4) + x_3(y_4 - y_2) + x_4(y_2 - y_3)]^n$$

$$A^n_b = \frac{1}{2}[x_1(y_2 - y_4) + x_2(y_4 - y_1) + x_4(y_1 - y_2)]^n$$

Conservation of mass requires that the mass of each zone remains constant during the motion. The density and specific volume of material in any zone are calculated from

$$\rho^n = \frac{m}{A^n} \qquad \text{and} \qquad V^n = \frac{1}{\rho^n}$$

where $A^n = A^n_a + A^n_b$.

Momentum Equations

The equations of motion, Eqs. (2.17), are given by

$$X\text{-momentum:} \qquad \rho\frac{Dv_x}{Dt} = \frac{\partial\sigma_{xx}}{\partial x} + \frac{\partial\sigma_{xy}}{\partial y} \tag{4.14}$$

$$Y\text{-momentum:} \qquad \rho\frac{Dv_y}{Dt} = \frac{\partial\sigma_{xy}}{\partial x} + \frac{\partial\sigma_{yy}}{\partial y} \tag{4.15}$$

The material derivatives at time n are approximated by

$$\frac{\mathrm{D}v_x}{\mathrm{D}t} = \frac{(v_x)^{n+1/2}_{j,k} - (v_x)^{n-1/2}_{j,k}}{\Delta t^n} \tag{4.16}$$

and a similar equation can be written for the material derivative of v_y.

The spatial derivatives are approximated by the so-called "integral method." Consider any continuous function $f(x, y)$ with continuous first-order partial derivatives. By integrating the x-derivative of f over the closed area A, indicated in Fig. 33, we obtain the relation

$$\int_A\!\!\int \frac{\partial f}{\partial x}\mathrm{d}x\ \mathrm{d}y = \int_c f\ \mathrm{d}y$$

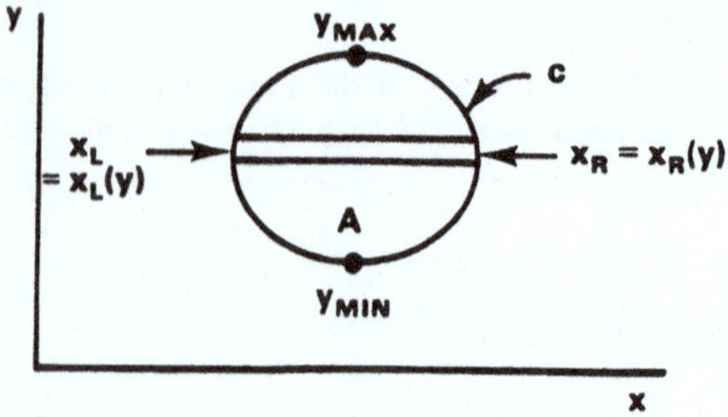

Fig. 33 Area of integration used to derive the formula for spatial derivatives.

where the integral on the right is the line integral along the boundary curve c. The area integral can also be expressed as

$$\int_A \int \frac{\partial f}{\partial x} \mathrm{d}x \ \mathrm{d}y = A\left(\frac{\partial f}{\partial x}\right)_{\mathrm{AVE}}$$

where the derivative on the right is the average value over the area A. From the preceding equations, we are led to the integral approximation to the derivative

$$\frac{\partial f}{\partial x} = \frac{1}{A}\int_c f \ \mathrm{d}y \tag{4.17}$$

The corresponding approximation to the y-derivative is

$$\frac{\partial f}{\partial y} = -\frac{1}{A}\int_c f \ \mathrm{d}x \tag{4.18}$$

By using the notation of Fig. 34, the derivatives of any dependent variable f are approximated at the node m by

$$\left(\frac{\partial f}{\partial x}\right)_m^n = \frac{1}{A^n}[f_1(y_{\mathrm{III}} - y_{\mathrm{II}}) + f_2(y_{\mathrm{IV}} - y_{\mathrm{III}}) + f_3(y_{\mathrm{I}} - y_{\mathrm{IV}}) + f_4(y_{\mathrm{II}} - y_{\mathrm{I}})]^n \tag{4.19}$$

and

$$\left(\frac{\partial f}{\partial y}\right)_m^n = \frac{-1}{A^n}[f_1(x_{\mathrm{III}} - x_{\mathrm{II}}) + f_2(x_{\mathrm{IV}} - x_{\mathrm{III}}) + f_3(x_{\mathrm{I}} - x_{\mathrm{IV}}) + f_4(x_{\mathrm{II}} - x_{\mathrm{I}})]^n \tag{4.20}$$

where f_1 is the value of f at zone center 1, and so on.

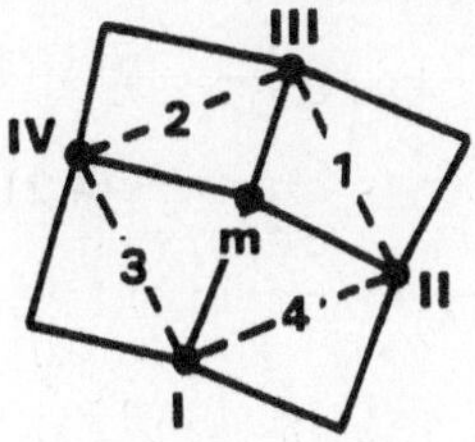

Fig. 34 Notation used in the finite difference formula for spatial derivatives at a mesh point.

These finite difference approximations, Eqs. (4.16), (4.19), and (4.20), can be used in Eqs. (4.14) and (4.15) to give equations for the velocity components at time step $n + 1/2$

$$(v_x)_{j,k}^{n+1/2} \qquad \text{and} \qquad (v_y)_{j,k}^{n+1/2}$$

in terms of known quantities at time step n. After these velocity components are determined, new values of the node coordinates of the Lagrangian grid can be calculated from

$$x_{j,k}^{n+1} = x_{j,k}^{n} + (v_x)_{j,k}^{n+1/2} \Delta t^{n+1/2}$$

and a similar equation for the y coordinate. With the grid established at time step $n + 1$, new values of density and specific volume can be calculated.

Deformation Rates

In two dimensions, the nonzero deformation rates evaluated at zone centers, as indicated by zone 1 of Fig. 35, are given by

$$(d_{xx})_1^{n+1/2} = \left(\frac{\partial v_x}{\partial x}\right)_1^{n+1/2}, \qquad (d_{yy})_1^{n+1/2} = \left(\frac{\partial v_y}{\partial y}\right)_1^{n+1/2}$$

and

$$(d_{xy})_1^{n+1/2} = \frac{1}{2}\left[\left(\frac{\partial v_y}{\partial x}\right)_1^{n+1/2} + \left(\frac{\partial v_x}{\partial y}\right)_1^{n+1/2}\right]$$

The finite difference approximations to the spatial derivatives are obtained by applying Eqs. (4.17) and (4.18) around the boundary of the zone. For example,

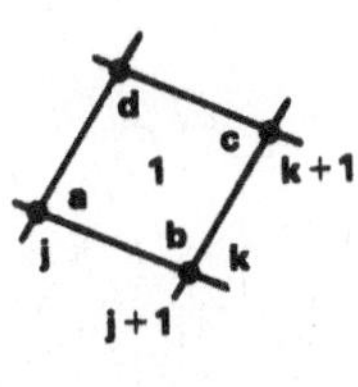

Fig. 35 Notation used for the difference approximation to spatial derivatives at a zone center.

$$\left(\frac{\partial v_x}{\partial x}\right)_1^{n+1/2} = \frac{1}{2A_1^{n+1/2}}[(v_{xb} - v_{xd})(y_c - y_a) + (v_{xa} - v_{xc})(y_b - y_d)]^{n+1/2} \tag{4.21}$$

where

$$A_1^{n+1/2} = \frac{1}{2}\left(A_1^{n+1} + A_1^n\right)$$
$$y_a^{n+1/2} = \frac{1}{2}\left(y_a^{n+1} + y_a^n\right)$$

and so on. The deviatoric components are then computed from

$$\left(\mathrm{d}'_{xx}\right)_1^{n+1/2} = (\mathrm{d}_{xx})_1^{n+1/2} - \frac{1}{3}\left(\frac{\dot{V}}{V}\right)_1^{n+1/2}$$

where

$$\left(\frac{\dot{V}}{V}\right)_1^{n+1/2} = \frac{V_1^{n+1} - V_1^n}{\frac{1}{2}\Delta t^{n+1/2}(V^{n+1} + V^n)}$$

In the preceding equation, the relationship $d_{xx} + d_{yy} + d_{zz} = \dot{V}/V$, obtained from the continuity equation, has been used. Similar expressions are used to calculate d'_{yy} and d'_{xy}. The deviator d'_{zz} can be calculated from $d'_{zz} = -d'_{xx} - d'_{yy}$.

Stress Rate-Deformation Rate Relations

For two-dimensional motion, ω_{xy} is the only nonvanishing component of spin; therefore, the stress rate-deformation rate relations, Eq. (2.37), can be written as

$$\frac{\mathrm{D}\sigma'_{xx}}{\mathrm{D}t} - 2\omega_{xy}\sigma'_{xy} = 2Gd'_{xx}$$
$$\frac{\mathrm{D}\sigma'_{yy}}{\mathrm{D}t} + 2\omega_{xy}\sigma'_{xy} = 2Gd'_{yy}$$
$$\frac{\mathrm{D}\sigma'_{xy}}{\mathrm{D}t} - \omega_{xy}\left(\sigma'_{yy} - \sigma'_{xx}\right) = 2Gd'_{xy}$$
$$\frac{\mathrm{D}\sigma'_{zz}}{\mathrm{D}t} = 2Gd'_{zz}$$

where

$$\omega_{xy} = \frac{1}{2}\left(\frac{\partial v_x}{\partial y} - \frac{\partial v_y}{\partial x}\right)$$

In these equations, the deformation is assumed to be elastic. Difference equations approximating these relations can be written for a typical zone, zone 1 of Fig. 35. For example, the first equation is approximated by

$$\left({}^{T}\sigma'_{xx}\right)_1^{n+1} = \left(\sigma'_{xx}\right)_1^n + 2\left(\omega_{xy}\right)_1^{n+1/2}\left(\sigma'_{xy}\right)_1^n \Delta t^{n+1/2} + 2G\left(d'_{xx}\right)_1^{n+1/2}\Delta t^{n+1/2} \tag{4.22}$$

These approximations determine the tentative values of stress at time $n+1$. To determine if the material is in an elastic or plastic state, the von Mises yield function is first evaluated from

$$f_1^{n+1} = \left({}^{T}J_2\right)_1^{n+1} - k^2$$

where

$$\left({}^{T}J_2\right)_1^{n+1} = \frac{1}{2}\left[\left({}^{T}\sigma'_{xx}\right)_1^2 + \left({}^{T}\sigma'_{yy}\right)_1^2 + \left({}^{T}\sigma'_{zz}\right)_1^2 + 2\left({}^{T}\sigma'_{xy}\right)_1^2\right]^{n+1}$$

If $f_1^{n+1} \le 0$, the deformation is elastic, and the tentative values of stress are correct. However, if $f_1^{n+1} > 0$, the tentative values of stress must be modified to satisfy the yield condition and flow rule since plastic flow has occurred during the time step.

When plastic flow occurs, the deformation rate deviators can be written as the sum of elastic and plastic parts. For example,

$$\left(d'_{xx}\right)_1^{n+1/2} = \left(d'^{E}_{xx}\right)_1^{n+1/2} + \left(d'^{P}_{xx}\right)_1^{n+1/2}$$

Only the elastic part should have been used in Eq. (4.22), that is,

$$\left(\sigma'_{xx}\right)_1^{n+1} = \left(\sigma'_{xx}\right)_1^n + 2\left(\omega_{xy}\right)^{n+1/2}\left(\sigma'_{xy}\right)_1^n \Delta t^{n+1/2} + 2G\left(d'^{E}_{xx}\right)_1^{n+1/2}\Delta t^{n+1/2} \tag{4.23}$$

By subtracting Eq. (4.23) from (4.22), we obtain

$$\left({}^{T}\sigma'_{xx}\right)_1^{n+1} - \left(\sigma'_{xx}\right)_1^{n+1} = 2G\left(d^{P}_{xx}\right)_1^{n+1/2}\Delta t^{n+1/2} \tag{4.24}$$

Similar equations can be derived for the other stress components.

The values of the stress deviator components must satisfy the flow rule

$$d^{P}_{ij} = \lambda\sigma'_{ij}$$

These relations can be written as

$$\frac{\sigma'_{xx}}{d^{P}_{xx}} = \frac{\sigma'_{yy}}{d^{P}_{yy}} = \frac{\sigma'_{zz}}{d^{P}_{zz}} = \frac{\sigma'_{xy}}{d^{P}_{xy}} = \frac{1}{\lambda}$$

By combining these relations with Eq. (4.24) and the similar equation for the σ'_{yy} component, we obtain the equation

$$\frac{{}^T\sigma'_{xx} - \sigma'_{xx}}{{}^T\sigma'_{yy} - \sigma'_{yy}} = \frac{d^P_{xx}}{d^P_{yy}} = \frac{\sigma'_{xx}}{\sigma'_{yy}} \tag{4.25}$$

We now assume that the true values of stress can be obtained by multiplying the tentative values of stress by constants of proportionality; therefore,

$$\begin{aligned} \left(\sigma'_{xx}\right)_1^{n+1} &= K_x\left({}^T\sigma'_{xx}\right)_1^{n+1}, \qquad \left(\sigma'_{yy}\right)_1^{n+1} = K_y\left({}^T\sigma'_{yy}\right)_1^{n+1} \\ \left(\sigma'_{zz}\right)_1^{n+1} &= K_z\left({}^T\sigma'_{zz}\right)_1^{n+1}, \qquad \left(\sigma'_{xy}\right)_1^{n+1} = K_{xy}\left({}^T\sigma'_{xy}\right)_1^{n+1} \end{aligned} \tag{4.26}$$

Combining the first two of Eqs. (4.26) with (4.25), we conclude that $K_x = K_y$. By repeating this analysis for the stress components σ'_{zz} and σ'_{xy}, we find that, to satisfy the flow rule, all constants of proportionality must be equal. Therefore,

$$K_x = K_y = K_z = K_{xy} = K$$

The final values of stress must also satisfy the yield condition

$$f_1^{n+1} = (J_2)_1^{n+1} - k^2 = 0$$

where

$$\begin{aligned} (J_2)_1^{n+1} = \frac{1}{2}\{&[K\left({}^T\sigma'_{xx}\right)_1^{n+1}]^2 + [K\left({}^T\sigma'_{yy}\right)_1^{n+1}]^2 + [K\left({}^T\sigma'_{zz}\right)_1^{n+1}]^2 \\ &+ 2[K\left({}^T\sigma'_{xy}\right)_1^{n+1}]^2\} = K^2\left({}^TJ_2\right)_1^{n+1} \end{aligned}$$

Therefore, to satisfy the yield condition, the value of the proportionality constant K must be

$$K = \frac{k}{\sqrt{({}^TJ_2)_1^{n+1}}}$$

With the value of K established for a typical zone, the final values of the stress deviator components are calculated from Eqs. (4.26). This technique of

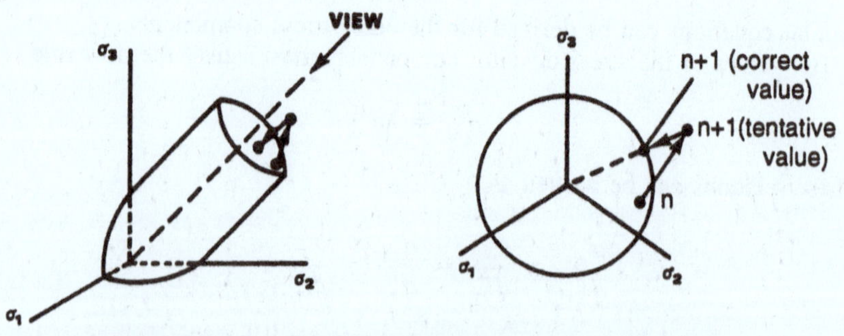

Fig. 36 Geometric illustration of the radial return method.

calculating the stress components is known as the *radial return method* and is used in most hydrocodes. It is geometrically illustrated in Fig. 36.

In Eq. (4.22), the stress component $(\sigma'_{xy})_1^n$ is not quite centered. To maintain second-order accuracy, the component should be $(\sigma'_{xy})_1^{n+1/2}$. Centering can be accomplished by a slightly more complicated analysis which is described in Ref. 26.

Artificial Viscosity

A viscous pressure must be used to simulate shock waves properly. A relation similar to Eq. (C.5) in Table 5, which was recommended for one-dimensional motion, is usually used. For two-dimensional motion, the viscous pressure can be computed from

$$q_1^{n+1/2} = B_2^2 \rho_1^{n+1/2} A_1^{n+1/2} \left[\left(\frac{\dot{V}}{V} \right)_1^{n+1/2} \right]^2 - B_1 \rho_1^{n+1/2} c_1^{n+1/2} \sqrt{A_1^{n+1/2}} \left(\frac{\dot{V}}{V} \right)_1^{n+1/2}$$

for

$$\left(\frac{\dot{V}}{V} \right)_1^{n+1/2} < 0 \text{ and } q_1^{n+1/2} = 0 \text{ for } \left(\frac{\dot{V}}{V} \right)_1^{n+1/2} \geq 0$$

In addition to this bulk viscosity, some codes also introduce deviator components of a tensor viscosity.[26]

Energy Equation

The new value of internal energy, for a typical zone, can be calculated from a difference equation similar to Eq. (4.1), which was used for one-dimensional motion. However, the work done by the deviator stress components, W_d, must be added to the numerator of Eq. (4.1). The quantity W_d is evaluated from

$$W_{\rm d} = \left[\sigma'_{xx}\, d_{xx} + \sigma'_{yy}\, d_{yy} + 2\sigma'_{xy}\, d_{xy} \right]_1^{n+1/2} V_1^{n+1/2} \Delta t^{n+1/2}$$

The pressure and sound speed calculations follow the method used for the one-dimensional case. The normal stresses are calculated from equations of the form

$$\sigma_{xx} = \sigma'_{xx} - P - q$$

$$\sigma_{yy} = \sigma'_{yy} - P - q$$

and the shear stress is

$$\sigma_{xy} = \sigma'_{xy}$$

The time step is calculated from an extension of the one-dimensional formula given in Eq (C.9). For example, the time step relation described in Ref. 26, for zones being compressed, is

$$\Delta t = \frac{KD_1}{B_1 c - B_2^2 D_2 \left(\frac{\dot{V}}{V}\right) + \left\{\left(B_1 c - B_2^2 D_2 \left(\frac{\dot{V}}{V}\right)\right)^2 + c^2\right\}^{1/2}}$$

where K is a constant, usually equal to 0.9, and D_1 and D_2 are the minimum and maximum distances between the nodes of the zone under consideration, respectively.

Hourglass Distortions

In Lagrangian calculations using quadrilateral zones, certain mesh distortions may result that produce no net restoring forces. These deformations are associated with the smallest resolvable wavelengths of the mesh ($2\Delta x$). The hourglass distortion pattern, also known as bowtie or keystone distortions, is illustrated in Fig. 37. Note that no volume change or deformation rates [see Eq. (4.31)], as calculated by the finite difference method, are associated with this nonphysical velocity field. To prevent these deformations from slowly degrading the solution, most Lagrangian hydrocodes introduce a small artificial resistance.

In the HEMP code,[18] an additional artificial viscosity is introduced that is a function of the rate of change of the angle between the grid lines. In the SALE code,[27] small additional accelerations, which are based on the local velocity field and tend to keep the velocities from deviating too strongly from their neighbors, are introduced at each node. This fourth order-coupling scheme is

$$v_o' = v_o + \frac{1}{4} a \left[2(v_1 + v_2 + v_3 + v_4) - v_5 - v_6 - v_7 - v_8 - 4v_o\right]$$

where v stands for either v_x or v_y and v_o' is the adjusted node velocity. The node numbering is indicated in Fig. 37, and a is a coefficient that controls the amount of coupling. A value of 0.05 or less is recommended in Ref. 27.

Boundary Conditions

Most Lagrangian codes include a variety of boundary conditions. Boundary conditions consist of the specification of velocities (or displacements), stresses, or a combination of these. The minimum boundary conditions required for warhead problems consist of free surfaces (applied boundary stress of zero) and planes of symmetry or an axis of symmetry for problems in cylindrical coordinates. In most codes, the boundary conditions are applied by assuming that an additional layer of zones of a *pseudo material* surrounds the object. The

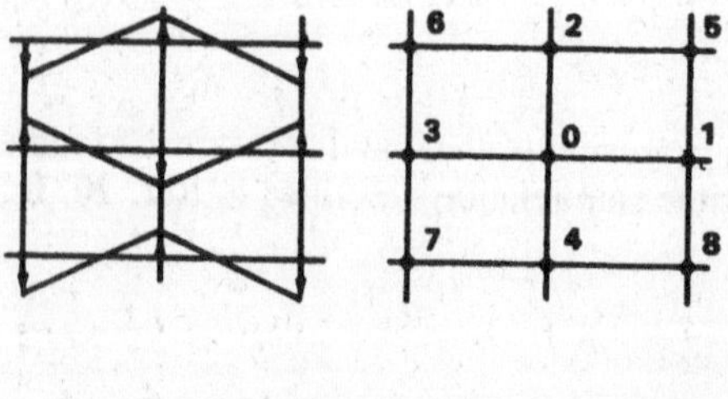

Fig. 37 Illustration of the hourglass or bowtie mesh distortion pattern.

variables in these zones are given values that cause the equations of motion to produce the desired behavior of the boundary points. For example, a free surface is simulated by allowing the pseudo material in the added zones to have zero stress and zero mass. The standard finite difference equations can then be used to compute the motion of the boundary points.

Sliding Interface

When two materials slide along each other, a decoupling of the Lagrangian grids at the interface must be provided to eliminate large zone deformations. The slide line calculational routine provides this decoupling. The geometry of the slide line (line K) is illustrated in Fig. 38 at time steps n and $n + 1$. Mesh points associated with the grid below the slide line are allowed to move along it. Point f is a typical sliding point and its position at time step n is denoted by P. The motion of point f is first calculated by assuming that the slide line is a fixed surface. The new location of point f, denoted by f^*, is determined by applying the equation of motion in the direction tangential to the slide line K^n. Next, the points on the slide line that are associated with the grid above it are moved by the usual equations of motion. However, the acceleration of these points (points a and b), is composed of components parallel and perpendicular to the slide line. The perpendicular component includes the appropriate mass of material below the slide line, whereas the parallel component does not. Node points not on the slide line are moved according

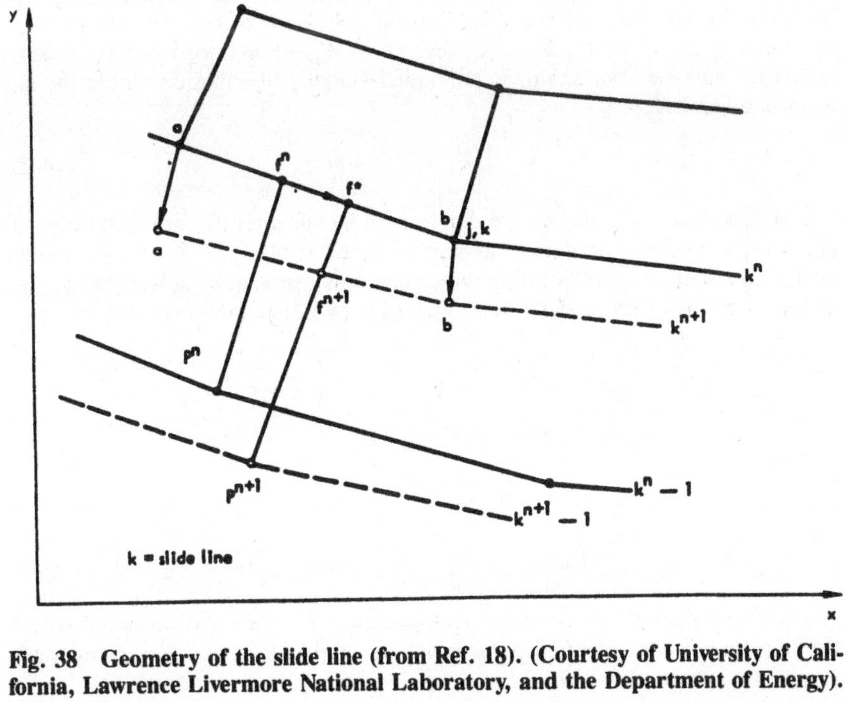

Fig. 38 Geometry of the slide line (from Ref. 18). (Courtesy of University of California, Lawrence Livermore National Laboratory, and the Department of Energy).

to the usual equations of motion. The final position of slide point f, designated as f^{n+1}, is located at the point of intersection between the new position of the slide line, K^{n+1}, and the line connecting point f^* and point P^{n+1}. This type of slide line calculation has been developed by Wilkins[18] for the case when the two materials remain in contact. This case is common for an explosive in contact with a metal surface. However, void-opening and void-closing at the slide line can also be treated by a slightly more complicated routine.

Three types of sliding interfaces are usually modeled. These types represent sliding without friction, sliding with friction, and tied slide lines that do not allow slip. Tied slide lines can be used to connect regions of different zone sizes.

4.5 Description of a Finite Element Calculation

The EPIC-2 code has been successfully applied to many problems in terminal ballistics. In this section, we outline the computational technique described by Johnson.[13]

The EPIC-2 code is based upon a Lagrangian finite element formulation where the equations of motion are integrated directly, rather than through the traditional stiffness matrix approach. As with most finite element formulations, complex geometries can be represented simply by providing an adequate assemblage of elements since an orderly arrangement of nodes is not required. The computational sequence essentially follows the method illustrated in Sec. 4.4.

The formulation is based upon triangular finite elements. Fig. 39 illustrates a typical axisymmetric element. In addition to motion in the rz-plane, a rotation about the axis of symmetry is also permitted. The formulation is still two-dimensional since all variables depend only on r and z. The three velocity components, $v_r = \dot{U}$, $v_z = \dot{V}$, and $v_\theta = \dot{W}$, are assumed to vary linearly within the element. For example, the radial velocity distribution within the element is given by

$$v_r = a_1 + a_2 r + a_3 z \tag{4.27}$$

The factors a_1, a_2, and a_3 are determined by substituting, one at a time, the radial velocities and coordinates at each of the three nodes, i, j, and m, into the preceding equation. The resulting three equations can then be solved for a_1, a_2, and a_3 in terms of the nodal values. Equation (4.27) can then be rewritten as

$$v_r = \frac{1}{2A}[(a_i + b_i r + c_i z) v_{ri} + (a_j + b_j r + c_j z) v_{rj} + (a_m + b_m r + c_m z) v_{rm}]$$

where

$$a_i = r_j z_m - r_m z_j, \qquad b_i = z_j - z_m, \qquad c_i = r_m - r_j$$

and similar equations for the i and m subscripts. A is the cross-sectional area of the element in the rz-plane. Similar equations apply for the velocity components v_z and v_θ, where $v_\theta = r\dot{\theta}$.

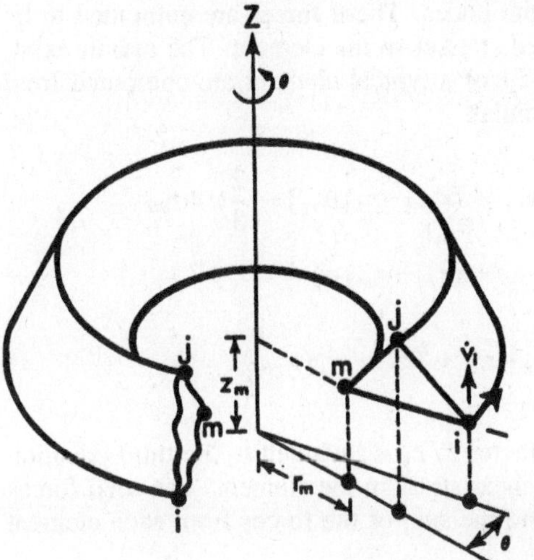

Fig. 39 Geometry of an axisymmetric finite element (from Ref. 13).

In cylindrical coordinates, the deformation rate components, for motions independent of the θ-coordinate, are given by

$$d_{rr} = \frac{\partial v_r}{\partial r}, \qquad d_{zz} = \frac{\partial v_z}{\partial z}, \qquad d_{\theta\theta} = \frac{v_r}{r}$$

$$d_{rz} = \frac{1}{2}\left(\frac{\partial v_r}{\partial z} + \frac{\partial v_z}{\partial r}\right), \quad d_{r\theta} = \frac{1}{2}\left(\frac{\partial v_\theta}{\partial r} - \frac{v_\theta}{r}\right), \quad d_{z\theta} = \frac{1}{2}\frac{\partial v_\theta}{\partial z}$$

The deformation rates within a typical element are determined from the expressions for the velocity distributions. All deformation rate components except $d_{\theta\theta}$ and $d_{r\theta}$ involve derivatives of the linear velocity distributions; therefore, these components are constants within the element. Since the expressions for $d_{\theta\theta}$ and $d_{r\theta}$ involve the terms v_r/r and v_θ/r, they will, in general, vary over the element. These terms are evaluated by using the average of the nodal values. For example,

$$d_{\theta\theta} = \frac{\overline{v}_r}{\overline{r}} = \frac{v_{ri} + v_{rj} + v_{rm}}{r_i + r_j + r_m}$$

The deviatoric components are determined from the deformation rate components in the usual manner.

Once the deviatoric deformation rate components are established at time step n, deviatoric stress components at time step $n+1$ can be calculated by differencing the stress rate-deformation rate relations. The stress calculation essentially follows the method described in Sec. 4.4.

In this technique, the mass of an element is assumed to be concentrated at the nodes by assigning one third of the element mass to each node. When a node i is associated with more than one element, the total mass of the node, $\overline{M}_i$ is equal to 1/3 the mass of all elements associated with the node. Concentrated forces are assumed to act on the mass at the nodes. These forces are computed to be statically equivalent to the distributed stresses in the element. The radial, axial, and tangential forces acting on node i of a typical element are computed from the stress in the element by the formulas

$$F_r^i = -\pi\overline{r}\,[(z_j - z_m)\sigma_{rr} + (r_m - r_j)\sigma_{rz}] - \frac{2}{3}\pi A\sigma_{\theta\theta}$$

$$F_z^i = -\pi\overline{r}\,[(r_m - r_j)\sigma_{zz} + (z_j - z_m)\sigma_{rz}]$$

$$F_\theta^i = -\pi\overline{r}\left[\frac{\overline{r}}{r_i}(z_j - z_m)\sigma_{r\theta} + (r_m - r_j)\sigma_{z\theta}\right]$$

where $\overline{r} = (r_i + r_j + r_m)/3$. The factor $\overline{r}/r_i$ is included in the third equation to represent the effect of a variable shear stress in the element. The total forces acting at node i, $\overline{F}_r^i$, $\overline{F}_z^i$, and $\overline{F}_\theta^i$, are the sum of the forces from each element that is associated with node i.

In cylindrical coordinates, the acceleration components in the radial, axial, and angular directions of a typical node are given by

$$a_r = \dot{v}_r - r\dot{\theta}^2, \qquad a_z = \dot{v}_z, \qquad a_\theta = \frac{1}{r}\frac{\mathrm{D}}{\mathrm{D}t}(r^2\dot{\theta})$$

A direct application of Newton's law of motion to the mass point $\overline{M}_i$ gives the equations of motion

$$(\dot{v}_r - r\dot{\theta}^2)_i^n = \left(\overline{F}_r^i\right)^n / \overline{M}_i$$

$$\dot{v}_{zi}^n = \left(\overline{F}_z^i\right)^n / \overline{M}_i$$

$$\left[\frac{1}{r}\frac{\mathrm{D}}{\mathrm{D}t}\left(r^2\dot{\theta}\right)\right]_i^n = \left(\overline{F}_\theta^i\right)^n / \overline{M}_i$$

These equations then determine the quantities

$$\dot{v}_r, \qquad \dot{v}_z, \qquad \text{and} \qquad \frac{\mathrm{D}}{\mathrm{D}t}(r^2\dot{\theta})$$

at time n. New values of velocities and displacements are obtained by an application of the central-difference approximation. For example, the new value of the radial velocity is obtained from

$$v_{ri}^{n+1/2} = v_{ri}^{n-1/2} + \dot{v}_{ri}^n \Delta t^n$$

and the new radial coordinate is obtained from

$$r_i^{n+1} = r_i^n + v_{ri}^{n+1/2} \Delta t^{n+1/2}$$

The EPIC-2 program uses an artificial viscosity q that is similar to the form given in Sec. 4.4. For numerical stability, the time step is limited by

$$\Delta t < \frac{h_{\min}}{\sqrt{g^2 + \sqrt{g^2 + c^2}}}$$

where $g^2 = 4q/\rho$ and $h_{\min}$ is the minimum altitude of the triangular finite element. A slide line technique based upon conservation of translational and rotational momenta is also incorporated into the program.

4.6 Three-Dimensional Lagrangian Calculations and Multidimensional Eulerian Calculations

Some problems in terminal ballistics cannot be adequately simulated by two-dimensional calculations. A notable example is the oblique impact problem. For these problems, three-dimensional continuum mechanics codes must be utilized. Many two-dimensional finite difference and finite element codes have three dimensional counterparts. The extension to three dimensions introduces increased geometric complexity, but no new physical or mathematical concepts. In the three-dimensional, Lagrangian finite difference code HEMP 3D,[28] the integral method is used to approximate spatial derivatives. The geometry involved in this calculation is illustrated in Fig. 40. To approximate spatial derivatives at point 0, three-dimensional analogs of Eqs. (4.17) and (4.18) are used. In three dimensions, these formulas involve surface integrals instead of line integrals. The surface integral is evaluated by summing the contributions from each of the eight triangular surfaces forming the octahedron about point 0. The three-dimensional technique employs a calculational sequence that is similar to the technique used with two-dimensional methods. A bulk viscous pressure is used to simulate shock waves, and a tensor viscosity is used to suppress nonphysical grid oscillations. Slide surfaces, analogous to slide lines used in two-dimensional calculations, are also incorporated into most three-dimensional Lagrangian codes.

The Eulerian calculational technique has been extensively used to solve problems of two- and three-dimensional motion where large deformations occur. The extension of the one-dimensional method to multidimensional motion is relatively straightforward because of the fixed orthogonal finite difference mesh. The conservative form of the governing equations for two-dimensional motion of an inviscid fluid in Cartesian coordinates, analogous to Eqs. (4.5–4.7), can be written as

$$\frac{\partial \rho}{\partial t} = -\frac{\partial}{\partial x}(\rho v_x) - \frac{\partial}{\partial y}(\rho v_y)$$

$$\frac{\partial}{\partial t}(\rho v_x) = -\frac{\partial}{\partial x}(\rho v_x^2) - \frac{\partial}{\partial y}(\rho v_x v_y) - \frac{\partial P}{\partial x}$$

$$\frac{\partial}{\partial t}(\rho v_y) = -\frac{\partial}{\partial x}(\rho v_y v_x) - \frac{\partial}{\partial y}(\rho v_y^2) - \frac{\partial P}{\partial y}$$

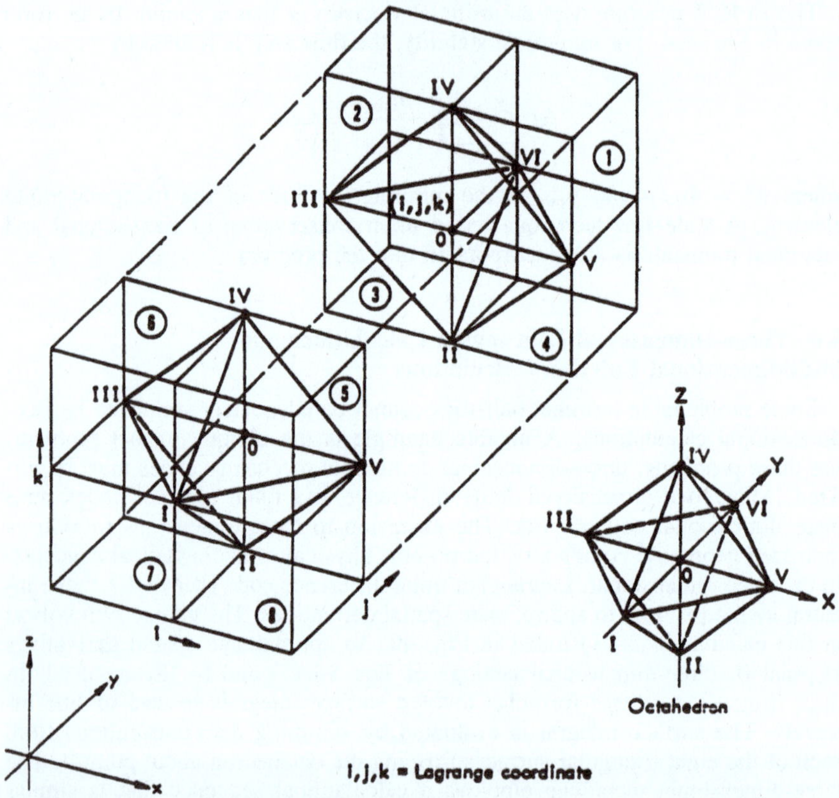

Fig. 40 Geometry of the three-dimensional mesh used to calculate spatial derivatives at point "0" (from Ref. 28). (Coutresy of University of California, Lawrence Livermore National Laboratory, and the Department of Energy).

$$\frac{\partial}{\partial t}(\rho e_t) = -\frac{\partial}{\partial x}(\rho e_t v_x) - \frac{\partial}{\partial y}(\rho e_t v_y) - \frac{\partial}{\partial x}(P v_x) - \frac{\partial}{\partial y}(P v_y)$$

The basic donor cell technique, illustrated in Sec. 4.3, can be directly applied to this system of equations. For example, the difference equation approximating the equation of motion in the x-direction takes the form

$$(\rho v_x)^{n+1}_{j,k} = (\rho v_x)^n_{j,k} - \frac{\Delta t}{\Delta x}\left[< \rho v_x >^n_{j+1/2,k}(v_x)^n_{j+1/2,k} - < \rho v_x >^n_{j-1/2,k}(v_x)^n_{j-1/2,k}\right]$$

$$-\frac{\Delta t}{\Delta y}\left[< \rho v_x >^n_{j,k+1/2}(v_y)^n_{j,k+1/2} - < \rho v_x >^n_{j,k-1/2}(v_y)^n_{j,k-1/2}\right]$$

$$-\frac{\Delta t}{\Delta x}\left[(P+q)^n_{j+1/2,k} - (P+q)^n_{j-1/2,k}\right]$$

where j and k indicate the center of a typical Eulerian zone bounded by the grid lines $x_{j-1/2}$, $x_{j+1/2}$, $y_{k-1/2}$, and $y_{k+1/2}$.

This difference equation is the two-dimensional analog of Eq. (4.9). The two-phase donor cell technique has been used in the HELP[14] code for two-dimensional calculations and in the METRIC[29] code for three-dimensional calculations. The two-step method of a Lagrangian phase followed by an Eulerian remap phase has been applied to two- and three-dimensional problems in the HULL[22] code.

A review of the application of hydrocodes to multidimensional problems in terminal ballistics has been given by Zukas in Chapters 10 and 11 of Ref. 30.

5. Illustrations of Hydrocode Results

In this section, several examples of results from hydrocode calculations of warhead problems are presented.

In Fig. 41, some results of the Lagrangian finite difference code HEMP are shown for the initial formation stage of a shaped charge device. This device, which was studied by Harrison,[31] consists of a hemispherical copper liner in contact with a cylindrical, explosive charge of octol. The velocity field is plotted above the horizontal axis of symmetry, and the current grid configuration is shown below. A slide line is used to allow the explosive gas to slide along the metal surface. Later stages of jet formation are shown in Fig. 42. In this figure, only the liner material is plotted. These graphs give the analyst a good qualitative picture of the motion and deformation developing in the liner. To obtain more quantitative information, contour plots (lines of constant values) of any of the dependent variables can be made. Figure 43 illustrates contour plots of density and pressure within the liner material at two different times. Time history plots of dependent variables at selected zones also provide useful quantitative information. For some problems, motion pictures of the deforming grid or other quantities have revealed significant features of the problem. Adequate graphical displays of the calculated results are a requirement for an effective hydrocode analysis.

At 35 μs, the liner material has experienced large deformations, as indicated by Fig. 42. Engineering strains in the axial direction for zones near the axis are over 2000%, and, to continue the solution, the liner should be rezoned to improve the aspect ratio of the elongated zones.

In Fig. 44 is illustrated the calculated formation of an explosively formed projectile (EFP). This calculation was performed by Bender[32] with the Lagrangian finite-element code EPIC-2. This axisymmetric device consists of a tantalum liner in contact with an LX-14 explosive charge that is confined by a steel casing and initiated on the axis at the rear. The complete device is shown initially in Fig. 44, but only the mesh corresponding to the liner is shown at subsequent times. After 60 μs, the mesh representing the explosive and casing was deleted from the remainder of the calculation. This deletion is justified because pressures in the explosive have become negligible and no longer affect the liner motion. This deletion technique shortens the calculation time and avoids eventual zone entanglement problems. The final calculated shape and velocity of the liner agree well with experimental results.

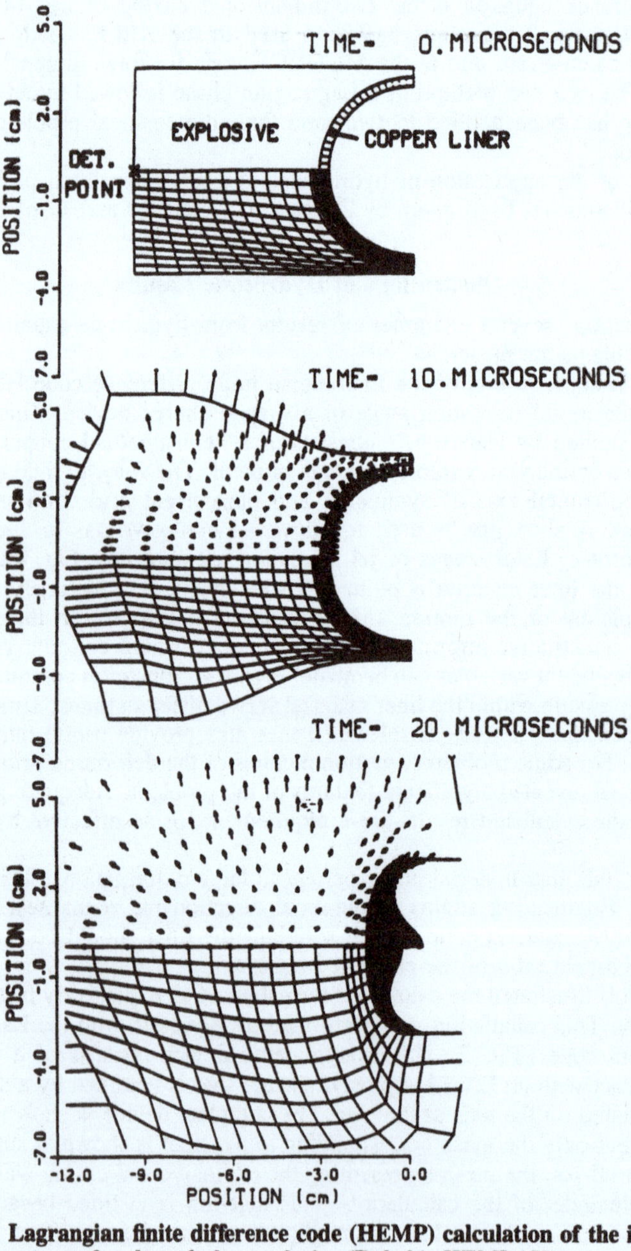

Fig. 41 Lagrangian finite difference code (HEMP) calculation of the initial jet formation stage of a shaped charge device (Ref. 31, HEMI 150).

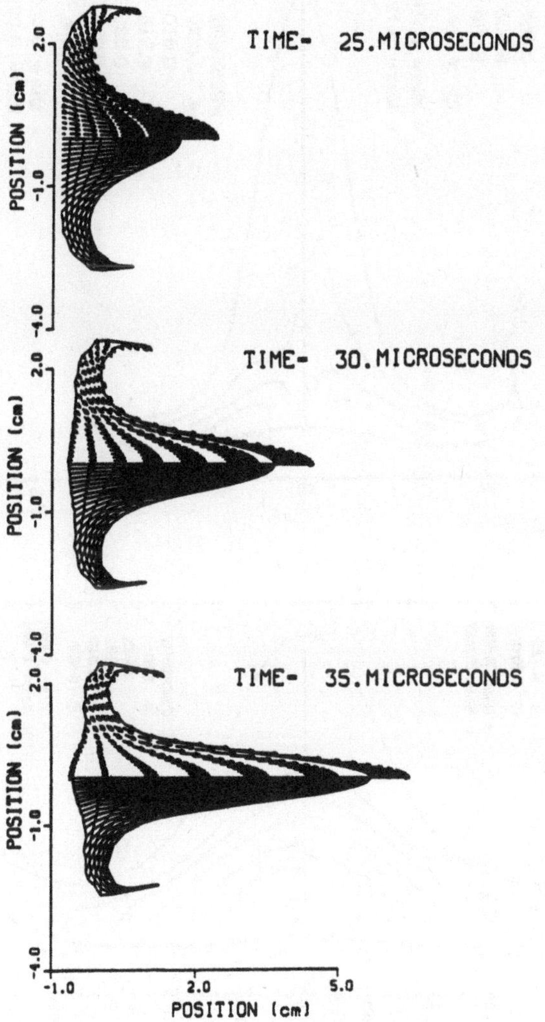

Fig. 42 Lagrangian finite difference code (HEMP) calculation of the intermediate jet formation stage of the shaped charge shown in Fig. 5.1.

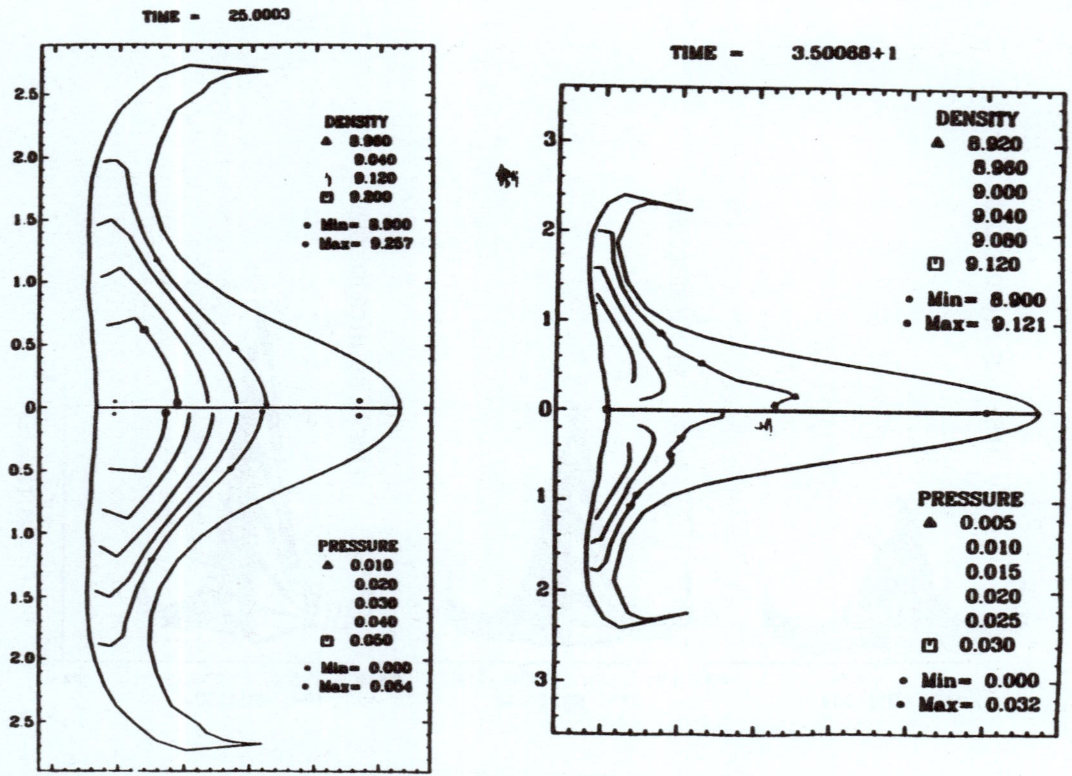

Fig. 43 Contour plots of density and pressure obtained from HEMP code calculations of the shaped charge shown in Fig. 5.1 (density in gm/cc and pressure in Mbar.)

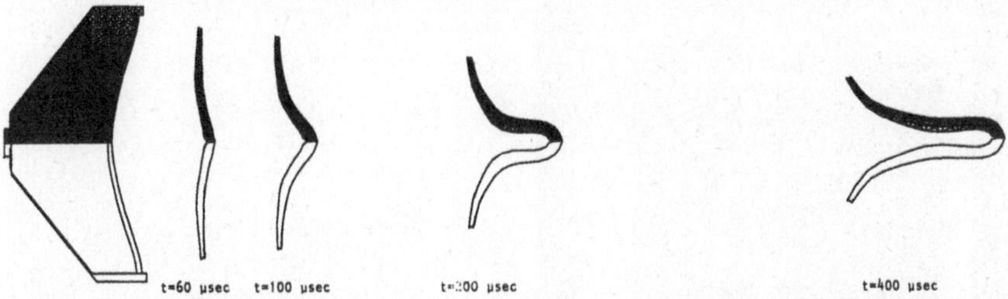

Fig. 44 Lagrangian finite element code (EPIC-2) calculation of the formation of an EFP (from Ref. 32).

In Fig. 45 the results of a deep penetration calculation performed with a modified version of the EPIC-2 code are illustrated. The problem consists of a thick steel target impacted by a long steel rod (L/D= 10) moving with an initial velocity of 3.114 km/s. In a typical Lagrangian calculation without rezoning, calculations of deep penetrations are usually not possible because of severe distortions of the grid. However, in this calculation, the sliding interface algorithm has been restructured to simulate contact surface erosion during impact. This method can be thought of as a localized rezoning technique where elements that have failed according to some criterion are removed from the remainder of the calculation. Details of this approximate technique, which is currently under active investigation, can be found in Ref. 33 and in the references cited there. In this calculation and

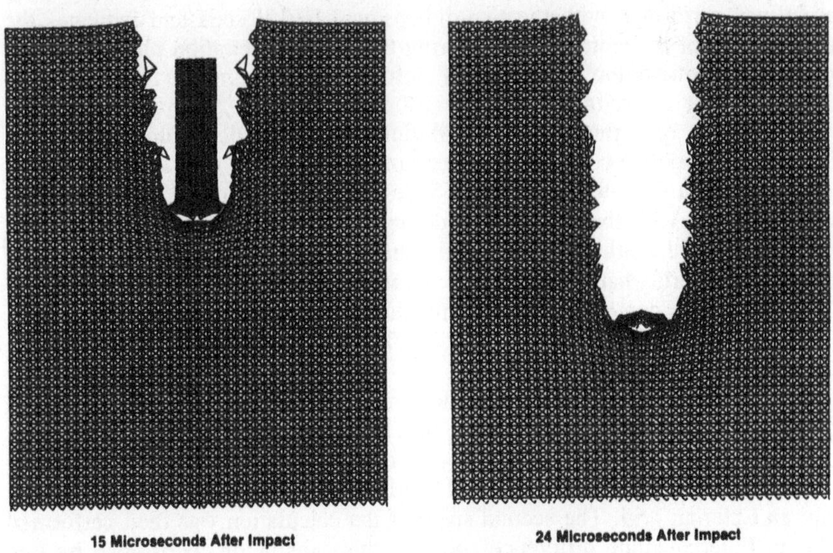

Fig. 45 Calculated penetration of a long rod into a thick armor target computed with a modified version of the EPIC-2 code (from Ref. 33).

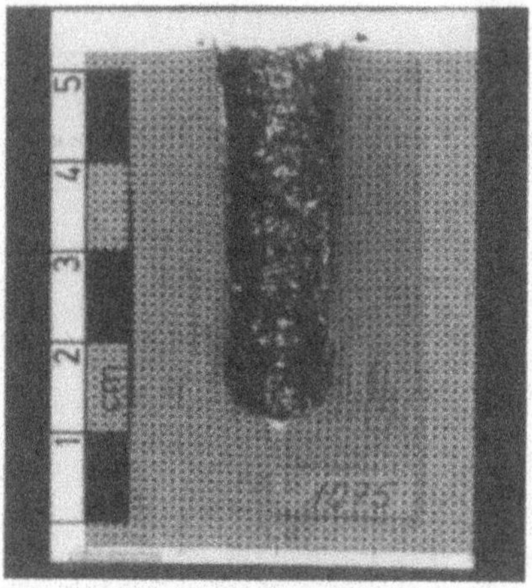

Fig. 46 Overlay of computed and experimental hole profiles for the calculation illustrated in Fig. 5.5 (from Ref. 33).

in the corresponding experiment, the rod material is completely eroded during the penetration process. A comparison between the calculated and experimental hole profiles is shown in Fig. 46.

The three-dimensional calculation of an oblique impact problem is illustrated in Fig. 47. The calculation was performed with the three-dimensional, finite-element code EPIC-3. At 23 μs after impact, the calculation indicates that the front portion of the rod ricochets from the surface of the plate.

In Fig. 48 is shown an Eulerian finite difference code calculation of the hypervelocity impact ($V = 5$ km/s) of an aluminum rod into water. The calculation was performed with the two-dimensional HELP code and indicates the rapid erosion of the aluminum rod during the initial penetration phase followed by a slower penetration of the highly deformed aluminum. At the end of the calculation, the cavity in the water is still expanding.

An example of a three-dimensional Eulerian code calculation is shown in Fig. 49. The problem consists of a 3 kg tungsten alloy rod with a hemispherical cap impacting a thin steel plate at a 65-deg angle of obliquity. The calculation was performed with the METRIC code and indicates perforation of the plate. Although the calculational grid is Eulerian, the plots shown in Fig. 49 are of a Lagrangian grid that indicates the deformation occurring in the materials. Lagrangian grids can be plotted by embedding tracer particles in the materials. These particles are moved with the local flow velocity and can be used as the nodes of a Lagrangian grid.

Figs. 50 and 51 illustrate a simulation of jet formation from a shaped charge that was calculated in two stages. The first stage consists of a Lagrangian finite-difference code calculation similar to the one illustrated in Figs. 41–43. After 15 μs of problem time, the results of the Lagrangian calculation were mapped onto an Eulerian grid. The second stage of the calculation was then performed with an Eulerian finite difference code. In this calculation, performed by Lee and Kwak,[36] the second stage was calculated with a second-order-accurate PIC code. This technique is efficient for problems involving thin liners. If an

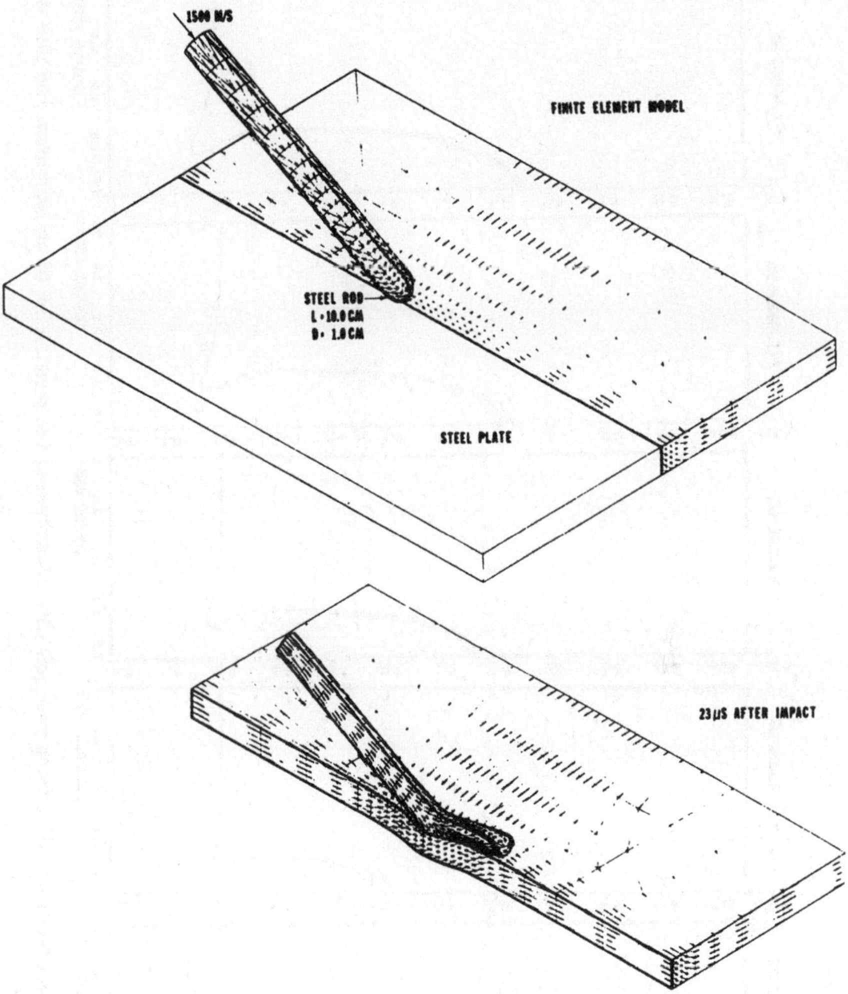

Fig. 47 Oblique impact (60 deg from normal) of a steel rod onto a steel plate at 1500 m/s calculated with the EPIC-3 code (from Ref. 34).

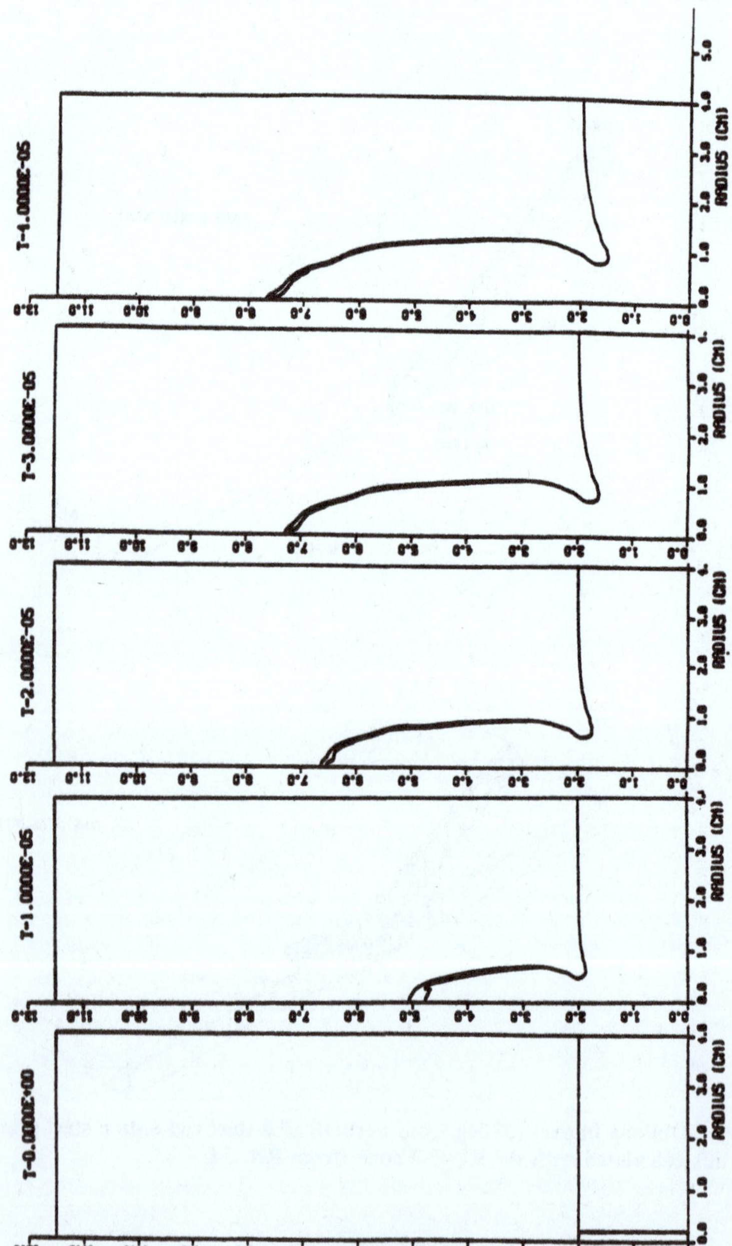

Fig. 48 Eulerian finite difference code (HELP) calculation of the penetration of an aluminum rod into water ($V = 5$ km/s).

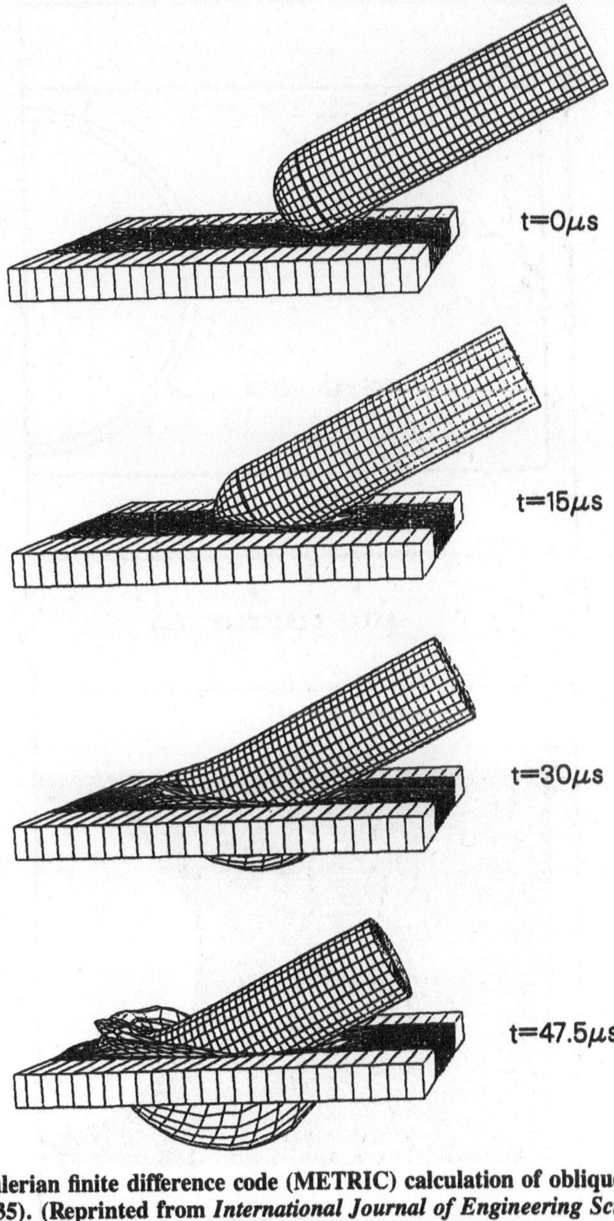

Fig. 49 Eulerian finite difference code (METRIC) calculation of oblique perforation (from Ref. 35). (Reprinted from *International Journal of Engineering Science*, Vol. 16, No. 7, R. T. Sedgwick, W. J. Hageman, R. G. Herrmann, and J. L. Waddell, "Numerical Investigations in Penetration Mechanics," pp. 859–869. Copyright 1978, with kind permission from Pergamon Press, Ltd, Headington Hill Hall, Oxford, OX3 OBW, UK).

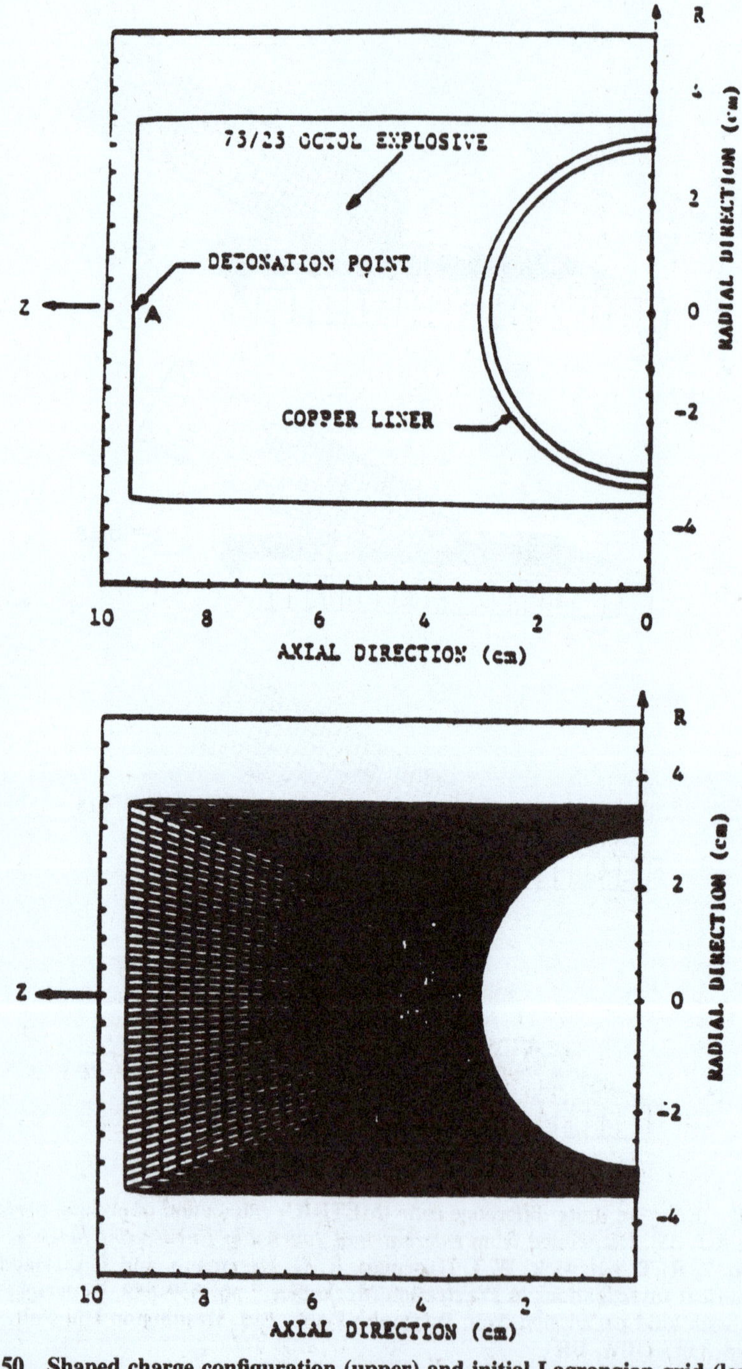

Fig. 50 Shaped charge configuration (upper) and initial Lagrangian grid (lower) for the two-stage Lagrangian/Eulerian calculation of jet formation (from Ref. 36).

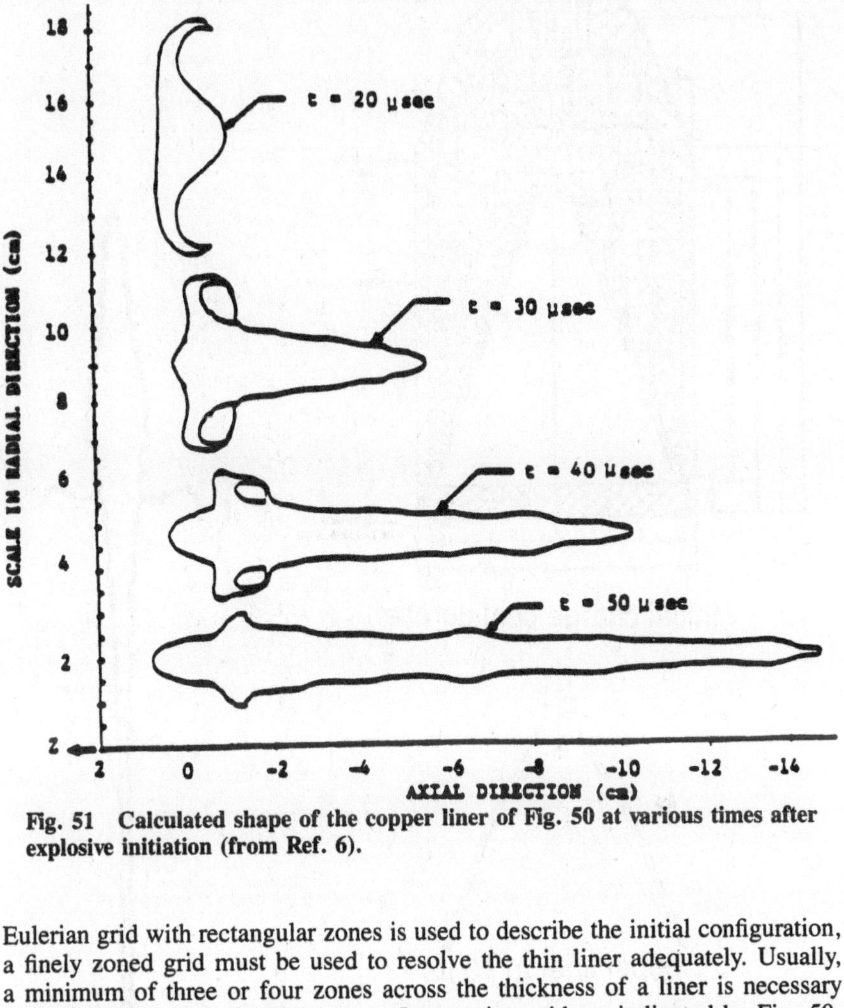

Fig. 51 Calculated shape of the copper liner of Fig. 50 at various times after explosive initiation (from Ref. 6).

Eulerian grid with rectangular zones is used to describe the initial configuration, a finely zoned grid must be used to resolve the thin liner adequately. Usually, a minimum of three or four zones across the thickness of a liner is necessary for reasonable accuracy. However, a Lagrangian grid, as indicated by Fig. 50, can be used to zone the problem efficiently since the grid conforms to the shape of the materials. The grid shown in Fig. 50 consists of eleven zones across the liner thickness and ninety zones normal to the liner thickness. With this zoning, the Lagrangian stage of the calculation should be quite accurate. Harrison has also developed a two-stage calculational technique for solving shaped charge problems.[31]

In Fig. 52 an illustration of the simulation of a conical shaped charge computed with the two-stage Lagrangian/Eulerian technique is shown.

Several additional examples of hydrocode simulations and a review of available three-dimensional production codes can be found in Ref. 30. A partial listing of hydrocodes that have been used on a production basis for warhead simulations is given in Table 7.

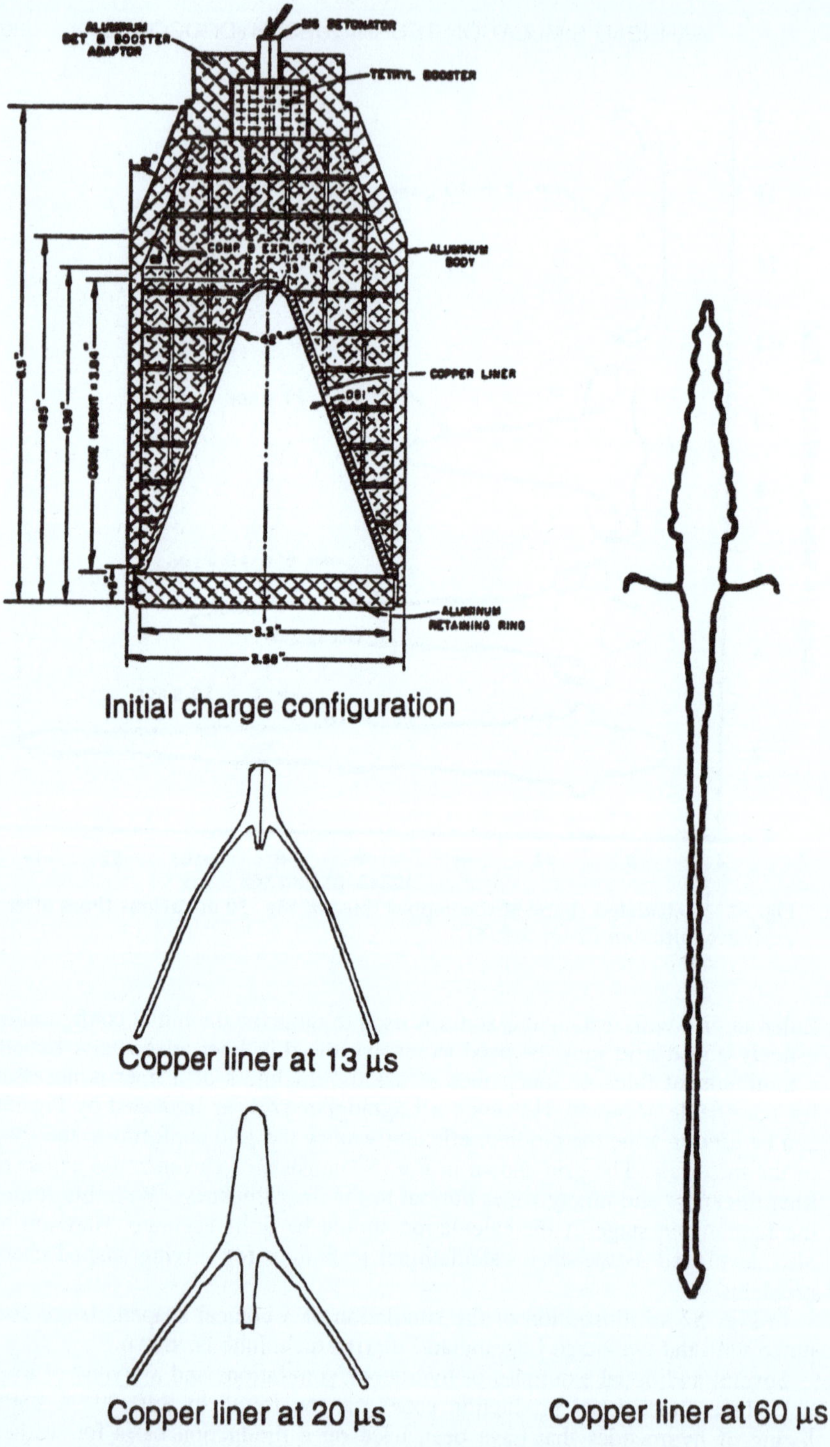

Fig. 52 Simulation of the jet formation process of the BRL precision shaped charge (from Wen Ho Lee, Los Alamos National Laboratory, private communication, 1988).

Table 7 Partial listing of hydrocodes

Code name	Dimension	Type	References	Developers
	Lagrangian codes			
KO	1-D	F-D[a]	18	M. L. Wilkins, et al., Lawrence Livermore National Laboratories, Livermore, CA
HEMP	2-D	F-D	37	
HEMPDS	2-D	F-D	38	
HEMP3D	3-D	F-D	28	
WONDY	1-D	F-D	19	B. J. Thorne, W. Herrmann, et al., Sandia Laboratories, Albuquerque, NM
TOODY	2-D	F-D	26	
TOOREZ	2-D	Rezoning Code	39	
THREEDY	3-D	F-D		
EPIC-2	2-D	F-E[b]	13	G. R. Johnson, Honeywell, Inc., Hopkins, MN
EPIC-3	3-D	F-E	40	
DYNA2D	2-D	F-E	41	J. O. Hallquist, Lawrence Livermore National Laboratories, Livermore, CA
DYNA3D	3-D	F-E	42	
DEFEL	2-D	F-E	43	W. J. Flis, Dyna East Corp., Philadelphia, PA
	Eulerian codes			
MESA	2-D	F-D	49	D. J. Cagliostro, et al., Los Alamos National lab, Los Alamos, NM
	3-D	F-D	49	

(continued on next page)

[a]F-D = finite difference method.

[b]F-E = finite element method.

[c]Proprietary to PISCES International.

Table 7(cont.) Partial listing of hydrocodes

Code name	Dimension	Type	References	Developers
Eulerian Codes				
CTH	2-D 3-D	F-D F-D	50 50	J. M. McGlaun, et al., Sandia National Lab, Albuquerque, NM
HULL	2-D 3-D	F-D F-D	22 22	D. A. Matuska, R. E. Durrett, Orlando Technology, Inc., Shalimar, FL
HELP METRIC	2-D 3-D	F-D F-D	14 29	L. J. Hageman, et al., Systems, Science and Software, Inc., La Jolla, CA.
CSQ	2-D	F-D	23	S. L. Thompson, Sandia Laboratories, Albuquerque, NM
JOY	3-D	F-D	44	R. Couch, Lawrence Livermore National Laboratories, Livermore, CA
Hybrid codes (codes using both Eulerian and Lagrangian grids				
HULL(Eulerian/Lagrangian) HULL/EPIC-3	2-D, 3-D 3-D	F-D F-D/F-E	22 45	D. A. Matuska, J. J. Osborn, Orlando Technology, Inc., Shalimar, FL
HOIL	2-D	F-D	31	J. T. Harrison, Army Ballistic Research Laboratories, Aberdeen Proving Ground, MD
PISCES-2DELK[c] PISCES-3DELK[c]	2-D 3-D	F-D F-D	46 47	S. L. Hancock, Physics International, San Leandro, CA

References

[1]Eringen, A. C., *Nonlinear Theory of Continuous Media*, McGraw-Hill, New York, 1962.

[2]Fung, Y. C., *Foundations of Solid Mechanics*, Prentice-Hall, Englewood Cliffs, NJ, 1965.

[3]Malvern, L. E., *Introduction to the Mechanics of a Continuous Medium*, Prentice-Hall, Englewood Cliffs, NJ, 1969.

[4]Aris, R., *Vectors, Tensors, and the Basic Equations of Fluid Mechanics*, Prentice-Hall, Englewood Cliffs, NJ, 1962.

[5]Lee, E. L., Hornig, H. C., and Kury, J. W., "Adiabatic Expansion of High Explosive Detonation Products," Lawrence Livermore Lab. Rept. UCRL-50422, 1968.

[6]Dobratz, B. M., and Crawford, P. C., "Properties of Chemical Explosives and Explosive Simulants," Lawrence Livermore National Lab. Explosives Handbook UCRL-52997 Change 2, Jan. 1985, pp. 8–21 to 8–23.

[7]McQueen, R. G., Marsh, S. P., Taylor, J. W., Fritz, J. N., and Carter, W. J., "The Equation of State of Solids from Shock Wave Measurements," *High-Velocity Impact*, edited by R. Kinslow, Academic Press, New York, 1970, Chap. 7.

[8]Tillotson, J. H., "Metallic Equations of State for Hypervelocity Impact," General Atomic Rept. GA-3216, July 1962.

[9]Richtmyer, R. D., and Morton, K. W., *Difference Methods for Initial-Value Problems*, Interscience, Wiley, New York, 1967.

[10]Karpp, R. R., and Chou, P. C., "The Method of Characteristics," *Dynamic Response of Materials to Intense Impulsive Loading*, Air Force Materials Laboratory, Wright Patterson AFB, Dayton, OH, 1972, Chap. 6.

[11]Lax, P. D., and Richtmyer, R. D., "Survey of the Stability of Linear Finite Difference Equations," *Communications of Pure and Applied Mathematics*, Vol. 9, No. 2, 1956, pp. 267–293.

[12]Zienkiewicz, O. C., and Morgan, K., *Finite Elements and Approximations*, Wiley, New York, 1983 pp. 272–275.

[13]Johnson, G. R., "EPIC-2, A Computer Program for Elastic-Plastic Impact Computations in 2 Dimensions Plus Spin," Ballistic Research Lab. ARBRL-CR-00373, Aberdeen Proving Ground, MD, June 1978.

[14]Hageman, L. J., Wilkins, D. E., Sedgwick, R. T., and Waddell, J. L., "HELP: A Multi-Material Eulerian Program for Compressible Fluid and Elastic-Plastic Flows in Two Space Dimensions and Time," Systems, Science and Software Rept. SSS-R-75-2654, La Jolla, CA, July 1975.

[15]Hirt, C. W., Amsden, A. A., and Cook, J. L., "An Arbitrary Lagrangian-Eulerian Computing Method for All Flow Speeds," *Journal of Computational Physics*, Vol. 14, No. 3, 1974, pp. 227–253.

[16]Crowley, W. P., "FLAG: A Free-Lagrangian Method for Numerically Simulating Hydrodynamic Flows in Two Dimensions," *Proceedings of the Second International Conference on Numerical Methods in Fluid Dynamics, Lecture Notes in Physics*, Vol. 8, Springer-Verlag, New York, 1971.

[17]Von Neumann, J., and Richtmyer, R. D., "A Method for the Numerical Calculation of Hydrodynamic Shocks," *Journal of Applied Physics*, Vol. 21, No. 3, 1950 pp. 232–237.

[18]Wilkins, M. L., "Calculation of Elastic-Plastic Flow," Lawrence Livermore Lab. Rept. UCRL-7322, rev. 1, 1969.

[19]Herrmann, W., Holzhauser, P., and Thompson, R.J., "WONDY, A Computer Program for Calculating Problems of Motion in One Dimension," Sandia Lab., Rept. SC-RR-66-601, Albuquerque, NM, Feb. 1967.

[20]Hirt, C. W., "Heuristic Stability Theory for Finite-Difference Equations," Journal of Computational Physics, Vol. 2, No. 4, 1968, 339–355.

[21]Kershner, J. D., and Mader, C. L. "2 DE: A Two-Dimensional Continuous Eulerian Hydrocode for Computing Multicomponent Reactive Hydrodynamic Problems," Los Alamos Scientific Lab. Rept. LA-4846, March 1972.

[22]Matuska, D. A., and Osborn, J. J., "Hull Documentation, Volume I, Technical Discussion," Orlando Technology, Inc., Shalimar, FL, 1985.

[23]Thompson, S. L., "CSQ—A Two-Dimensional Hydrodynamic Program with Energy Flow and Material Strength," Sandia Lab. Rept. SAND 74-0122, 1975.

[24]Evans, M. W., and Harlow, F. H. "The Particle-in-Cell Method for Hydrodynamic Calculations," Los Alamos Scientific Lab. Rept. LA-2139, 1957.

[25]Van Leer, B., "Towards the Ultimate Conservative Difference Scheme. IV. A New Approach to Numerical Convection," Journal of Computational Physics, Vol. 23, No. 3, 1977, pp. 276–299.

[26]Thorne, B. J. and Herrmann, W., "Toody, A Computer Program for Calculating Problems of Motion in Two Dimensions," Sandia Lab., SC-RR-66-602, July 1967; and Swegle, J. E., "Toody IV—A Computer Program for Two-Dimensional Wave Propagation," Sandia Lab. Rept. SAND-78-0552, Albuquerue, NM, Sept. 1978.

[27]Amsden, A. A., Ruppel, H. M., and Hirt, C. W., "Sale: A Simplified ALE Computer Program for Fluid Flow at All Speeds," Los Alamos National Lab. Rept. LA-8095, June 1980.

[28]Wilkins, M. L., Blum, R. E., Cronshagen, E., and Grantham, P., "A Method for Computer Simulation of Problems in Solid Mechanics and Gas Dynamics in Three Dimensions and Time," Lawrence Livermore National Lab. Rept. UCRL-51574 rev. 1, May 1975.

[29]Hageman, L. J., and Lee, E. P., "Development of an Ordnance Velocity Multimaterial 3-D Perforation Code for Finite Plates," Systems, Science and Software, La Jolla, CA, SSS-R-76-2973, Feb. 14, 1976.

[30]Zukas, J. A., *Impact Dynamics*, Wiley, New York, 1982.

[31]Harrison, J. T., "A Two Stage, Hydrodynamic, Numerical Technique, 'Hoil' and an Analysis of the Results of a Hemispherical, Shaped Charge Liner Collapse," Ballistic Research Lab. Memorandum Rept. BRL-MR-3494, Aberdeen Proving Ground, MD, Jan. 1986.

[32]Bender, D., Private Communication, Aerojet Ordnance Company, Tustin, CA, 1986.

[33]Kimsey, K. D., and Zukas, J. A., "Contact Surface Erosion for Hypervelocity Problems," Ballistic Research Lab. Memorandum Rept. BRL-MR-3495, Aberdeen Proving Ground, MD, Feb. 1986.

[34]Johnson, G. R., "A New Computational Technique for Intense Impulsive Loads," Proceedings of the Third International Symposium on Ballistics, Karlsruhe, Germany, March 1977.

[35]Sedgwick, R. T., Hageman, L. J., Herrmann, R. G., and Waddell, J. L., "Numerical Investigations in Penetration Mechanics," *International Journal of Engineering Science*, Vol. 16, No. 7, 1978, pp. 859–869.

[36]Lee, W. H. and Kwak, D., "On the Pic Method for Modelling the Shaped-Charge Problems," Proceedings of the Ninth International Symposium on Ballistics, Shrivenham, UK, April 1986, pp. 2-253 to 2-262.

[37]Giroux, E. D., "Hemp User's Manual," Lawrence Livermore National Lab. Rept. UCRL-51079 rev. 1, 1973.

[38]Warren, K. H., "HEMPDS User's Manual," Lawrence Livermore National Lab. Rept. UCID-18075 rev. 1, 1983.

[39]Thorne, B. J. and Holdridge, D. B., "The Toorez Lagrangian Rezoning Code," Sandia Lab. Rept. SLA-1057, 1974.

[40]Johnson, G. R., "EPIC-3, a Computer Program for Elastic-Plastic Impact Calculations in 3 Dimensions," Ballistic Research Lab. CR No. 343, Aberdeen Proving Ground, MD, 1977.

[41]Hallquist, J. O., "User's Manual for DYNA2D—An Explicit Two-Dimensional Hydrodynamic Finite Element Code with Interactive Rezoning," Lawrence Livermore Lab. Rept. UCID-18756, rev. 2, 1984.

[42]Hallquist, J. O., "User's Manual for DYNA3D and DYNAP (Nonlinear Dynamic Analysis of Solids in Three Dimensions)," Lawrence Livermore Lab. Rept. UCID-19156, Livermore, CA, 1981.

[43]Flis, W. J., Miller, S., and Clark, W. J., "DEFEL: A Finite-Element Hydrodynamic Code," Dyna East Corp. No. DE-TR-84-05, Philadelphia, PA, 1984.

[44]Couch, R., Albright, E. and Alexander, N., "The Joy Computer Code," Lawrence Livermore National Lab. Rept. UCID-19688, Livermore, CA, 1983.

[45]Matuska, D. A., and Osborn, J. J., "HULL/EPIC-3 Linked Eulerian/Lagrangian Calculation in Three Dimensions," Ballistic Research Lab. ABRL-CR-00467, Aberdeen Proving Ground, MD, 1981.

[46]Hancock, S. L., "Pisces-2Delk User's Manual," Physics International, San Leandro, CA, 1985.

[47]Cowler, M., "Pisces-3Delk User's Manual," Physics International, San Leandro, CA, 1983.

[48]Lee, W. H., private communication, Los Alamos National Laboratory, Los Alamos, NM, 1988.

[49]Cagliostro, D. J., Mandell, D. A., Schwalbe, L. A., Adams, T. F., and Chapyak, E. J., "MESA 3-D Calculations of Armor Penetration by Projectiles with Combined Obliquity and Yaw," *International Journal of Impact Engineering*, Vol. 10, No. 1-4, 1990, pp. 81–92.

[50]McGlaun, J. M., Thompson, S. L., and Elrick M. G., "CTH: A Three-Dimensional Shock Wave Physics Code," *International Journal of Impact Engineering*, Vol. 10, No. 1-4, 1990, pp. 351–360.

Chapter 6

Mechanics of Shaped Charges

Joseph Carleone*
Aerojet General Corporation, Azusa, California 91702

THIS chapter will focus upon shaped charge mechanisms and properties of the resulting jet. In this chapter the term *shaped charge* will refer to all jet-forming warheads with liner shapes of many types. The presentation assumes some familiarity with the classical shaped charge and will not address the historical background of the device, which has been adequately covered in the excellent text by Kennedy[1] and the book on fundamentals of shaped charges by Walters and Zukas.[2] Emphasis in this chapter will be placed on the developments in shaped charge mechanics since 1970. It is intended to be the extension of and amplification to the theory of shaped charge mechanics presented by Walters and Zukas.[2] Discussion includes generally shaped liners, improved collapse formulas including acceleration of liner elements, formulas for the estimation of jet radius and jet stretching rate, and jet particulation or breakup formulas.

An engineering discussion of shaped charge configuration is first given. Each of the key shaped charge components is discussed, that is, liner configuration, explosive configuration, method of initiation, case configuration, and assembly. A generalized jet formation theory is then presented. This theory extends former theories to include a number of effects, such as finite acceleration time for liner elements, as well as formulas for the jet properties (mass, velocity, strain, etc.) and criteria for jet tip formation.

A jet particulation model is presented in detail. This model demonstrates that particulation is controlled by a plastic instability driven by the jet material's strength. Formulas usable in a comprehensive shaped charge model are also developed, including jet particulation time and a method to compute the final number of jet segments.

The chapter concludes with a series of parametric calculations examining the effects of liner wall thickness and cone angle on the ensuing jet properties.

*Vice-President, Armament Programs, Electronic Systems Division.

1. Shaped Charge Configurations

Various liner and explosive charge configurations have evolved since the invention of the shaped charge. It is important to point out that there is no one optimum design but that the shaped charge must be optimized within the constraints and requirements of the particular weapon system. In this section we will briefly discuss the more common warhead configurations and define terminology. The rationale for these different shapes will become more apparent after a detailed discussion of jet formation is presented. Although we made a clear distinction in the Introduction between a shaped charge with a conical liner and a hemispherical liner, for the purposes of this chapter we will refer to them each as special cases of the family of shaped charges.

1.1 Liner Configuration

The early work on shaped charges led to the cone as the most reliable liner shape, although many other shapes had been pursued. With the advent of improved manufacturing and inspection technology, which enabled the fabrication of high-precision complex shapes, and the introduction of computer technology, which enabled the exploration of many configurations without the cost of testing them all, more complex liners became more prevalent in the 1970s. The more common shapes are shown in Fig. 1.

The biconic design evolved as an attempt to make more efficient use of the apex region of the standard cone. Carleone et al. found that approximately 40% of the cone height of a 42 deg copper conical liner was responsible for the jet tip particle and did not contribute to the remainder of the jet.[3] Therefore, by redesigning this apex region the jet mass and energy could be more efficiently used. The biconic liner is one attempt to use the region near the apex more effectively. It produces an elongated tip region of the jet and with the proper wall thickness the biconic design can produce jet tip speeds close to 10 km/s. It also produces a somewhat nonlinear velocity gradient that stretches the rear section of the jet at a faster rate and tends to enhance the penetration of the jet.

The biconic design naturally leads to a trumpet configuration, which consists of a continuously varying liner angle. This angle is small (typically 30 deg) toward the axis of the charge and gradually gets larger toward the outer diameter of the charge. The trumpet gives the designer more freedom than the biconic to optimize the jet properties for high tip speed and large stretching rate to achieve maximum depth of penetration in homogeneous armor.

An alternative approach is the tulip liner, also shown in Fig. 1. This design first evolved as an approach to use the explosive energy more efficiently in collapsing the liner.

In the apex region of a standard cone, say 42 deg, there is not enough distance to the axis to allow full acceleration of the liner elements. Therefore, by moving these elements away from the axis, they will arrive at the axis with higher velocity. Also, less explosive will be used in the process. Thus, the tulip concept began as an approach to maximize jet energy for a given amount of explosive mass. This approach tends to yield a hole diameter that is somewhat larger for a given amount of explosive mass.[4] More recently, the tulip liner concept has been

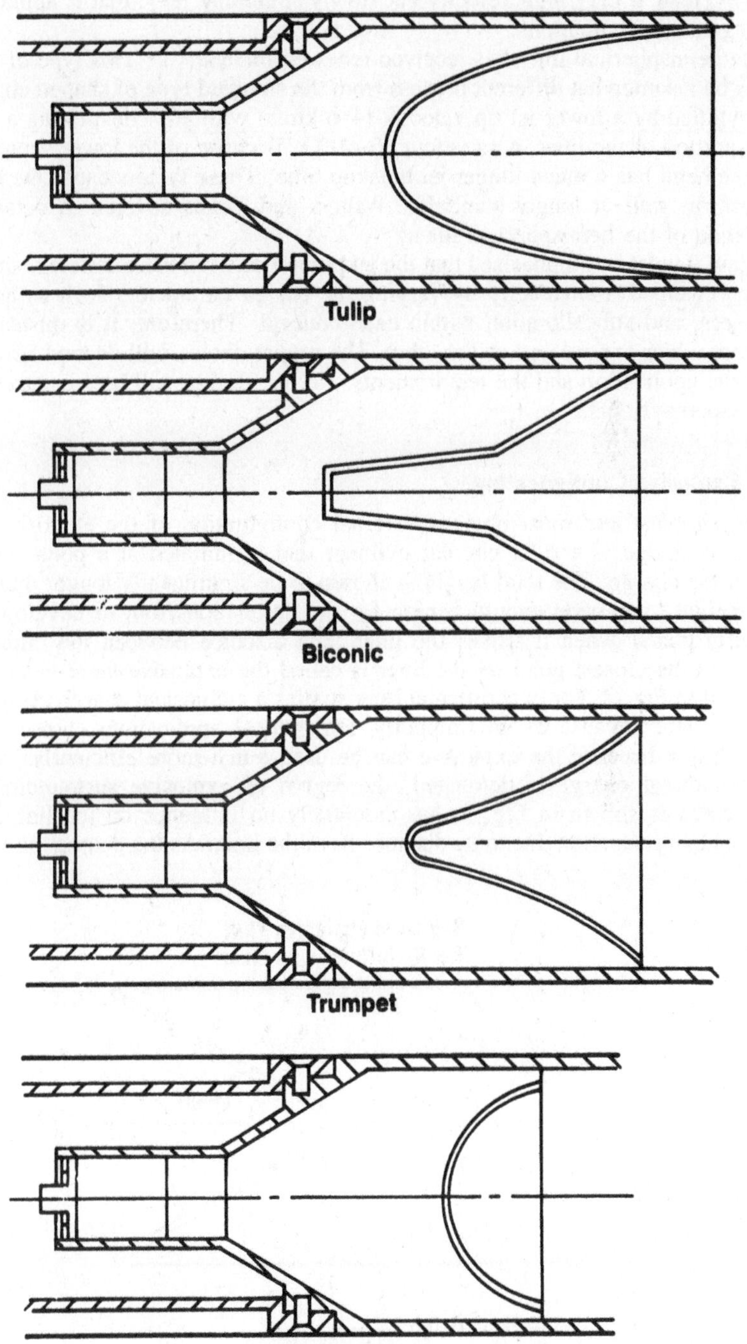

Fig. 1 Common warhead liner design concepts.

used to create a very high velocity but slowly stretching jet[5,6] that is applicable to long standoff situations.

The hemispherical liner has received renewed interest.[7–12] This type of liner forms by a somewhat different process from the standard type of shaped charge. It is typified by a lower jet tip velocity (4–6 km/s) with jets constituting a very large portion of the liner mass (about 60–80%). Because of the lower stretching rate the hemi has a much longer jet breakup time. These factors cause the hemi to perform well at longer standoffs. Walters and Zukas covered in detail jet formation of the hemispherical liner.[2]

Again it must be emphasized that the jet properties from these different shapes can be changed dramatically by varying the design parameters such as angle, thickness, and subcalibration within each concept. Therefore, it is misleading to adopt a bias toward one or the other. The proper choice will depend strongly upon the application and the requirements. No one design will be a panacea for all weapon systems.

1.2 Explosive Configuration

The simplest and most obvious external configuration of the explosive for a shaped charge is a right circular cylinder that is initiated at a point on the axis of the charge. The total height is chosen to be significantly longer than the liner height to provide enough length for the detonation front to develop and be fairly planar when it strikes the liner. The distance between the initiation point and the closest point on the liner is called the explosive *head height*, as indicated in Fig. 2. For experimental liner studies a cylindrical charge geometry is convenient because of its simplicity. For tactical applications, however, it is too heavy because the explosive can be used much more efficiently. When the cylindrical charge is detonated, the region of explosive surrounding an inner core, as shown in Fig. 2, has essentially no influence on the liner collapse and jet formation due to its distance from the liner. As the detonation wave

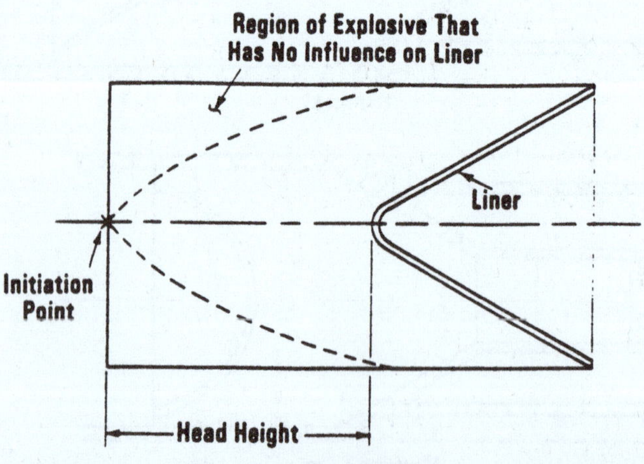

Fig. 2 External configuration of an explosive for a shaped charge.

reaches the liner, it applies a pressure pulse and begins to collapse the liner. The region in question will only change the impulse delivered to the liner at a time when release waves from the boundary reach the liner region. Therefore, if we remove the region shown, release waves will arrive at the liner somewhat sooner, but only after the liner has collapsed and the jet formed. Therefore, a region of explosive mass can be removed as long as the pressure pulse resulting from the new configuration is essentially the same as the pulse in the original configuration at early times and differs only at times after liner collapse.

For these reasons, most tactical shaped charge warheads do not have a right circular configuration, but the initiation end has a conical or contoured surface. This shape is often called a *boat-tail*.

The length of the shaped charge can also be reduced. At a certain point, performance of the jet is reduced. By introducing inert barriers in the explosive not only can performance be regained but also enhancements can be achieved. A typical charge with a barrier is shown in Fig. 3.[13] In principle, the barrier directs the detonation wave to the periphery of the charge. It then moves forward in the form of a torus upon the liner, causing two major effects. The detonation sweep speed across the liner is increased, and the liner elements near the axis are accelerated at a greater rate because of the more normal impact of the detonation wave. This greater acceleration yields higher tip velocities and a faster stretching rate and hence a longer jet.

The use of such a barrier is often called *waveshaping*, and the barrier is usually called a *waveshaper*. This terminology is not truly precise. The barrier does not shape the wave but merely directs its path. There are techniques to shape waves by purposely varying the explosive composition. These techniques have been applied to shaped charges but are not in wide use and will not be covered in this chapter.

A similar effect, that is, *peripheral initiation*, can be achieved by the use of a multi-point initiator (MPI). In its basic form, an MPI is an explosive network embedded in a plastic block that accepts an initiation pulse at one point from a

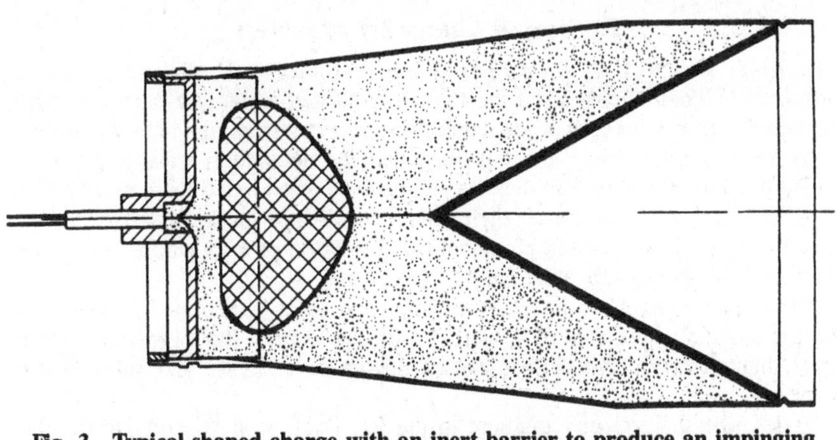

Fig. 3 Typical shaped charge with an inert barrier to produce an impinging toroidal detonation wave on the liner (design reproduced from Ref. 13).

detonator. Connected to this point are a number of channels filled with explosive whose detonation wave propagates at a very precise detonation rate. The lengths of the explosive filaments are also very precise. The explosive filaments lead to a number of booster pellets. Thus, a single point initiation can be expanded into a multiple point initiation. MPI devices can take many forms. For peripheral initiation the exposive filaments can expand radially outward to a number of boosters at the periphery. Such a device, with 20 peripheral initiation points, is shown in Fig. 4. This is a basic configuration. Very often, redundant paths are used to increase the reliability of the MPI device.

As in the case of the liner configuration, there is no single explosive configuration that works best for all applications. It is highly system dependent and must meet the overall requirements of the weapon system.

1.3 Case Configuration and Assembly

In most applications the shaped charge will have a casing. The casing serves to protect the explosive during handling and provides a structural member for assembly with the rest of the system. In some instances, the case will provide fragmentation capability as well. For cast explosives and some pressed explosives, the casing serves as the mold during the explosive loading operation. In any event, the casing will affect the operation and functioning of the shaped charge and must be considered in the design.

In general, the casing must be kept axisymmetric. Attachment points should be well aft and forward of the liner. For cast explosive charges the liner often has a precision flange that seats on a step in the case and is held in place by a retaining ring (other methods have also been developed). The metal parts align the liner axis to the case axis and the explosive is cast into the subassembly. This type of assembly is shown in Fig. 5 for the U.S. Army Ballistics Research Lab (BRL) 81.3-mm precision charge.[14] When pressed explosives are used, the liner is usually inserted directly into the explosive. A ring or other mechanical fastener is used to ensure positive mechanical retention of the liner.

2. Shaped Charge Jet Formation

In 1948, Birkhoff et al. published the first theory of shaped charge jet formation.[15] Their theory used the fact that the detonation pressure was large compared with the strength of the metal liner and, therefore, the liner was treated as an inviscid fluid. They also assumed that the detonation wave impulsively loaded the liner elements and thus instantaneously accelerated them to their final collapse velocity. Furthermore, their theory was based on steady-state, incompressible hydrodynamics and predicted the formation of a jet whose length was constant and equal to the slant height of the cone.

It has been experimentally observed, however, that jets formed from shaped charges contain a spectrum of velocities with the front of the jet traveling much faster than the rear. This effect causes the jet to elongate and subsequently break up.

To describe this velocity gradient in the jet, Pugh et al.[16] modified the theory of Ref. 15. This modified theory, known as the *nonsteady* theory, is based on

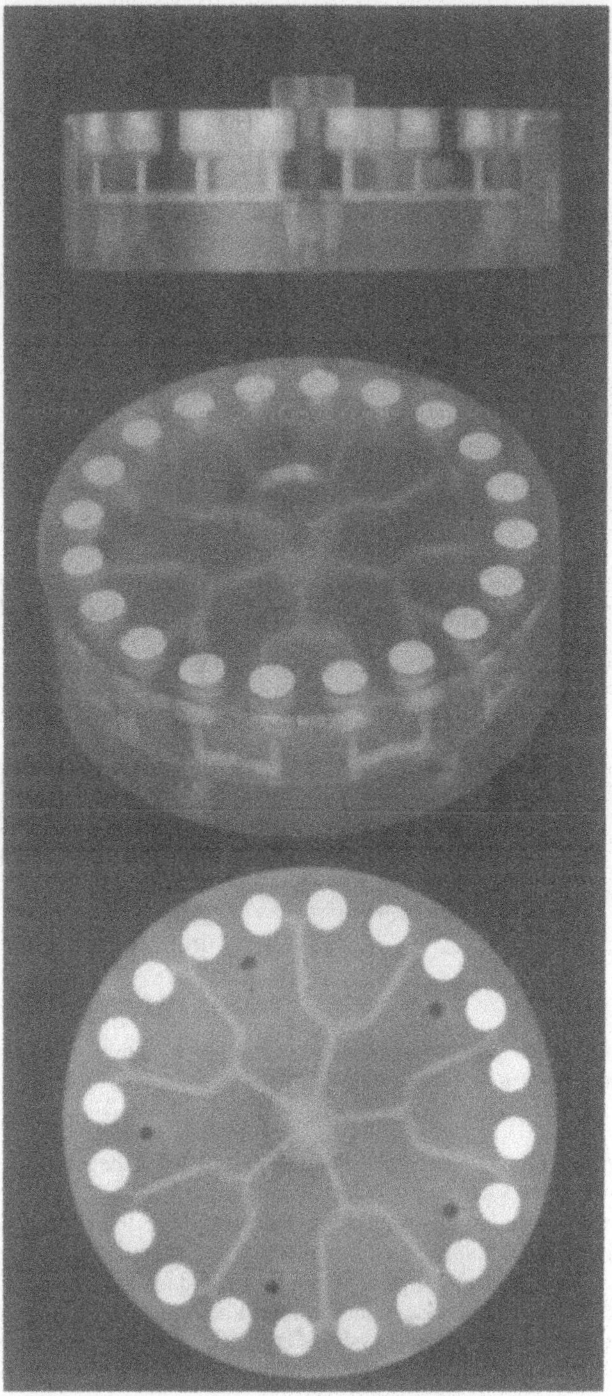

Fig. 4 Peripheral initiation device.

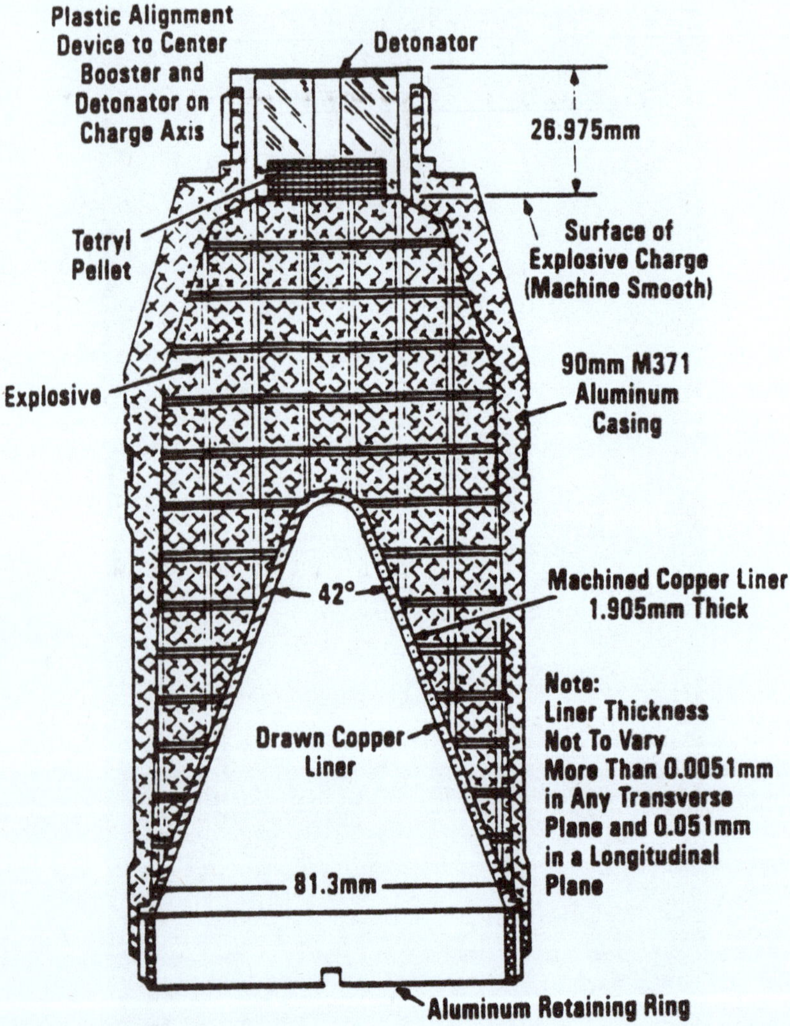

Fig. 5 U.S. Army Ballistics Research Laboratory 81.3-mm precision charge (reproduced from Ref. 14).

the same hydrodynamic concepts as the original theory, except that the velocities with which various elements collapse are not the same for all elements but depend upon the original position of the element in the liner. According to this modified theory, the collapse velocity decreases continuously from the cone apex to the base. This effect produces a significant elongation of the jet.

Several attempts have been made to verify the nonsteady theory experimentally. Eichelberger and Pugh demonstrated the qualitative accuracy of the theory.[17] In 1954, Eichelberger showed evidence of a systematic discrepancy between theoretical and experimental results.[18] Subsequently, Bryan et al. applied a radioactive tracer technique to shaped charges and obtained jet velocity as a function of original position in the liner.[19] In 1962, Allison and Vitali, using a more accurate radioactive tracer technique and a larger 105-mm shaped charge, obtained excellent agreement between theory and experiment.[20]

Carleone and Chou extended the theory of Ref. 16 to describe the jet properties including jet radius and elongation. These studies considered only a conical liner and a plane detonation wave.[21] In this section we will present a generalized theory of jet formation for shaped charges having a general liner and explosive configuration. It will also include the more recent collapse formulas presented in a previous chapter.

2.1 Brief Review of the PER Jet Formation Theory

In this section we will briefly review the basic assumptions and equations of the Pugh, Eichelberger, and Rostoker (PER) theory. For specific details see Ref. 16.

A diagram of the cone collapse and geometry is shown in Fig. 6. Consider a detonation wave that is normal to the axis of the cone and passes from P to Q in unit time. When the detonation wave reaches Q, let us say, the element originally at P collapses in a direction making an angle δ with the normal to the liner to point J on the axis. The contour of the collapsing liner is then given by QMJ, as shown. In Ref. 16, the authors comment that, although the changing velocities rule out the possibility of a true steady state, each zonal element of the liner can be considered separately in its appropriate constant velocity

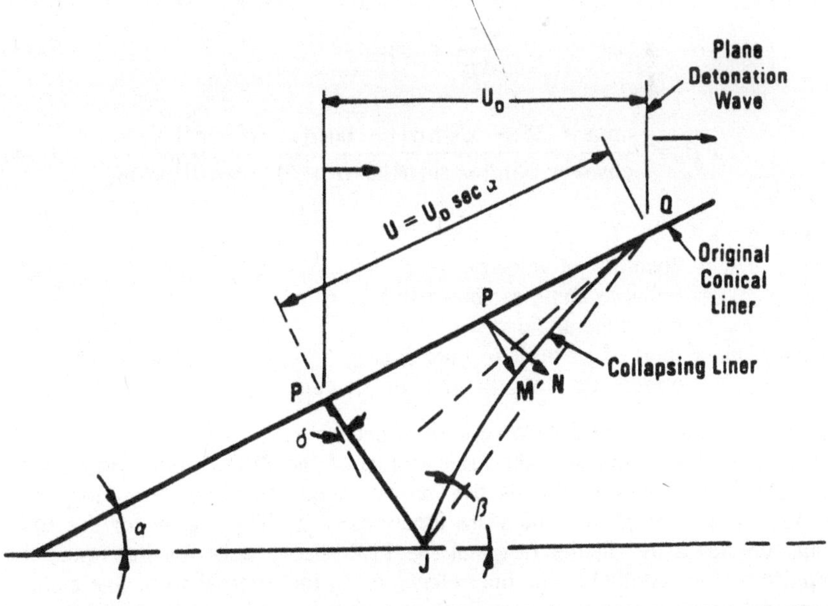

Fig. 6 Cross section showing collapse process geometry (reproduced from Ref. 16 by permission of American Institute of Physics).

coordinate system. This consideration is valid since one assumes that there is no interaction among liner elements during collapse. The detonation speed U_D is assumed constant, and the speed with which the wave sweeps across the liner is given by $U = U_D \sec \alpha$. By considering the detonation wave to supply only an instantaneous impulse to the liner elements, we conclude that, as observed in a system of coordinates moving along the liner from P to Q at a velocity of $U_D \sec \alpha$, a liner element changes only its direction when acted upon by the detonation wave, not its velocity. This conclusion leads to the following relationship between the angle α and the absolute collapse velocity V_0:

$$\sin \delta = \frac{V_0}{2U} \tag{1}$$

This expression often referred to as Taylor's relation, was originally developed for exploding cylinders.[22] Now consider the point J where the ring element collapses on the axis and splits into two segments, one of which is called the jet, the other the slug. In a system of coordinates moving with the point J, the flow enters the region and splits with flow in both directions. It is assumed that the velocity relative to the moving coordinates entering this stagnation region is equal to the relative velocity of both the jet and the slug.[16] Under this assumption and, of course, the assumption of incompressible flow, the following results are obtained:

$$V_j = V_0 \csc \frac{\beta}{2} \cos\left(\alpha + \delta - \frac{\beta}{2}\right) \tag{2}$$

$$\frac{\mathrm{d}m_j}{\mathrm{d}m} = \sin^2 \frac{\beta}{2} \tag{3}$$

$$\tan \beta = \frac{\sin(\alpha + 2\delta) - x \sin \alpha [1 - \tan \delta \tan(\alpha + \delta)] V_0' / V_0}{\cos(\alpha + 2\delta) + x \sin \alpha [\tan(\alpha + \delta) + \tan \delta] V_0' / V_0} \tag{4}$$

where

V_j = absolute jet velocity
β = collapse angle as shown in Fig. 6
α = half cone angle
$\mathrm{d}m_j/\mathrm{d}m$ = ratio of jet mass to liner mass
x = coordinate along the axis of the cone

The prime indicates the derivative with respect to x.

Equations (1–4) represent the basic results of the PER theory. Examination of these four equations reveals five unknown quantities, namely, $\mathrm{d}m_j/\mathrm{d}m$, β, V_j, V_0, and δ, all of which are functions of x. The reason for the additional unknown lies in the fact that the PER theory does not determine the impulse that is supplied to the liner elements. In the original work, the authors proposed measuring at least one of the quantities experimentally and thus providing a complete description of the process.[16] The obvious drawback to this technique is that an experimental testing program would have to be conducted for each different charge. More recently, additional approaches to obtain a complete description of the process have been suggested. Kiwan and Wisniewski take into account the properties of the explosive by using an empirical pressure function for the explosive products and then integrating the equations of motion of the liner element to obtain the collapse velocity.[23] Karpp has used two-dimensional computer code calculations to model the explosive-metal interaction problem.[24] He obtains the collapse velocity and angle from this approach and uses them to compute V_j from the PER theory. This approach has produced very encouraging results. Defourneaux has proposed the use of an empirical equation that relates the angle of projection to the explosive and liner properties.[25] The equation is of the form

$$\frac{1}{2\delta} = \frac{1}{\phi_0} + K\rho\left(\frac{\epsilon}{e}\right) \tag{5}$$

where K and ϕ_0 are empirical constants that depend on the explosive and the angle at which the detonation wave intersects the liner; ϵ is the liner thickness; ρ is the liner density; and e is the thickness of explosive measured normal to the liner.

Other approaches use a Gurney relationship for V_0 in terms of the explosive charge-to-mass ratio μ and the Gurney constant $\sqrt{2E}$.

2.2 Generalized Jet Formation Theory

Since 1970 various fundamental improvements to the theory of shaped charge jet formation have been made in the published and unpublished literature. They include the following: generally shaped liner; spherical or toroidal detonation wave (in addition to plane detonation); improved explosive-metal interaction formulas for liner collapse; finite acceleration of liner elements during collapse; determination of jet tip origin by supersonic criterion or inverse velocity gradient; and formulas for jet characteristics (radius, stretching rate). A comprehensive theory that incorporates these improvements will be presented next.

2.2.1 *Liner Geometry and Collapse*

A diagram of the general charge geometry and liner collapse is shown in Fig. 7. We consider the explosive to be detonated at D. If D is on the axis the wave will be spherical; if it is not the wave will be toroidal. The coordinate x will always refer to the original axial position of the elements on the liner. As the detonation wave passes P, the liner element originally at P begins to collapse in a direction that makes an angle δ with the normal to the original liner at P. We assume that the element accelerates from zero velocity to some final velocity V_0 in a finite time. This assumption differs from the classical PER theory, which assumes that the element instantaneously obtains the velocity V_0 when the detonation wave reaches the element.[16] We will discuss this acceleration in more detail later.

As in the PER theory, we will assume that there is no physical interaction among liner elements during collapse and that each zonal element of the

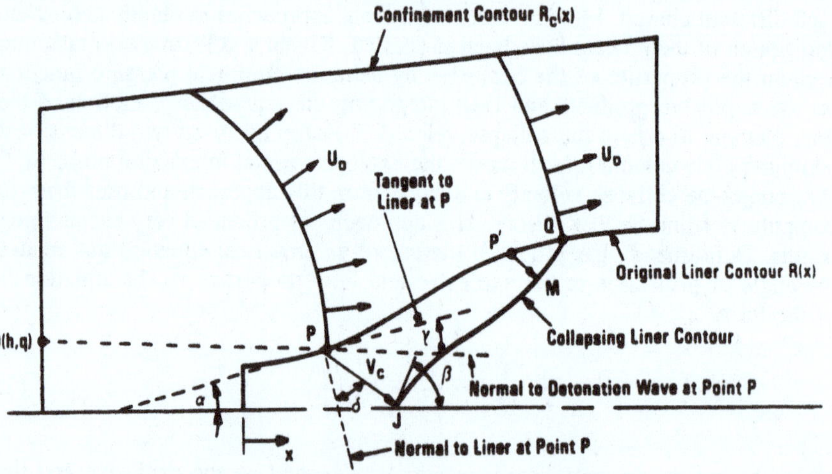

Fig. 7 Charge geometry and collapse.

liner may be considered separately in its appropriate constant-velocity coordinate system. The detonation wave speed U_D is considered constant normal to the wavefront. The velocity with which the wave sweeps the liner is not constant, however, and is given by the formula

$$U(x) = \frac{U_D}{\cos\gamma} \tag{6}$$

where $\gamma = \gamma(x)$ is the angle between the normal to the detonation wave at P and the tangent to the liner at P.

In the classical PER theory, by considering the detonation wave only to supply an impulse normal to the liner elements, one concludes that, as observed in a system of coordinates moving along the liner at the sweeping velocity U, a liner element changes only its direction when acted upon by the detonation wave, not its velocity. This conclusion leads to the Taylor relationship between the angle δ and the absolute collapse velocity V_0, [Eq. (1)]. The Taylor relation assumes a steady-state process and strictly applies to simple geometries like the uniformly expanding cylinder detonated at one end. Chou et al. extended the Taylor projection angle formula to the unsteady case,[26] which is given by

$$\delta = \frac{V_0}{2U} - \frac{1}{2}\tau V_0' + \frac{1}{4}\tau' V_0 \tag{7}$$

where the prime indicates differentiation with respect to the Lagrangian liner coordinate x. The symbol τ is a time constant related to the acceleration of the liner.

In the present approach, either Eq. (1) or Eq. (7) may be used to provide the direction of each liner element collapse velocity. The liner elements in the classical theory were assumed to reach collapse velocity V_0 instantaneously. The first level of refinement assumes the liner acceleration to be constant over a finite period of time. Thus, the velocity increases linearly over a short period until it reaches final velocity V_0 or collapses on the axis. These two approaches are depicted graphically in Figs. 8a and 8b. Another velocity history that assumes an exponential form was proposed by Randers-Pehrson [27]

$$V(t) = V_0\left[1 - \exp\left(-\frac{t-T}{\tau}\right)\right] \tag{8}$$

and is also shown graphically in Fig. 8c. Carleone and Flis showed that this form fits experimental data for expanding cylinders and hydrocode calculations for imploding cylinders[28] and have suggested the following form for the time constant τ:

$$\tau = A_1\frac{mV_0}{P_{CJ}} + A_2 \tag{9}$$

where

m = initial mass per unit area of the liner
P_{CJ} = Chapman-Jouget pressure
A_1, A_2 = constants

The final liner collapse velocity V_0 can be given by the proper Gurney equation, previously discussed. The alternative method is to combine the Taylor relation with an empirical formula for projection angle.

In the present formulation, we will develop the equations using the constant acceleration in time and, of course, allow the acceleration to vary for each element, that is, $a = a(x)$. It is a simple exercise to implement the exponential forms if so desired. The time base used is $t = 0$ when the detonation wave is at $x = 0$ on the liner. Therefore, a given liner element will have an absolute collapse velocity given by

$$V_c(x,t) = a(t-T) \qquad t \leq t_m \tag{10}$$

$$= V_0 \qquad t \geq t_m \tag{11}$$

and $T = T(x)$ is the time when the detonation wave reaches the element x on the liner. We may now solve for the time t_m when the acceleration ceases by the condition $V_c(x, t_m) = V_0$. This expression yields

$$t_m = \frac{2U \sin\delta}{a} + T \tag{12}$$

We now seek to develop the contour of the collapsing liner and to find β, the angle of impact of the liner with the axis. We can do this by writing the coordinates of a general point, say P' in Fig. 7, as it collapses. Suppose at time

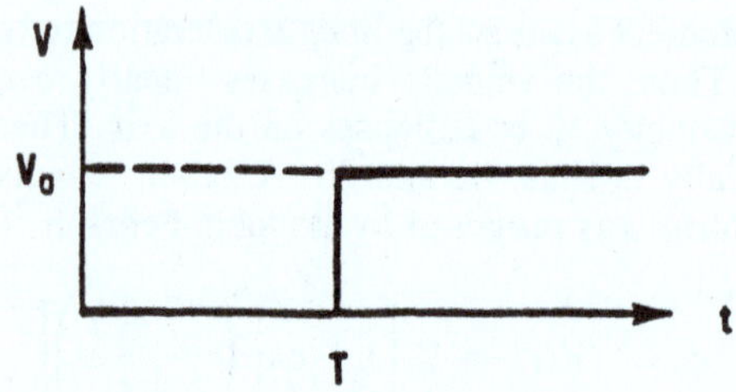

A. Instantaneous Acceleration

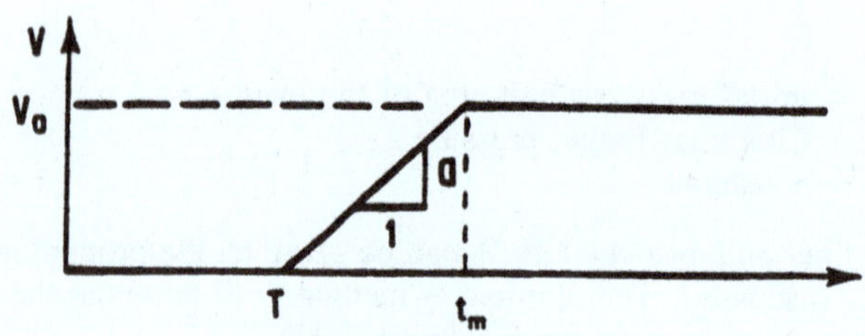

B. Constant Acceleration

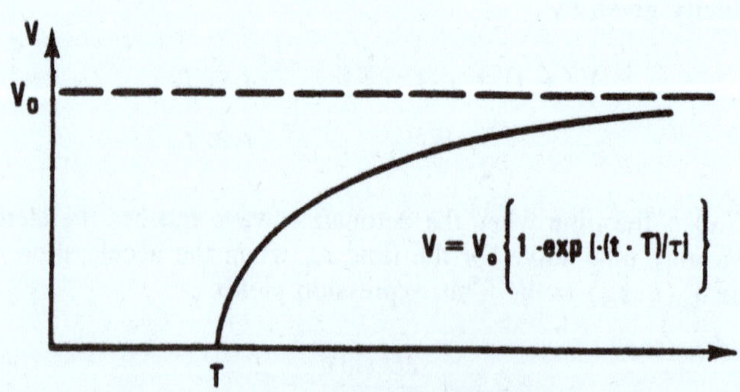

C. Exponential Acceleration

Fig. 8 Liner collapse velocity.

t the element originally at P' is now at point M. The coordinates of point M may be written as

$$z = x + \ell(x, t)\sin(\alpha + \delta) \tag{13}$$

$$r = R(x) - \ell(x, t)\cos(\alpha + \delta) \tag{14}$$

where z is the axial coordinate, r is the radial coordinate, R is the original liner radius, and $\ell(x, t)$ is the distance the element has traveled from P' to M, given by

$$\ell(x, t) = \begin{cases} \frac{a(t-T)^2}{2}; & t \leq t_m \\ \frac{a(t_m-T)^2}{2} + 2U(t - t_m)\sin\delta; & t > t_m \end{cases} \tag{15}$$

For all x, Eqs. (13) and (14) represent the contour of the liner at any time t.

To compute the impacting angle β we need the slope of the collapsing contour. This can be evaluated as follows:

$$\frac{\partial r}{\partial z} = \frac{\partial x}{\partial z}\frac{\partial r}{\partial x}$$

Performing the differentiations, we obtain

$$\frac{\partial r}{\partial z} = \frac{R' - \dfrac{\partial \ell(x, t)}{\partial x}\cos A + \ell(x, t)A'\sin A}{1 + \dfrac{\partial \ell(x, t)}{\partial x}\sin A + \ell(x, t)A'\cos A} \tag{16}$$

where $A = \alpha + \delta$, the prime indicates differentiation with respect to x, and the quantity $\partial\ell(x, t)/\partial x$ may be obtained by differentiating Eq. (15). Now to find β, we must evaluate this slope when each element impacts the axis. Let us denote the time of impact as $t_c = t_c(x)$. The distance traveled by the liner element in reaching the axis may be found by setting $r = 0$ in Eq. (14), or

$$\ell(x, t_c) = \frac{R(x)}{\cos A} \tag{17}$$

Combining Eq. (17) and Eq. (15), we may solve for the time of impact t_c as

$$t_c = \begin{cases} \left[\dfrac{2R}{a\cos A}\right]^{1/2} + T; & V_c < V_0 \\ \dfrac{1}{2U\sin\delta}\left[\dfrac{R}{\cos A} - \dfrac{a(t_m - T)^2}{2}\right] + t_m; & V_c = V_0 \end{cases} \tag{18}$$

Thus, the angle β is given by

$$\tan\beta = \left.\frac{\partial r}{\partial z}\right|_{t=t_c}$$

or

$$\tan\beta = \frac{R' - \dfrac{\partial \ell_n}{\partial x}(x,t_c)\cos A + RA'\tan A}{1 + \dfrac{\partial \ell_n}{\partial x}(x,t_c)\sin A + RA'} \tag{19}$$

where the derivative $\partial\ell/\partial x$ is evaluated at the proper impact time given by Eq. (18).

2.2.2 *Jet Formation*

Consider the point of impact on the axis shown as J in Fig. 7. The development in this region follows identically that of the PER theory. Figure 9 shows the collision region referred to laboratory coordinates as well as referred to coordinates moving at the juncture velocity V_1. Note that V_s, V_c, V_1, and V_j are velocities referred to laboratory coordinates, and v, v_j, and v_s are velocities relative to the moving point J.

Thus, an observer moving with the juncture will observe an element moving toward the juncture with a velocity v that may be computed vectorially from a simple coordinate transformation as

$$\boldsymbol{v} = \boldsymbol{V}_c - \boldsymbol{V}_1 \tag{20}$$

where here and in the remainder of this report the symbol $\boldsymbol{V}_c$ will refer to $\boldsymbol{V}_c(x, t_c)$. This vector relationship is shown diagrammatically in Fig. 10. From this triangle, we readily obtain

$$v = V_c\ \cos(\alpha + \delta)\ \csc\beta \tag{21}$$

and

$$V_1 = V_c\ \cos(\beta - \alpha - \delta)\ \csc\beta \tag{22}$$

Again considering the flow relative to the juncture J, we assume that the magnitude of the impinging velocity v remains the same after splitting at the juncture, which yields

$$v_j = -v_s = v \tag{23}$$

This expression is identical to the assumption made by the PER theory. Note that if the flow were truly steady and incompressible, this relationship would be exact.

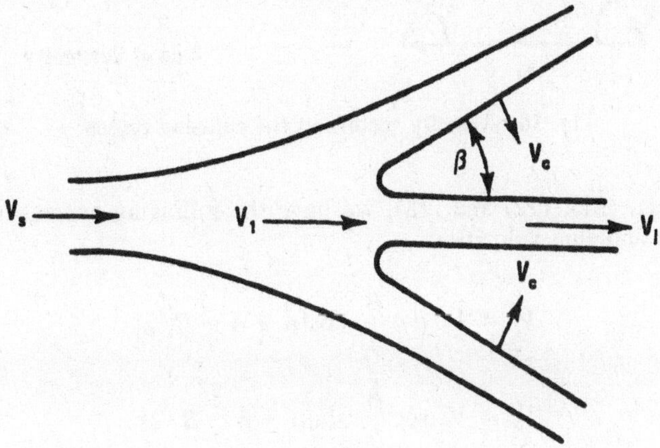

A. Velocities Referred to a Laboratory System of Coordinates

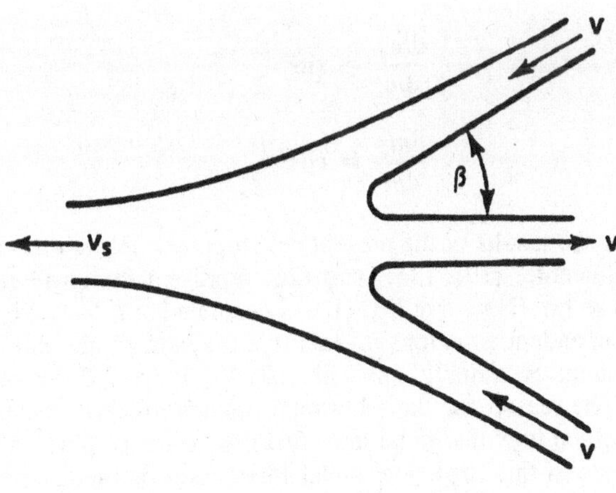

B. Velocities Referred to a System of Coordinates Moving at Velocity V_1

Fig. 9 Region of jet formation.

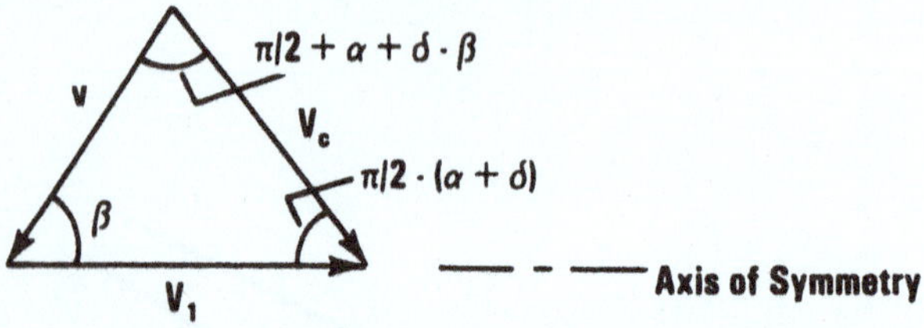

Fig. 10 Velocity vectors at the collision region.

Using Eqs. (21), (22) and (23), we have the following expressions for the absolute jet and slug velocities:

$$V_j = V_c \csc\frac{\beta}{2}\cos[\alpha + \delta - \beta/2] \tag{24}$$

$$V_s = V_c \sec\frac{\beta}{2}\sin[\alpha + \delta - \beta/2] \tag{25}$$

Let us now describe how the mass of the liner splits into the jet and slug. Consider a ring element of the liner of mass dm. As this element enters the collision region, it splits into an element of jet of mass dm_j and an element of slug dm_s. If we now write the conservation of mass and conservation of horizontal momentum equations with respect to a system of coordinates moving with the juncture J, we find

$$\frac{\mathrm{d}m_j}{\mathrm{d}m} = \sin^2\frac{\beta}{2} \tag{26}$$

$$\frac{\mathrm{d}m_s}{\mathrm{d}m} = \cos^2\frac{\beta}{2} \tag{27}$$

Let us summarize the results of the present development. Equations (10) or (11), whichever is applicable, (19), (24), and (26) represent the basic equations of theory presented so far. (Note that Eq. (10) is evaluated at $t = t_c$.) Examination of these four independent equations reveals that they are expressible in terms of five unknown quantities, namely, d$m_j/$dm, β, V_j, V_c, and δ, all of which are functions of x. The reason for the additional unknown lies in the fact that we have not yet specified how the metal liner and the explosive physically interact. Several approaches to this explosive-metal interaction problem were discussed in the previous section.

2.2.3 Jet Position and Radius

The theory just described develops the collapse and jet formation. We now desire to trace the location of the jet elements and study their stretching after formation. The coordinate system conventions are shown in Fig. 11. We will continue to use x as the coordinate along the axis of the cone that defines the original position of the liner elements (Lagrangian coordinate). The position coordinate in the jet is measured from the original position of the cone apex and is designated as $\xi(x, t)$. The position of an element at the time it just reaches the axis is denoted by the symbol $\overline{z}(x)$. From the basic collapse geometry, (Fig. 7), this position may be calculated as

$$\overline{z}(x) = x + R\tan(\alpha + \delta) \tag{28}$$

Recall the time that the element just reaches the liner is given by t_c, from Eq. (18). Therefore, at any time $t > t_c$, a portion of the element originally at x will be in the jet and a portion will be in the slug. Assuming that each element of the jet portion travels at a constant velocity $V_j(x)$ immediately after it is formed (this is consistent with the original PER assumption of no interaction among liner elements), we may write the position of the jet portion of the x element at time t as

$$\xi(x, t) = \overline{z} + (t - t_c)V_j, \qquad t \geq t_c \tag{29}$$

We may now define the one-dimensional extension of the jet elements in the following manner. Consider two points x_1 and x_2 on the liner separated by a distance Δx as shown in Fig. 12. Let us follow the motion of these points as they collapse and proceed to form the jet.

We will not discuss the slug here. During the collapse process, the point x_1 first reaches the axis and then proceeds to jet along the axis. Meanwhile, x_2 has begun to collapse and reaches the axis at time $t_0 = t_c(x_2)$. At this time t_0, jetting material originally at x_1 is located at $\xi(x_1, t_0)$ and x_2 has just reached

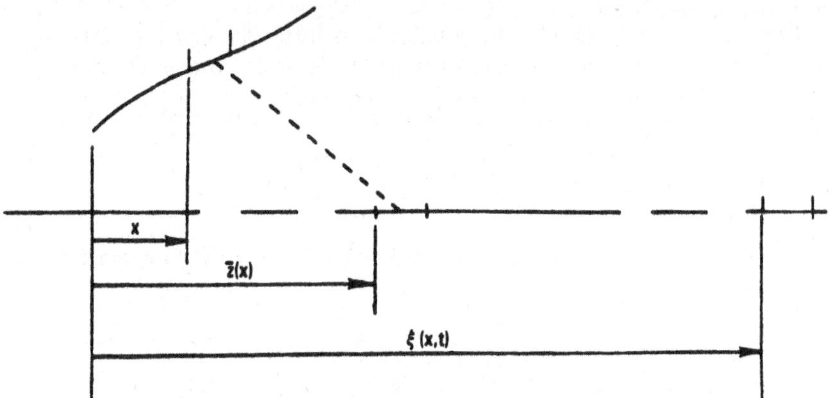

Fig. 11 Description of coordinates x, z, and ξ.

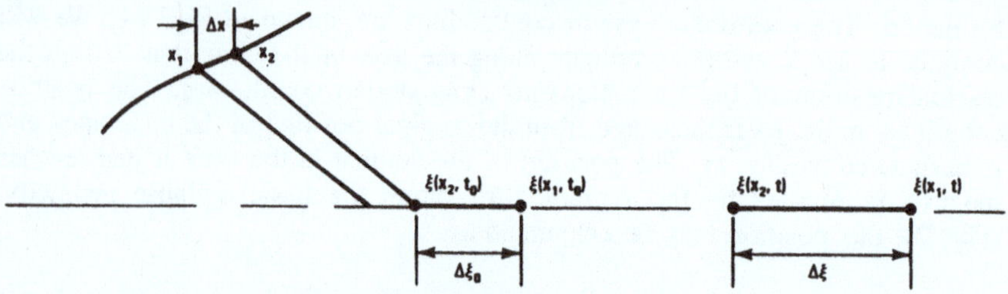

Fig. 12 Schematic diagram showing stretching of a jet element.

the axis and is located at $\xi(x_2, t_0)$, as shown in Fig. 12. At some arbitrary later time t, their positions are $\xi(x_1, t)$ and $\xi(x_2, t)$, respectively. We now define the one-dimensional jet extension (simple strain) as the ratio of the increase in length of a jet element to its length when first formed. Mathematically, we may express this in the limit as

$$E \equiv \lim_{x_2 \to x_1} \frac{\Delta\xi - \Delta\xi_0}{\Delta\xi_0} \equiv \lim_{x_2 \to x_1} \left[\frac{\xi(x_2, t) - \xi(x_1, t)}{\xi(x_2, t_0) - \xi(x_1, t_0)} \right] - 1$$

$$= \left[\frac{\partial \xi}{\partial x}(x_1, t) \Big/ \frac{\partial \xi}{\partial x}(x_1, t_0) \right] - 1$$

where in the limit $t_0 = t_c(x_1)$. For any x, we therefore write

$$E(x, t) = \left[\frac{\partial \xi}{\partial x}(x, t) \Big/ \frac{\partial \xi}{\partial x}(x, t_0) \right] - 1 \tag{30}$$

where $t_0 = t_c(x)$.

The derivative $\partial\xi/\partial x$ may be computed by differentiating Eq. (29). In doing this, many subsequent differentiations are needed. After performing these differentiations and making substitutions, we obtain

$$E(x, t) = \frac{(t - t_c)V_j'}{\overline{z}' - t_c' V_j} \tag{31}$$

where the primes indicate differentiation with respect to x. The Lagrangian strain rate is, therefore,

$$\dot{E}(x, t) = \frac{V_j'}{\overline{z}' - t_c' V_j} \tag{32}$$

Let us now compute the radius of the jet under the assumptions that the cross section of the jet is circular and the flow is steady and incompressible. The mass of a jet element per unit length of original cone may be given by

$$\frac{\mathrm{d}m_j}{\mathrm{d}x} = \frac{2\pi\rho\epsilon R}{\cos\alpha}\sin^2\frac{\beta}{2} \tag{33}$$

Now, assuming a circular jet cross section and constant density, we may express the mass of a jet element that is $\mathrm{d}\xi$ in length by

$$\mathrm{d}m_j = \pi\rho\, r_j^2\, \mathrm{d}\xi$$

or

$$\frac{\mathrm{d}m_j}{\mathrm{d}x} = \pi\rho\, r_j^2 \left|\frac{\partial\xi}{\partial x}\right| \tag{34}$$

where r_j is the radius of the jet. The absolute value of $\partial\xi/\partial x$ is used since $\mathrm{d}\xi$ has a sense opposite to the original $\mathrm{d}x$. Now combining Eqs. (33) and (34) and solving for the radius of the jet, we have

$$r_j = \left[\frac{2R\epsilon}{\cos\alpha}\,\frac{\sin^2(\beta/2)}{|\partial\xi/\partial x|}\right]^{1/2} \tag{35}$$

2.2.4 *Jet Tip Formation*

In many cases, because of the finite acceleration of liner elements, the liner has a region where the elements do not reach the final collapse velocity. In this region, normally close to the charge axis, an element having a jetting velocity V_{j1} may be followed by an element with a jetting velocity that is greater than this velocity, that is, $V_{j2} > V_{j1}$. This inverse velocity situation usually continues throughout this region. Thus, interference of the jet elements takes place and the mass piles up and usually forms the jet tip. This effect, first pointed out by Kiwan and Wisniewski,[23] is known as the inverse-velocity gradient. Carleone et al. verified this effect experimentally.[3] For conventional conical charges, the first 30 to 40% of the liner from the theoretical apex forms the tip, which has a much larger radius than the rest of the jet.

In the present theory, we assume that the elements experience a perfectly plastic impact and that the elements *pile up* to form a tip particle whose velocity may be predicted by conservation of linear momentum. We consider each element to impact until we find the first jetting element whose velocity is less than the velocity of the combined tip particle. Then conventional jetting ensues.

In mathematical form, this may be calculated as

$$\overline{V}_j(x_{\mathrm{tip}}) = \frac{\displaystyle\int_0^{x_{\mathrm{tip}}} V_j(x)\frac{\mathrm{d}m_j}{\mathrm{d}x}\mathrm{d}x}{\displaystyle\int_0^{x_{\mathrm{tip}}} \frac{\mathrm{d}m_j}{\mathrm{d}x}\mathrm{d}x}\,, \tag{36}$$

where $\overline{V}_j$ indicates the velocity of the combined tip particle. In actuality this expression is integrated step by step until a point x_{tip} is found such that

$$V_j(x_{\text{tip}}) \leq \overline{V}_j(x_{\text{tip}}) \tag{37}$$

Then this value of x_{tip} is considered that point on the liner that distinguishes where the formation of the tip stops and where normal jetting begins. This value of x_{tip} is depicted graphically in Fig. 13. The expression $\mathrm{d}m_j/\mathrm{d}x$ may be evaluated as follows:

$$\frac{\mathrm{d}m_j}{\mathrm{d}x} = \frac{\mathrm{d}m}{\mathrm{d}x}\frac{\mathrm{d}m_j}{\mathrm{d}m} = \frac{\mathrm{d}m}{\mathrm{d}x}\sin^2\frac{\beta}{2}$$

From geometry

$$\frac{\mathrm{d}m}{\mathrm{d}x} = \frac{2\pi\rho\epsilon R}{\cos\alpha} \tag{38}$$

Equation (38) is exact for a conical cone with constant wall thickness. It is also a very good approximation for most practical liners.

Thus, we have

$$\frac{\mathrm{d}m_j}{\mathrm{d}x} = \frac{2\pi\rho\epsilon R}{\cos\alpha}\sin^2\frac{\beta}{2} \tag{39}$$

Note also that the mass of the tip particle may be calculated from the denominator of Eq. (36), or

$$m_{\text{tip}} = \int_0^{x_{\text{tip}}} \frac{\mathrm{d}m_j}{\mathrm{d}x}\mathrm{d}x \tag{40}$$

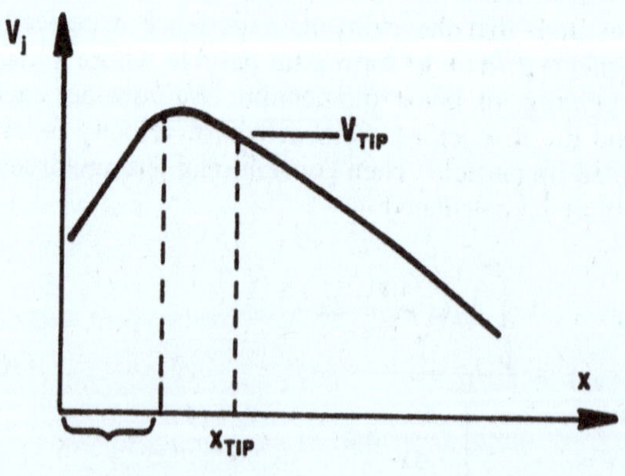

Fig. 13 Typical jet velocity distribution curve.

2.2.5 *Coherent Jet Formation*

When the jet collapse velocity exceeds a certain threshold, one finds that a radially dispersed jet is formed. This result has often been termed an *overdriven jet*. This effect is governed by the compressibility of the liner material and by the shock that occurs in the collapsing liner under supersonic flow conditions. The effect of compressibility on jet formation for two impinging plates (plane strain) has been examined by several investigators (e.g., see Ref. 29). These studies addressed the problem of explosive welding and were primarily concerned with conditions of jet formation.

Walsh et al. concluded that jetting always occurs if the fluid is incompressible or if the collision velocity v, as defined in Figs. 7, 8, and 9, relative to the moving coordinates is subsonic.[29] In supersonic collisions, jetting occurs if $\beta > \beta_c$, where β_c is the critical turning angle for an attached oblique shock at the velocity v.

In Refs. 30 and 31, jetting criteria for both plane and axisymmetric cases are proposed that not only provide limiting conditions on jetting but also predict the quality of the jet (i.e., whether a coherent jet forms). These criteria are based on extensive experimental results and hydrocode simulations. The jetting criteria may be stated as follows:

1) For subsonic collisions, $v < c$, a solid coherent jet always occurs.

2) For supersonic collisions, $v > c$, jetting occurs if $\beta > \beta_c$ for a given v; but the jet is not coherent.

3) For supersonic collisions in which $\beta < \beta_c$, no jet will be formed.

In the preceding, c is the bulk sound speed of the liner material, and β_c is the maximum angle that an attached shock wave can form at a prescribed supersonic velocity v.

For shaped charge applications, the most important criterion is that the formation must be subsonic, otherwise the jet will be noncoherent and disperse radially. Since the value of the sound speed under the very high pressure and temperature of the formation region is not known exactly, the subsonic collision condition is useful as a general principle. In practice, the value of the critical Mach number based on the bulk sound speed of the liner is about 1.2. This approach gives reasonable agreement with experimental observations of shaped charge jets. Figure 14 shows radiographs of a coherent and a noncoherent jet.

3. Jet Particulation

The jet formed for most liner designs possesses a large velocity gradient, causing the jet to stretch to great lengths and making it capable of penetrating deeply into solid targets. The length, and hence penetration capability, is limited, however, by the eventual axial breakup of the jet into many constant-length segments. Chou and Carleone have shown that shaped charge breakup is controlled by a plastic instability, driven by the material strength or flow stress.[32] This instability causes a series of necks to develop in the jet that eventually fail. This is different from the instability in liquid jets, which is caused by surface tension.

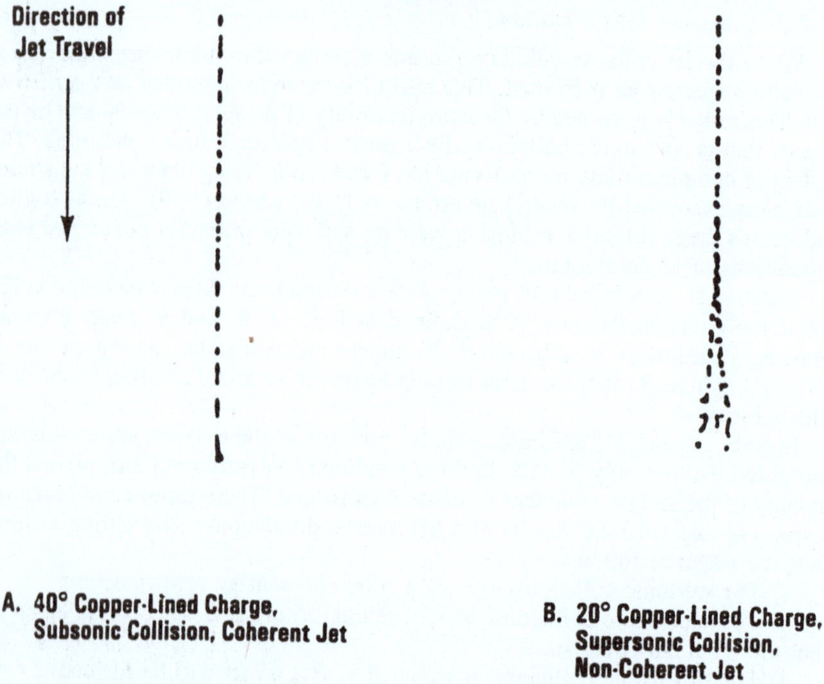

Fig. 14 Radiograph of jets from two typical conical charges.

In this section a comprehensive analytical model of shaped charge jet instability and breakup will be presented. On the basis of this model, a design curve for the jet breakup time is presented, as well as a method to calculate the number of jet segments. The results of the model agree well with both two-dimensional hydrocode calculations and experimental data.

Historically, it appears that Karpp and Simon provided the first numerical simulation of this problem[33] and applied the two-dimensional hydrocode HEMP[34] to a stretching jet with a surface perturbation and found that the perturbation led to necking of the jet and eventual particulation at each neck. They showed this for a range of surface wavelengths. Subsequent to this work, Refs. 32, 35, and 36 have shown that a critical or predominant unstable wavelength exists that grows faster than all other surface perturbations. At this point, let us offer a simplified physical explanation of the stable and unstable surface disturbances.

Consider a stretching jet having a surface disturbance with a very small wavelength. In the valleys or necks, the average axial stress is higher than the nominal flow stress Y because of the three-dimensional stress distribution that exists there (see Bridgman[37]). In fact, Bridgman showed the smaller the radius of curvature of the neck, the higher the stress. Therefore, for a small wavelength disturbance, the net force on an element of jet is toward the neck causing it to "heal" and thus stabilize the jet. For a long-wavelength disturbance, on the other hand, the stress concentration at the neck is much closer to the flow stress. Since the cross-sectional jet area is greater away from the neck, the net force is away from the neck. This condition causes the material to move away from the neck and, hence, results in continuous cross-sectional area reduction until breakup.

Of course, the exact phenomenon is more complicated. Large inertial forces exist since the jet is stretching at a very high rate. These inertial forces tend to stabilize the jet. In our numerical computations, we find the critical wavelength to be a function of the stretching rate of the jet. Work hardening and viscoplastic effects will also tend to delay the breakup of the jet; however, excellent agreement with experimental data has been obtained with a perfectly plastic model.

Chou and Carleone have also shown that perturbations in jet strength (flow stress) or velocity will also cause the plastic instability to occur.[32] In fact, the critical wavelength appears independent of the physical quantity used to trigger the instability. Furthermore, Miller performed two-dimensional hydrocode calculations of a jet with distributed cross-sectional area and compressed tip region.[38] His calculation used a constitutive model developed by Steinberg et al. that accounts for strain hardening and thermal softening.[39] Apparently, the variable distribution from this constitutive law and geometry was sufficient to trigger the plastic instability. Again, the proper wavelength resulted. Osborn has also applied two-dimensional code calculations to the jet breakup problem.[40]

The source of such perturbations in shaped charges may occur from the metallurgical grain structure or perhaps even the machine marks that exist on the surface of the liner. In any event, some type of mechanical or metallurgical perturbation seems always to exist, causing the eventual breakup of the jet. The identification and control of these factors appear to be an important fruitful area for research.

The first analytical treatment of the shaped charge jet stability problem was presented by Chou and Carleone.[32] This early one-dimensional model showed that the value of the ratio of jet flow stress to jet density Y/ρ controlled the growth of the instability. A low value of this ratio causes the jet to neck down more slowly than a material with a large value, all other conditions being equal. This early version of the theory, however, cannot predict the critical wavelength. In Carleone et al.,[36] the early theory of Ref. 32 was extended to include stress concentration at the jet necks. The resulting equations clearly showed the existence of a critical wavelength as described above. The linearized solution to the theory in Ref. 36 did not show the critical wavelength to be a function of the jet stretching rate, but, in fact, showed the critical wavelength λ_{ocrit} to be 2.22 times the jet diameter d_0 at the time of onset of the instability. Detailed two-dimensional hydrocode simulations, also presented in Ref. 36, clearly showed the critical wavelength to be a function of jet stretching rate (axial strain rate η_0). Nonlinear solutions of the one-dimensional theory presented in this section and in Refs. 41 and 42 show the proper dependence of critical wavelengths on jet stretching rate. On the basis of these solutions, a breakup time design curve and simple formula were established. The proper critical wavelength also allowed the calculation of the number of jet segments.

Hirsch has presented a phenomenological formula for the jet breakup time.[43,44] In this formula the breakup time is related to the smallest characteristic length (original liner thickness) and an empirical constant V_p. This formula gives reasonable breakup times but is not so general as the model presented here. Pfeffer has also presented a formula for jet breakup time.[45] He performed two-dimensional hydrocode calculations of jet stability and fit a simple formula to the results. His formulas correctly show the critical wavelength to be approximately independent of jet diameter (as does the model presented here), but incorrectly show the breakup time to be independent of jet strength (flow stress). A comparison of the present model with Hirsch's results and Pfeffer's results are presented in Refs. 41 and 42, respectively. Walsh recently published a jet necking model based on the work of Refs. 32, 35, 36, 41, and 42 but used different nondimensional groups in his formulation.[46] As a final note, Taylor et al. have developed a surface instability approach[47] similar to the one used in this paper to study the problem of stretching sheets of metal.

In Sec. 3.1, a concise formulation of one-dimensional stability theory that evolved from the work in Refs. 32, 35, 37, 41, and 42 is presented. Nondimensional groups are defined to simplify the presentation of results. A linearized solution to the theory is first presented in Sec. 3.2 to show more easily the physical interdependence between strength, density, radius, stretching rate, and wavelength. Nonlinear solutions are also presented and compared with two-dimensional hydrocode simulations in Sec. 3.2. A final jet breakup model including design curves is presented in Sec. 3.3.

3.1 Formulation of the One-Dimensional Jet Stability Model

3.1.1 Conservation Equations

Consider a section of stretching jet, as shown in Fig. 15. For the general case we assume that there is some initial disturbance of the surface and that an initial velocity distribution exists in the jet. Detailed boundary and initial conditions will be discussed below. We first define a material coordinate a that assigns an initial position to each material particle of jet and remains with that particle, or

$$a = \frac{1}{\rho_0}\int_{x(0,t)}^{x(a,t)} \frac{\rho A}{A_I} \mathrm{d}x \tag{41}$$

where
ρ = density of the jet
A = cross-sectional area
ρ_0 = initial density
A_I = initial cross-sectional area (with surface disturbance)
x = spatial coordinate along the axis of symmetry fixed to the laboratory in the Eulerian sense

The continuity equation follows directly from Eq. (41) as

$$x_a = A_I/A \tag{42}$$

where we have assumed the jet to be incompressible, $\rho = \rho_0$, and the subscript

a indicates differentiation with respect to the material coordinate. Conservation of axial momentum of a jet with variable cross-sectional area may be written as

$$\rho_0 A_I x_{tt} = (\sigma A)_a \tag{43}$$

where σ is the axial stress averaged over the jet cross-sectional area, and the subscript t indicates differentiation with respect to time. Equations (42) and (43) represent the conservation equations of the jet.

3.1.2 Definition of Strain and Strain Rate

We now choose the logarithmic definition of strain (see Ref. 37)

$$\bar{\epsilon} = \ell n(A_I/A) \tag{44}$$

where we have used the initial cross-sectional area A_I as a basis so that $\epsilon = 0$ initially. The strain rate may be obtained by direct differentiation as

$$\bar{\epsilon}_t = -A_t/A \tag{45}$$

Using continuity Eq. (42), we rewrite Eq. (45) as

$$\bar{\epsilon}_t = (A/A_I)v_a \tag{46}$$

where we have defined $v \equiv x_t$.

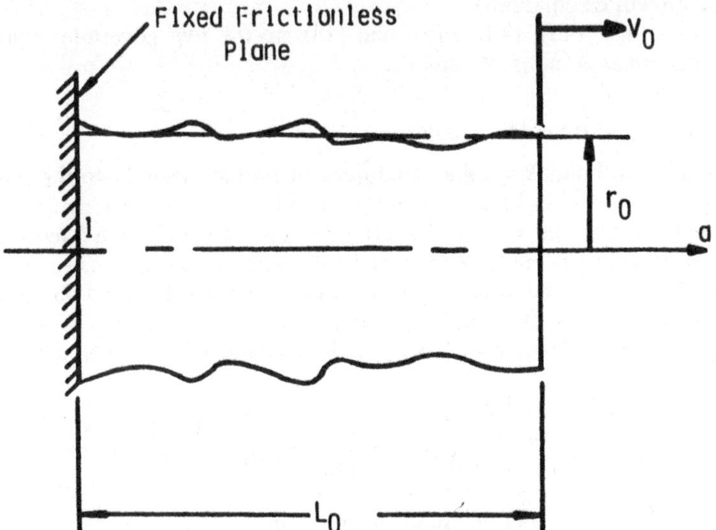

Fig. 15 Geometry of a section of stretching jet with surface perturbation.

3.1.3 Constitutive Equation

For a one-dimensional plastic formulation, it is sufficient to prescribe a uniaxial stress-strain curve as a description of the material properties. Let us consider a universal stress-strain curve given in functional form as

$$T = T(e, \dot{e}) \tag{47}$$

where T is the effective stress, e the effective strain, and $\dot{e}$ the effective rate of strain (see Malvern[48] for definitions of effective stress, strain, and strain rate). For this essentially one-dimensional problem, we assume that the effective strain and strain rates are given by the one-dimensional quantities of $e = \bar{\epsilon}, \dot{e} = \bar{\epsilon}_t$.

Once a surface disturbance occurs in the jet, the stress distribution becomes truly three dimensional. The average axial stress σ varies along the axis of the jet. For a static tensile test, Bridgman showed that the axial stress at the neck is equal to the flow stress (effective stress) at the free surface and that it increased to a maximum on the axis.[37] Thus, the average axial stress at the neck is higher than the flow stress. Let us use the Bridgman result to provide an estimate of the stress concentration effect for our dynamic stretching jet problem. If we assume small surface disturbances the average axial stress is given by

$$\sigma = T\left(1 + \frac{r}{2}\frac{\partial^2 r}{\partial x^2}\right) \tag{48}$$

where r is the radius of the jet assuming a circular cross section, or $r = \sqrt{A}/\pi$, and T is given by Eq. (47). (This assumption is valid since the onset of the instability occurs when the surface disturbance is small. After the jet becomes unstable, the value of the stress concentration is not important to the final disturbance growth calculation.)

Equations (42), (43), (44), (46), and (48) are the five governing equations for the unknowns A, x, σ, $\bar{\epsilon}$, and $\bar{\epsilon}_t$.

3.1.4 Initial and Boundary Conditions

There are two boundary-value conditions of primary interest to the problem of jet breakup, fixed-velocity end conditions and free end conditions. For a long jet, we may assume that any small interior section stretches with a fixed velocity difference between the ends. For a more fundamental understanding of the entire process, the free end solution is useful. In fact, with the free end solution, we can explain under which conditions the interior sections stretch with a fixed velocity difference. These two cases may be mathematically stated as follows:

Fixed-velocity end conditions

$$x_t = 0 \qquad \text{at} \qquad a = 0 \tag{49}$$

$$x_t = \eta_0 L_0 \qquad \text{at} \qquad a = L_0 \tag{50}$$

where η_0 is the initial strain (stretching) rate, defined by the velocity at the moving end V_0, divided by the initial length of jet L_0, $\eta_0 = V_0/L_0$.

Free end conditions

$$x_t = 0 \qquad \text{at} \qquad a = 0 \tag{51}$$

$$\sigma = 0 \qquad \text{at} \qquad a = L_0 \tag{52}$$

For a complete formulation of the problem, we specify the initial velocity and initial perturbed cross-sectional area. In general, these are as follows:

Initial conditions

$$x_t = V_I(a) \qquad \text{at} \qquad t = 0 \tag{53}$$

$$A = A_I(a) \qquad \text{at} \qquad t = 0 \tag{54}$$

Therefore, the five governing equations [Eqs. (42), (43), (44), (46), and (48)] together with the boundary conditions [Eqs. (49) and (50) or (51) and (52)] and the initial conditions [Eqs. (53) and (54)] provide a complete formulation of the stretching jet problem.

3.1.5 Nondimensional Quantities

For convenience, it will be useful to introduce the following nondimensional quantities:

$$\bar{x} = x/r_0, \qquad \bar{a} = a/r_0, \qquad \bar{r} = r/r_0$$

$$\bar{L} = L_0/r_0, \qquad \bar{\lambda}_0 = \lambda_0/r_0, \qquad \bar{t} = C_p t/r_0 \tag{55}$$

$$\bar{\eta}_0 = \eta_0 r_0/C_p, \qquad \bar{\sigma} = \sigma/(\rho_0 C_p^2)$$

where r_0 is the initial undisturbed radius of the jet, C_p is defined by

$$C_p = \sqrt{Y/\rho_0} \tag{56}$$

and Y is the dynamic flow stress in simple tension.

Using these quantities and expressing the cross-sectional area in terms of the radius, we may formulate the governing equations and boundary conditions in nondimensional form. Doing this will make the final numerical solution much more generally applicable as well as provide insight and simplifications to final interpretation of results. In the interest of space, we will not present the equations in nondimensional form, but the reader may refer to Carleone et al.[42] for these equations.

3.2 Solutions

In this section we shall present linearized and complete nonlinear solutions to the jet stability theory just formulated. Only solutions for a perfectly plastic material will be presented here. In this case, Eq. (47) becomes

$$T = Y \tag{57}$$

where Y is taken as a constant for a given material. We see that under this condition, Eq. (44) and (46) drop out of the formulation and only three governing equations are required.

A linearized solution for a simple strain-hardening material has been presented in Ref. 36. A computer program for a nonlinear finite-difference numerical solution including strain-hardening and viscoplastic effects has been formulated in Ref. 42. Results show that strain-hardening and viscous effects tend to retard the growth in amplitude of surface perturbations.

For the development of a generalized breakup model for shaped charge jets, we have found that a perfectly plastic constitutive law appears adequate. In addition, this simple material description requires specification of only one empirical quantity Y for each jet material. Because of the dynamic nature of these problems, static measurements of Y cannot be used; instead, values of Y must be estimated from shaped charge experiments.

3.2.1 Linearized Solution

By introducing a sinusoidal surface disturbance of the form

$$A(a,0) = A_I = A_0(1 + f_0 \cos\, ba) \tag{58}$$

and a linear velocity gradient

$$x_t(a,0) = V_I = \eta_0 a \tag{59}$$

with the fixed end boundary conditions Eqs. (49) and (50) where $b = 2\pi/\lambda_0$, and by assuming small disturbances and neglecting higher-order terms in the governing equations, we find a solution for A and x of the form

$$A(a,t) = \frac{A_0(1 + f \cos\, ba)}{\eta_0 t + 1} \tag{60}$$

$$x(a,t) = (\eta_0 t + 1)[a - (f/b)\sin\, ba] \tag{61}$$

where $f = f(t)$ is a function governing the amplitude of the disturbance term and $f(0) = f_0$. The solution for $f(t)$ is

$$f(t) = Q_1(\eta_0 t + 1)^{S_1} + Q_2(\eta_0 t + 1)^{S_2} \tag{62}$$

where Q_1 and Q_2 are constants to be determined from initial conditions and

$$S_1 = \frac{1}{2}(-1 + \sqrt{1 + 4k}) \tag{63a}$$

$$S_2 = \frac{1}{2}(-1 - \sqrt{1 + 4k}) \tag{63b}$$

$$K = \frac{4Y}{\rho_0}\left(\frac{\pi}{\eta_0 \lambda_0}\right)^2 \left[1 - \left(\frac{\pi r_0}{\lambda_0}\right)^2\right] \tag{64}$$

When the exponents of Eq. (62) are negative, the amplitude of the surface disturbance will decay and the jet is stable. However, when S_1 becomes positive, the amplitude becomes unstable and grows without bound. From Eqs. (63) and (64), we see that the jet is stable if

$$\frac{\lambda_0}{2r_0} < \frac{\pi}{2} \tag{65}$$

Otherwise the jet is unstable.

We also note from Eqs. (62) and (64) the existence of a critical wavelength. That is, there exists a wavelength that maximizes the value of S_1 and causes the disturbance amplitude to grow faster than all others. By differentiating Eq. (63) and setting the result equal to zero, we find the critical wavelength as

$$\left(\frac{\lambda_0}{2r_0}\right)_{\text{crit}} = \frac{\pi}{\sqrt{2}} \qquad \text{(linearized solution)} \tag{66}$$

Two-dimensional code calculations show the critical wavelength to be a function of initial stretching rate. We must find a nonlinear solution to obtain such a result from the one-dimensional model.

Note that by introducing the nondimensional quantities defined by Eqs. (55) and (56), as shown in Ref. 41, the linearized solution can be expressed in functional form as

$$f/f_0 = g(\bar{t}, \bar{\eta}_0, \bar{\lambda}_0) \tag{67}$$

This fact will be important in choosing the final form of the breakup curve.

3.2.2 Nonlinear Solution

The preceding linearized solution applies only to short times after the jet becomes unstable. This was extended to longer times in Ref. 41 by using an incremental method in time, but it still retained the restrictions that the amplitude of the disturbance remain small, the disturbances be sinusoidal, and the velocity difference between the ends remain fixed. In this section, we will present results of a finite-difference solution in which all of these restrictions may be relaxed. The specific differencing approach and detailed finite-difference equations are presented in Ref. 42.

A comparison of the one-dimensional nonlinear solution, the two-dimensional HEMP code solution[41] and the linearized solution (incremental method of Ref. 41) is shown in Fig. 16. A basic sinusoidal surface disturbance, fixed velocity end conditions, and initial linear velocity distribution were used. Shown in Fig. 16 is the amplitude growth as a function of time. The nonlinear solution agrees much better with the HEMP simulation, indicating the importance of including the nonlinear terms for large amounts of necking and longer times.

The surface shape of the one-dimensional finite-difference solution is compared to the HEMP solution at two times in Figs. 17 and 18.

Next, let us demonstrate the concept of critical wavelength from this solution. One method to do this is trial and error. That is, one can calculate a parametric series, varying the initial wavelength and finding the wavelength that causes the surface amplitude to grow the fastest. This was the approach used with the two-dimensional code calculations of Refs. 32 and 35. The current work developed another approach that was found to be more useful and more efficient. If an irregular or random surface disturbance is introduced, we find that

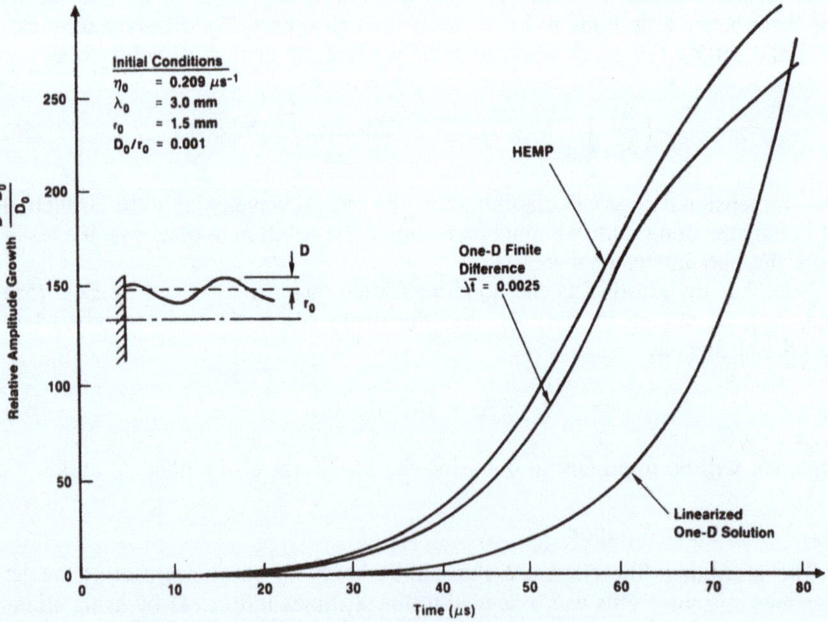

Fig. 16 Comparison of two-dimensional HEMP, one-dimensional finite-difference nonlinear, and one-dimensional linearized solutions.

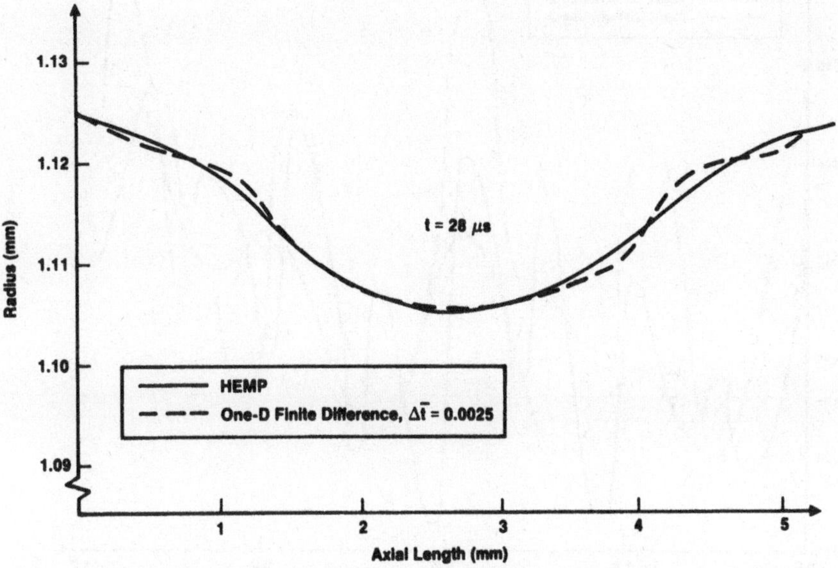

Fig. 17 Surface shape comparison at $t = 28 \mu s$.

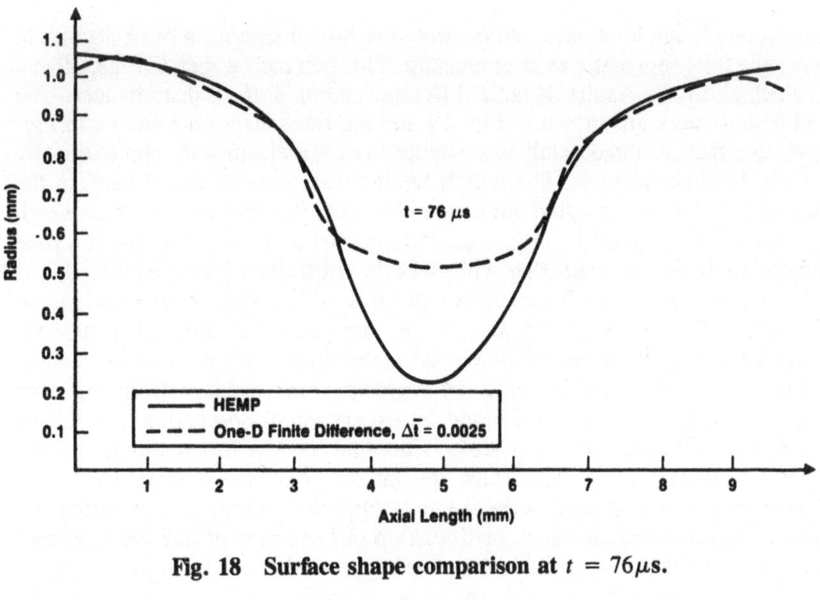

Fig. 18 Surface shape comparison at $t = 76 \mu s$.

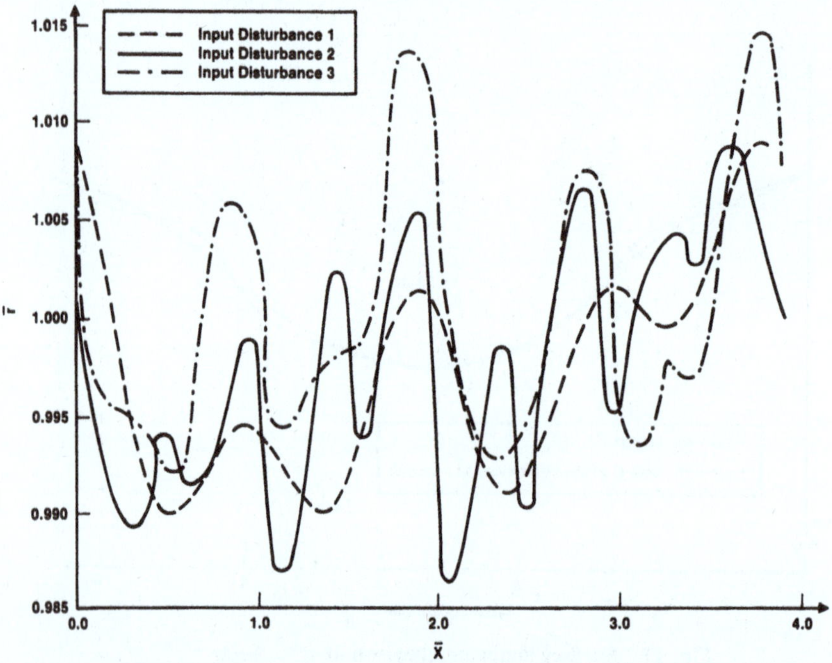

Fig. 19 Three different initial surface disturbances for $\overline{\eta} = 2.0$.

one wavelength predominates and controls the final distribution of necks and in essence the final segment size after breakup. This fact can be seen in Figs. 19 and 20, which show the results of three different random surface disturbances. The initial disturbances are shown in Fig. 19 and the final shapes are shown in Fig. 20. We see that all three result in a simple final waveform with approximately the same final wavelength. The length of this wave can be traced back to the initial time by using the stretching rate. We call this the critical wavelength $\overline{\lambda}_{ocrit}$. For practical shaped charge calculations, this is the wavelength, at a time when the jet is first formed, that will yield the most disturbance growth.

We have performed such computations for a wide range of nondimensional strain rates $\overline{\eta}_0$. The results of $\overline{\lambda}_{ocrit}$ vs $\overline{\eta}_0$ are plotted in Fig. 21. Throughout the foregoing development of the jet stability theory, we have considered fixed-velocity end conditions on a jet segment some distance from the ends of the jet. This calculation was based on experimental observations of the jet radiographs. A general question arose: since the jet was in free flight and not subject to external forces and since the jet material was flowing plastically and absorbing strain energy, were we not supplying energy to the system by imposing constant velocity end conditions on our segment of jet? We answered this objection with the hypothesis that, for a typical shaped charge jet, the large amount of stretching far exceeds the jet's capability to absorb strain energy;

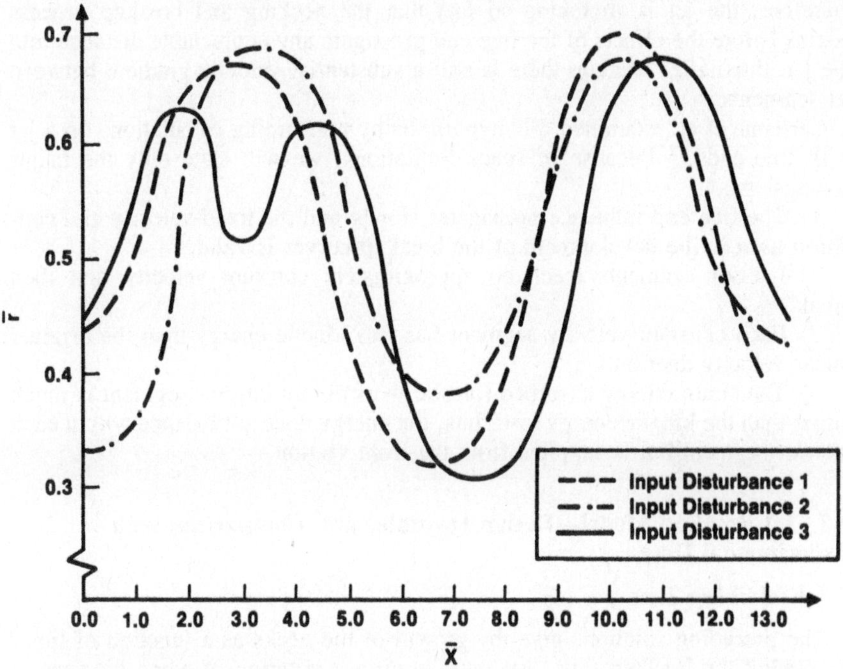

Fig. 20 Resulting disturbance shape at f = 0.66 for each different initial disturbance.

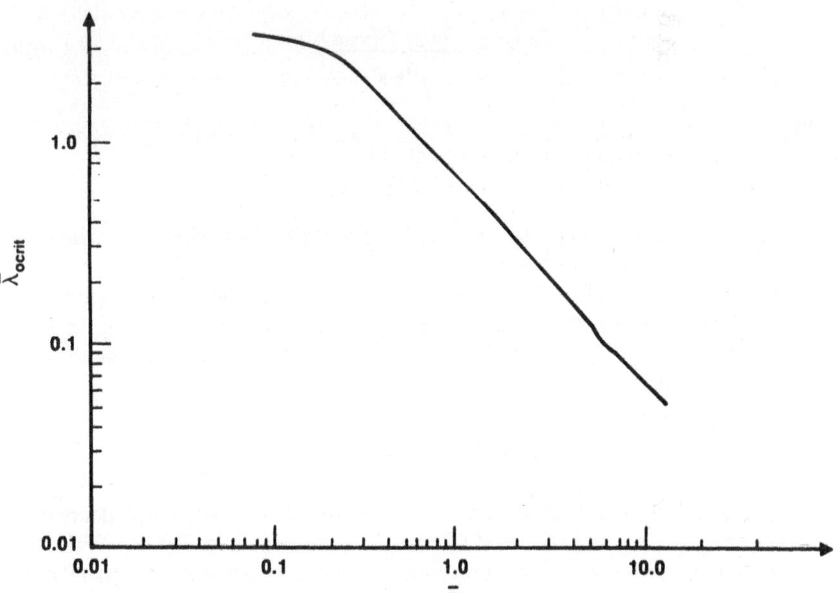

Fig. 21 Plot of critical wavelength $\overline{\lambda}_{ocrit}$ as a function of strain rate $\overline{\eta}_0$.

therefore, the jet is stretching so fast that the necking and breakup process begins before the effects of the free end propagate any appreciable distance into the jet; thus, after breakup there is still a substantial velocity gradient between jet segments.

Carleone et al. examined this hypothesis by performing calculations on a jet with free ends.[42] Because of space limitations, we will state only the major conclusions:

1) The free end influence propagates slowly and the fixed-velocity end condition used in the development of the breakup curves is valid.

2) Interior segments reach an approximately constant velocity first then break.

3) Each constant velocity segment has less kinetic energy than the original linear velocity distribution.

4) The strain energy absorbed (plastic work) in an interior segment is much larger than the kinetic energy lost; thus, the energy does not balance within each interior segment but is supplied from the front section.

3.3 Jet Breakup Model—Design Formulas and Comparison with Experimental Data

3.3.1 Breakup Time

The preceding solutions give the growth of the necks as a function of time. To predict the breakup time, we must impose a criterion of when breakup or segment separation occurs. Consider the amplitude growth function f defined in Eq. (60). Physically, this quantity may be interpreted as

$$f = \frac{A_{\text{avg}} - A_{\text{min}}}{A_{\text{avg}'}} \tag{68}$$

where

A_{avg} = current average cross-sectional area
A_{min} = current minimum cross-sectional area

Let us assume that the neck breaks when f reaches a given value f_s, which we shall call the dynamic ductility.

The time when f reaches f_s is relatively insensitive to the initial amplitude.[42] In general, the breakup time will, therefore, be a function of $\overline{\eta}_0$, $\overline{\lambda}_0$, and f_s. Since we need only to consider the critical wavelength, which is a function of $\overline{\eta}_0$, we may, therefore, assume

$$\bar{t}_b = \bar{t}_b(\overline{\eta}_0, f_s) \tag{69}$$

Plotting the solution indicated, we find $\bar{t}_b$ to be a monotonically decreasing function of $\overline{\eta}_0$ for $\overline{\eta}_0 > \frac{1}{2}$. Several such curves are shown in Fig. 22. Values of $\bar{t}_b$ vs $\overline{\eta}_0$ for experimental data from conical shaped charges are given in Refs. 41 and 42.

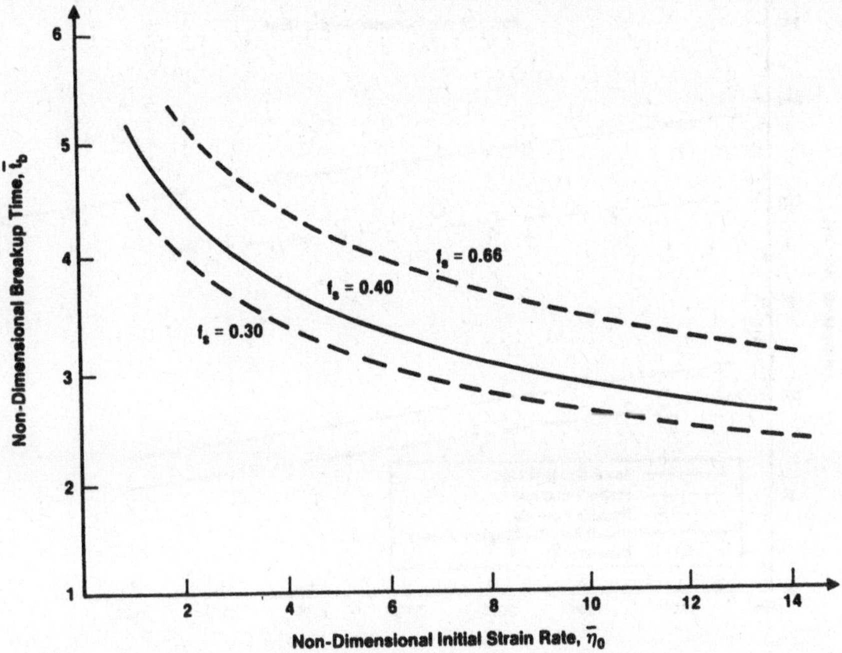

Fig. 22 Breakup curves from finite-difference numerical computations.

We have found that the final comparison of breakup time for all ductile jets can be approximated by the formula

$$\bar{t}_b = \gamma \bar{\eta}_0^{-\delta} - \left(\frac{1}{\bar{\eta}_0}\right) \tag{70}$$

where $\gamma = 5.4427$ and $\delta = 0.2953$. To place the results in dimensional form, we need the proper value of Y/ρ. For copper jets, we have found that $\sqrt{Y/\rho} = 0.15$ km/s is a reasonable value.

The complete jet breakup time distribution for two different copper-lined shaped charges is shown in Figs. 23 and 24. Measured jet breakup time is plotted vs jet segment position at a convenient time. A least-squares straight line is plotted to show the breakup trend. The results of the one-dimensional model are also compared and show excellent agreement. For the case of a 60° copper cone, we have plotted the results calculated by Pfeffer[45] and Hirsch[43] as presented in their papers. (In a private communication, Hirsch demonstrated that by varying the parameter $V_{p\ell}$, results can be matched more closely. However, then $V_{p\ell}$ will vary for a given material.) In some cases, these formulas agree more closely with experiments; however, each cannot describe the full range of geometries and materials often encountered. See Refs. 41 and 42 for more detailed comparisons.

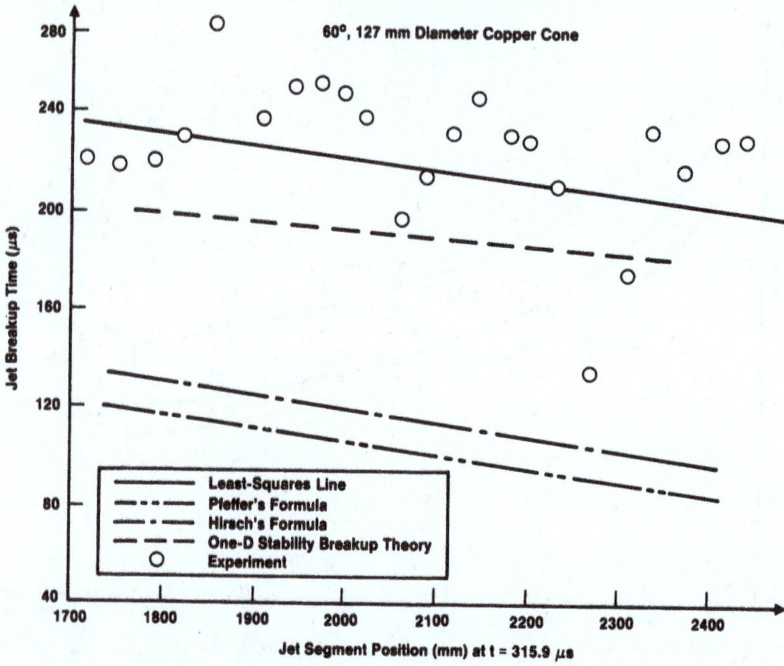

Fig. 23 Comparison of Hirsch's formula, Pfeffer's formula, present theory, and experiment for 127-mm, 60 deg, copper-lined charge.

3.3.2 Number of Jet Segments

Using the critical wavelength λ_{ocrit} given by the one-dimensional stability theory, we can predict the number of jet segments for shaped charge jets. Conceptually, we can explain this method as follows. By definition, the critical wavelength is that surface disturbance wavelength at the formation of the jet element that is *most unstable* and will predominate among all other wavelengths. The surface disturbance amplitude of this wavelength will grow and reach the breaking value before all others. Each full cycle on the surface will correspond to a jet segment. Thus, by dividing each jet element length as it is formed by the critical wavelength for that element and summing over all elements, the number of eventual segments can be predicted. Mathematically, we can express this as follows. Let $d\xi_0$ be the jet element length as it is formed. Then the number of critical waves in this element is given by $d\xi_0/\lambda_{ocrit}$. The total number of waves or jet segments can then be obtained by

$$N = \int_{\text{all jet elements}} \frac{1}{\lambda_{ocrit}} d\xi_0 \tag{71}$$

where N is the total number of jet segments. The increment $d\xi_0$ and the integration indicated by Eq. (71) may be calculated by a one-dimensional jet formation

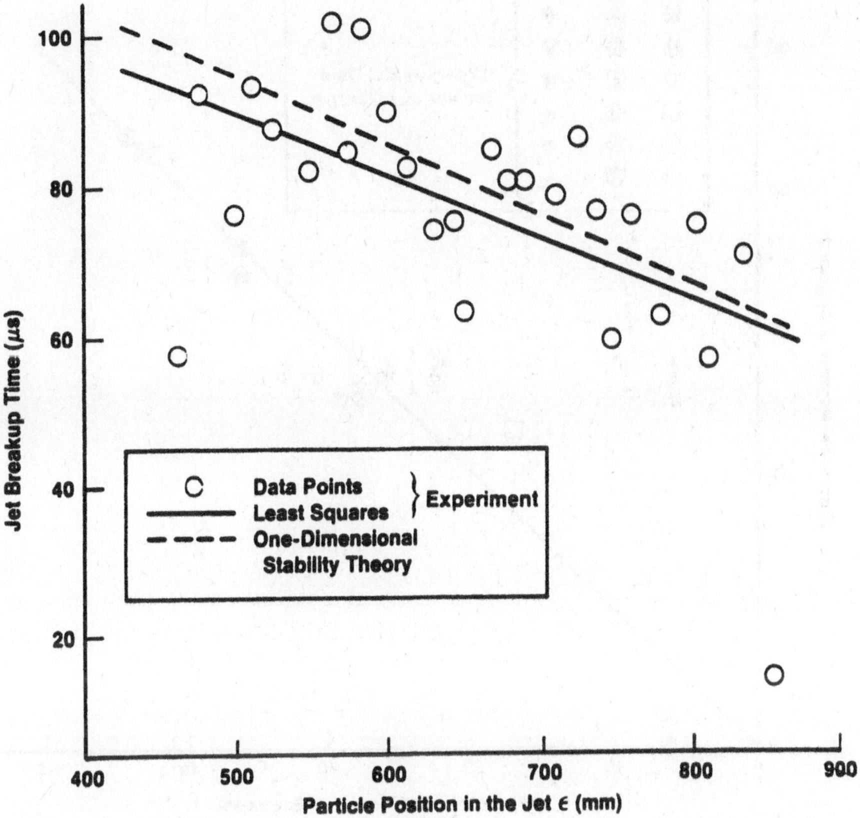

Fig. 24 Jet breakup time vs jet position at $t = 130.7\mu s$ for the 50.8-mm, 42 deg, 2.54-mm wall, copper-lined charge.

code. For convenience, we note that for $\overline{\eta}_0 > 0.03$, we can express the curve for λ_{ocrit} shown in Fig. 21 by the function

$$\overline{\lambda}_{ocrit} = \beta \overline{\eta}_0^{\alpha} \tag{72}$$

where $\beta = 0.6807$ and $\alpha = -0.9879$. By using a one-dimensional jet formation code, the calculated number of segments given by Eq. (71) is compared to the experimentally observed number of segments for a large number of different charges in Fig. 25. An exact agreement between theory and experiment falls on a 45 deg line through the origin. We see excellent agreement between theory and experiment. Furthermore, we can observe how the number of jet segments varies with cone angle in Fig. 26. Also, the effect of liner wall thickness on the number of jet segments is shown in Fig. 27.

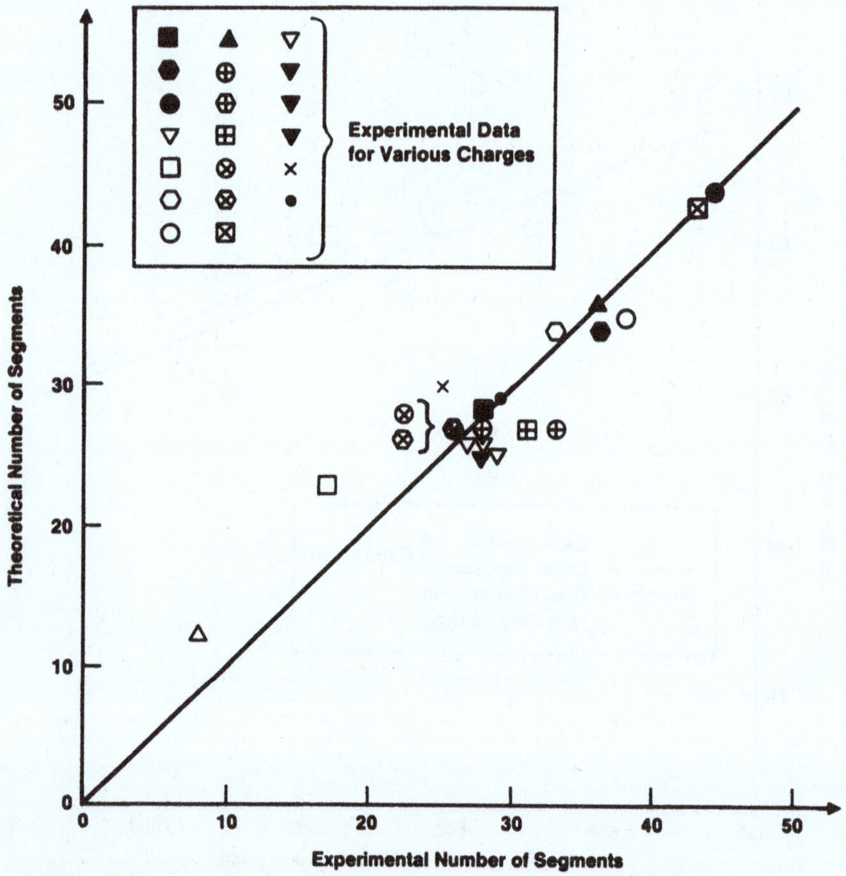

Fig. 25 Comparison of theoretical and experimental number of jet segments.

Finally, it is interesting to note that Eq. (72) may be useful in estimating $\sqrt{Y/\rho}$ from the observed velocity difference between segments after breakup. Since $\alpha \cong -1$, we may rewrite Eq. (72) in dimensional form as

$$\eta_0 \lambda_{\text{ocrit}} = \beta (Y/\rho)^{\frac{1}{2}} \tag{73}$$

The quantity $\eta_0 \lambda_{\text{ocrit}}$ represents the velocity difference between segments, which can easily be measured from radiographic data of jets in free flight.

3.4 Summary of the Jet Particulation Phenomenon

A one-dimensional theory of shaped charge jet instability and breakup has been presented that identifies the primary factors in the jet breakup process.

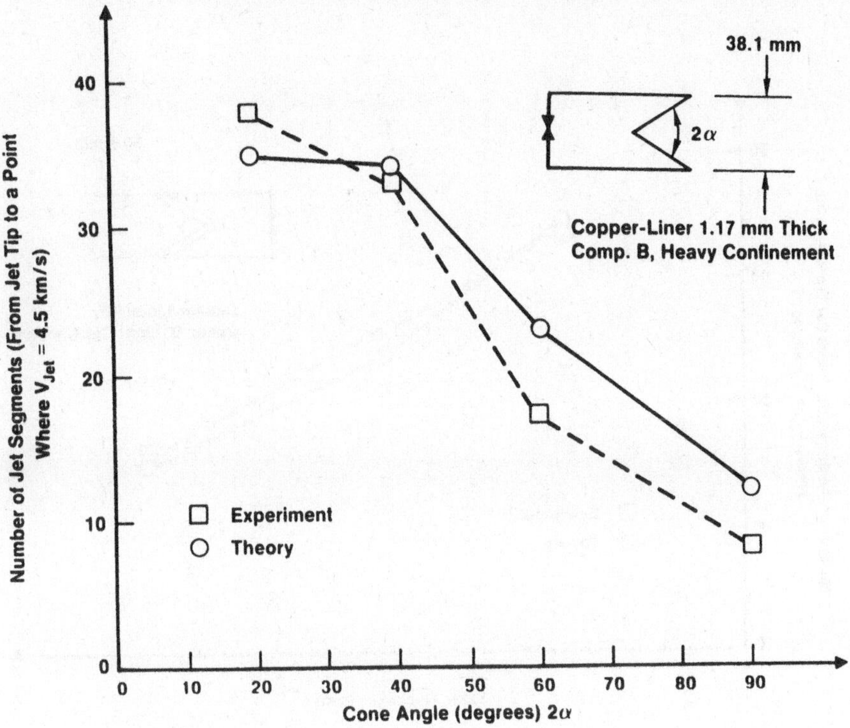

Fig. 26 Comparison of theoretical and experimental number of jet segments for a series of charges with different cone angles.

The model uses the concept of a plastic instability triggered by very small surface perturbations. The one-dimensional theory shows that a critical wavelength exists; the amplitude of this critical wavelength grows faster than that of all other wavelengths. This wavelength determines the number of segments after jet breakup.

The final breakup is a function of two material parameters: strength-to-density ratio Y/ρ and dynamic ductility. It also depends primarily on the initial jet radius r_0 and stretching rate η_0. The final growth appears insensitive to the initial amplitude as long as it is relatively small ($f_0 < 0.002$), although more work is required on this topic to identify the relationship between relative magnitudes of potential sources of input disturbances.

The model shows that *breakup time can be increased by* the following: increasing jet radius, decreasing stretching rate, decreasing jet strength, increasing density, and increasing ductility. The final model is embodied in a simple formula that relates the key parameters. It is easy to apply and has been incorporated into simplified one-dimensional jet formation computer programs.

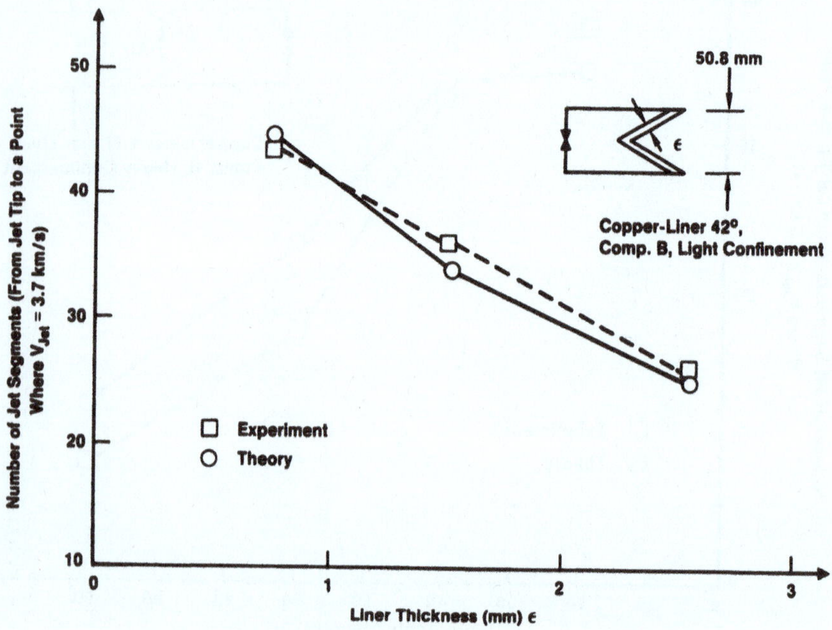

Fig. 27 Effect of wall thickness on number of jet segments.

2α (degrees)	E (% CD)
30	2.0
60	1.5
60	2.0
60	2.5
90	2.0

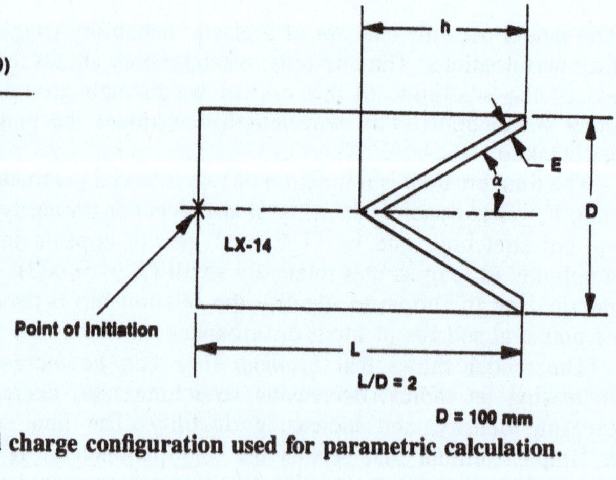

Fig. 28 Shaped charge configuration used for parametric calculation.

4. Parametric Calculations

A complete theory for shaped charge jet formation and jet breakup has been presented in the foregoing sections. One of the major uses of such a theory, or model, is to provide a tool that can be used for examining trends and sensitivities of various shaped charge design parameters or characteristics. In this section, we present a series of parametric calculations to exemplify the use of this theory. The analysis will be restricted to copper conical liners of constant wall thickness. The effect of varying the magnitude of wall thickness and cone angle will be examined. The basic configurations used in the computations are shown in Fig. 28. The explosive used in the calculations was LX-14 with detonation velocity of 8.8 km/s and a Chapman-Jouget (C-J) pressure of 0.370 Mbar. The calculations used the jet formation theory described in Sec. 2.2 with the linear acceleration model. Also the Defourneaux relation, Eq. (5), was used to provide the explosive-metal interaction model.

Figure 29 shows the jet velocity as a function of the original position on the liner of each liner element. The position is normalized by the liner height and denoted by $\overline{x}$. In all cases considered, the jet formation was calculated to be in the subsonic regime. Note the inverse velocity gradient near the apex of the cone. The jet tip speed, as calculated by Eq. (36), is noted by the cross located on each of these curves.

Note that the tip particle of the jet, which is formed from the inverse velocity gradient region, for all three cone angles is formed from approximately the same percentage of liner length. That is, the apex portion of the cone with height equal to 30–35% of the total liner height forms the tip particle. Normal jetting ensues from the rest of the liner. Also, for conical shaped charges with heavy-wall steel confinement, the apex region that forms the jet tip can extend to 45–50% of the liner height.[3]

It is interesting to note the sensitivity of jet tip velocity to the cone angle of the shaped charge liner. This is exemplified in Fig. 30, which shows plots of jet tip velocity as a function of cone angle for a fixed wall thickness (2% of the charge diameter) as well as a function of wall thickness for a fixed cone angle (60 deg). The plots show that jet tip velocity is much less sensitive to wall thickness changes in the practical region of interest (1.5–3% charge diameter) but is strongly tied to cone angle.

The mass distribution in the jet as a function of the velocity is addressed in Fig. 31. The mass is expressed as the cumulative mass starting from the mass at the tip and summing rearward toward the aft section of the jet. For example, the mass associated with the highest velocity on the curve for a given design represents the mass of the jet tip particle. At a lower velocity, we read the mass axis as the total jet mass from the tip to the section in the jet traveling at that lower velocity.

It is interesting to point out that, for the examples chosen, the accumulated jet mass at 4 km/s is approximately the same for all three designs. This is an extremely important observation because, for many applications, the most useful portion of the jet is the forward portion. Elements of the rear portion of the jet ($<$ 3 km/s) often do not contribute to the penetration depth because they often interact with the side of the target hole or with other jet particles that have

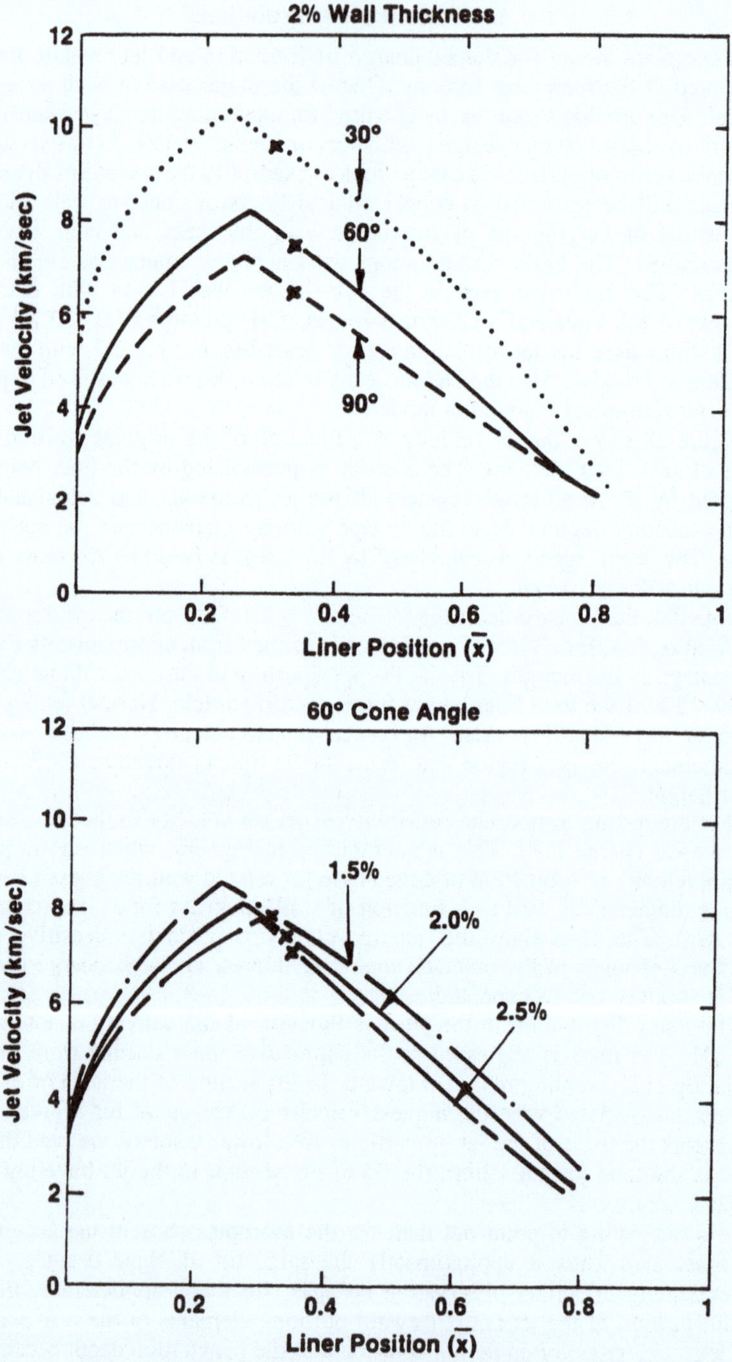

Fig. 29 Jet velocity distribution as a function of position along the liner.

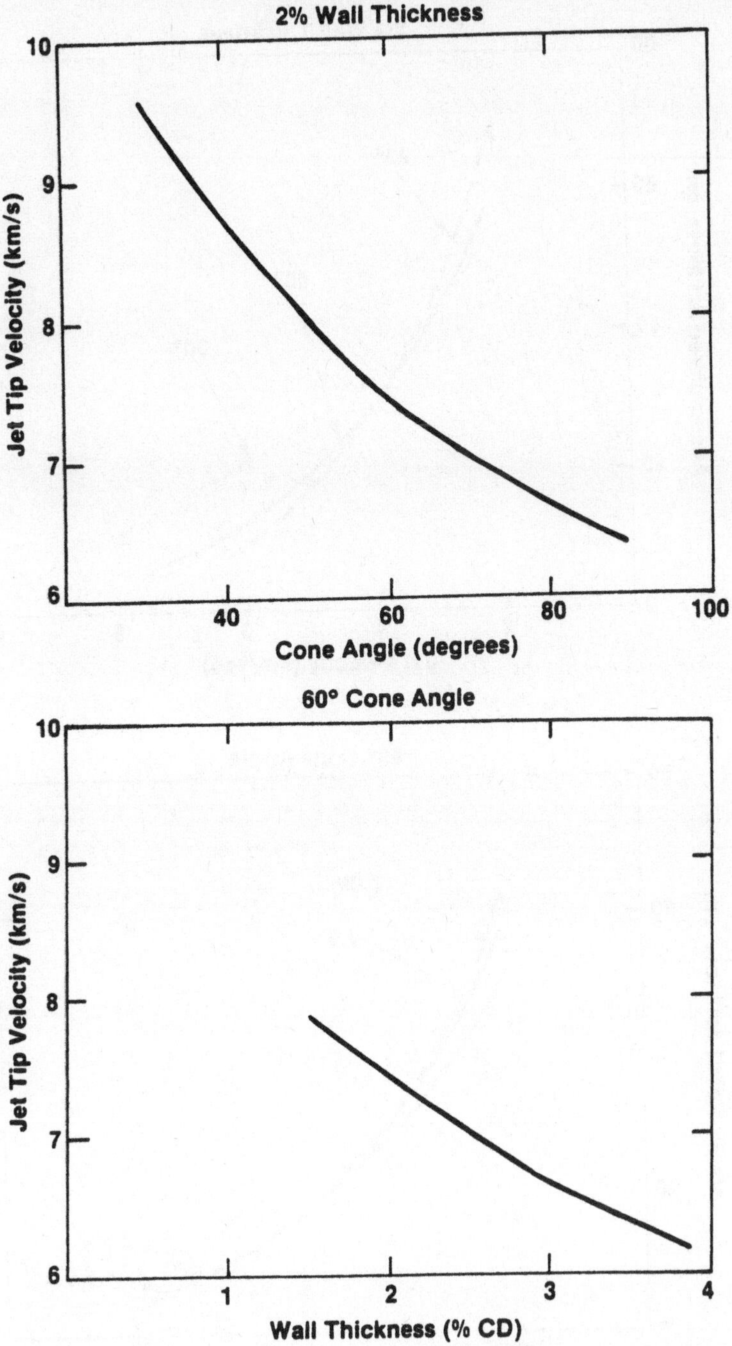

Fig. 30 Jet tip velocity as a function of cone angle and wall thickness.

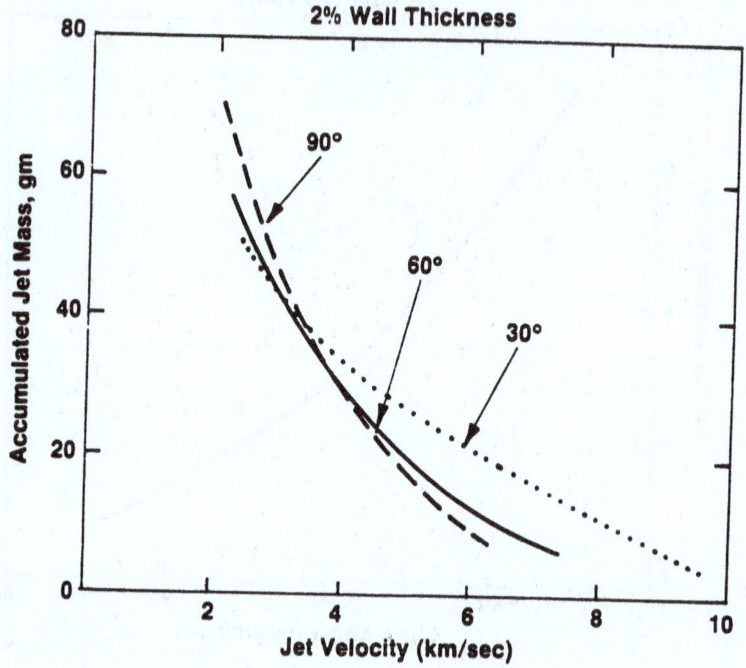

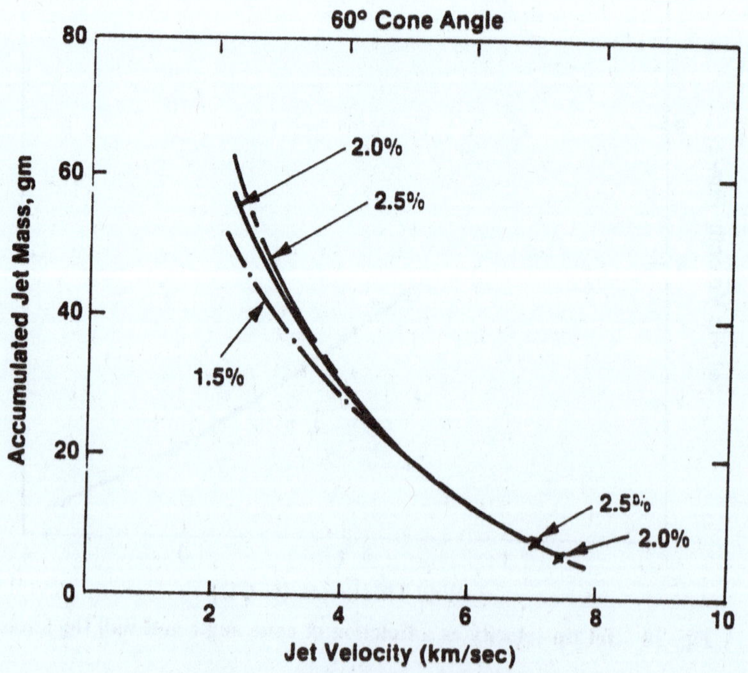

Fig. 31 Jet mass distribution.

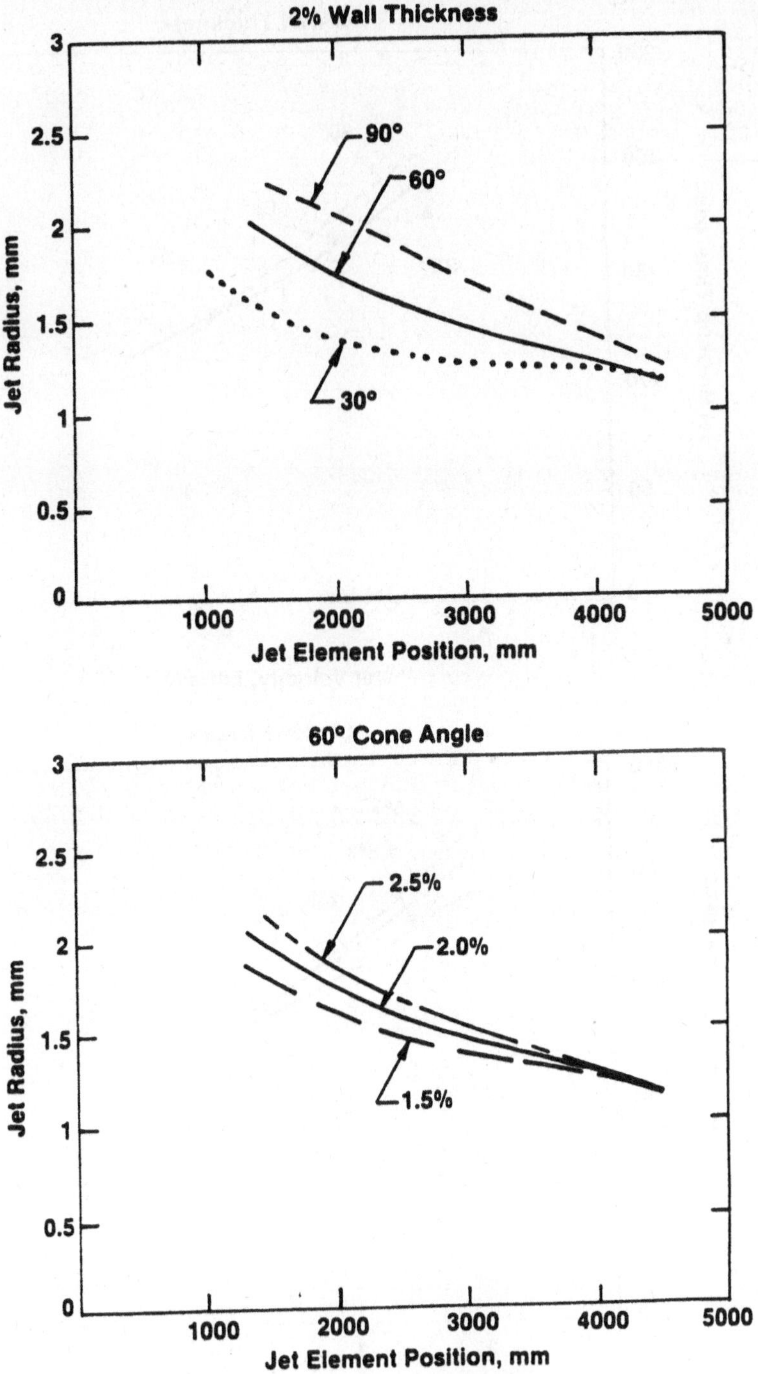

Fig. 32 Jet radius distribution.

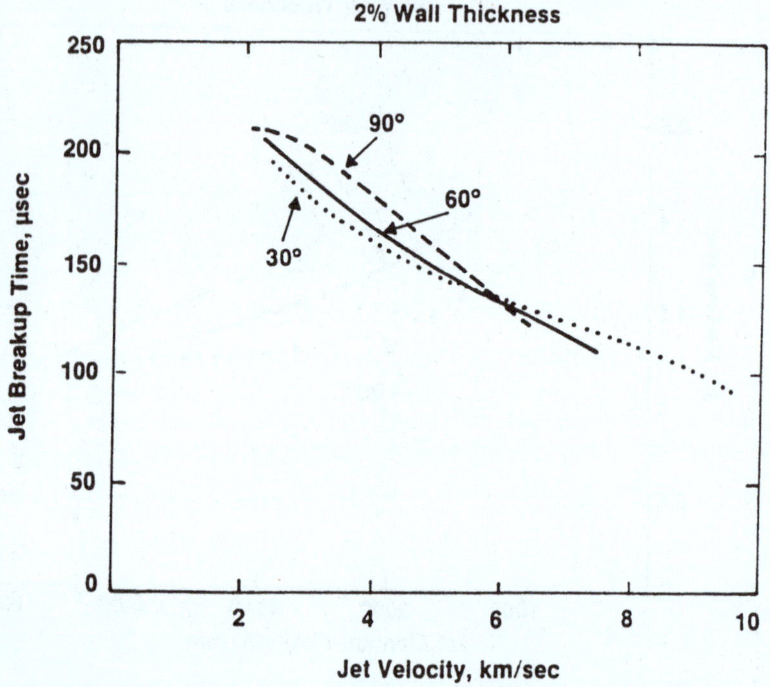

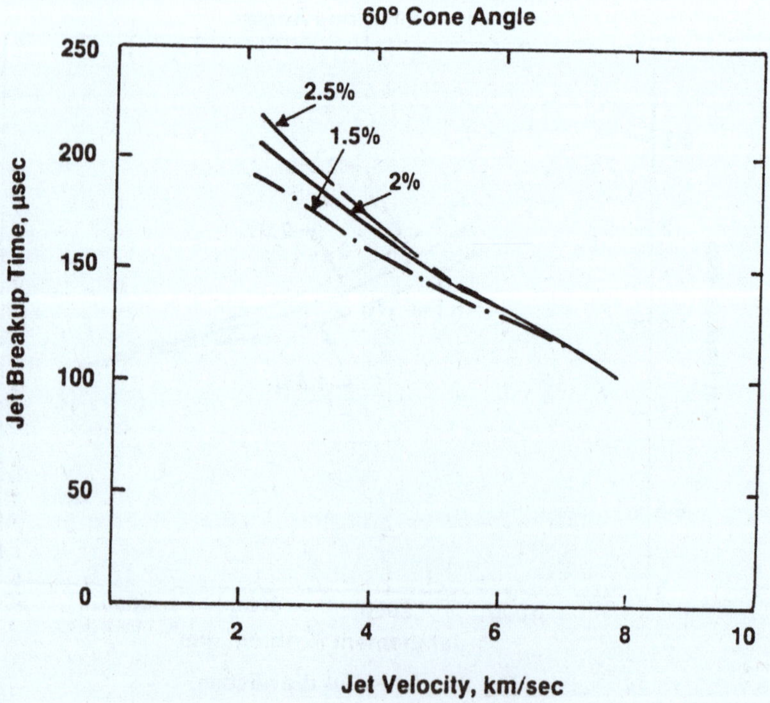

Fig. 33 Jet breakup time distribution.

been slowed or tumbled by noncollinear interactions with the target. Therefore, placing more mass in the forward portion of the jet at the expense of the rear portion of the jet is usually a good trade.

A comparison of jet radius distribution is made in Fig. 32. The radius is plotted as a function of position along the jets when the tip of each jet reaches 4500 mm from the cone apex. Such a long distance was chosen for convenience, since the entire jet is certainly particulated and the jet radius is no longer changing with time. Obviously, because the tip velocity of each design is different, this comparison is made at different times after initiation. We see from Fig. 32 that at a given standoff, the jet radius very near the tip is approximately equal for all the designs, but at the aft end of the larger cone angle (90 deg) gives a significantly larger radius jet. Similarly, the thicker liner gives a larger radius jet as we proceed aft from the tip. Again, jet radius is much more sensitive to cone angle than to liner thickness over the typical range of each variable.

Jet breakup time is plotted in Fig. 33 as a function of jet velocity. Here, we see that the breakup time is about 15% greater at the aft end of the jet from the 90 deg cone compared to the 30 deg cone. The liner thickness appears to affect breakup time at the aft end of the jet.

The power of such a model is in examining trends from one configuration to another. Absolute values given by these models are less important. However, by calibrating such a model using experimental jet measurements for a given liner material, explosive, initiation techniques, and a given overall liner shape, very accurate estimates of jet properties can be made for that family of shaped charges. The model may give erroneous results if it is applied to a configuration that is significantly different from the configuration for which the model is calibrated. When applying the model to a new family of designs, we recommend that some test data be generated to recalibrate the model.

Various computer codes have been developed to facilitate such calculations using the analytical models described in this chapter or variations of them. Several examples are BASC (Harrison[49]), DESC (Carleone et al.[50]), BRIGS (Backofen[51]), SC1D (Buckley[52]), and ASC (Conner and Quong[53]).

References

[1]Kennedy, D. R., "History of the Shaped Charge Effect: The First 100 Years," D. R. Kennedy and Associates, P.O. Box 4003, Mountain View, CA 1983.

[2]Walters, W. P., and Zukas, J. A., *Fundamentals of Shaped Charges*, Wiley, New York, 1989.

[3]Carleone, J., Jameson, R., and Chou, P. C., "The Tip Origin of a Shaped Charge Jet," *Propellants and Explosives*, Vol. 2, 1977, pp. 126–130.

[4]Behrmann, L., private communication, 1980.

[5]Chanteret, P. Y., and Jamet, F., "Quasi Non-Stretching Hypervelocity Jets," *Proceedings of the Eighth International Symposium on Ballistics*, Orlando, FL, Oct. 23–25, 1984.

[6]Arvidsson, B., and Gustafsson, H., "A Comparative Investigation of Various Analytical and Numerical Shaped-Charge Liner Collapse Models," *Proceedings of the Ninth International Symposium on Ballistics*, Shrivenham, UK, April–May 1, 1986.

[7]Aseltine, C. L., Walters, W. P., and Clifford, A. L., "Hemispherical Shaped Charges Utilizing Tapered Liners," *Proceedings of the Fourth International Symposium on Ballistics*, Monterey, CA, Oct. 17–19, 1978.

[8]Chou, P. C., Ciccarelli, R. D., and Walters, W. P., "The Formation of Jets from Hemispherical Liner Warheads," *Proceedings of the Seventh International Symposium on Ballistics*, The Hague, The Netherlands, April 19–21, 1983.

[9]Grace, F. I., Golaski, S. K., and Scott, B. R., "The Nature of Jets from Hemispherical Lined Explosive Charges," *The Proceedings of the Eighth International Symposium on Ballistics*, Orlando, FL, Oct. 23–25, 1984.

[10]Harrison, J. T., "A Two Staged, Hydrodynamic, Numerical Technique and an Analysis of a Hemispherical Shaped-Charge Liner Collapse," *Proceedings of the Eighth International Symposium on Ballistics*, Orlando, FL, Oct. 23–25, 1984.

[11]Bransky, I., "Investigation of Shaped Charges with Conical and Hemispherical Liners of Tungsten Alloys," *Proceedings of the Ninth International Symposium on Ballistics*, Shrivenham, UK, April 29–May 1, 1986.

[12]Chou, P. C., "A Simplified Model of Jet Formation in Hemispherical Shaped Charges," *Proceedings of the Ninth International Symposium on Ballistics*, Shrivenham, UK, April 29–May 1, 1986.

[13]Held, M., "Shaped Charge Measuring Techniques," presented at Fundamental Aspects of Hypervelocity Impact and Shaped-Change Phenomena," Baltimore, MD, April 7–11, 1986.

[14]Aseltine, C. L., "Analytical Predictions of the Effect of Warhead Asymmetries on Shaped-Charge Performance," *Proceedings of the Fourth International Symposium on Ballistics*, Monterey, CA, Oct. 17–19, 1978.

[15]Birkhoff, G., MacDougall, D. P., Pugh, E. M., and Taylor, G., "Explosives with Lined Cavities," *Journal of Applied Physics*, Vol. 19, June 1948, pp. 563–582.

[16]Pugh, E. M., Eichelberger, R. J., and Rostoker, N., "Theory of Jet Formation by Charges with Lined Conical Cavities," *Journal of Applied Physics*, Vol. 23, No. 5, May 1952, pp. 532–536.

[17]Eichelberger, R. J., and Pugh, E. M., "Experimental Verification of the Theory of Jet Formation by Charges with Lined Conical Cavities," *Journal of Applied Physics*, Vol. 23, No. 5, May 1952, pp. 537–542.

[18]Eichelberger, R. J., "Re-examination of the Nonsteady Theory of Formation by Lined Cavity Charges," *Journal of Applied Physics*, Vol. 26, No. 4, April 1955, pp. 398–402.

[19]Bryan, G. M., Eichelberger, R. J., MacDonald, D., and Zigman, P. E., "Application of Radioactive Tracers to the Study of Shaped Charge Phenomena," *Journal of Applied Physics*, Vol. 28, No. 10, Oct. 1957, pp. 1152–1155.

[20]Allison, F. E., and Vitali, R., "A New Method of Computing Variables for Shaped Charge Jets," U. S. Army Ballistic Research Lab, BRL Report No. 1184, Aberdeen Proving Ground, MD, Jan. 1963.

[21]Carleone, J., and Chou, P. C., "A One-Dimensional Theory to Predict the Strain and Radius of Shaped-Charge Jets," *Proceedings of the First International Symposium on Ballistics*, Orlando, FL, Nov. 13–15, 1974.

[22]Taylor, G. I., "Analysis of the Explosion of a Long Cylindrical Bomb Detonated at One End," (1941), *Scientific Papers of G. I. Taylor*, Vol. III, Cambridge Univ. Press, Cambridge, England, 1963, pp. 277–286.

[23]Kiwan, A. R., and Wisniewski, H., "Theory and Computations of Collapse and Jet Velocities of Metallic Shaped Charge Liners," U.S. Army Ballistic Research Lab, BRL Report No. 1620, Aberdeen Proving Ground, MD, Nov. 1972.

[24]Karpp, R., Los Alamos National Laboratory, Los Alamos, NM, private communication, 1974.

[25]Defourneaux, M., "Theorie Hydrodynamique des Charges Creuses," *Memorial de l' Artillerie Française,* T. 44, 2^e Fasc., 1970, pp. 293–334.

[26]Chou, P. C., Carleone, J., Flis, W. J., Ciccarelli, R. D., and Hirsch, E., "Improved Formulas for Velocity, Acceleration, and Projection Angle of Explosively Driven Liners," *Propellants, Explosives and Pyrotechnics*, Vol. 8, 1983, pp. 175–183.

[27]Randers-Pehrson, G., "An Improved Equation for Calculating Fragment Projection Angle," *Proceedings of the Second International Symposium on Ballistics,* Daytona Beach, FL, March 9–11, 1976.

[28]Carleone, J., and Flis, W. J., "Improved Formulas for the Motion of Explosively Driven Liners," Air Force Armament Lab. Rep. AFATL-TR-82-01, Jan. 1982.

[29]Walsh, J. M., Shreffler, R. G., and Willig, F. J., "Limiting Conditions for Jet Formation in High Velocity Collisions," *Journal of Applied Physics,* Vol. 24, No. 3, March 1953, pp. 349–359.

[30]Chou, P. C., Carleone, J., and Karpp, R. R., "The Effect of Compressibility on the Formation of Shaped Charge Jets," *Proceedings of the First International Symposium on Ballistics,* Orlando, FL, Nov. 13–15, 1974.

[31]Chou, P. C., Carleone, J., and Karpp, R. R., "Criteria for Jet Formation from Impinging Shells and Plates," *Journal of Applied Physics,* Vol. 47, No. 7, July 1976, pp. 2975–2981.

[32]Chou, P. C. and Carleone, J., "The Stability of Shaped Charge Jets," *Journal of Applied Physics,* Vol. 48, No. 10, Oct. 1977, pp. 4187–4195.

[33]Karpp, R. R., and Simon, J., "An Estimate of the Strength of a Copper Shaped Charge Jet and the Effect of Strength on the Breakup of a Stretching Jet," U.S. Army Ballistic Research Labs, BRL Report No. 1893, June 1976.

[34]Wilkins, M. L., "Calculation of Elastic-Plastic Flow," University of California, Lawrence Livermore Laboratory, Rept. UCRL-7322, rev. 1, Jan. 24, 1969.

[35]Chou, P. C., Carleone, J., Tanzio, C. A., and Ciccarelli, R. D., "Shaped Charge Jet Breakup Studies Using Radiograph Measurement and Surface Instability Calculations," U.S. Army Ballistic Research Lab BRLCR No. 337, April 1977.

[36]Carleone, J., Chou, P. C., and Ciccarelli, R. D., "Shaped Charge Jet Stability and Penetration Calculations," U.S. Army Ballistic Research Lab BRLCR No. 351, Sept. 1977.

[37]Bridgman, P. W., *Studies in Large Plastic Flow and Fracture,* Harvard Univ. Press, Cambridge, MA, 1964.

[38]Miller, C. W., "A New Approach to the Numerical Analysis of Shaped Charge Jets," *Proceedings of the Sixth International Symposium on Ballistics*, Orlando, FL, Oct. 1980.

[39]Steinberg, D. J., Cochran, S. G., and Guinan, M. W., "A Constitutive Model for Metals Applicable at High-String Rate," *Journal of Applied Physics,* Vol. 51, No.3, March 1980, pp. 1498–1504.

[40]Osborn, J. J., "Effects of Material Properties on the Deformation and Necking of Rods Stretching under Large Velocity Gradients," Seventh Annual Tech. Meeting U.S.-German DEA-AF-71-F/G-7304, Lawrence Livermore Lab, Oct. 21–23, 1981.

[41]Carleone, J., Chou, P. C., Walters, W., Jameson, R., and Ciccarelli, R. D., "Prediction of Shaped-Charge Jet Breakup Time and Calculation of Penetration," U.S. Army Ballistic Research Lab ARBRL-CR-00396, April 1979.

[42]Carleone, J., Chou, P, C., and Walters, W. P., "An Improved Shaped-Charge Jet Breakup Calculation and Measurement of Jet Properties," U.S. Army Ballistic Research Lab. ARBRL-CR-00445, Jan. 1981.

[43]Hirsch, E., "A Formula for the Shaped Charge Jet Breakup Time," *Propellants and Explosives,* Vol. 4, 1979, pp. 89–94.

[44]Hirsch, E., "A Model Explaining the Rule for Calculating the Breakup Time of Homogeneous Ductile Metals," *Propellants and Explosives,* Vol. 6, 1981, pp. 11–14.

[45]Pfeffer, G., "Determination par Simulations Numeriques de l'État et des Lois de Fragmentation des Jets de Charges Creuses," *Proceedings of the Fifth Symposium on Ballistics*, Toulouse, France, April 16–18, 1980.

[46]Walsh, J. M., "Plastic Instability and Particulation in Stretching Metal Jets," *Journal of Applied Physics*, Vol. 56, No. 7, 1984, pp. 1997–2006.

[47]Taylor, J. W., Harlow, F. H., and Amsden, A. A., "Dynamic Plastic Instabilities in Stretching Plates and Shells," *Journal of Applied Mechanics,* Vol. 45, No. 1, 1978, pp. 105–110.

[48]Malvern, L. E., *Introduction to the Mechanics of a Continuous Medium,* Prentice-Hall, Englewood Cliffs, NJ, 1969.

[49]Harrison, J. T., "BASC, an Analytical Code for Calculating Shaped Charge Properties," *Proceedings of the Sixth International Symposium on Ballistics,* Orlando, FL, October 27–29, 1981.

[50]Carleone, J., Chou, P. C., and Tanzio, C. A., "User's Manual for DESC-1, A One-Dimensional Computer Code to Model Shaped Charge Liner Collapse, Jet Formation, and Jet Properties," Dyna East Corporation, No. DE-TR-75-4, Dec. 1975.

[51]Backofen, J. E., Jr., BRIGS Co., Herndon, VA 22071.

[52]Buckley, P., PMC Inc., 1222 Apache Drive, Socorro, NM 87801.

[53]Conner, J., and Quong, A., "Aerojet Shaped-Charge Code (ASC)," Aerojet Electronic Systems Division, Azusa, CA 91702.

Chapter 7

Explosively Formed Projectiles

David Bender* and Joseph Carleone†
Aerojet General Corporation, Azusa, California 91702

1. Concept Origin

THE concept of using explosive energy to deform a metal plate into a coherent penetrator while simultaneously accelerating it to extremely high velocities offers a unique method of employing a kinetic energy penetrator without the use of a large gun. However, the concept was primarily used in mines until the mid-1970s when top attack systems capable of standing off from the target and addressing the more vulnerable target tops were introduced. The explosively formed projectile (EFP) is a natural warhead selection for these systems. This chapter will point out that many differently shaped projectiles can be formed, and warheads can be sized to fit most weapon systems.

A typical EFP warhead is shown in Fig 1. Also known as an *explosively formed penetrator, self-forging fragment, Misznay-Schardin charge, mass focus device,* and the *P-charge,* the EFP warhead is made up of a metallic liner and case, an explosive section, and an initiation train. Very often there is also a retaining ring to position and hold the liner-explosive subassembly in place. After detonation, the explosive products create enormous pressures that accelerate the liner while simultaneously reshaping it into a rod or some other desired shape. The EFP then hits the target at speeds in excess of 2000 m/s, delivering billions of watts of mechanical power.

The EFP concept is certainly not new. In fact, the earliest published demonstration of such a device was by R. W. Wood in 1936.[1] Wood experimented with detonators having shallow cavities in the explosive. The cavities at the end of the detonators were lined with copper. Wood demonstrated that after detonation the copper in each cavity formed a pellet that was projected at a very high velocity and that traveled long distances in a coherent fashion.

The explosive formation of a projectile later became known as the Misznay-Schardin effect. Misznay demonstrated this effect in Germany with a 400-mm diameter charge and a 300-mm diameter iron liner in July 1944. According to

*Program manager.

†Vice-president, Armament Programs, Electronic Systems Division.

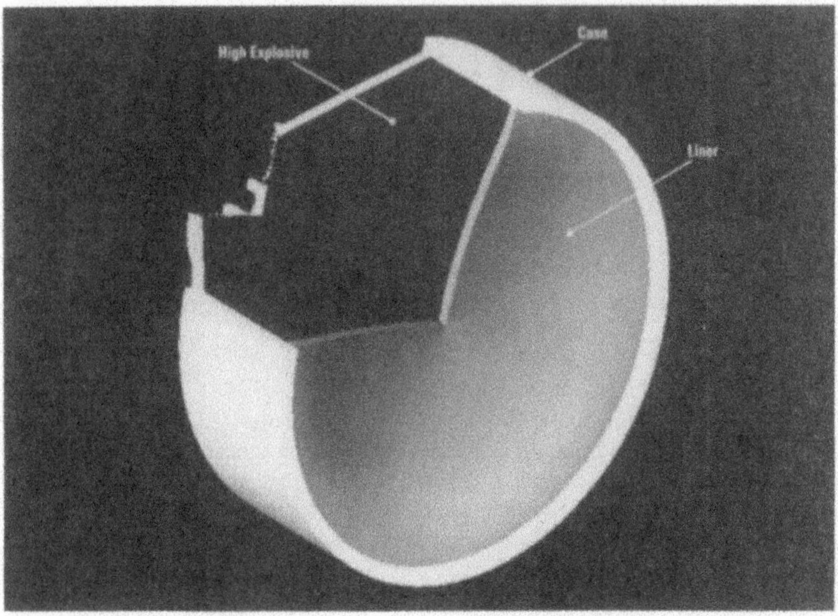

Fig. 1 Typical EFP warhead configuration.

Held, this demonstration was conducted before W. Trinks of the Wehrmacht Army Ordnance Office and H. Schardin and G. Thomer of the Luftwaffe Technical Academy, Berlin-Gatow.[2] Schardin later published technical data for these types of charges.[3,4] These charges were relatively simple compared with today's design; the liners typically were constant-thickness plates with constant curvature, as have since been used in mine systems to attack the belly of tanks. Such fundamental designs are still called Misznay-Schardin charges, although the term is often applied to more advanced designs as well.

It was found that constant-wall-thickness Misznay-Schardin devices did not work well at very long standoff (i.e., distance from warhead to target) since the penetrator typically fractured a meter or two away from the original warhead position. As Hermann et al. indicated, Kronman of the Army Ballistic Research Laboratory developed a coherent, bullet-shaped projectile by tapering the liner of a Misznay-Schardin charge.[5] Held investigated wide-angle conical liners with tapered wall thickness.[6] He called these P-charges, short for projectile-charge, since they also formed a coherent projectile capable of traveling long distances to defeat armor.

During the early to mid-1970s, EFP technology escalated significantly, primarily because of three simultaneous developments: 1) the success of hydrocode simulation techniques to model the EFP device, 2) the progress in high-precision computer numerical control (CNC) manufacturing techniques, and 3) a number of system concepts sponsored by both the U.S. Army and U.S. Air Force that used EFP technology. The first gave designers the ability to achieve improved EFP shapes rapidly, the second allowed complex liner shapes to be made with precision, and the system projects provided the funding base and a specific requirement to allow a significant focus on EFP technology. This work demonstrated that virtually all practical shapes can be explosively formed by varying the contour and thickness profile of the metal liner in the proper way. In fact, many shapes, such as compact spheroids, long rods, hollow cups, and rods with a conical flare have been demonstrated. Table 1 summarizes the key origins of the EFP concept that have contributed to its evolution.

Two major applications have evolved since 1975 for EFP warheads, namely, long-standoff sensor-fuzed submunitions and medium-standoff, close-overflight missiles, as depicted in Fig. 2. The former application, which is the more traditional one, requires the formation of a single-piece EFP capable of flying in a stable fashion to the target. This refinement has led to the flared EFP rod and, more recently, to the finned EFP rod designs.

For the medium- or short-standoff applications, a new type of EFP was developed. The need for an aerodynamic shape is not necessary for these applications because of the short distance the EFP must travel, hence, the length of the rod was increased and the flared tail was eliminated from the design. In fact, some of these rods are purposely stretched beyond their breaking point and fracture into several pieces resulting in greater total length. This type of EFP was pioneered during the Anti-armor Cluster Munition (ACM) program and was reported by Carleone and Johnson.[7]

2. Mechanics of Explosively Formed Projectile Formation

To place the EFP technology in perspective, it is helpful to examine its liner collapse in the context of the more familiar shaped charge warhead. A shaped

Table 1 Evolution of the explosively formed projectile

Investigator(s)	Event	Date
Wood[1]	Detonators with shallow cavities	1936
Misznay[3]	Curved iron plate (300-mm diam) in a charge (400-mm diam)	1944
Schardin[4]	Published studies of shallow cavity-lined charges	1954–1956
Kronman[5]	Tested "ballistic disc" (tapered constant curvature liner)	1956–1958
Held[2]	Wide-angle cones with tapered walls	1965–1975
Karpp[15]	Applied HEMP code to simulate ballistic disc	
Hermann, Randers-Pehrson, and Berus[5]	Hyperbolic liner to form compact ball projectile and applied hydrocode analysis	1975–1977
Many investigators	Flared EFPs	1980
Bender, Carleone, and Singer[10]	Finned EFPs	1988
Carleone and Johnson,[7] Bender and Archibald[16]	Long-rod EFPs	1982–1990

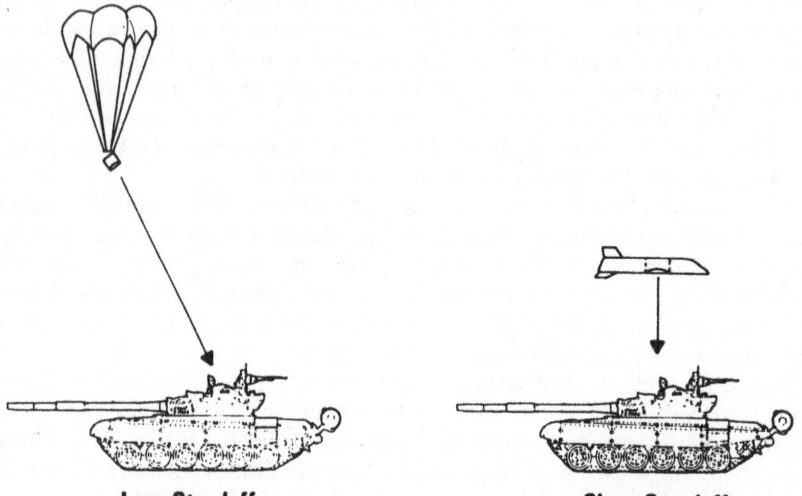

Fig. 2 Two modern applications of EFP technology.

charge is an explosive device that has a deep metal-lined cavity that is typically conical. As shown in Fig. 3, the shaped charge liner normally splits into a very fast jet and a much slower-moving slug, which eventually separate from each other. The jet makes up about 15 percent of the liner mass, and the rest remains in the slug. As the half-cone angle is increased, the inward collapse becomes less violent and there is correspondingly less of a velocity difference between the jet and slug. In fact, as shown in Fig. 4, Held found that for α approaching 75 deg, the jet and slug approach the same velocity and become indistinguishable.[2] Hence, at these angles, an EFP is formed.

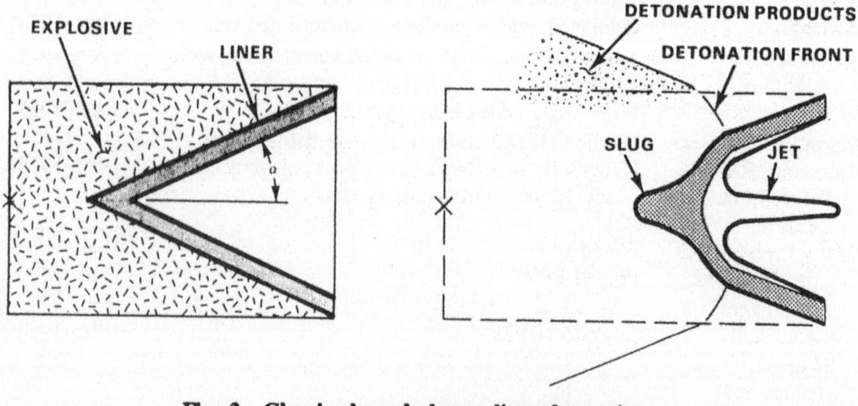

Fig. 3 Classic shaped-charge liner formation.

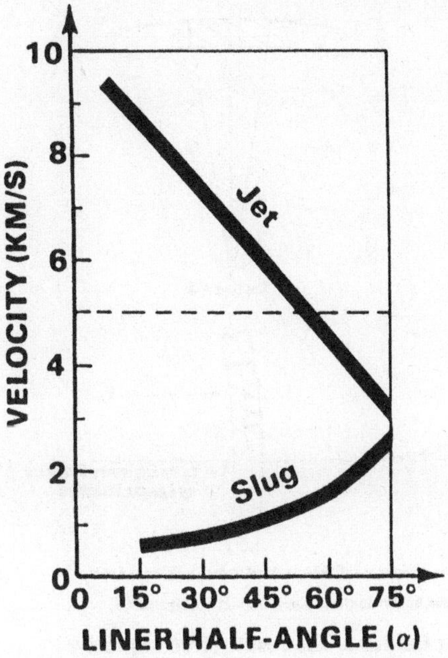

Fig. 4 Jet and slug velocity as a function of cone angle.

On the basis of fundamental momentum principles, let us consider the interaction between the explosive and the liner. After detonation, a shock wave passes through the explosive charge and impinges upon the liner. Behind this shock wave are the gas products of the explosion, which are at very high pressures. Now consider an element of the liner, as shown in Fig. 5, with thickness h and surface area dA given by

$$\mathrm{d}A = r_\theta r_\phi \boldsymbol{n}\, \mathrm{d}\theta \mathrm{d}\phi \tag{1}$$

where r_θ and r_ϕ are the local radii of curvature of the element and $\boldsymbol{n}$ is its unit normal vector.

As the shock wave passes over this element, the gaseous products push on this element with a pressure $p = p(t)$, causing the element to accelerate. The total time of effective interaction is short, on the order of tens of microseconds in most cases, because the pressure in the products decays very rapidly. This interaction results in an impulse $\boldsymbol{P}$ that is delivered to the liner element causing it to attain a final velocity $\boldsymbol{V}$, or

$$\boldsymbol{P} = \int p\, \mathrm{d}\boldsymbol{A}\, \mathrm{d}t = \boldsymbol{V}\, \mathrm{d}m \tag{2}$$

where dm is the element's mass given by $\mathrm{d}m = \rho h\ \mathrm{d}\boldsymbol{A} \cdot \boldsymbol{n}$. For simplicity of discussion let us consider axisymmetric or two-dimensional liner surfaces.

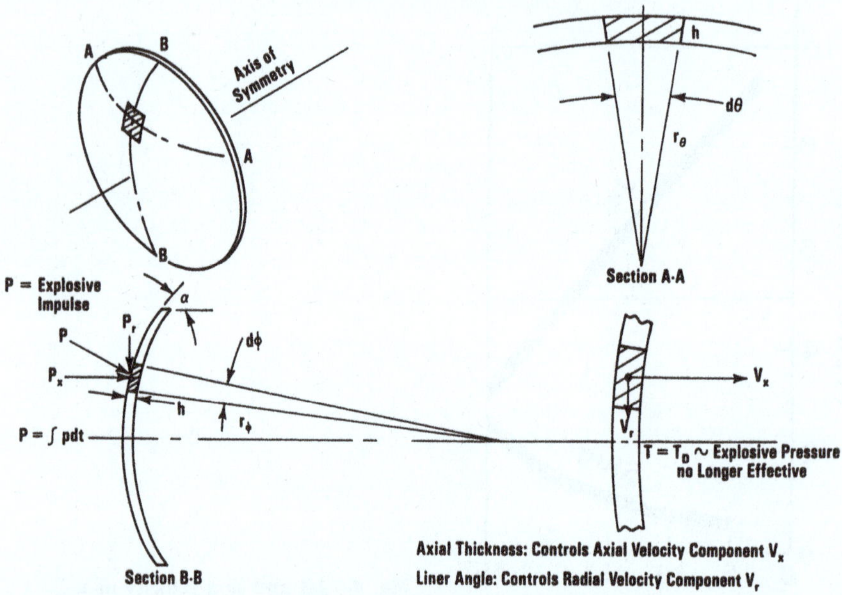

Fig. 5 Elements of the EFP liner that influence the result of an explosive impulse.

The fundamental rules that will be established will also apply, or can easily be extended, to three-dimensional liners as well. The fundamental vectors can then be written in component form as

$$\begin{aligned} \boldsymbol{P} &= P_x\boldsymbol{i} + P_r\boldsymbol{j} \\ \boldsymbol{V} &= V_x\boldsymbol{i} + V_r\boldsymbol{j} \\ \boldsymbol{n} &= \boldsymbol{i}\ \sin\alpha + \boldsymbol{j}\ \cos\alpha \end{aligned} \tag{3}$$

where x and r represent the axial and radial components and $\boldsymbol{i}$ and $\boldsymbol{j}$ are the unit vectors in the x and r direction, respectively.

The individual velocity components are, therefore, given by

$$\begin{aligned} V_x &= P \sin\alpha / \rho h dA \\ V_r &= P \cos\alpha / \rho h dA \end{aligned} \tag{4}$$

The final mapping of all of the liner elements, that is, the final EFP shape, is governed by the distribution of these velocity components along the liner. Therefore, given an initial liner configuration that produces a certain EFP shape, one can modify the liner and produce a different or modified shape by changing the liner angle α and thickness h in accordance with Eq. (4).

3. Warhead Configurations and Penetrator Shapes

In developing a warhead configuration that meets system constraints and also meets performance requirements, several basic parameters in the warhead configuration must be adjusted until the optimum configuration is obtained. The mode of liner formation must be chosen and the liner diameter optimized for the given target requirements. The casing must be optimized to provide sufficient performance, and symmetry should be maintained. Also, if the liner must form through an obstruction, the mass of this obstruction should be minimized.

Several warhead configurations have been developed over the years to accommodate varying system constraints; however, the basic components are essentially the same. Figure 1 shows a typical EFP warhead configuration comprised of a steel case, high-explosive charge, and metallic liner.

The case provides a protective housing for the explosive and liner, but more important, the mass of the case provides confinement for the explosive. The addition of mass around the explosive and liner increases the duration of the explosive impulse and hence the total energy delivered to the liner. The material of choice for the case is typically steel because of its low cost, high strength (if desired), and density. However, most materials can be used, as long as the mass is sufficient to provide the necessary confinement.

In designing an EFP warhead, symmetry must be maintained. If the amount of mass confinement provided by the casing varies about the axis of symmetry, an unbalanced venting of explosive products will occur and an asymmetric explosive impulse will result. This asymmetric explosive impulse will cause the liner to form asymmetrically. This consideration is especially critical in forming an aerodynamically stable rod since an asymmetric flare will more than likely cause the rod to tumble or diverge greatly from the planned trajectory. Yiu studied the effects of asymmetric confinement on the formation of EFPs.[8] Using a relatively simple liner design, he examined the effects of axisymmetric explosive confinement. The explosive and liner were surrounded by a two-piece case, 180 deg of steel and 180 deg of epoxy. When the warhead was initiated, the unbalanced explosive impulse caused by the uneven venting of explosive products produced a severely distorted EFP. As another example of changing the mass confining the explosive, a series of two-dimensional hydrocode simulations of the same liner design were conducted. The steel case wall thickness was simulated at 10 mm and 5 mm. The results are illustrated in Fig. 6. In the 10-mm-thick configuration, the liner formed an EFP that is relatively compact and has a velocity of 2.57 km/s. In the case with a wall thickness of 5 mm, the same liner design formed an EFP that is relatively more hollow and has a velocity of 2.43 km/s, 5.4% less than that in the 10-mm-thick wall configuration. Hence, the liner design is highly dependent on the surrounding mass and is very sensitive to mass asymmetries.

The density and the physical dimensions of the explosive charge are also of major importance. The symmetry of explosive properties has significant influence on the symmetry of the formed projectile. If the explosive density was higher over one half of the liner relative to the other, it would have much the same effect as the asymmetric confinement described above, with an unbalanced

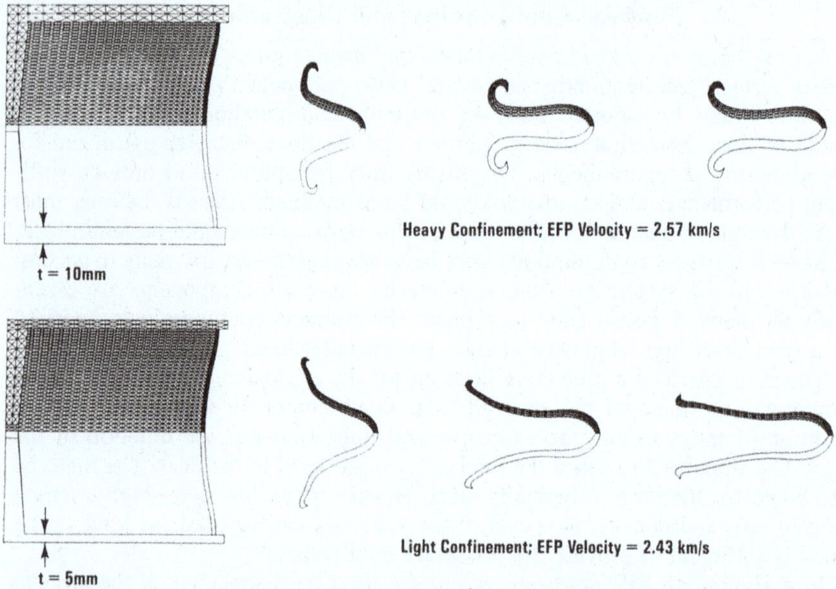

Fig. 6 Final geometry of EFP design as a function of variations in the explosive confinement.

detonation velocity causing the liner to be distorted. Furthermore, if voids were present in the explosive fill, the liner might be distorted or even torn. The length-to-diameter ratio of the explosive charge, L/D, is also of importance. Because of overall system constraints, the explosive charge head height or length is usually limited. As the explosive charge length-to-diameter ratio L/D is increased, the kinetic energy of the EFP increases until a point of diminishing returns is reached. As an example, let us consider a 117-mm charge diameter warhead with a copper liner. Figure 7 is a graph of explosive charge L/D vs EFP kinetic energy. As the explosive charge L/D increases, so does the EFP kinetic energy, until an L/D of approximately 1.5 is reached. The curve then begins to flatten out as the warhead weight continues to increase, which suggests a design with an L/D of approximately 1.5 is best. However, if there is not enough room for an explosive charge of $L/D = 1.5$, empirical data from several researchers have shown an L/D of about 0.75 to be sufficient.

The selection of the liner configuration and formation mode is based primarily on the target requirements and the overall system mission. Various EFP shapes can be formed by changing the thickness and contour of the liner. The three basic shapes that have evolved are the compact ball, the long rod, and the flared rod. Early EFP designs focused on forming compact ball shapes. Because of increased target requirements, however, these are no longer of primary interest against heavy armor but are still very effective against lightly armored targets.

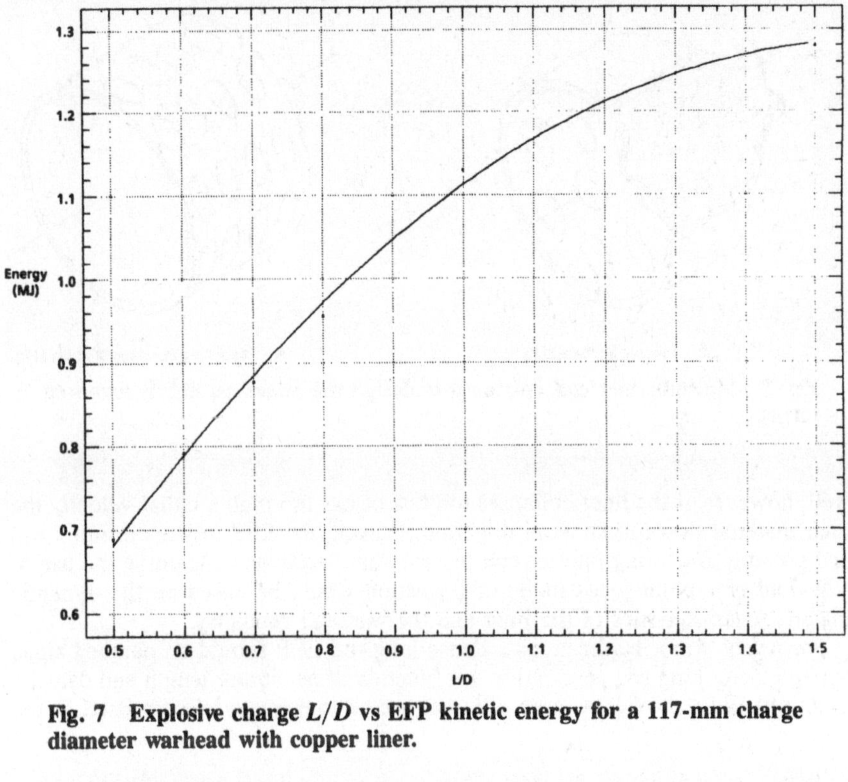

Fig. 7 Explosive charge L/D vs EFP kinetic energy for a 117-mm charge diameter warhead with copper liner.

In fact, ball-shaped EFPs are still very effective in a multiple-fragment application. In this type of application, it is desirable to fire several small EFPs at a soft or thin target and control the fragment spray pattern much like a shotgun. The result is a higher lethality compared to firing a single, much larger, EFP at a thin target. This is accomplished by simply replacing the large single liner with several small liners that form into balls, as seen in Fig. 8a. Another application for ball-shaped fragments would be to fire several EFPs at thin targets omnidirectionally and eliminate the fragment spray pattern control. This is accomplished by fabricating the entire case out of small liners that form into balls, as seen in Fig. 8b.

Compact balls can be formed by two methods. The first, *point focus*, focuses all the liner material toward a common point. Figures 9 and 10 illustrate this method as shown by Kivity and Tzur with a hydrocode simulation using the EPIC code and flash radiograph of the EFP.[9] The second method of forming a ball is the W-fold. Hermann et al. showed that compact shapes could be obtained by designing the liner such that its cross section forms a W-shape and gently collapses upon itself, as seen in Figs. 11 and 12 with an EPIC hydrocode simulation and flash radiograph of the EFP.[5] Both of these methods work equally

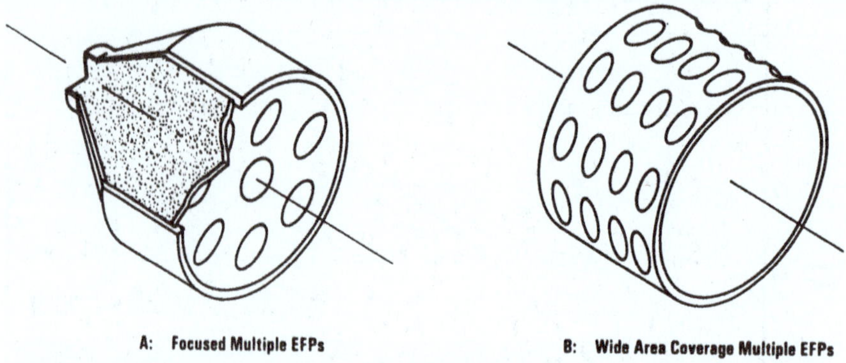

Fig. 8 Multiple liner concepts used to design the attack on lightly armored targets.

well; however, if the liner collapses too fast or has too high a radial velocity the liner material flows in an axial direction, causing the EFP to stretch into a rod and possibly fracturing into several pieces. Any decisions concerning the use of a W-fold or a point focus mode of formation would be based on the dynamic material characteristics of the liner and the warhead geometry.

For heavy armor applications a dense long-rod EFP should be pursued since hypervelocity long-rod penetration is a function of penetrator length and density. Long-rod EFPs provide a more efficient use of liner material compared to the

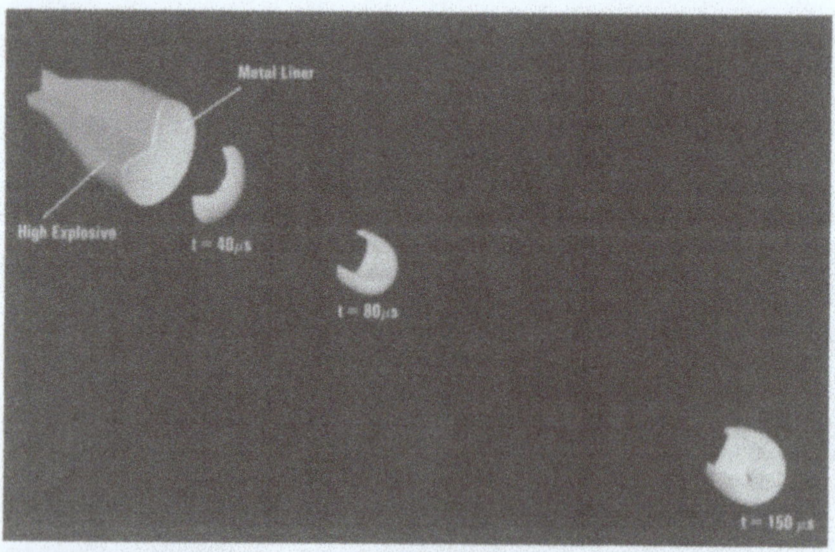

Fig. 9 EPIC code simulation of a ball-shaped EFP using the point focus approach.

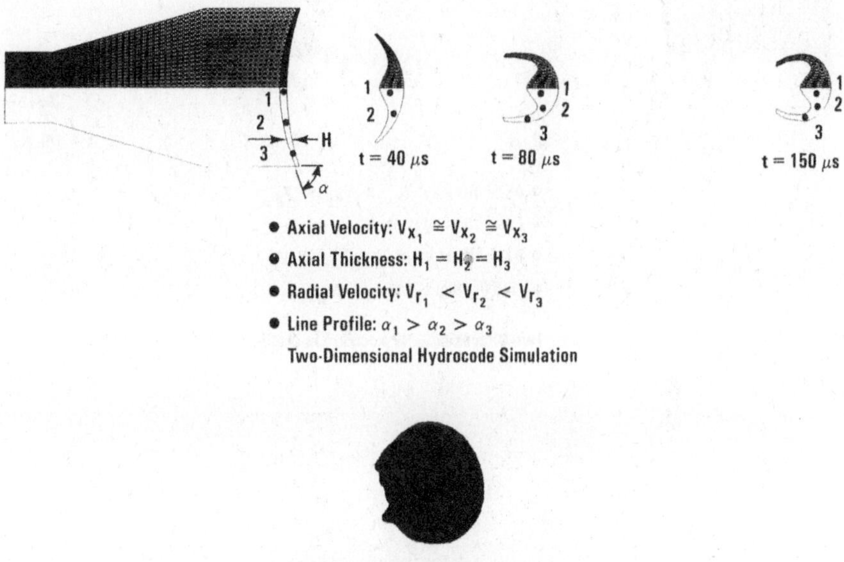

Fig. 10 Detailed representation of the point focus formation method comparing an EPIC hydrocode simulation with a flash radiograph of the fully formed EFP.

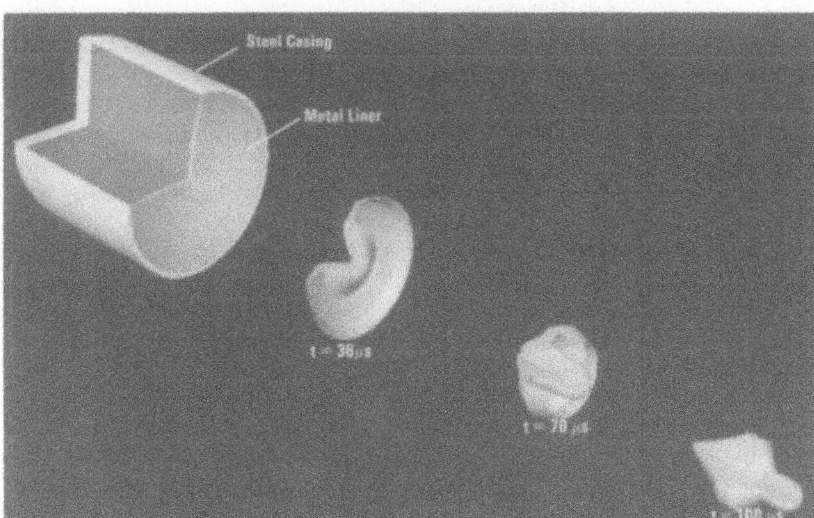

Fig. 11 EPIC code simulation of a ball-shaped EFP using the W-fold approach.

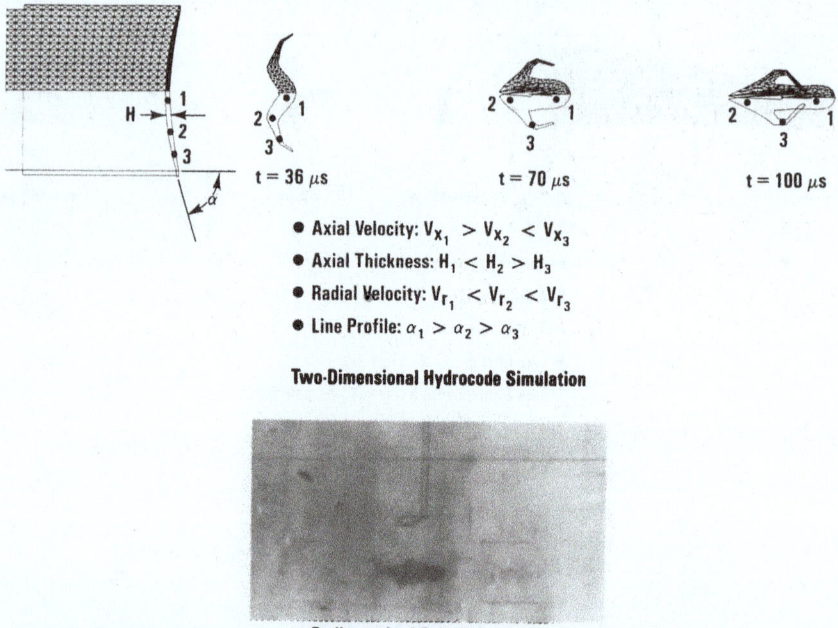

Fig. 12 Detailed representation of the W-fold formation method comparing an EPIC code simulation with a flash radiograph of the fully formed EFP.

ball-shaped EFP; and if designed with a stabilizing flare the EFP can fly very long distances to the target without tumbling.

Rod-shaped EFPs can also be formed by two formation methods, the forward and rearward fold. In the forward-folding mode the edge of the liner is accelerated ahead of the center and at the same time the liner is gently driven toward the axis of symmetry. The edge of the liner consequently becomes the tip of the rod and the center becomes the tail.

In general, rod-shaped EFPs without a stabilizing flare on the tail are formed in the forward-folding manner. The forward-folding mode produces very compact and very long EFPs, but maintaining a single-piece penetrator for more than a meter or two is very difficult as a result of high axial stretching and radial collapse, as shown by Carleone and Johnson.[7]

In the rearward-folding mode the center of the liner is accelerated ahead of the edge while at the same time the liner is driven toward the axis of symmetry, hence, the liner inverts. Long-rod EFPs with and without stabilizing flares can be formed in this manner. The rearward-folding mode is very well suited for forming aerostable EFPs since the tip is usually very symmetric and the flare is hollow, forcing the center of gravity forward. Also a single-piece penetrator is much more likely to be formed since the radial collapse is usually not so high as forward-folding EFP. However, as a result the rearward-folding mode tends to leave voids at the axis of symmetry, decreasing the target penetration capability of the EFP.

4. Aerodynamics of Explosively Formed Projectiles

In many EFP systems, the objective of the EFP warhead is to form an aerodynamically stable penetrator shape that travels to distant targets. Early designs were ball-shaped, and as target requirements grew increasingly difficult the warhead designer was forced to evolve the ball shape into longrods with stabilizing flares. More recently, considerable effort has been expended on the development of EFPs with fins replacing the flare.[10] The replacement of the flare with fins has a three-fold effect. First, aerodynamic drag is reduced. Second, the sometimes hollow flare becomes more compact, resulting in a greater effective penetrator length, and third, repeatable rod and tail formation becomes more reliable as a preferred mode of buckling is forced.

The problem of aerodynamic stability of the EFP is very complex. In general the typical hypersonic flow pattern around a flared EFP is very complicated, as is illustrated in Fig. 13.[11] This sketch illustrates the shock wave generated by the nose of the EFP as well as the shock waves produced by the beginning of the separation region and the reattachment point. The separated flow in the recirculation area as well as the flow turning angle into the wake are governed by the boundary-layer conditions as well as the pressure fields generated by the shock waves. These interactions are nonlinear functions of the detailed EFP geometry as well as the angle of attack of the EFP and its flight environment.

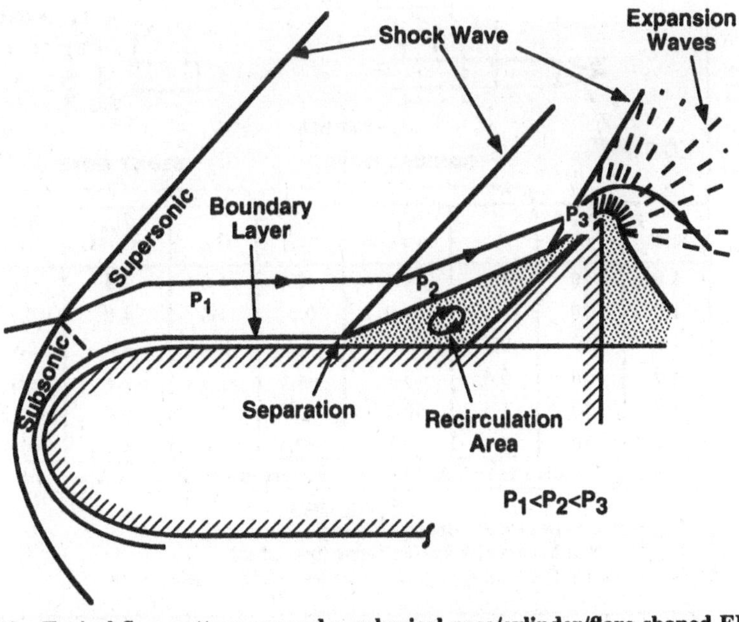

Fig. 13 Typical flow pattern around a spherical nose/cylinder/flare shaped EFP.

Two major areas of interest are the extent of the separated flow region, which is a strong function of Mach number and angle of attack, and the dynamic pressure of the flow at the flare, which is strongly influenced by the strength and interaction of the shock waves through which the air must pass.

Figure 14 was developed using models from Refs. 12 and 13 and serves as an example of the effects of these factors on the drag and stability of two classes of EFP geometry. Extrapolation beyond the geometries considered in the table is risky because of the nonlinear characteristics of the interaction. The figure compares the aerodynamic characteristics of a sharp cone-cylinder-flare for cylinder-length-to-cylinder-diameter ratios of 1.0 and 4.0. The data are presented as a function of flare angle, assuming a constant ratio of 8.62 for flare surface area to reference area and for a Mach number of 6.0.

There are two principal considerations in the ultimate selection of a given design. The first one is to select a configuration having low drag, and the second is to choose a configuration with adequate static and dynamic stability to minimize the sensitivity to initial pitch and yaw rates as well as asymmetries in EFP shape. As seen from the data, increasing the flare angle increases the drag coefficient at zero angle of attack, increases the normal force coefficient derivative, and moves the center of pressure rearward. In both cases of cylinder

Mass Properties

$M = 600$ g $l = 10.0$cm $d_C = 2.6$ cm

$I_X = 2400$ g/cm² $I_Y = 7900$g/cm² $d_F = 7.6$ cm

Aerodynamic Data Comparison[a]

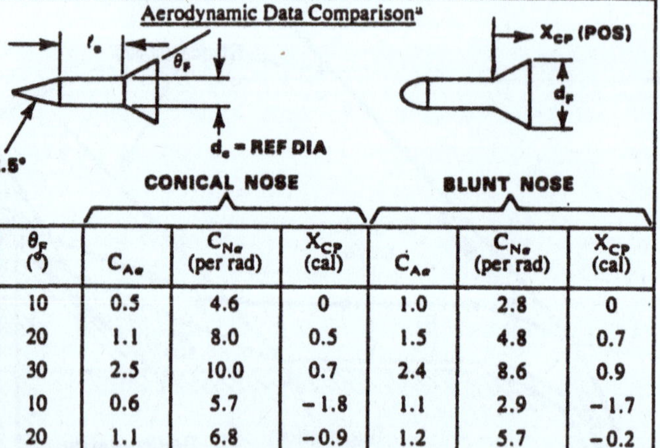

		Conical nose			Blunt nose		
lc/dc	θ_F (°)	C_{A_0}	C_{N_α} (per rad)	X_{CP} (cal)	C_{A_0}	C_{N_α} (per rad)	X_{CP} (cal)
1.0	10	0.5	4.6	0	1.0	2.8	0
1.0	20	1.1	8.0	0.5	1.5	4.8	0.7
1.0	30	2.5	10.0	0.7	2.4	8.6	0.9
4.0	10	0.6	5.7	−1.8	1.1	2.9	−1.7
4.0	20	1.1	6.8	−0.9	1.2	5.7	−0.2
4.0	30	2.4	9.1	−0.6	2.0	8.2	+0.4

[a] Zero angle of attack and constant flare surface area to reference area ratio of 8.62

Nomenclature

C_{A_0} = zero angle of attack drag coefficient
C_{N_α} = normal force coefficient derivative (per radian)
X_{CP} = normal force center of pressure calibers aft of reference point

Fig. 14 Typical EFP mass properties and aerodynamic data.

length-to-diameter ratio, the conical forebody exhibits better drag and stability values than the blunt nose shape for flare angles of 10deg and 20deg. However, for a flare angle of 30 deg, the blunt nose aerodynamic characteristics are preferred.

In general, the forces and moments of flare-stabilized bodies are affected by 1) boundary layer separation due to axial shape variation and flow characteristics, 2) bow shock influence on the dynamic pressure near the flare, and 3) the bow shock-flare/shock intersections effects. Experimental data have shown that nose blunting increases the axial force contribution of the nose and tends to decrease the axial force contribution of the flare, owing to the reduction of dynamic pressure near the body. Radially, however, the dynamic pressure gradient is increased over that of the conical shape; and as the flare angle is increased, the flare emerges from being embedded in a near-constant dynamic pressure region to a higher-energy state, as illustrated by the higher drag for the 30 deg flare angle in Fig. 14.

To complicate matters further, the flare of virtually every EFP buckles to some degree, causing asymmetries that, if severe enough, result in trim angles other than zero and cause the EFP to diverge greatly from its planned trajectory or even tumble. Hence, total target impact error is two-fold, resulting from initial disturbances such as aiming error that increase linearly with time and from asymmetries that increase quadratically with time, as seen in Fig. 15. Therefore, the flare should be designed properly to provide stability, and asymmetries should be kept to a minimum.

5. Dynamic Materials Properties for Explosively Formed Projectile Designs

During the formation of an explosively formed penetrator, the liner experiences a wide range of conditions—for example, temperatures ranging from ambient to 1000K, strains up to 300%, and strain rates up to 1.0×10^5/s all

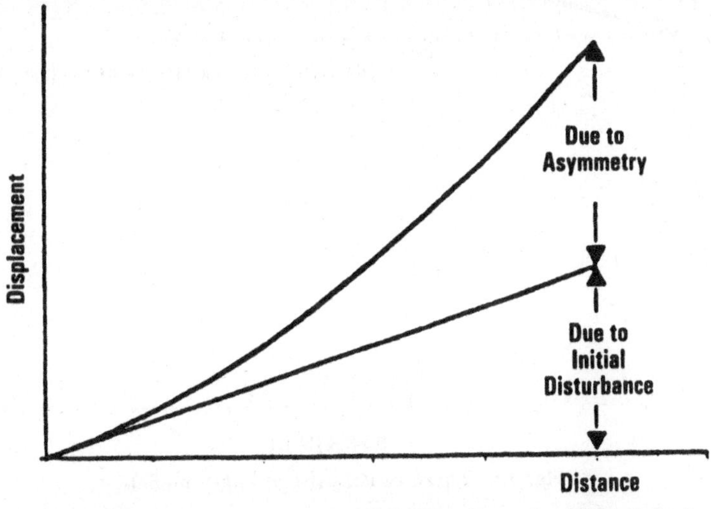

Fig. 15 Effects of symmetries and initial disturbance on flight trajectory.

within the first 400 to 500 μs of the liner formation. Hence, to model the formation process with any degree of accuracy requires complicated constitutive relationships. Recently, extensive effort has been placed on dynamic materials characterization. In the past, a simple elastic perfectly plastic material model was used in most hydrocodes. This model produced simulations, with reasonable accuracy, at early stages of the formation process only.

As warhead performance requirements increased, the need for a better materials model was obvious. The elastic perfectly plastic model evolved into the linear work-hardening model, and then the strain-and-strain-rate hardening, thermal-softening model. In Fig. 16 flow stress is plotted as a function of strain for the three types of constitutive models just mentioned, illustrating the difference in calculated material response. Johnson and Cook developed a constitutive model that expresses the Von Mises flow stress as a function of strain, strain-rate, and temperature,[14] or

$$\sigma = (A + B\epsilon^n)(1 + C \ln \dot{\epsilon}*)(1 - T*^m)$$

where ϵ is the equivalent plastic strain, $\dot{\epsilon}* = \dot{\epsilon}/\dot{\epsilon}_0$ is the dimensionless plastic strain rate for $\dot{\epsilon}_0 = 1.0\ \text{s}^{-1}$, and $T*$ is the homologous temperature, which is expressed as

$$T* = \frac{(T - T_{\text{initial}})}{(T_{\text{melt}} - T_{\text{initial}})}$$

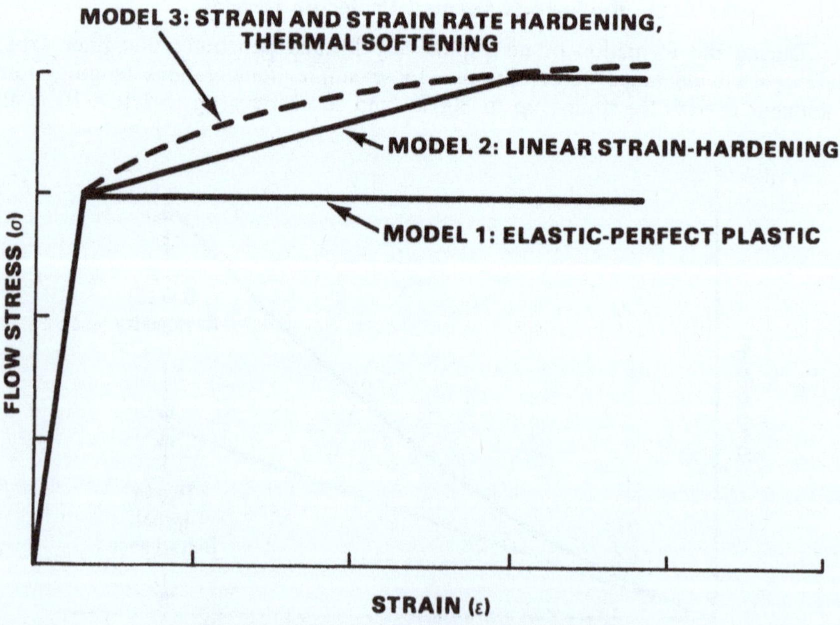

Fig. 16 Three constitutive strength models.

The expression in the first set of brackets gives the stress as a function of strain; the expression in the second gives stress as a function of strain rate; and the last expression gives stress as a function of temperature. A, B, C, n, and m are the material constants.

To determine values for the five material constants, a series of laboratory tests must be conducted. The constants in the first set of brackets—A, B, and n, where A is the yield stress and B and n are strain-hardening coefficients—are determined by conducting high strain rate tensile tests at room temperature and at a strain rate of $\dot{\epsilon}* = 1.0$, hence eliminating rate effects and temperature effects other than adiabatic heating. The temperature effect exponent m is determined by repeating the high strain rate tensile test with $\dot{\epsilon}* = 1.0$ at elevated temperatures, hence, rate effects are still eliminated but strain and temperature effects are included. Finally, the strain rate coefficient c is determined by conducting high strain rate tests such as torsional Hopkinson bar. Torsional tests can be conducted at room and elevated temperatures to include the effect of thermal softening.

After all the coefficients have been determined, a check of the basic coefficients is made by conducting Taylor tests. In a Taylor test, a cylindrical sample of the material to be used for liner fabrication is fired against a rigid surface at various impact velocities and temperatures. This test provides strains in excess of 200% and strain rates in excess of 10^5 s^{-1}. Hydrocode simulations are conducted to model the Taylor test, and the plastically deformed test samples are compared to the simulation. If other than minor deviations are present, the constants can be varied until the two match, hence, the constants are fine tuned. Figure 17 shows results of copper samples compared to EPIC-2 hydrocode simulations. Once the material constants have been determined, the design of the EFP may proceed with fairly high confidence. Figure 18 illustrates a comparison of a flash radiograph of a fully formed copper EFP with EPIC hydrocode simulations employing a calibrated constitutive model yielding favorable results.

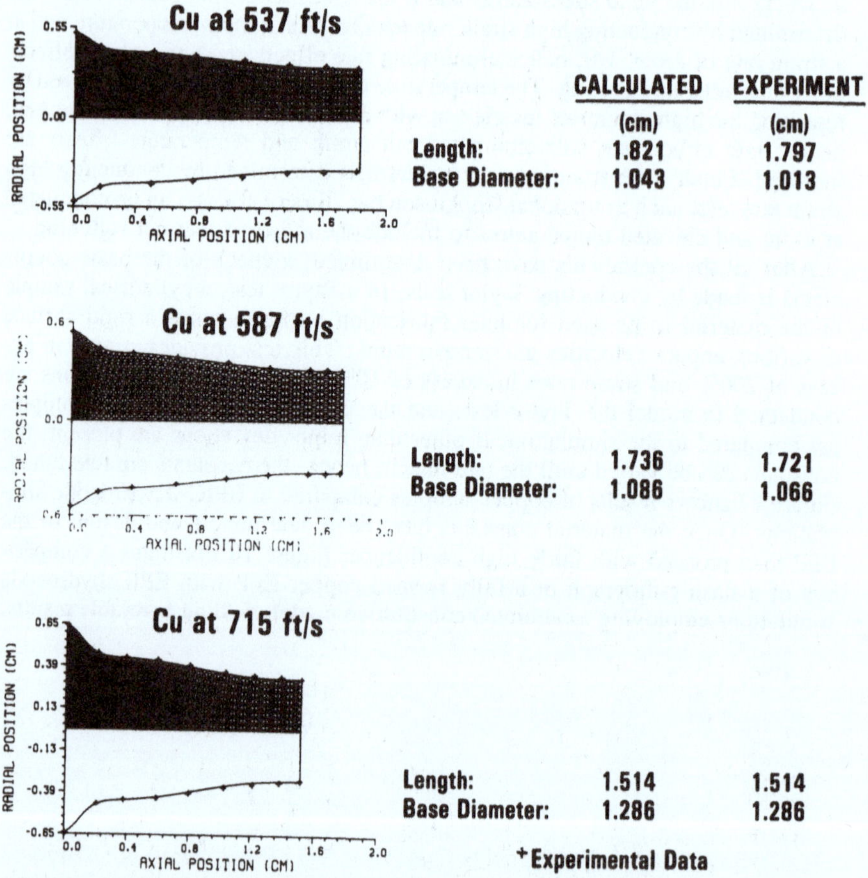

Fig. 17 Copper Taylor anvil samples compared to EPIC hydrocode simulation with calibrated constitutive model.

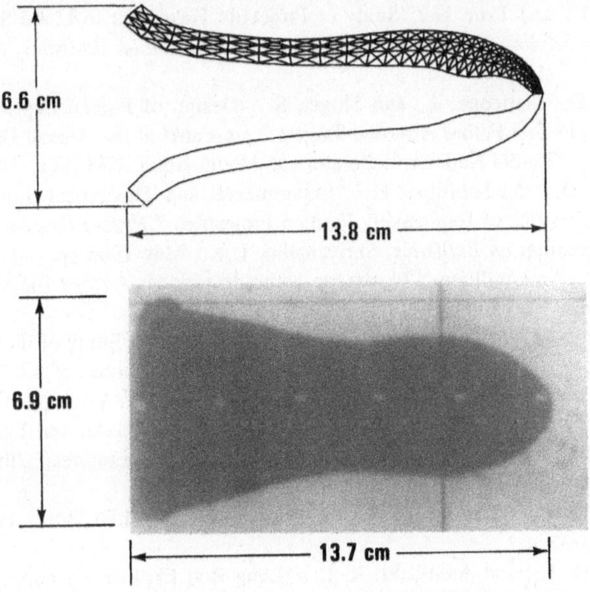

Fig. 18 Comparison of EPIC hydrocode simulation and flash radiograph of fully formed copper EFP.

References

[1]Wood, R. W., "Optical and Physical Effects of High Explosives," *Proceedings of The Royal Society of London*, Vol. 157, 1936, pp. 249–261.

[2]Held, M., "Self-Forging Fragmentation Charges," presented at the 3rd International Symposium on Ballistics, Karlsruhe, Germany, March 23–25, 1977.

[3]Schardin, H., "Über das Wesen der Hohlladung," Wehrtechnische Monatshefte, Vol. 51, 1954, pp. 97–120.

[4]Schardin, H., "Über das Wesen der Hohlladung," VDI-Zeitschrift, Vol. 98, 1956, pp. 1837–1842.

[5]Hermann, J. W., Jr., Randers-Pehrson, G., Berus, E. R., "Experimental and Analytical Investigation of Self-Forging Fragments for the Defeat of Armor at Extremely Long Standoff," *Proceedings of the 3rd International Symposium on Ballistics*, Karlsruhe, Germany, March 23–25, 1977.

[6]Held, M., "Air Target Warheads," *International Defense Review,* Vol. 5, No. 5, 1975, pp. 123–128.

[7]Carleone, J., and Johnson, G. R., "Multi-Slug Self-Forging-Fragment Warheads," *Proceedings of the First Self-Forging Fragment Conference*, Ballistics Research Laboratory, Aberdeen Proving Ground, MD, Oct. 13–14, 1982.

[8]Yiu, S. K., "Experimental Study of the Effects of Asymmetric Variations of an Explosive Device," Chung Shan Institute, Taiwan. *Proceedings of the Eighth International Symposium on Ballistics*, Orlando, FL, 1984, pp. VI-41 to VI-52.

[9]Kivity, Y., and Tzur, D., "Study of Projectile Formation in Lined Shallow Cavity Charges," *Proceedings of the 4th International Symposium on Ballistics*, Monterey, CA, 1978.

[10]Bender, D., Carleone, J., and Singer, S., "Design of Finned Explosively Formed Penetrators to Defeat Future Armored Targets," presented at the Annual Bomb and Warheads Meeting, Sandia National Laboratories, Albuquerque, NM, May 18–19, 1988.

[11]Warken, D., and Schilling, H., "Experimental and Theoretical Simulation of the Flight Characteristics of Explosively Formed Projectiles," *Proceedings of the 9th International Symposium on Ballistics*, Shrivenahm, U.K., May 1986 pp. 2-177 to 2-185.

[12]Jones, R., and Williams, T., private communications, Aerojet Electronic Systems Division, Azusa, CA, 1987–1993.

[13]Ashley, G. C., Jr., and Cary, A. M., Jr., "A Parametric Study of the Aerodynamic Characteristics of Nose-Cylinder-Flare Bodies at a Mach Number of 6.0," Langley Research Center, NASA IND.-2854, Langley Station, Hampton, VA, June 1965.

[14]Johnson, G. R., and Cook, W. H., "A Constitutive Model and Data for Metals Subjected to Large Strains, High Strain Rates and High Temperatures," 7th International Symposium on Ballistics, The Netherlands, April 1983.

[15]Karpp, R. R., "Accuracy of HEMP Code Solutions," BRL Memorandum Report 2268, Jan. 1973.

[16]Bender, D. E., and Archibald, R. L., "Long Rod Explosively Formed Penetrators to Defeat Future Armored Targets," 40th Annual American Defense Preparedness Association Bomb and Warhead Technical Meeting, Eglin Air Force Base, Ft. Walton Beach, FL, May 16–18, 1990.

Chapter 8

Fragmentation Warheads

Manfred Held*
MBB Dynamics, Schrobenhausen, Germany

1. Introduction

PERHAPS it may be asked why an explosive charge is usually surrounded by a fragmenting casing or a preformed fragment casing instead of merely allowing the blast wave to impact directly on the target. Blast warheads exploit the destructive effect of the shock of blast waves produced by the detonation of the explosive charge. Within some ten microseconds, about 1000 liters of extremely highly compressed gas (at a pressure of some 100,000 bars) is produced per kilogram of explosive. The expansion force of these highly compressed gases produces the gas *stream* that pushes the air before it as a shock wave. The casing certainly absorbs part of the energy that would otherwise be available for the blast wave.

The question is answered by the fact that the destructive capability of the blast wave from a conventional high-explosive (HE) charge decreases considerably in proportion to the distance from the detonation point. On the other hand, the fragments absorb a portion of the energy from the explosive charge in the form of kinetic energy, which carries them over greater distances to the target. It is for this reason that the lethal zone is not so evenly swept by the fragments as by the volumetrically expanding blast wave, since the fragments can only expend their energy in the direction of their flight path.

A fragmenting warhead can produce either many small fragments of relatively low penetration efficiency or a reduced number of fragments that have a greater penetration effect. The armor protection and size of the target plus the required lethal zone of the warhead determine the optimum size of the fragments for a given size of warhead.

There exist a variety of warheads with different fragment formations (Fig. 1). The main groups are natural fragmentation, controlled fragmentation, and preformed fragments. The controlled fragment formation can be split into four

*Professor, Manager of Technology Warheads.

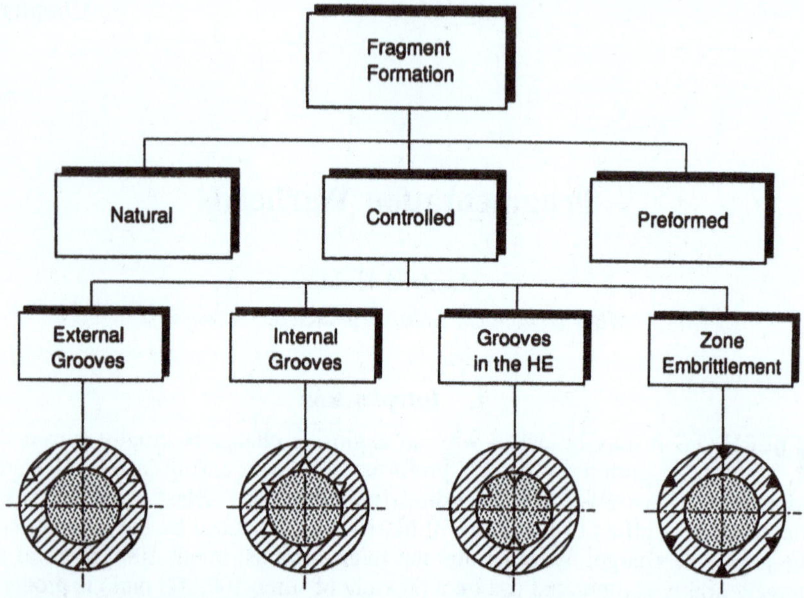

Fig. 1 Ways of generating controlled fragments, compared to natural and preformed fragments.

main technologies, namely, grooves on the outside or inside, or zone embrittlement in the casing, or grooves in the high explosive charge. All of these types will be discussed in the following chapters, together with the special fragmenting warhead types such as multi-P charge, multishaped charge, continuous rod, polygon, and aimable warheads.

2. Classification of Fragmenting Warheads by Fragment Speed

Fragmenting warheads can be classified according to their design or the theoretical behavior of their fragment velocities. The term fragment in this context covers a wide range of types, but is here taken to include the following:

1) Natural fragments, which result from the disintegration of the warhead casing when the explosive charge is detonated

2) Preformed fragments (e.g. balls, cubes, rods, etc.), which are accelerated to high speeds by the detonation of the charge

3) Projectiles (P), which are explosively formed from the saucer-shaped inserts of a multi-P warhead when it is detonated

4) Particles which are produced from the shaped charge jet

5) The linked rods of a continuous-rod warhead

Depending on the design of the warhead casing, fragments can have various velocities and velocity distribution patterns. In evaluating the hit probability against targets, these various velocities must be known and taken account of in

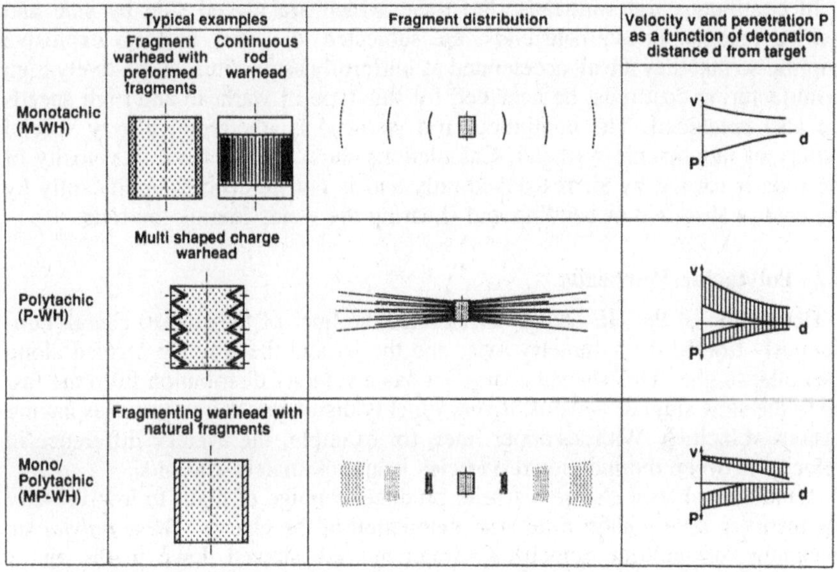

Fig. 2 Classification of antiair warheads according to their typical fragment velocities.

the calculations. For this reason warheads are classified according to the velocity spectrum of their fragments into three main groups: monotachic, polytachic, and mono/polytachic (Fig. 2).[1]

2.1 Monotachic Warheads

In the monotachic (coming from the Greek and meaning "of uniform velocity") warhead group, a first approximation assumes that all fragments have the same velocity. In this group, the mass of the warhead casing is homogeneous (the same portion of explosive lies behind each element in the fragmentation casing) and a uniform explosive impulse (integral of pressure over time) is applied to every point of the casing. In practice, this ideal situation can only be achieved at the cost of great complexity, because of the influence of initiation paths from the detonation position and the faster reduction of pressure at the front end of the charge relative to the center of the charge resulting from pressure release waves.

If, however, the fragments all have the same size, which occurs in preformed fragmentation warheads (e.g., uniform ball-shaped fragments embedded in a plastic casing), then the ideal case is very closely achieved in practice. These similar-sized fragments or balls all have a similar velocity pattern (i.e., their velocity decreases uniformly with range as a function of drag resistance) and follow an exponential curve.

The penetration or perforation capability of a fragment against the target is approximately proportional to its impact velocity. Performance falls exponentially with range because of the decrease in velocity.

In continuous rod warheads, the rods, which are placed side by side and welded together at alternate ends, are subjected to a very uniform explosive impulse so that they are all accelerated as uniformly as possible. A relatively high manufacturing cost must be accepted for this type of warhead and high speeds are also precluded. The continuous rod warhead is, therefore, a very special variety of monotachic warhead. Calculations show that the drop in velocity of the rods is caused by air resistance only and is not influenced significantly by the energy absorbed in bending and shearing the welds joining the rods.

2.2 Polytachic Warheads

Detonation of the HE charge accelerates the liner of the shaped charge continuously toward the symmetry axis, and the jet and the slug are formed along the collapse line. This shaped charge jet has a velocity distribution from the fast tip to the slow slug; as a result, it very quickly disintegrates into particles having various velocities. With a copper liner, for example, the average difference in velocity between the individual particles is approximately 100 m/s.

A multiple shaped charge warhead produces a range of high- to low-velocity fragments in a very short time after detonation of the charge. These *polytachic* (meaning "of multiple velocities") fragments are slowed down in the air or ablated at the tip as a result of their high speed through the atmosphere. These fragments also have different penetration and perforation capabilities, depending on the various flight speeds, impact velocities, and fragment sizes.

2.3 Mono/Polytachic Warheads

When a homogeneous casing is accelerated by the detonation of a charge, all the case elements of constant mass attain approximately the same velocity. Under the effect of radial expansion, the casing splits with shear or radial fractures into a large number of different-sized pieces or natural fragments. The fragmentation is a function of the brittleness, toughness, and crystal structure of the material, and the magnitude of the explosive impulse and temperature. These different-sized fragments are rapidly retarded by atmospheric drag, depending on fragment size, with the large and heavy fragments flying ahead of the smaller ones. Initially monotachic, the swarm of fragments becomes polytachic. Because of their different masses and velocities, these fragments also possess different penetration or perforation capabilities, even if their external shape is not taken into account. The irregular geometry and different cross-sectional area of the fragments further enlarges the dispersion zone of the fragment penetration capability. The penetration power of small fragments declines more rapidly with range than the larger fragments as a result of the greater atmospheric retardation.

3. Natural Fragmentation

3.1 Introduction

If the explosive charge is detonated in a casing made of homogenous material, a certain velocity will be imparted to the casing depending on the ratio of the casing mass to the mass of the high explosive charge. Radial expansion causes the case to split into various sized *natural* fragments through shear and/or radial fractures depending on the toughness, brittleness, and grain size of the material, and on the power of the shock wave, magnitude of the explosive impulse, and initial temperature of the warhead.

The mass distribution of shells or warheads with natural fragmentation is generally described in the literature by the Mott formula.[2] However, it gives a limited description that excludes the smaller fragments. It will be shown that another type of description developed by Held[3–6] is more general. With two parameters and the total mass M_0 (or a best fit for the total mass $M_{0,\text{Best}}$) it gives an excellent description of the experimentally found mass distributions of natural fragmenting warheads. Note that the mass distribution is a fit to the experimentally found mass distributions. The mass distribution depends on the geometry, such as diameter and length of the fragmenting charge, casing thickness, cover plate thicknesses, and on the material behavior of the casing and the type of high explosive.

3.2 Distribution Function

The Weibull distribution[2] can be applied to a great variety of technical problems. The distribution density of the three-parameter Weibull distribution is as follows:

$$f(x) = \frac{\lambda(x-\mu)^{\lambda-1}}{\delta^{\lambda}} \cdot e^{-(x/\delta)^{\lambda}}, \qquad x > 0 \tag{1}$$

where

$$\delta = \text{scale parameter}$$
$$\lambda = \text{shape parameter}$$
$$\mu = \text{location parameter}$$

Equation (1) reduces to a two-parameter Weibull distribution when the location parameter μ is set equal to 0; this step is equivalent to a transformation to the new variable $x - \mu$, or

$$f(x) = \frac{\lambda x^{\lambda-1}}{\delta^{\lambda}} \cdot e^{-(x/\delta)^{\lambda}}, \qquad x > 0 \tag{2}$$

The two-parameter Weibull distribution follows from Eq. (2) by an integration, that is,

$$F(x) = 1 - e^{-(x/\delta)^{\lambda}}, \qquad x > 0 \tag{3}$$

In some papers on fragment mass distribution, the two-parameter Weibull distribution, Eq. (3), is referred to as Rosin-Rammler-Sperrling (RRS) distribution, which goes back to the description of the grain size distribution in grinding processes.

In the context of fragmentation and the distribution of fragment sizes, the Weibull distribution, Eq. (3), has entered fragmentation ballistics since Mott (see Ref. 2). With regard to this particular application it is, therefore, often referred to as Mott distribution, and it is usually taken to describe the distribution of the *number of fragments*.

On the basis of flash x-ray pictures that permit analyzing smaller and finer fragments, Held[3,7] has developed an experimental approach to represent the fragment mass distribution as a function of the number of fragments. For comparison, the formulas are given here:

$$\text{RRS}: \quad M(m) = M_0 e^{-(m/m_R)^{\lambda_R}} \tag{4}$$

$$\text{MOTT}: \quad N(m) = N_0 e^{-(m/m_M)^{\lambda_M}} \tag{5}$$

$$\text{HELD}: \quad M(n) = M_0(1 - e^{-Bn^{\lambda H}}) \tag{6}$$

where

M_0	= total mass of all fragments
$M(m)$	= cumulative fragment mass, that is, overall mass of all fragments whose mass is greater than or equal to a given mass m
$M(n)$	= cumulative fragment mass, that is, overall mass of the fragment number n, beginning with the largest fragment mass of the nth fragment
m	= mass of the nth fragment
N_0	= total number of fragments
$N(m)$	= cumulative fragment number, that is, number of all fragments whose mass is greater than or equal to a given fragment mass m
n	= cumulative fragment number, beginning with the heaviest fragment
m, λ, B	= (with subscripts R, M, H) constants.

The RRS formula is not related to the number of fragments at all, and the Mott formula requires a given number N_0 of fragments, whereas Held's formula does not need the number of fragments. Any number of fine fragments may be added, even if they contribute virtually nothing to the overall mass M_0.

3.3 Description of Mass Distribution with Held's Formula

The method of analyzing mass distributions by means of Held's formula[4,5,7] is explained as follows. This formula, when applied correctly, gives an excellent description of the mass distribution of the natural fragments generated in the detonation of all high-explosive projectiles examined, even when filled with various types of explosives, and also of secondary fragments behind a target plate, or even behind a set of multiple spaced target plates. The simple equation for this relation is given by

$$M(n) = M_0(1 - e^{-Bn^{\lambda}}) \tag{7}$$

This equation is shown schematically in Fig. 3.

The constants B and λ in equation (7) are readily found by isolating the exponential term

$$\frac{M_0 - M(n)}{M_0} = e^{-Bn^\lambda} \tag{8}$$

and then taking the natural logarithm of Eq. (8):

$$\ln \frac{M_0 - M(n)}{M_0} = -Bn^\lambda \tag{9}$$

To determine the values of B and λ it is convenient again to take the logarithm of Eq. (9) so that in a logarithmic representation the point of intersection $n = 1$, or $\log n = 0$, will give the constant $\log B$ directly on the ordinate axis, and the exponent λ can be determined from the slope of the straight line (Fig. 4). That is,

$$\log\left(-\ln \frac{M_0 - M(n)}{M_0}\right) = \log\left(\ln \frac{M_0}{M_0 - M(n)}\right) = \log B + \lambda \cdot \log n \tag{10}$$

To this end, the value of $M(n)$ must be computed with the associated cumulative number of fragments n, beginning with the largest fragment. This value must be then subtracted from the total fragment mass M_0, and then be divided by M_0. The corresponding logarithms can then be plotted in a log-log diagram.

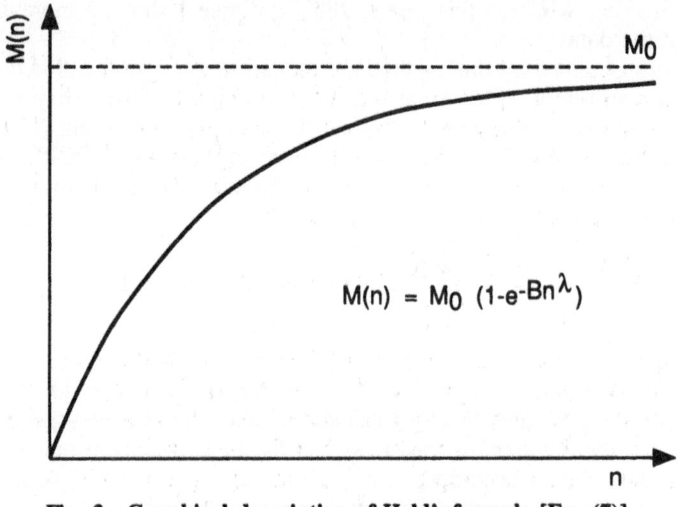

Fig. 3 Graphical description of Held's formula [Eq. (7)].

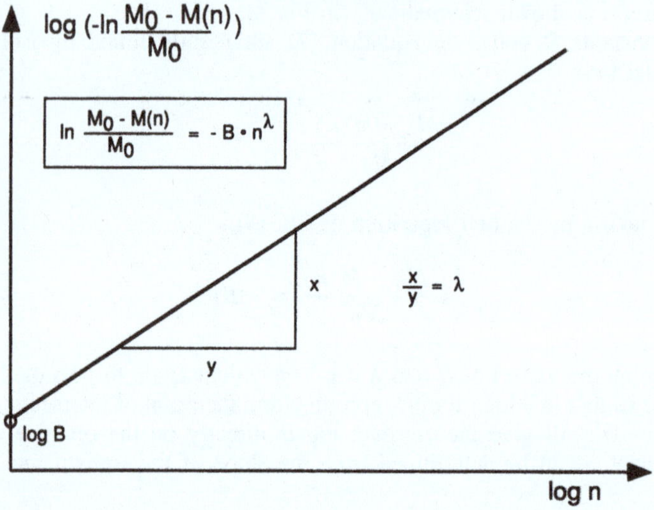

Fig. 4 Easy determination of the constants *B* and λ from the log-log plot.

3.4 Fragment Mass Distribution of High-Explosive-Filled Projectiles

Example No. 1: Fragment mass distribution of the 30 mm × 170 UIA HE projectile. In Fig. 5 are shown the natural fragments generated by the detonation of the HISPANO SUIZO 30-mm × 170 UIA incendiary HE projectile, classified by fragment mass.[7] For Held's approach, the fragment masses $M(n)$ must be summed up over the corresponding numbers, beginning with the largest fragment (in each class), that is, beginning with the top line. The result must then be evaluated with Eq. (9). The total mass M_0 of the fragments is the sum of all partial masses, which in this case is 288.7 g. Table 1 shows a more detailed analysis of the data.

The values obtained by this method are plotted in a diagram called the fragment mass distribution (FMD) log-log diagram (Fig. 6, left). They show an excellent agreement with a best-fit straight line and give a constant B of 0.511 and an exponent λ of 0.7318 with a correlation coefficient of 0.9999.

Taking the derivative of Eq. (7) with respect to n yields Eq. (11) for the mass m of the nth fragment,

$$m = \frac{\mathrm{d}M(n)}{\mathrm{d}n} = M_0 B \lambda n^{\lambda - 1} e^{-Bn^\lambda} \tag{11}$$

This equation, when plotted in a diagram of mean fragment mass as a function of the cumulated number n or, for short, an MFM diagram, with the given M_0 and with the constants B and λ calculated, also shows a very good agreement between the numbers of fragments and the mean fragment masses in the individual mass classes according to Fig. 5 and Eq. (11) (see Fig. 6, right).

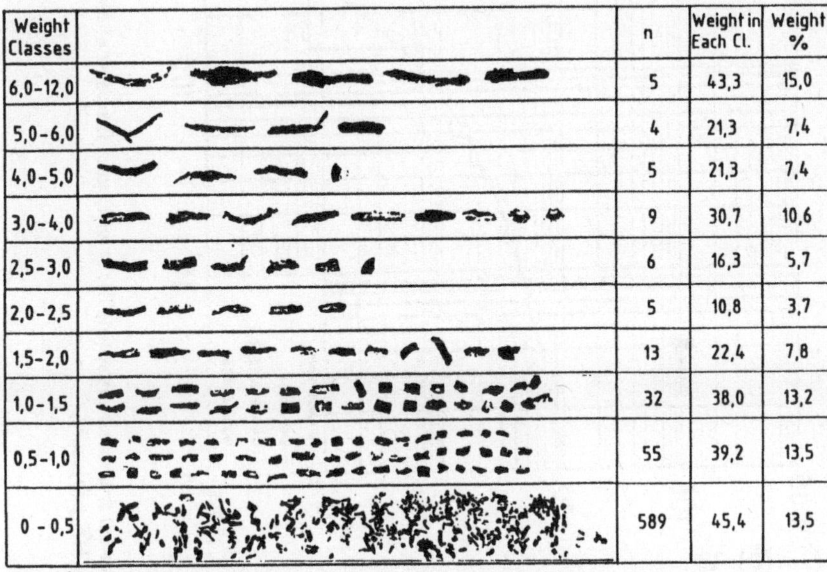

Fig. 5 Fragment mass distribution of the HISPANO SUIZA 30-mm × 170 UIA incendiary HE projectile.

Table 1 Fragmentation classification for Example 1

Fragment classes, g	Number of fragments, n	Weight of fragment in each class, g	Σn	$M(n)$, g	x_1	x_2
6–12	5	43.3	5	43.3	0.850	0.162
5–6	4	21.3	9	64.	0.776	0.253
4–5	5	21.3	14	85.9	0.702	0.353
3–4	9	30.7	23	115.5	0.596	0.517
2.5–3	6	16.3	29	132.9	0.540	0.617
2.0–2.0	5	10.8	34	143.7	0.502	0.689
1.5–2.0	13	22.4	47	166.1	0.425	0.856
1–1.5	12	38.0	79	204.1	0.293	1.227
0.5–1.0	55	39.2	134	243.3	0.157	1.850
0–0.5	589	45.4	723	288.7	0	—

$$x_1 = \frac{M_0 - M(n)}{M_0} \qquad x_2 = \ln \frac{M_0 - M(n)}{M_0}$$

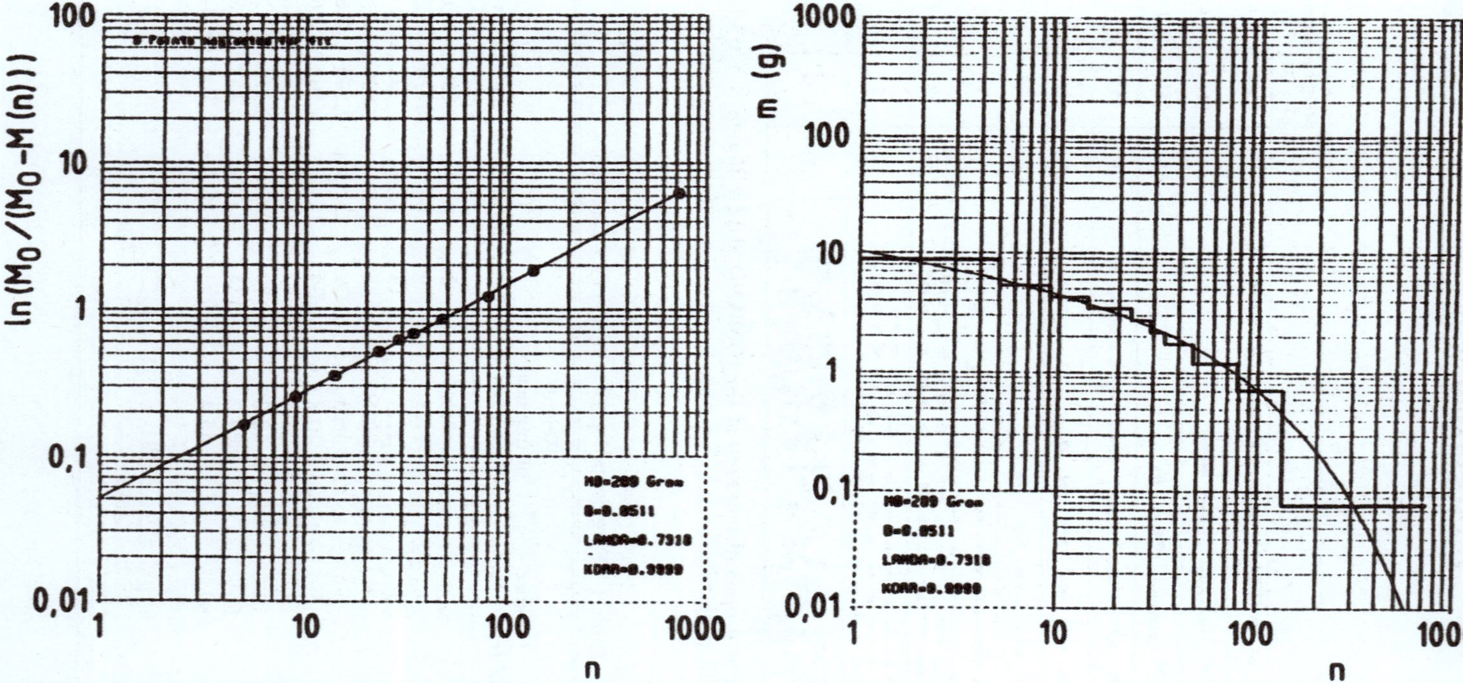

Fig. 6 Fragment mass distribution (FMD) log-log diagram (left) and mean fragment mass (MFM) (right) as a function of the cumulative fragment number *n* for the HISPANO SUIZA 30-mm $\times 170$ incendiary HE projectile.

Example No. 2: Fragment mass distribution of the Oerlikon 35-mm × 225 MSB/K HE projectile. The method will be demonstrated by a second example, where the ad hoc agreement between the given overall mass M_0 and Eq. (7) is not quite so good as in Example 1. However, the agreement can be improved by adapting the overall mass M_0 as well as by neglecting some of the largest fragments, which do not correlate with the fragment mass distribution of the shell casing, because they originate from the end plate and from the fuze adapter flange. The example considered here is the measured fragment mass distribution of the 35-mm × 225 MSB/K HE explosive projectile manufactured by Oerlikon (Fig. 7).

In the log-log diagram, the straight line does not fit the measured data very well (Fig. 8). The points representing $n > 100$ fragments, in particular, deviate from this straight line. The resulting constants are $B = 0.0182$ and $\lambda = 0.7945$, with a correlation of 0.9977. Either not all of the fragments have been taken into account for the overall mass M_0 or the given total mass M_0 is not the optimum mass M_0 for the fragment mass distribution. The individual fragment masses found experimentally cannot be described very well with Eq. (11) when these constants are inserted (Fig. 8, right).

Using the constants B and λ as originally determined, one can now calculate an optimum mass M_{OB}, that is, a total mass M_{OB} that best fits this set of equations, as

$$M_{OB} = \frac{M(n)}{1 - e^{-Bn^{\lambda}}} \tag{12}$$

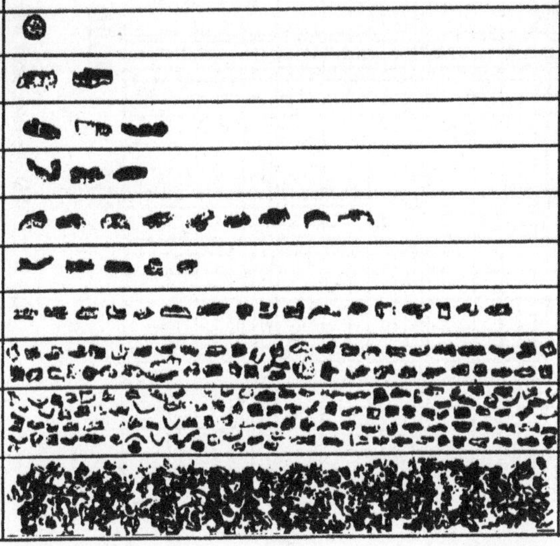

Weight Classes		n	Weight in Each Cl.	Weight %
6,0-8,0		1	6,7	1,7
5,0-6,0		2	11,1	2,8
4,0-5,0		3	12,7	3,3
3,0-4,0		3	10,0	2,6
2,5-3,0		9	24,6	6,4
2,0-2,5		6	13,0	3,4
1,5-2,0		17	29,1	7,5
1,0-1,5		50	59,0	15,3
0,5-1,0		135	96,2	25,0
0 - 1,0		1118	123,2	32,0

Fig. 7 FMD of the Oerlikon 35-mm × 225 MSB/K HE projectile.

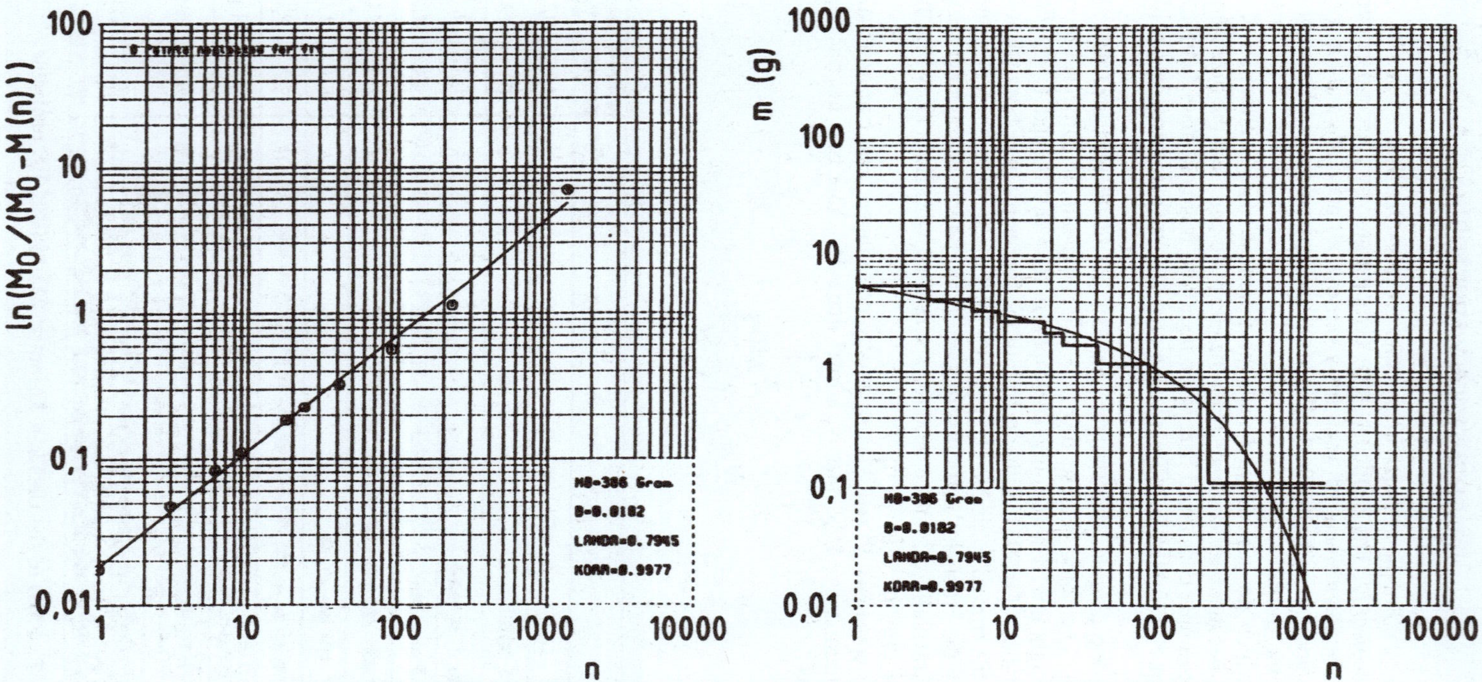

Fig. 8 FMD log-log and MFM diagrams for the Oerlikon 35-mm × 225 MSB/K HE projectile.

The new constants B_B and λ_B are now determined with this new total mass M_{0B} as

$$M(n) = M_{0B} \cdot (1 - e^{-B_B \cdot n^{\lambda_B}}) \tag{13}$$

With this new total mass $M_{0B} = 390$ g instead of $M_0 = 386$ g, which in the example given means 1% more mass. Thus, the experimental data are much better described by the fitting of a straight line. The new constants (Fig. 9, left) are now $B_B = 0.0198$ (instead of $B = 0.0182$) and $\lambda_B = 0.7526$ (instead of $\lambda = 0.7945$), with a correlation coefficient of 0.9992 (instead of r = 0.9977). As one can see in Fig. 9 (right), the cumulative number n of fragments can now be described much better as a function of the mass classes when these constants are used.

However, in Fig. 9 (left) we see that the individual points still fluctuate about the best-fit straight line, with the first two points—that is, the three largest fragments—deviating even more than the rest. These three fragments, with their random masses, must not be relevant to the fragment mass distribution.

When the first two points, corresponding to these three fragments, are omitted in this example, optimizing the mass M_0 leads to a value of $M_{0B} = 394$ g, and the constants become $B_B = 0.0226$ and $\lambda_B = 0.7128$, with a correlation coefficient of 0.9999. As can be seen in Fig. 10 (left), all points—except for the two that have been previously omitted—fit the calculated straight line rather well. Of course, the fragment mass equation with m as a function of the cumulative number of particles n averages the experimental values particularly well (Fig. 10, right).

Example No. 3: Antiship warhead, including fragment mass distribution of secondary fragments. Detailed results are available from firing tests of the KORMORAN anti-ship warhead in a special arena setup.[8] The KORMORAN warhead is a so-called multi-P warhead, where the detonating high-explosive charge generates fast, *explosive-formed* projectiles from a set of *liner* inserts, and natural fragments from the remainder of the casing structure. The fact that the steel insert liners (for the projectiles) and the remaining structure, as well as the target plates, are all different types of steel makes it possible to distinguish between fragments from the structural casing, fragments from the projectiles that break up to some degree on perforating the target, and secondary fragments from the spaced target plates.

The trials were made against various numbers of spaced plates of 6-mm thick St-50-grade steel. The primary fragments from the structure and from the insert liners and the secondary fragments from the spaced target plates were all recovered from so-called strawboards and were then carefully analyzed for their mass distribution (Fig. 11).

As an example, let us consider fragments from the strawboard in the firing direction S1, where a projectile had been aimed at the middle of a single 6-mm thick St-50-grade steel plate; this projectile, of course, broke up into various fragments as it perforated the target plate. The fragments recovered from the strawboard have been designated in the following way:

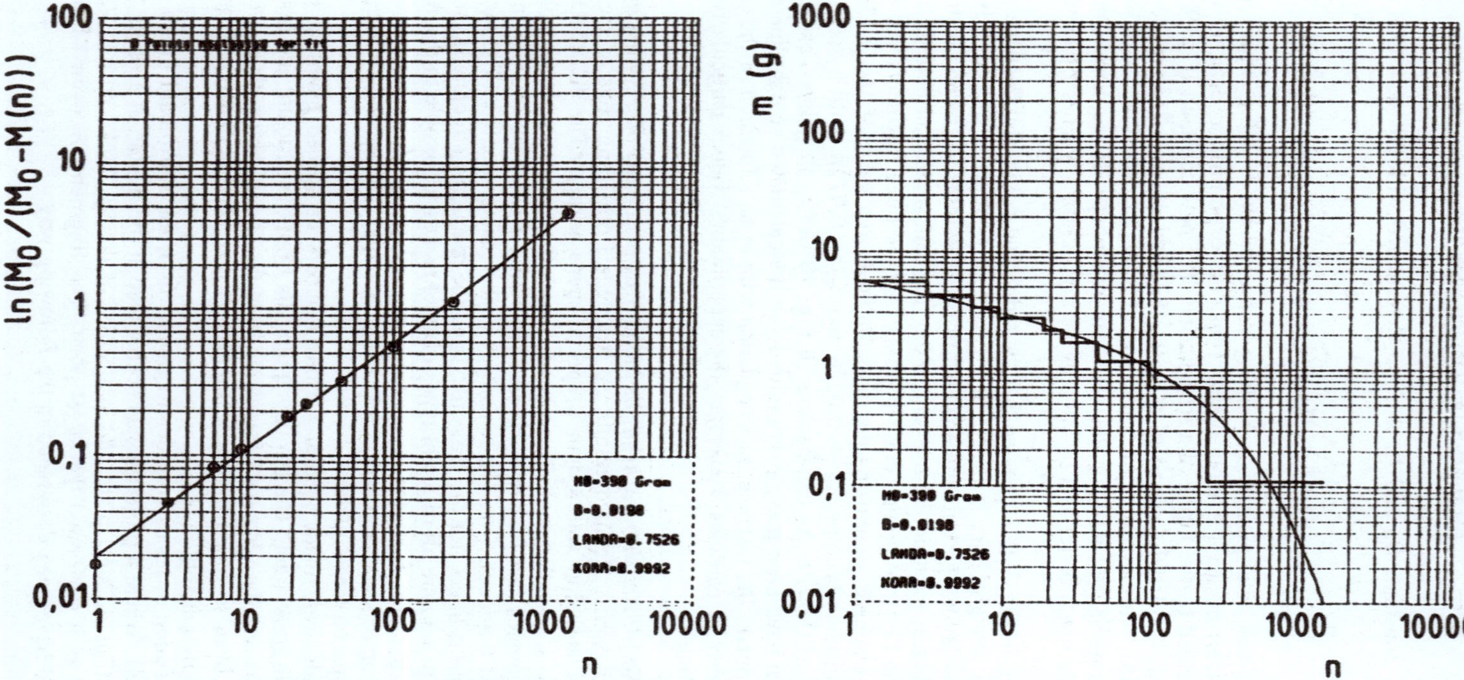

Fig. 9 FMD log-log and MFM diagram for the 35-mm × 225 MSB/K HE projectile with corrected mass.

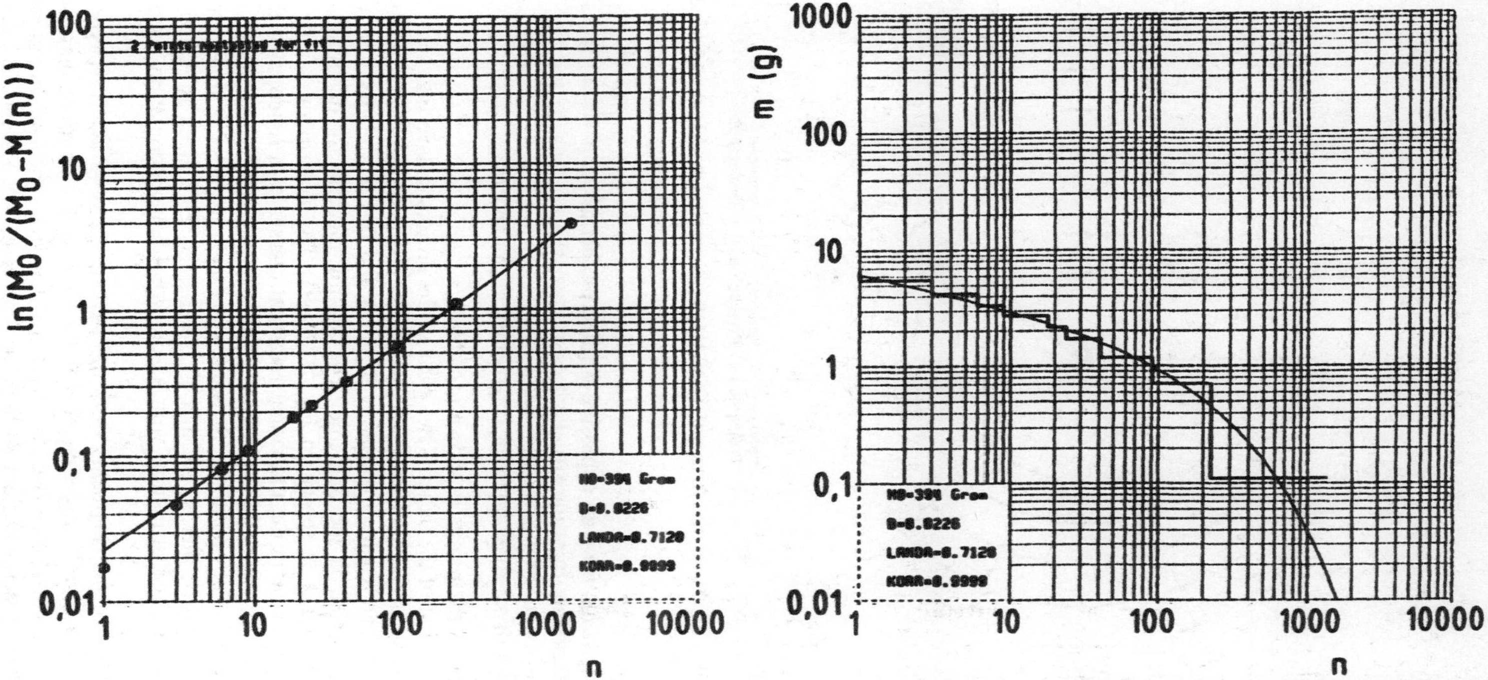

Fig. 10 FMD log-log and MFM diagrams for the 35-mm × 225 MSB/K HE projectile with corrected mass, and neglecting the first two points (equivalent to the first three fragments).

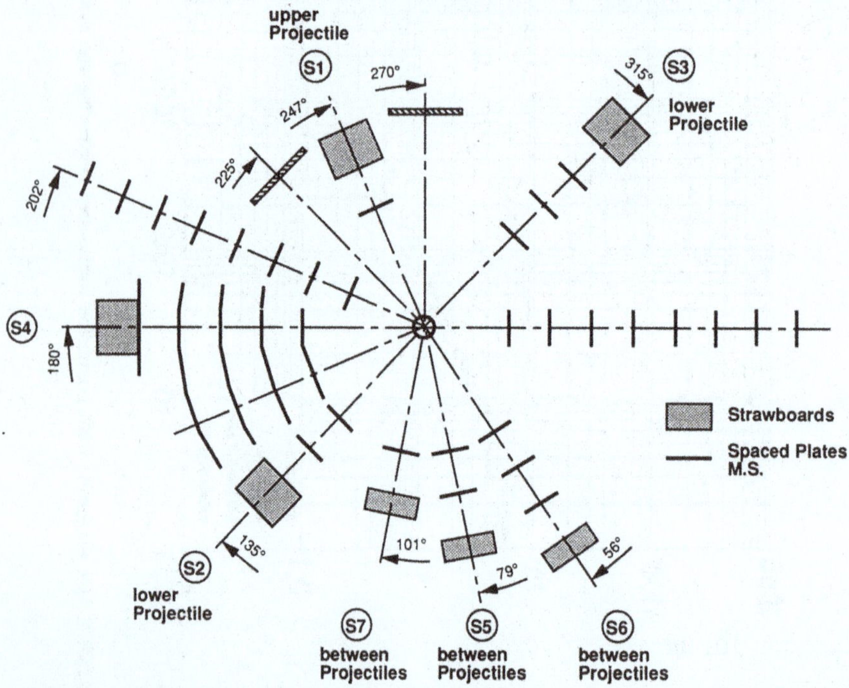

Fig. 11 Test arena for the firing of KORMORAN warhead with strawboards for fragment recovery.

P = fragments from an explosive-formed projectile
H = fragments from the structural casing
S = fragments from the target plate
T = sum of all three types of fragments, P, S, and H

A summary of the results is shown in Table 2.

On perforation of the target plate, parts of the projectile have fractured. These fragments have been collected as P-type fragments, and their overall mass adds up to 683 g. The resulting constants B and λ and their correlation coefficients have been entered in the first line. The log-log diagram for determining the constants B and λ is shown in Fig. 12 (left). The description of the mean fragment masses m in the individual fragment classes as a function of the cumulative number n with these constants is not too bad (see Fig. 12, right). It can still be improved by optimizing the total mass and, at the same time, omitting the first two points, that is, the first six fragments. This gives an equation for the fragment mass distribution with a somewhat better correlation coefficient (Fig. 13).

An analysis of the fragments from the structural casing alone (H-type fragments) gives a good correlation of 0.9863 even without further optimization (Fig. 14). The correlation becomes very good indeed when an optimum fragment mass is chosen and the two heaviest fragments—one point—are omitted.

Table 2 Results of the analysis of the total masses M_0, the calculated constants B and λ, and the correlation coefficients for the test against target S1

Fragment type	Optimum	M_B, g	B	λ	C
P	—	683	0.1682	0.6391	0.9909
	M_{B+P2}	685	0.2613	0.5270	0.9966
H	—	566	0.1234	0.6199	0.9863
	M_{B+P1}	578	0.1922	0.4797	0.9999
S	—	293	0.1103	0.6357	0.9221
	M_{B+P1}	340	0.1010	0.5087	0.9973
T	—	1542	0.0726	0.6224	0.9865
	M_{B+P2}	1590	0.1484	0.4477	0.9996

In this case, with a mass M_{0B} = 578 g (as compared to 566 g previously), the correlation is 0.9999 (Fig. 15).

The mass distribution of the secondary fragments from the 6-mm thick St-50 plate, that is, of the S-type fragments, shows a particularly great deviation of the best-fit straight line from the experimental points (Fig. 16).

Optimizing the fragment mass M_0 from 293 g to only 340 g gives a fairly good fit of the straight line, with a correlation coefficient of 0.9973 (Fig. 17, left), and a good representation of the mean fragment masses m in the individual mass classes as a function of the cumulative number n (Fig. 17, right).

This formula also gives a very good description of the mass distribution of all primary and secondary fragments that have been generated from the target plate. The large number of fragments gives good statistics a priori, and so a good description of the fragment mass distribution of all fragments is possible without any correction or adjustment, as shown by the correlation of 0.9865 (Fig. 18). When the first two measured points, that is, the first nine fragments, are omitted, a correlation coefficient of 0.9996 follows with the optimum mass of 1590 g (as compared to 1542 g), which means that the measured points from a number of fragments of 10 to over 1000 fit rather well to a straight line (Fig. 19).

4. Controlled Fragmentation

4.1 Introduction

Naturally fragmenting warheads will produce many fragments that will be too large to achieve an optimum destructive effect on a given target. On the other hand, a large number of fragments will be too small, so that their penetration power for a successful attack on the target is too low. Although there are many small fragments, they usually only account for a small portion of the total weight. In contrast, the lesser number of large fragments constitutes the major portion of the warhead casing mass. This pattern reduces the number of effective fragments; consequently, the probability of hitting the target is low.

For attacking a target with a given warhead there is an optimum fragment size that, in addition to providing an adequate destructive effect, gives a sufficient number of fragments to guarantee a high hit probability. Only a high number

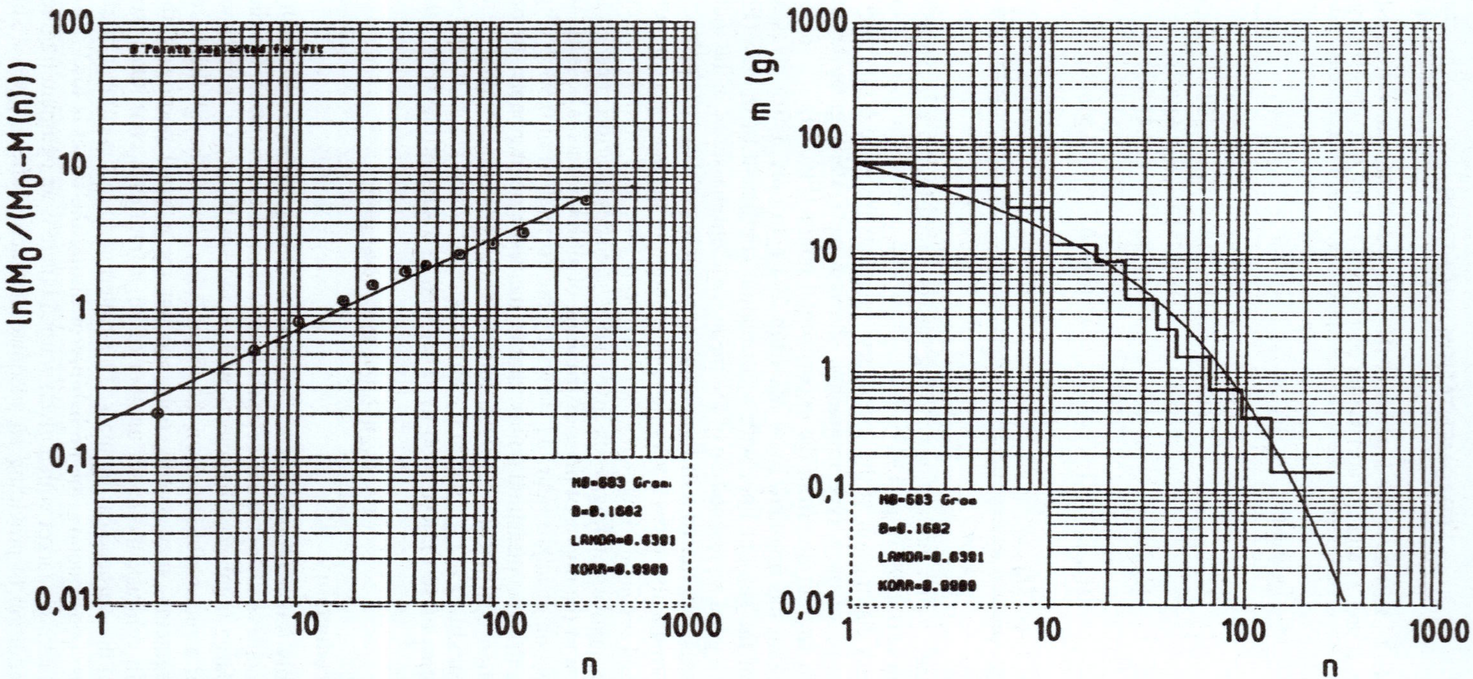

Fig. 12 FMD log-log and MFM diagrams for the fragments mass distribution of the explosive-formed projectile without any correction.

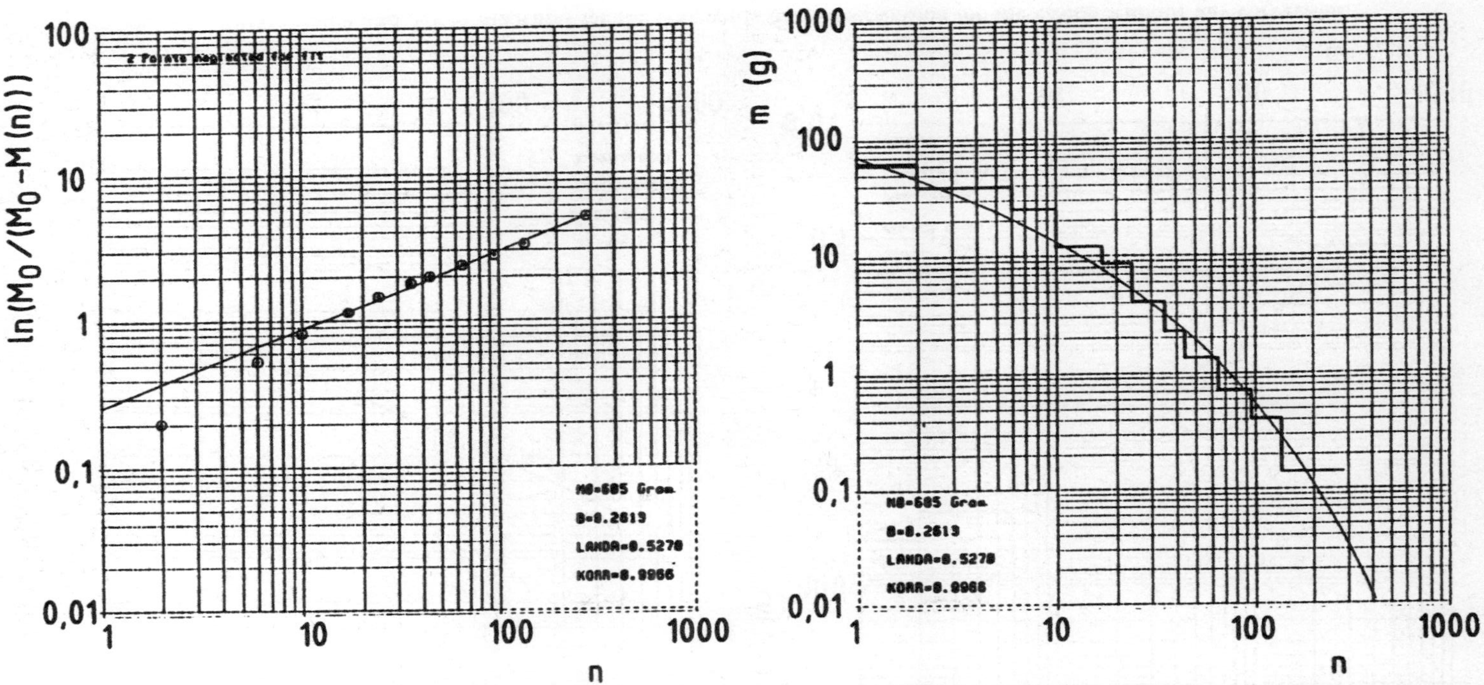

Fig. 13 FMD log-log and MFM diagrams for the fragments mass distribution of the explosive formed projectile with mass correction and two points (equal to six fragments) omitted.

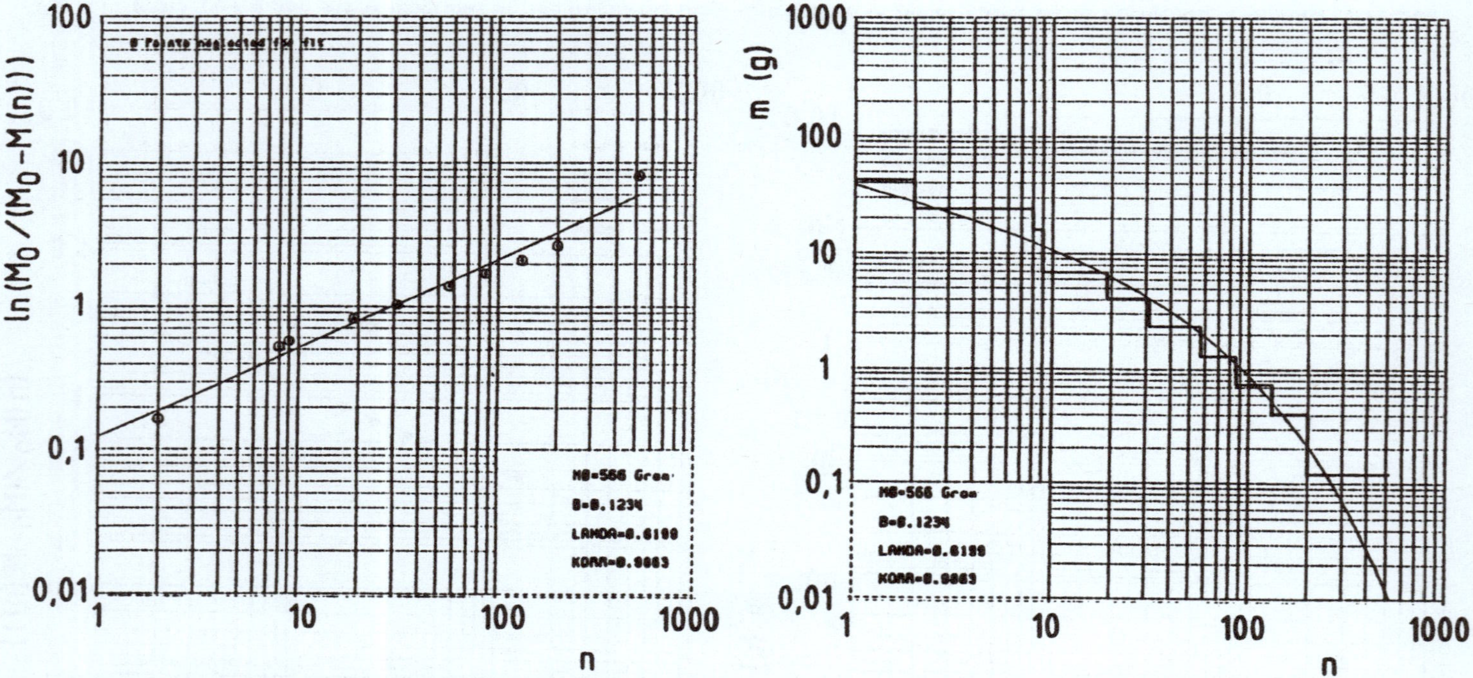

Fig. 14 FMD log-log and MFM diagrams for the fragments mass distribution for the casing without any correction.

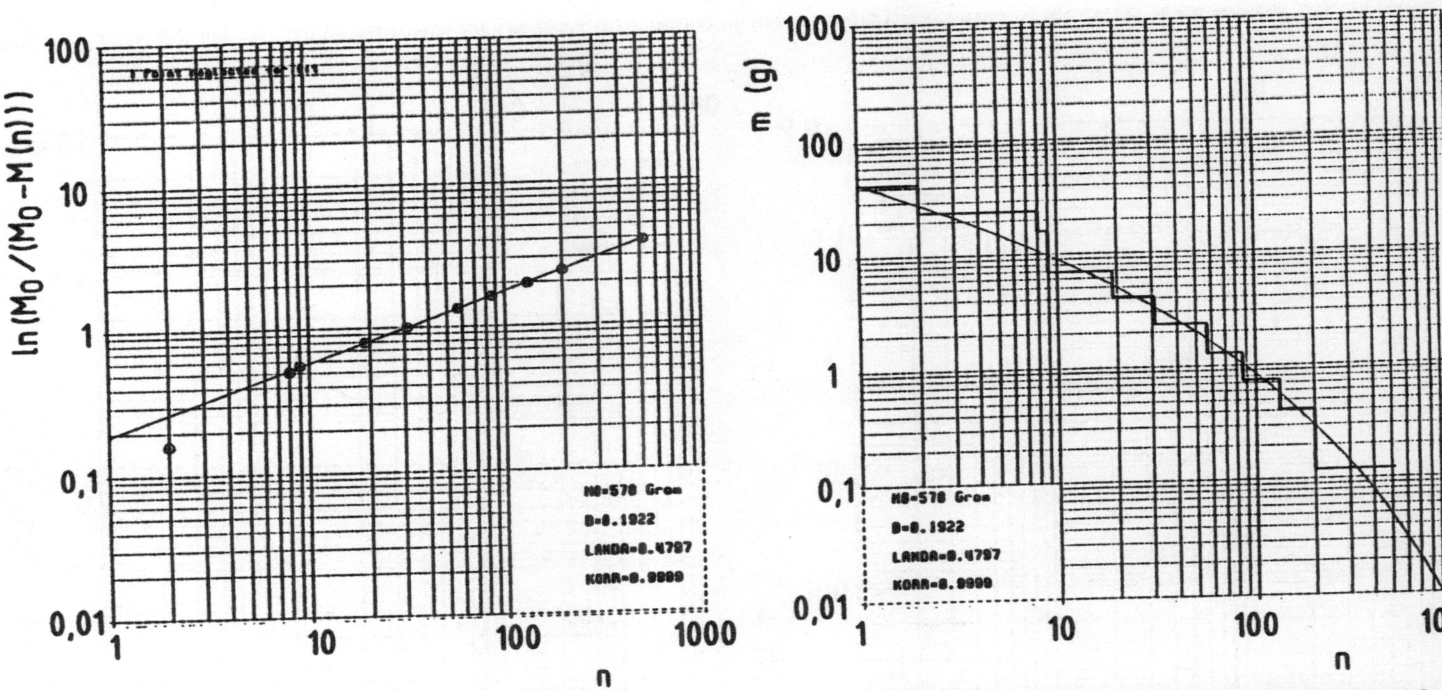

Fig. 15 FMD log-log and MFM diagrams for the fragments mass distribution of the casing with mass correction and one point (equal to two fragments) omitted.

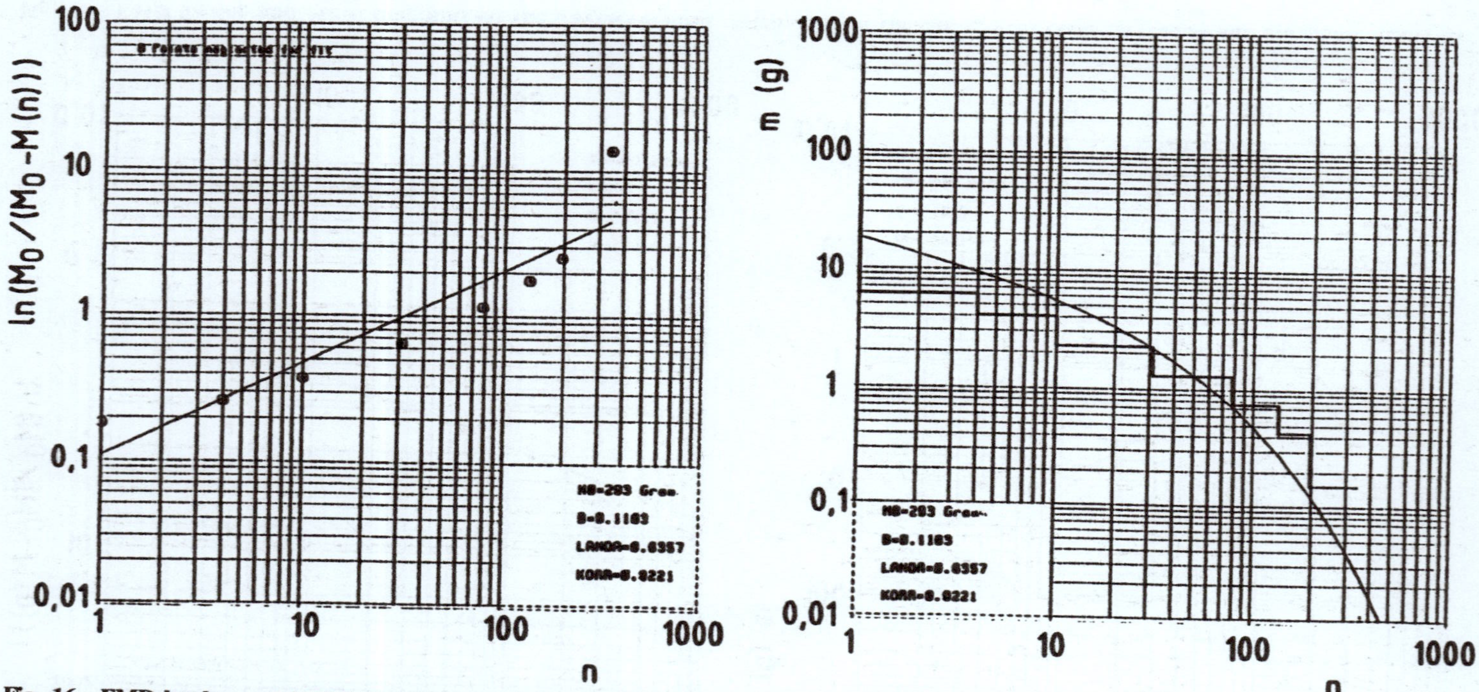

Fig. 16 FMD log-log and MFM diagrams for the mass distribution of the secondary fragments of the target plate without any correction.

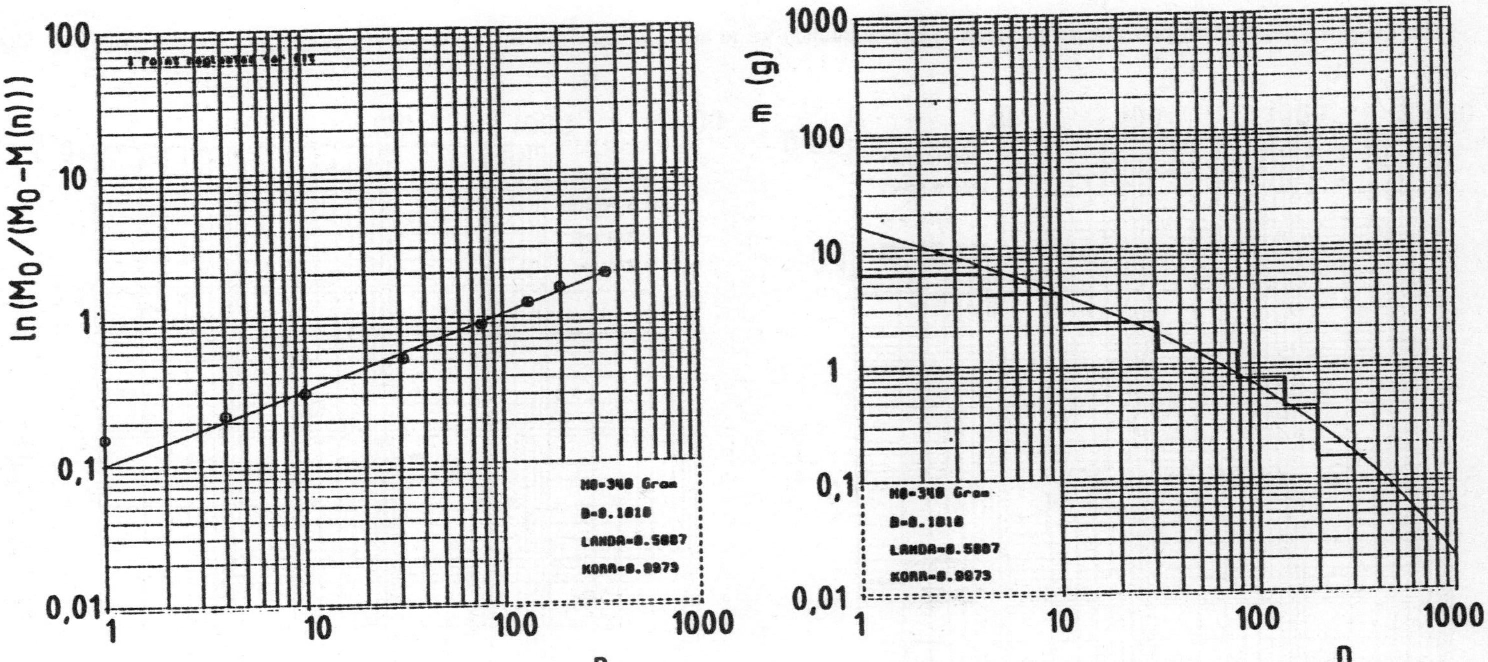

Fig. 17 FMD log-log and MFM diagrams for the mass distribution of the secondary fragments of the target plate with mass correction and one point (equal to one fragment) omitted.

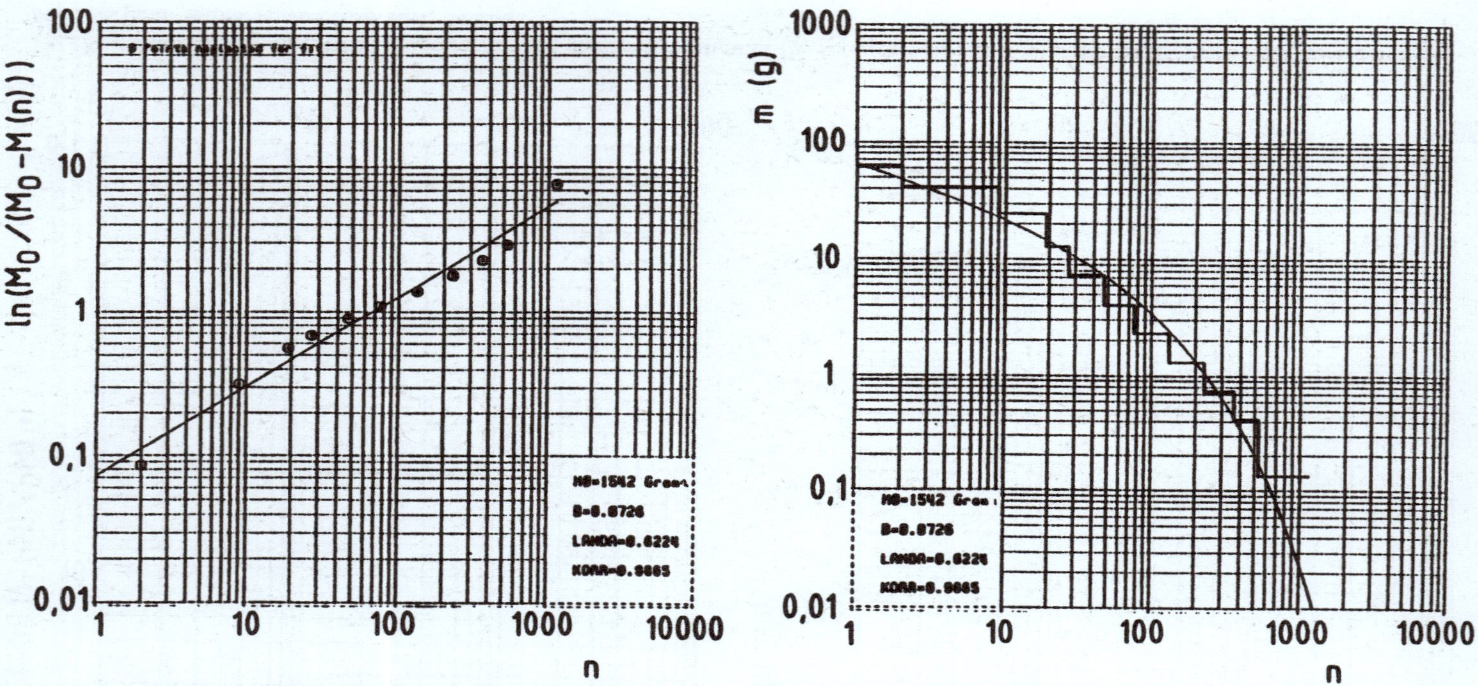

Fig. 18 FMD log-log and MFM diagrams for the mass distribution of all fragments (EFP, warhead casing, and target plate) without any correction.

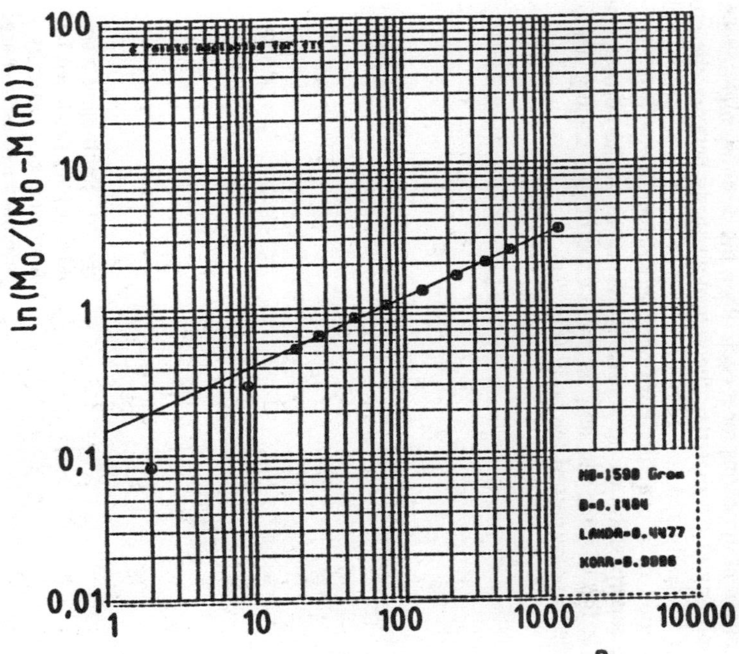

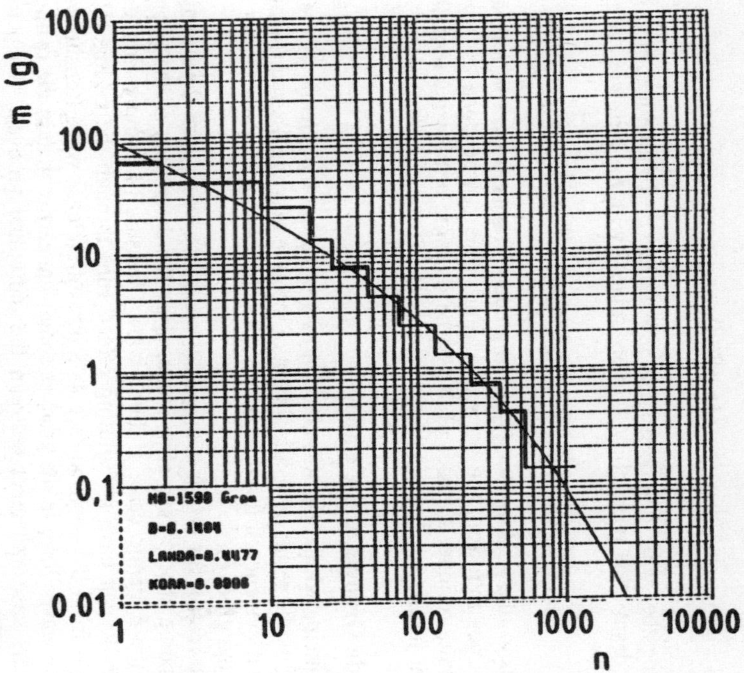

Fig. 19 FMD log-log and MFM diagrams for the mass distribution of all fragments with mass correction and two points (equal to nine fragments) omitted.

of hits n, together with high destructive probability p of individual hits, give a high target destruction probability Z in accordance with the following equation:

$$Z = 1 - (1 - p)^n \tag{14}$$

For this reason, the goal has long been to produce fragment sizes as regular and effective as possible from the warhead casing. This can be done by grooving either the outer or inner surfaces of the warhead casing or by using roof-shaped or triangular inserts.

4.2 Outside Grooves in the Casing

It was anticipated that the introduction of grooves on the outside of the warhead casing would result in disintegration of the case into equal-sized fragments. Experimental evidence showed, however, that the casing does not in fact often split along the grooves.

The basic reason for this is a quasistatic pressure loading on the casing and not the dynamic shock wave effects caused by the detonation of the explosive charge. In the dynamic case, a shock wave (pressure or compression wave), which naturally can have no splitting effect, runs through the casing. This is then reflected from the outer surface as a pressure release wave (tensile wave). These tensile waves form in the grooves somewhat earlier than in the cylindrical surface of the case and meet between the individual grooves. When tensile waves from different directions meet each other, the resulting high tensile stress produces a splitting of the material, theoretically exactly between the grooves.

This effect is clearly apparent in the splitting of a six-sided prism, produced by the detonation of a centrally-positioned explosive charge (Fig. 20). As a result of the superposition of the tensile stress waves at the corners, this six-sided splits at the corners and not, as static experiments usually lead one to believe, at the thinnest part of the body (Fig. 21).

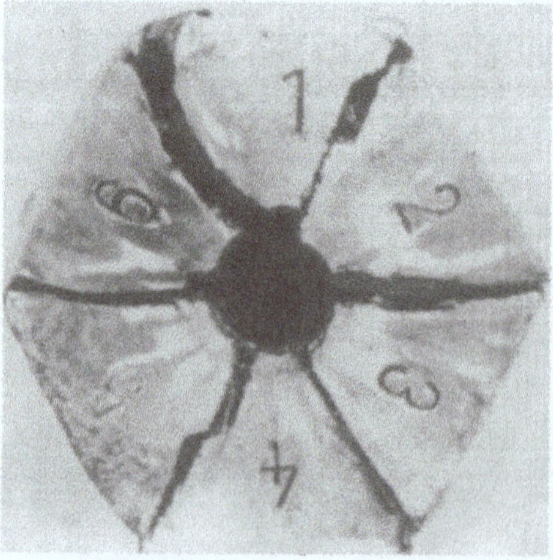

Fig. 20 Splitting of a six-sided prism, produced by detonation of a centrally positioned explosive charge.

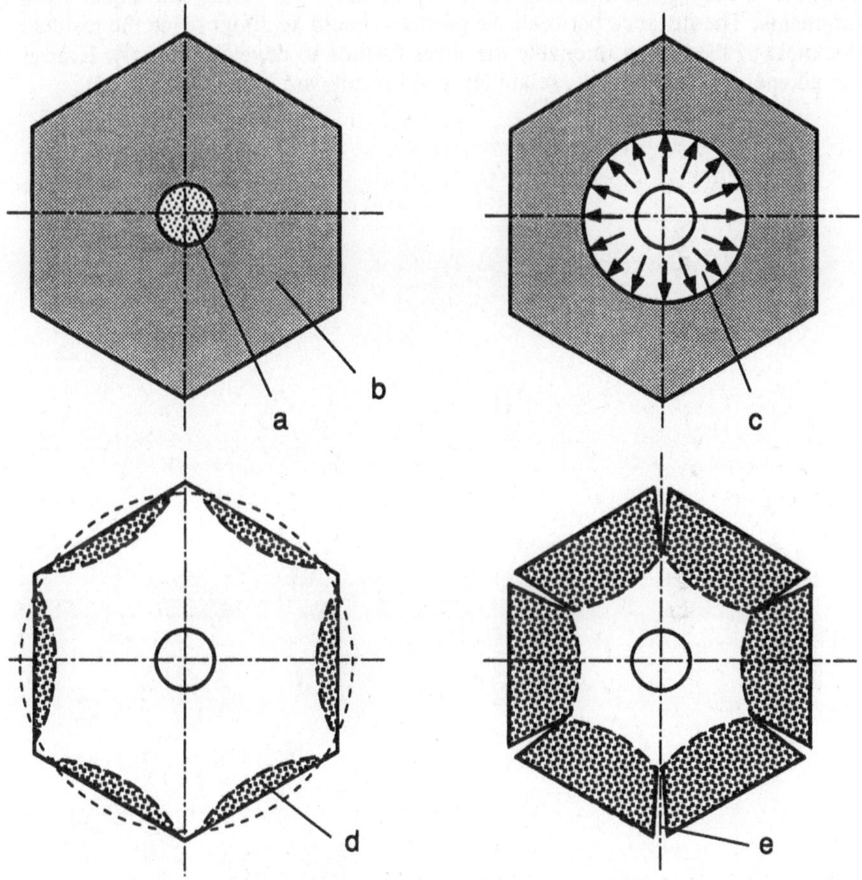

Fig. 21 Diagram illustrating why a six-sided prism subjected to detonation of a centrally positioned explosive charge splits at the corners a) HE - Charge b) Steel prism c) Shock wave d) Release or rarefraction wave e) Interference of release waves with crack formation.

4.3 Inside Grooves in the Casing

4.3.1 Introduction

Warhead casings with grooves on the inside surface do not experience these release wave effects. The grooves can, therefore, be regarded as preformed weak points of the casing. The depth, shape, and spacing of the grooves have, however, a very great influence on the splitting of the casing into equal-sized fragments. The distance between the grooves should be about twice the residual thickness of the casing to enable the shear fracture to develop properly. Even if the geometry is not perfect, relatively good results are possible (Fig. 22).

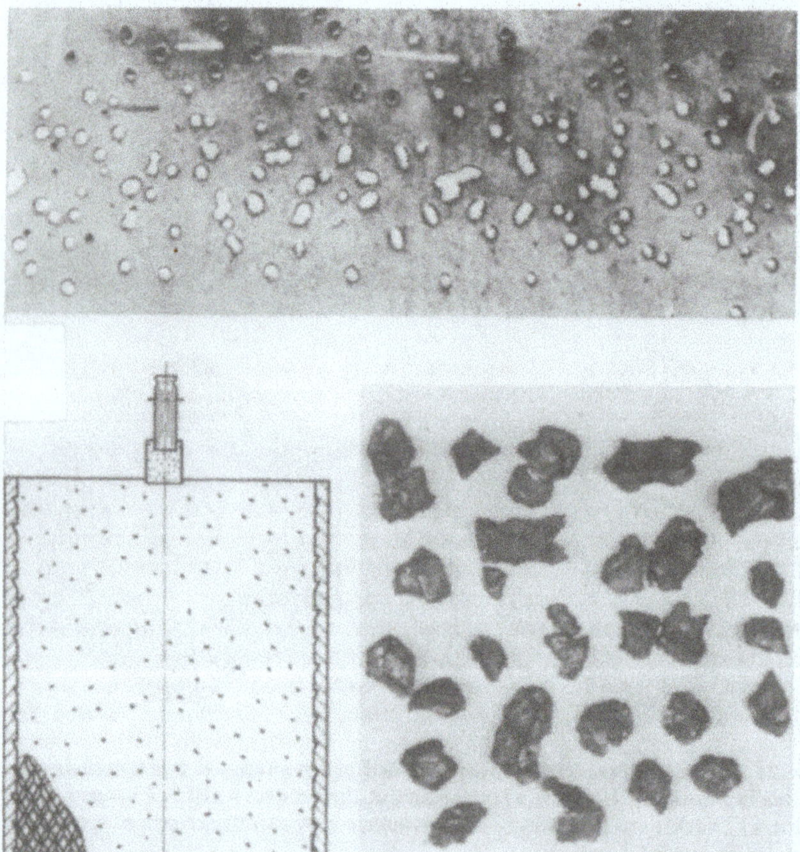

Fig. 22 Fragments produced from fragmentation warheads with grooves on the inside of the casing. The upper photo of the group shows the fragment perforation pattern (at a range of 1 m) of the warhead with internal grooving shown at lower left (casing thickness 4 mm, groove depth 0.6 mm), whereas the photo lower right shows typical fragments collected from the test.

The following sections give a detailed description of the shear-control method, which uses an inside grid system for warhead fragmentation. It addresses grid profile, grid spacing, orientation of detonation front, material properties, type of high explosive, and so on.

The shear-control method of fragmentation, patented by Pearson[9,10] is a highly effective method for controlling the fragmentation of a warhead case into fragments of predetermined size and shape. The shear-control method derives its name from the ability to control both the initiation locations of shear fractures in the warhead case and the orientation of the planes along which the fractures propagate. This method of warhead fragmentation uses families of mechanical stress raisers in the form of a grid system that is usually machined or formed into the inner surface of the warhead case. The elements of the grid system control the initiation of shear fractures at the root of each grid element, and the shear fractures then propagate along fracture trajectories established by the stress fields existing in the warhead case during the initial phase of case expansion. The control grid is designed to match the geometry of this stress field and utilize the principal strains in the metal to activate only specific families of fracture trajectories and thus produce fragments of a desired size and shape. This method can be used with many warhead shapes, wherever there is sufficient ductile expansion of the case prior to fragmentation, and the geometry of the strain field is compatible with control requirements of the method.

Although the theoretical network for all of the shear fracture trajectories that can exist in a warhead case for a specific grid geometry can be predetermined, actual fracture activation will occur only on a certain percentage of these trajectories. The actual number of trajectories activated, and the relative orientation of the trajectories, will determine the size and shape of the fragments produced and the resulting signature of the fragment mass distribution plot for that warhead.

Some of the more important parameters that influence trajectory activation and that need to be considered in the design of a warhead include 1) the properties of the warhead case material, 2) the basic grid geometry, 3) the cross-sectional profile of the grid elements, 4) the orientation of the detonation front, 5) the type of explosive, 6) design considerations such as the presence of buffers between the explosive and case, and 7) possible design variables associated with the type of manufacturing process used to produce the grids. A brief discussion of each of these parameters follows, in order to show its relationship to the overall fragmentation process.

The most important single factor in the use of the shear-control method is the dynamic property of the case material. Since this method is based on shear fracture as the primary mode of failure, the method works best with ductile steels and is at its worst when used with very brittle steels. A good guide in this respect is the normal fragmentation characteristics of a given case material with no control involved. If the uncontrolled case fractures completely through the wall in shear, or essentially so, then that material should work well with this method. If, on the other hand, the case fractures in a predominantly brittle manner, that metal is generally not suited to this application.

The basic configuration of the fragment is determined by the geometry of the grid system used on the inner surface of the warhead case. For example, a family of parallel, longitudinal grooves will produce rodlike fragments, whereas a diamond pattern will produce smaller fragments of predetermined shape and size. The actual size of the fragments resulting from a specific grid pattern will depend mainly on the wall thickness and the distance between the grid lines.

The shear trajectories emanate in mutually orthogonal pairs from the root of each element in the grid system. Whether the controlled fractures tend to propagate along both trajectories or are restricted to only certain trajectories of a specific orientation is governed strongly by the cross-sectional profile of the grid element. Symmetrical profiles, such as sharp V-shaped grooves, tend to activate fractures along both trajectories. Nonsymmetrical profiles, such as a sawtooth configuration, tend to restrict the fractures to only one orientation of the trajectory pair. Thus, by changing the shape of the grid element profile, different fragment mass distributions are available to the warhead designer.

The orientation of the detonation front relative to the shape of the grid element profile in a diamond pattern system is normally considered only when a nonsymmetrical profile is employed. Then the relative orientation between the detonation front and the profile configuration affects the number of preferred and nonpreferred trajectories that are activated. Therefore, the orientation of an asymmetric grid relative to the detonation wave of a warhead is important for the quality of controlled fragmentation.

The type of explosive is related to the fragmentation behavior of the warhead case through what might be most simply termed its brisance characteristic. Having fixed the preceding parameters for a given warhead case, the fragmentation characteristics will vary to a lesser degree with the type of explosive used. This variation will be determined by the relative number of fracture trajectories activated in the entire case, and for a nonsymmetrical profile by the relative activation of trajectories of preferred and nonpreferred orientation. In general, the greater the brisance of the explosive, the larger the total number of trajectories activated, and the increased tendency to activate trajectories with a less preferred orientation. Within the range of military explosives studied with this method, however, fragmentation variations due to changing the explosive were much less important than the variations associated with changes in the other parameters.

The presence of a buffer material between the explosive and the warhead case, such as a thin metal liner or sleeve, affects the activation of fracture trajectories. It can reduce the total number of trajectories activated, and for the nonsymmetrical profile type of grid it can affect the activation ratio between preferred and nonpreferred trajectories. The effect of the buffer will vary considerably, depending upon its properties and thickness. For example, the use of a thin, hot-melt layer (cavity paint) such as is used with some of the cast explosives appears to have little effect, while a $\frac{1}{16}$-inch thick steel sleeve can cause a marked difference in the fragmentation behavior of the case.

Numerous types of manufacturing processes have been tried to produce control grids for experimental warheads. In general, these processes can be separated into two categories: 1) machining operations, where the metal is actually removed, and 2) embossing operations (deformation processing), where the metal is displaced. Differences in hardness and microstructure (grain deformation) may occur locally in the metal at the root of the grid element for these two categories of operations. Small geometrical differences may also occur in the grid that are associated with metal flow and types of tooling. However, a review of many such operations by the author indicates that satisfactory results have been obtained for most types of grid production methods, and that the type of manufacturing process employed should be determined mainly by the economics of the problem rather than by possible minor variations in case behavior.

4.3.2 Geometry of Grid System

The geometry of the grid system used on the inner surface of the warhead case determines the basic shape of the fragments. For a cylindrical case, a family of parallel, longitudinal grooves produces long rodlike fragments, whereas a diamond pattern grid produces small fragments of controlled shape and size. Once initiated at the root of the grid element, the shear fractures follow maximum shear trajectories through the wall. The fragments produced in this manner are formed either by the propagation of fractures completely through the wall, or by the intersection of multiple fractures within the wall.

The major considerations in planning the geometry of a grid system are 1) to design the grid in such a way that it utilizes the maximum strain conditions that exist in the warhead case and 2) to use a grid geometry that maintains symmetry with respect to the strain field. Violation of either of these considerations will result in a fragmentation system that is less than satisfactory.

During the deformation and expansion of a cylindrical warhead case, the maximum strain ϵ_θ occurs in the circumferential direction whereas the axial or longitudinal strain ϵ_z has a much smaller value. Therefore, the grid design should be based primarily on the action of the circumferential rather than the axial strain. Thus, a grid system of parallel longitudinal grooves that utilizes the action of ϵ_θ is highly effective in forming rodlike fragments. On the other hand, a system that uses parallel circumferential grooves and relies on the action of ϵ_z normally would be ineffective. In Fig. 23 is a cross-sectional view of a case having a parallel grid system. The dashed lines emanating from the roots of the grids represent the paths the shear fractures will follow. Because of the geometry of this fracture field, rods from the inner and outer portions of the case will have different cross-sectional shapes.

To form small fragments, a diamond pattern is extremely effective. The grid should be designed to utilize the action of the circumferential strain effectively; accordingly, a pattern with diamonds elongated in the axial direction (Fig. 24a) is far more effective than a pattern with the diamonds elongated in the circumferential direction (Fig. 24b). The most desirable angle to use for the diamond pattern (β in Fig. 24a) becomes a compromise between having the angle small enough to get effective control and at the same time having the angle large enough to form fragments with a desirable shape. An angle of 60 degrees for β has proven to be an excellent choice and has been widely used in warhead programs.

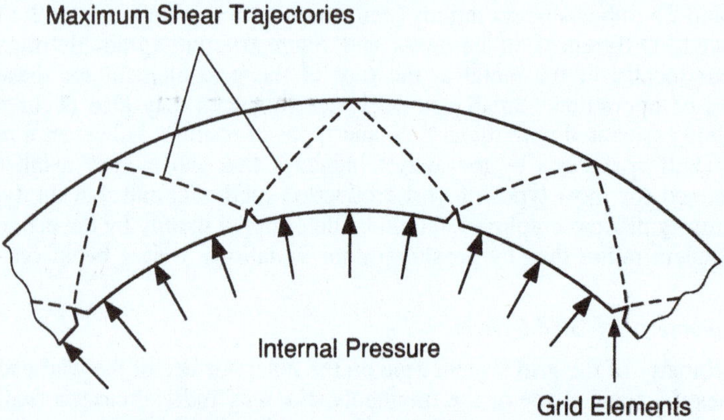

Fig. 23 Cross-sectional view of warhead case showing grid of parallel circumferential grooves with maximum shear trajectories.

The fragmentation behavior of the warhead case is reasonably sensitive to the symmetry of the grid system with respect to the strain field. If all elements of the grid system use the action of the strain field to an equal degree, each element can be expected to participate in a comparable fashion in the fragmentation of the case. If, however, the grid design is not symmetrical with respect to the strain field there is imbalance in the utilization of the grid elements, and so those elements of the grid are most preferentially oriented will predominate in controlling the formation of fragments.

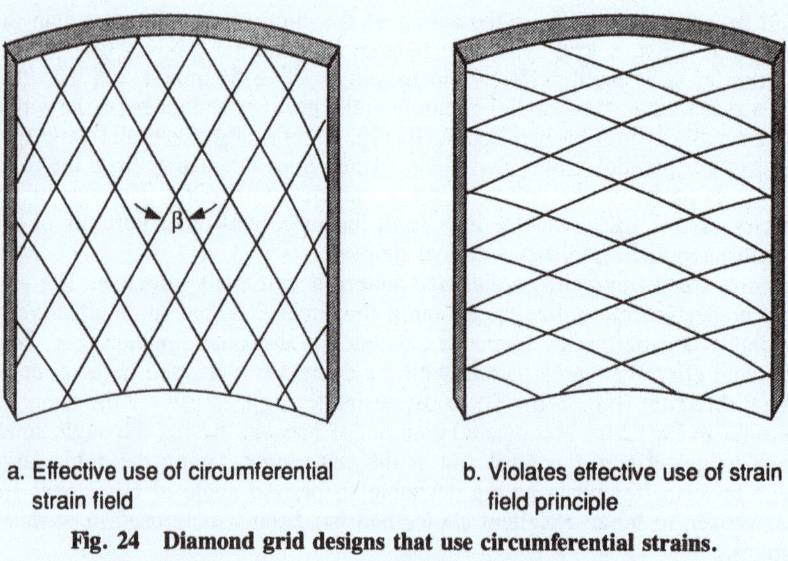

a. Effective use of circumferential strain field

b. Violates effective use of strain field principle

Fig. 24 Diamond grid designs that use circumferential strains.

The diamond pattern system of Fig. 24a is symmetrical with respect to the longitudinal axis of the cylinder and, hence, with respect to the field of strain. Accordingly, each line element in the grid system participates equally in the fragmentation process.

4.3.3 Grid Profile Configuration

Shear-control grids can be used in terms of either semiprecision or precision control. The difference between the two types of control concepts is based on the use of the grid element profile. A brief review of each control type is given here based on the use of the grid element profiles shown in Fig. 25.

Semiprecision control involves the use of grid elements with symmetrical profiles, such as the V notch configuration in Fig. 25a. The maximum shear trajectories occur in orthogonal pairs, and for the symmetrical profile the shear fractures may propagate along either trajectory. In some instances the fractures will propagate along both trajectories simultaneously. In Fig. 25a both trajectories are classed as primary fracture trajectories since both trajectories support fracture to the same degree. When used in a diamond pattern grid, this type of grid element profile will produce fragments in a variety of shapes and sizes, but with the bulk of the fragment mass clustered in a relatively narrow range in the fragment mass distribution plot.

Precision control involves the use of grid elements with nonsymmetrical profiles such as the one shown in Figure 25b. For this configuration of profile the trajectory pair emanating from the root of the grid no longer support fracture on an equal basis. Rather, there is now a preferred orientation for fracture activation, with the orthogonal pair separated into primary and secondary trajectories. For this profile, the primary trajectory supports fracture to a much greater degree than does the secondary trajectory. The orientation of these two trajectories relative to the shape of the profile is a constant feature and, as shown in Fig. 25b, the primary trajectory is always located over the slant face of the profile. Since fracture activation is now restricted to only one trajectory of a known orientation, the precise shape and size of the fragments can be predetermined.

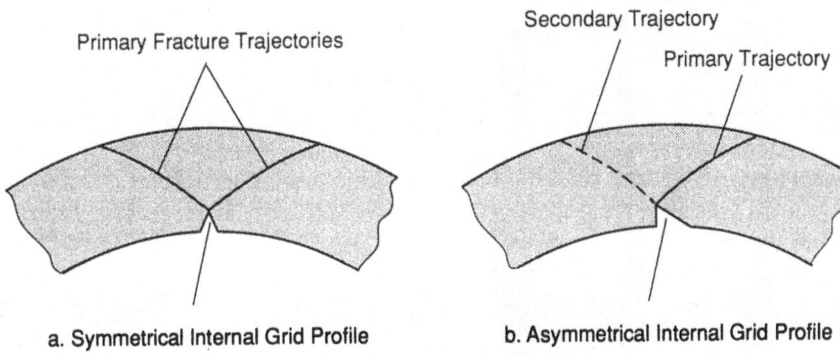

Fig. 25 Grid element profiles.

Through appropriate use of case material properties and grid design, the warhead designer normally tries to maximize the number of primary trajectories activated and to minimize activation of the secondary trajectories.

4.3.4 Formation of Rods

The simplest application for visualizing the differences associated with the two types of profiles is the parallel grid used to form rods. In Fig. 26 a parallel system of equally spaced grooves with nonsymmetrical profiles is illustrated. In such a system only the primary trajectories are activated, and the result is the formation of rodlike fragments all having the same cross-sectional shape and size. The formation of rods by means of a parallel grid system with equally spaced symmetrical V grooves involves the possible activation of all trajectories as shown in Fig. 27. Three different situations are considered in this figure, based on the grid spacing distance relative to the thickness of the case.

In Fig. 27a, the spacing between the parallel grooves is such that the right-hand and left-hand trajectories intersect at the outer surface of the case. Such a system produces two different families of fragments (A and B) with cross-sectional shapes that are approximately triangular.

If the parallel grooves have a spacing distance such that the fracture trajectories do not intersect, the fragments will have cross-sectional shapes as shown in Fig. 27b. Since the shapes of the trajectories are constant, the cross-sectional area of the shear fragments (B) depends only on the wall thickness, whereas the cross-sectional area of the main fragments (A) depends on both the case thickness and the distance between the grooves. This particular concept is of considerable value if families of rods with two different cross-sectional areas are required, since the ratio of the areas can be precisely controlled by the distance between the grid lines.

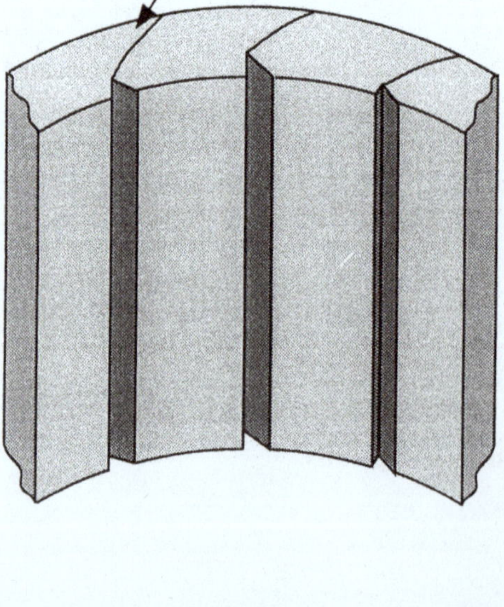

Fig. 26 Fracture trajectories for parallel grid with asymmetrical element profiles.

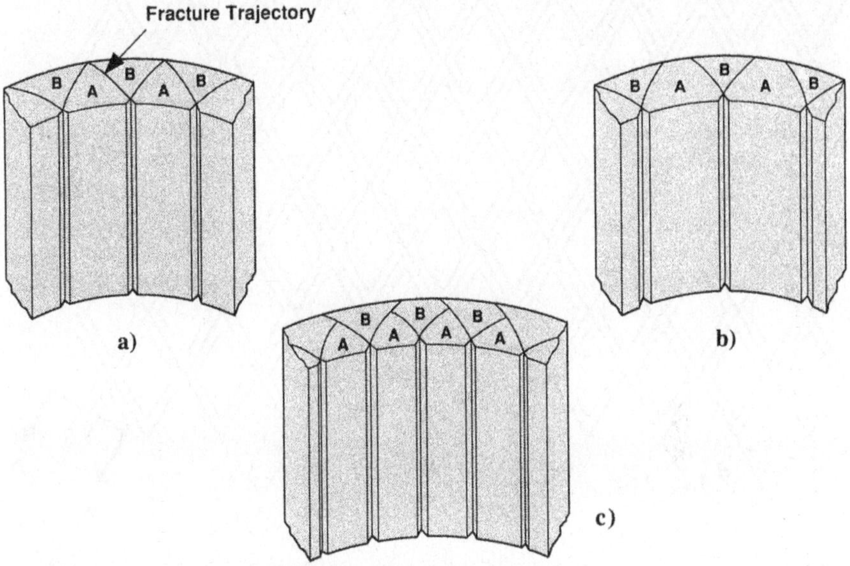

Fig. 27 Fracture trajectories for parallel grids with symmetrical grid element profiles (A: main fragments; B: shear fragments).

A third possibility is illustrated in Fig. 27c, where the spacing is such that all of the fracture trajectories intersect within the wall of the case. The result is the formation of a family of shear fragments (A) with a triangular cross-sectional configuration of uniform shape and size. The main fragments (B), however, will have cross-sectional configurations in a variety of shapes and sizes because of the unpredictable manner in which the trajectory intersections will occur.

4.3.5 Formation of Small Fragments

If small fragments are required, a diamond pattern grid is normally used. The concepts expressed in the previous section on rod formation also apply to the formation of small fragments. Since each diamond in the grid is bounded by four straight sides, the shape and size of the fragment produced by a given diamond in the grid will depend on the activation of shear fractures from four grid elements (the sides of the diamond).

In Figs. 28 and 29 are shown diamond pattern grid designs using symmetrical and asymmetrical profiles. The symmetrical profile grid (Fig. 28) will produce fragments in a variety of shapes and sizes, but with the bulk of the fragment mass clustered in a relatively narrow mass range. The asymmetrical profile grid (Fig. 29) tends to produce fragments of one uniform size and shape. As indicated by the small arrows, each fragment is formed by two wing fractures (diverging arrows) and two undercut fractures (converging arrows), producing a fragment shape as shown. A view of such a grid in a cylindrical casing is shown in Fig. 30.

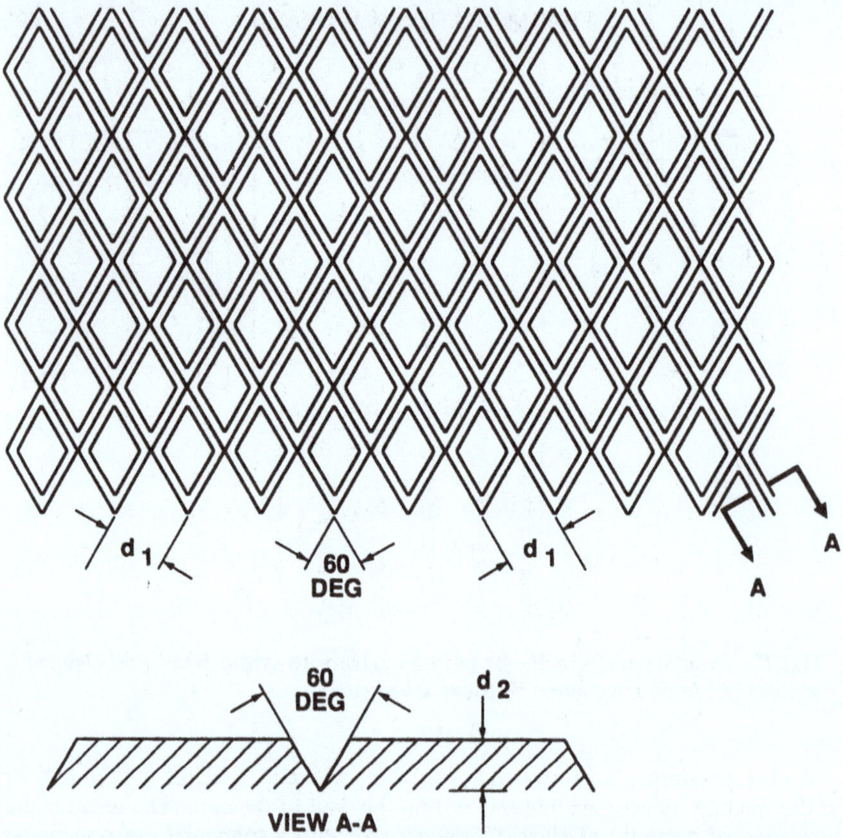

Fig. 28 Diamond grid design with symmetrical profile.

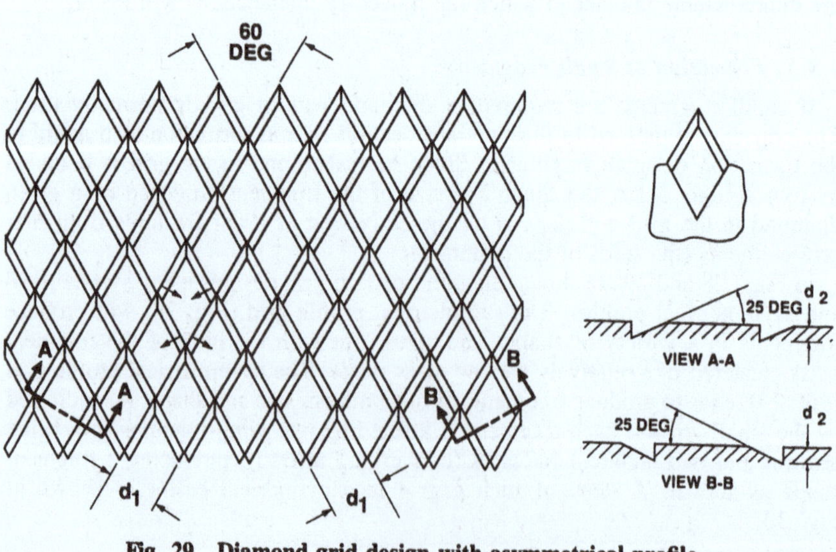

Fig. 29 Diamond grid design with asymmetrical profile.

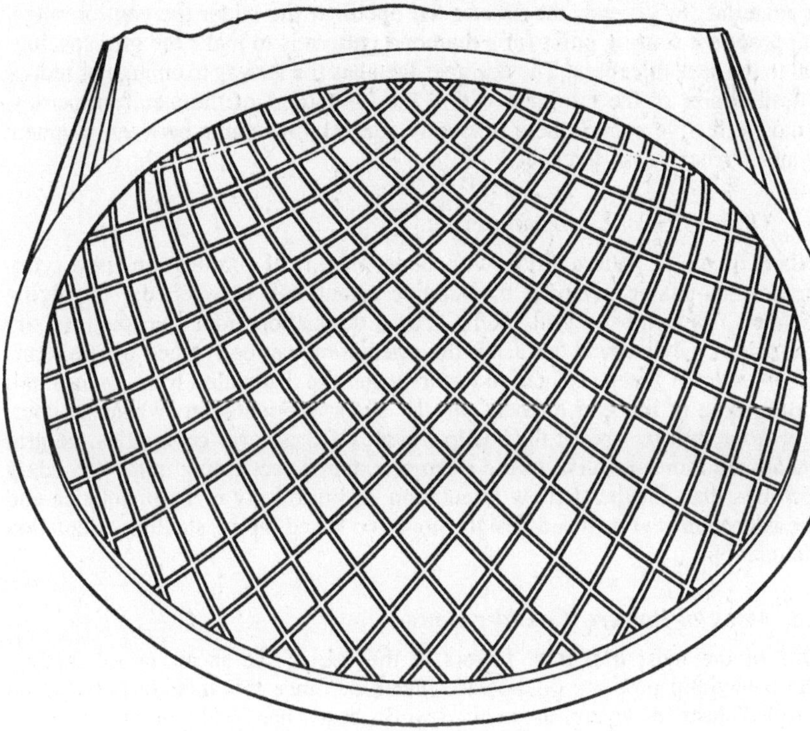

Fig. 30 Cylindrical casing with an internal diamond grid.

4.3.6 Grid Spacing Requirements

In designing a precision control grid of the type shown in Fig. 29, an additional factor to consider is the required size of the fragment relative to the case thickness. For a cylindrical warhead case of a given size and material, there is normally a range of fragment sizes that can be effectively obtained by the shear-control method. There are also limitations in terms of how large or how small the fragment size control can be effectively exercised. For a diamond pattern grid, these limitations are directly related to the distance between the grid lines. If the grid lines are too close together relative to the case thickness, fragments with multiple diamonds are formed. At the other extreme, if the grid lines are too far apart, the large fragments produced by the control method will, in turn, be broken into smaller pieces by the natural fragmentation process. There are no sharp lines of demarcation to establish these boundaries. Rather, it is a gradual transition process in both directions.

Between these two extremes is a range of effective fragment sizes that can be obtained through the appropriate spacing of the grid lines. The width of this effective control range can be expected to vary with the properties of the case material. In general, the greater the ductility, the wider the control range. An appropriate starting point for a diamond pattern is to make the grid spacings equal to the case thickness. The designer then has the leeway to enlarge or reduce the desired size of the fragment within the limitations of the metal properties. For most effective control there is usually a general matching between fragment size and warhead case thickness.

4.3.7 Orientation of Detonation Front

When diamond pattern grids with nonsymmetrical profiles are used, consideration should be given to the relative orientation between the impinging detonation front and the grid profile. For a detonation front sweeping axially the length of a cylindrical warhead, the orientation can be typified by the terms *steep* or *shallow*. Steep orientation occurs when the detonation front sweeps into the steep side of the grid element profile; shallow orientation, when the detonation front sweeps against the shallow side. Of these two geometries the grid elements are more sensitive to the steep orientation, activating more secondary trajectories than for the shallow orientation. If uniformity of fragment size and shape is desired, the warhead design should be based on the shallow orientation of detonation.

4.3.8 Material Property Considerations

One of the most important factors in the use of the shear-control method is the behavioral property of the case material. Since this method is based on controlled shear fracturing, the method works best when used with ductile steels and is least effective with brittle steels. A good guide to the effectiveness of the method with any specific steel is the normal fragmentation characteristics of the case material when no control feature is used. If the warhead case fractures completely through the wall in shear, or essentially so, then that case material should work well with this method for a warhead of that size. If, on the other hand, the case fractures predominately in tension, that metal is generally not so well suited to this type of control.

Most steels will behave somewhere between the two extremes of completely ductile and complete brittle behavior, with the case fracturing through in some ratio, say 80/20, where the inner 80% of the wall thickness fractures in shear, and the outer 20% fractures in tension. Although such a material does not behave so well as the 100% shear type for precision control, nevertheless, it should be acceptable for most applications. At what ratio the metal would be deemed not acceptable to use with this control method is a figure that will vary with the requirements of the weapon.

A variety of conditions can influence the ductility of the warhead case material and, hence, the degree of control obtained by the shear-control method. Three areas where ductility effects need to be considered are 1) where the ductility of a specific steel is changed through heat treatment, 2) where the choice of different higher-strength steels offers different ductility behavior, and 3) where the ductility may be changed by lowering the temperature.

Most of the warhead programs that have considered the use of the shear-control method have not required high-strength characteristics in the case material. Accordingly, the material most commonly chosen has been one of the plain, low-carbon steels. These steels are low in cost, readily available, and easy to machine and fabricate. In addition, their ductility characteristics make them admirably suited to this method of fragmentation.

The shear-control method has also been considered for use with weapons that require higher strength characteristics in the warhead case, either for penetration effects or to sustain the forces associated with the gun launching of projectiles. Accordingly, a number of high-strength, heat-treatable steels have been studied in view of their potential application. In general, each of the heat-treatable steels can be treated to produce a fairly wide range of hardness and strength properties. The heat treatment usually represents a trade-off between increased strength and reduced ductility. The problem that faces the designer is to pick a type of steel that can be heat-treated to give the required strength and still maintain sufficient ductility to work with the control method. Most of the heat-treatable steels studied behaved in an acceptable manner in their lower and middle hardness ranges.[9] Near the higher limits of their hardness ranges it is expected that many of them will not be satisfactory. A general guide, based on limited studies, indicates that most of these steels should be generally acceptable at least up to a hardness value of about $R_C 40$.[10]

A change in behavior, associated with the hardening and strengthening of the steel, appears to be representative of the heat-treatable steels in general. However, the quantitative results should vary with different steels, since they may show different elongation characteristics at a given hardness value.

4.3.9 Effect of Low Temperatures

The effect of decreasing temperature on the shear-control process was, for example, studied for a plain, low-carbon steel at temperatures ranging from +80 to −110°F.[10] Although there was some change in the signature of the fragment mass distribution plots at the different temperatures, the shear-control method was quite effective even at the lowest temperatures studied. These tests demonstrated that for the type of steel studied the lowest service temperatures that normally may be encountered in the use of a controlled fragmentation warhead should present no problem in the use of this control method.

4.3.10 Types of Explosives

The shear-control method is used effectively with a variety of military explosives, including such general types as hand-loaded (plastic), cast, and pressed. More specifically, satisfactory tests have been conducted with such explosives as Composition C-3, Composition C-4, Composition A-3, Composition B, Octol, H-6, PBX C-12, and PBX C-108. Most of these explosives have been used with different warhead designs so that a controlled comparison is not readily available.

It is known, however, that some variation in the fragment mass distribution occurs when different explosives are used with a given warhead case design.

In general, it is thought that for a given case design an increase in the power (brisant) characteristics of the explosive tends to activate a somewhat greater percentage of the total fracture trajectories for a symmetrical profile and a somewhat greater percentage of secondary trajectories for a nonsymmetrical profile.

In general, any of the recognized military explosives commonly used for fragmentation warheads will work well with the shear-control method. Control variations that may occur with the use of these different explosives should be relatively minor and should not be so important as those variations that can occur as a result of changes in other warhead design parameters.

4.3.11 Methods for Producing Grid Systems

The methods used to produce grid systems on the inner surface of a warhead case fabricated from tubular or flat stock can be separated into two categories: 1) machining operations where the metal is actually removed, and 2) deformation processing operations where the metal is displaced. For both areas of operations, many of the processes have been applied directly to the cylindrical case, whereas others have been used to produce the grids on flat stock that was subsequently formed into the tubular shape. The grids produced by these various processes have functioned effectively in terms of controlling the fragmentation, although each type of process may produce minor differences in the grid, which can be considered a signature of the respective process.

Machining operations are normally used to manufacture small numbers of experimental units, since the cost of tooling for a deformation processing operation would be prohibitive, especially where the grid parameters might be changed from unit to unit. Machining operations on cylinders have been performed mainly with the use of a lathe or milling machine, whereas for flat plate, the grids have been machined on a shaper.

Deformation processing operations are used to form the grids when actual warhead cases are produced in large numbers. Such operations have included 1) stamping or roll embossing of flat plate, which was then formed into tubular sections; 2) the use of internal expansion tools and multiroller heads for cylinders; and 3) the employment of male dies during the sizing operation in the hot-cup, cold-draw process. All such operations have worked well.

A general review of many operations based on both machining and deformation processing shows that satisfactory results have been obtained for a large variety of grid production methods. This would indicate that the type of manufacturing process employed should be determined by the economics of the situation rather than by possible minor variations in case behavior.

4.4 Zone Embrittlement

Another highly effective method of controlling the fragmentation of a warhead case into fragments of predetermined size and shape is zone embrittlement of the casing, either by special chemical treatment or by laser- or electron-beam welding processes.

Figure 31 is shown as an example of a casing with electron-beam welded zones parallel to the axis, with different depths of embrittlement in the wall. The types of fragments found experimentally are also illustrated.

With an embrittlement about 25% to 50% deep into the wall from the outside, a mix is found of *main fragments*, which are generated by radial cracking that initiates at the embrittled zones, and *shear fragments*, from cracks that extend from the end of each radial crack. When the embrittlement is as deep as 50% to 75% of the wall thickness, the shear fragments become smaller and smaller until, finally, no shear fragments are formed at all when the embrittled zone goes all the way through the wall of the metal casing. In Fig. 32 the impacts of rods formed in such a controlled manner are shown in a steel witness plate. The left-hand side of this figure shows *main fragments* and the relatively large *shear fragments* from a casing that had been embrittled to a depth of 25% to 50%, and the right-hand side shows the larger main fragments and the smaller shear fragments for a case where the depth of embrittlement had been about 75% of the wall thickness.

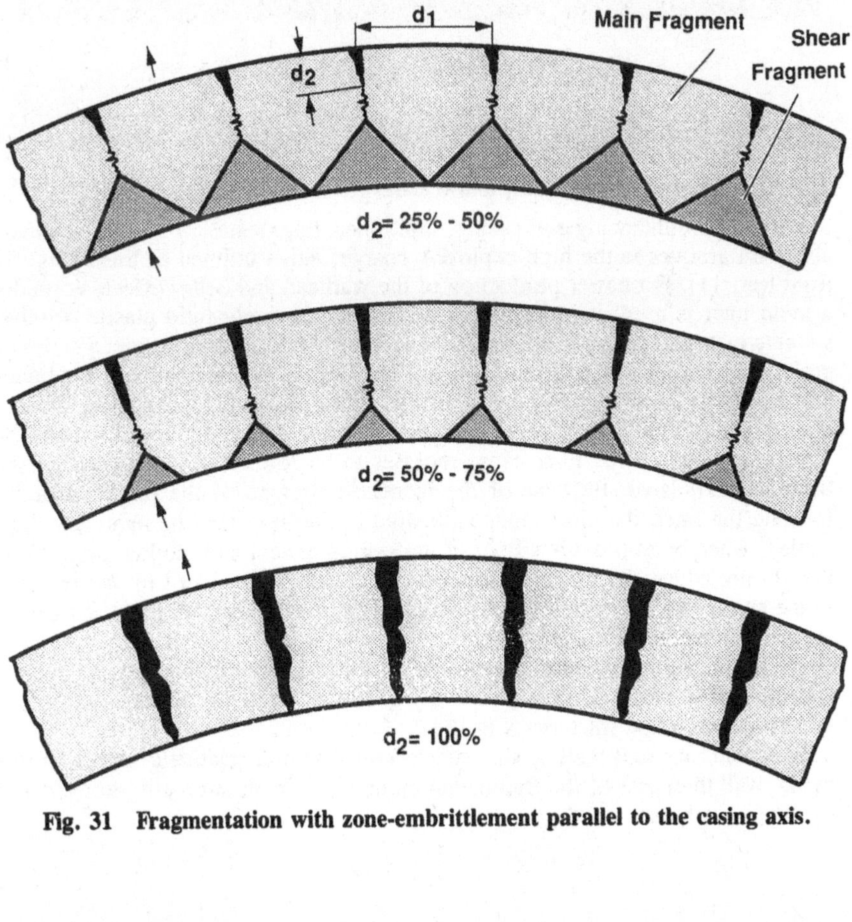

Fig. 31 Fragmentation with zone-embrittlement parallel to the casing axis.

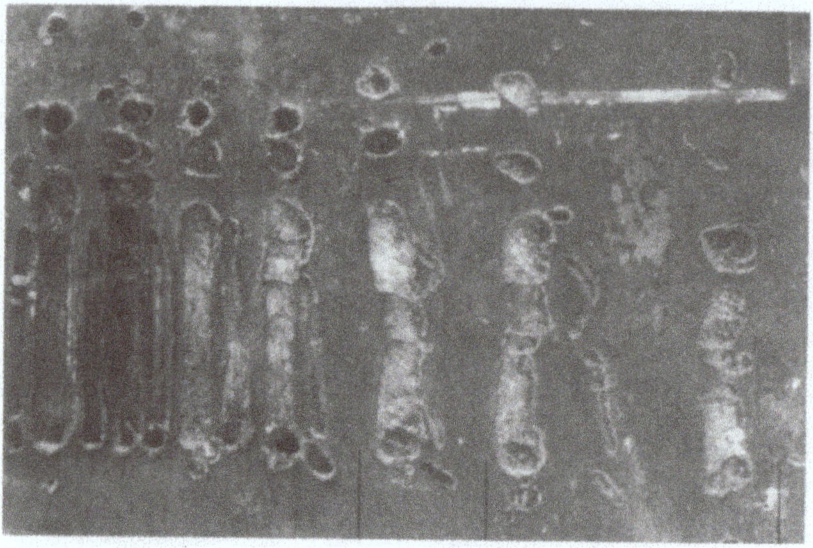

4.5 Grooves in the High-Explosive Charge

Instead of influencing the casing, controlled fragmentation can be achieved also with grooves in the high-explosive charge, either unlined or lined (Fig. 33 from Ref. 11). For easier production of the warhead and better effect, normally a mold liner is used; it is formed of a sheet of relatively rigid plastic or other suitable material, which is embossed in a manner to form on one side; a series of projections arranged in a desired pattern. The liner is then formed into a cylinder and placed in the metallic casing constituting the shell of the explosive device, after which a liquified explosive is poured into the liner and allowed to solidify. The projections on the liner cause recesses to be formed in the surface of the solidified explosive. Because of the increased strength of the plastic material forming the liner, the projections embossed in the lines may be relatively thin-walled, whereby to provide a liner of uniform thickness with hollow projections that do not adversely affect the shaped-charge effects produced by the recesses in the surface of the explosive. Accordingly, it is not necessary that the liner be removed from the explosive charge.

By suitably dimensioning the profile of these liners (angle of roof, length, spacing, offset, etc.), a correspondingly good fragmentation pattern can be obtained by controlled fractures (Fig. 34).

Fragment size and weight are predetermined within relatively narrow limits by the wall thickness of the fragmentation casing in both internally-grooved and roof-type warheads.

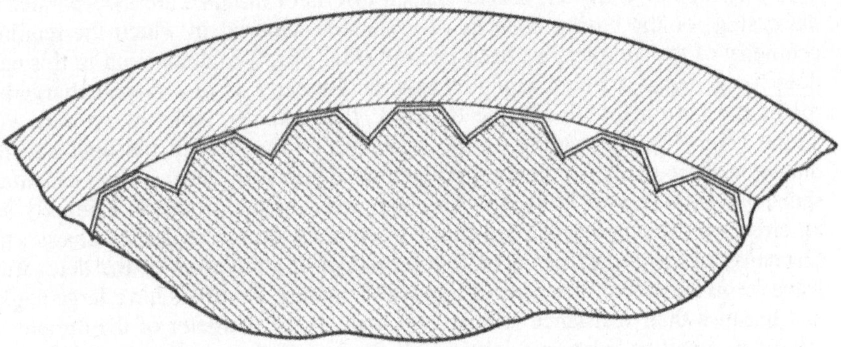

Fig. 33 Controlled fragmentation resulting from lined grooves in the high-explosive charge.

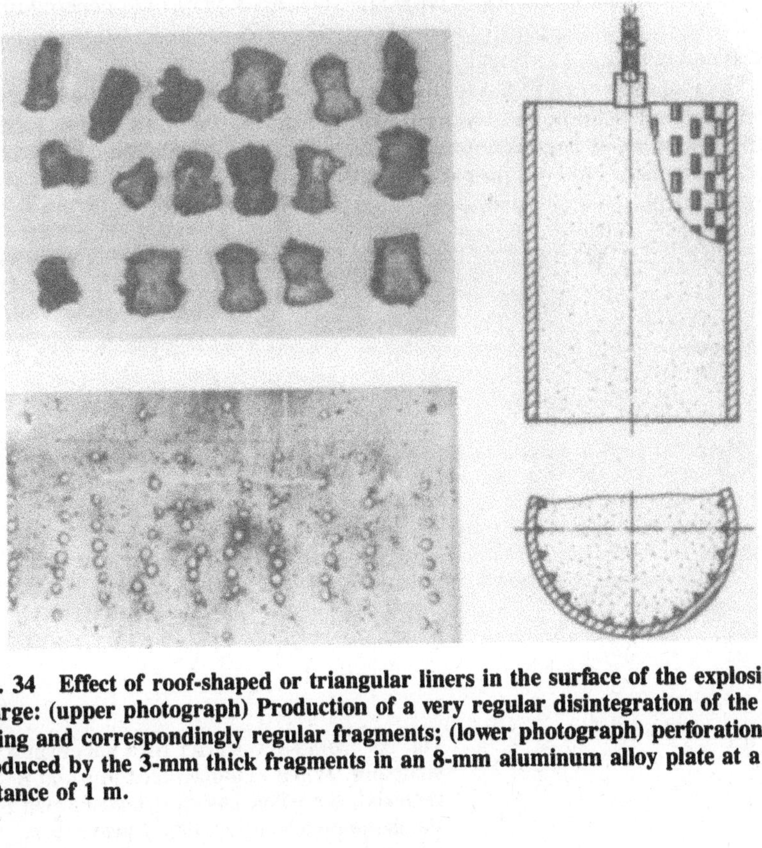

Fig. 34 Effect of roof-shaped or triangular liners in the surface of the explosive charge: (upper photograph) Production of a very regular disintegration of the casing and correspondingly regular fragments; (lower photograph) perforations produced by the 3-mm thick fragments in an 8-mm aluminum alloy plate at a distance of 1 m.

5. Multi-P Charge Warhead

In this type of warhead, saucer-shaped inserts or dimples are incorporated in the casing, or the casing consists of a suitable material in which the required geometry of the dimples has been pressed (Fig. 35). The detonation in this case does not collapse the liner along its axis of symmetry as in a shaped charge but forces the bottom of the dimple outward. It eventually looks like a dart with a core (Fig. 36). Depending on the resistance of the wall of the liner and the diameter of the explosive charge, projectiles of relatively great weight can attain speeds of 1600 m/s to 2200 m/s. The large mass of the projectile evidently has an effect on the number of projectiles. In comparison to the shaped charge, the formation of the projectiles does not depend on the direction of the detonation wave or on the size of the various impulses, because the liners have large angles and because their resistance is relatively high. If the diameter of the dimples is relatively small in relation to their wall thickness, then perfect projectiles are obtained, even if the detonation wave only grazes them.

If larger and heavier projectiles are required, then larger liners are required in the casing. But the detonation wave must then hit the liner almost perpendicularly or almost symmetrically to the axis of the liners. As a result of their greater mass and speed, these projectiles possess a very high penetration capability, even against spaced plate targets (Fig. 37). For steel targets, they can penetrate between a half and a full diameter of the liner.

The inconvenience of the multi-P charge warhead is that dimples of this type can only be placed in one layer around the warhead casing; therefore, their size and penetration power cannot be increased except by reducing their number, and consequently the hit probability of the warhead is also reduced. As a result of their increased mass, their penetration power remains practically unchanged as

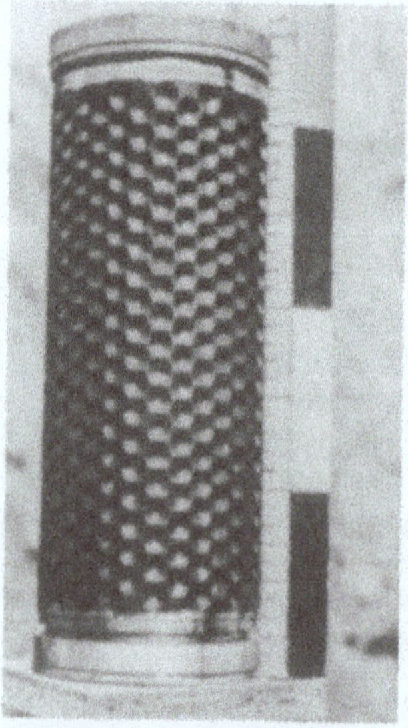

Fig. 35 Multi-P warhead with honeycomb structure. When manufactured in suitable material, the whole casing is transformed by the detonation into individual projectiles.

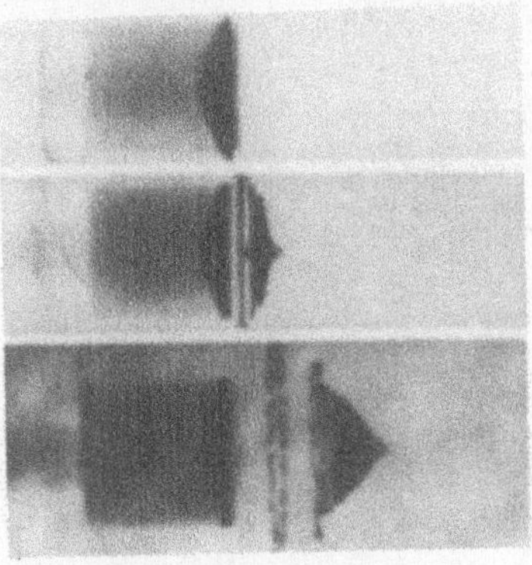

Fig. 36 Flash X-ray photograph of the transformation of a dimple into an explosive-formed projectile after 4, 10, and 25 microseconds.

a function of range in the zones normally of interest for air target warheads. The hit probability and the number of hits alone determine the level of effectiveness of the multi-P warhead. Because of its high penetration power, it is capable of cutting through or damaging even strong structures (the warheads of ballistic missiles, for example) and to detonate their high-explosive charge.

6. Multishaped Charge Warhead

The axis of symmetry of these liners normally intersects the axis of the warhead, and the liners are symmetrically distributed around the circumference of the charge. The liners can be conical, hemispherical, in the form of a segment of a sphere in shape, or of any desired form that can be arranged symmetrically with the axis of rotation (an example is shown in Fig. 38). The liner material can be copper, iron, or aluminum depending on the operational role.

It is known that the detonative transformation of the high-explosive charge accelerates the liner along its axis, causing the elements of the inserts to merge with each other along the collapse line.[12] In accordance with the laws of hydrodynamics, the formation of a jet and a slug then takes place, with the jet exhibiting a velocity gradient between its tip and the slug. This causes the jet to break up into individual particles or fragments.

If the liner is of copper, there will be a velocity difference of about 100 m/s between individual particles, regardless of the initial included angle, wall thickness, and basic diameter of the liner. A jet tip velocity of 7000 m/s—which corresponds to liner angle of 60 degrees—produces approximately 60 particles or fragments.

The hollow charge jets of this type of liner only form precisely along the axis of the liner if the detonation wave arrives on the liner normal to the rotation axis and runs from the tip of the cone to its base or if each circular segment of the casing simultaneously receives pressure impulses of the same magnitude and duration. In general, however, these conditions are not fulfilled in multishaped charge warheads. Multishaped charge warheads are usually detonated on one side only, so that in the various levels of liners the detonation wave has

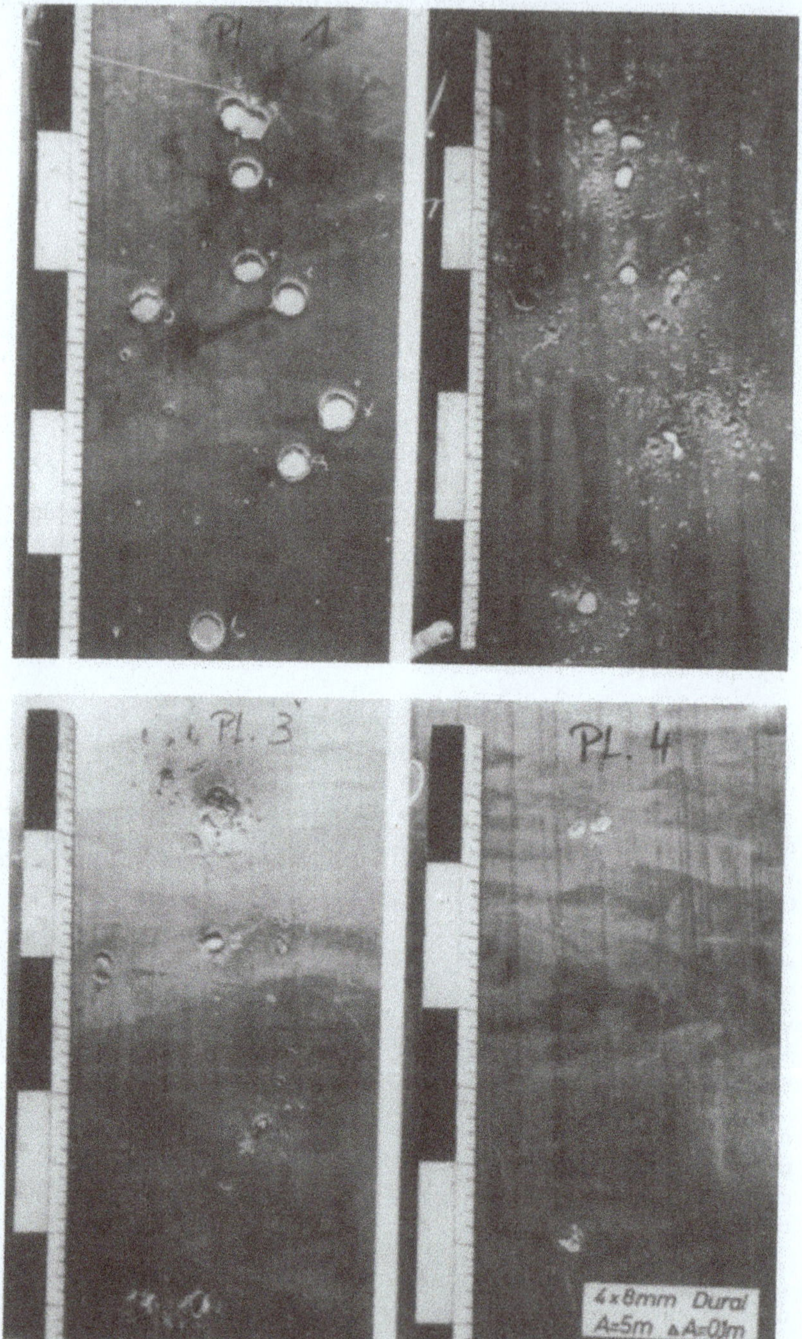

Fig. 37 Spaced plate penetration capability of multi-P charge fragments (5-m distance, 8-mm aluminum alloy, spaced 100 mm).

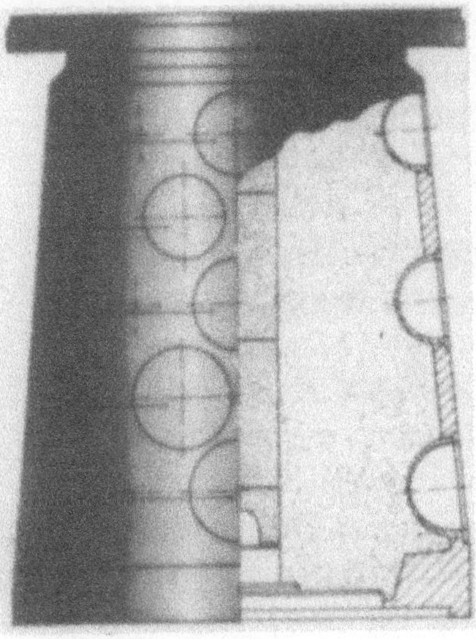

Fig. 38 Example of a multishaped charge warhead.

various angles with the axis of the liners, with the result that shaped charge jets are formed in various different directions. If a hollow charge is initiated eccentrically, a hollow charge jet is formed, the tip of which subtends a greater angle to the charge axis than the remainder of the jet. Then as the impulses are balanced out, the slug of the jet can even move into an exactly opposite angular direction. The individual particles of a hollow charge jet do not then strike a spot on the target, but make impact along a line, thus making a cut in the target area instead of a deep penetration. The curved jet of the shaped charge is clearly shown in a flash x-ray photograph of this phenomenon, though of course all the particles forming the jet in fact fly in a straight line from the point at which they were formed and thus produce a cutting or slicing action on the target plate (Fig. 39).

Since it is not possible to initiate an explosive charge at a precise spot, accurate initiation on the warhead axis always produces an elliptical detonation wave, this wave forming an intersection curve of the fourth order with the liner. The jet formation in this case is similar to that of an eccentric initiation at two points (Fig. 40).

Furthermore, if several liners are positioned in the casing of a multishaped charge warhead, no uniformity of detonation wave impulse on the elements in one row of liners is obtained. The duration of the pressure on the sides of the liners lying closest to the neighboring liners is less than on the other parts, since the mass of explosive is less at this point. This configuration corresponds to an irregular confinement of a hollow charge but is symmetrical about the axis of rotation. It is clearly seen in a flash x-ray photograph that, in a charge of this type, the hollow charge jet forms across the plane of the confinement, in other words, vertically (Fig. 41).

Fig. 39 Flash x ray of a *curved* jet, and the associated cut.

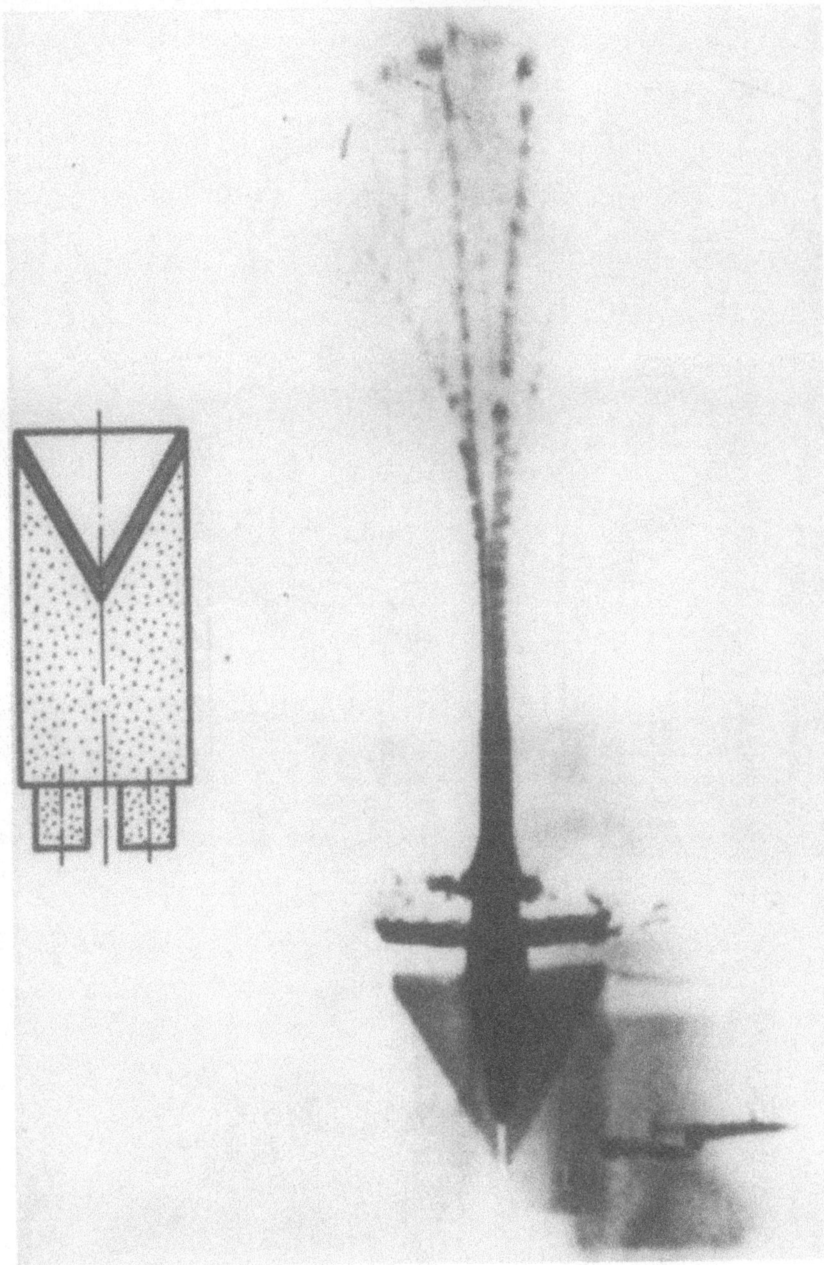

Fig. 40 Flash x ray of a jet splitting in its tip region as a result of a double eccentric initiation of the shaped charge.

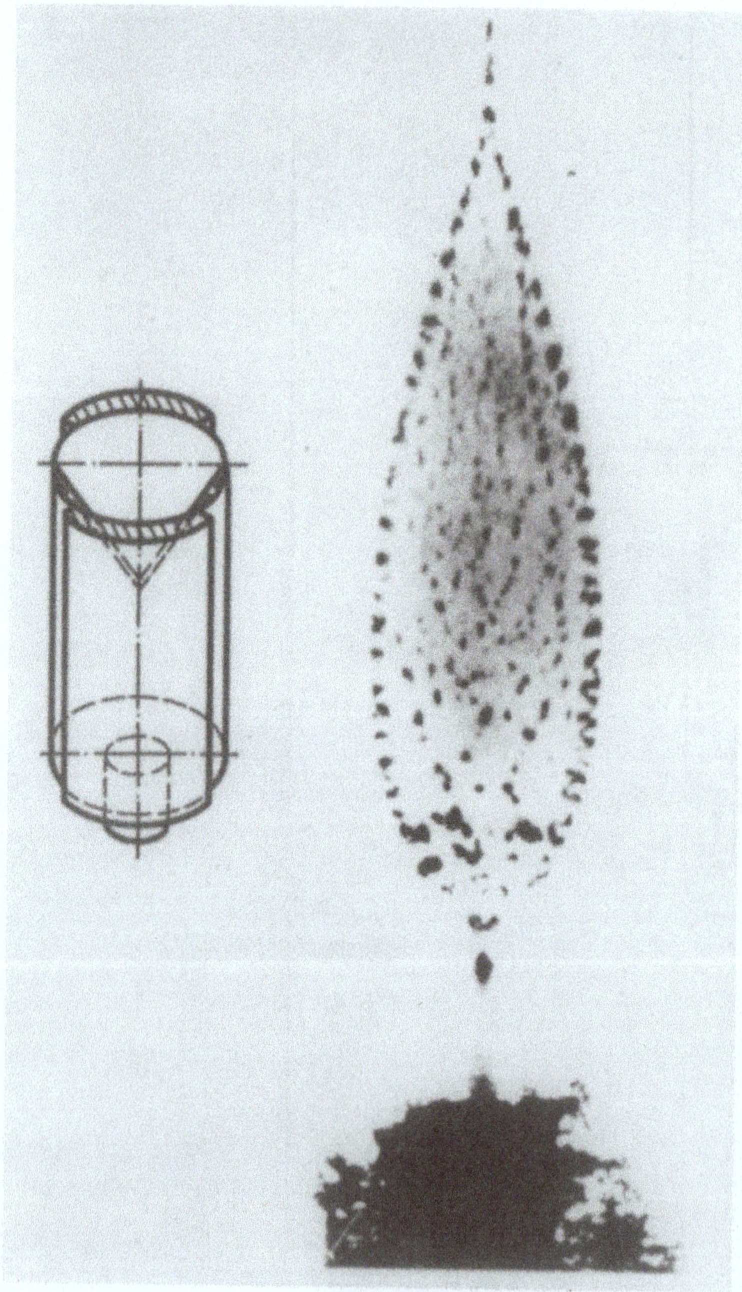

Fig. 41 Flash x ray of a jet splitting in the rear portion as a result of a shaped charge being confined on two opposite sides over a 90°-sector each.

The following effects always lead to a large deviation of the jet particles relative to the axis of the liners.

- Detonation wave hitting the liners asymmetrically
- Detonation wave unsymmetrical with the axis of rotation
- Different impulse effects on the liners in one plane

The radial disintegration effect in particular increases the number of fragments, thereby reducing their individual mass.

An advantage of multishaped charge warheads is the high tip velocity of the fragments resulting from the particulation of the jet. The tip velocities are in the range of 5000 m/s. This also causes partial ablation in the air and creates brightly luminous traces (Fig. 42). The fast-moving particles characteristically flatten out when they hit the target and knock off numerous secondary fragments from the target (Fig. 43) and so on, all of which have considerable destructive capability,

Fig. 42 Detonation of a multishaped charge warhead during firing trials with the ROLAND missile at the Centre d'Essais, France. Note the radiating trajectories of the individual shaped charge fragments and the CT-20 target drone just visible in the upper part of the "chrysanthemum."

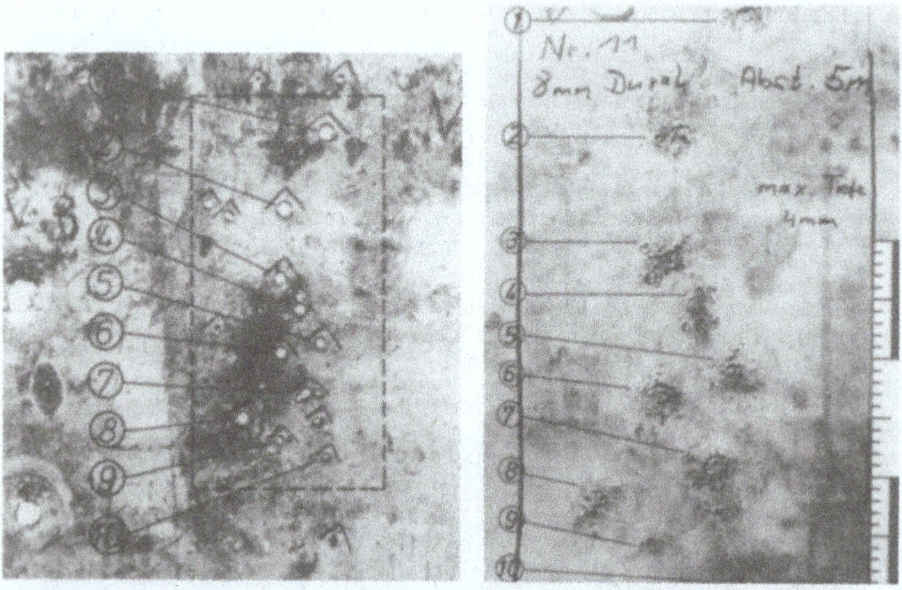

Fig. 43 Multishaped charge warhead striking a primary and a backup aluminum alloy plate. (Left photograph) Perforations produced in an 8-mm thick aluminum alloy plate at 5 m distance by the jet of a shaped charge that is already particulated fragments. The high velocity at which the jet fragments have hit the plate has both resulted in the fragments of the shaped charge and also caused the target material to disintegrate into a large number of very small fragments, which have penetrated a second dural plate 60 mm behind the first (right photograph) to a maximum depth of 4 mm. The perforations in the first plate and the corresponding effect on the second are numbered from 1 to 10.

particularly against electric cable runs, hydraulic lines, (Fig. 43). This high-speed impact causes target material at the point of impact to vaporize or to form very fine particles, so that with aluminum plates in particular, explosive-type secondary reactions with atmospheric oxygen take place. These explosions and reactions of the aluminum clouds with oxygen in the air have the capability to ignite gasoline and oil.

The disadvantage of the particles from multishaped charge warheads, however, is that in the case of moving airborne targets they do not impact at the same spot but cover a wider area, depending on the crossing speed and distance from the target. Even when hollow charge jets form and move exactly along the liner axis, they only produce a cut in the crossing target.

If a small shaped charge of 32-mm diameter, positioned at a distance of 800 mm laterally from the flight path of a projectile moving at a velocity of 500 m/s, is fired just as the projectile flies by, the particulated jet with the decreasing velocity from the tip of the jet to the slug is very clearly demonstrated (Fig. 44).

Fig. 44 Verification of the trajectory of fragments (particles) and the efficiency of individual fragments of a shaped charge fired against a fast-moving target. (upper photograph) The target selected was a fin-stabilized 106-mm round fired from a recoilless gun with a velocity of 500 m/sec that on hitting the contact sheet seen at right, caused the 32-mm shaped charge mounted on the support in the background, 800-mm to the side of the target's trajectory, to fire. The effect shows as a smooth cut from the jet tip region, followed by the increasingly discrete impacts from the jet particles that get slower and scarcer toward the tail end of the jet (lower photograph)!

It has not been possible to date to compensate for this spreading of the particles by a corresponding deviation of the jet, obtained, for example, by an oblique propagation of the detonation wave.

7. Preformed Fragments

Considerable freedom in the choice of the number and size of fragments is provided by preformed fragments in the shape of balls, cubes, rods, and so on. These fragments can also be arranged in a number of layers, when the weight of the warhead and the diameter of the charge permit this.

Freedom exists also in the fragment material. Instead of steel, it is also convenient to use the following:

- Heavier fragments for higher perforation potential, like tungsten-sintered alloys or depleted uranium (DU)
- Lighter fragments for higher velocities, like titanium
- Incendiary fragments for higher burning probability from materials like zirconium, coruscatives[13], DU

The initial velocity of the fragments, of course, decreases in relation to the increase in fragment weight. Generally speaking, the preformed fragments are embedded in a resin matrix and these casings, therefore, have a low structural rigidity. It is for this reason that it is necessary to incorporate in addition an inner and/or an outer casing for structural strength in warheads of this type when used for missiles, to take the tensile bending and torsional loads or the inertial forces during the launch phase.

By embedding the fragments in an aluminum or steel matrix, the structural rigidity of the casing can be significantly improved. As a further step, attempts to produce a rigid fragmentation casing, mainly in the form of a hexagonal fragment structure, by sintering together the individual fragments are proceeding very successfully.

A special preformed fragment warhead type is the *multirod* or *tumbling-rod* warhead. This warhead type projects many discrete rods, normally packed longitudinally in the warhead casing, as shown in Fig. 45. The main feature of this warhead is the potential for high penetration capability, which requires weight and velocity. The number of rods is normally limited by the available surface area.

Summary

A disadvantage of warheads with external or internal grooving, and particularly those with preformed fragments, is their higher cost in comparison to homogeneous casings. The introduction of metal inserts makes little difference to fabrication costs compared to the smooth casing and very good results can be achieved by this method. This conclusion is only valid, however, if casing wall thickness and the desired fragment size are compatible, and not, for example,

Fig. 45 Tumbling rod warhead projecting a small number of large high-velocity fragments.

when a large number of small fragments, which can only be arranged in multiple layers, are required.

8. Continuous Rod Warhead

Continuous rod warheads as the name implies, incorporate a series of rods that are positioned alongside each other around an extremely homogeneous charge and are welded together at alternate ends (Fig. 46). The detonation of the charge provides an impulse of a magnitude exactly sufficient to project the rods radially outward, the welds at the ends being bent until a ring of rods is formed that does not break up until it attains about 80% of the theoretical maximum perimeter. The velocity of the ring and its relatively heavy rods confer a certain destructive effect, particularly against aircraft (Fig. 47). The explosive impulse is purposely limited to obtain the necessary homogeneity of the charge. As a result, the initial velocity of the rods is approximately 1000 to 1400 m/s, roughly half the initial velocity of the fragments of a fragmentation or multi-P warhead. The restricted effective area of the continuous rod warhead, perpendicular to the axis of the charge, also limits the hit probability against air targets approaching at various angles. There is also some doubt whether the structure of a modern aircraft could actually be seriously damaged by a hit from a continuous rod of this type.

As the radius increases, the ring, of course, disintegrates and the individual rods turn and tumble in various directions. They can, therefore, penetrate a target with a considerable transverse force. In general, however, the efficacy of a continuous rod circle after breakage is considered to be very small or even negligible.

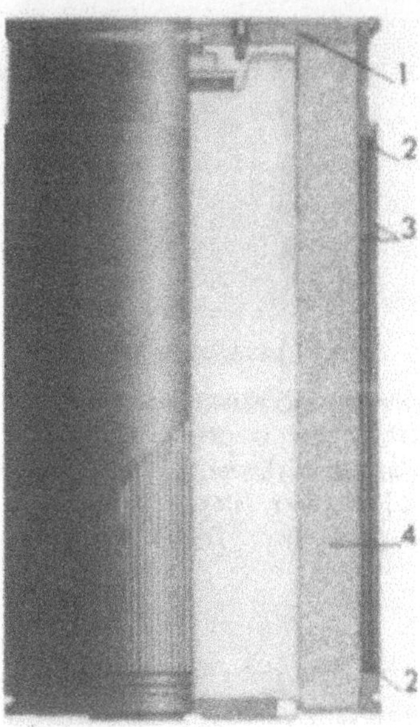

Fig. 46 Continuous rod warhead. The two layers of rods are positioned around a relatively thin layer of explosive, which provides a homogeneous accelerating impulse. Key: 1—ring initiation; 2—welds; 3—rods in layers; 4—high explosive charge.

Fig. 47 Effects of a continuous rod warhead: (upper photograph) rods of a continuous rod warhead pierce the semicircle of plates at left in photo, while the slightly more distant plates at right remain undamaged; (lower photo) Effect on target plates pierced by the rods while still linked together. (Reproduced by permission of the Direction des Constructions et Armes Navales, Toulon, France, 1974.)

9. Linear Shaped Charge or Polygon Warhead

Instead of having liners in the casing arranged symmetrically around the axis of rotation, the polygon warhead features a casing with angular corrugations around the circumference. These angular corrugations, the tips of which have an included angle of, for example, 60 or 90 degrees, are combined with relatively thin casing walls and act as linear cutting charges, the jet and slug being formed in the longitudinal plane, with the slugs in the shape of rods. If the tip angles are more open, the peak velocity of the jet decreases whereas that of the slug increases, until at an angle of 130 to 150 degrees both have practically equal velocity and constitute one single linear projectile (Fig. 48).[12]

Depending on their direction of impact and their cross-sectional loading, the rods possess various penetration capabilities and, if the impact is in a favorable position, they can successfully perforate thick plates. As a result of the restricted number of grooved corrugations, the hit probability of such warheads is limited. This type of warhead is, therefore, generally used against stronger and larger targets like tanks, amored personnel carriers, and supporting beams or pillars of bridges.

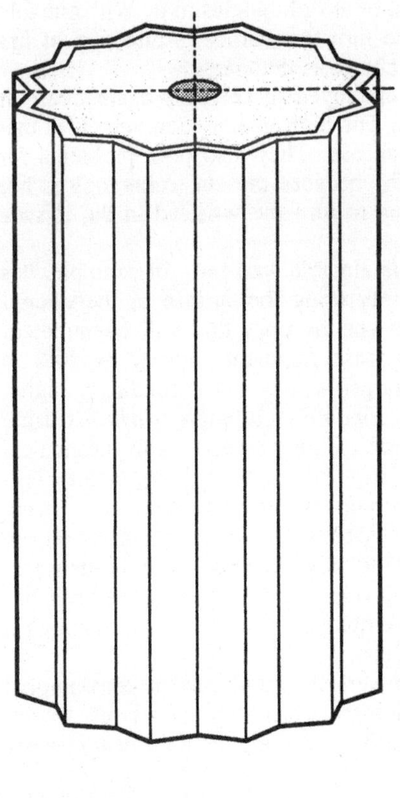

Fig. 48 Polygon warhead with a series of linear cutting charges designed to generate a number of linear projectiles.

10. Aimable Warheads

Warheads can differ in their fragment beam width (see Fig. 49).[14] The fragment group often has a very wide angle of spread, either to ensure that a larger area of the target is hit, or to allow for variations in the point at which the warhead is detonated. Nevertheless, many fragments from a warhead will not hit the target. The percentage that miss will depend on the distance between the warhead detonation point and the target, and on the size of the target itself.

To achieve greater cumulative effects, designers try to produce narrow fragment distributions or even parallel fragment paths. These will increase the cumulative and synergistic effects, at least within certain distances.

The narrowest fragment distribution is, of course, produced with the continuous rod warhead, in which suitably sized rods are joined together at their ends in zigzag fashion such that a hoop is formed when the charge is detonated. Use of this type of warhead is gradually being discontinued because of the hoop's limited speed of expansion and the fact that a *structure kill* can no longer be obtained on some areas of air targets (e.g., bomb bays that have been armored).

Higher hit densities (i.e., more fragments per unit of surface area) with less dependence on the distance between warhead and target can be achieved by using so-called aimable warheads (Fig. 50). These are not the same as *directional* or *aimed* warheads. A directional warhead has one defined direction of attack, be it along the longitudinal axis of the missile or at right angles to it. With aimable warheads, various devices are available to turn the warhead's direction of fire toward the target immediately before the charge is detonated.

Directional warheads can no longer be employed successfully against current highly maneuverable aircraft and missiles. The high interception velocities that are now possible also militate against such use. They also pose problems for aimable mechanically rearranged warheads, because present technology is not equal to the task of last-minute *fine tuning* to turn the warhead in the missile toward the target sufficiently rapidly.

The *detonation wave aiming* or *steerable* aimable warhead, for example, has in every quadrant a row of initiation points along the surface of the casing. The detonation points opposite the target are the ones that will be initiated. These shape the detonation front and increase fragment velocity by 20% to 35%. Also the energy brought onto the target will be correspondingly higher (Fig. 51). There are two further types of *aimable deformable warheads* using either mechanical or detonative deformation of the warhead. With mechanical deformation, the warheads are unfolded explosively toward the target. This type of aiming, also referred to as a quadrant opening warhead, takes many milliseconds, often too long for practical purposes. Detonative deformation, on the other hand, requires about one millisecond. Significantly higher hit densities can be achieved,[15] although the cost of developing this type of warhead has been shown to be high.[15] The technology employed is well known, although it requires adaptation for any particular warhead.

The author expects no significant improvements from current conventional fragmentation warheads,[15] since the technology is probably at its limit. Development of current warheads will take the form of adapting them for new missiles or for attacking new types of target.

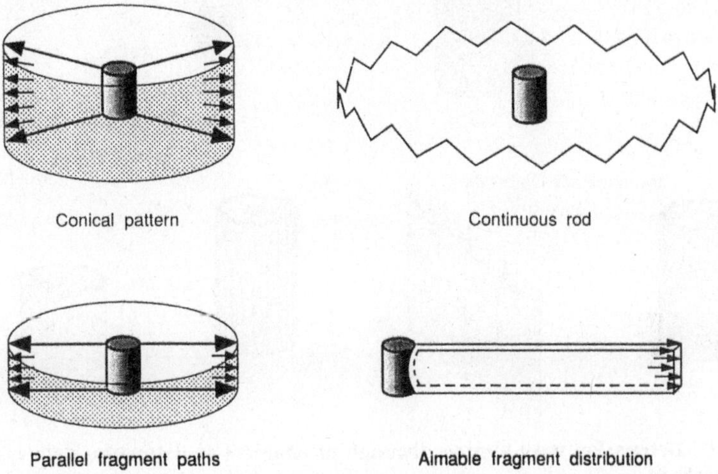

Fig. 49 Hit density as a function of fragment distribution pattern and standoff distance. With a conical pattern, the density n decreases as the square of the standoff distance R ($n \sim \Lambda/R^2$). With parallel fragment paths the decrease is linear up to the stand off distance ($n \sim \Lambda/R$). For an aimable warhead with exactly parallel fragment paths, hit density would not depend on standoff distance, that is, n would be constant.

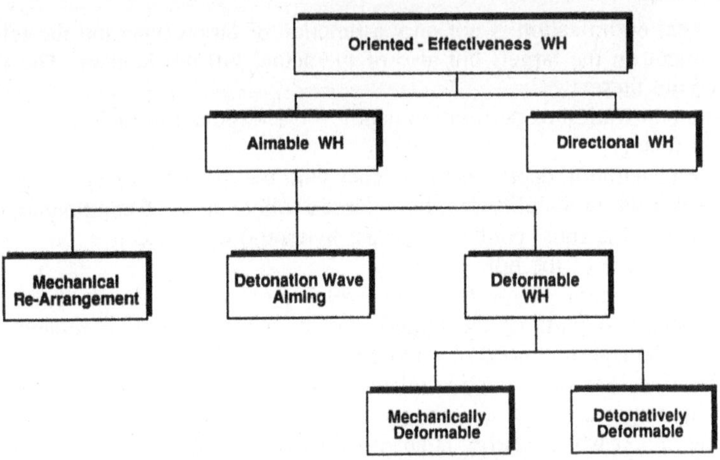

Fig. 50 Definition of directional and aimable warheads.

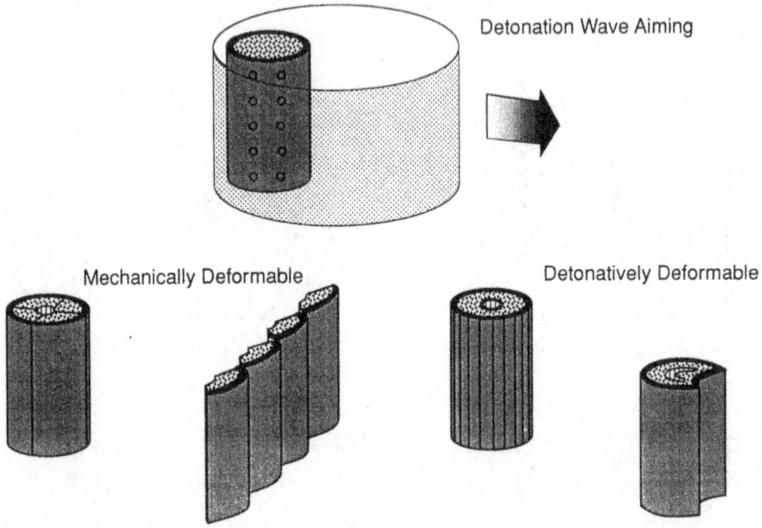

Fig. 51 Detonation wave aiming, through mechanical or detonative deformation of a warhead.

There is considerable potential, however, in aimable warheads, and there are no apparent technological limits to this area of development, which could lead to considerable progress in warhead technology to engage airborne targets.

11. Fragment Kill Mechanisms

11.1 Introduction

Fragment optimization is not only a function of target type and the effect of the fragment on the target, but also of the actual kill mechanism. The factors involved are these:

- The penetration or perforation depth of the fragment (which is regarded as the main factor)
- The penetration depth taken together with the diameter of the hole (which would affect the probability of a hit against a whole array of components in the target, such cable runs, control linkages, hydraulic lines, etc.)
- The volume of the hole produced
- The momentum of the fragment (which provides a measure of the deformation caused to parts of the target) and its energy (which is related to the volume of damage produced in a large target)
- The *structure kill* (which is related to the fragment energy and the square root of the target area that absorbs it)
- The rate at which energy is absorbed by the target

In general terms, destructive effect is produced by a number of fragments hitting the target and not just one. The preceding kill mechanisms must, therefore, be seen in connection with hit density, that is, the number of hits in a given area.

11.2 Cumulative or Synergistic Effects

11.2.1 General

To increase the probability of destroying a target, it is necessary to increase the energy acting on that target. Using larger and heavier fragmentation warheads will achieve this end, since the kinetic energy of each fragment is proportional to its weight and to the square of its speed.

The effect on the target can also be increased by achieving more hits on the target and also by cumulative or synergistic effects. A synergistic effect is one where the overall effect is greater than the sum of the individual effects.

When many fragments hit a target, especially at close distance (i.e., at high hit density and in rapid succession or simultaneously), the effects are very specific, but various effects may overlap one another. At present, there are no clear differentiations for these individual effects. They are very generally designated as *high density effect* or *multiple fragment effect* or *vapor* or *vaporific effect.*

Let us consider the effects of different loads on the targets and the related effects of fragments at a high hit density or fragments at a high hit density in rapid bursts. To do this, let us fix clearly the definitions for the different phenomena.

11.2.2 Additive Effects

A component is certainly more greatly weakened by many closely situated fragment perforations or penetrations, even when they arrive in succession, than by a wide spread of fragments at the target. It is obvious that the strength of a support is more substantially reduced, for example, by closely spaced penetrations than by a wider spread of target hits on the support structure. The increased effect of a large terminal density becomes clearly apparent in this case. It is unimportant for the remaining strength of this support whether the holes are made simultaneously or in succession. This specific weakening of a structure or component by fragments of a high terminal density is defined as an *additive effect*. This additive effect is to be the dominant concept for all further definitions to be presented as they relate to impacting targets with *fragment showers* (Fig. 52).

11.2.3 Cumulative Effects

Special effects are observed at the target not only because of this high fragment density, but also because the fragments hit and penetrate the target within a short period of time or simultaneously. The additional phenomena that occur because of this are defined as *cumulative effect*.

Compared to the pure additive effect, in the cumulative effect, a rapid time sequence is required in addition to a high fragment hit density. The mechanisms that can occur with a cumulative effect follow (Fig. 52).

11.2.3.1 Cumulative Mechanical Effect. Pressure and shock waves, possibly overlapping, are created by the impact of fragments on a structure, causing special damage due to the overlapping of these pressure and shock waves when

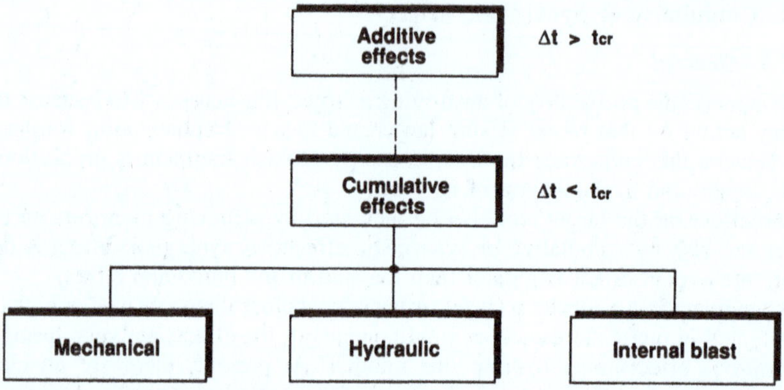

Fig. 52 Additive and cumulative effects. Cumulative effects occur when the time interval between hits is shorter than t_{cr} and include mechanical, hydraulic, and internal blast effects.

many fragments impact almost simultaneously. This mechanical effect can be observed, for example, when a large hole is knocked out of a target plate upon high hit density of fragments. Figure 53 is a flash x-ray exposure of the fragment shower at a distance of approximately 2 m from the place of detonation, together with the effect on a spaced plate arrangement of 1.5-mm thick mild steel plates 100 mm apart, positioned at a distance of 10 m. The photographs show the two spaced plates bent by the force of the fragments, with the middle completely punched away. The mechanical stress was apparently too great at the center of fragment hit density for the plates between the individual points of impact to withstand the load. The flash x-ray exposure proves that the fragments did not impact simultaneously on the target plates, but in short intervals. It is easy to imagine that more complex structures would be substantially damaged by such fragment densities because of the overlap of the considerable mechanical stresses.

11.2.3.2 Cumulative Hydraulic Effect. When a fragment enters a liquid, even at a residual velocity of only several hundred meters per second, it generates, in addition to the shock wave, a cavitation bubble with specific pressure effects on the housing, connection lines, and valves. In this case the diameter of the cavitation bubble is a multiple of the fragment diameter. This effect is multiplied when several fragments enter water or a tank filled with fuel simultaneously or in rapid succession.

11.2.3.3 Cumulative Internal Blast Effect. The cumulative internal blast effect is caused by the interaction of projectile and target material with material derived from either the target or the projectile. A differentiation therefore has to be made between these two possibilities.

In Fig. 54 a steel ball of 10-mm diameter is shown perforating a through 2-mm thick dural plate. The exposures are copies of every fifth individual frame of a film series, made at 10^6 frames per second with the aid of a rotating mirror framing camera. Time delay 0 μs shows the pellet coming from the left at 1500

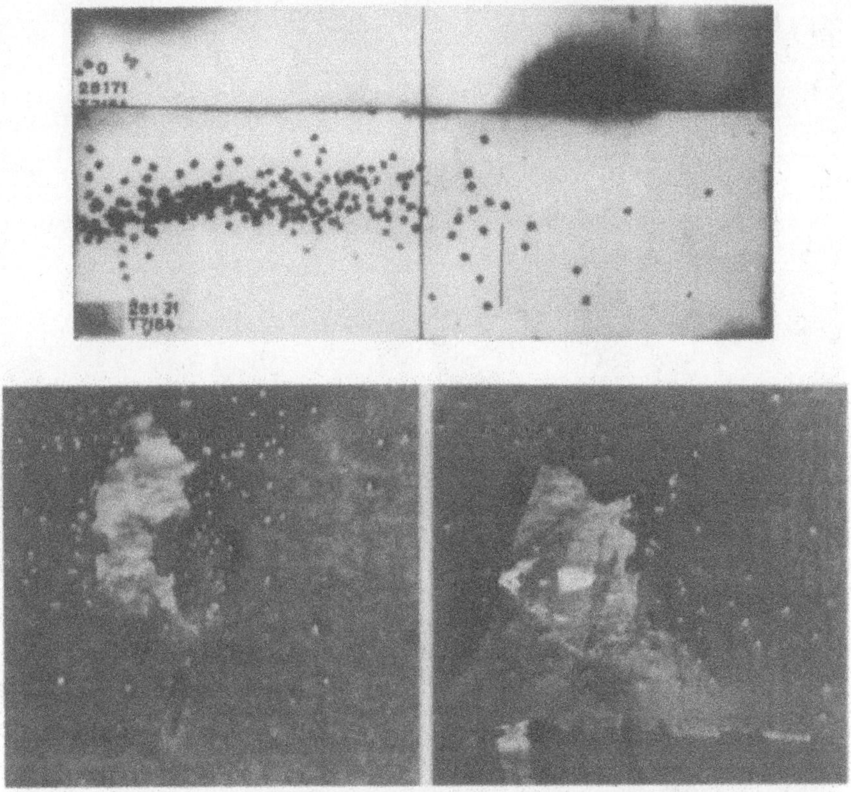

Fig. 53 Example of a mechanical cumulative effect. The upper photograph shows a flash x-ray image of a fragment shower. The lower ones show the effect of the shower of fragments on a target consisting of two plates of 4-mm aluminum alloy, spaced 100 mm apart. The first plate to be hit is on the left, the second is on the right.

m/s in front of the 2-mm thick dural plate. Time delay 5 μs already shows the beginning of the material cloud created by the *shaped charge effect* on the entrance side, spreading conically in the opposite direction to the projectile.

This material cloud has already spread further in time delay 10 μs. It must be assumed in this case that this cloud consists of fine particles from both the target and the projectile material and that it reacts with oxygen in the air. This reaction cloud makes the space in front of the target plate almost opaque in the further time delay against the argon flash bomb background lighting. In the time delay 25 μs to 55 μs the fragment, greatly deformed by the impact, is clearly visible together with a fragment cloud, which probably also consists of material from the target plate as well as from the fragment.

The material cloud stemming from the shaped charge effect, especially on the entry side, probably causes the smoky traces on the entry side, often focusing on alumninum target plates.

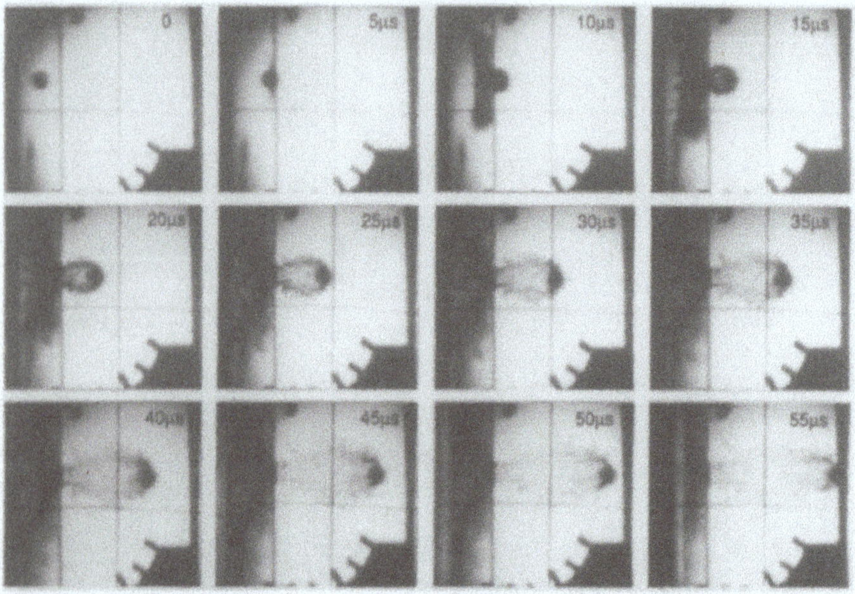

Fig. 54 Perforation of a 2-mm aluminum plate by a 10-mm steel sphere at 1500 m/s velocity. Special attention should be given to the cloud formation on the impact side.

11.3 Reaction Mechanisms

11.3.1 Reactions of the Target Material

When fragments penetrate different aluminum alloys at velocities generally above 1000 m/s, reactions of the aluminum surfaces exposed or of the aluminum vapor take place with the oxygen in the air behind the target material. The air near the penetration hole is heated by these reactions. In the case of flying targets or aircraft there is often a small volume in a more or less enclosed space. The heating of the air in this volume results in a *quasistatic overpressure*.

An individual fragment usually generates only a relatively small amount of heat, which could not appreciably increase pressure to destroy the components by means of this quasistatic inner pressure; however, if *many fragments* impact at a rapid rate or simultaneously at a high density before the pressure is reduced, the cumulative blast effect may occur, creating an inner pressure due to the multiple reaction heat of the fragment penetrations and damaging the components of the target. In this instance damaging means ripping the weakest point, in the case of an aircraft usually identical to ripping the outer hull. More target material is divided into fine fragments with increasing fragment velocity. The reaction with oxygen in the air is then substantially increased. High-velocity fragments are found in multi shaped charge warheads. The range of fragment impact velocities is 3–5 km/s. The target material is completely broken apart and divided into fine particles, especially at these high impact velocities, so that especially strong subsequent reactions occur in this case with oxygen in the air.

Shaped charge fragments do not impact simultaneously, but in such rapid succession (that a cumulative blast effect occurs).

11.3.2 Reaction of the Fragment Material with Oxygen in the Air

In the penetration of steel fragments through aluminum, the fragments are deformed and broken apart so that new free surfaces are created on the fragment material, which may react with oxygen in the air. Generally, however, steel has a relatively low reaction capacity, that is, it is not particularly pyrophoric. For this reason attempts are made at employing fragments with a high capacity for reaction with oxygen in the air, that is, they are extremely pyrophoric.

In addition preformed fragments can be produced for antiair warheads in such a manner that they burst into many small reactive fragments upon penetrating the outer covering. Fragment materials that do not break apart to any great extent and still exhibit good pyrophoric characteristics are, for example, zirconium or hafnium.

Fragments that are already breaking apart upon penetration even of thin plates may consist of metal mixtures, produced from appropriate materials by pressing at several tons per square centimeter. Such fragments, however, may only be accelerated to the velocity necessary for large penetration capacity under certain conditions by means of an explosive charge. On the other hand, these provide extremely high heating values after penetrating even thin bulkheads, since they disintegrate into many small fragments and react violently with the oxygen in the air, creating a considerable internal overpressure due to many fragment impacts in a closed volume, for example, in an aircraft.

11.3.3 Reactions of the Fragment Material with Itself

Various studies[13] are also being conducted on the production of fragments that are not dependent on ambient oxygen for reaction but that carry the reaction partner themselves. The reaction is not necessarily conducted with oxygen in this case, but it may also take place in intermetallic reactions, such as in the so-called coruscative materials, in which considerable heat energies may also be released.

11.4 Time Intervals

The time interval between hits that is required to produce cumulative and synergistic effects varies according to the medium in which they occur (see Fig. 55). Whereas cumulative internal blast or overpressure effects almost always occur, whatever the hit density, superimposed hydraulic shock wave effects require time intervals between hits that are relatively shorter. Cumulative mechanical effects in metals, for example, require fragments to impact on the target almost simultaneously.

If we assume that the distance between fragment impacts is 50 mm, and that the propagation speed of the shock wave in metal is 5 mm/μs, the critical time interval between fragment impacts will be about 10 μs. For a fluid in which the

- **Shock Wave in Metal**

  $\Delta t < 10\ \mu sec \qquad (50\ mm : 5\ \frac{mm}{\mu sec})$

- **Shock Wave in Liquids**

  $\Delta t < 100\ \mu sec \qquad (150\ mm : 5\ \frac{mm}{\mu sec})$

- **Internal Blast (vapourific effect)**

  $\Delta t < 1\ msec$

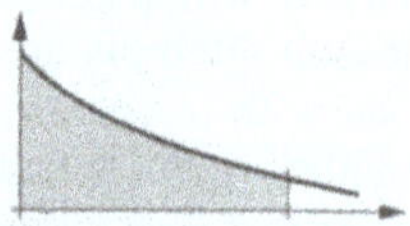

Fig. 55 Critical time intervals for cumulative effects.

shock wave propagates at 1.5 mm/μs, and the fragments impact 150 mm apart, the critical time between impacts to induce superimposed shock waves can be calculated as 100 μs. For fragments penetrating an enclosed space to induce a quasistatic overpressure, the critical time between impacts is about 1 ms.

12. Airborne Target Warheads

The air threat comes in a large variety of shapes, sizes, and speeds. The selection of warheads with which to cope is equally varied, and many of the types available have already been discussed in these pages.

There are essentially two ways of destroying an airborne target with a projectile—by a direct hit or by the standoff effect generated by a near miss.[17] Small-caliber projectiles and cannon shells are not generally equipped with proximity fuzes to produce near-miss effects. They depend on a direct hit. The high rate of fire of modern automatic weapons gives a good chance of a direct hit on an airborne target, but such weapons have limited range.

Guided missiles generally have a lower probability of gaining a direct hit—especially over the longer ranges for which they are designed. Command guidance and seeker accuracy are not generally sufficient for a direct hit at long range if the target engaged is highly maneuverable. The missile usually flies past the target at a short distance, and with a target the size of a ground-attack aircraft—20 to 30 m long with a wing span of between 5 and 10 m—missing by a few meters is almost the same as a direct hit.

With warheads that require a direct hit on the target, there is a school of thought that strongly believes that premature detonation of the warhead is to be avoided at all costs. However, if the proximity fuze in such a warhead does detonate before reaching the target, some sort of hit is still likely to be obtained. Parts of the missile such as the motor and guidance section might be expected to continue for several meters along roughly the same flight path. If such debris hits the air target, it will still inflict considerable damage. Given the greater probability of a near miss, however, the fact remains that an air target can only be engaged effectively if the energy stored in the warhead is released at the right time.

Quite generally, the source of energy is the chemical energy stored in the high-explosive charge. The particular performance of a high-explosive charge is not due to its extraordinarily high energy content, which is of the order of 5000 kJ/kg and, therefore, is only about one-tenth of that of gasoline or coal, which use the oxygen of the air for combustion. Instead, explosives and propellants have the oxygen necessary for the reaction (detonation) stored in a molecular form. The specific feature of high explosives is their extremely high reaction rate, their detonation velocity. This means that their *power* and gas production per unit time is greater by a factor of about 10^5 than any other quick chemical reaction.

The reason for this is that the reaction is set off and is maintained by a shock wave, namely, the detonation wave, which in high-energy military explosives propagates at 8000 to 9000 m/s instead of at only a few tenths of meters per second, as in the case of the combustion of rocket propellants.

The energy content of military high explosives for blast and fragment acceleration is today, in the author's opinion, largely extracted to the ultimate. In the near future there will be no explosives with a remarkably higher power content than RDX and HMX, which at present are the most widely used explosives.

On the other hand, research and development for so-called insensitive high explosives has made considerable progress. Today high explosives are available that resist much higher temperatures, namely, 200 to 300°C, than do RDX (about 140°C) or HMX (160–180°C).

In particular, the hitherto preferred *binding agent* TNT, which melts at about 80°C, is being replaced by plastic binders. Such explosives are now mainly used in filling the warheads of air-to-air missiles, which are likely to attain enhanced temperatures by aerodynamic heating. Moreover, insensitive explosive mixtures are being developed to resist bullet attack and fuel fire hazards.

All warheads for standoff effect are filled with a high-explosive charge. On detonation of the latter, the latent chemical energy is released producing—in the case of military high explosives—about 1000 liters of high-pressure high-temperature gas (3000—5000 K) per kilogram of explosive. Pressures are approximately 30 GPa. The gas attempts to expand, thus compressing the surrounding air that it pushes away. This effect results in what is known as a blast wave.

In general, the blast effect is considered to be easy to calculate. However, going into greater detail, one will discover the complexity of the interaction between a blast wave (which is characterized by peak overpressure p_{max}, positive phase duration t_+, and positive phase impulse I_+) and the target structure (which may consist of many components having different strengths and natural frequencies of vibration). For a rough estimate, a square-root law can be assumed for the distance D vs the HE weight w, that is,

$$D = K\sqrt{W} \tag{15}$$

where the value of K is in the range of 0.3 to 0.5 m/kg$^{1/2}$ depending on the type of air target. For smaller charges, however, the distance D from the air target must not be measured from a center of mass, but from a line or surface of reference (Fig. 56).

The high-pressure gases from a detonating high-explosive charge are capable of accelerating the casing in which they are enclosed, and to impart to it a high velocity that depends on the charge-to-casing mass ratio. A smooth casing will be broken up into so-called natural fragments, that is, into both small and large fragments. For a most efficient attack on an air target, however, a favorable size of fragment is wanted, which can be obtained by controlled fragmentation or by using preformed fragments. Fast and heavy fragments, though few in number, can be produced from multi-P warheads. Especially fast fragments can be obtained from the so-called multi-shaped charge warheads with tip velocities of up to 4–5 km/s.

A continuous, expanding rod can be obtained by a cleverly engineered arrangement of rods that surround the charge and that are welded together at their ends. Particularly high hit densities can be reached with aimable warheads (Fig. 57).

Whereas a blast wave continuously covers the entire space, thus having a quite limited range of action, a detonating high-explosive charge is able to bring a fragment casing to a high velocity so that the fragments transmit concentrated energy over long distances; however, this is done only along the discrete trajectories of the fragments between the warhead and the target (Fig. 58).

13. Fragment Acceleration

13.1 Acceleration of Individual Fragments

The acceleration of single fragments by a high-explosive charge is in certain respects completely different from the acceleration of fragment array or a closed casing or sheath on a charge. Yet, an examination of the acceleration of single fragments leads to interesting information on the interaction between a detonating high-explosive charge and the acceleration of fragments.

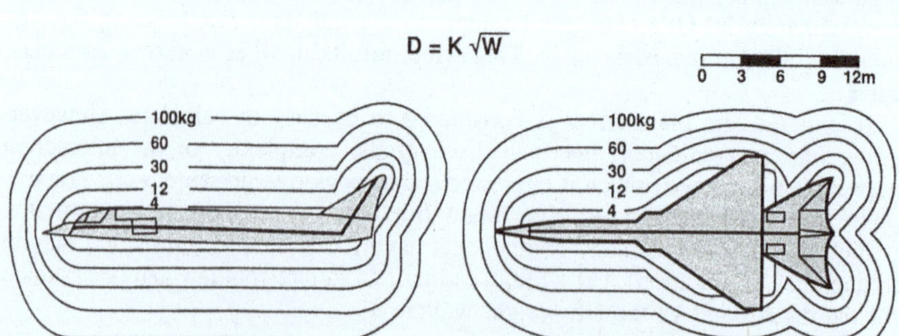

Fig. 56 Blast wave effect against an air target.

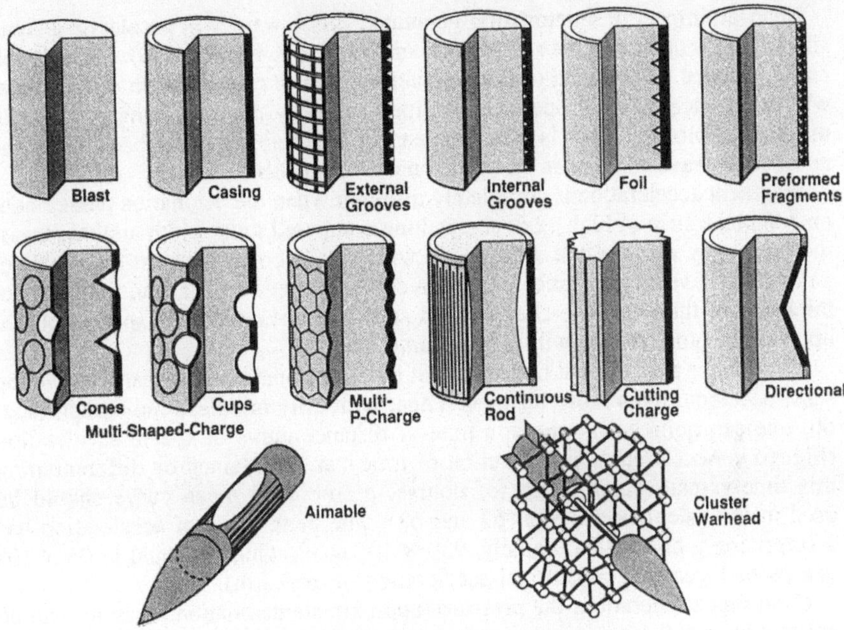

Fig. 57 Warheads against air targets.

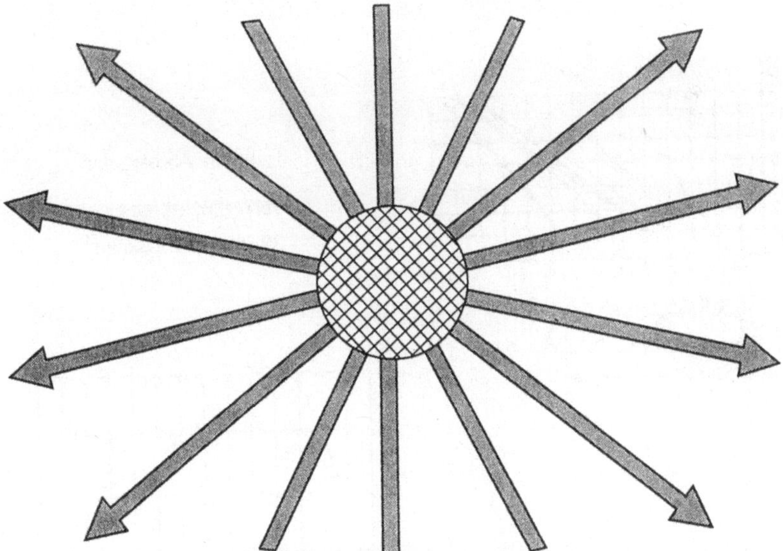

Fig. 58 Transportation of high energies by fragments along discrete trajectories to large distances, compared to the damage radius of a blast wave.

Head-on impact of a detonation-generated shock wave will accelerate a small single fragment in less than a microsecond (Fig. 59, from Ref. 3). After acceleration caused by the front of the detonation wave, a closed sheath of fragments will experience an additional acceleration caused by the expanding products of the high explosive. This is not the case for a single fragment because of the rarefaction wave running in from the charge front face.

Fragment acceleration is completely different when the detonation passes side-on.[3] When a strip of high explosive is long compared to its width and thickness, then one may assume that an impulse will have become constant after a detonation has traveled a distance about three to five times longer than the width or thickness of the explosive charge. This statement means that all spheres placed upon the explosive strip will get the same impulse.

Thus, those spheres that have been hit by the detonation wave earlier have the same acceleration as those hit later. Thus, with only one flash radiograph (Fig. 60) one can quite easily obtain a time-vs-distance curve of sphere acceleration (Fig. 61). Acceleration as a function of time can be obtained by differentiating this time-vs-path curve twice (of course, a smoothed mean curve should be used in the calculation) (Figs. 62 and 63). The peak value of acceleration has a surprisingly high value, namely, 930×10^6 m/s^2, which is equal to 93×10^6 gee (with 1 gee = gravitational acceleration of the earth).

From this acceleration, the pressure (approximate detonation pressure) can be calculated as follows:

$$p = \frac{F}{A} = \frac{m \cdot a}{A} \tag{16}$$

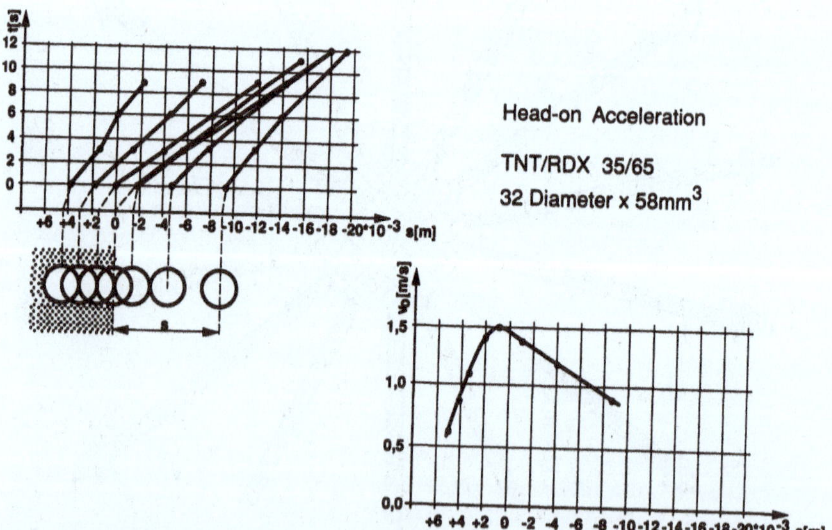

Fig. 59 Distance/time curves, or initial velocity v_o as a function of the depth of embedding of 3-mm steel spheres for head-on or perpendicular impact of the detonation wave.

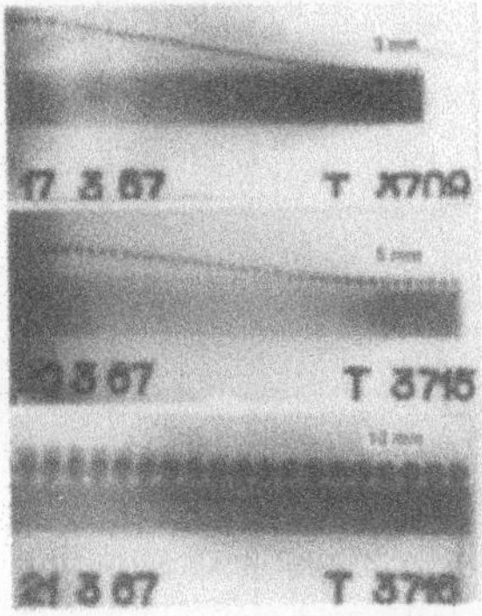

Fig. 60 X-ray flash radiographs of side-on detonation wave with 3-, 5- and 10-mm spheres. A diminished preliminary flash maintains the charge and the sphere on the same film prior to the detonation. Charge: 200-mm long, 50-mm wide, 20-mm high.

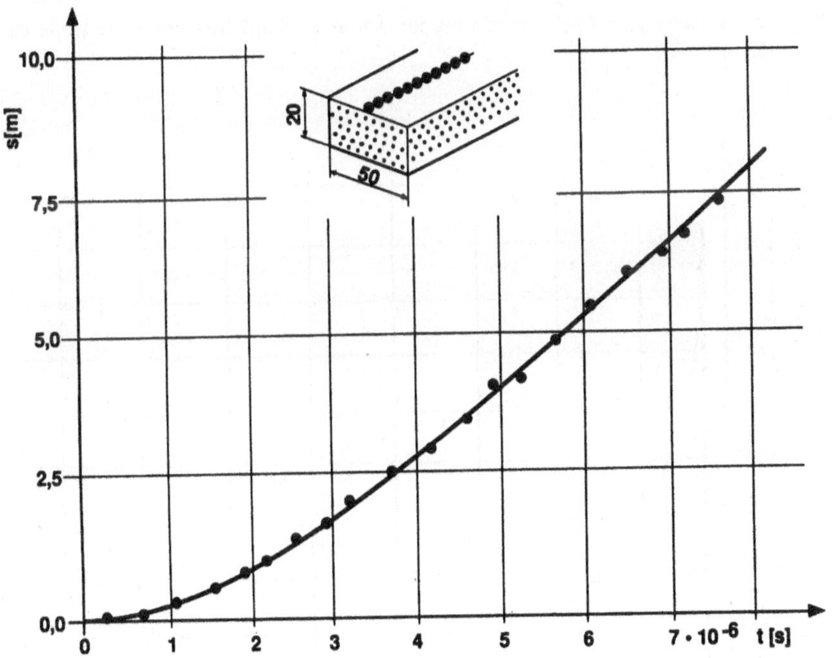

Fig. 61 Distance/time curves for 3-mm steel spheres placed on an explosive slab of 200-mm length × 50-mm wide × 20-mm thick for a side-on detonation wave.

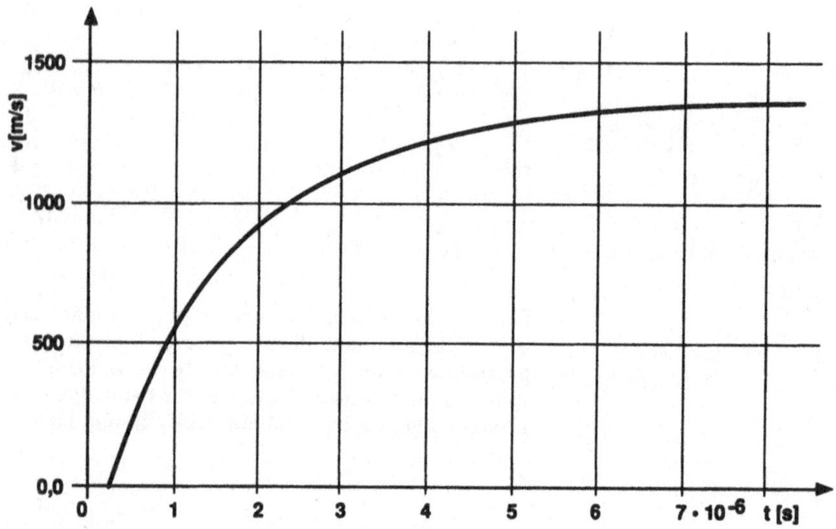

Fig. 62 Velocity as a function of time for 3-mm steel spheres (data the same as in Figs. 60 and 61).

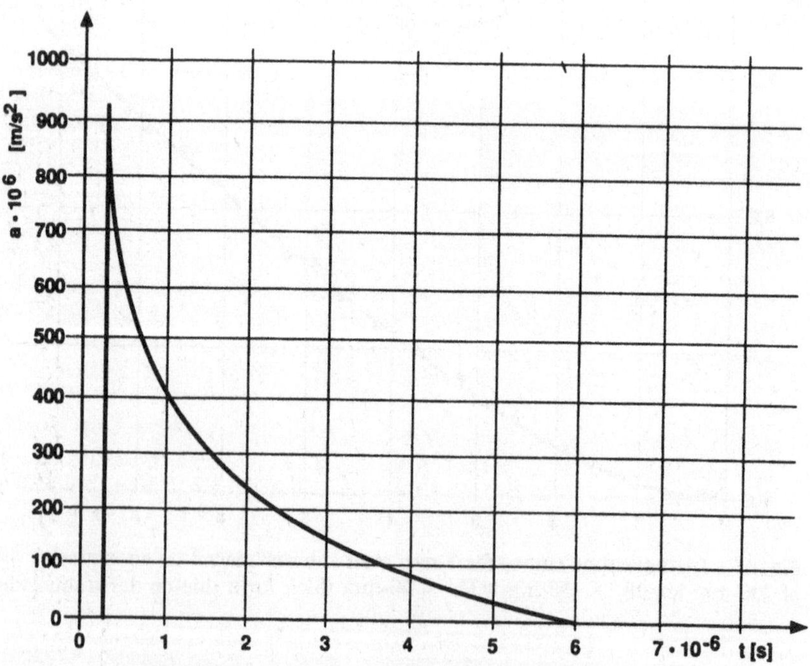

Fig. 63 Acceleration as a function of time (data the same as in Figs. 60–62).

where

p = pressure
F = force
A = area of the fragment
m = mass of the fragment
a = the acceleration measured

The ratio of m to A of a sphere with diameter D follows from

$$\frac{m}{A} = \frac{(D^3/6) \cdot \pi \cdot \rho_s}{\frac{D^2}{4} \cdot \pi} = \frac{2}{3} \cdot D \cdot \rho_s$$

whence

$$p = \frac{2}{3} \cdot D \cdot \rho_s \cdot a \tag{17}$$

The 3-mm steel spheres experienced an acceleration of 930×10^6 m/s, whence the pressure can be calculated

$$p = \frac{2}{3} \cdot 3 \cdot 10^{-3} \cdot 7.85 \cdot 10^3 \cdot 930 \cdot 10^6 (\mathrm{m} \cdot \frac{\mathrm{kg}}{\mathrm{m}^3} \cdot \frac{\mathrm{m}}{\mathrm{s}^2}) \tag{18}$$

A pressure of 14.6 GPa or 146 kbar is of the order of magnitude that corresponds to the detonation pressure of this type of high explosive.

The force that accelerates the sphere at the first moment follows from

$$F = ma$$

$$F = \frac{D^3}{6} \cdot \rho_S \cdot \pi \cdot a \tag{19}$$

The force F acting on the 3-mm steel sphere is thus

$$F = \frac{(3 \cdot 10^{-3})^3}{6} \cdot 7.85 \cdot 10^3 \cdot \pi \cdot 930 \cdot 10^6 (\mathrm{m}^3 \cdot \frac{\mathrm{kg}}{\mathrm{m}^3} \cdot \frac{\mathrm{m}}{\mathrm{s}^2}) \tag{20}$$

$$F = 103,208 \ N = 10.5 \text{ metric tons}$$

Thus, a 3-mm sphere will, at the first moment, be pushed off the surface with a force of 10 tons.

13.2 Acceleration of Casings

First, the fragments are accelerated by the shock wave that is induced in the casing, which entails a material velocity of the casing. Second, they suffer an after acceleration by shock reverberations and by the expanding reaction gasses of the detonation (Fig. 64).

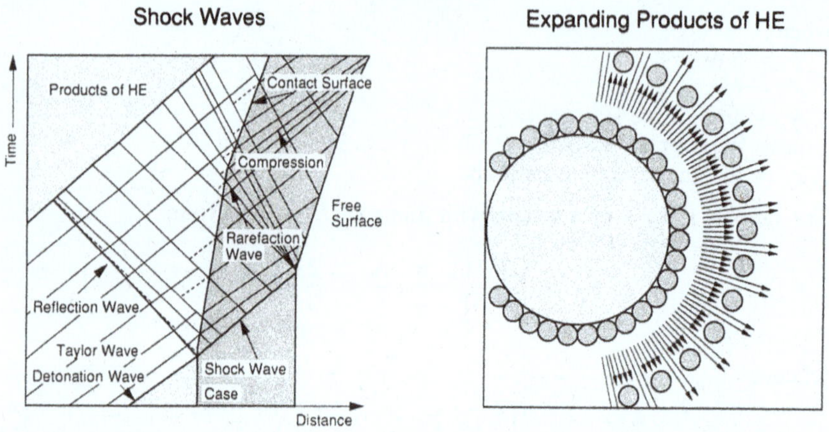

Fig. 64 Fragment acceleration by shock waves and by expanding products of high explosives.

A good engineering approach for calculating the initial velocity of fragments is by the Gurney formulas for different geometries. Without doubt, the Gurney formula for a cylindrical charge gives a rough prediction of the velocities of real warhead fragments—apart from the ratio of the mass of the metal casing m to that of the high-explosive charge c and the so-called Gurney constant $\sqrt{2e}$. But the following parameters have to be considered:

- Ratio of length to diameter of the warhead
- Mass of the end plates
- Types of fragmentation (preformed fragments are much less effectively accelerated than controlled fragments or natural fragments)

Once a warhead has been fired and the final velocity distribution of the fragments' along the warhead axis has been established, then the prediction of the velocity distribution of similar warheads is for engineering purposes, fairly precise if the diameter or casing thickness and so on are not too different. At this point these general remarks should be sufficient.

The Gurney equations have been described in detail in Chapter 4. However, a brief derivation will be given of a formula after Lukanow and Molitz,[16] which had been established independently of Gurney and which includes the use of material data on the casing.

Lukanow and Molitz[16] used a more general energy equation. The kinetic energy of the casing m and of a certain fraction ϵ of the high-explosive charge c plus the work $f \cdot m$ done in breaking up the casing into fragments, is equal to the chemical detonation energy $e \cdot c$ of the high-explosive charge multiplied by a certain efficiency factor η. Thus, the equation is written as

$$\underbrace{\tfrac{1}{2} \cdot (m + \epsilon c) v_o^2}_{\text{kinetic energy portion}} + \underbrace{f \cdot m}_{\text{work for deformation and breakup}} = \underbrace{\eta \cdot e^1 \cdot c}_{\text{effective used explosive energy}} \qquad (21)$$

where

v_o = initial velocity of the casing
m = weight of casing
c = charge weight
f = specific work for deformation and breakup of casing
e^1 = specific detonation energy of the high explosive
ϵ = fraction of reaction products (scaled to v_o)
η = HE-charge efficiency

The formula can also be written as

$$v_o^2 = \frac{2 \cdot (\eta e^1 c - am)}{m + \epsilon c}$$

$$v_o = \sqrt{2 \cdot \frac{(\eta e^1 - am/c)}{m/c + \epsilon}} \tag{22}$$

The authors found good agreement with their tests[16] for values of $\epsilon = 0.75$ and $a = 0$ for brittle materials (glass) and $a = 0.151$ J/kg for ductile steel. The experimental tests were done with large m/c ratios, which do not necessarily apply to military warheads.

If we make the following assumptions,

$$a = 0$$

$$\epsilon = 0.5$$

$$\eta \cdot e^1 = e$$

one obtains this expression from Eq. (22)

$$v_o = \sqrt{2e}\sqrt{\frac{1}{m/c + 0.5}} \tag{23}$$

which is precisely the Gurney equation for cylindrical warheads.

14. Aerodynamic Considerations for Fragments

This section is concerned with the behavior of a fragment during its flight from the point of acceleration up to its impact upon the target. Individual fragments are decelerated through the air according to the well-known laws of air resistance. The first phase of the fragment movement, caused by the detonation acceleration while the fragments are still within the cloud of gaseous combustion products, will be ignored. On the one hand very little is known about it and on the other hand this distance, depending on the charge size, is generally small.

The force F acting upon a moving body in air is

$$F = \frac{1}{2} c_w \rho_{\text{air}} A v^2 \tag{24}$$

c_w = drag coefficient, a dimensionless number
ρ_{air} = air density
A = cross-sectional area of the body at right angles to the flight direction
v = relative of the body velocity to the surrounding air

If a force F acts upon a body, it will be either accelerated or decelerated (here the latter) so that the following equation applies:

$$F = -ma \tag{25}$$

Further,

$$ma = m\frac{dv}{dt} = m\frac{dv}{dx}\frac{dx}{dt} = m\frac{dv}{dx}v \tag{26}$$

so that the two equations can be combined

$$m\frac{dv}{dx}v = -\frac{1}{2}c_w \rho_a A v^2 \tag{27}$$

and after separation of the variables

$$\frac{dv}{v} = -\frac{c_w \rho_a A}{2m}dx \tag{28}$$

The initial conditions for the integration of this equation are at $x = 0$, $v = v_o$ and at $x = x$, $v = v$, so that

$$\int_{v_o}^{v} \frac{dv}{v} = \int_{o}^{x} \frac{c_w \rho_a A}{2m}dx \tag{29}$$

After integration, this equation becomes

$$\ln \frac{v}{v_o} = \frac{c_w \rho_a A}{2m}x \tag{30}$$

or

$$v = v_o e^{-\frac{c_w \rho_a A}{2m}x} \tag{31}$$

For spheres, A is always $D^2\pi/4$; for non-preformed fragments, that is, originating from the blowing up of a casing, A is variable because the fragment is constantly revolving and A can vary between A_{min} and A_{max}. For the special case of a sphere of diameter D and density ρ_m,

$$A = \frac{D^2\pi}{4} \tag{32}$$

$$m = \rho_m \frac{D^3 \pi}{6} \tag{33}$$

and

$$\frac{A}{m} = \frac{3}{2\rho_m D} \tag{34}$$

so that the Eq. (31) for v is written

$$v = v_o e^{\frac{0.75 c_w \rho_a}{\rho_m} \cdot \frac{x}{D}} \tag{35}$$

For steel spheres the following equation is obtained if $\rho_{\text{air}} = 1.28\ \text{kg/m}^3$ is introduced:

$$v = v_o e^{-c_w 1.22 \times 10^{-4} \frac{x}{D}} \tag{36}$$

From the *explosively deformed steel sphere* tests, c_w came out to 0.97,[3] and so the equation for steel spheres is

$$v = v_o e^{-1.18 \times 10^{-4} \frac{x}{D}} \tag{37}$$

Thus, Eq. (35) states that, say, a 1-mm steel sphere has only about 1/3 (precisely, 0.368) of its initial velocity v_o after 9.5 m.

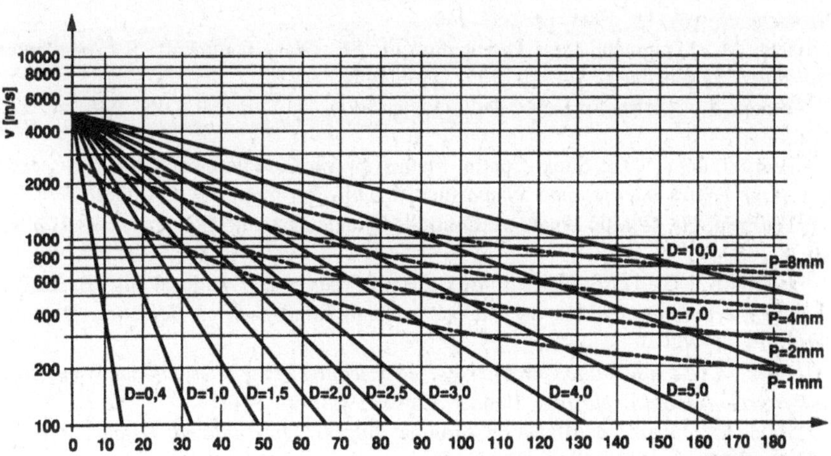

Fig. 65 Decrease in velocity v of steel spheres of diameter D (mm) as a function of the distance x. The decrease in velocity can be read from the diagram for any initial velocity at any distance interval. The dotted lines show roughly the penetration depth P into aluminum alloy as a function of the initial velocity.

In Fig. 65, Eq. (37) for steel spheres is shown as full-line graphs. Using a log/linear representation, straight lines are obtained for the different sphere diameters. The origin, with $v = 5000$ m/s, is chosen arbitrarily. We can start with any initial velocity and choose the path difference so that the corresponding velocity decrease along this distance is obtained.

The broken lines indicate the penetration depths P of steel spheres into dural plates. A higher velocity is required for a small sphere than with a larger sphere in order to obtain a penetration depth of, say, 2 mm.

References

[1]Held, M., "Air Target Warheads," *International Defense Review*, Vol. 8, 1975, pp. 719–724.

[2]Heiser, R., "Gesetze für die Massen- und Formverteilung von naturlichen Splittern," *Handbuch der Munitionsbewertung*, BWB, WM VI 2, 1979, pp. 203–312.

[3]Held, M., "Splitterballistik," *Explosivstoffe*, Vol. 15, 1967, pp. 265–274; Vol. 16, 1968, pp. 73–78. These translations are available: "Fragmentation Ballistics," Atomic Weapon Research Establishment (AWRE) Translation No. 64, UK, May 1972. "Ballistics of Projectile Fragments," translated by R. F. Brinkley, Bureau of Mines, Pittsburgh, PA, Oct. 1968.

[4]Held, M., "Berechnung der Splittermassenverteilung von Splittermunition," *Explosivstoffe*, Vol. 16, 1968, pp. 241–244.

[5]Held, M., "Consideration of the Mass Distribution of Fragments by Natural Fragmentation in Combination with Preformed Fragments," *Propellants and Explosives*, 1979, vol 1, pp. 20–23.

[6]Oerlikon Taschenbuch, Para. 6.2.1.3, "Splittermassenverteilung," 1981, pp. 210–216.

[7]Held, M., "Fragment Mass Distribution of HE Projectiles," *Propellants, Explosives, Pyrotechnics*, Vol. 15, 1990, pp. 254–260.

[8]Held, M., "Fragment Mass Distribution of 'Secondary Fragments,'" *Propellants, Explosives, Pyrotechnics*, Vol. 16, 1991, pp. 21–26.

[9]Pearson, J., "Means for Controlled Fragmentation," U.S. Patent 4 068 590, Jan. 17, 1978.

[10]Pearson, John, "The Shear-Control Method of Warhead Fragmentation," *Proceedings of the Fourth International Symposium on Ballistics*, Monterey, CA, Oct. 1978.

[11]Fountain, G., "Plastic Liners for Controlled Fragmentation," U.S. Patent 3 491 694, Jan. 27, 1970.

[12]Held, M., "The Performance of the Different Types of Conventional High Explosive Charges," *Proceedings of the Second International Symposium on Ballistics*, Daytona Beach, FL, 1976.

[13]Held, M., "A Study into Coruscatives," *Proceedings of the International Symposium on Pyrotechnics and Explosives*, Beijing, China, 1987, pp. 34–41.

[14]Held, M., "Warhead Effects on Airborne Targets," *International Defense Review*, Vol. 19, 1986, pp. 343–347.

[15]Held, M., "Aimable Fragmenting Warhead," *Proceedings of Thirteenth International Symposium on Ballistics*, Stockholm, Sweden, 1992, Vol. 2, pp. 539–548.

[16]Lukanow, H., und Molitz, H., "Über die Zerlegung von Sprenggeschossen," *Wehrtechnische Monatshefte*, 1959, Vol. 56, pp. 334–343.

[17]Ball, R. E., *The Fundamentals of Aircraft Combat Survivability Analysis and Design*, AIAA Education Series, AIAA, New York, 1985.

Chapter 9
Part A

Target Interaction: Fundamentals of Penetration Mechanics—Jets and Rods

David Chi*
Aerojet General Corporation, Azusa, California 91702

Introduction

The primary purpose of the tactical warheads discussed in this book is to penetrate or damage a specific target, usually a vehicle or structure of some type. The warhead causes damage to the target by depositing a large amount of kinetic energy over a relatively small area by means of a high-speed penetration, typically a jet or rod of metal. This high-speed collision causes interaction pressures between the target and penetrator that result in stresses that exceed the strength of those materials. For metallic targets, the material flows plastically away from the impact area causing a cavity to be formed. The penetrator is also eroded by plastic flow during this process. The target cavity continues to grow in depth until the penetrator is totally consumed or the target structure is perforated. This chapter presents the fundamentals of penetration mechanics for stretching jets and constant-velocity rods.

Obviously, the penetration mechanics of a directed energy warhead depends on the characteristics of penetrators formed by the warhead. Although clear distinction is not possible, directed energy warheads may be conveniently divided into shaped charge, hemispherical charge, and explosively formed penetrators (EFPs).

A shaped charge warhead forms a metallic jet from a conical, trumpet, or tulip liner. The liner material collapses on the axis of the charge to form the jet that constitutes about 15% of the liner mass. The remainder of the liner mass forms the slower-moving slug. The velocity typically varies between 10 and 2 km/s within the jet from tip to tail. The jet stretches during its flight to the target because of its velocity gradient until it eventually breaks axially into

*Electronic Systems Division, currently Manager of Design Analysis Group, SPARTA, Inc., Brea, CA 92621.

a number of particles. The broken jet particles drift away from the shot line, and tumbling moment is imparted to these particles at breakup. The penetration of the jet is reduced significantly by the lateral dispersion due to the drift and the rotational motion induced by the tumbling moment.

An explosively formed penetrator warhead, on the other hand, forms a constant-velocity long-rod penetrator from a disk-shaped metal liner. Nearly all of the liner mass remains in the EFP. Characteristic velocities of these projectiles vary between 1.5 and 3 km/s. The velocity does not vary significantly along the EFP, and consequently classical EFPs remain in one piece. Therefore, EFPs can be used over very long standoff distances of 150 m or more.

A hemispherical warhead forms a jet from a hemispherical liner. The jet formation process is significantly different from that of the shaped charge or EFP warheads. The tip velocity the hemispherical charge jet is about 5 or 6 km/s compared to 10 km/s for the shaped charge jets. This smaller velocity gradient within the jet results in a larger breakup time, making the hemispherical jet more efficient at long standoffs than the shaped charge jet. However, since jets from hemispherical warheads exhibit characteristics similar to the shaped charge after formation, the same approach can be used to model the penetration for both the hemispherical and shaped charge jets.

During the past 40 years, much effort has been made to develop analytical models for the prediction of the penetration of jets and rods. During World War II, an analytical model of the penetration of a constant-velocity jet or rod based on one-dimensional, incompressible hydrodynamic theory was developed.[1] Since then, the hydrodynamic model has evolved along two lines: nonuniform velocity, or stretching, jets for shaped charges and constant velocity rods for EFPs. An analytical approach has been followed for the modeling of the nonuniform velocity jets. On the other hand, for the constant-velocity rods, empirical models were developed first and analytical models were developed later. Penetration mechanics models for these types of penetrators will be presented in the following two sections. Examples will also be discussed.

2. Shaped Charge Jets

2.1 Hydrodynamic Theory

Early work of Birkhoff et al.[1] described an analytical model of jet penetration. Assuming that both the target and the jet behave as incompressible fluids, they proposed the simple Bernoulli relation for the jet penetration depicted in Fig. 1:

$$\frac{1}{2}\rho_j(V-U)^2 = \frac{1}{2}\rho_t U^2 \tag{1}$$

where V is the jet velocity, U is the penetration velocity, and ρ_j and ρ_t denote the densities of the jet and target, respectively. Equation (1) describes the penetration process with reasonable accuracy for the penetration of very-high-velocity jets into metallic targets, that is, for velocities yielding a flow process in which strength becomes a negligible factor. For jets on homogeneous steel armor, this is typically a velocity of 4000 m/s or greater. Most of subsequent analytical models are forms of Eq. (1) modified to include such effects as segmentation

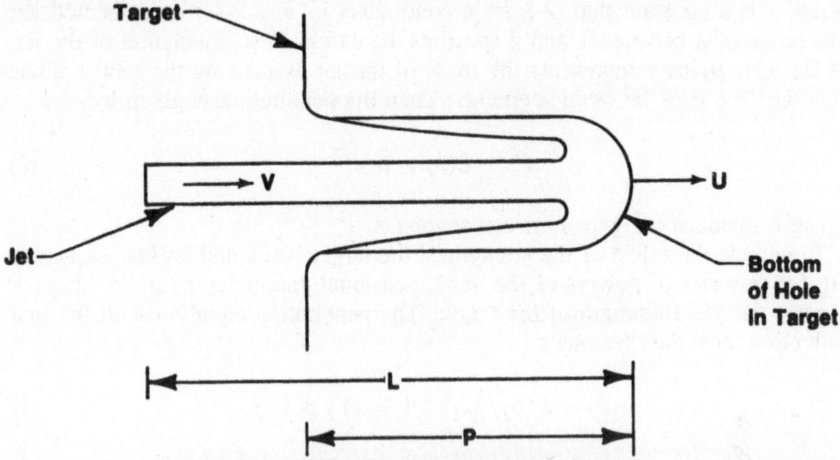

Fig. 1 Shaped charge jet penetration.

of the jet and the strengths of the target and the jet. Solving Eq. (1) for the penetration velocity, that is, the velocity with which the bottom of the hole moves during the penetration process, gives

$$U = V/\left(1 + \sqrt{\rho_t/\rho_j}\right) \tag{2}$$

Then the penetration P as a function of time t is given by

$$P(t) = \int_0^t U\,\mathrm{d}t \tag{3}$$

For the jet of constant velocity V and constant length L, assuming that the penetration stops when the jet length is consumed, that is, there is no afterflow effect, the duration of the penetration is $L/(V - U)$. The penetration is then given by

$$P = L(\rho_j/\rho_t)^{1/2} \tag{4}$$

This is usually referred to as the *density law*. It is interesting to note that the total penetration for a constant-velocity jet depends only on the length of the jet and the density ratio of the jet and the target. The depth of penetration by a given jet varies inversely with the square root of the density of the target. The lack of dependence on the jet velocity holds for velocities great enough to produce pressures far above the yield strength of the target.

Pack and Evans[2] extended Eq. (1) to include segmented jets. The penetration by a segmented jet is then governed by

$$\lambda\rho_j(V - U)^2 = \rho_t U^2 \tag{5}$$

where λ is a constant that is 1 for a continuous jet and 2 for a segmented jet. The range of λ between 1 and 2 specifies the extent of segmentation of the jet. In Eq. (5), ρ_j now represents the mass of the jet divided by the total volume including the gaps between segments. Then the penetration is given by

$$P = L(\lambda \rho_j / \rho_t)^{1/2} \tag{6}$$

where L includes the gaps between segments.

To include the effect of the strength of the target, Pack and Evans[2] expanded Eq. (4) in series of powers of the nondimensional parameter $Y/\rho_j V^2$ where Y denotes the yield strength of the target. The penetration equation with the first correction term then becomes

$$P = L(\rho_j / \rho_t)^{1/2}(1 - \alpha Y / \rho_j V^2) \tag{7}$$

where α is a parameter that is a function of the densities of the jet and the target.

From experimental measurements of the jet velocity and penetration velocity as a function of the depth of penetration, Eichelberger[3] concluded that though the hydrodynamic formula, Eq. (5), is accurate for early stages of the penetration process, the strengths of the target and jet play an important role in later stages of the penetration when the jet velocity falls to a relatively low value. The strength effects are accounted for by adding a strength term to Eq. (5):

$$\lambda \rho_j (V - U)^2 = \rho_t U^2 + 2\sigma \tag{8}$$

where σ is the difference between σ_t and σ_j, which represent the resistance to plastic deformation of the target and the jet, respectively, usually taken as one to three times the yield strength of the target and jet.

2.2 Nonuniform Velocity Jets

For nonuniform velocity jets, though the penetration can be expressed by Eq. (3), an explicit formula is not readily obtainable because the penetration velocity is a function of time. Abrahamson and Goodier[4] derived an explicit formula for the penetration of a continuous, stretching jet of nonuniform velocity. However, jet length is assumed to be known at a certain distance from the target. Since the relation between the jet length and distance from the charge is not given, their results are not suitable for direct application to shaped charge performance estimates.

Allison and Vitali[5] made the basic theory practical by assuming the existence of a virtual origin. The virtual origin approximation is derived from the observation that the velocity of a given jet particle remains fairly constant and the spatial distribution of velocity is linear. The virtual origin is a mathematical convenience that defines the point source of the entire jet at which the jet length is zero. The accuracy of the virtual origin approximation in representing a variety of shaped charge jets was reviewed in detail by Carleone et al.[6] Neglecting the strengths of the jet and target materials, they obtained the penetration for a continuous jet as a function of time as

$$P(t) = V_0 t(t_0/t)^{\gamma/(1+\gamma)} - s \tag{9}$$

where $\gamma = (\rho_t/\rho_j)^{1/2}$, V_0 is the tip velocity of the jet, t_0 is the time for the tip to travel from the virtual origin to the target, and s is the distance from the virtual origin to the target surface. When the jet breaks up at time t_1, the penetration process changes from one involving a continuous jet to one involving a jet of discrete particles. Assuming that breakup occurs simultaneously throughout the remaining jet and that each particle of a broken jet behaves as a short steady-state jet, the total penetration P_T is

$$P_T = P(t_1) + [s + P(t_1) - t_1 V_{\min}]/\gamma \tag{10}$$

where $V_{\min}$ is the minimum jet velocity capable of penetrating the target material and $P(t_1)$ is the penetration at time t_1, given by Eq. (9).

DiPersio and Simon[7] presented explicit formulas, based on Allison and Vitali's theory, for three cases: 1) penetration completed before jet breakup, 2) jet breakup occurring during penetration, and 3) jet breakup occurring before the start of penetration. For case 1, the total penetration is

$$P_T = s\left[(V_0/V_{\min})^{1/\gamma} - 1\right] \tag{11}$$

For case 2,

$$P_T = \{[(1+\gamma)(V_0 t_1)^{1/(1+\gamma)} s^{\gamma/(1+\gamma)} - V_{\min} t_1]/\gamma\} - s \tag{12}$$

For case 3,

$$P_T = (V_0 - V_{\min}) t_1/\gamma \tag{13}$$

These penetration equations based on $V_{\min}$ provide useful practical formulas, but the physical phenomena that cause the decrease in jet penetration with standoff are not directly modeled in the formulas. These phenomena, which include jet segmentation and wavering and tumbling of segmented jet particles, have been the focus of several recent studies.

An approximate method for determining the decrease in penetration with standoff was introduced by Carleone et al.[8] in a one-dimensional jet formation code, DESC. The breakup time of each portion of the jet is calculated using the theory of Chou and Carleone.[9,10] If a jet particle reaches the target before its computed breakup time, its penetration is calculated from Eq. (1) for a continuous jet. If the jet particle has broken, however, its actual increment of penetration dP' is reduced from the continuous-jet penetration dP' by

$$dP' = dP(1 - g/g_0) \tag{14}$$

where g is the distance between jet particles, or gap, nondimensionalized with respect to the jet particle length, and g_0 is an empirical constant determined by fitting to experimental data.

Observation that the penetration process of a jet is stochastic in nature led Smith[11] to apply the Monte Carlo approach to breakup, tumbling, and wavering of the jet particles in a computer code, PENSO. The parameters for these processes are assumed to follow prescribed probability distributions, and random values of the parameters consistent with these distributions are assigned to each jet particle. Proper selection of the statistical parameters such as mean and standard deviations describing the distribution of tumbling rates, dispersion angle, and breakup time yields good agreement with measured penetrations. However, the large number of Monte Carlo iterations involved makes the computation time high.

In addition to applying the Monte Carlo algorithm to the drift velocity and tumbling rate, Chi et al.[12] assumed that the tumbling of the broken jet particles follows the equation of motion based on an aerodynamic model for cylindrical rods. They found this aerodynamic effect on the particles to be very important in providing a consistent penetration model for the entire range of shaped charge standoffs. Given an initial angular velocity, the motion of the particle is governed by the equation derived from the conservation of angular momentum

$$\ddot{\theta} = \beta \sin^2 \theta \cos \theta \tag{15}$$

where θ is the angle of rotation counterclockwise from the flight axis, and

$$\beta = C_D \frac{1}{4} \rho_{\text{air}} V^2 L^2 D / I_y \tag{16}$$

where C_D is the drag coefficient at 90 deg crossflow, ρ_{air} is the density of air, and D is the particle diameter. In Eq. (16), I_y is the transverse moment of inertia of a cylindrical rod given as

$$I_y = m\left[(D^2/16) + (L^2/12)\right] \tag{17}$$

where m denotes the mass of the particle. The initial conditions at breakup are

$$\theta = 0 \tag{18}$$

$$\dot{\theta} = \dot{\theta}_i \tag{19}$$

Since this second-order, nonlinear differential equation has no available closed-form solution, it is solved numerically.

The model computes the hole profile in an incremental fashion whereby jet particles contribute sequentially as they arrive at the hole. Some jet particles may arrive at the bottom of the hole resulting in increased penetration depth. Others may impact at an intermediate hole depth because of the random drift velocity. These two situations are illustrated in Fig. 2.

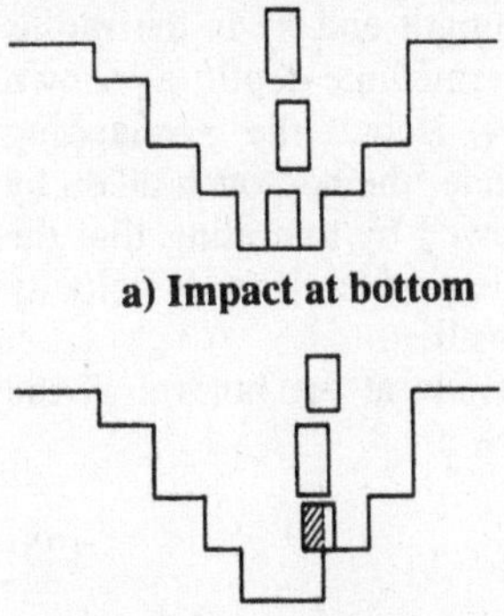

Fig. 2 Impact of jet particles at the hole profile.

The penetration increment, ΔP, contributed by a jet particle of length ΔL is given by

$$\Delta P = U\,[\Delta L/(V-U)] \tag{20}$$

where V is taken as the local velocity of each element. From Eq. (8) with $\lambda = 1$, U is given as

$$U = \frac{1}{1-(\rho_t/\rho_j)}\left\{V - \sqrt{\frac{\rho_t}{\rho_j}\left[V^2 + \frac{2\sigma}{\rho_t}\left(1-\frac{\rho_t}{\rho_j}\right)\right]}\right\} \tag{21}$$

When $\rho_j = \rho_t$, Eq. (21) becomes

$$U = (V/2) - (\sigma/\rho_j V) \tag{22}$$

The hydrodynamic cutoff velocity, V_{cutoff}, below which the jet will not penetrate the target hydrodynamically is determined by setting $U = 0$ in Eq. (21):

$$V_{\text{cutoff}} = \sqrt{2\sigma/\rho_j} \tag{23}$$

The unsteady hole radius, r_H, as a function of time after the impact of a jet particle on the hole is given by Szendrei[13] as

$$r_H^2 = (A/B) - \left[\sqrt{(A/B) - r_j^2} - t\sqrt{(\sigma/\rho_t)}\right]^2 \tag{24}$$

where

$$A = \rho_j r_j^2 V^2 / 2\rho_t (1 + \mathrm{d}P/\mathrm{d}L)^2$$
$$B = \sigma/\rho_t$$

Here dP/dL is the rate of penetration per unit jet length and r_j is the radius of the jet particle. If the jet particle impacts at an intermediate depth, as shown in Fig. 2b, the effective jet radius used in Eq. (24) is half the overlapping region. For the nonoverlapping portion of the jet particle, the approach taken by Szendrei[13] and Golesworthy and Townsend[14] is followed by assuming that the nonoverlapping portion does not contribute to the hole profile since its velocity is significantly reduced by interaction with the hole wall.

Tumbling reduces the effective length of a jet particle at the bottom of the hole. The effective length of the particle is then given as

$$\Delta L_{\text{eff}} = \Delta L \cos\theta \tag{25}$$

Although all of the characteristic jet properties such as velocity, breakup time, and particle radius are known to vary, the predominant factor determining the mean penetration depth for precision charges at a given standoff is jet particle dispersion. Therefore, in an effort to reduce the computation time, Chi et al.[12] applied the Monte Carlo algorithm to the drift velocity and tumbling only, the parameters that would contribute to the jet dispersion. Thus, this model accounts for only a few of many stochastic processes. However, they have shown that this is sufficient to match the penetration-standoff data for precision shaped charges.

Excellent agreement is shown in Fig. 3 where the predicted penetration-standoff curve of the 81-mm precision charge is compared with the experimental values. The calculations were performed with the jet characterized by two

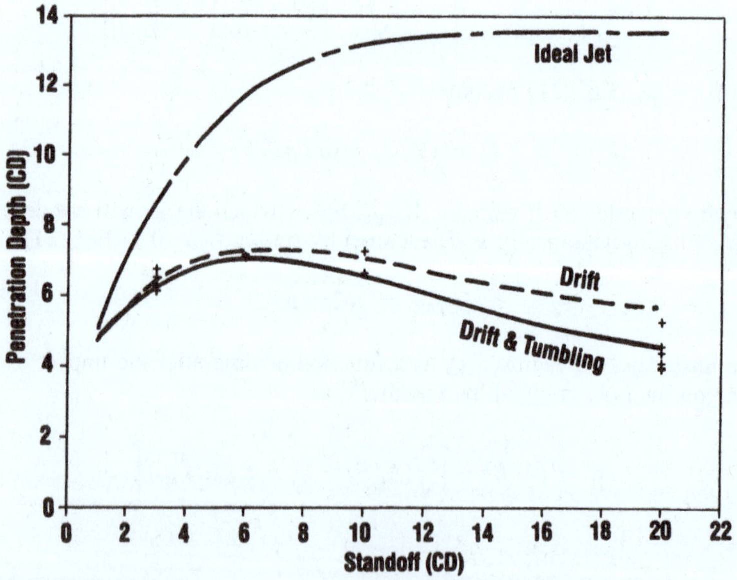

Fig. 3 Predicted penetration-standoff curves for the BRL 81-mm precision charge. (+ represents experimental data.)

Table 1 Jet parameters for the BRL 81-mm precision charge

Jet parameter	Segment 1	Segment 2
Tip velocity (km/s)	9.25	6.08
Tail velocity (km/s)	6.08	2.20
Maximum drift velocity (km/s)	0.017	0.06
Maximum initial angular velocity (rad/s)	4500	28,000
Virtual origin (mm)	5.51	2.33
Breakup time (μs)	105	143
Radius at breakup (mm)	1	1
Particle length at breakup (mm)	8.75	13.48

virtual origins with the parameters listed in Table 1. Also shown in Fig. 3 are the predictions with the drift only and the ideal jet predictions. If only drift is assumed, the predicted values are always larger than the experimental values, and greater difference is observed for large standoffs. It can also be seen that, for the ideal jet, the penetration does not decrease with the standoff, but the penetration remains constant after all jet particles are broken. The model predictions agree well with the test data for profile hole diameter and penetration depth as a function of time during the penetration process as shown in Figs. 4, 5, and 6. The abbreviation CD indicates the explosive charge diameter.

The effect of variable tumbling rate on the penetration depth is shown in Fig. 7. The predictions with variable tumbling rate match with the experimental values, whereas the predicted penetrations with constant tumbling rate are larger than the experimental values at long standoffs. Hence, the variable tumbling rate based on the aerodynamic model improves the predictions at long standoffs.

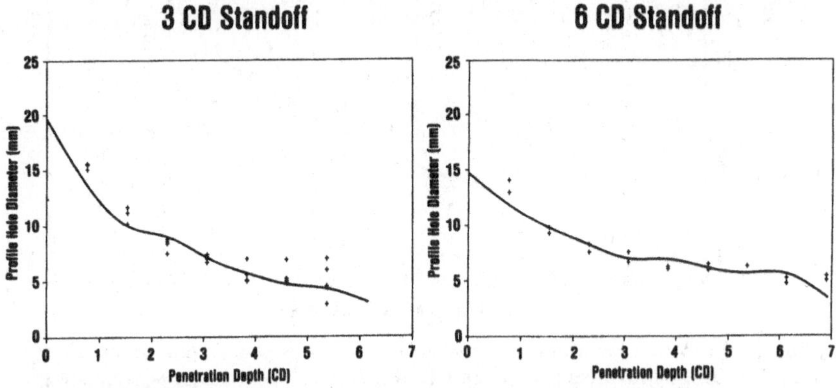

Fig. 4 Predicted variation of profile hole diameter with penetration depth for the precision charge into steel armor, 3 CD and 6 CD standoffs. (+ represents experimental data.)

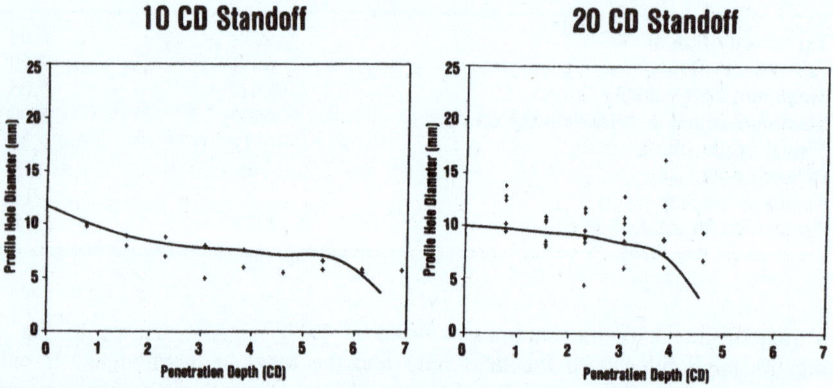

Fig. 5 Predicted variation of profile hole diameter with penetration depth for the precision charge into steel armor, 10 CD and 20 CD standoffs. (+ represents experimental data.)

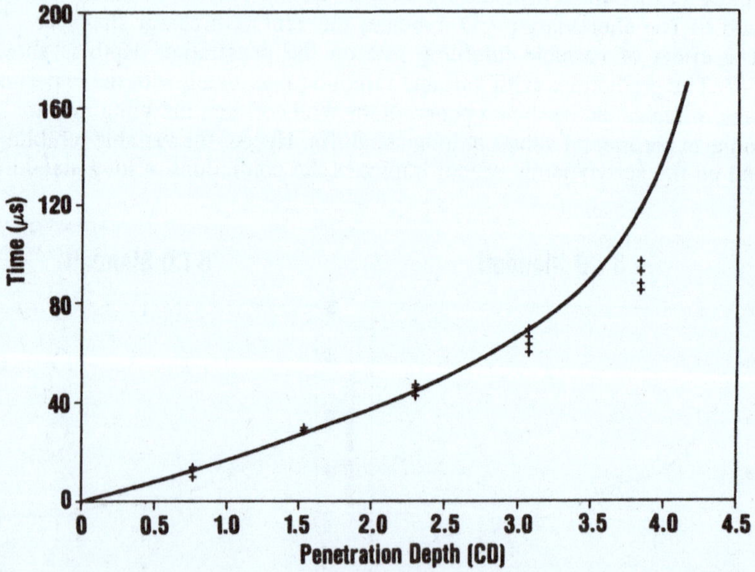

Fig. 6 Predicted variation of penetration time with penetration depth for the precise charge into steel armor, 20 CD standoff. (+ represents experimental data.)

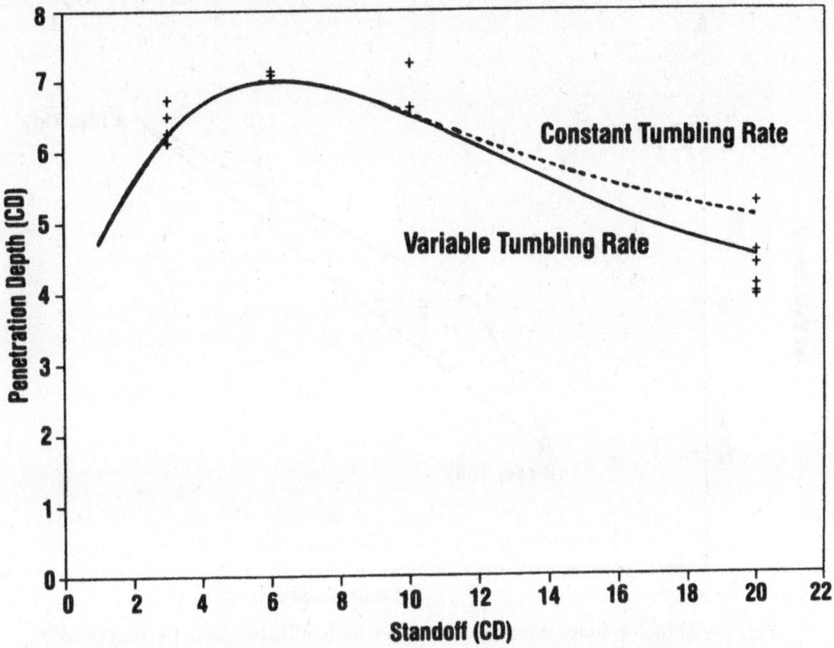

Fig. 7 Effect of variable tumbling rate on the penetration-standoff curve for the precision charge.

An integral approach was used by Walters and Majerus[15] to develop a model for penetration and hole growth based on the conservation of mass and momentum. The penetration process is divided into two regions, the undeformed region of the penetrator and the interaction region between deforming target and penetrator material at the bottom of the crater. For each region, global equations of motion are derived in terms of normal and shear stresses, and the resulting system of ordinary differential equations is solved numerically. With proper selection of target strength and viscosity, good agreement with measured penetration was achieved. A broken jet is first approximated by a continuous jet of equal length with a density equal to the density of the broken jet. This treatment of the broken jet is similar to that of Allison and Vitali.[5] Then the continuous jet is shortened to the sum of the lengths of the particles in the original jet, and the jet radius is also adjusted so that mass is conserved. Though this method yields satisfactory results, it does not give insight into the mechanism of penetration by broken jets.

2.3 Nonlinear Velocity Jets

The penetration equation for a jet of nonlinear velocity distribution was developed by Chou and Foster.[16] They used a bilinear velocity distribution to represent the nonlinear velocity distribution of the jet, as shown in Fig. 8 where ξ represents the space variable and z is the distance between C and D. Two

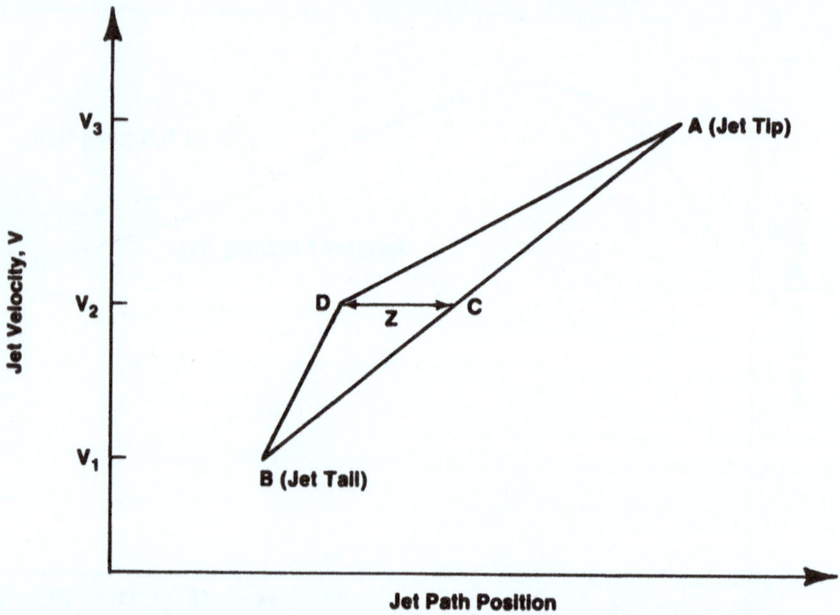

Fig. 8 Velocity-space diagram for a jet with bilinear velocity distribution.

straight line segments, AD and DB, represent the bilinear velocity distribution, and the line ACB represents the linear velocity distribution. The points C and D have the same velocity V_2. The velocity at A is the tip velocity of the jet, V_1, and the velocity at B is the tail velocity, V_3. The total penetration is given as

$$P = s\left(\frac{V_1^{1/\gamma}}{V_3^{1/\gamma}} - 1\right) + z\left[\frac{V_1}{V_3^{1/\gamma}}\left(\frac{V_1^{1/\gamma} - V_2^{1/\gamma}}{V_1 - V_2}\right) - \frac{V_3}{V_3^{1/\gamma}}\left(\frac{V_2^{1/\gamma} - V_3^{1/\gamma}}{V_2 - V_3}\right)\right] \quad (26)$$

The first term in Eq. (26) is the penetration of a linear jet. Additional penetration given by the second term is gained by the bilinear velocity distribution.

In Fig. 9, the ratio between the penetration of a bilinear velocity jet and that of a linear velocity jet is plotted against standoff. The bilinear velocity distribution is from a biconic copper liner reported by Foster.[17] For this biconic charge, $V_1 = 10$ km/s, $V_2 = 7.6$ km/s, $V_3 = 2$ km/s, and $z = 12.5$ cm. At the short standoff of 12.5 cm, the penetration of the bilinear jet is about twice that of the linear jet. As the standoff increases, the ratio between the penetrations of the bilinear jet and the linear jet decreases because the first term in Eq. (26) increases with the standoff while the second term remains constant.

2.4 Compressibility Effects

Allison and Vitali[5] studied the effects of compressibility using the compressible Bernoulli equation with the equation-of-state data available to them. Their

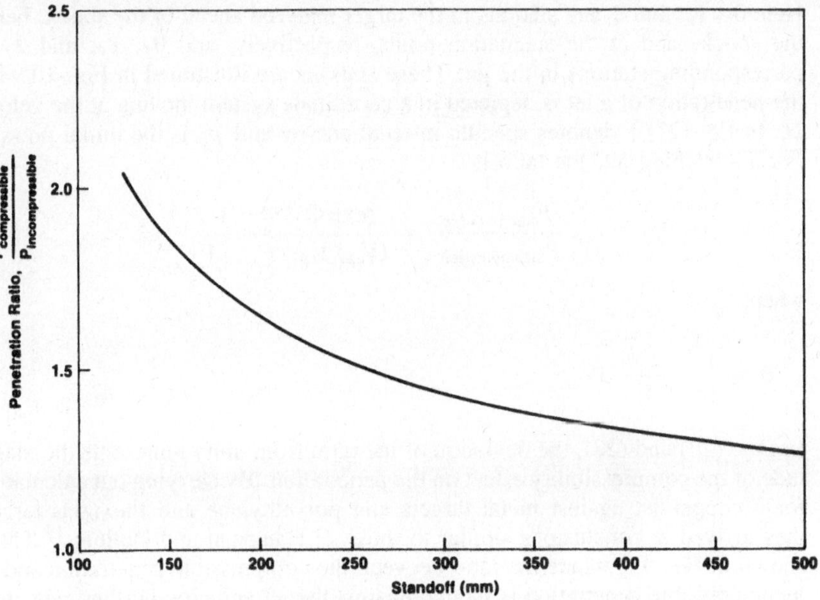

Fig. 9 Effect of nonlinear velocity distribution on the penetration of the jet from a biconic copper liner into a steel target.

work concluded that for metal jets against metal targets the compressibility effect is small. This conclusion is also confirmed by Harlow and Pracht[18] in two-dimensional modeling of an aluminum jet against an aluminum target and an iron jet against an iron target.

However, for nonmetallic targets such as plastics and liquids there can be a substantial reduction in penetration. Haugstad and Dullum[19] predicted a 35% reduction in penetration for Plexiglas targets including the effect of volume compression work and stationary shocks. Flis and Chou[20] later extended the work of Haugstad and Dullum[19] to include a generalized equation of state. For a uniform-velocity jet, they obtained the ratio between the penetration predicted by the compressible model and that by the incompressible model to be

$$\frac{P_{\text{compressible}}}{P_{\text{incompressible}}} = \sqrt{\psi} \tag{27}$$

where

$$\psi = \left\{ \beta + \frac{(\beta - 1)p_0}{(\rho_{0t}U^2)/2} + \frac{\rho_{0j}(e_{2j} - e_{0j}) - \beta\rho_{0t}(e_{2t} - e_{0t})}{(\rho_{0t}U^2)/2} \right\}^{-1}$$

$$\beta = (\rho_{2t}/\rho_{2j})\,(\rho_{0j}/\rho_{0t})$$

Here 0_t, 1_t, and 2_t are stations in the target material ahead of the shock, behind the shock, and at the stagnation point, respectively, and 0_j, 1_j, and 2_j are corresponding stations in the jet. These stations are illustrated in Fig. 10 where the penetration of a jet is depicted in a coordinate system moving at the velocity U. In Eq. (27) e denotes specific internal energy and p_0 is the initial pressure. For a stretching jet, the ratio is

$$\frac{P_{\text{compressible}}}{P_{\text{incompressible}}} = \frac{(\exp \Phi)^{1/\gamma} - 1}{(V_{\text{tip}}/V_{\text{tail}})^{1/\gamma} - 1} \tag{28}$$

where

$$\Phi = \int_{V_{\text{tail}}}^{V_{\text{tip}}} \frac{\sqrt{\psi}}{V}\, dV$$

In Eqs. (27) and (28), the deviation of the ratio from unity represents the magnitude of the compressibility effect on the penetration. By carrying out calculations for a copper jet against metal targets and polyethylene and Plexiglas targets, they arrived at conclusions similar to those of Haugstad and Dullum.[19] This is shown in Fig. 11, where the ratio between the compressible penetration and the incompressible penetration is plotted against the jet velocity. Furthermore, for a stretching jet, their calculated results indicate that the effect of compressibility is amplified, because if the penetration of the front portion of the jet is degraded even a little, the rear will have stretched less by the time it reaches the bottom of the crater.

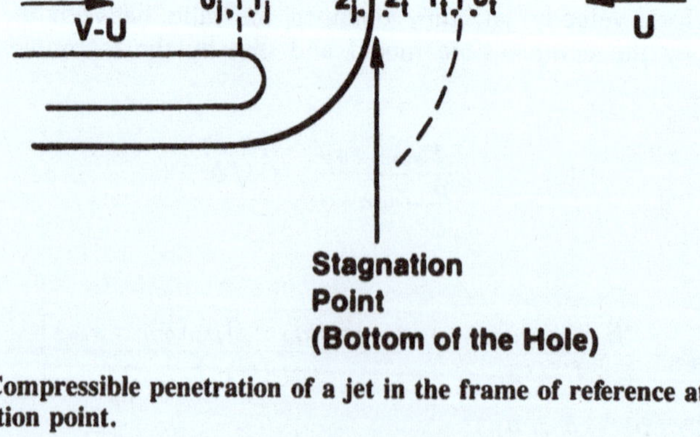

Fig. 10 Compressible penetration of a jet in the frame of reference attached to the stagnation point.

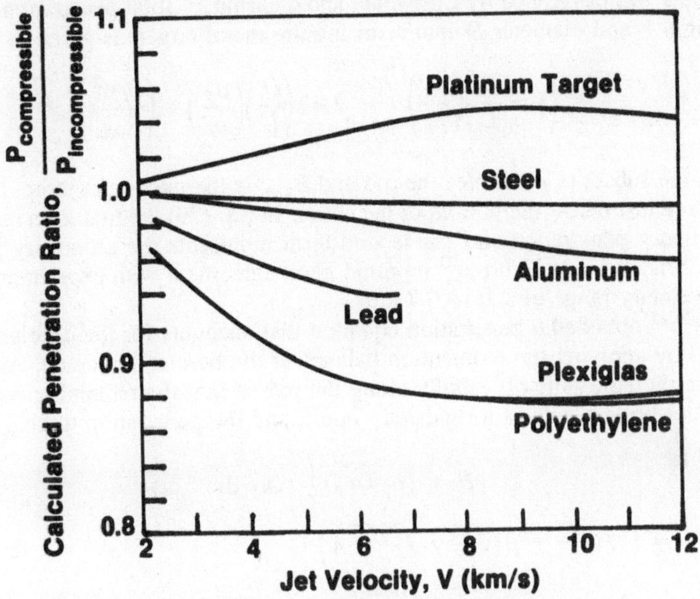

Fig. 11 Compressibility effect of a uniform-velocity copper jet against several materials predicted with the Mie-Gruneisen equation of state.[20]

Chick et al.[21] later compared the experimental penetration velocity of a copper jet in Plexiglas with that predicted by the theory developed by Flis and Chou.[20] They obtained reasonably good agreement between the measured penetration velocity and those calculated using the compressible theory with apparent origins of jet particles based on the data giving the trajectories of the particles of the jet after breakup.

3. Constant-Velocity Rods

Explosively formed penetrators are basically constant-velocity rods with length-to-diameter ratios of 4 to 8. Impact velocities generally range from 1.5 to 3 km/s. Because of the lower impact velocities, material strength will exert a strong influence on the penetration process. Consequently, deceleration of the projectile must be included in the penetration analysis. Furthermore, "secondary penetration" arises when a high-density rod penetrates a low-density target, as observed by Allen and Rogers[22] for a gold rod against an aluminum target. In this instance, after the rod is used up in the penetration, the penetrator material that has been reflected from the crater bottom still has a net velocity into the target and thus causes additional penetration.

An empirical penetration model for a constant-velocity rod based on experimental data of aluminum and steel rods against metallic and nonmetallic target materials was developed by Christman and Gehring.[23] Total penetration of a rod of length L and diameter D into semi-infinite metal targets is given as

$$\frac{P}{L} = \left(1 - \frac{D}{L}\right)\left(\frac{\rho_p}{\rho_t}\right)^{1/2} + 2.42\left(\frac{D}{L}\right)\left(\frac{\rho_p}{\rho_t}\right)^{2/3}\left(\frac{\rho_t V^2}{B_{\max}}\right)^{1/3} \tag{29}$$

where the subscript p denotes the rod and $B_{\max}$ is the maximum target hardness measured just below the bottom of the crater. In Eq. (29) the first term represents the primary penetration and the second term represents the secondary penetration. Christman and Gehring[23] reported good agreement with experimental data for a velocity range of 2.0 to 6.7 km/s.

Tate[24,25] obtained a penetration equation that accounts for the deceleration of the rod by applying the momentum balance at the base of the crater. Assuming that the pressure falls off rapidly along the rod so that the remaining rod above the crater bottom acts as a rigid rod, Tate found the penetration to be given by

$$P = (\rho_p/\sigma_p)\int_v^V u\ell \, dv \tag{30}$$

where

$$u = [1/(1 - \gamma^2)]\left(v - \gamma\sqrt{v^2 + A}\right) \tag{31}$$

$$\frac{\ell}{L} = \left\{\frac{v + \sqrt{v^2 + A}}{V + \sqrt{V^2 + A}}\right\}^{[(\sigma_t - \sigma_p)/\gamma\sigma_p]}$$

$$\times \exp\left[\frac{\gamma\rho_p}{2(1 - \gamma^2)\sigma_p}\left\{[v\sqrt{v^2 + A} - \gamma v^2] - [V\sqrt{V^2 + A} - \gamma V^2]\right\}\right] \tag{32}$$

In Eqs. (31) and (32),

$$A = 2(\sigma_t - \sigma_p)(1 - \gamma^2)/\rho_t \tag{33}$$

Termination of the penetration process was classified into three cases based upon the relative magnitude of σ_t and σ_p. When $\sigma_t > \sigma_p$ (case 1), the rod flows hydrodynamically until penetration ceases, and this occurs when the velocity of the back end of the rod, v, is

$$v = [2(\sigma_t - \sigma_p)/\rho_p]^{1/2} \tag{34}$$

When $\sigma_t < \sigma_p$ (case 2), hydrodynamic penetration continues until $v = u$, or

$$v = [2(\sigma_p - \sigma_t)/\rho_t]^{1/2} \tag{35}$$

and from this point until it is brought to rest the rod behaves as a rigid body. Finally, when $\sigma_t = \sigma_p$ (case 3), the rod always behaves as a liquid and penetration ceases when v and u are both zero. For each of the three cases, v defined by Eq. (34), Eq. (35), or zero represents the lower limit of integration in Eq.

(30). Equation (30) with u and ℓ substituted by Eqs. (31) and (32) is evaluated numerically for cases 1 and 2. For case 3, analytical integration yields

$$(P/L) = (1/\gamma)\{1 - \exp[-B(V^2 - v^2)]\} \tag{36}$$

where

$$B = \gamma\rho_p/2(1+\gamma)\sigma_p \tag{37}$$

The deceleration of the rod as it penetrates the target is illustrated in Fig. 12, where the velocity of the back end of the rod is plotted against the depth of

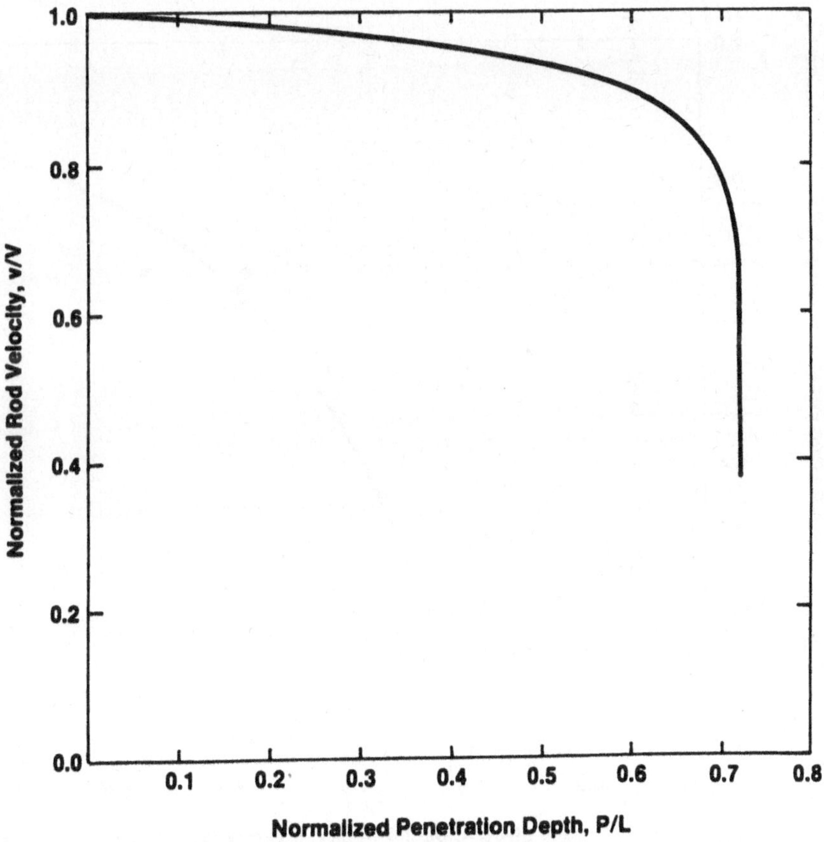

Fig. 12 Variation of the velocity of the back end of the rod with depth of penetration for a steel rod impacting on a steel target. The velocity is normalized to the initial impact velocity $V = 2$ km/s and the penetration is normalized to the initial rod length $L = 6.35$ cm. The strength values of $\sigma p = 11$ kbar and $\sigma t = 33$ kbar are used.

penetration for a steel rod penetrating a steel target. The velocity is normalized to the initial impact velocity, $V = 2$ km/s, and the penetration is normalized to the initial rod length, $L = 6.35$ cm. The strength values of $\sigma_p = 11$ kbar and $\sigma_t = 33$ kbar are used in the calculation. The velocity decreases slowly with the depth of penetration during most of the penetration process but falls off rapidly toward the end of the penetration process. In Fig. 13, the predicted penetration is compared with the experimental data given by Tate[24] as a function of the initial impact velocity for the penetration of the rod described above. Although significant scatter exists in the data for the lower impact velocities, the agreement between the predicted and experimental penetration is good.

For high-density rods impacting low-density targets, the secondary penetration exists as observed by Allen and Rogers.[22] When the rod is used up in the penetration, the rod material reflected from the bottom of the crater still has

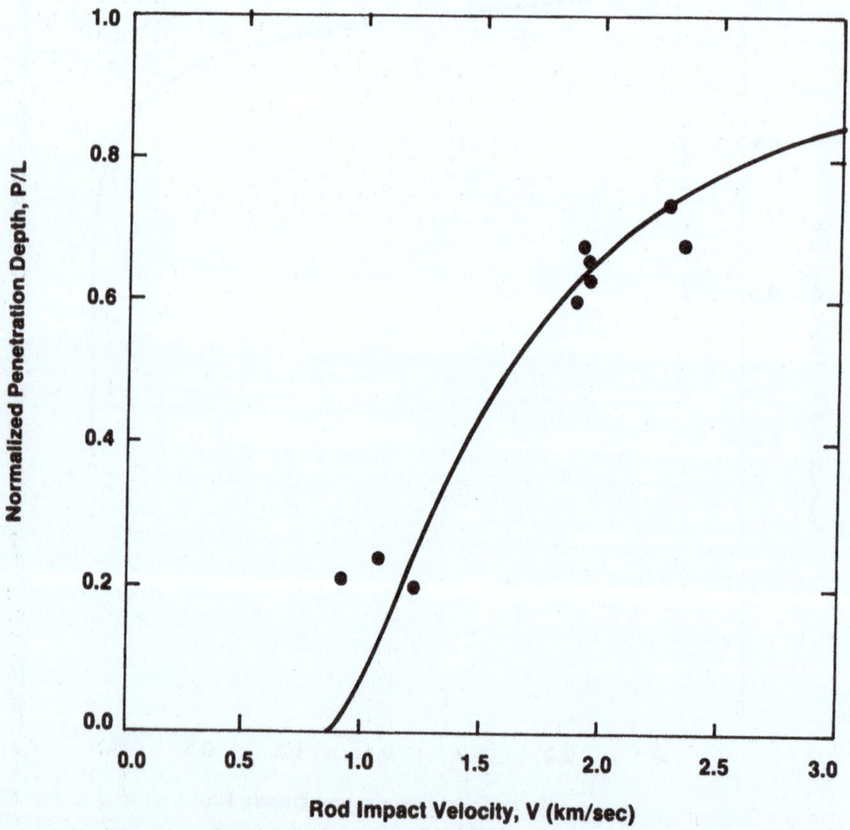

Fig. 13 Variation of the penetration with the impact velocity for a steel rod impacting on a steel target. The penetration is normalized to the initial rod length of 6.35 cm. The points shown are taken from the experimental data presented by Tate.[24]

the velocity toward the target as shown in Fig. 14. In a coordinate system that moves with the velocity U, the jet velocity is $V - U$. Since the speed of the free streamline is not affected by a change in direction, the velocity of the reflected rod material is given by $U - V$. In a coordinate system fixed with the target, the residual velocity V_r is

$$V_r = 2U - V \tag{38}$$

Substitution of U in Eq. (38) with Eq. (21), with ρ_j replaced by ρ_p, gives the residual velocity in terms of the impact velocity, densities of the rod and jet, and relative resistance.

In Fig. 15, the predicted penetration of a gold rod into 7075-T6 aluminum is compared to the experimental data reported by Allen and Rogers.[22] The penetration is nondimensionalized with respect to $(\rho_p/\rho_t)^{1/2}L$, and $\sigma = 23$ kbar is used in the calculation. Good agreement exists between the experimental data and the predictions, including both primary and secondary penetrations. The primary penetration alone is less than the experimental penetration for higher rod velocities for which the secondary penetration becomes significant.

4. Concluding Remarks

The analytical models that evolved from the hydrodynamic theory provide reasonably accurate predictions for the penetration of a jet or rod into homogeneous metallic targets. For the jets, efforts have been directed toward developing models that describe the degradation in penetration after breakup. Some advances have been made to include the compressibility effect, which significantly reduces penetration against compressible targets such as plastics and liquids. For the rods, analytical models have been developed to include deceleration of the rods and the secondary penetration that has been observed when high-density rods penetrate low-density targets.

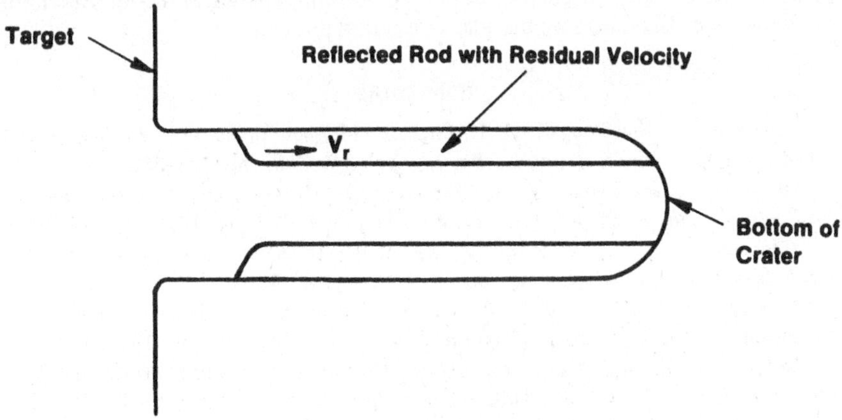

Fig. 14 Secondary penetration of a jet.

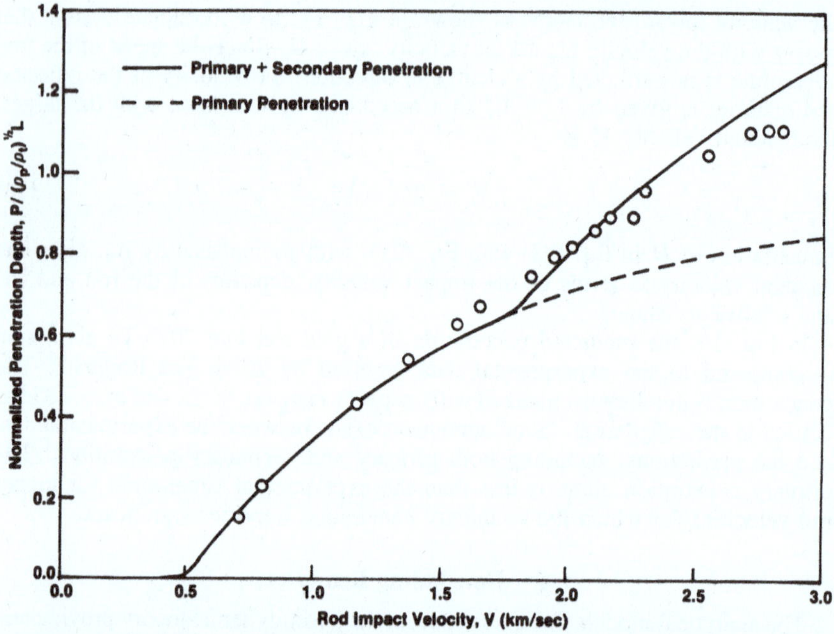

Fig. 15 Predicted penetration of a gold rod into 7075-T-6 aluminum, normalized to $(\rho p/\rho t)^{1/2L}$. The circles represent the experimental data presented by Allen and Roger.[21] The strength value of σ = 23 kbar is used.

However, with the advent of composite armors with glass and ceramics, the existing models became inadequate to predict performance. The primary mechanism of jet disruption by confined glass or ceramic layers is considered to be the rebounding of debris into the path of the jet. Though some efforts been made,[26] more work is needed to develop a working analytical model describing the penetrator disruption by the glass or ceramic layers.

References

[1]Birkhoff, G., MacDougall, D. P., Pugh, E. M., and Taylor, G., "Explosives with Lined Cavities," *Journal of Applied Physics,* Vol. 19, 1948, pp. 563–582.

[2]Pack, D. C., and Evans, W. M., "Penetration by High-Velocity ('Munroe') Jets: I," *Proceedings of the Physical Society, London, Section B,* Vol. 64, 1951, pp. 298–302.

[3]Eichelberger, R. J., "Experimental Test of the Theory of Penetration by Metallic Jets," *Journal of Applied Physics,* Vol. 27, 1956, pp. 63–68.

[4]Abrahamson, G. R., and Goodier, J. N., "Penetration by Shaped Charge Jet of Nonuniform Velocity," *Journal of Applied Physics,* Vol. 34, 1963, pp. 195–199.

[5]Allison, F. E., and Vitali, R., *A New Method of Computing Penetration Variables for Shaped-Charge Jets,* U.S. Army Ballistic Research Lab., Rept. 1184, Jan. 1963.

[6]Carleone, J., Chou, P. C., and Walters, W. P., *The Virtual Origin Approximation in Hemispherical Charges and Shaped Charges,* U.S. Army Ballistic Research Lab., Rept. ARBRL-CR-00460, July 1981.

[7]DiPersio, R., and Simon, J., *The Penetration-Standoff Relation for Idealized Shaped-Charge Jets,* U.S. Army Ballistic Research Lab., MR 1542, Feb. 1964.

[8]Carleone, J., Chou, P. C., Flis, W. J., and Tanzio, C., *User's Manual for DESC-2, a One-Dimensional Computer Code to Model Shaped Charge Liner Collapse, Jet Formation, and Penetration,* Dyna East Corp., Philadelphia, PA, Rept. DE-TR-75-4, Dec. 1975.

[9]Chou, P. C., and Carleone, J., "Breakup of Shaped-Charge Jets," *Proceedings of the 2nd International Symposium on Ballistics,* Daytona Beach, FL, 1976.

[10]Chou, P. C., and Carleone, J., "The Stability of Shaped Charge Jets," *Journal of Applied Physics,* Vol. 48, 1977, pp. 4187–4195.

[11]Smith, J. A., *Shaped Charge Penetration at Long Standoff,* Air Force Armament Laboratory Rept. No. AFATL-TR-81-25, Feb. 1981.

[12]Chi, D. B., Conner, J., and Jones, R., "Computational Model for the Penetration of Precision Shaped Charge Warhead," *Proceedings of the 11th International Symposium on Ballistics,* Brussels, Belgium, 1989.

[13]Szendrei, T., "Analytical Model of Crater Formation by Jet Impact and Its Application to Calculation of Penetration Curves and Hole Profiles," *Proceedings of the 7th International Symposium on Ballistics,* The Hague, Netherlands, 1983.

[14]Golesworthy, R. C., and Townsend, I., "Analytical Model of Shaped Charge Penetration in the Direct Overflying Top Attack Modes," *Proceedings of the 8th International Symposium on Ballistics,* Orlando, FL, 1984.

[15]Walters, W. P., and Majerus, J. N., "Impact Models for Penetration and Hole Growth," U.S. Army Ballistic Research Lab., Rept. ARBRL-TR-02069, Oct. 1980.

[16]Chou, P. C., and Foster, J. C., "Theory of Penetration by Jets of Non-linear Velocity and in Layered Targets," *Proceedings of the 10th International Symposium on Ballistics,* San Diego, CA, 1987.

[17]Foster, J., "Integrated Flash Radiograph Analysis—an Approach to Studying Time-Dependent Phenomena in the Explosive Formation and Projection of Metals," *Proceedings of the 7th Annual Technical Meeting on Physics of Explosives,* Livermore, CA, 1981.

[18]Harlow, F. H., and Pracht, W. E., "Formation and Penetration of High-Speed Collapse Jets," *Physics of Fluids,* Vol. 9, 1966, pp. 1951–1959.

[19]Haugstad, B. S., and Dullum, O. S., "Finite Compressibility in Shaped Charge Jet and Long Rod Penetration—the Effect of Shocks," *Journal of Applied Physics,* Vol. 52, 1981, pp. 5066–5071.

[20]Flis, W. J., and Chou, P. C., "Penetration of Compressible Materials by Shaped-Charge Jets," *Proceedings of the 7th International Symposium on Ballistics,* The Hague, Netherlands, 1983.

[21]Chick, M. C., Frey, R. B., Trimble, J. J., and Bines, A., "Jet Penetration in Plexiglas," *Proceedings of the 8th International Symposium on Ballistics,* Orlando, FL, 1984.

[22]Allen, W. A., and Rogers, J. W., "Penetration of a Rod into a Semi-Infinite Target," *Journal of the Franklin Institute,* Vol. 272, 1961, pp. 275–284.

[23]Christman, D. R., and Gehring, J. W., "Analysis of High-Velocity Projectile Penetration Mechanics," *Journal of Applied Physics,* Vol. 37, 1966, pp. 1579–1587.

[24]Tate, A., "A Theory for the Deceleration of Long Rods after Impact," *Journal of the Mechanics and Physics of Solids,* Vol. 15, 1967, pp. 387–399.

[25]Tate, A., "Further Results in the Theory of Long Rod Penetration," *Journal of the Mechanics and Physics of Solids,* Vol. 17, 1969, pp. 141–150.

[26]Eichelberger, R. J., "Defeat of Penetrators by Ceramics," *Proceedings of the 12th International Symposium on Ballistics,* San Antonio, TX, 1990.

Chapter 9
Part B

Target Interaction: Penetration Mechanism of the Shaped Charge Jet

Robert C. Golesworthy* and Ian Townsend†
Impact Sciences Limited, Chesham, Bucks, United Kingdom

1. Introduction

THE practical motivation for studying the penetration mechanisms of shaped charge jets is to acquire the means of assessing and extending the effectiveness of the shaped charge or, taking protection as the objective, assessing and extending the effectiveness of the armor. One outcome of such study is the ability to simulate real devices and explore their performance to the point where a direct experiment can be performed to validate the modeling and advance the design capability. Proven methods of modeling shaped charge jet interactions with target structures therefore provide an essential link between chemical energy warhead technology and armor technology. Both technologies have advanced substantially in recent years. Until the late 1960s, armor had been almost exclusively homogeneous and of a thickness generally within the defeat capabilities of warhead designs of the time. Therefore, a simple hydrodynamic penetration law and a linear scaling law were sufficient to determine the relationship between warhead caliber and jet penetration depth. Residual, behind-armor penetrative capabilities could be taken as the algebraic difference between warhead penetration capability and armor thickness in the line of attack. This residual penetration could, in turn, be used directly in an assessment of behind-armor effects and lethality.

For armor design, the main consideration is often protection against long-rod kinetic energy penetrators, but in improving that aspect of protection, there is usually a greater improvement in the protection achieved against the shaped charge jet. A series of new challenges came to antiarmor shaped charge technology, first with spaced armor and laminated armor and then, with the development of reactive armor, in nonexplosive and explosive variations.

*Managing Director.

†Technical Director.

These reactive armor systems have a nonisotropic, nonlinear response to attack, which is also time-dependent; that is, after the response is stimulated by the attack, it develops, propagates in the armor structure, and then decays. It has therefore become more important to understand the penetrative behavior of the shaped charge jet, in space and time frames, in order to understand its interactions with these armor systems.

At this time numerical finite difference modeling methods, as they are embodied in the many available hydrocodes, do not provide the means of resolving the dynamic and structural complexity of real warhead jets, modern armors, or unconventional attack modes, such as dive or fly-over–shoot-down attack. There is no doubt that as computer power is increased, it will become practical and economical to use hydrocodes at this level. Meanwhile, work in these areas has progressed mainly with analytical methods, by building on the classical hydrodynamic theory of penetration, as it was originally applied to ideal shaped charge jets and semi-infinite homogeneous targets. The methodology for the problem has been to consider the ideal jet, use classical hydrodynamic penetration theory, and then to examine how a real jet deviates from ideality and how nonideal aspects may be incorporated in a practical modeling framework for homogeneous targets. From here, the modeling may be extended to address the effects of lateral translation velocities between jets and targets and to address explosive reactive armor, where such lateral translations are an important part of the jet/target interaction.

2. Ideal Jets and Homogeneous Targets

At the impact velocities achieved with shaped charge jets, the mechanical strength of the jet material and the target material are greatly exceeded by the strain levels and strain rates induced by the impact. It has therefore been accepted practice, in first approximations, to neglect strength effects altogether. A stagnation volume is created at the jet/target interface, which moves into the target with a velocity U once a local steady state has been established. The stagnation pressure p, in this condition, applies to both the jet material and the target material in the stagnation volume. The Bernoulli equation for incompressible flow gives the following equalities:

$$
\begin{aligned}
p &= \frac{1}{2}\rho_j(V_j - U)^2 \\
p &= \frac{1}{2}\rho_t U^2
\end{aligned}
\tag{1}
$$

where ρ_j and ρ_t are the densities of the jet material and target material and V_j is the jet velocity approaching the stagnation volume. If it is assumed that the jet velocity V_j is constant for a length of jet L, then penetration velocity U in homogeneous targets is also constant:

$$U = V_j/\sqrt{(1 + \rho_j/\rho_t)} \tag{2}$$

and the penetration depth P resulting from the impact of that length of jet L is

$$P = \frac{LU}{V_j - U} = L\sqrt{(\rho_j/\rho_t)} \tag{3}$$

These equations may be used to provide an approximation of the penetration depth and penetration-time history of an ideal jet impacting a homogeneous target.

In this context, an ideal jet may be specified as originating from a single point on the charge axis as a single event in time. It is also specified as perfectly collinear with the charge axis and as stretching continuously, with a linear velocity gradient, either without breakup or with a single instantaneous breakup throughout its length.

All of the above characteristics attributed to an ideal jet have limited validity. Hydrocode models and flash radiography reveal that the collapse of a typical conical liner occurs over tens of microseconds, during which time the collapse stagnation volume has traveled typically tens of centimeters down the charge axis. Nonlinear jet velocity gradients are common; indeed, many warheads are now designed to produce multiple or nonlinear jet velocity gradients to achieve specific effects. For example, most high-penetration-performance warheads have lower velocity gradients in the front of the jet and steeper velocity gradients at the rear. Jet breakup is more likely to be progressive rather than instantaneous and to take place as the local stress in the stretching tapered jet exceeds the local strength. Exact collinearity with the jet axis is achieved only with stringent design and quality control, which may not be appropriate in some applications.

These features are not easy to reconcile with the concept of an ideal jet; even so, approximations based on linear velocity gradients, instantaneous breakup, and a virtual jet origin have offered an acceptable modeling basis for some recent work[1,2] and still provide an acceptable simplification for jets that show a high degree of particle alignment, or when the warhead-target standoff distance of interest is short. The original research[3,4,5] performed at the U.S. Army Ballistics Research Laboratory provided models for the prediction of ideal jet penetration as early as 28 years ago. It is worth examining a more recent interpretation[6] of that work.

Taking the concept of an ideal jet originating at a virtual origin, having a linear velocity gradient, and having an instantaneous breakup, we may assign the following variables:

Z_0 = the distance from the jet virtual origin to the target front plane
V_0 = the jet tip velocity
t_0 = the time of flight of the jet tip to travel from the virtual origin, at time zero, to the target front plane
t_b = the time after its formation that the jet is fully broken up
V_r = the limiting velocity of the fastest residual jet element that is just incapable of adding to hydrodynamic penetration at the crater base
$\gamma = \sqrt{(\rho_t/\rho_j)}$ = the square root of the ratio of target to jet densities

The penetration depths $P(t)$ achieved as a function of time t will be characteristic of three regimes:

1) At short standoff distances when the jet is continuous, the time-dependent penetration depth is

$$P(t) = Z_0[(t/t_0)^{1/(1+\gamma)} - 1] \tag{4}$$

2) In the intermediate standoff regime, the initial penetration is generated by a continuous stretching jet, but after time t_b, it continues with the particulated jet, in which the individual jet particles no longer stretch. The time-dependent penetration is then

$$P(t) = Z_0[(1-\gamma)(t_b/t_0)^{1/(1+\gamma)}t/(t_b+\gamma t) - 1] \tag{5}$$

3) Finally, in the long standoff regime, where all the penetration is generated by the particulated jet, the time-dependent penetration is given by

$$P(t) = Z_0[(t/t_0) - 1]t_b/(t_0 + \gamma t) \tag{6}$$

A limiting penetration $P(V_r)$ may be calculated for each standoff condition by solving Eqs. (4–6), using the identity

$$t = (P + Z_0)/V_r \tag{7}$$

to provide the expected maximum penetration depths $P(V_r)$ from within the same three standoff regimes:

1) Close-in, continuous jet:

$$P(V_r) = Z_0(V_0/V_r)^{1/\gamma} - 1 \tag{8}$$

2) Intermediate standoff, part continuous, part particulated jet:

$$P(V_r) = \left\{[(1+\gamma)(V_0 t_b)^{1/(1+\gamma)} - Z_0^{\gamma(1+\gamma)} - V_r t_b]/\gamma\right\} - Z_0 \tag{9}$$

3) Long standoff, fully particulated jet:

$$P(V_r) = (V_0 - V_r)t_b/\gamma \tag{10}$$

A set of ideal penetration standoff curves may be constructed for any jet or target using V_r, the limiting residual jet velocity, as a parameter. For example, for rolled homogenous armor (RHA), penetration[6] by a precision shaped charge that produces a copper jet with a jet tip velocity $V_0 = 8.5$ mm/μs and a nominal jet breakup time $t_b = 130\mu$s. The value of V_r may be chosen to range between 2 and 8 mm/μs. The family of curves that results is shown in Fig. 1, in which an experimental penetration standoff curve for the same warhead is also included.

Ideal jet penetration increases with increasing standoff distance until the time of flight of the residual jet tip, from the virtual origin to the base of the

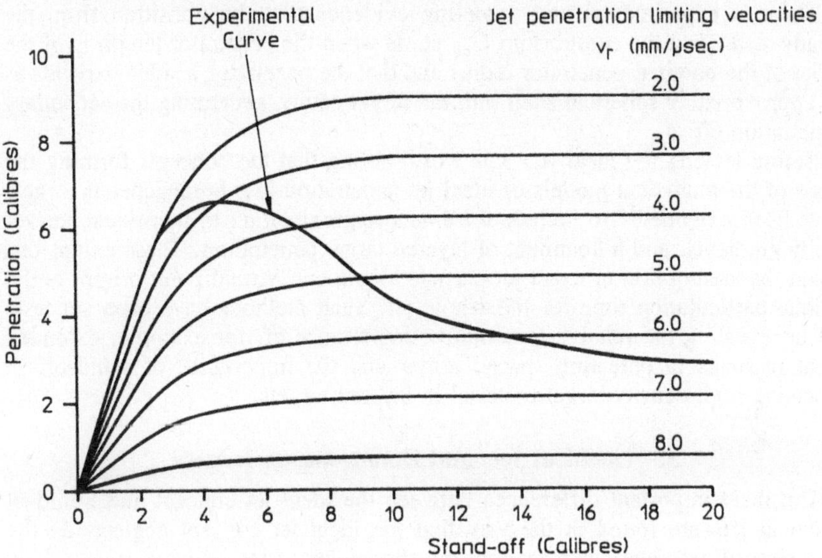

Fig. 1 Ideal penetration standoff curves as a function of limiting velocity.

penetration crater, has equaled the jet breakup time. From then on, no further extension of the jet material is modeled; that is, total cumulative residual jet element length is constant and the cumulative penetration is independent of target stand-off. However, the experimental curve in Fig. 1 shows a progressive negative deviation from the form of the ideal curve long before jet breakup is complete, and the apparent limiting jet velocity for continued penetration, V_r, is rising as stand-off increases.

Particulation of the jet does not itself lead to reduction in penetration performance. On the contrary there is a body of experimental and theoretical work,[7–10] showing that sequential multiple impacts at the base of a target crater are capable of greater cumulative penetration depth than that resulting from the impact of an equivalent length of continuous jet. The phenomenon is described as *afterflow*, or *secondary penetration*. An impacting jet particle establishes a steady-state penetration velocity U [Eq. (2)]. As the particle becomes expended in the crater, the penetration velocity decays from the steady-state value. The crater boundary continues to expand toward a new equilibrium condition in the target material such that the local induced stress is equal to the target dynamic yield stress Y_t. The limiting penetration crater base velocity U_{eq} at which the equilibrium crater boundary is established is

$$U_{eq} = \sqrt{2Y_t/\rho_t} \tag{11}$$

Substituting the appropriate values for RHA gives a value of 450 m/s.

There is experimental and modeling evidence that the transition from the steady state U to the equilibrium U_{eq} starts when the penetrator length is of the order of the nominal penetrator radius and that the penetrator residue expands as an approximately spherical shell into the target cavity, generating the secondary penetration effect.

Before leaving the ideal jet, it is worth noting that the concepts forming the basis of the analytical models of ideal jet penetration into homogeneous targets have been extended[11] to include a bilinear approximation to non-linear jet velocity gradients, and a treatment of layered target penetration. These extensions retain the assumption of exact jet particle alignment, virtual point origin, and a unique particulation time for the whole jet. Such methods have been successful in revealing the relative quantitative importance of, for example, extended breakup times in defeating spaced armor and the importance of stand-off in defeating high-density target material in layered targets.

3. Nonideal Jets and Homogeneous Targets

The most important differences between the ideal jet concept and nonideal practical jets are found in the way that the ideal jet concept neglects 1) the treatment of systematic and random misalignments of jet velocity vectors from the nominal jet axis, 2) the treatment of nonlinearity of the jet longitudinal velocity profile, and 3) the treatment of the time dependence of breakup as a function of position in the jet.

This section develops a method, based on experimental jet data, that leads to a predictive model of nonideal jet penetration of homogeneous targets. Because it is based on experimental jet data, it carries none of the assumptions of ideality. Once the model is established, it is then quite feasible to construct a hypothetical nonideal jet and to vary its characteristics parametrically by way of examining its performance potential.

It has been shown experimentally by many workers using flash x-ray radiography that the shaped charge jet comprises one or more elongated, stretching components that undergo successive breakup in flight to form a sequence of smaller particles, which eventually show no further tendency to breakup. Flash x-ray radiography of jets shows that, for a given warhead design, made to good standards of precision and consistency, the number, the velocities, and even the particle masses and shapes of the jet particle sequences are remarkably reproducible from round to round.

For a given particle in a jet, some longitudinal oscillation may persist, but in general, jet particles in a free-flight condition maintain constant lengths and constant motions. The observed motions include tumbling (rotation) and lateral motion away from the mean longitudinal axis of the jet. Two types of lateral motion are discernible, systematic and random. Figure 2 gives a simple representation of this condition, compared with the straight, well-ordered particulated jet. Off-axis asymmetries in warhead initiation or liner misalignment[2] invariably produce systematic jet deflections. Random transverse particle velocities may be superimposed on systematic deflections; these tend to occur if the quality of manufacture has been relaxed, with respect to the metallurgical condition

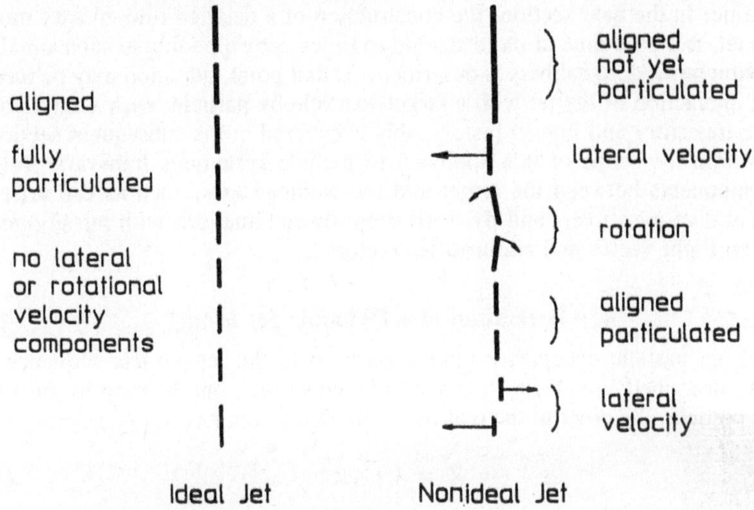

Fig. 2 Qualitative comparison of ideal and nonideal jets.

of the charge liner and the granularity and homogeneity of the explosive filling.

The experimentally observed systematic lateral deflections that can arise in jets may be simply characterized by an angular standard deviation of velocity vectors of the order 0.1 mrad to 0.3 mrad, measured from the nominal jet axis and originating from within the jet formation zone. Strictly, the systematic particle deflections are not random Gaussian and should be defined explicitly as a function of position in the jet.

The random angular dispersions of individual particle velocities after breakup are likely to conform to a standard deviation of 0.2 mrad or more, again referred to the jet formation zone. Jet particle tumbling motions are seen in comparatively few of the particle trajectories. Where they arise they introduce angular rates having a standard deviation up to 4 mrad/μs about the jet particle centroid.

It is evident that jet particles that are misaligned from the jet axis, to the extent indicated here, may experience premature impacts with the penetration crater wall, before they have a chance of impacting with the crater base. In this case they will fail to contribute to the extension of the crater depth but may contribute to the increase of crater volume. The form and motions of such particles are important for the modeling process. It is therefore important that the consistency of jet radiographic data from consistent hardware extends to this level of detail in the jet structure and particle motions.

Given these round-to-round consistencies, accurate radiographic data from one or more similar jets may be used to construct a dynamic three-dimensional jet model, valid for that design of warhead. The method makes use of sequential orthogonal radiograph pairs that reveal the spatial structure of the jet at two instants in time; the sequential radiograph pairs reveal the particle movement with time. In the next section, the construction of a detailed time-history model of the jet, from the time of its formation to times corresponding to substantial jet extension beyond 20 calibers is described. At that point, attention may be turned to the interaction of the jet with a target, particle by particle, each with its own unique trajectory and impact history; this is covered in the subsequent sections. It is within the scope of this approach to include systematic transverse velocity components between the target and the warhead axis, such as can arise in terminal dive maneuvers and fly-over–shoot-down situations with misalignment between flight vector and warhead jet vectors.

4. Derivation of a Dynamic Jet Model

Each jet particle occupies a unique position in the jet particle sequence. It may be described in velocity terms by two equations, one relating to the front of the particle, the other to the rear of the particle, each having the general form

$$Z = Vt + C \tag{12}$$

where Z is the position of the particle reference point at time t derived from an orthogonal radiograph pair; V is the velocity of the particle reference point (assumed constant once the part of the jet containing the particle has been formed at the origin); t is the time, measured from charge initiation; and C is an axial coordinate identifying the particle virtual origin.

From two sequential radiograph pairs, obtained after complete breakup, the position of each particle may be determined at two times. A series of pairs of simultaneous equations in V and C with known values of Z and t may be generated for the front and the rear of each particle. When solved, they provide the velocity equations, which are needed to generate and map the complete jet extension and particulation history. Particle volumes may be calculated to a first approximation by assuming that particles are cylindrical in form. Particles that are found to be tumbling or reverberating may be processed more simply by assuming that the attributed volume is the mean of the measured volumes, at the two measurement times, taking particle geometry to be cylindrical and correcting for imaging distortions. Particle volume is

$$\text{Volume} = (\pi/8)\left[d_1^2(Z_{1,f} - Z_{1,r}) - d_2^2(Z_{2,f} - Z_{2,r})\right] \tag{13}$$

where d_1 and d_2 are the particle diameters measured at the time of the first and second radiographs, respectively, $Z_{1,f}$ is the axial coordinate of the front of the particle at the time of the first radiograph, measured from the base of the charge liner; $Z_{1,r}$ is the axial coordinate of the rear of the particle at the time of the first radiograph, measured from the base of the charge liner; and $Z_{2,f}$ and $Z_{z,r}$ are the front and rear axial coordinates, measured from the base of the charge liner, at the time of the second radiograph.

Positions are calculated from the radiograph measurements with respect to the warhead axis at the base of the charge liner, taking radiographic scaling factors into account. The breakup time between two particles within the jet is obtained from the particle velocity equations by solving the equations for adjacent front and rear reference points of neighboring particles:

$$V_r(i)t_b + C_r(i) = V_f(i+1)t_b + C_f(i+1) \tag{14}$$

where $V_r(i)$ and $C_r(i)$ are the velocity and virtual origin for the rear of the leading particle, $V_f(i+1)$ and $C_f(i+1)$ are the velocity and virtual origin for the front of the following particle, and t_b is the required breakup time for that particle pair.

In the event that the condition $V_r(i) = V_f(i+1)$ arises from the radiograph measurements, the breakup time is arbitrarily set to zero; that is, the particles are treated as having been separated since formation.

The reconstruction of the jet breakup process from the experimental velocity equations may be achieved by progressively recovering earlier breakup times and jet structures starting from the last breakup and working backward through time. As each breakup event is encountered in the regression from complete breakup, there will be one less particle in the jet. Each parent particle, which in the forward sequence gives rise to two daughter particles, is attributed with the rear velocity of the rear particle and the front velocity of the front particle. Its volume is the sum of the daughter particle volumes. The reverse time sequence procedure is continued until there is a minimum number of particles for times greater than or equal to zero. Because the experimental orthogonal radiographic positional data for the fully particulated jet permit measurement of displacements from the geometrical warhead axis, the orthogonal transverse velocity components attributable to each particle may also be determined in a manner similar to the longitudinal velocities.

The simplest instrumentation needed to acquire this data uses only two orthogonal x-ray sources and sequences them to trigger at times t_1 and t_2, which both exceed the time at which the experimental jet is fully particulated. This minimum level of instrumentation is based on the assumption that the particles started from the geometrical charge axis; with the two orthogonal pairs of x-ray sources, this constraint is removed.

The measured displacements from the charge axis for the ith particle at t_1 and t_2 are measured as $D_1(i)$ and $D_2(i)$. The component transverse velocities V_T are then

$$V_{T,1}(i) = D_1(i)/t_1 \qquad V_{T,2}(i) = D_2(i)/t_2 \tag{15}$$

The resultant is

$$V_T(i) = \sqrt{V_{T,1}^2(i) + V_{T,2}^2(i)} \tag{16}$$

which subtends an angle $\phi(i)$ to the reference image plane of the second x-ray source

$$\phi(i) = \tan^{-1}[(V_{T,2}(i)/V_{T,1}(i))] \tag{17}$$

5. Jet/Target Interaction

The jet model generated from jet radiographs in the manner described above provides for a full three-dimensional mapping of individual particles and their impact histories in target space. Some combinations of target construction, attack geometry, and dynamics of the interaction require a full three-dimensional representation, but most jet/target interactions of interest may be resolved into a plane that expresses the dominant effects of energy and momentum transfer. A rotation of jet particle transverse velocity vectors into such a reference modeling plane retains actual particle displacement magnitudes and does not lead directly to an overestimate of particle impacts with the penetration crater base. A second-order effect may arise in the form of an overestimation of the upper-crater volume from impacts attributed to earlier particles in the jet sequence.

For simplicity, each jet particle may be regarded as having cylindrical geometry; it will be stretching or dimensionally stable, depending on the local condition of the jet at the time of its impact with the target. Because a particle impact may be oblique or grazing, it will not necessarily involve the whole particle in a consistent manner. It is therefore convenient to consider each particle as comprising a number of axial modeling elements. It is convenient to keep the total number of elements constant for a given jet. If, for example, it is decided that the unit modeling dimension for the fully extended jet will be 1 mm, then if $L(i)$ is the length of the ith particle in millimeters, at the time of the last particle-particle breakup the number of modeling elements in that particle may be defined as follows:

$$N(i) = \text{Integer part of } (1 + L(i)) \tag{18}$$

$N(i)$ will be constant with time.

This process may be applied to each particle in sequence from the jet tip to the jet tail to produce a jet element number sequence, which overlays the jet particle sequence and which may be used in determining the crater impact sequence. It is to be expected that some later particles will contain elements that impact the crater wall earlier than elements ahead of them in the same or earlier particles because of transverse deflections and/or tumbling motions.

The penetration hole depth L_h contributed by the full impact of an element of length L_e will be, according to Eq. (3),

$$L_h = L_e \sqrt{(\rho_j / \rho_t)} \tag{19}$$

depending on the extension of the particle containing that element, so L_e and the corresponding element diameter D_e will vary. The volume of any element is conserved in the development and extension of the jet.

Penetration hole volume will be proportional to the kinetic energy of the impacting element, for full impact:

$$\text{Vol} = K_i (M \; V_e^2)/2 \tag{20}$$

where

V_e = the jet element longitudinal velocity
M = the jet element mass
= $(\pi/4)D_e^2 L_e \rho_j$, where D_e is the cylindrical jet element diameter, derived from the corresponding particle diameter
K_i = constant for a given jet/target combination

A further approximation may be made; the target penetration volume contributed by each element can be assumed to be cylindrical. This will generally[13] not be the case for the leading and trailing elements of the particle. A prolate hemispherical hole profile would be more accurate for a jet element of higher density than the target and a hemispherical hole for a jet element of lower density than the target. The experimental work[14] in this field also shows that hole volume increases with penetrator density for a given target material and penetrator kinetic energy, but this dependence of hole volume on penetrator density is not constant for all penetrator velocities. Both of these factors could be added to the hole development model in a straightforward manner, if required.

Hole volume may also be defined in terms of a diameter D_h and a length L_h:

$$\text{Vol} = (\pi/4)D_h^2 L_h = (\pi/8)K_i D_e^2 L_e \rho_j V_e^2 \tag{21}$$

or, making the substitution,

$$K = (K_i/2)^{1/2}(\rho_j/\rho_t)^{1/4} \tag{22}$$

$$D_h = K D_e V_e \tag{23}$$

and K is a hole profile constant for cylindrical geometry.

The ratio of jet element kinetic energy to penetration hole volume is also a function of the target strength, as well as the relative densities of jet and target. A hydrodynamic model[15] presented by Szendrei has provided the relationship following:

$$\frac{\text{Jet energy}}{\text{Hole volume}} = \frac{[1 + \sqrt{\rho_j/\rho_t}]^2 \sigma}{\sqrt{(\rho_j/\rho_t)}} \tag{24}$$

The strength term, σ, may be related to the Brinnel Hardness Number (BHN), that is, $\sigma = 0.63$ BHN. Expressing σ in dynes/cm^2 and by substitution of the right-hand side of Eq. (24) in Eq. (20),

$$K_i = \sqrt{(\rho_j/\rho_t)}/[1 + \sqrt{(\rho_j/\rho_t)}]^2 0.0063 \text{ BHN} \tag{25}$$

$$K = (\rho_j/\{2[1 + \sqrt{(\rho_j/\rho_t)}]^2 0.0063 \text{ BHN}\})^{1/2} \tag{26}$$

The next aspect of penetration by nonideal jets, that has to be accounted for is the impact of later jet particles with the crater walls formed from earlier impacts.

For a known warhead to target standoff distance and the current penetration crater profile, the impact of an element within a jet particle with that crater profile may be predicted from the particle velocity equations [Eqs. (12) and (16)] and a knowledge of the diameter D_e of the element.

The criterion for impact is

$$X_d - (D_e/2) < X_c \tag{27}$$

where X_d is the transverse coordinate of the center of the leading element face and X_c is the transverse coordinate of the penetration crater profile in the same reference plane and at the same longitudinal depth in the target space. A simplified two-dimensional representation of this geometry is shown in Fig. 3.

If the standoff distance from the origin of the kth jet element and the front face of the target is $S(k)$ and the current depth of the element in the penetration crater is $Z(k)$, then the transverse displacement of that element in the crater, $X_d(k)$, is

$$X_d(k) = V_t(k)[S(k) + Z(k)]/V(k) \tag{28}$$

where

$V_t(k)$ = the transverse velocity of the kth jet element [Eq. (16)]
$V(k)$ = the longitudinal velocity of the kth jet element

A further modeling simplification may be introduced to define the penetrative interaction between an element within a particle that satisfies Eq. (27) at the target. The part of the presented element cross section that may be pro-

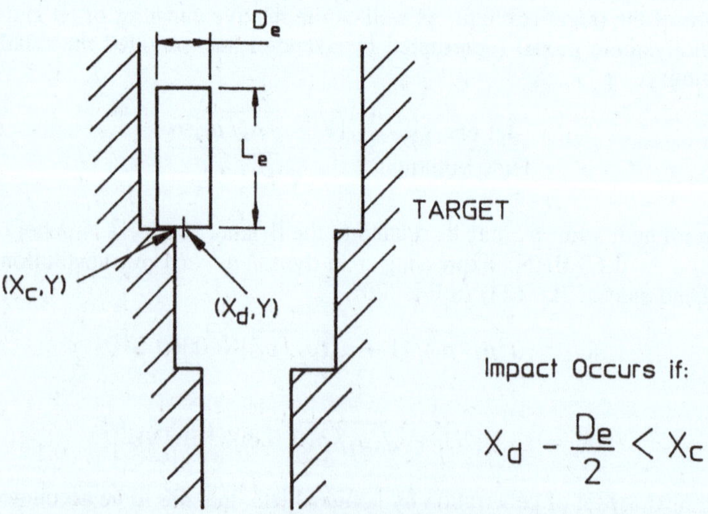

Fig. 3 Geometry of partial jet element impact upon the penetration crater wall.

jected into the target is assumed to be equated to an equivalent element with reduced diameter. The impacting kth element of the jet of nominal diameter $D(k)$ is assigned an effective diameter $D'(k)$:

$$D'(k) = X_d(k) + [D(k)/2] - X_c \tag{29}$$

No account is taken of the asymmetric pressure loading on the particle beyond acknowledging that nonoverlapping portions of the particle will ricochet from the crater wall and will not add to systematic crater enlargement in the line of the original particle velocity vector. Equation (23) is used with $D'(k)$ and $V(k)$ to calculate an effective crater diameter to be added asymmetrically to the previous crater profile.

This modeling approach was first validated[12] against experimental data derived from a standard, nonprecision shaped charge that produced a distinctly nonideal jet; this enabled the jet misalignments to be measured unambiguously. The predicted penetration standoff curve and the mean experimental curve are reproduced in Fig. 4, and a comparison of a predicted penetration crater cross section is shown in Fig. 5.

It is a simple step to add a constant translational velocity to the particles of a jet, such as might be induced by a misalignment of a missile flight vector from the warhead axis. In the extreme case, when the warhead axis is at 60° or more below the flight vector, we have the so-called fly-over–shoot-down condition. The simulation can proceed as before. The systematic translational velocity component is summed with the inherent transverse velocity of each element in turn to produce the net value for $V_t(k)$ in Eq. (28).

Comparisons between predicted and experimental results,[12] for the same type of warhead as was used in static tests are shown here in Figs. 6 and 7. The warheads were fired normal to the RHA target but with a velocity of 150 m/s or

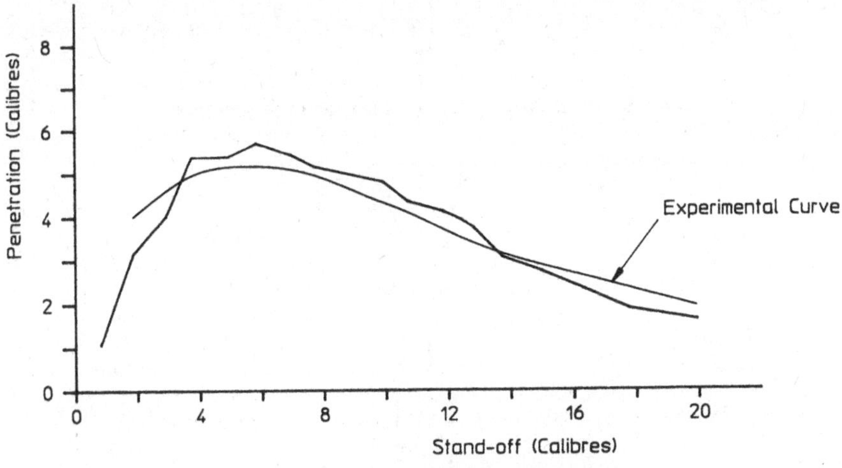

Fig. 4 Predicted penetration standoff curve.

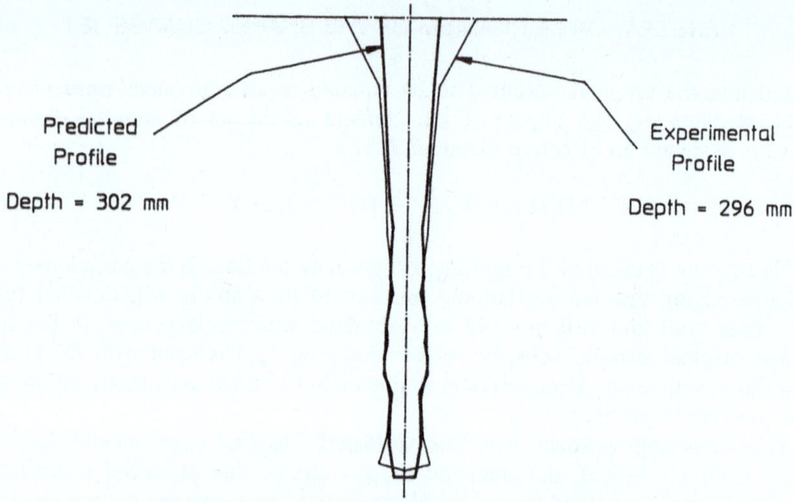

Fig. 5 Predicted penetration crater profile.

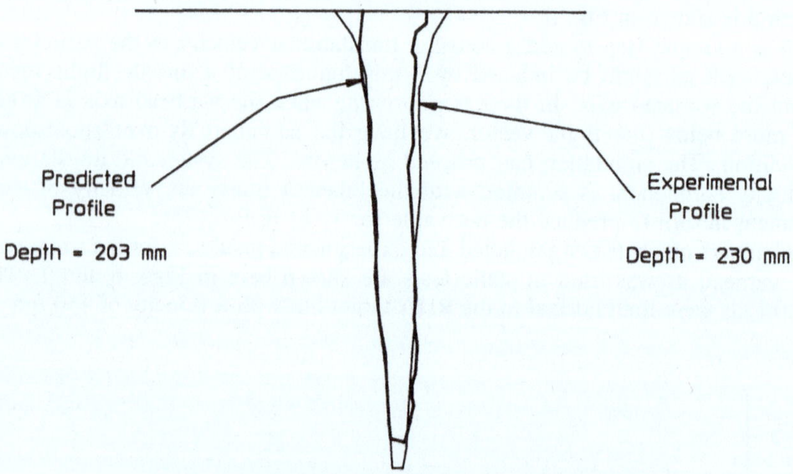

Fig. 6 Predicted dynamic penetration at 150 m/s.

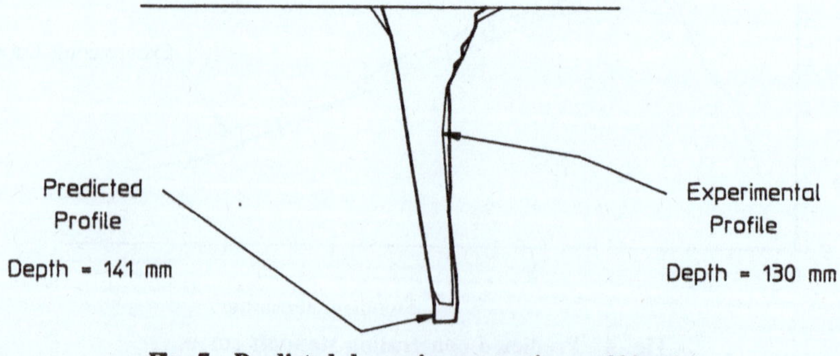

Fig. 7 Predicted dynamic penetration at 300 m/s.

300 m/s parallel to the target surface. The agreement between the penetration depths and crater profiles is sufficient to recommend this modeling approach for comparative assessment and design applications. The next section addresses an extension of this approach, to the attack of explosive reactive armor, ERA.

6. Jet Interaction with ERA

In the early 1970s, a patent[16] was claimed relating to a reactive armor element comprising two metal plates forming a sandwich with an explosive layer between them. As a shaped charge jet penetrates the sandwich, the explosive layer is initiated. The metal plates are accelerated apart, normal to the plane of the sandwich.

If the attacking jet axis is oblique to the plane of the sandwich, then the metal plates present a continuous transfer of material and momentum into the path of the oncoming jet. There are distinct similarities between this condition and the fly-over–shoot-down situation, previously discussed, except that some additional factors come into play:

1) For a single simple sandwich, there is one advancing and one retreating plate.

2) The plate spacing and hole alignment is changing with time; therefore, ricochet at the first plate can contribute to increased jet impact with the second plate.

3) Gas flow in the explosive detonation products can induce deflections and instabilities in the jet particles.

Some simplified theoretical models of these phenomena have been presented[17] that are supported by experimental data obtained from flash x-ray radiographs. The typical sequence of events is illustrated in Fig. 8.

Erosion of jet length by full impact between a jet element and front plate material can be mapped and calculated as described in previous sections, once the plate movement history is known. In the absence of experimental data on plate acceleration, hydrocode simulations or an empirical model can be used.[18] The hydrocode model has the advantage of providing a velocity time curve; this is most important in dealing with the initiation phase and the plate acceleration from rest, during which time the plate material flux into and across the path of the jet is changing. A typical plate velocity time curve is shown in Fig. 9.

For a grazing incidence on the edge of the front plate crater, the overlap criteria of Eq. (27) may be used and the effective penetrating element diameter $D'(k)$ extracted with Eq. (29). The deflected portion of the element may then be tracked to the receding plane of the back plate. The mass flux ratio[17] for the jet and the front plate is

$$\mathring{m}_j / \mathring{m}_p = K_2\{(V_j/V_p)\tan\phi - [1/\cos\phi]\} \tag{30}$$

where

$K_2 = \rho_j \pi[D(k) - D'(k)]/(4\rho_p h)$
$\mathring{m}_j$ = the local mass flux of the jet interacting with the plate
$\mathring{m}_p$ = the local mass flux of the plate

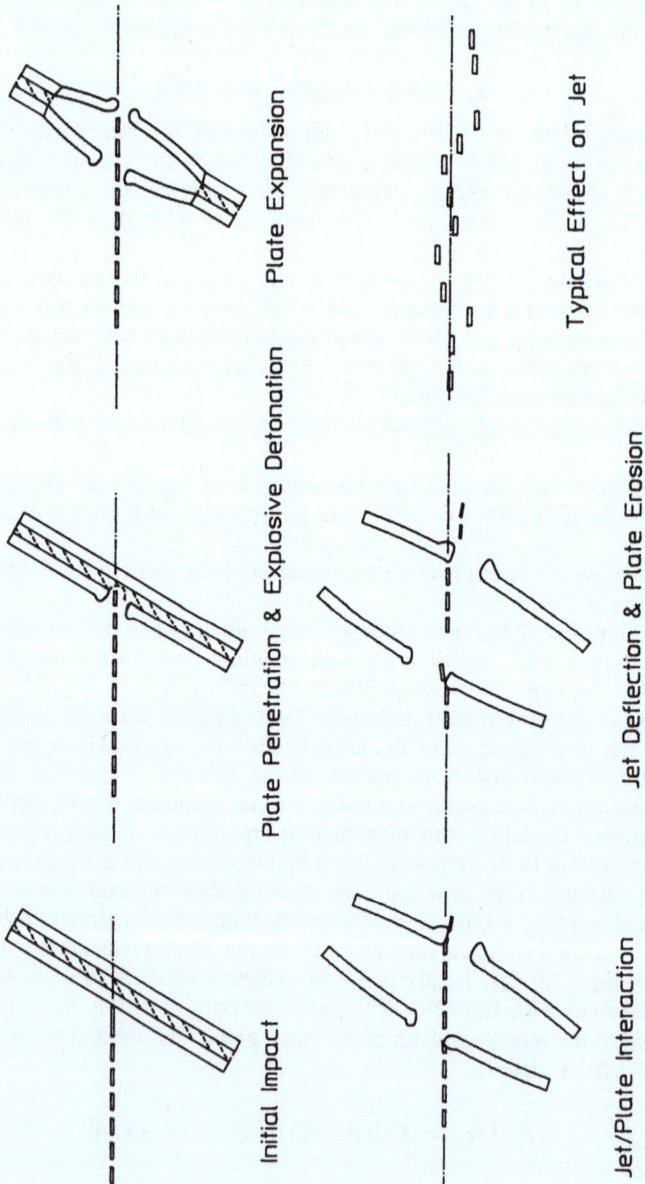

Fig. 8 Interaction between jet and ERA sandwich.

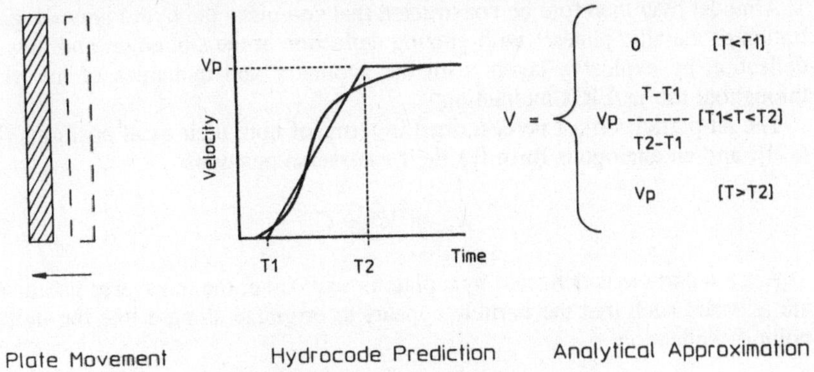

Fig. 9 Modeling of ERA plate movement.

ρ_p = the density of the plate
h = the plate thickness
ϕ = the angle between the jet axis and plate surface

If it is assumed that the jet particle and the local plate material undergo a fully plastic collision, then they share a common deflection angle of α given by

$$\frac{1}{\tan\alpha} = +\left[\frac{V_j}{V_p}\frac{K_2}{\cos\phi}\left(\frac{V_j}{V_p}\tan\phi + \frac{1}{\cos\phi}\right) - \tan\phi\right] \tag{31}$$

For an interaction with the back plate, the +/− signs in Eqs. (30) and (31) are interchanged, giving both a higher relative mass flux and a higher deflection. This means that the front plate of the sandwich will have a lesser effect on the jet than will the back plate. Some experimental results confirm this.[17]

Those jet particles that pass through the front plate, either through the penetration slot or by deflection off its sides, are then subject to a turbulent pressure effect from the gaseous detonation products of the explosive layer. This effect may be approximated by a smooth pressure lasting for between 10 and 40 μs, depending on the sandwich geometry and materials. This pressure will tend to exaggerate any off-axis velocity of the jet particle. Assuming that the particle is approximately cylindrical, the incremental transverse velocity, ΔV_t, added to the particle during each modeling time step, Δt, will be approximately

$$\Delta V_t = 2p\Delta t\cos\phi/\pi R_j\rho_j \tag{32}$$

where p is the instantaneous pressure and R_j is the jet particle radius.

For a multiple sandwich ERA, the possibility exists that the back plate of one sandwich and the front plate of the adjacent sandwich will collide. At typical plate velocities of 200 m/s to 700 m/s, it is likely that the plates will become welded together and thereafter behave as a single plate at a lower velocity derived from conservation of their net momentum.

A model may therefore be constructed that combines the hydrodynamic penetration of metallic plates[12] with grazing deflection at the slot edges and pressure deflection by explosive layers with the geometry and dynamics of the plates throughout the jet/ERA interaction.

The jet particles must be described in terms of both their axial positions [Eq. (12)], and an analogous form for their transverse positions.

$$X_d = V_t t + C_t \tag{33}$$

When a particle is deflected by a plate or explosive, the transverse parameters are adjusted such that the particle appears to originate along a line through the point of deflection:

$$V_t' = V_t + \Delta V_t \tag{34}$$

$$C_t' = (V_t - V_t')t + C_t \tag{35}$$

As a good approximation for the central portions of the plates that are likely to be impacted, they may be assumed to maintain their rectangular profiles and fly parallel to their original positions. Each plate may then be described as a simple quadrilateral, with a multipoint polygon describing the penetration slot. This polygon will initially have zero area but will increase as the penetration slot is opened by successive jet particle impacts while remaining within the bounds of the plate coordinates.

Each particle coordinate position (X, Z) at a given time may be checked for overlap with the target plates, and according to the overlap position, it will be deflected, consumed by penetrating the plate surface or consumed by widening the penetration slot. After the entire jet has passed the ERA plates, the residual particles may then be tracked into a backing armor, using the penetration model described previously.

The results of a typical simulation are shown in Fig. 10. A shaped charge is fired at 8 charge diameters (CD) standoff into an RHA target. The standard penetration model predicts 5.78 CD penetration. When an ERA sandwich of 3-mm mild steel/6-mm PETN/13-mm RHA is placed at 6 CD stand-off from the charge, the predicted penetration into the RHA backing armor, still at 8 CD standoff from the charge, is reduced to 1.44 CD; this is due to fast portions of the jet near the jet tip. Asymmetric slots are cut in the plates as they move apart.

As a particle impacts a plate, it forms a rapidly expanding crater, wider than the particle. Hence, a following particle can pass through this hole before the plate has moved enough to present fresh material to the jet path. The jet will emerge from this process as a series of minimally affected particles with gaps where others have been consumed or deflected by the plates; this has been described as the "pebblestone" effect.[17] It will be most notable near the faster jet tip, whereas the slower particles toward the jet tail will allow the plates time to move into their path; although they will tend to be more massive and hence less deflected by grazing impacts.

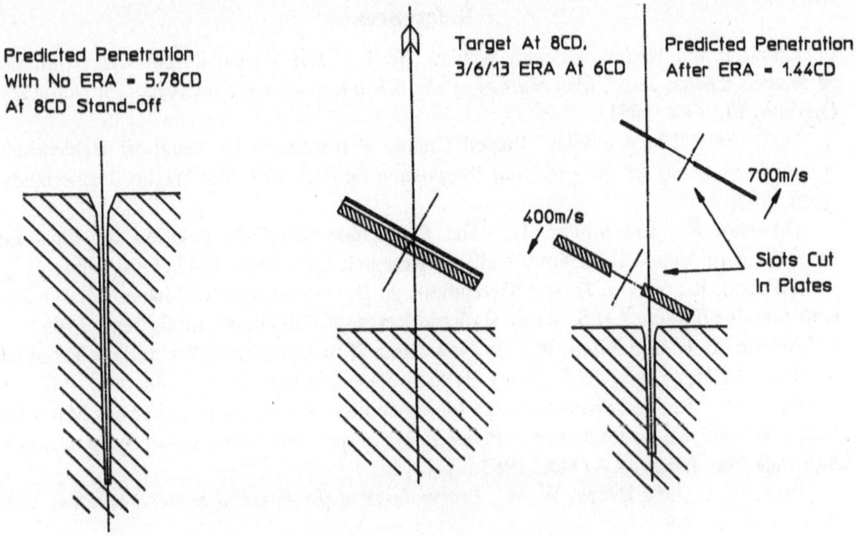

Fig. 10 ERA penetration model predictions.

There are a number of factors to be considered in both the design of a jet to defeat a given ERA combination, and the design of an ERA to defeat a given jet. These include the jet velocity gradient, the plate attack angles, and velocities and interactions between plates in adjacent sandwiches of an ERA.

In general a fast jet will have a better chance of defeating a single sandwich than a slower one. This arises from a combination of the finite explosive response time and a reduced pebblestone effect. The explosive takes a few microseconds to respond to the jet impact, and the plates may take up to 30 μs to accelerate to their final velocity. During this time, a number of jet particles may pass through the plates virtually unaffected, before the plates start to deflect the jet. For a fast jet, particularly if that jet has a low velocity gradient, a much larger number of particles may get through than for a slower jet; in addition, faster particles create larger holes in the target plates. A very fast pseudo long-rod penetrator would constitute an efficient attack mechanism for defeat of lightweight ERA, as considered here.

7. Conclusion

The modeling methods described here have found a wide application in practical design and assessment problems. In some applications, the degree of agreement between test and prediction can be very close; in others, where substantial simplifying assumptions are needed to reduce problem complexity, the methods still remain valuable for comparative assessments. It is to be expected that substantial further modeling refinement and extension will be achieved in this area, if for no other reason than that the methods are generally compatible with the ever-more-powerful desktop computer, and they satisfy a growing need for fast parametric evaluation of design tradeoffs.

References

[1]Chou, P. C., Hirsch, E., and Walters, W. P., "The Virtual Origin Approximation of Shaped Charge Jets," *Proceedings of the 6th International Symposium on Ballistics*, Orlando, FL, Oct. 1981.

[2]Held, M., "Characterising Shaped Charge Performance by Stand-off Behaviour," *Proceedings of the 7th International Symposium on Ballistics*, The Hague, Netherlands, April 1983.

[3]DiPersio, R., and Simon, J., "The Penetration-Stand-Off Relation for Idealised Shaped-Charge Jets," U.S. Army Ballistic Research Lab, Rept. 1542, Feb. 1964.

[4]DiPersio, R., Simon, J., and Merendino, A. B., "Penetration of Shaped Charge Jets with Metallic Targets," U.S. Army Ballistic Research Lab, Rept. 1296, Sept. 1965.

[5]Allison, F. E. and Vitali, R., "A New Method of Computing Penetration Variables for Shaped-Charge Jets," U.S. Army Ballistic Research Lab, Rept. 1184, Jan. 1963.

[6]Held, M., "Refined Shaped Charge Jet Penetration Measurement in Various Materials Versus Hydrodynamic Prediction," *Proceedings of the 10th International Symposium on Ballistics*, San Diego, CA, Oct. 1987.

[7]Pack, D. C., and Evans, W. M., *Proceedings of the Physical Society, London*, Vol. B64, 1958, p. 298.

[8]deRosset, W., "Multiple Impacts on Monolithic Steel," *Proceeding of the 6th International Symposium on Ballistics*, Orlando, FL, Oct. 1981.

[9]Christian, D., and Gehring, J., "Analysis of High Velocity Projectile Penetration Mechanics," *Journal of Applied Physics*, Vol. 37, No. 4, 1966, p. 1579.

[10]Cullis, I. G., and Nash, M. A., "The Importance of Afterflow in High Impedance Mismatch Penetration," *Proceedings of the 9th International Symposium on Ballistics*, Shrivenham, England, April–May 1986.

[11]Chou, P. C., and Foster, J. C., Jr., "Theory of Penetration by Jets of Non-Linear Velocity and in Layered Targets," *Proceedings of the 10th International Symposium on Ballistics*, San Diego, CA, Oct. 1987.

[12]Golesworthy, R. C., and Townsend, I., "Analytical Model of Shaped Charge Penetration in the Direct and Overflying Top-Attack Modes," *Proceedings of the 8th International Symposium on Ballistics*, Orlando, FL, Oct. 1984.

[13]Engel, O. G., "Crater Depths in Fluid Impacts," *Journal of Applied Physics*, Vol. 37, No. 4, 1966, pp. 1798-1808.

[14]Christman, D. R., Gehring, J. W., and McMillan, A. R., "Hypervelocity Impact Studies on Aerospace Materials and Structures," General Motors Defense Research Laboratories, Rept. TM63 205A, 1963.

[15]Szendrei, T., "Analytical Model of Crater Formation by Jet Impact and its Application to Calculation of Penetration, and Hole Profiles," *Proceedings of the 7th International Symposium on Ballistics*, The Hague, Netherlands, April 1983.

[16]Held, M., British Patent No. 1581125, 1974.

[17]Mayseless, M., Erlich, Y., Falcovitz, Y., and Rosenberg, G., "Interaction of Shaped-Charge Jets with Reactive Armour," *Proceedings of the 8th International Symposium on Ballistics*, Orlando, FL, Oct. 1984.

[18]Fucke, W., Bol, J., and St. Schuman, "Velocity of Sandwich Plates Driven by Thin H. E. Layers," *Proceedings of the 10th International Symposium on Ballistics*, San Diego, CA, Oct. 1987.

Chapter 9
Part C

Target Interaction: Penetration of Thin Targets by Fragments

Rodney F. Recht*
Ballistic Impact Dynamics, Englewood, Colorado 80110

1. Introduction

BETTER engineering methods are being sought that can be used with confidence to control and improve fragment lethality with respect to various air targets under various encounter conditions. A first step is to consider how ballistic impact damages targets and target components. To reach components that are shielded, fragments must penetrate into the target. To be effective, they must retain sufficient lethality to cause component malfunction or to increase malfunction probability after penetrating the shielding. A single fragment may disintegrate into a number of smaller fragments and/or generate secondary ballistic particles derived from target components. Fragments may act individually or collectively to penetrate and/or damage targets. To be effective in penetrating multiple element targets, good penetrators impart relatively small impulsive loads to elements (i.e., these penetrators retain relatively high momentum). Depending upon the target component, penetration may or may not contribute to the lethality of a fragment; impulsive loading may be much more damaging (particularly in situations involving multiple fragment impacts). Large impulsive loads are produced by fragments that are not good penetrators. Hence, it is often desirable for a fragment to be a good penetrator during penetration through shielding and then to become a poor penetrator.

This section addresses the interaction of targets with fragments. The presentation focuses upon a single fragment impacting upon thin plates. The fragment configuration, mass, and velocity are characteristic of those that are propelled by a fragmentation warhead; and the target materials and thicknesses, for the most part, are characteristic of aircraft targets and light vehicle targets.

*Proprietor, Ballistic Impact Dynamics.

2. Momentum and Energy Balances

Penetration behavior of a fragment in a multiple element target is defined principally by residual velocity V_r and residual mass M_r at points along the path of penetration. Impulsive loading I is associated with the changes in velocity and mass at each impact. Given that appropriate models exist for computing the residual mass of a fragment after each impact as a function of impact characteristics, the characteristics of the penetration process and the impulsive loading of target elements are tied together completely by the equations of motion.

As applied to a single fragment impacting a single plate, the conservation of momentum (a vector equation) and the conservation of energy (a scalar equation) may be written precisely as follows:

$$M_o V_o = M_{\mathrm{rp}} V_{\mathrm{rp}} + m V_{\mathrm{r\,m}} + I \tag{1}$$

$$\frac{M_o V_o^2}{2} = \frac{M_{\mathrm{rp}} V_{\mathrm{rp}}^2}{2} + \frac{m V_{\mathrm{r\,m}}^2}{2} + \frac{(M_o - M_{\mathrm{rp}}) V_o^2}{2} + E_f + W_s \tag{2}$$

where

M_o = impacting fragment mass
M_{rp} = all residual fragment mass that passes through the target element
m = total mass driven from the target element
V_{rp} = residual velocity of mass center of residual pieces of fragment that pass through the element
$V_{\mathrm{r\,m}}$ = residual velocity of mass center of m
I = impulse transmitted to target element due to the shear strength of the target material and the fragment material stopped by the element
E_f = plastic deformation energy associated with the impact of two *free* masses of mass M_o and mass m (as if m were not attached to plate)
W_s = all plastic deformation energy due to m being attached to a target element made of a material that has shear strength

Added together, E_f and W_s represent all of the kinetic energy that is transformed into fragment and target deformation energy. The term $(M_o - M_{\mathrm{rp}})V_o^2/2$ represents the kinetic energy of the material stripped from the fragment that does not pass through the element.

These equations pertain to oblique plate perforation, where fragment direction changes and the direction of the mass center of m is different from that of the fragment pieces. Three-dimensional equations have been developed. However, for purposes of this illustrative discussion, normal perforation will be assumed so that all motion is unidirectional.

A precise equation can be written for E_f by considering the elastic-plastic impact of the two free masses M_o and m. This equation has the form

$$E_f = [m/(M_{\mathrm{rp}} + m)]\frac{M_{\mathrm{rp}} V_o^2}{2} - m(\Delta V_{\mathrm{r\,m}})^2 \frac{(M_{\mathrm{rp}} + m)}{2M_{\mathrm{rp}}} \tag{3}$$

where ΔV_{rm} is the increase of velocity in m due to postperforation elastic decompression and conservation of momentum, and

$$V_{rm} = V_r + \Delta V_{rm} \tag{4}$$

$$V_{rp} = V_r - (m/M_{rp})\Delta V_{rm} \tag{5}$$

Here V_r = residual velocity of mass center of all pieces of both M_{rp} and m.

Similarly, a precise equation may be written for W_s by incorporating E_f in Eq. (2) and solving for W_s; thus,

$$W_s = \{[M_{rp}/(M_{rp} + m)]\, M_{rp}V_0^2/2\} - [(M_{rp} + m)\; V_r^2/2] \tag{6}$$

When $V_r = 0$, then the impact velocity required for penetration is V_{50} (the ballistic limit velocity) and the associated value of W_s, defined as $(W_s)_{50}$, is

$$(W_s)_{50} = [M_{rp}(M_{rp} + m)]\; M_{rp}V_{50}^2/2 \tag{7}$$

This unique value is precise. A similar equation can be written for W_s by replacing V_{50} with V_x, an unknown characteristic velocity with a value defined as that required to make the equation true; thus,

$$W_s = [M_{rp}/(M_{rp} + M)]\; M_{rp}V_x^2/2 \tag{8}$$

3. Residual Velocity

If Eqs. (3–5), and (8) are substituted into Eq. (2) and the resulting equation is resolved in favor of V_r, the ΔV_{rm} terms cancel conveniently and the equation for V_r is

$$V_r = \sqrt{V_o^2 - V_x^2}/(1 + m/M_{rp}) \tag{9}$$

Figure 1 is an illustration of how well Eq. (9) correlates residual velocity data, even when the variable V_x is replaced by the constant V_{50}. Since the value of V_x is not usually very sensitive to changes in V and since the influence of V_x on V_r diminishes rapidly as V_o rises above V_x, this replacement is almost always permissible.[2]

4. Impulse Transmitted to a Plate

The impulse I transmitted to the element can be written in terms of V_r by using Eqs. (1), (4), and (5). Then Eq. (9) can be substituted for V_r so as to yield I as a function of V_x. Normalized with respect to initial momentum M_oV_o, this equation becomes

$$I/M_oV_o = 1 - (M_{rp}/M_o)\sqrt{1 - (V_x/V_o)^2} \tag{10}$$

In Fig. 2 is shown how the impulse I (now normalized with respect to the momentum M_oV_{50} at the ballistic limit velocity rather than M_oV_o) rises in

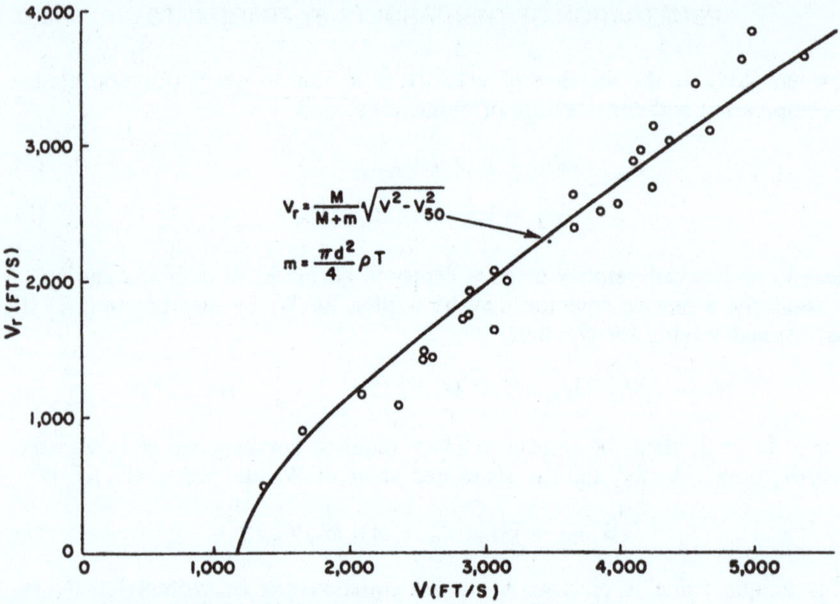

Fig. 1 Typical residual velocity data correlation assuming $V_x = V_{50}$ in equations; blunt cylindrical projectiles (44 grain), mild steel plate (1/8 in.).

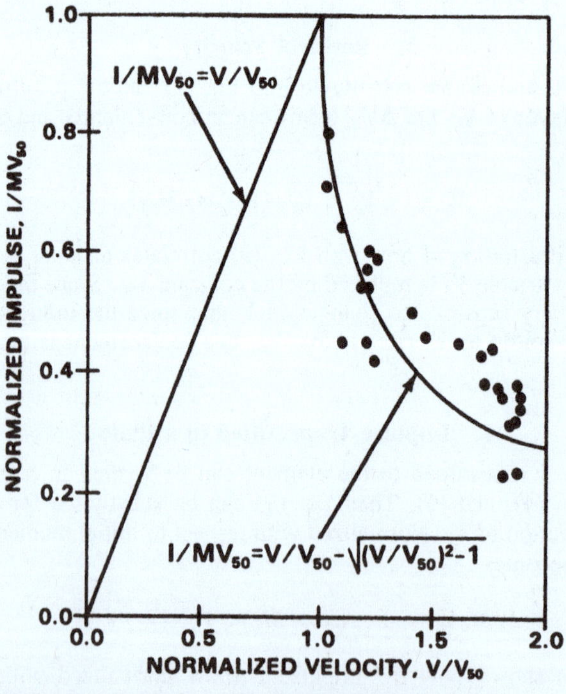

Fig. 2 Impulse transmitted to plate during ballistic perforation; steel cubes (0.25 g), aluminum alloy plate (4.8 mm), V_{50} = 850 m/s.

proportion to V_o below $V_o = V_{50}$ (i.e., during ricochet impacts) and then rapidly diminishes as V_o/V_{50} increases above 1.0 (perforations). Thus, fragments are designed for penetration by keeping V_{50} as small as possible. Conversely, to produce high-impulse loads, V_{50} values must be high.

Experiments (such as those described in Sec. 9.2 and in Ref. 10) have demonstrated how a fragment that breaks up during perforation of a first plate (where V_o/V_{50} is much greater than 1.0) imparts very little impulse to the first plate (as indicated by Fig. 2) and makes a small perforation. Resulting smaller fragments that encounter a second identical plate possess much smaller V_o/V_{50} values (often less than 1.0), and all such fragments contribute relatively large amounts of their momentum to impulsive loading of the second plate and may, acting in unison, produce a large perforation.

5. Ballistic Penetration Resistance and Its Measurement

5.1 Use of a Ballistic Pendulum to Measure Impulse

Equation (10) can be rearranged to give an expression for V_x as a function of I. Fragments can be fired through plates mounted in pendulums to measure I and thereby learn empirically how V_x changes with impact velocity in a given encounter situation. In Fig. 3 are presented typical experimental curves of V_x vs V_o. The intersections of these curves with the straight line represent $V_o = V_{50}$;

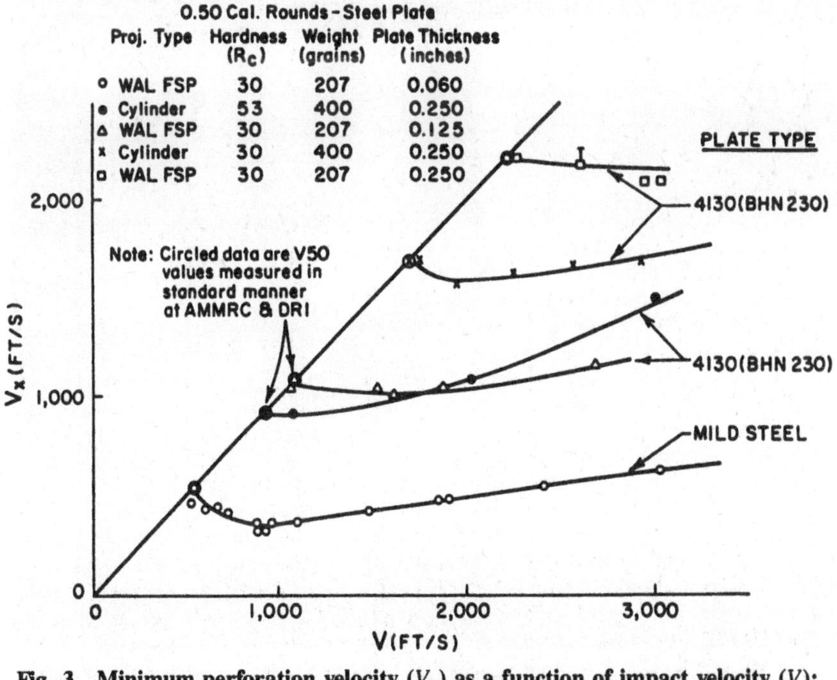

Fig. 3 Minimum perforation velocity (V_x) as a function of impact velocity (V); caliber 0.50 cylindrical rounds perforating steel plate at normal incidence.

V_{50} data plotted on the line were obtained by standard means and illustrate the correspondence between pendulum results and standard methods.[2]

5.2 Interpretation of Pendulum Data

Note that V_x is not very sensitive to V_o in Fig. 3. Downward trends relate primarily to plate bending near the limit velocity; upward trends relate primarily to fragment deformation. Data are presented in Ref. 2 that do show more sensitivity for certain situations; however, the curves shown are typical, particularly with regard to air target encounters. Thus, in both Eqs. (9) and (10), V_x can usually be replaced by V_{50} without producing significant errors in predictions of V_r and I values. In both cases, as the ratio V_o/V_{50} increases, the effects of differences between V_x and V_{50} on prediction errors of V_r and I values diminish rapidly.

In Figs. 4 and 5 we see how useful the pendulum is for obtaining V_{50} values for oddly shaped fragments. Points plotted concern impacts of the elongated triangular fragment where the principal axis of the fragment was parallel to the plate surface. Open data points pertain to coplanar impact of a flat side with the plate surface; closed points pertain to edge impact as was experienced by the fragment shown in Fig. 4. The pendulum measures the impulse I for each

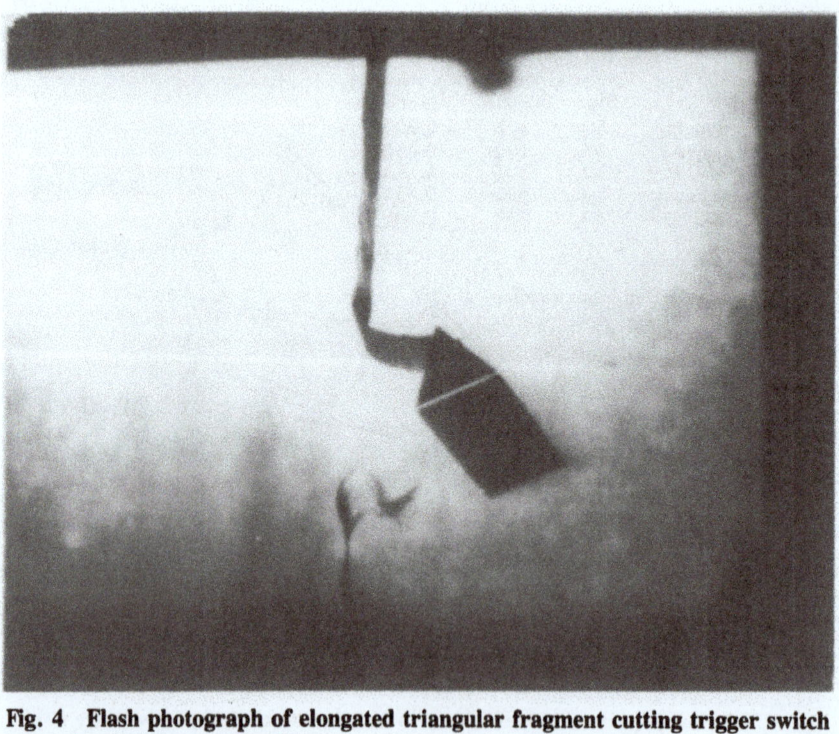

Fig. 4 Flash photograph of elongated triangular fragment cutting trigger switch just prior to impact with plate.

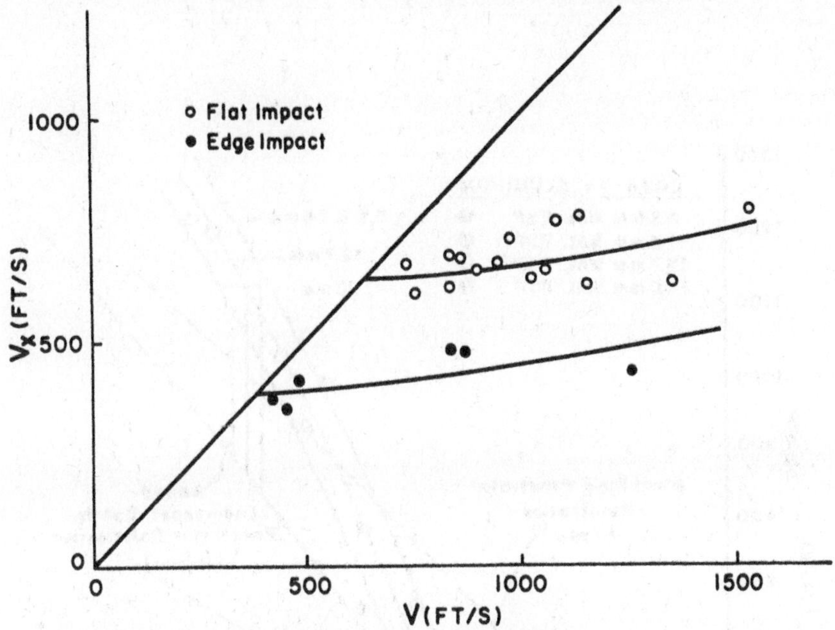

Fig. 5 Minimum perforation velocity (V_x) as a function of impact velocity (V); elongated triangular fragment perforating steel plate.

perforation that is converted to a V_x value. Curves plotted through the data intersect the straight line very near the limit velocity for each situation.

6. Disassembling the Ballistic Limit Velocity

6.1 Components of the Ballistic Limit Velocity

An explanation of the means by which the pendulum can be used to disassemble the limit velocity into its components is presented in Ref. 3. Normal ballistic limit data (V_{50n}) for four fragment simulators impacting aluminum alloy plate are plotted in Fig. 6. If plate perforation by these steel cylinders were accomplished by simple shear plugging (symbolized by τ), without any plate deformation external to the perforation and without fragment deformation, then the limit velocity would be represented by the lower dashed curve. This curve is true only in the region where plates are relatively stiff without being thick and velocities are low enough that the fragments do not deform (i.e., from a plate thickness to penetrator diameter ratio of $T/d = 0.6$ to $T/d = 1.0$). At $T/d < 0.6$, plate response (dishing and bulging, symbolized by δ) adds a velocity component $\Delta V_{50\delta}$ to the simple shear component $\Delta V_{50\tau}$. Where $T/d > 1.0$, the perforation process involves both cratering penetration (symbolized by p) and plugging (symbolized by d), and the steel fragments deform. These two components, and ΔV_{50p} and ΔV_{50d}, act to increase the limit velocity greatly.

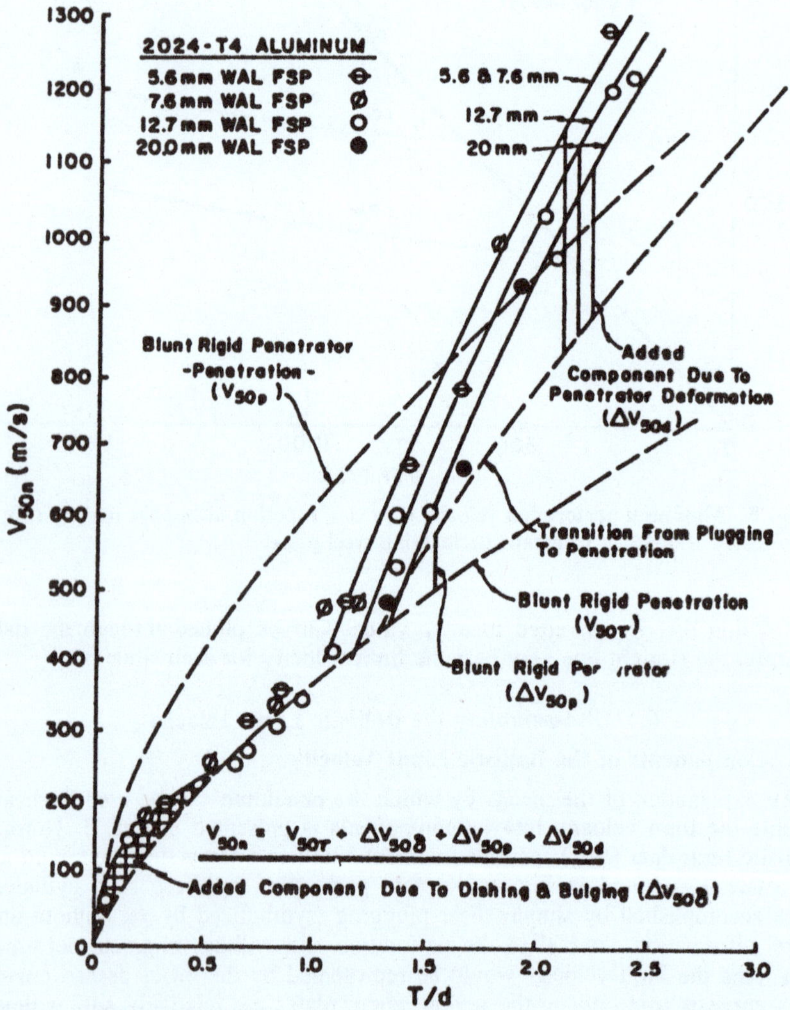

Fig. 6 Components of the ballistic limit velocity; blunt compact fragments vs aluminum plate.

6.2 Contrasting Rigid-Sharp and Rigid-Blunt Fragments

Sharp penetrators that do not change shape penetrate heavy plates by pushing material aside, which is a less efficient process than plate plugging. A comparison of the normal ballistic limit velocities for a rigid-blunt cylinder of L/d (L = penetrator length, d = penetrator diameter) = 2 and a rigid armor-piercing (AP) projectile of the same weight and caliber [steel plate; 230 Brinell hardness (BHN)][3] is presented in Fig. 7. The rigid-sharp (penetration) curve represents V_{50n} for the AP projectile, except in the plate response region where petalling modifies the curve downward. The plug-shear curve represents the V_{50n} for the cylinder, modified upward in the plate response region by bulging and dishing and at higher T/d values by penetration. At low T/d values, the AP projectile is superior; at moderate T/d values, the blunt fragment is superior; and at high T/d values, the AP projectile regains its superiority. Deformation of either penetrator would modify the curves upward, of course.

6.3 Effects of Fragment Deformation

A dramatic example of the influence of fragment deformation upon perforation is illustrated by the photographs shown in Fig. 8 and related test data.[3] Except

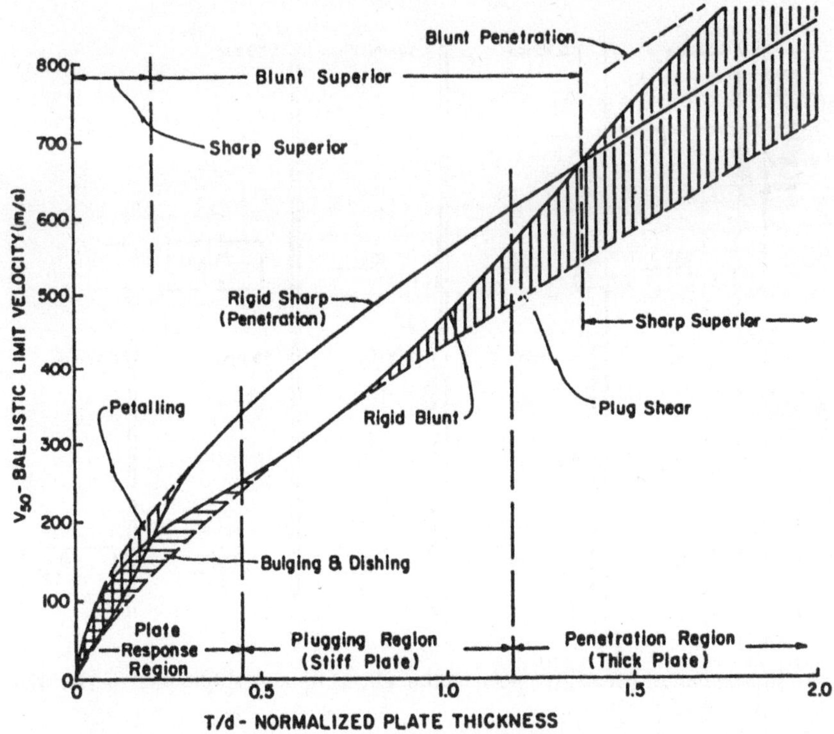

Fig. 7 Influence of penetrator and target characteristics upon ballistic perforation resistance; steel plate.

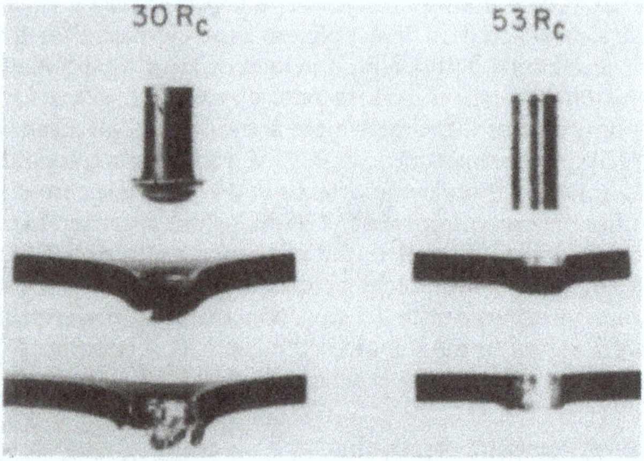

Fig. 8 Effects of penetrator deformation upon perforation characteristics (R_c = Rockwell Hardness No.).

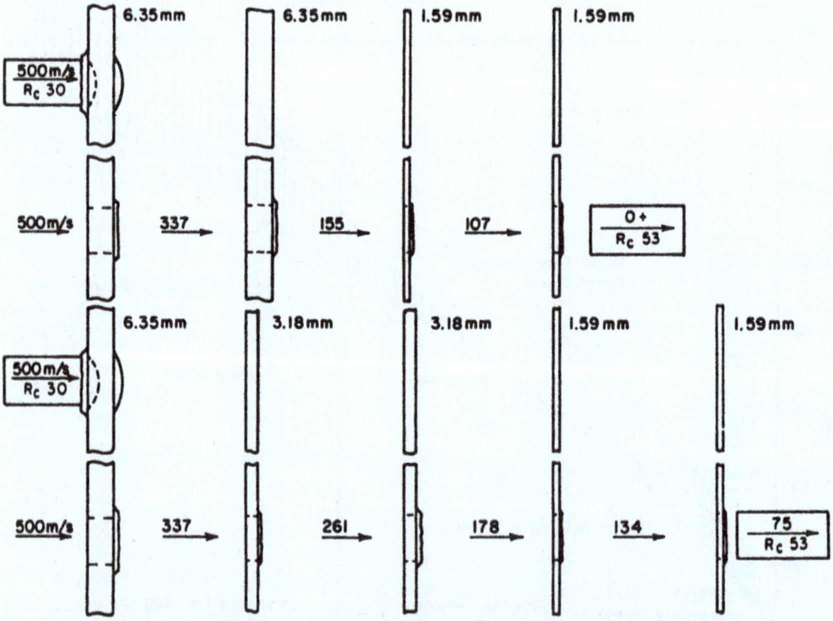

Fig. 9 Illustration of penetrator deformation effects upon multiple plate perforation.

for hardness, the two cylindrical fragments were identical prior to impact.[3] Each fragment impacted a single steel plate at a velocity near the limit velocity. As shown by the postimpact photographs, the hard fragment did not deform plastically. The two plates below each fragment show plate deformation just below and above the limit velocity for each case. As a result of the deformation experienced by the soft fragment, shear stress gradients in the plate are low and the limit velocity was 510 m/s, whereas the limit velocity for the hard fragment was only 275 m/s. Plate bulging, dishing, and petalling were involved in the first instance; simple plug shear was the principal perforation mechanism in the second instance.

Although the increase in V_{50} due to deformation was dramatic in itself, we see in Fig. 9 what this can mean with respect to multiple plate targets. Two different plate arrays are presented, each of which involves 6.35-mm steel plate as a first element. As shown, this element stops the soft fragment that impacts at 500 m/s. However, the hard fragment (also impacting at 500 m/s) perforates the four elements of the first array retaining a small residual velocity; it perforates the second (five-element) array retaining a velocity of 75 m/s. The significance of fragment deformation upon multiple plate perforation is clearly demonstrated.

6.4 Fragments of Arbitrary Shape

Reference 3 presents ballistic limit models for fragments of arbitrary shape. Experimental limit velocities for the fragments shown in Fig. 10 are compared with the model predictions shown in Fig. 11. A number of the limit velocities were obtained using the pendulum.

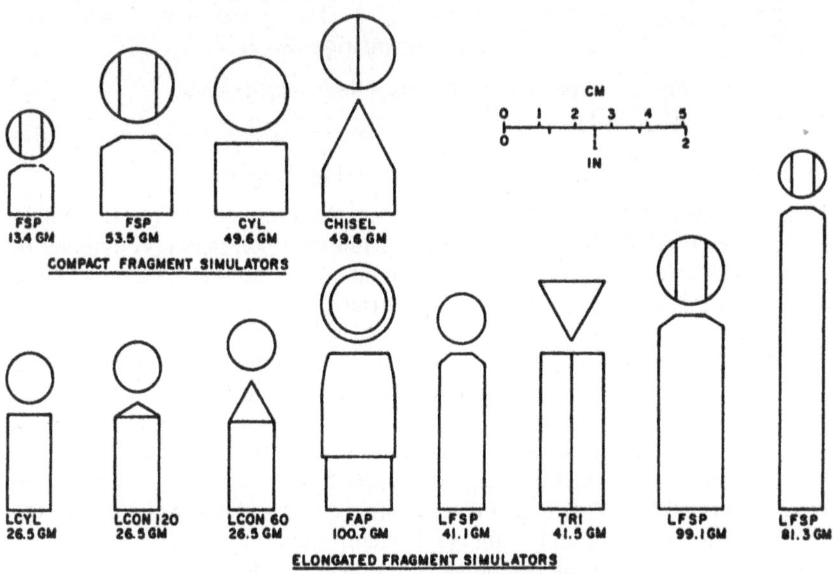

Fig. 10 Steel fragment simulating projectiles.

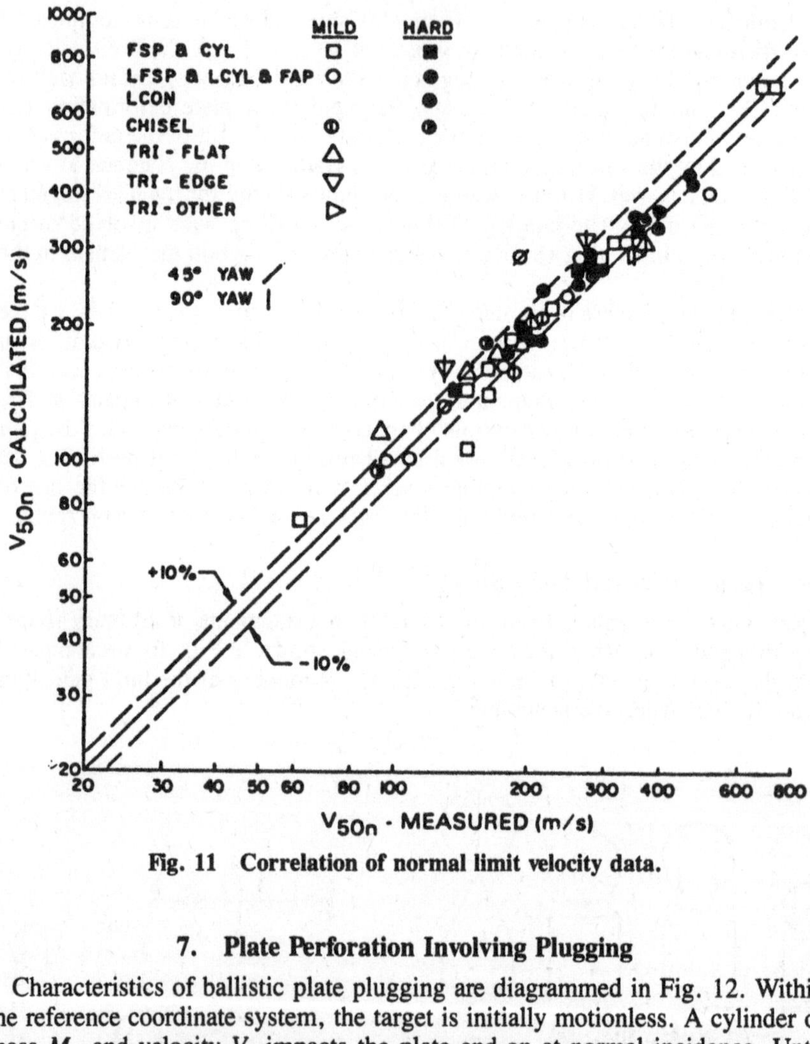

Fig. 11 Correlation of normal limit velocity data.

7. Plate Perforation Involving Plugging

Characteristics of ballistic plate plugging are diagrammed in Fig. 12. Within the reference coordinate system, the target is initially motionless. A cylinder of mass M_o and velocity V_o impacts the plate end-on at normal incidence. Until stress waves reflect back to the impact interface from the rear surface of the plate, the plate responds as a half-space. As long as the undeformed remainder of the cylinder has a velocity $U > kU_c$ (i.e., relative velocity of the cylinder with respect to the impact interface is greater than U_c, where U_c is the plastic wave velocity in the penetrator and k is an impedance matching factor), a shock wave stands above the impact interface and material passing through the shock is eroded away; melting produces impact flash. Cratering of the plate surface occurs. When V falls to kU_c, erosion ceases and the remainder of the cylinder begins to mushroom in the Taylor deformation modeling mode.[4] Cratering, bulging, and dishing responses of the plate produce plate deformations and affect the motion of the impact interface. A plug begins to shear from the

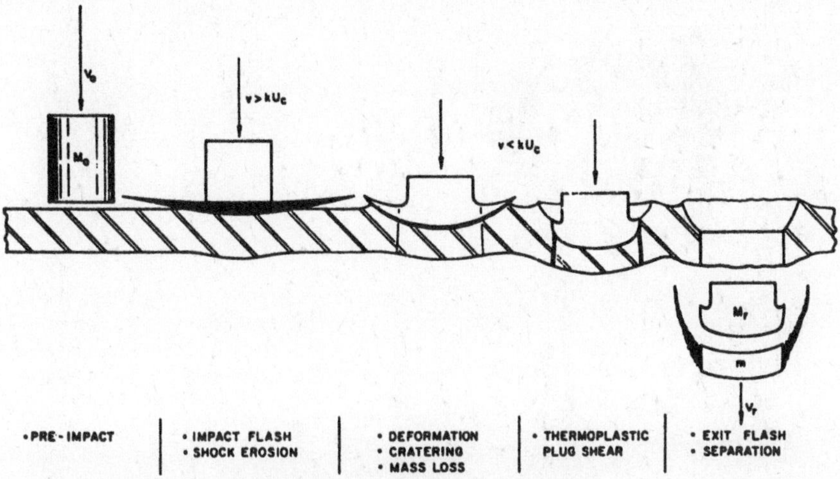

Fig. 12 Ballistic plate plugging.

plate; cylinder material that has been extruded beyond the plug radius is sheared from the deformed cylinder. As the plug shear progresses, heating occurs in the shear zone. At low velocities, the cool material on either side of the shear zone acts to quench the deforming material. At higher velocities, heating approaches being adiabatic and molten material is often rejected as exit flash. Subsequent to perforation, stored elastic energy is released, which causes separation of the two ballistic masses.

Photographic and radiographic inserts on the residual velocity curve of Fig. 13 illustrate typical effects of velocity on the plate plugging process. At low velocity (1600 ft/s), the cylinder suffers little deformation and the plate plug has the same diameter as the cylinder. At higher velocity (3000 ft/s), deformation increases the presented area of the fragment, which is reflected in plug size. At 5500 ft/s, deformation shear and/or shatter processes have greatly reduced the residual mass of the penetrator; the plate plug has broken up. Fragments behind the plate form a hollow, expanding ellipsoidal pattern. Near 900 ft/s this ellipsoidal pattern is fully developed; the largest residual remnant of the cylinder is at the front of the expanding cloud. Obviously, during plate perforation, the cylinder has been transformed into a ballistic impact damage agent of quite different characteristics. Both cylinder and plate fragments are now represented in the ballistic impact threat to subsequent target components that are subjected to multiple impacts.

8. Fragment Deformation and Mass Loss

8.1 Erosion Mass Loss Mode

As explained in the previous section, when a blunt fragment impacts a plate, material erodes from the front of the fragment until the relative velocity between the remaining fragment and the impact interface diminishes to the plastic wave

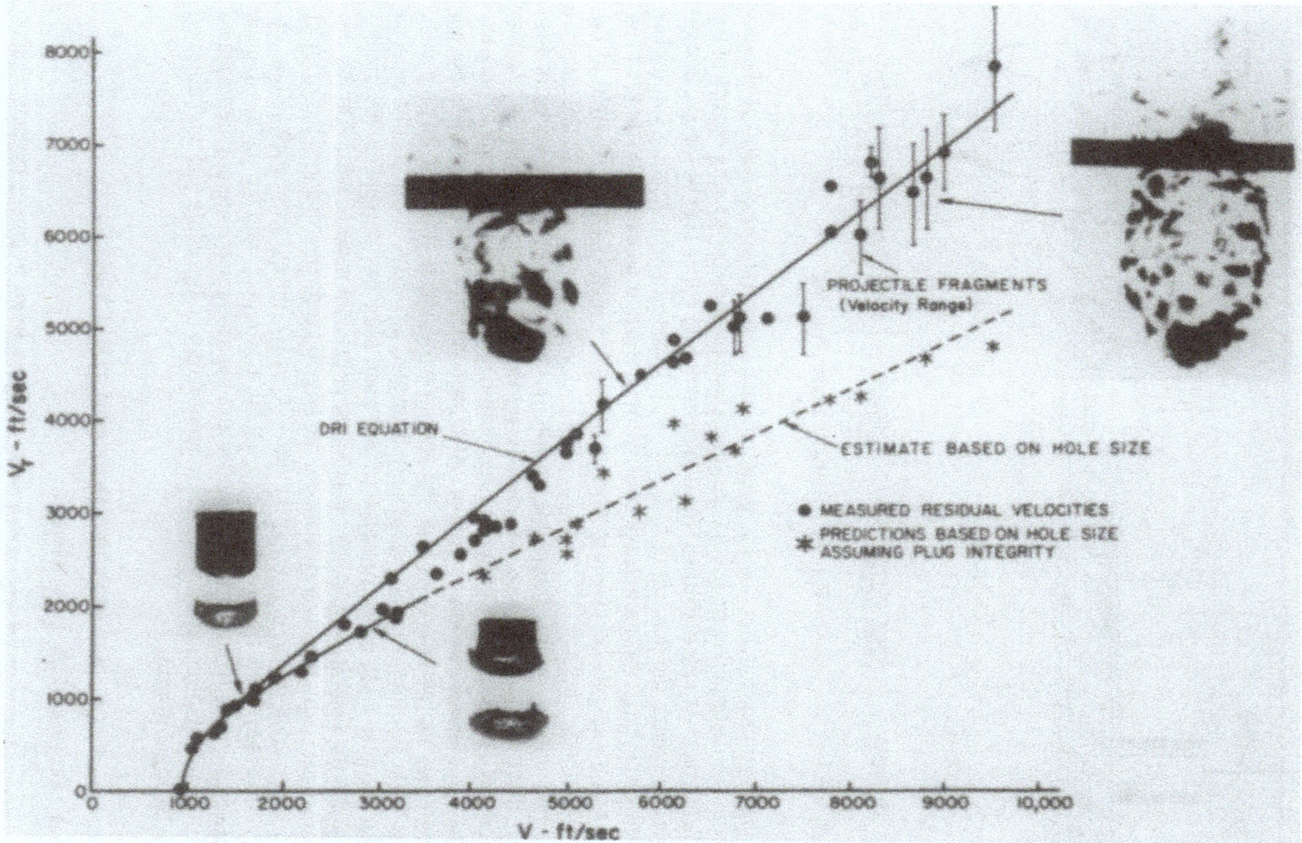

Fig. 13 **Effects of deformation and breakup on residual velocity prediction; *Watertown National Laboratory* fragment simulating projectiles perforating thin mild steel plate.**

velocity U_c in the fragment material. The following equation (derived in Ref. 4) defines the remaining fragment mass M_{re} (normalized with respect to initial mass M_o) subsequent to erosion for the plate plugging situation:

$$M_{re}/M_o = 1 + (m/2M_o)\ln\{(1 + 1/Q)/[1 + (V_o/U_c)^2/Q]\} \qquad (11)$$

where

M = plate plug mass
V_o = initial impact velocity
$Q = \sigma_e/\rho_c U_c^2$ (dimensionless)
σ_e = dynamic yield strength of fragment material
ρ_c = mass density of fragment material

8.2 Taylor Deformation

Displayed in Fig. 14 are top and side views of cylindrical steel fragments that impacted the surface of a hardened steel block at the indicated velocities. Below 410 m/s, the block experienced little deformation, but was cratered at higher velocities. Two photographs are reproduced in Fig. 15; the superimposed outlines are Taylor model predictions of final shape.[4] The dashed line within each outline is the boundary above which no plastic deformation was predicted. The dashed lines below the photographs indicate the front of the fragment prior to deformation. Predictions of final length and the plastic deformation boundary correspond very well to the observed deformations. Shape is also well predicted; upturning of the extruded material in the lower photograph was a result of surface cratering (not considered by the model).

8.3 Deformation-Shear Mass Loss Mode

During plate perforation, some of the extruded material (such as that shown in Fig. 15b) is sheared from the fragment. Mass loss by both erosion and deformation-shear mass loss modes is illustrated in Fig. 16. Given the long cylinder of indicated initial length, the portion diagrammed as a cross-hatched rectangle is lost to shock erosion (illustrated in Fig. 12 and described in Sec. 8.1). Then, the remaining undeformed cylinder deforms in the Taylor deformation mode to the shape indicated by the solid outline. As it passes through the plate, the material external to the assumed shear surface is sheared from the fragment (diameter of the shear surface is typically about 25% greater than the diameter of the cylinder). The final size and shape of the fragment are predicted to be as defined by the solid outlines, less the material outside the shear surface. Model predictions are compared with final shapes of cylinders that perforated plates in Fig. 17.

8.4 Shatter Mass Loss Mode (Critical Velocity, Orientation)

The two radiographs at the left of Fig. 18 show orthogonal views of a cubical steel fragment just prior to impacting the first of two spaced steel plates at a velocity of 5000 ft/s. Impact orientation can be characterized as a *corner* impact.

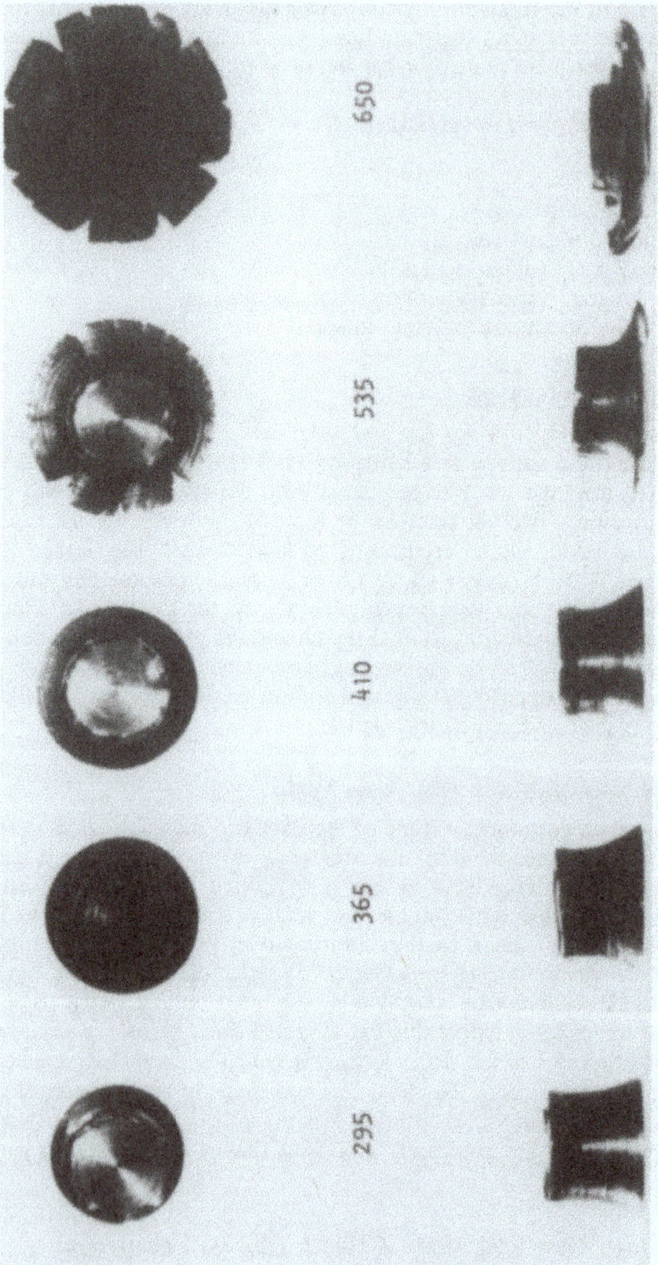

Fig. 14 Deformation of cylinders due to ballistic impact against the surface of a hardened steel block.

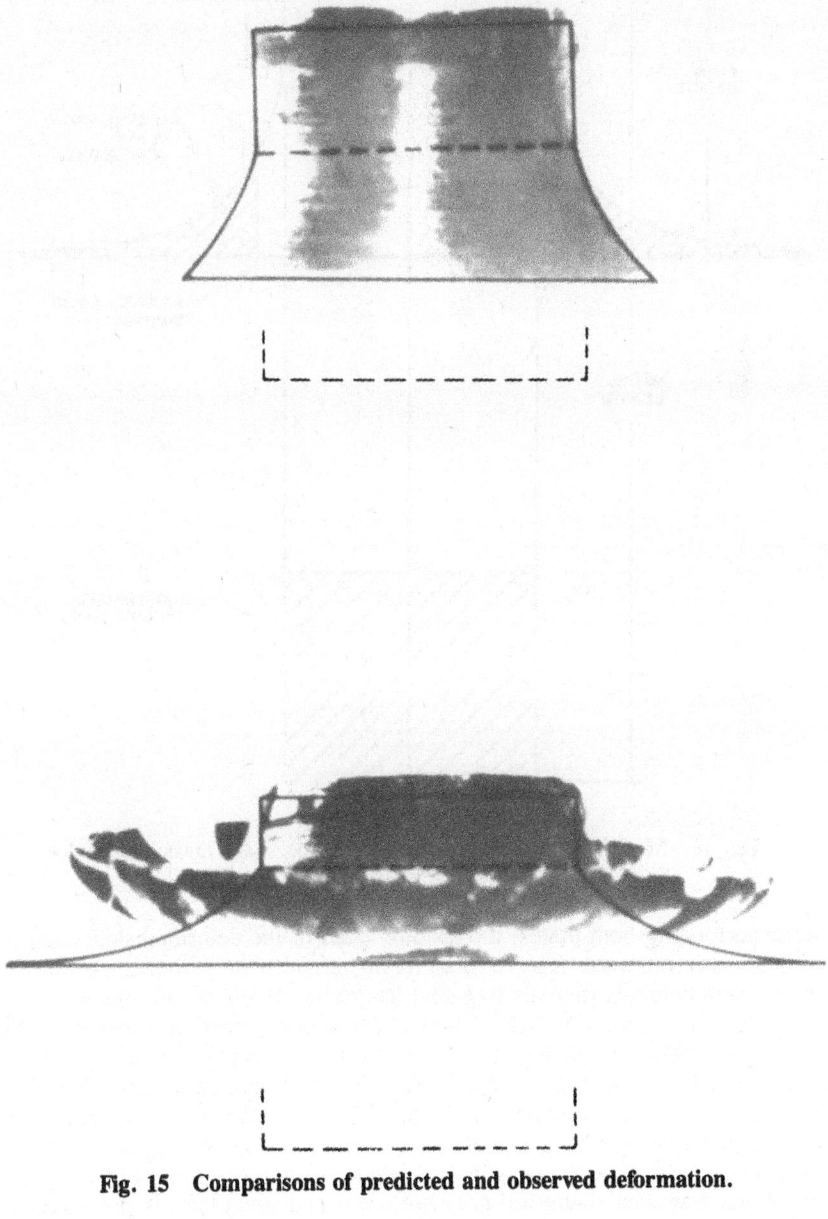

Fig. 15 Comparisons of predicted and observed deformation.

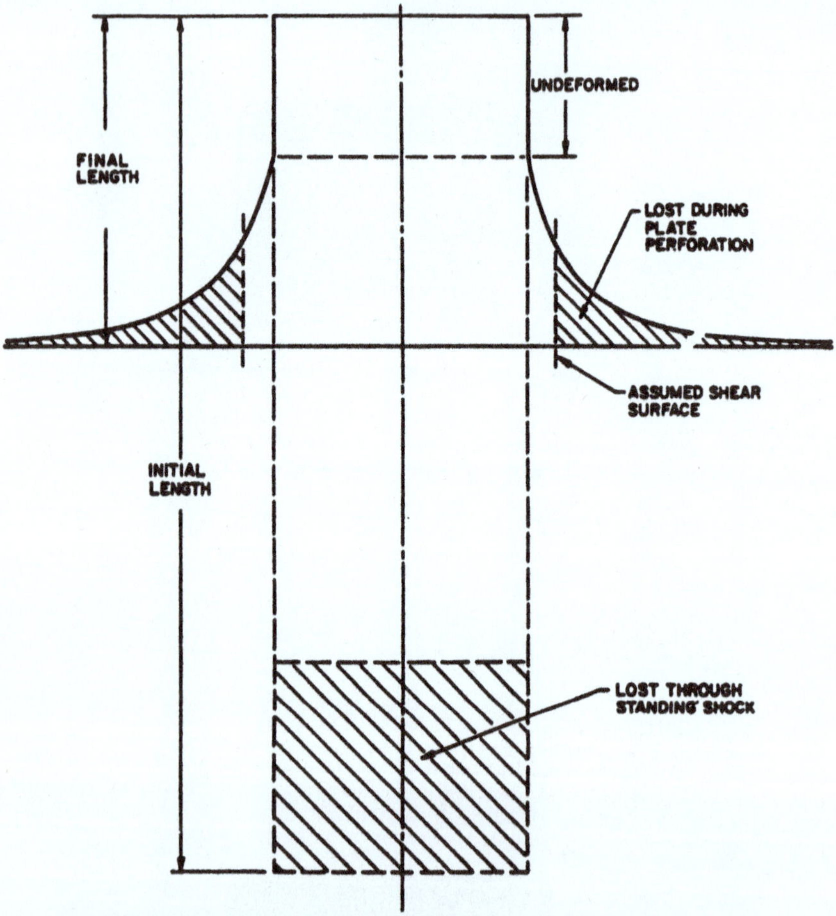

Fig. 16 Mass loss by erosion and deformation shear mass loss modes.

After perforating both plates, the residual mass of the deformed steel cube was 92% of its initial mass. The residual fragment and plate particles are shown in the large radiograph after the fragment has passed through both plates.

Except for impact orientation, the ballistic impact situation displayed in Fig. 19 is the same as that in Fig. 18. As shown by the orthogonal radiographs, the fragment orientation in this instance may be characterized as flat. The largest residual piece of the fragment was only 24% of initial mass. Because of the extreme pressures produced by flat impact at this velocity, the fragment shattered.

Below a critical velocity, which depends upon properties of the fragment and plate, fragment shattering does not occur and, therefore, is not a function of impact orientation. Above the critical velocity, a fragment will shatter if the impact can be characterized as flat; otherwise, the fragment will not shatter.

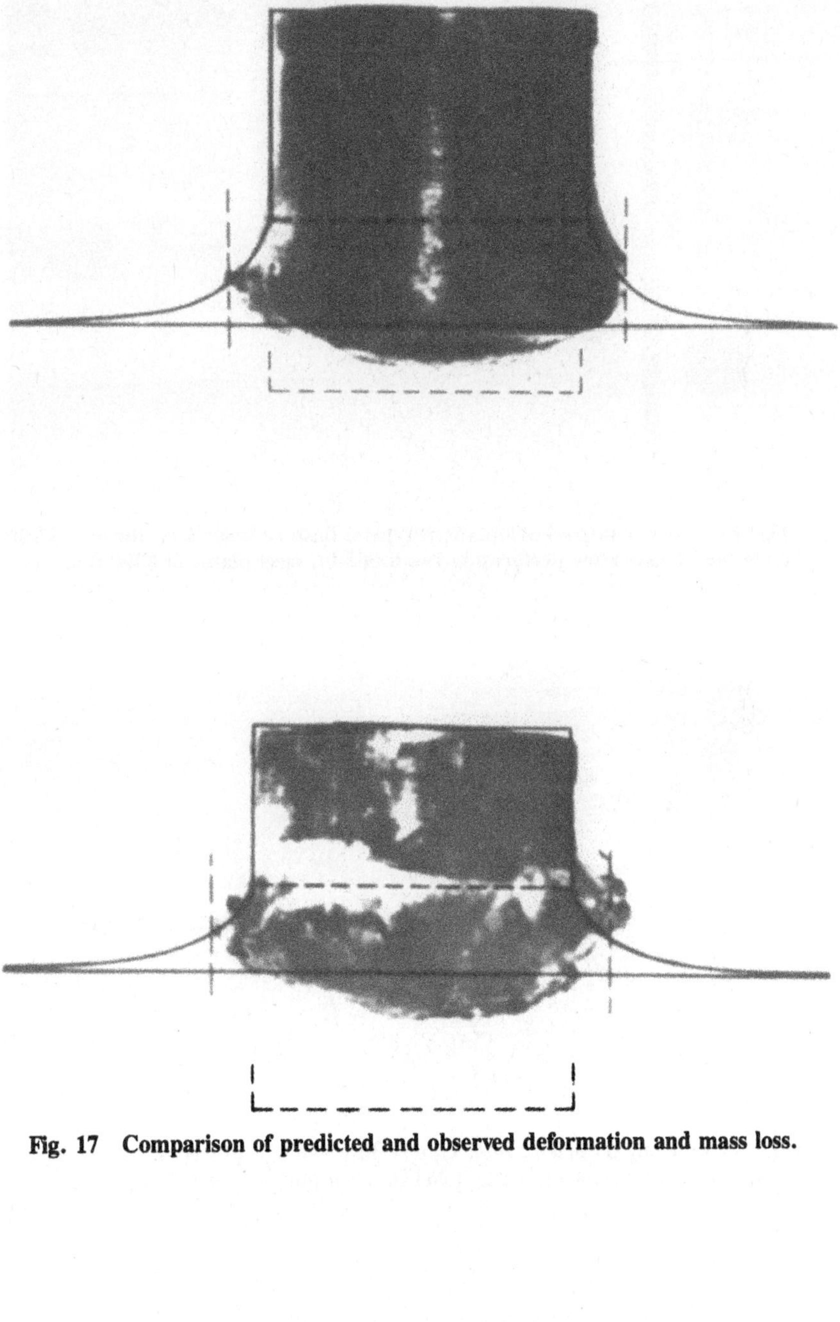

Fig. 17 Comparison of predicted and observed deformation and mass loss.

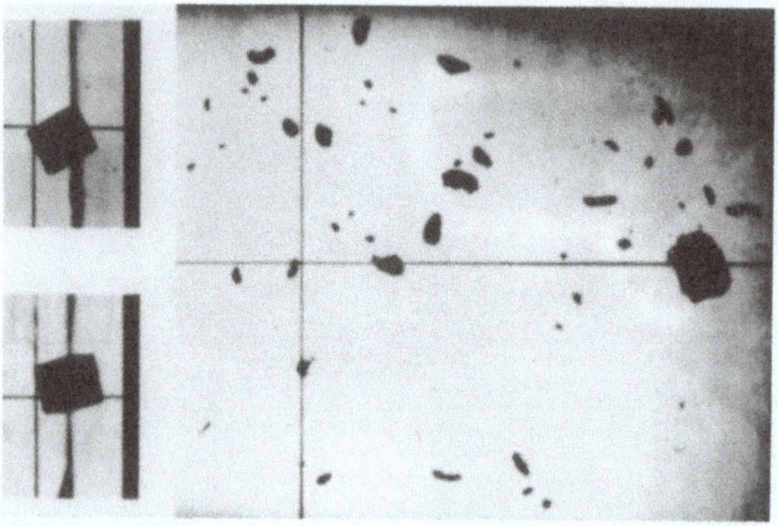

Fig. 18 Corner impact orientation; typical flash radiographs showing 3/8-in. cube before and after perforating two 0.063-in. steel plates at 5000 ft/s.

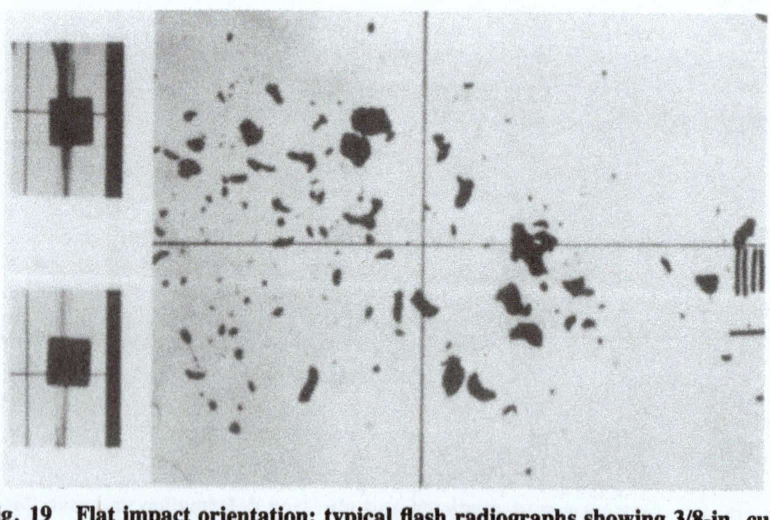

Fig. 19 Flat impact orientation; typical flash radiographs showing 3/8-in. cube before and after perforating two 0.063-in. steel plates at 5000 ft/s.

If the fragment does not shatter, mass loss is due to the erosion and deformation-shear mechanisms only.

In Ref. 5, the critical velocity V_c for steel fragments having a hardness near R_c30 is defined by

$$V_c = 2000(1 + (\rho U)_{st}/\rho U) \sec\theta, (\text{ft/s}) \tag{12}$$

where

U = Hugoniot stress wave velocity
ρ = density or specific weight
θ = impact obliquity angle

and the subscript *st* indicates steel; variables without subscripts refer to plate material. In the same reference, flatness is defined by a critical angle u_c. If u [the angle between the impacting surfaces (fragment and plate)] is equal to or less than u_c, then the impact is characterized as flat. The critical angle is defined by

$$u_c = \arcsin (V_0 \cos\theta/C) \tag{13}$$

where

V_0 = impact velocity
C = sonic wave velocity in the fragment

8.5 Mass Loss (Compact Fragments, Long Rods)

The mass loss equations presented in Ref. 5, as used to predict the results of the situations depicted by Figs. 18 and 19, are represented by the curves shown on the middle graph of Fig. 20. Also shown (solid curve) is the Thor equation prediction, which makes no distinction between corner and flat impacts. The Thor empirical equations were devised by Project Thor at the Institute of Cooperative Research, The Johns Hopkins University, to correlate fragment residual velocity and mass data.[5] The upper broken curve represents residual mass predictions for all impacts below the critical velocity V_c, which is about 4000 ft/s; all data below the critical velocity are in accord with this curve. The upper broken curve also represents residual mass predictions for fragments that have corner impact orientations at both plates. The middle broken curve represents predictions for fragments that impact the first plate flat, and the lower (dashed) curve represents predictions for fragments that experience two flat impacts. Open data points were known to be corner impacts and closed data points were known to be flat impacts [as defined by Eq. (13)]. Half-closed data points were of unknown orientation. All known corner impacts cluster about the upper curve, and all known flat impacts cluster about the lower curves, except one. It is reassuring to know that the undecided impact that produced the 70-grain residual fragment was associated with a flat orientation u that was very close to the critical value.

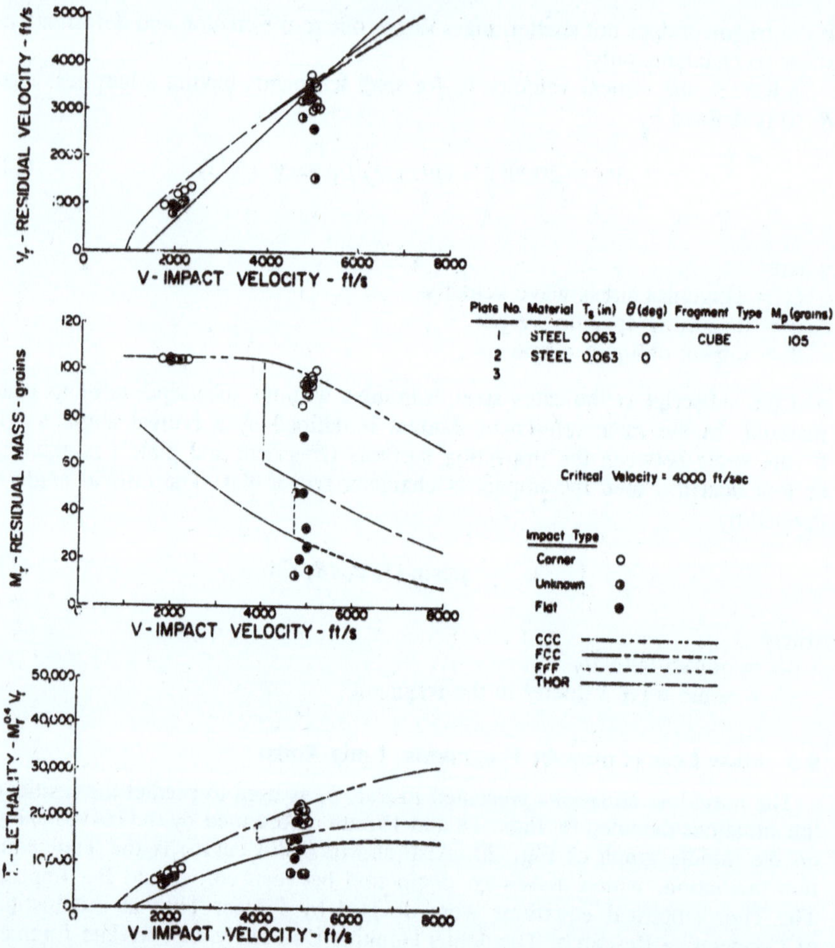

Fig. 20 Residual velocity/residual mass predictions; multiple spaced steel plates; normal incidence.

The upper graph on Fig. 20 compares Thor equation and Denver Research Institute (DRI) residual velocity predictions with data.[5] As previously discussed, component malfunction due to ballistic impact can often be related to a lethality parameter $M^a V$, where M is fragment mass, V is velocity, and a is an empirical constant often having a value near 0.4. The lower graph of Fig. 20 plots predicted lethality curves for the largest residual fragments that perforated the two plates.

The lower image of Fig. 21 (from Ref. 6) is a PHERMEX flash radiograph that shows the progress of a long rod that is penetrating a plate obliquely (at 25.6 microseconds after penetration began). The PHERMEX facility at Los Alamos

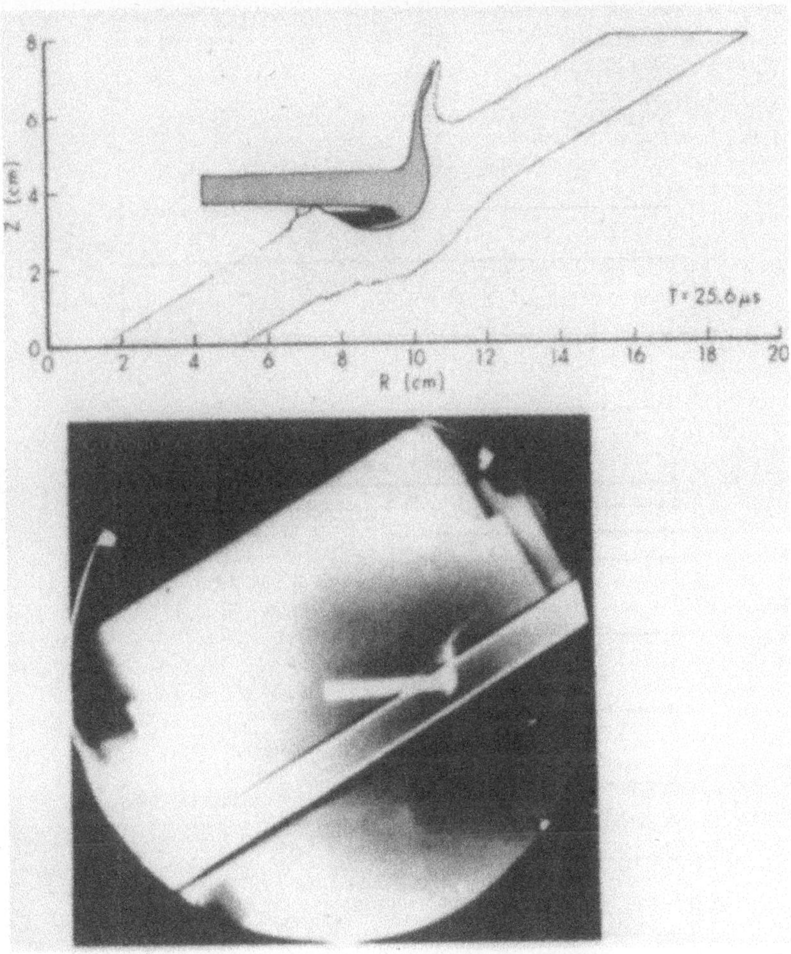

Fig. 21 Comparisons of plane strain (two-dimensional) computations and experimental results at 25.6 microseconds.

National Laboratories is a 6 MeV x-ray source capable of shining through six inches of steel. This invaluable tool reveals deformation and orientation of the penetrator within the target. The upper diagram is a computer code plane-strain (two-dimensional) approximate simulation of the three-dimensional event.

Taylor model predictions of erosion and deformation-shear mass loss and the final shape of steel rods perforating steel plates at 3800 m/s are shown in Figs. 22 and 23.[4] Rod length data obtained by Baker at the Naval Research Laboratory are plotted in Fig. 23 and may be compared with Baker's model as well as with the Taylor model predictions.[7]

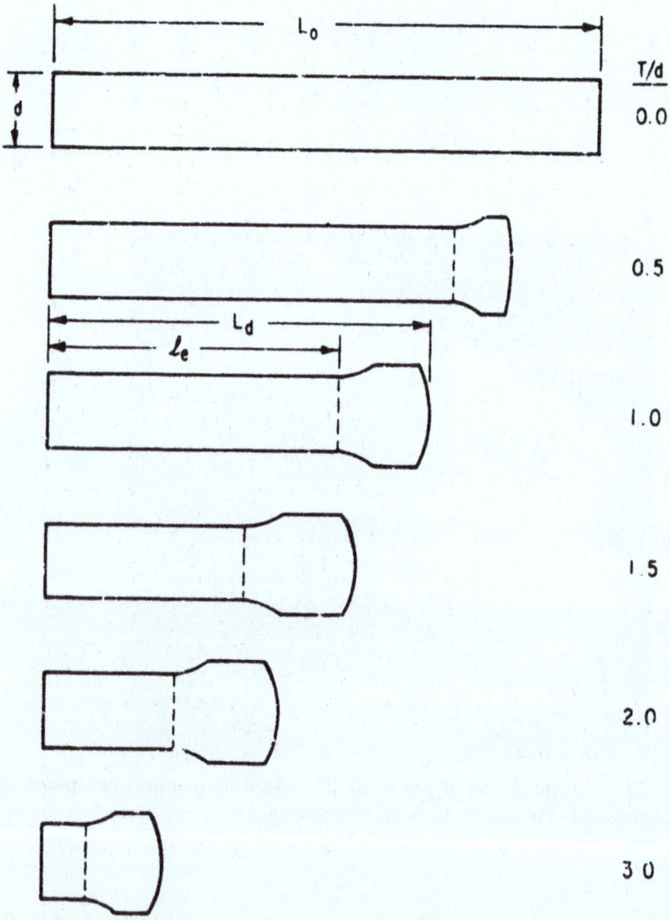

Fig. 22 Predictions of final size and shape of steel rods perforating steel plates at 3800 m/s.

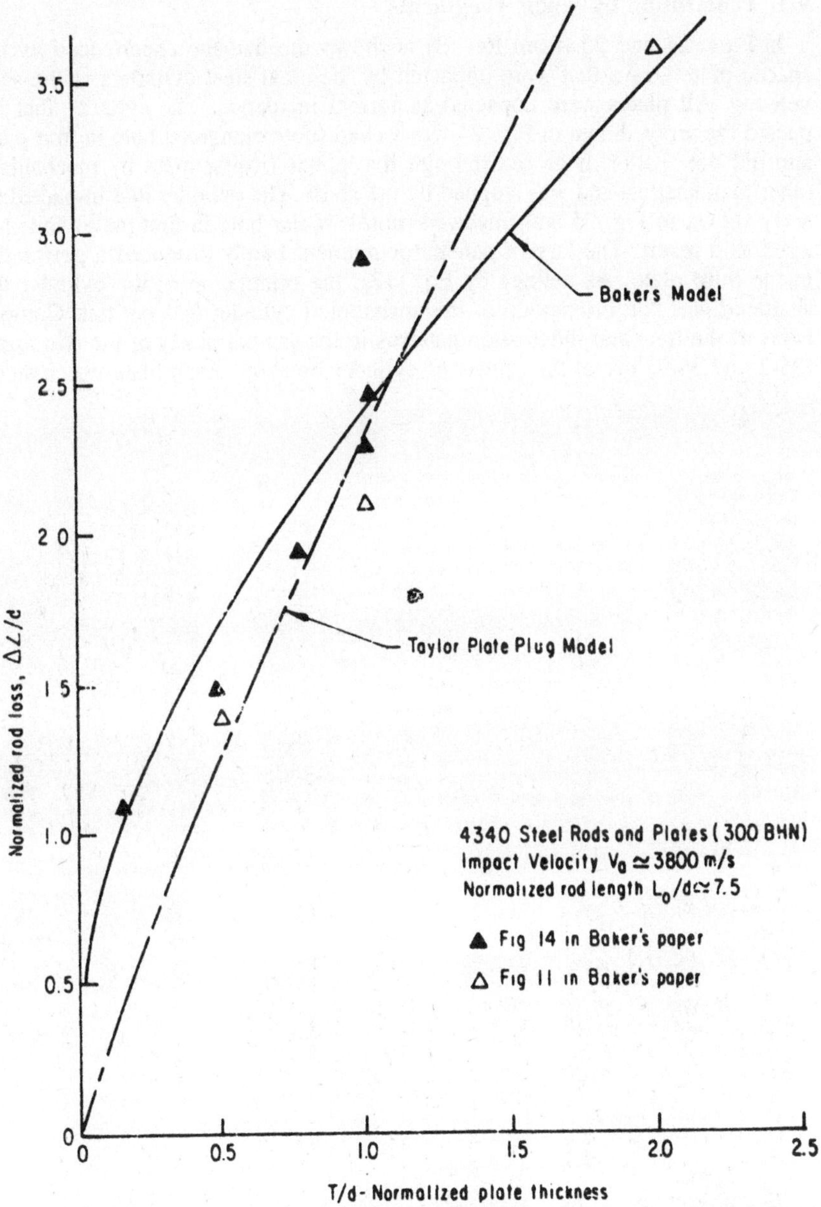

Fig. 23 Residual rod length after perforating plates at a very high velocity.

9. Multiple Plate Penetration Behavior of Fragments

9.1 Penetration by Single Fragments

In Figs. 24 and 25 (from Ref. 3) is shown the damage experienced by two spaced-plate arrays that were impacted by identical steel cylinders at the same velocity. All plates were impacted at normal incidence. The cylinder that impacted the array shown in Fig. 24 was yawed (note elongated hole in first plate) and did not shatter. It passed through five plates (losing mass by mechanisms other than shatter) and was stopped by the sixth. The cylinder that impacted the array shown in Fig. 25 was unyawed (note circular hole in first plate) and shattered as a result. The largest penetrator remnant barely produced a perforation in the third plate. As defined by Eq. (13), the orientation of the cylinder that shattered was flat; orientation of the unshattered cylinder was not flat. Comparisons of the hole and indentation patterns in the second plates of the two arrays (25-2 and 50-2) reveal the effects of cylinder breakup. Each plate was cratered

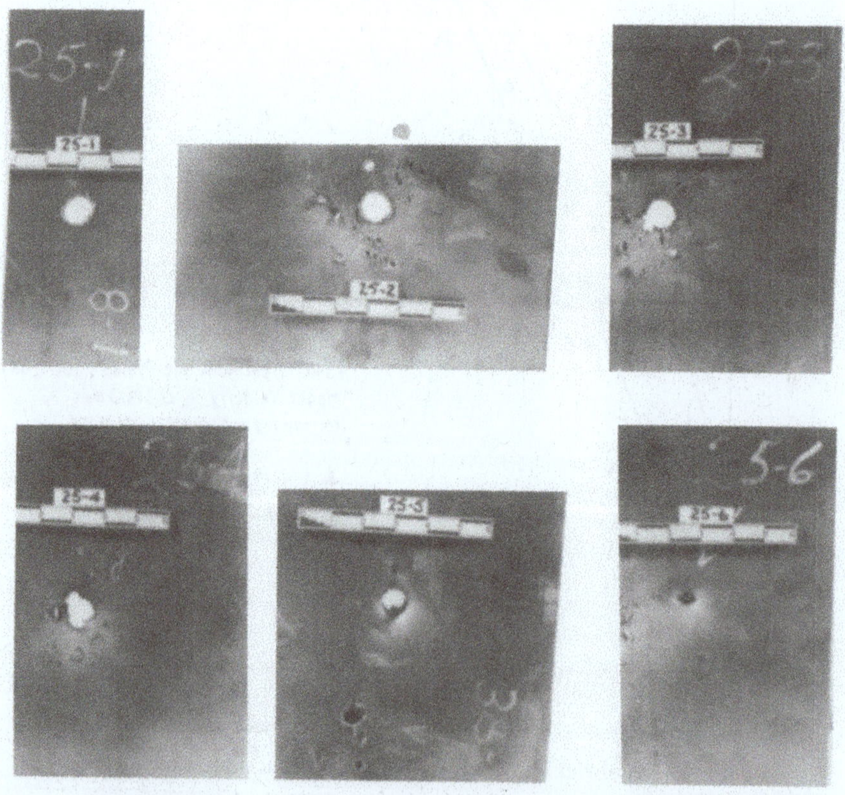

Fig. 24 Multiple plate perforation; 3.18-mm mild steel plate (120 BHN), 13-g compact steel cylinder (R_C 29) 100-mm spacing, 1,900 m/s velocity (yawed impact; penetrator did not shatter).

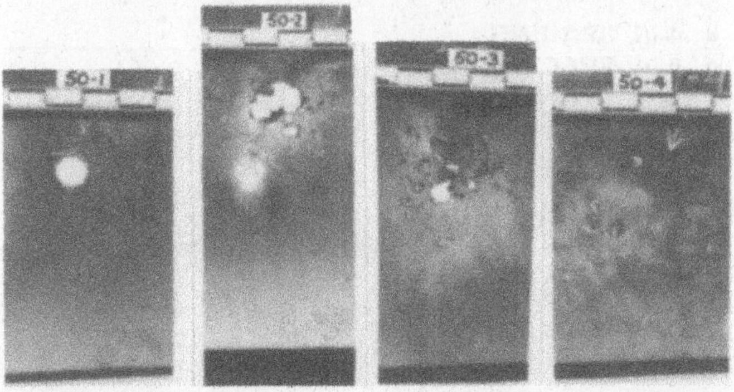

Fig. 25 Multiple plate perforation; 3.18-mm mild steel plate (120 BHN), 13-g compact steel cylinder (R_C 29); 100-mm spacing, 1,900 m/s velocity (unyawed impact; penetrator shattered).

and perforated by plate particles driven from the first plate. Plate 50-2 shows that the cylinder was broken into at least four pieces, whereas plate 25-2 shows that the other cylinder was unbroken prior to impact.

The data plotted in Figs. 26 and 27 indicate the number of plates that were perforated in each test of the entire series of experiments represented by Figs. 23, 24, and 25.[3] The superimposed stair-step and curve predictions were made

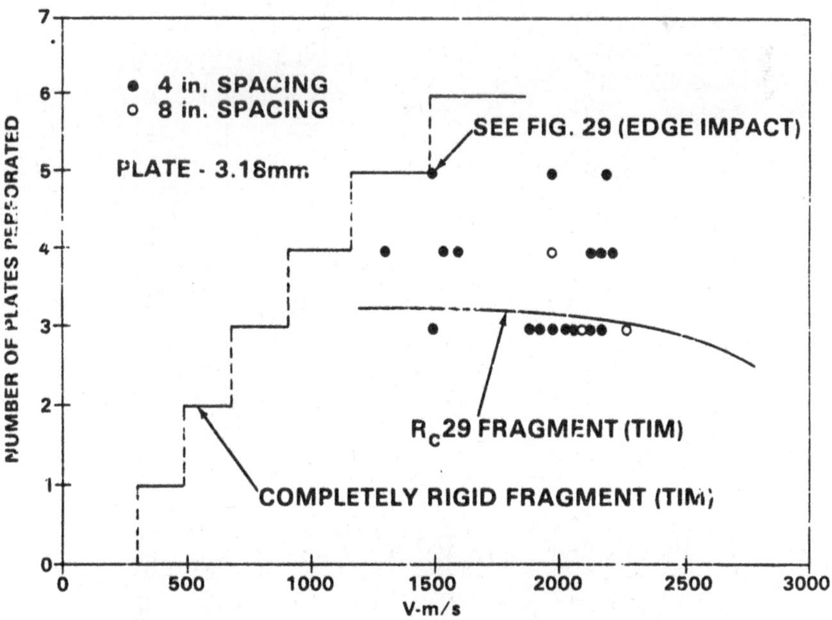

Fig. 26 Multiple plate perforation experiments; 13-g fragment, mild steel plate.

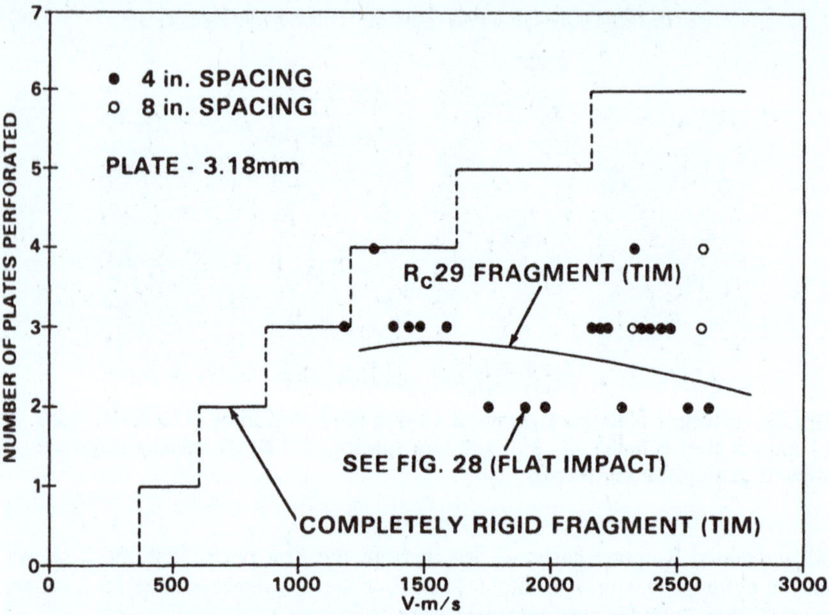

Fig. 27 Multiple plate perforation experiments; 6.5-g fragment, mild steel plate.

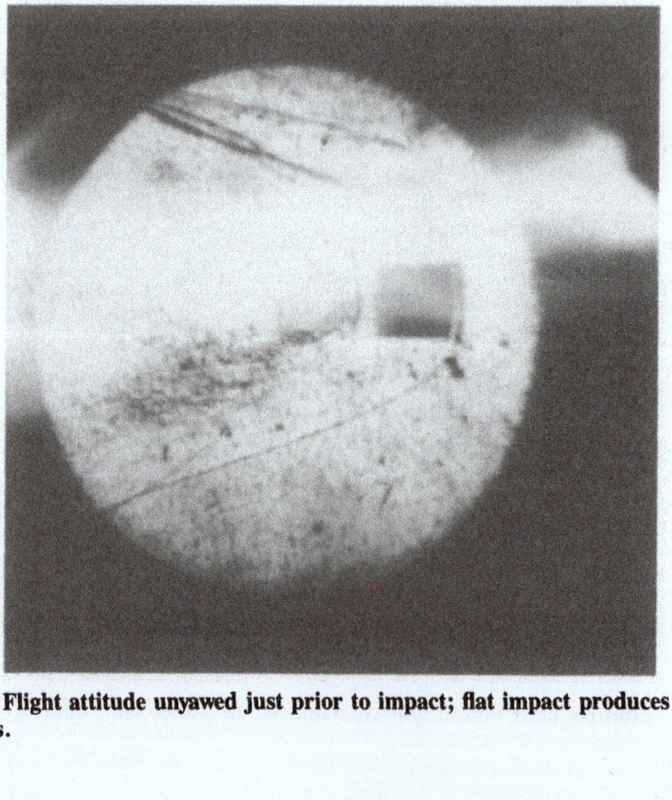

Fig. 28 Flight attitude unyawed just prior to impact; flat impact produces shatter pressures.

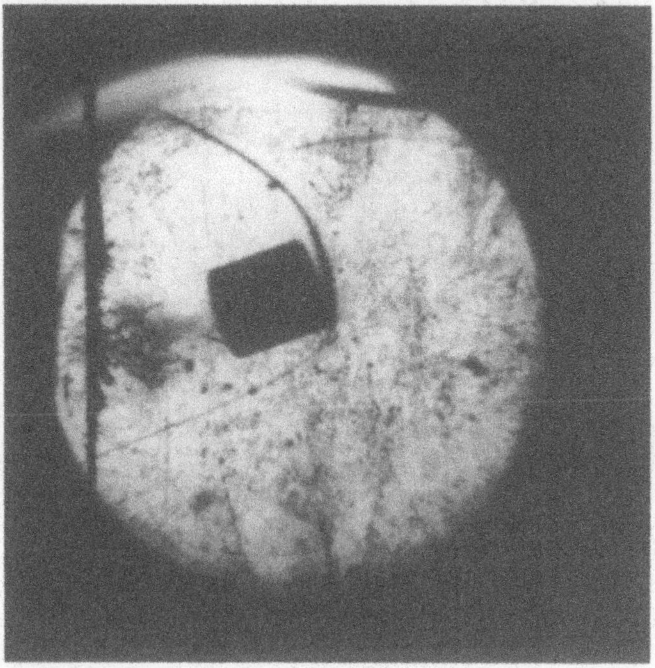

Fig. 29 Flight attitude yawed just prior to impact; edge impact does not produce shatter pressures.

using the Terminal Interaction Model (TIM) documented in Ref. 8. To create the stair-step predictions, it was assumed that the fragment retained its initial mass and shape. The curves consider the effects of fragment deformation and breakup and lie closest to the predominant mode of behavior (as defined by number of plates perforated). Data scatter is attributed principally to impact orientation, as supported by Figs. 28 and 29; these photographs show the cylindrical fragments in flight just prior to impacting the plates. The shock waves being generated in the air by the cylinders are indicated by the lines on the figures (see corresponding data points in Figs. 26 and 27).

Very high velocity experiments performed for the Naval Surface Weapons Center by the Naval Research Laboratories (NRL) involved single fragment penetration of the multiple aluminum plate array shown in Fig. 30.[9] Flat impact by a cube on the first plate at 8390 ft/s produced a damage pattern in the second plate that took the form of a cross (arms are perpendicular to the flat sides of the cube). Figures 31 and 32 show this type of breakup behavior and the resulting pattern on the second plate, which involves an impact velocity at 7240 ft/s. Figure 31 is a double exposure that shows the expansion of the fragment cloud moving right to left. Figures 33 and 34 represent the same situation, except that the velocity is 15,400 ft/s. Even though the steel projectile particles are smaller

Fig. 30 Multiple aluminum plate array used for very high velocity penetration experiments.

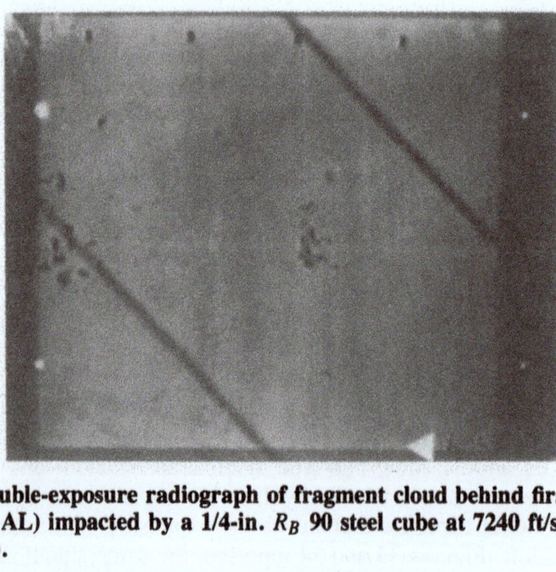

Fig. 31 Double-exposure radiograph of fragment cloud behind first plate (1/16 in. 2024-T3 AL) impacted by a 1/4-in. R_B 90 steel cube at 7240 ft/s (motion is right to left).

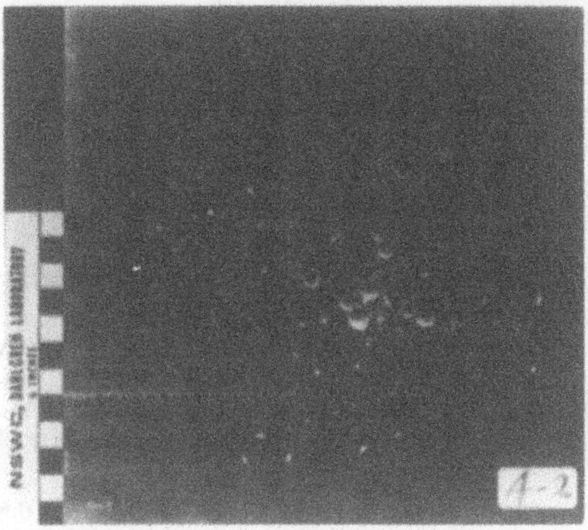

Fig. 32 Second plate damage (1/8 in. 2024-T3 AL) caused by fragments shown in Fig. 31.

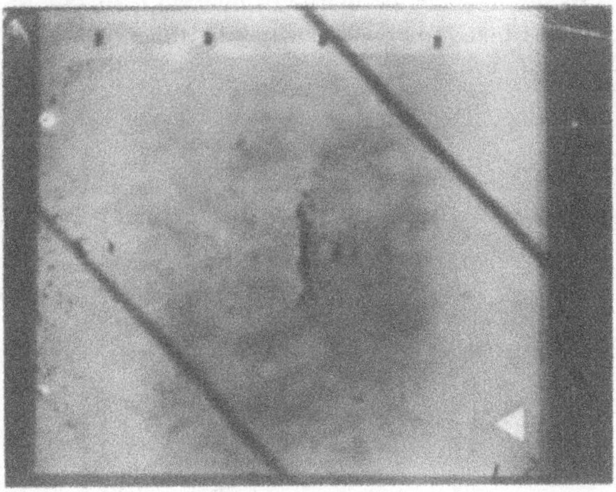

Fig. 33 Double-exposure radiograph of fragment cloud behind first plate (1/16 in. 2024-T3 AL) impacted by 1/4-in. R_B 90 steel cube at 15,400 ft/s (motion is right to left).

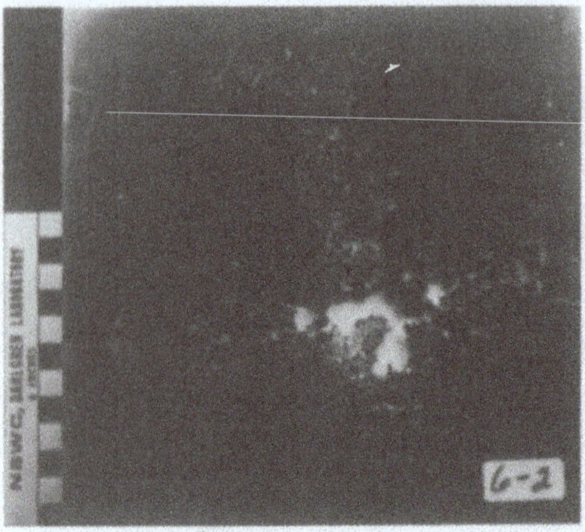

Fig. 34 Second plate damage (1/8 in. 2024-T3 AL) caused by fragments shown in Fig. 33.

in this case and none are capable (individually) of perforating the second plate, when acting impulsively together, they produce large perforations in the second plate.

In Fig. 35 is illustrated the nature of the model of multiple plate penetration by a single fragment (developed in Ref. 9). The angles ϕ_c and ϕ_s subtend the projectile fragment and the plate fragment dispersion patterns, respectively. Among other things, the model predicts the mass lost by each plate due to perforation. In Fig. 36 we see how breakup affects penetration capability adversely. Also a comparison of predictions with data for several impact velocities is presented in the figure.

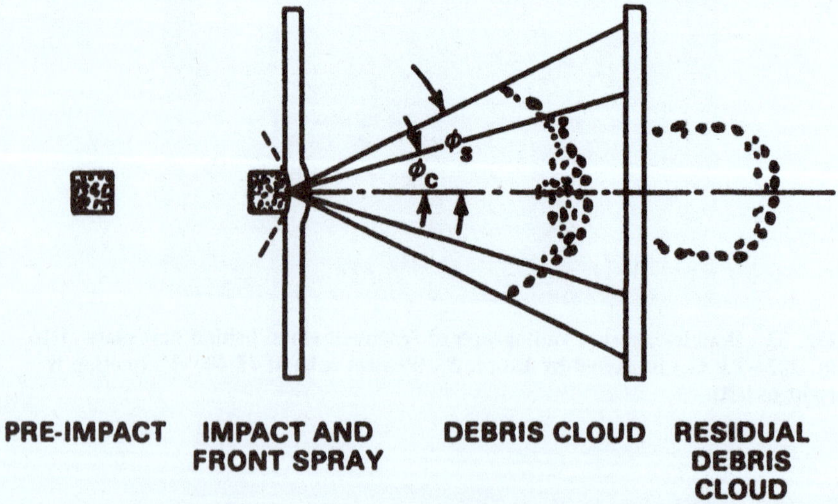

Fig. 35 Typical high velocity multiple plate penetration.

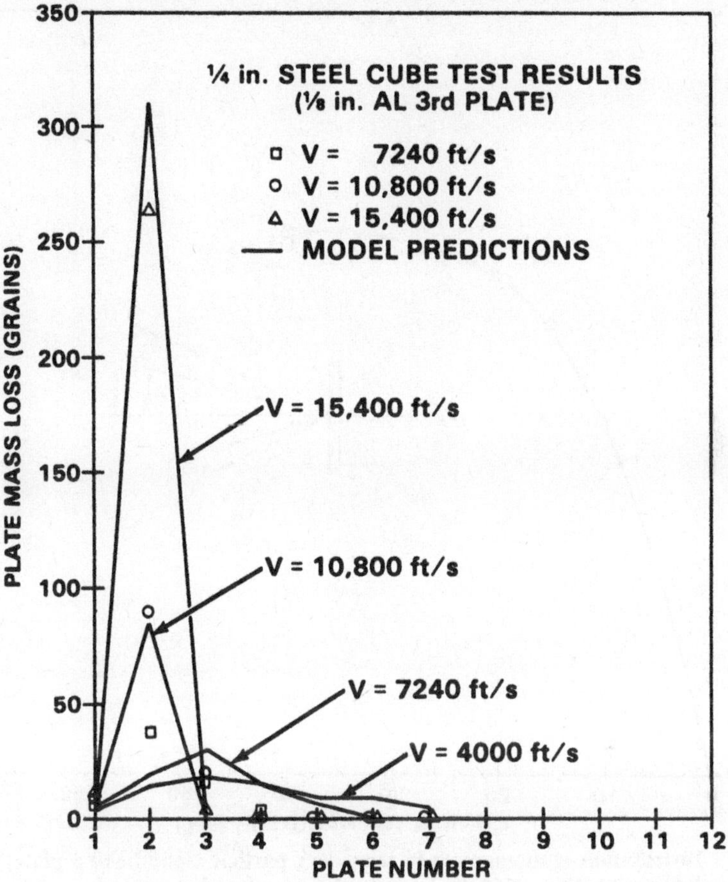

Fig. 36 Plate mass loss vs plate number for three impact velocities.

The Denver Research Institute of the University of Denver (DRI) determined the distribution of momentum in particle clouds behind plates by passing a portion of the cloud through a heavy stripper plate that incorporates holes of different sizes. The portion that passes through the stripper plate is caught by a ballistic pendulum that measures its momentum (impulse). An example of such a distribution appears in Fig. 37.

9.2 Penetration by Multiple Fragments

Characteristics of multiple fragment penetration are illustrated in Fig. 38. The lead shot produced individual holes in the first plate, breaking up in the process. As indicated by the bar graphs, relatively little impulse was transmitted to the first plate and damage was slight (as indicated by residual strength). At the second plate, individual particles were unable to perforate the plate; however,

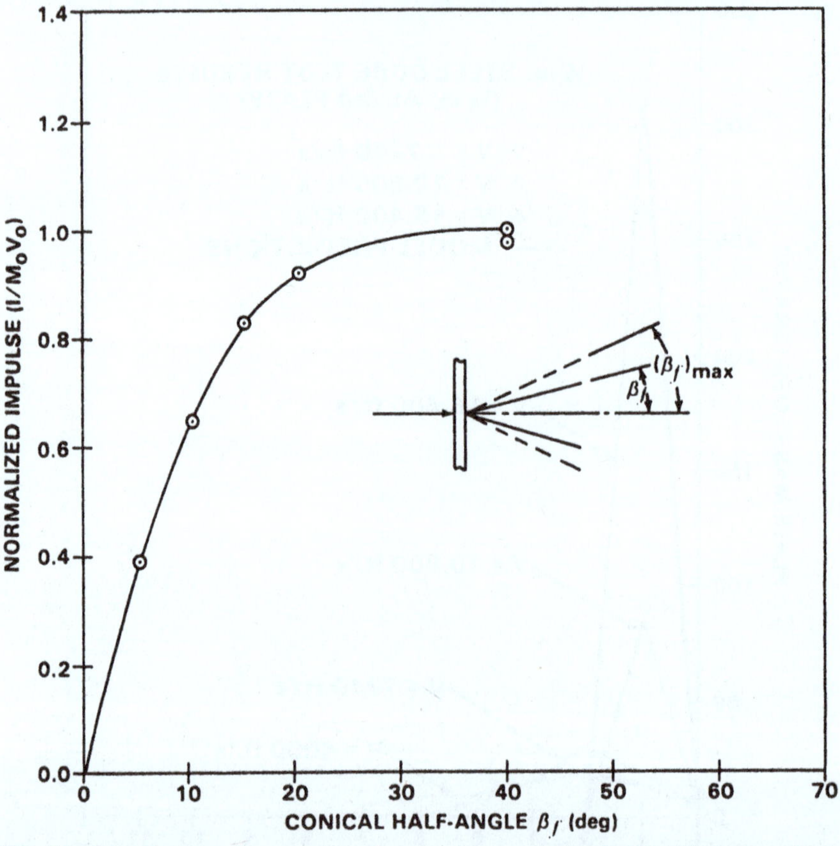

Fig. 37 Distribution of momentum in secondary particle cloud behind plate; high velocity plate perforation (2750 m/s), steel sphere perforating steel plate.

the total impulsive load was sufficient to tear a large hole in the plate and damage was great. Having spent most of the initial momentum, the impulse on the third plate was sufficient only to produce a small hole at the center of the pattern.

A series of high-velocity multiple fragment experiments involving penetration of multiple plate arrays is described and documented in great detail in Ref. 10. Clusters of $\frac{1}{16}$-in. spheres (aluminum and steel), rods, and platelets were fired at velocities ranging from 1300 to 2200 m/s. Plates in the array were mainly 0.32-in. aluminum spaced at 0.075, 1.5, 3, and 6 in. Total masses of typical clusters were 0.5, 1.0, and 2.0 g. High-speed photographs (Fig. 39) reveal some interesting features of the penetration process. For instance, off-axis particles in the cluster have lower velocities, impact the plates at higher obliquities, and perforate fewer plates.

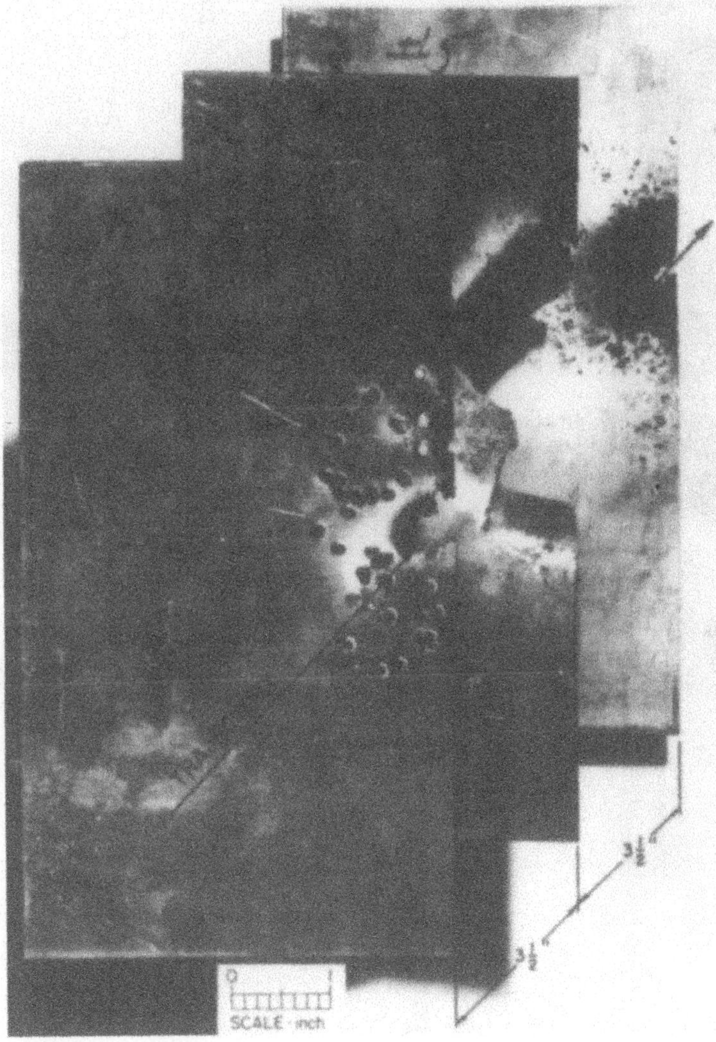

Fig. 38 Impulsive loading and damage resulting from multiple fragment penetration of multiple plates.

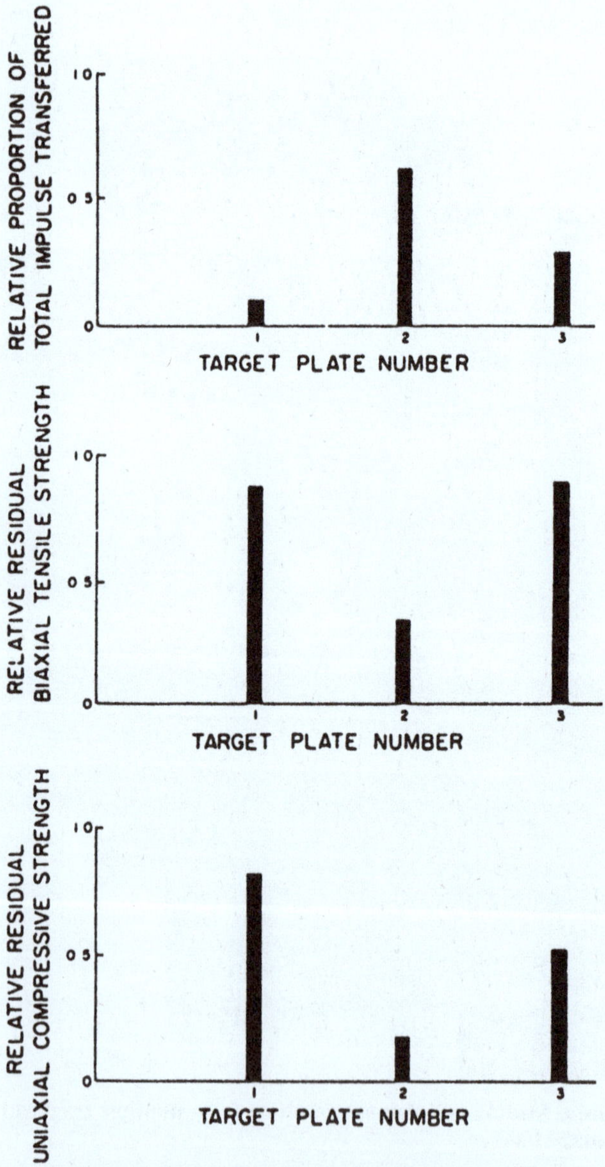

Fig. 38 (continued) Impulsive loading and damage resulting from multiple fragment penetration of multiple plates.

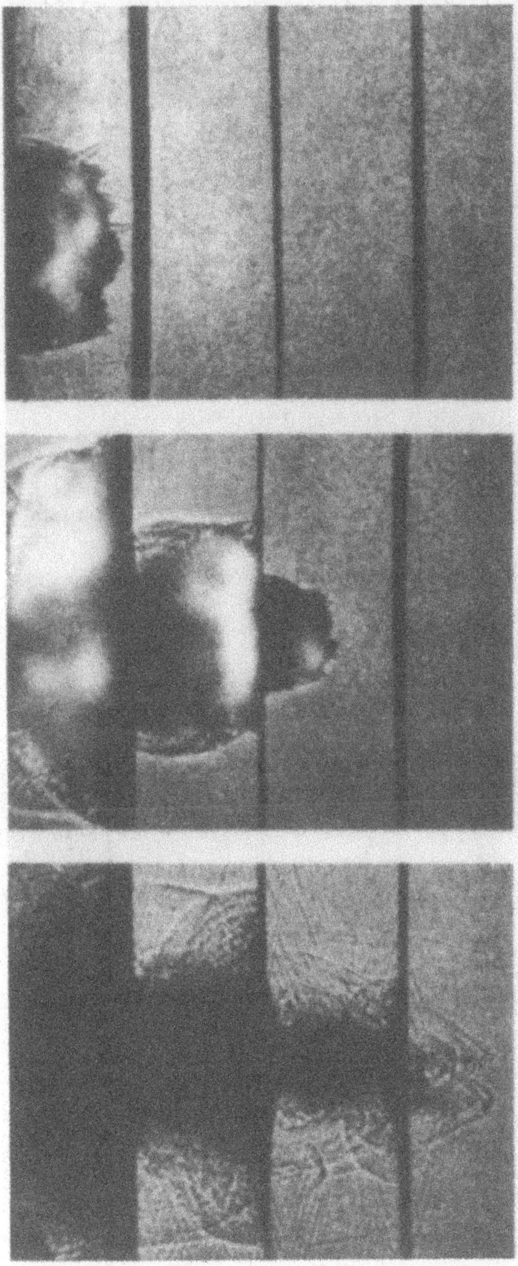

a. Fragment beam enters compartment number 1 (between plates 1 and 2).

b. Fragment beam in compartments 1, 2 and 3 (between plates 1 and 4).

c. Fragment beam in compartments 4, 5, 6 and 7 (between plates 4 and 8).

Fig. 39 High-speed photographs of the fragment beam within the target array.

Typical momentum and energy expenditures within the target array are shown in Figs. 40 and 41. Corresponding rupture diameter profiles are shown on Fig. 42. Invasion depth as a function of cluster mass (g) and impact velocity is compared with the penetration capability of a single $\frac{1}{16}$-in. steel sphere (0.0164 g) in Fig. 43.

Tests were performed on both confined and unconfined plate arrays to determine effects of pressure buildup in the atmosphere between plates. Sides of the array were left open for the unconfined geometry and were closed for the confined geometry so as to form compartments. As shown by the deformation (confined) curve in Fig. 44, the first two plates were bulged by pressure buildup in the direction opposite to the direction of motion and the third plate was straightened up. All of the curves provide an easily interpreted representation of target loading and response.

10. Effects of Fragment Material Properties

Material properties that influence fragment deformation and fracture include the yield strength σ_e, ductility, and toughness represent by fracture stress σ_f and fracture strain ϵ_f or fracture energy sonic velocity C (Young's modulus E and density ρ) and plastic wave velocity U_c. Dynamic values of these factors as related to uniaxial stress and uniaxial strain states in predominantly

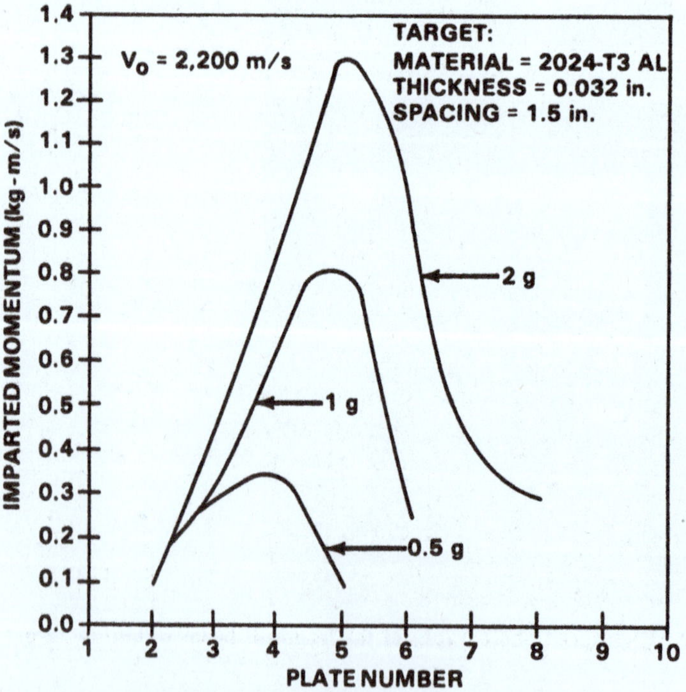

Fig. 40 Momentum imparted to target.

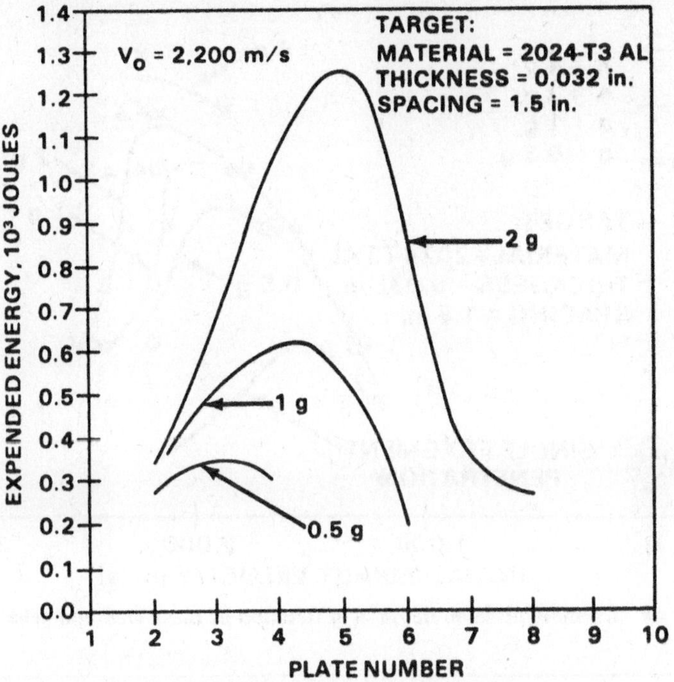

Fig. 41 Energy expended within target system per plate interval.

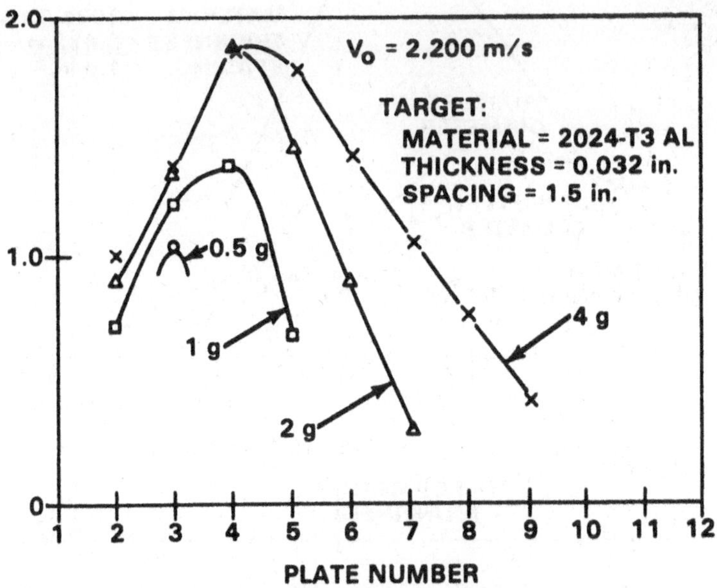

Fig. 42 Rupture damage profiles for the high velocity case.

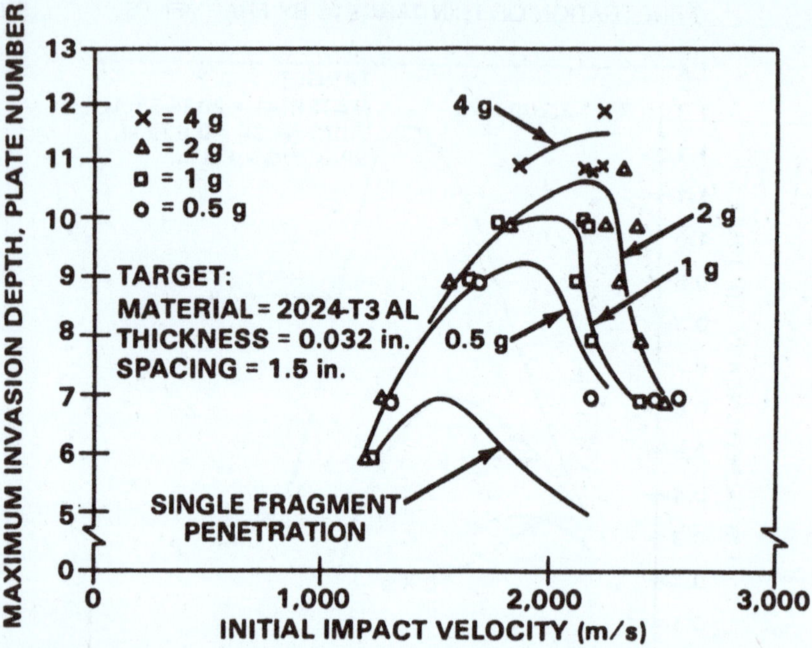

Fig. 43 Ultimate invasion depth as a function of mass load and velocity.

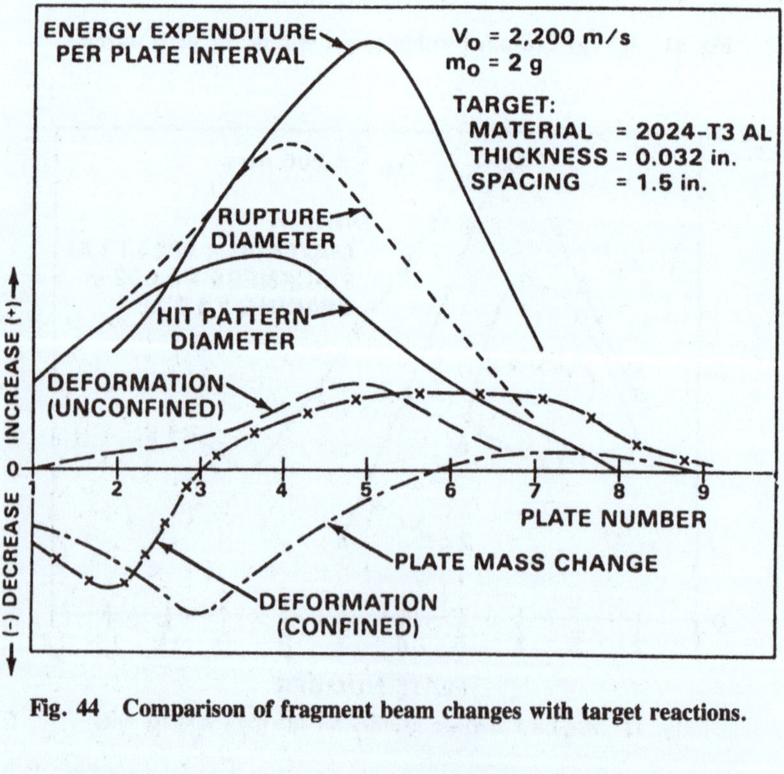

Fig. 44 Comparison of fragment beam changes with target reactions.

compressive environments are pertinent. Higher-strength materials possess higher plastic wave velocities ($U_c \simeq 118\sqrt{(\text{BHN})}$ in ft/s; BHN is Brinell Hardness No.) and, therefore, for given impact conditions, will erode less and deform less. However, the higher-strength materials also are more brittle and typically exhibit lower fracture toughness; hence, high-strength fragments will fracture at lower deformation levels and are more susceptible to shatter.

Most ballistic impact and penetration data concern cylindrical steel fragments having a Rockwell hardness near R_c 30. Very little has been done to examine the effects of material properties on penetration behavior. As a consequence, most estimates pertain to R_c 30 fragments and are used without considering actual hardness.

A series of experiments is described in Ref. 3 that involve the firing of 13-grain fragments of two hardnesses at radar van walls. Fragment hardnesses were R_b 93 and R_c 44 (200 and 415 BHN, respectively). The van wall sections consisted of a 1.25-mm mild steel sheet followed by 32 mm of fiberglass insulation and backed by a standard single-ply ($\frac{1}{4}$-in.) sheet of plywood. Velocities ranged up to 2500 m/s and impact obliquities were 0 and 60 deg (a few tests involved obliquities of 30 deg).

Residual velocity data are plotted in Fig. 45 along with prediction curves for undeformed fragments impacting at $\theta = 0$ deg and $\theta = 60$ deg. The degree to which the data fall below the curves reflects the effects of deformation and mass loss. Note that at higher velocities, the harder fragments had lower residual velocities (the result of shatter).

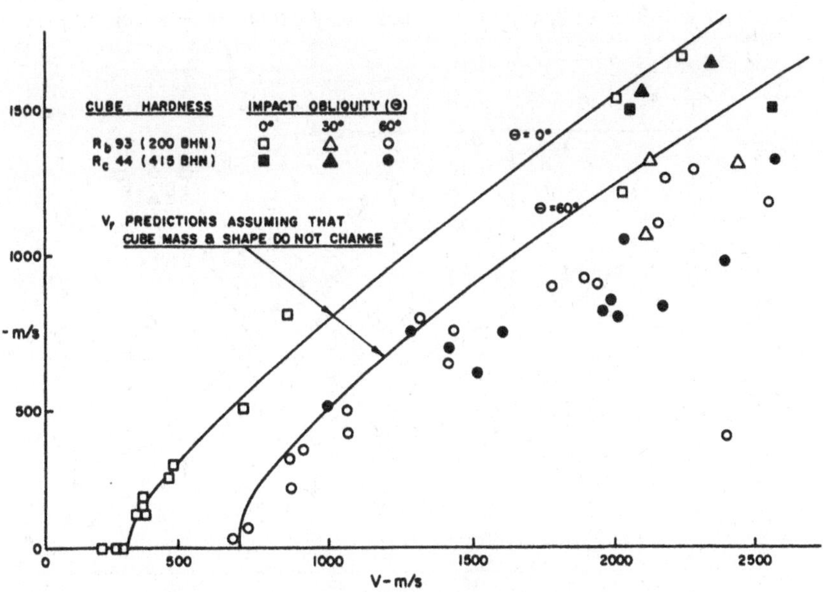

Fig. 45 Residual velocity data and predictions.

Residual mass data and predictions for R_c 30 fragments are plotted in Fig. 46. (See Fig. 20 and related discussion for interpretation of the curves: V_η values indicate the velocity associated with incipient deformation-shear mass loss; V_c values are critical velocities for shatter given that impact orientation is flat; shatter drops residual mass from the deformation mass-loss curve to the shatter mass-loss curve.) Recognizing that the deformation mass loss curve for R_b 93 cubes would be lower and the R_c 44 curve would be higher than the R_c 30 curve, we see that deformation-shear mass loss is the predominant mode for both fragments below 1800 m/s. Above the shatter velocity for the R_c 44 fragments, it appears that shatter is predominant in the harder fragments. Above 2000 m/s, the softer fragments lose less mass because they do not shatter.

In Fig. 47 the residual kinetic energies of the two fragments are compared. One can also compare these values with the residual kinetic energy that a fragment would have if it did not lose mass or change shape during perforation at a 60-deg obliquity. Both fragments suffer great losses of energy at the higher velocity because of deformation and mass loss. Up to 1500 m/s, the harder fragment is more effective as a penetrator, then it becomes inferior.

These experiments explore the effects of material properties only very superficially. Much more needs to be done.

11. Damage Mechanisms and Damage Interpretation

Ballistic impact imparts an impulsive load to target components. The fragment may penetrate into a component or ricochet. In either case, the component is subjected to shock loading, fractures may be initiated, and cracks may propagate. Penetration of a fluid produces hydrodynamic ram pressures on hardware in contact with the fluid; ruptures of tanks may occur. Perforations and ruptures

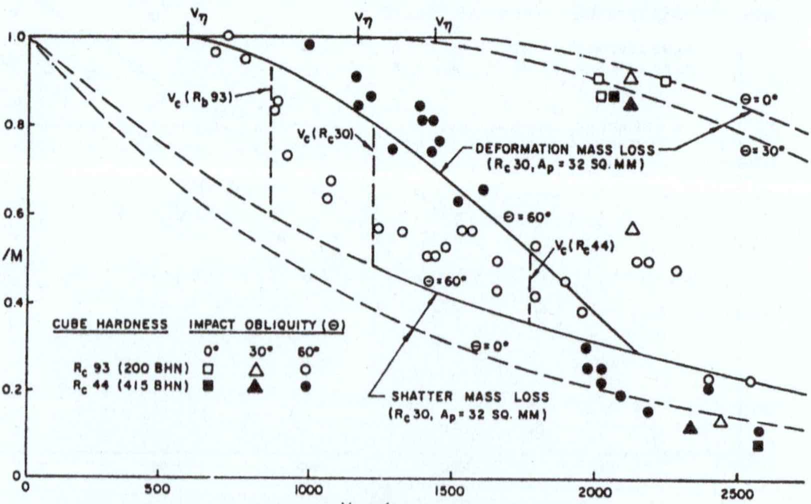

Fig. 46 Residual mass data and predictions.

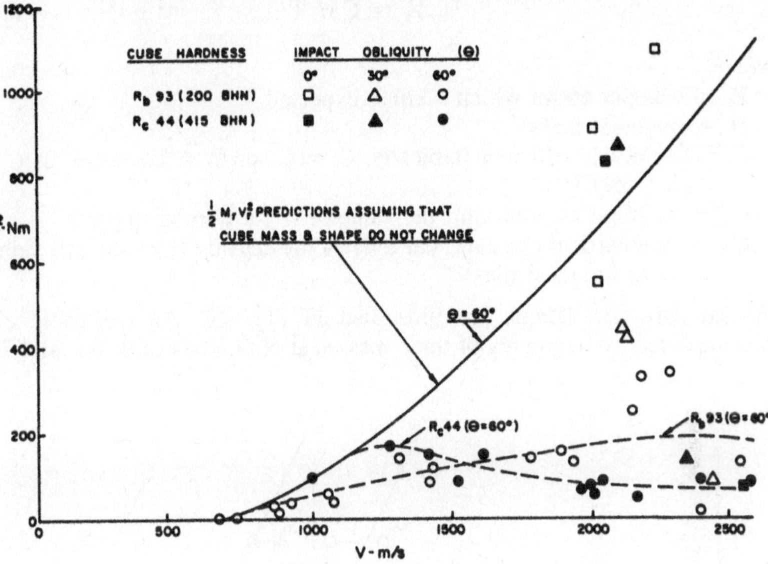

Fig. 47 Residual kinetic energy data and predictions for a cube that does not deform.

may represent leaks that may lead to hydraulic pressure losses or fuel fires. Propulsion, control, and guidance components may be damaged so as to suffer mechanical, electrical, or electronic malfunction. Reactions may be initiated in explosives, propellants, and fuels by shock loading or impact flash. Interconnecting wires, cables, and waveguide can be damaged so as to modify or disrupt transmission. As considered in Sec. 9, multiple fragment clouds can greatly increase component damage due primarily to impulsive loading.

Such is the nature of ballistic impact damage. Whatever impulsive loading and deformation a component experiences, unless this physical damage is interpreted in terms of functional damage, little can be said about operational malfunction. Once operational malfunction has been related successfully to ballistic impact damage, then the effects of operational malfunction on target system performance can be evaluated. Obviously, it is better actually to observe the effects of ballistic impact damage upon operational malfunction. Often, however, it is not practical to expose operating components to fragment impact. Instead, inoperative components are exposed and an estimate of operational malfunction is inferred from the physical damage. This area of investigation deserves a good deal of organized and careful analytical and experimental work.

12. Component Response to Ballistic Impact

12.1 Single Fragments

Fragment lethality with regard to a specific air target component or element threat can very often be expressed by the following equation[5]:

$$V_k = CM^{-C_2} \tag{14}$$

where

V_k = velocity above which a kill is expected
M = fragment mass
C = hardness coefficient [typically, $C = C$, or $C = C_1 \sec\theta$, or $C = C_1 (1 + \sec\theta)/2$]
C_1 = an empirical constant, the value of C for normal impacts
C_2 = an empirical constant that reflects the sensitivity of the kill velocity V_k to fragment mass

Typical data correlations are illustrated in Fig. 48. An electronic element was impacted by fragments of three masses at obliquities of 0, 60, and 75 deg.

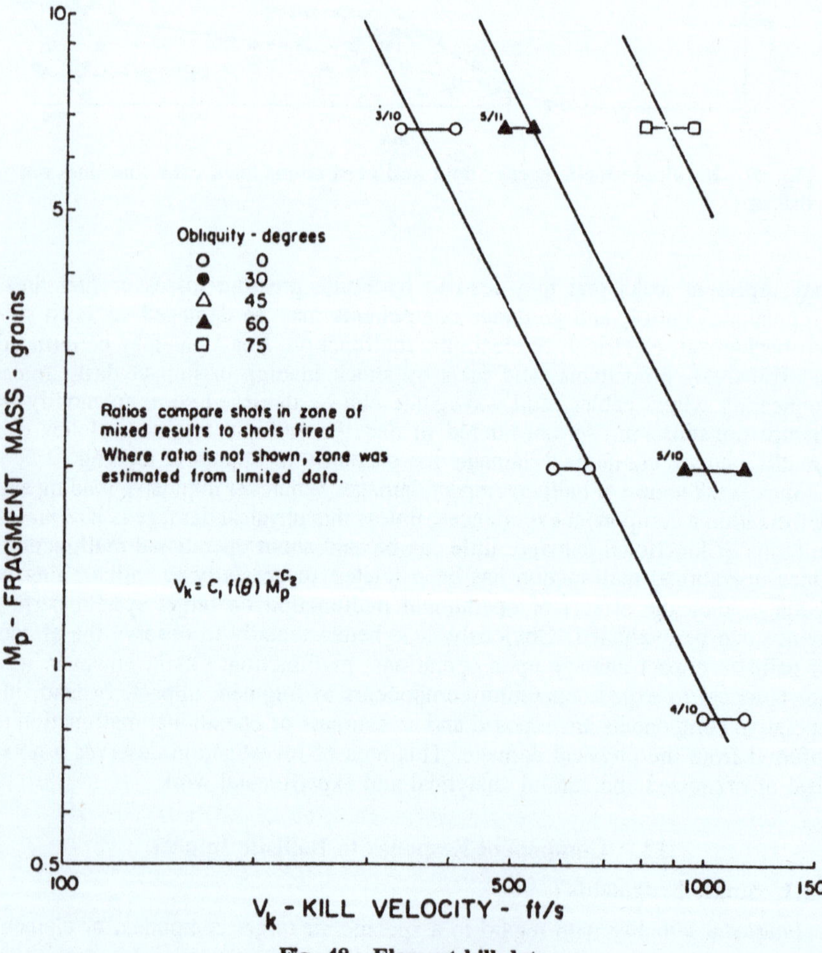

Fig. 48 Element kill data.

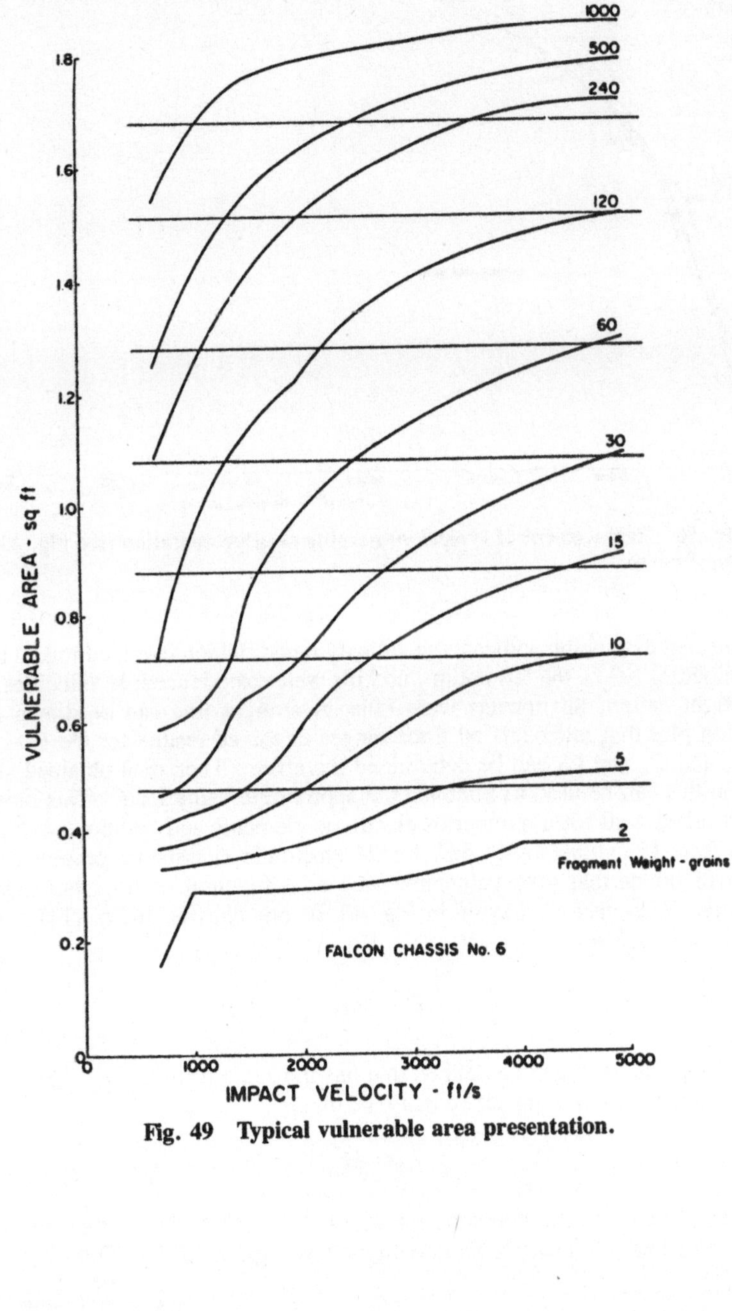

Fig. 49 Typical vulnerable area presentation.

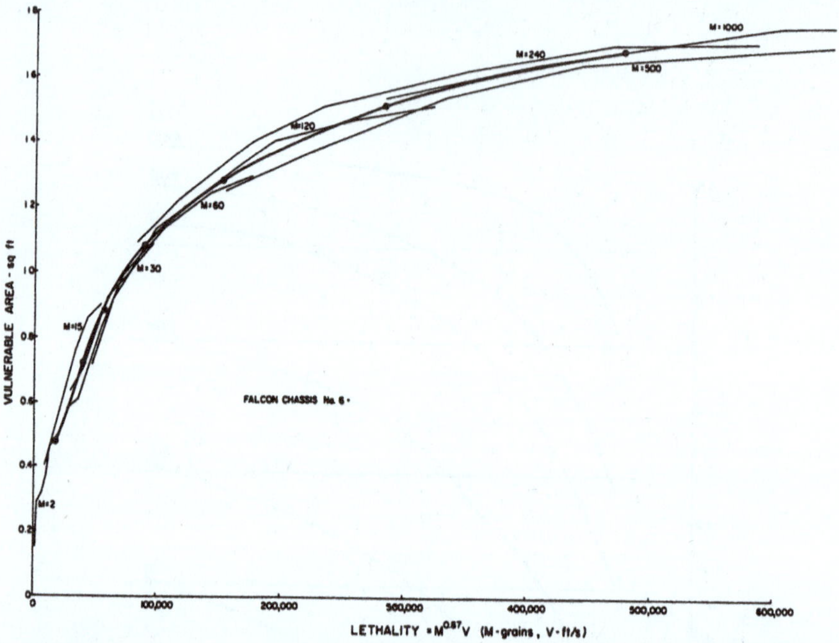

Fig. 50 Replacement of typical vulnerable area presentation (see Fig. 49) by single curve.

Connected data points indicate the velocity range, which involved mixed results; at velocities below the left datum, no kills were experienced; at velocities above the right datum, all impacts were kills. A straight line can be drawn on the log-log plot that intercepts all three ranges of mixed results for the $\theta = 0$ deg data, and C_1 and C_2 can be determined therefrom. Then data obtained at other obliquities can be used to establish the appropriate value for C. This procedure has worked well for a number of electronic elements and components.

In Ref. 11, vulnerability data for 24 electronic chassis are presented in the form of tables that give vulnerable area as a function of fragment mass and velocity. Such data are plotted in Fig. 49 for one chassis. If Eq. (14) is solved for C, it becomes

$$C = M^{C_2} V_k \tag{15}$$

A lethality factor $\overline{L}$ can be defined that has the same form as Eq. (15), wherein V_k is replaced by the impact velocity V; thus,

$$\overline{L} = M^{C_2} V \tag{16}$$

Mass and velocity combinations for which $\overline{L} \geq C$ kill the element or component. This lethality factor is plotted on the lower graph of Fig. 20 for $C_2 = 0.4$.

When the several curves of Fig. 49 are plotted vs $\overline{L}$ for the chassis, they define the locus of a single curve (Fig. 50). Thus, using the lethality factor concept, it is possible to reduce the complexity of vulnerability analysis by replacing a set of vulnerability curves by just one curve.

12.2 Multiple Fragments

To date, most multiple fragment component damage data pertain to plate arrays. This type of damage is described in some detail in a previous section. Although some other types of components have been exposed to multiple fragment impacts, interpretations of data have been qualitative rather than quantitative for the most part.

13. Suggested General Reference

High Velocity Impact Dynamics, Ref. 12, is recommended as a comprehensive general reference on ballistic impact dynamics and target penetration. Its 15 chapters were written by 10 contributors.

Acknowledgment

This work was published initially as the technical rationale of a Navy technical report (TR), Ref. 1; permission was obtained from the Department of Navy, Naval Surface Weapons Center (since renamed the Naval Surface Warfare Center), Dahlgren, VA, to publish the text of this portion of the TR, and related illustrations, herein.

References

[1]Recht, R. F., "Fragment Lethality," Naval Surface Weapons Center, NSWC TR 83-165, Dahlgren, VA, Aug. 1983.

[2]Ipson, T. W., and Recht, R. F., "Ballistic Penetration Resistance and Its Measurement," Naval Weapons Center, NWC TP 5648, China Lake, CA, April 1974.

[3]Ipson, T. W., and Recht, R. F., "Ballistic Perforation by Fragments of Arbitrary Shape," Naval Weapons Center NWC TP 5927, China Lake, CA, Feb. 1977.

[4]Recht, R. F., "Taylor Ballistic Impact Modelling Applied to Deformation and Mass Loss Determinations," *International Journal of Engineering Science*, Vol. 16, 1978, pp. 809–827.

[5]Ipson, T. W., Wittrock, E. P., and Recht, R. F., "Transformation of Terminal Ballistic Threat Definitions into Vital Component Malfunction Predictions," Naval Weapons Center NWC TP 4871, China Lake, CA, Dec. 1969.

[6]Jonas, G. H., and Zukas, J. A., "Mechanics of Penetration: Analysis and Experiment," Ballistic Research Laboratory ARBRL-TR-02137, Aberdeen Proving Ground, MD, Feb. 1979.

[7]Baker, J. R., "Rod Lethality Studies," Naval Research Laboratory Rept. No. 6920, Washington, DC, July 1969.

[8]Rolens, D. K., Grubin, E. S., and Recht, R. F., "Effects of G Loading on Aircraft Vulnerability," Ballistics Research Laboratory BRL CR 293, Aberdeen Proving Ground, MD, March 1976.

[9]Yatteau, J. D., "High Velocity Multiple Plate Penetration Model," Naval Surface Weapons Center, NSWC TR 82-123 Dahlgren, VA, Feb. 1982.

[10]Alexander, M. D., and Finnegan, S. A., "Interaction of Multiple Fragment Systems with Plate Array Targets," Naval Weapons Center TP 6058, China Lake, CA, Jan. 1979.

[11]Flint, J. B., "A Vulnerability Analysis of 24 Electronic Chassis Found in a Tube-Type Radar System (U)," Naval Weapons Center NWC TP 5034, China Lake, CA, Jan. 1971.

[12]Zukas, J. A. (ed.), *High Velocity Impact Dynamics*, John Wiley & Sons, New York, 1990.

Chapter 10

Flash Radiography

Manfred Held*

MBB Dynamics, Schrobenhausen, Germany

1. Flash Radiography in Ballistics

THE first quantitative measurements of particles moving at high speeds were made by Carl Cranz by means of electric discharge sparks in air. The electric spark was bright and short enough to produce shadowgraph records of the motion of projectiles and fragments. Even today, the spark approach is the basis of many short-exposure photographic procedures, but with significant improvements in brightness and shortened exposures.

A special type of spark is the *vacuum spark,* that is, an electric discharge in a vacuum in which the electrons are set free by field emission from needles or sharp edges. A strong electric field accelerates the electrons from the cathode toward the anode where they impact at a high velocity. X rays are emitted when these electrons are slowed down in the anode. Owing to the short duration of vacuum breakdown, the duration of x-ray emission is also very short. In this sense one may speak of an *x-ray flash*.

Flash radiography is an important technique in all fields of ballistic experimentation, with the main emphasis undoubtedly on terminal ballistics. In what follows, we shall not dwell in scientific thoroughness on the fundamentals of flash radiography with all the related theories and details. For such details the reader is referred to an excellent book by Francis Jamet and Gustav Thomer[1] and an article by Francis Charbonnier.[2] Both contain extensive lists of references. To complement these references, the present contribution will demonstrate, by means of examples, why x-ray flash radiography holds such a preeminent position in the entire field of ballistics and particularly, in high-explosive warheads (see also Ref. 3).

Flash radiography provides a very special element—time resolution or stop motion—that is not present in conventional radiography. This special capability, however, carries some limitations[2]:

1) The exposure time and radiation intensity must generally be preset, and there is no opportunity to change them while the exposure is in progress.

---

*Professor, Manager of Technology Warheads.

2) Relatively high voltages are required to achieve useful x-ray intensities during a very short pulse. Hence, image contrast is more limited than in conventional radiography.

3) Even at high voltages, the total radiation intensity per pulse is relatively low, typically an incident dose of 1 to 50 milliroentgens (mR) at 3 meters from the x-ray source; the intensity is much less after penetration of a thick object. Consequently, very fast film/screen combinations must be used, resulting in a loss of sharpness and an increase in quantum noise.

4) Because the target cannot be cooled effectively during the submicrosecond pulse, the x-ray target must be physically large (one to several millimeters) so as to absorb the electron beam energy. The large focal spot places an additional limitation on the sharpness of the radiographic image.

5) During observation of very violent events, such as large explosive detonations or impact phenomena, suitable shields and a substantial physical distance from the event must be used to protect the film from damage. This further degrades the contrast and sharpness of the radiographic image.

The ability to control and optimize techniques is much more limited in flash radiography than it is in conventional static radiography, and the same image quality cannot generally be attained. However, in many situations involving high-speed events, flash radiography represents the only available technique for imaging or observation, and its limitations are tolerated because of its unique ability to freeze motion and provide time-resolved information.

1.1 Flash X-Ray Equipment

Events in ballistics are, in general, very rapid. They proceed at velocities of several hundred meters per second, for example, bullets from pistols or rifles, up to 10 km/s, for example, detonations or shaped charge jets. To minimize motion blur, a very short duration x-ray flash must be used. Light sparks usually have breakdown durations of the order of microseconds. By using high voltage and breakdown in a vacuum, durations much shorter than a microsecond are easily achieved. Modern units usually have an effective x-ray flash duration of 20–30 nanoseconds. In general, a voltage of several hundred kilovolts is applied in order to obtain a sufficiently high x-ray intensity. Commercially available x-ray flash units usually operate at voltages of from 150 to 600 kV. But there are also commercial units at 1 to 2 million volts. In addition to a high voltage to produce hard x-rays, a very high current is also required to achieve a sufficient x-ray total intensity current. Values are in the range of several kiloamperes, and the electric energy of the discharge is in the range of 10 to 1000 joules.

In principle, there are several ways of producing short-duration high-current high-voltage pulses. The generally accepted technique, however, is a discharge of capacitors in a Marx high-voltage generator. It permits achieving an extremely short high-current discharge with a duration of 20 ns and output voltages from a few hundred kilovolts up to several megavolts, with charging voltages in the range of only 30 kV. The pulse capacitors C are connected through resistors and charged in parallel up to the charging voltage U_+, for example, to 30 kV (Fig. 1). By triggering spark gap F_1, the ground potential of the second capacitor C_2 is raised to the plus potential of the first capacitor, for which only a recharge of

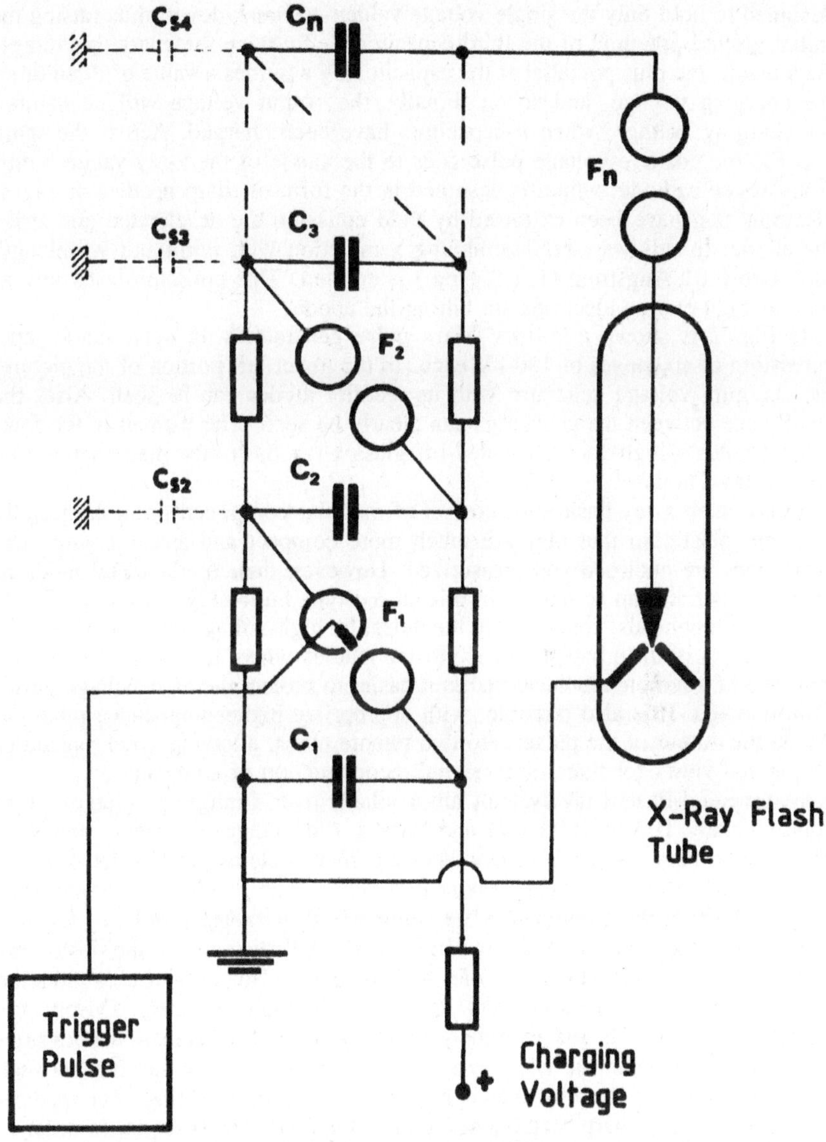

Fig. 1 Schematic circuit diagram of a Marx pulse generator, in which the capacitors C_1 through C_n are charged in parallel and are then discharged in series across the x-ray flash tube, via the spark gaps F_1 through F_n.

the stray capacitance C_{S2} at the electrode of the second capacitor C_2 is required (drawn by dotted lines in the circuit diagram). In this way, twice the charging voltage appears at the second spark gap F_2, which causes this spark gap—designed to hold only the single voltage value—to break down, thus raising the initial ground potential of the third capacitor C_3 to twice the charging voltage. As a result, the plus potential at the capacitor C_3 assumes a value of three times the charging voltage, and so on. Finally, the output voltage will be n-times the charging voltage, when n-capacitors have been charged. Across the spark gap F_n, the positive voltage pulse goes to the anode of the x-ray vacuum tube from whose cathode, which is designed in the form of sharp needles or edges, electrons that have been extracted by field emission are accelerated and strike the anode. In this way, bremsstrahlung x radiation with minimum wavelength of 0.1 to 0.01 Angstrom (1.0–0.1 pm) is emitted. This emission lasts only as long as high-energy electrons are hitting the anode.

In Fig. 2 is shown a 900-kV Marx pulse generator with open spark gaps, consisting of six stages of 150 kV each. In the lower left portion of the picture, the charging voltage generator with its rectifier diodes can be seen. Also, the spark gaps between the capacitors can clearly be seen. The woman in the foreground holds a high-vacuum sealed-off glass x-ray flash tube manufactured by Dr. Gustav Thomer.[1]

Today, such x-ray flash units consist of modular components that are usually cast into plastic so that they are much more compact and space saving. The spark gaps are enclosed and pressurized. However, their fundamental mode of operation is not seen as readily in this closed-type build (Fig. 3a and Fig. 3b).

Remote tubeheads, connected to the pulser by high-voltage coaxial cables, are particularly useful in ballistic or explosive studies where the event is very violent; use of a remote tubehead makes it easier to protect the high-voltage pulser from damage. It is also possible, with appropriate higher-impedance tubes, to divide the output of the pulser into two remote tubes, allowing synchronized or orthogonal views for three-dimensional reconstruction of an event.

Commercial flash x-ray systems are available over a range of voltages (from approximately 100 to 2000 kV) and with a wide range of output capability. The two primary companies manufacturing such systems are Hewlett-Packard in the United States and Scanditronix in Sweden. The published characteristics of systems from these companies are summarized in Tables 1 and 2.

Similar to the spark method used in the early ballistic studies, the flash x-ray method produces a shadowgraph of a ballistic event. However, it uses radiation in the x-ray spectrum rather than in the visible light spectrum. Despite the use of intensifier foils and extremely sensitive films, high radiation intensities are needed, owing to the short exposure times and the long spacing between the x-ray tube and the film that is required for warhead testing. For medical applications, very sharp pictures are usually required, for example, in order to be able to recognize bone fractures. In ballistics, such sharp pictures are usually not required. For this reason, intensifier foils made of calcium tungstate, which are much more efficient but have less resolution, can be used; the film is placed between two of these foils (Fig. 4). Specific cassettes have been developed that contain intensifier foils and ensure that the film, when inserted, is in close contact with the foils.

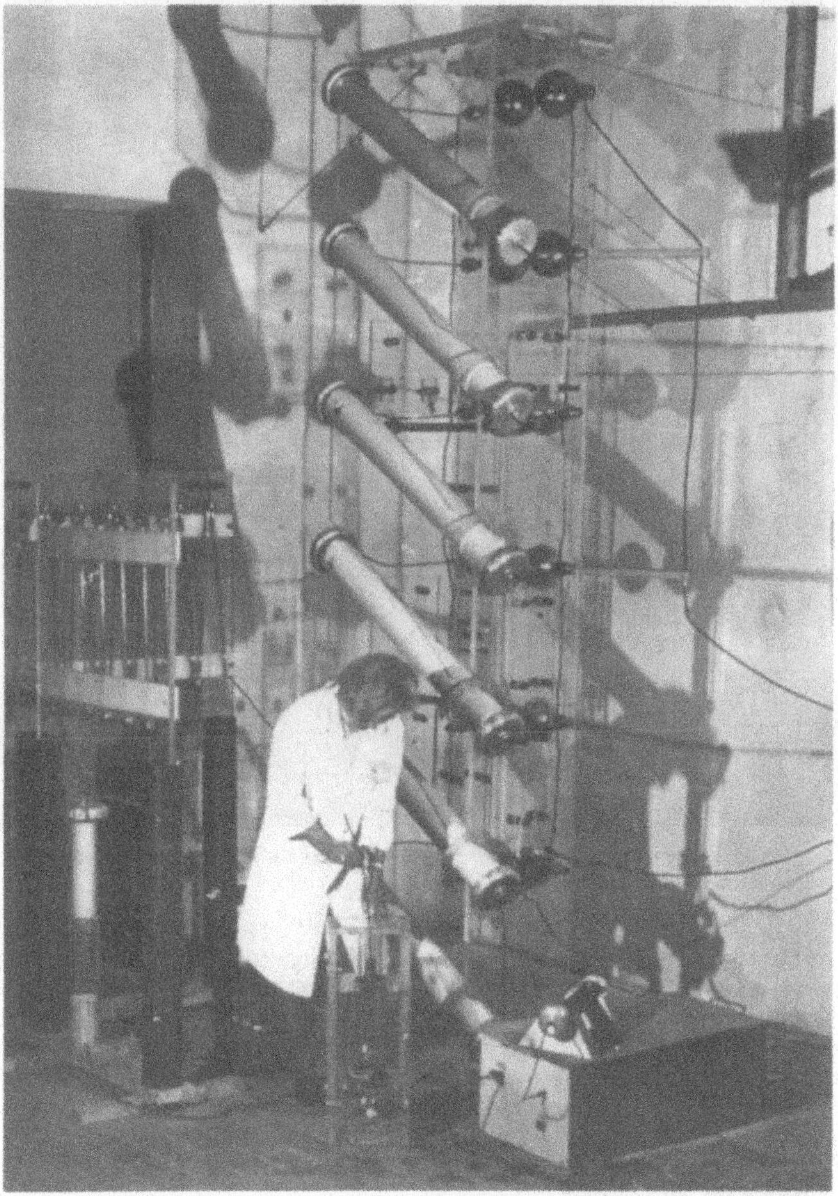

Fig. 2 900-kV x-ray flash unit, built by the author and in use for many years.

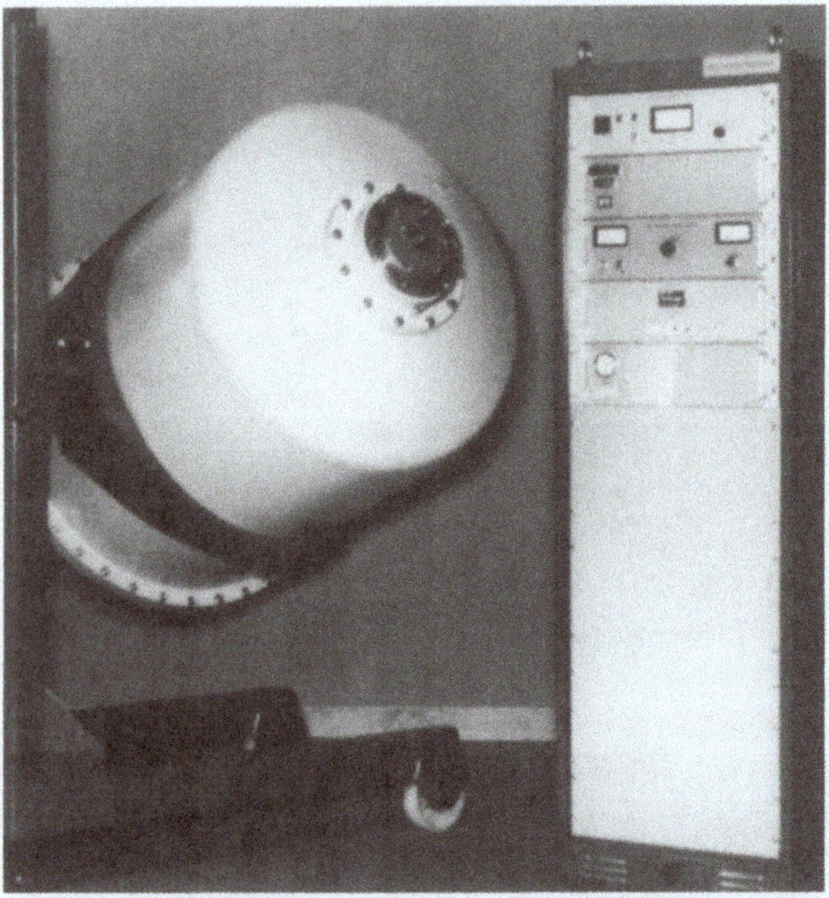

Fig. 3a Commercially available 450-kV unit by the Scandidron Company, Sweden.

The film cassette is, in turn, put into a protective cassette to shield it from the blast wave of exploding charge or from the fragments or other debris from detonating warheads. The protective cassette is in general very tough so that it can withstand even the detonation of a high-explosive charge weighing several kilograms and detonating at a distance as short as 0.3–0.5 m away. In general, the cover plate is a 10- to 20-mm thick high-grade aluminum alloy plate. Some space is left free between the cover plate and the film cassette and is lined with plastic foam so that, under the pressure loading, the cover can bulge without damaging the film cassette.

The x-ray flash recording is a shadowgraph of an event situated between the x-ray flash tube and the film. A general-purpose testing unit, which is particularly useful in terminal ballistic investigations and which can be used in a great variety of experiments, is shown in Fig. 6. The remotely controlled pulse generator and

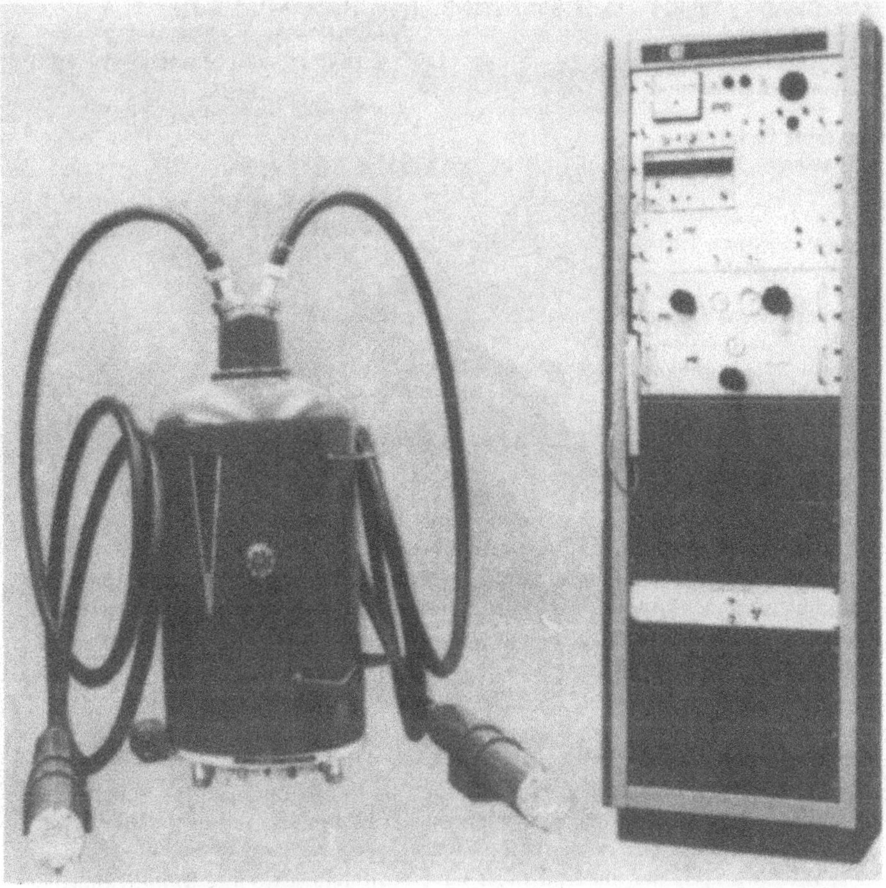

Fig. 3b 300-kV x-ray flash unit for two simultaneous x-ray flashes, by the Hewlett-Packard Company.

the x-ray tube are housed in a protective bunker. The latter is closed by a window of, for example, two thick polyurethane plates. Polyurethane has a particularly low x-ray absorption. For protective purposes, the window area can be kept relatively small, because the x-ray flash tube is located immediately behind the window; still, a sufficiently wide aperture angle of the x-radiation must be provided for. The x-ray film between the intensifier foils of the film cassette, all inside the protective cassette, is located at a distance of about 3 to 4 m from the window. Such a long distance can be chosen when only shadowgraphs of fragments or of a shaped charge jet are to be recorded. The distance must be shorter when additional material is to be penetrated by the x-rays.

To evaluate the position of a projectile or the straightness of a shaped charge jet in space, two simultaneous x-ray flashes at right angles to one another are needed. In Fig. 7 is shown a schematic drawing of a setup designed to record

Table 1 Flash x-ray systems from Hewlett-Packard[4]

Nominal peak voltage, kv	150	300	450	1000	2300
Nominal pulse duration, ns	70	30	30	30	30
Available Outputs					
Standard x rays	x	x	x	x	x
Soft x rays	x	x	x		
Remote tubeheads (stero)	x	x	x	x	x
Dual tubeheads (stero)	x	x	x		
Electron beam		x	x	x	x
Standard X-ray Output					
Shadowgraph distance, m	4	6	10	20	40
Penetration in aluminum at 1 m, cm	2	6	10	17	24
Penetration in steel at 2 m, cm	—	1	2.3	5	8
X-ray dose per pulse at 1 m, mR	1.6	6	20	55	500

two images at 90 deg to one another. In the figure the firing vessel appears, with one flash tube arranged vertically and one horizontally. The advantage of this arrangement is that the events proceed horizontally and can therefore easily be adjusted to various standoffs. The corresponding pulse generator units are located in a separate room. The entire unit is operated from a safe shelter. The *aperture technique* must be used to avoid multiple superimposed exposures of the film, in particular when more than two flashes are used (Fig. 8).

1.2 Design of Experiments

To size a flash x-ray system properly for a particular purpose, the dimensions of the object to be observed—its thickness d_v, length ℓ_v and/or height h_v, or, generally speaking, the shape of the object—must be known. For the purpose of a shadowgraph, the material must have sufficient x-ray absorption, that is, thickness times density, so that a contrast can be obtained. Conversely, the material to be penetrated by the x rays must not be so thick or dense that the rays cannot penetrate with the dimensions given.

Furthermore, it is important to estimate the rates at which the event will occur. If detonating high-explosive charges are to be included, their weight and mass

Table 2 Flash x-ray systems from Scandiflash AB[5]

Nominal peak voltage, kv	150	300	450	600	1200
Nominal pulse duration, ns	35	20	20	20	20
Remote tubehead	yes	yes	yes	yes	no
Dual tubehead	yes	yes	yes	yes	no
Dose per pulse at 1 m, mR	1,6	9	20	30	65
Penetration in steel at 2 m, cm	0,7	2,1	3,6	4,2	6,8

(1)Remote tube head can be used up to 450kV.
(2)Dual tube heads can be used up to 450kV.
(3)Penetration measured with Kodak XAR-5 film and Du Pont NDT 9 intensifying screens.

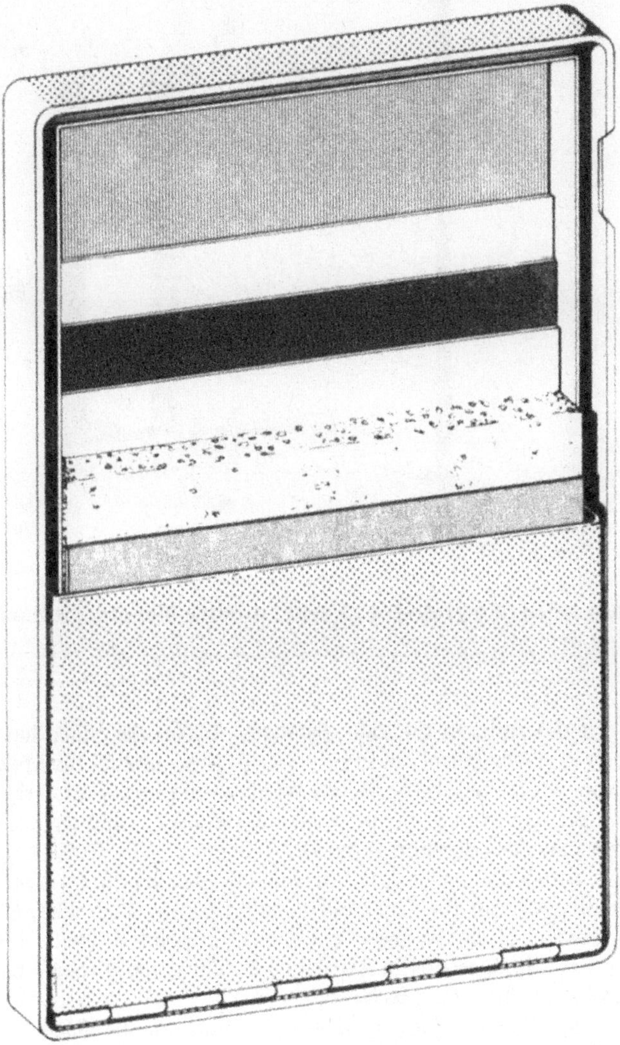

Fig. 4 Cutaway view of a typical film cassette with intensifier foils.

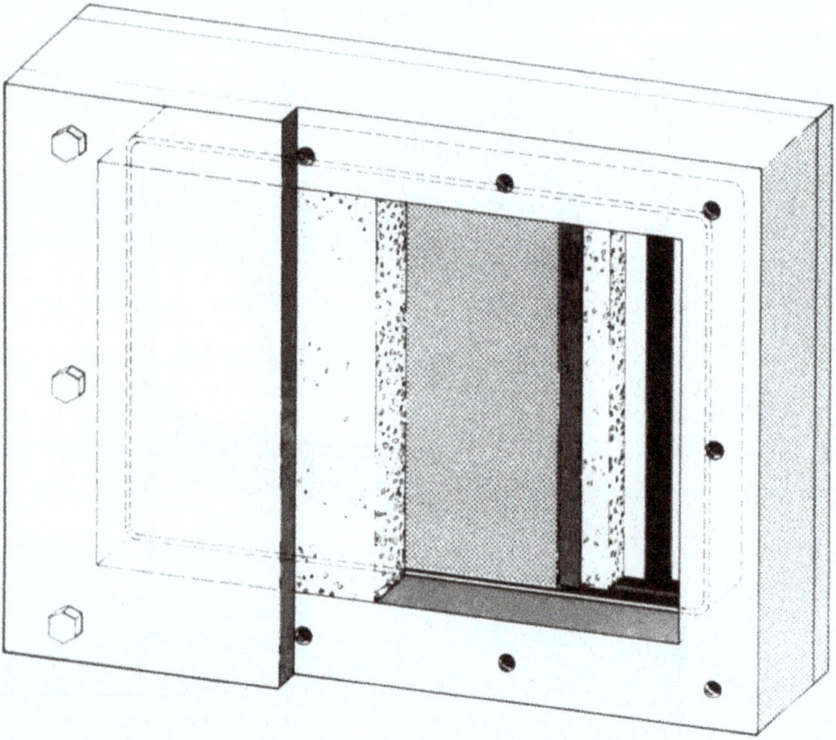

Fig. 5 Construction of a protective cassette, with the film cassette inserted at the bottom.

are needed for determining the safe distances. If, on the other hand, thicker materials such as a rifle barrel are to be x-rayed, this material thickness must be considered, too. Moreover, the protective windows in front of the x-ray tubes and the protective plate in front of the protective cassette, d_{s1} and d_{s2}, respectively, must be taken into consideration (Fig. 9).

From these considerations follow, in general, the maximum necessary distances of the unit between the film cassette and the x-ray flash tube, $E = a + b$. For various reasons of safety, the event should in general not be too close to the film cassette, which results in a distance b. Thus, the distance between the object and the x-ray flash tube is $E - b = a$.

The maximum distance $E_{max} = a + b$ depends on the flash unit, on the maximum pulse voltage and/or energy as well as on the film sensitivity with intensifier foils and the material between, such as, for example, the barrel of a rifle and/or protective plates for a fragmentation charge. The length and the height of the maximum possible field of view are limited by the aperture angle of the x-ray flash tube and/or sometimes by the windows of the unit. Several film cassettes can be arranged side by side so that the maximum possible field

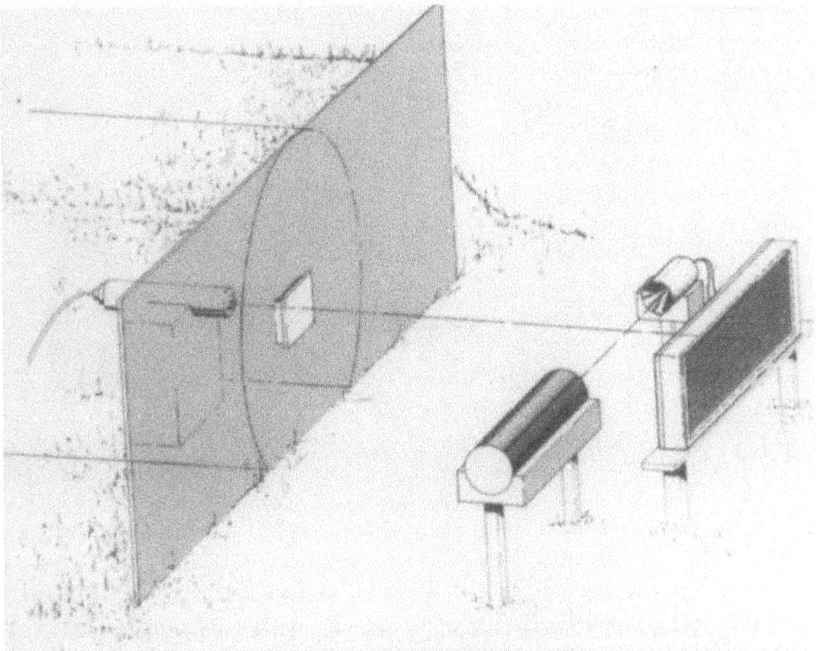

Fig. 6 Simplest arrangement of an x-ray flash unit for general applications in ballistics. Only the pulse generator with x-ray flash tube is located in a protective bunker. The radiation can emerge through a suitable window covered by 10 to 20 mm of a high-grade aluminum or magnesium alloy.

illuminated by the x-ray flash unit is utilized to its full extent. The radiation intensity is strongest along the axis and will be denoted by 100%; then, in an angular range of ±40°, the flash intensity will in general still be about 75%.

X-ray flash units can quite easily be started, that is, triggered, by the event. They require no advance time at all, once the capacitors have been charged. They can be switched through, so to say, at any time with submicrosecond accuracy and can be started by a trigger signal either directly or, after an adjustable time difference, by means of a time delay generator. This is the ideal mode of operation for a device that is intended for recording and keeping track of high-speed events.

Even though the sharpness of x-ray flash radiographs is relatively limited for several reasons, quite valuable information can be extracted from the recordings, as the examples below will show (Fig. 10). Motion blur will in general play the least significant part. The effective time of exposure of x-radiation is some 20 to 30 ns, that is, it is so short that a projectile moving at 1600 m/s will travel for only 0.05 mm within this time of exposure. A more critical point is, in general, the penumbra that results from the fact that the focal spot of the tube has a size of several millimeters, and that for various technical reasons the event cannot proceed immediately

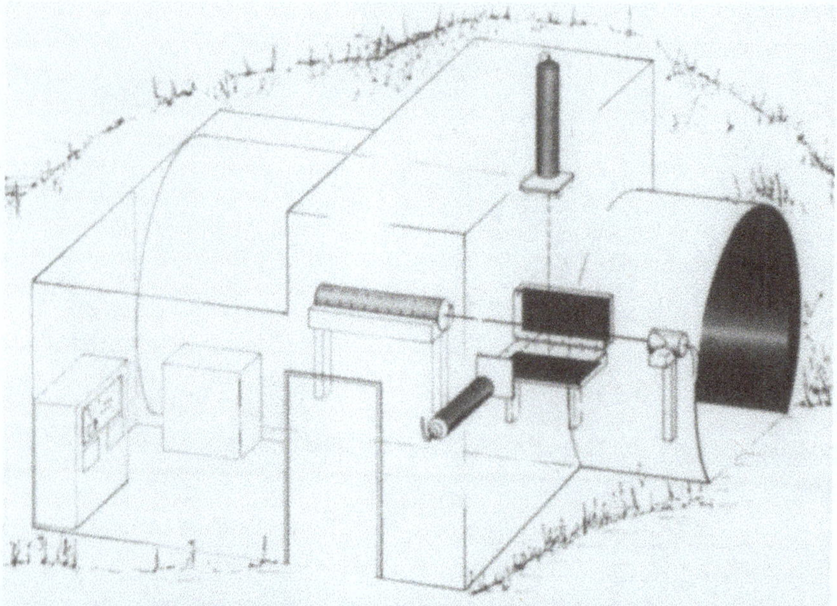

Fig. 7 X-ray flash unit for stereographic recording. That the tests can be made horizontally permits performing a great variety of different types of experiment and experimental arrangements, like both shooting and detonative firing.

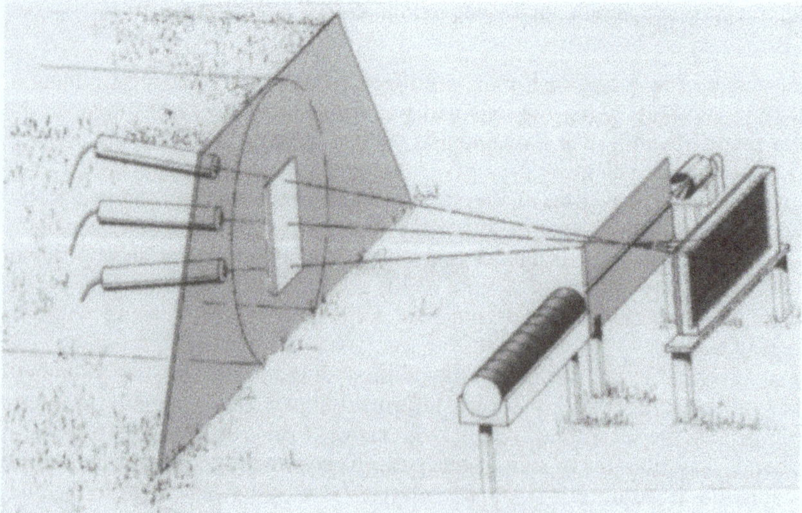

Fig. 8 Multiple x-ray flash array in which the individual pictures of the horizontally progressing process are screened out on the film with appropriate steel or lead masks.

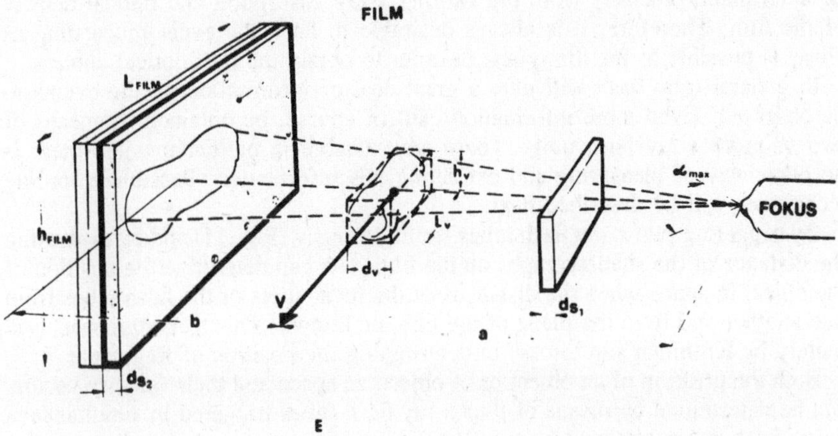

Fig. 9 Schematic of the process of observing an event at an x-ray flash unit.

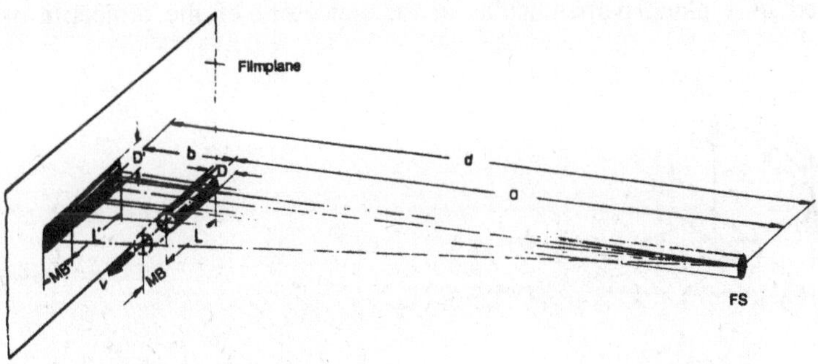

Magnification $M = \frac{d}{a} = \frac{a+b}{a} = 1 + \frac{b}{a}$

$D' = M\,D$ (not correct, because of varying thickness of cylinder)

$L' = M \cdot L$

$PU = \frac{b}{a} \cdot FS$

$MB = v \cdot t_B$

M = magnification
D = diameter
FS = focal spot
PU = penumbra
MB = motion blur
v = velocity
t_B = duration of flash

Fig. 10 Illustration of the penumbra HS given by the imaging scale or the magnification *V* of the shadowgraph and by the finite size of the focal spot BF, and of the motion blur BU given by the duration of the x-ray flash.

in front of the plane of the film. Moreover, the absorption by the object is not uniform so that, in particular in the case of round objects, the contour cannot be determined precisely from the varying x-ray absorption and optical density of the film. Therefore, it is always desirable to have the event proceeding as close as possible to the film plane in order to obtain the best optical contrast.

In general, one flash will give a great deal of information on the events to be observed. Even more information can, of course, be obtained by means of two or more x-ray flash units. These are valuable in particular when there is no other way of measuring and extracting this information. Possible recording techniques will be described next.

By triggering two x-ray flash tubes simultaneously (Fig. 11) and by evaluating the distance of the shadowgraphs on the film, one can determine the position of the object in space when the distances of the focal spots of the flash tubes from one another and from the plane of the film are known. This technique was used mainly by Kronman and Gross[6] in determining the position of fragments.

Both the position of an object or of objects in space and their velocity vectors can be determined by means of four x-ray flash tubes triggered in simultaneous pairs. Such an arrangement was used to determine the starting point of the natural fragments from a projectile shell and their velocities even at a considerable distance from the detonating projectile.[6]

Velocities can be determined in particular by means of two x-ray flash tubes arranged in a plane perpendicular to the trajectory of the projectile or frag-

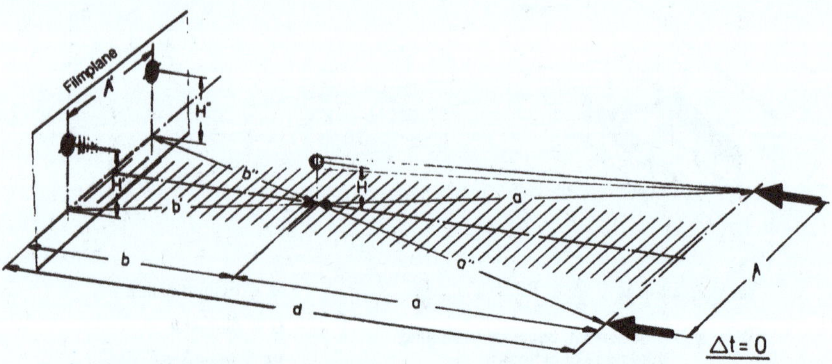

Find b or a and H

$$\frac{b}{a} = \frac{A'}{A} \qquad b = \frac{A' \cdot d}{A + A'} \qquad a = \frac{A \cdot d}{A + A'}$$

$$\frac{H}{H'} = \frac{a}{d} \qquad H = \frac{A \cdot H'}{A + A'}$$

Fig. 11 An exact determination of the position of an object can be made by two x-ray flash tubes that record the object simultaneously, that is, with $\Delta t = 0$. This technique was used by Kronman and Gross[6] in particular for determining the position of fragments.

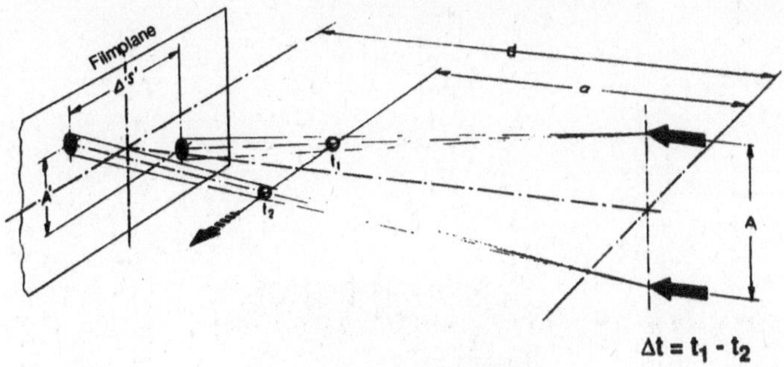

$$v = \frac{\Delta s'}{M\,\Delta t}$$

$$M = \frac{d}{a} = \left(1 + \frac{A'}{A}\right)$$

$$v = \frac{\Delta s'}{\Delta t\left(1 + \frac{A'}{A}\right)}$$

conditions:
trajectory parallel to film cassette and perpendicular to A

Fig. 12 An exact determination of velocity can be made by means of two x-ray flashes emitted at different times in a plane perpendicular to the direction of motion. The position in space can also be determined if the velocity vector is known.

ment (Fig. 12). When the trajectory is exactly perpendicular to the x-radiation and parallel to the plane of the film, the position of the object can also be determined.

1.3 Applications

Flash radiography has been adopted in the entire field of ballistics. By means of refined experimental techniques in *interior* ballistics, even an observation of the burning and of the motion of propellent charges in the barrel of a weapon is possible (Fig. 13).

The position of a tubular propellent charge in a booster motor can also be recorded by x-ray flash radiography shortly before cutoff (Fig. 14).

The separation of a triple projectile, still inside a rifle barrel, can be observed only by means of x-ray flash recording (Fig. 15). The figures on the picture give the separately measured individual velocities of the projectiles.

As a supplement to other measurement techniques, extremely soft x-radiation has been used in photographing the gas flow in the field of *intermediate* or *muzzle* ballistics. The processes occurring at the weapon muzzle during separation and discard of the sabot from the core in subcaliber munitions cannot be observed by optical means. The escaping propellant gases, which in part react with the oxygen of the air, are not optically transparent. These processes in the immediate vicinity of the muzzle can be recorded only by means of flash radiography (Fig. 16).

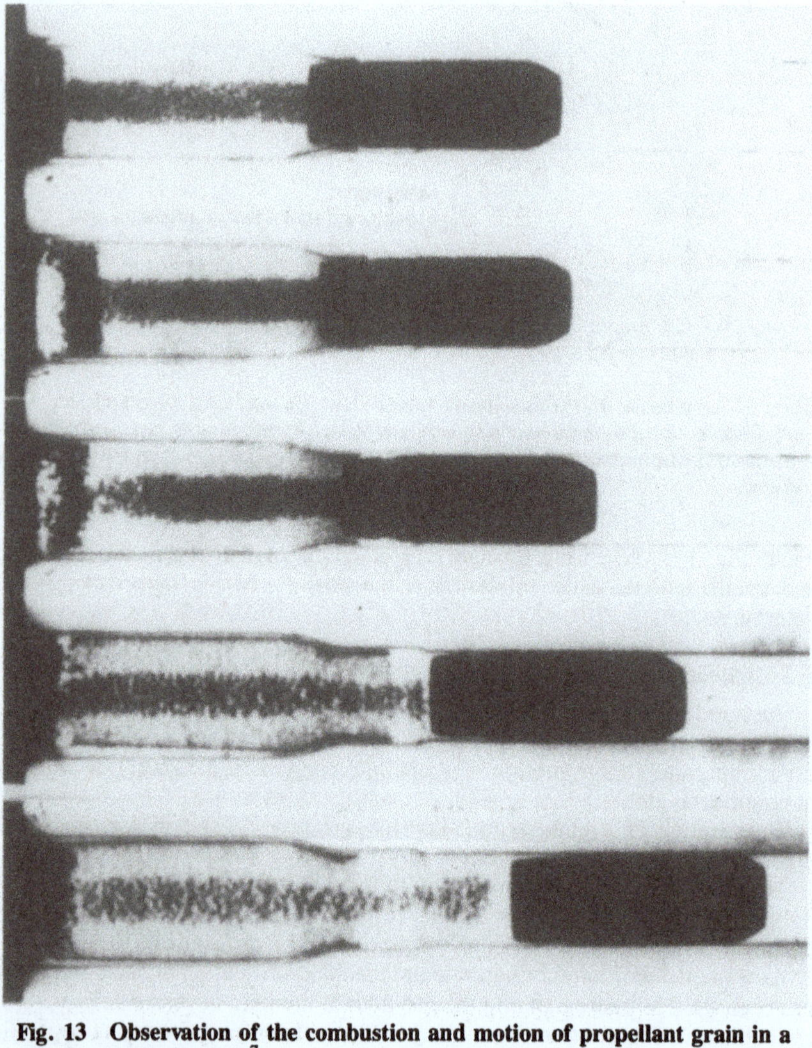

Fig. 13 Observation of the combustion and motion of propellant grain in a specific weapon barrel.[7]

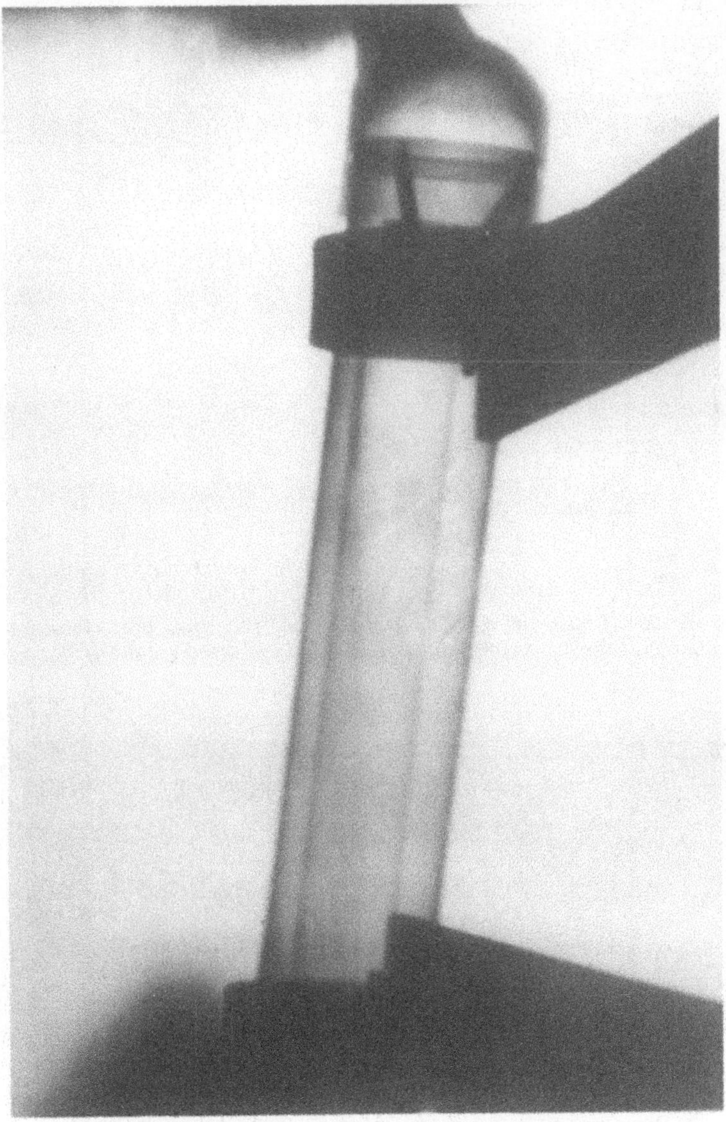

Fig. 14 Propellant charge of the tubular burner of the COBRA rocket motor; holder lost shortly before cutoff.

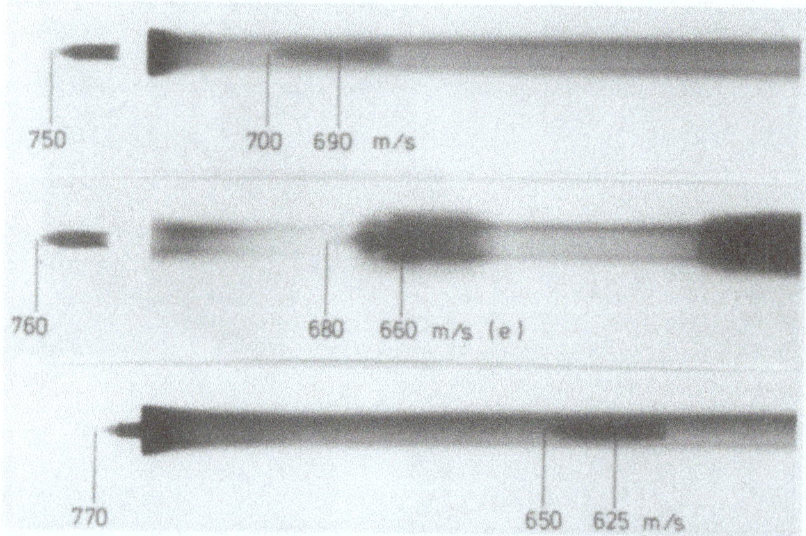

Fig. 15 X-ray flash radiographs of three tests with triple projectiles in a 7.62-mm barrel, showing insufficient separation of the second and third projectile.

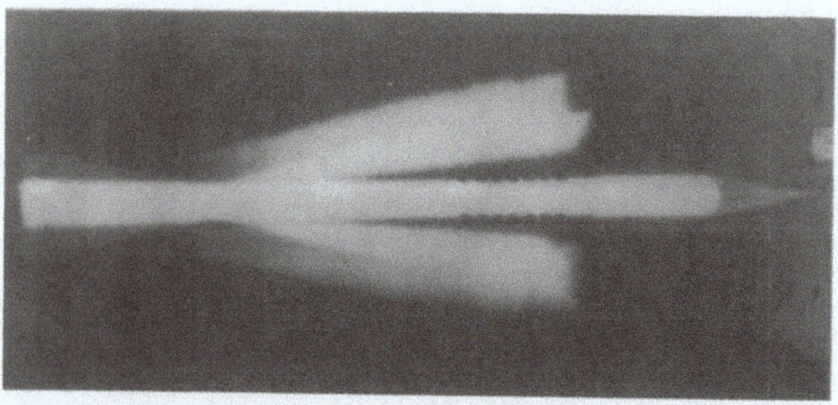

Fig. 16 X-ray flash radiograph of the incipient separation of the sabot parts of a high-velocity projectile, recorded by Schmidt, BRL.

X-ray flash radiography is not quite so essential in the field of *exterior ballistics* as it is in terminal ballistics. Projectile velocities, fragment velocities, questions of stability, and ablation problems can be treated in most cases just as well by means of other recording techniques, even though x-ray flash radiography has the advantage that a dark firing channel is not needed because the film is always packed into a lightproof cassette. From an experimental point of view, x-ray flash units have certain advantages in this respect, but usually they are not frequently used because of the large quantity (Fig. 17) of film materials needed and high processing costs.

Fuze operation of flying projectiles is often observed by means of x-ray flash radiography because the arming process on firing can be observed unambiguously only by radiating through the fuze casing (Fig. 18).

The principal domain of x-ray flash radiography is undoubtedly in the field of *terminal ballistics*. It is suitable for radiating through a high-explosive charge as well as through the reaction products of detonation of a high explosive, enabling one to observe all processes occurring inside, such as detonation waves, shock waves, compressions, material velocity, fragmentation, fragment acceleration, shaped charge jetting, or projectile formation in EFP charges. These processes are hardly accessible to any other measurement technique.

The compression of materials by shock waves, known as Hugoniot data, can be measured quantitatively. Apart from determining the shock or detonation

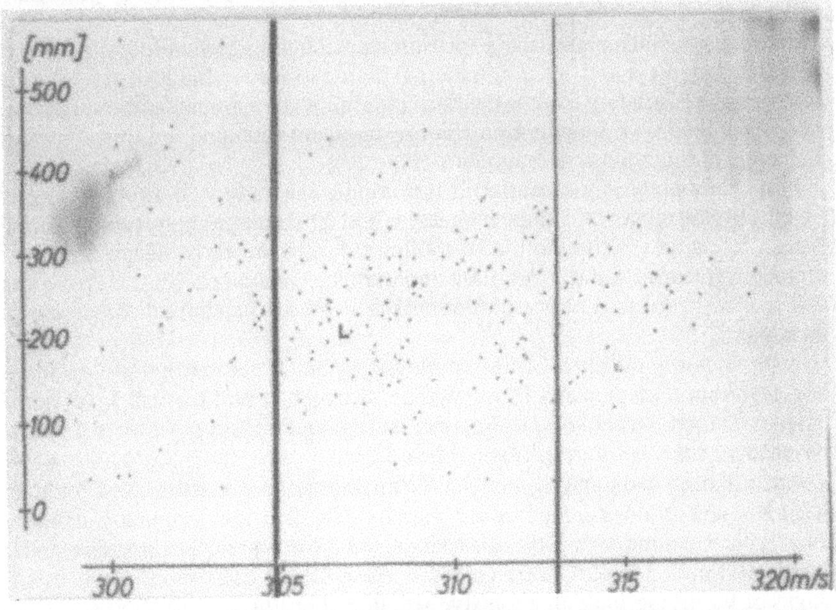

Fig. 17 X-ray flash radiograph of the distribution of a 3-mm diameter buckshot spray at 1/2 choke at 10 m distance from the muzzle, used for determining the spatial and the velocity distribution. The direction of the shot is from left to right.

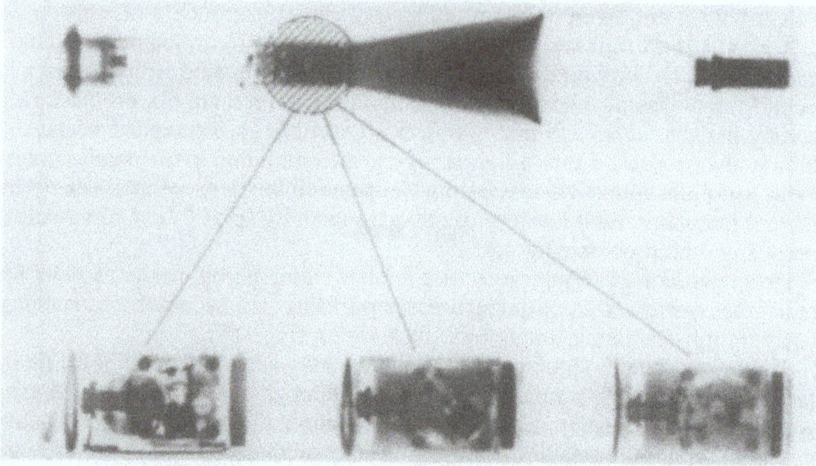

Fig. 18 X-ray flash radiographs of a partially armed safety-and-arming unit (middle), and a completely armed one (right picture), which can be recognized by the dark spot of lead azide in the detonator compared with the still picture (left) with the detonator in the safe position.

fronts, one can also determine the flow of the reaction products by embedding *contrast foils* (Fig. 19).

Particularly convenient is the measurement of the acceleration and velocity of fragments projected by a detonating high-explosive charge; here, one can determine immediately from x-ray flash radiographs the fragmentation, fragment velocities, angles of fragment projection, fragment rotation, and to some extent the shape and the mass of fragments (Fig. 20).

Flash radiography is essential in developing the family of *shaped charges,* such as hollow charges, flat cone charges, and EFP charges (explosively formed projectile charges). The motion of the liner during the early stages can be observed by means of the x-ray flash radiographic technique (Fig. 21). To date this has been the only successful approach in observing liner collapse and jet formation.

A particular application of flash radiography is the observation of penetration and perforation phenomena of projectiles, fragments, and the jets from shaped charges. Target deformation during penetration and cratering can be determined by means of the x-ray flash (Figs. 22 and 23).

Once a flash radiography unit has been installed, it can be used for many purposes and problems far beyond that for which it was originally intended. Two typical examples are the measurement of the response sensitivity and of the delay time of impact fuzes (see Sec. 3) and the observation of the *take off* angle of the flying plate in explosive welding (Fig. 24).

A great number of applications are given in the literature.[1,2] A few more, which are outside the field of ballistics, are the observation of the operation of

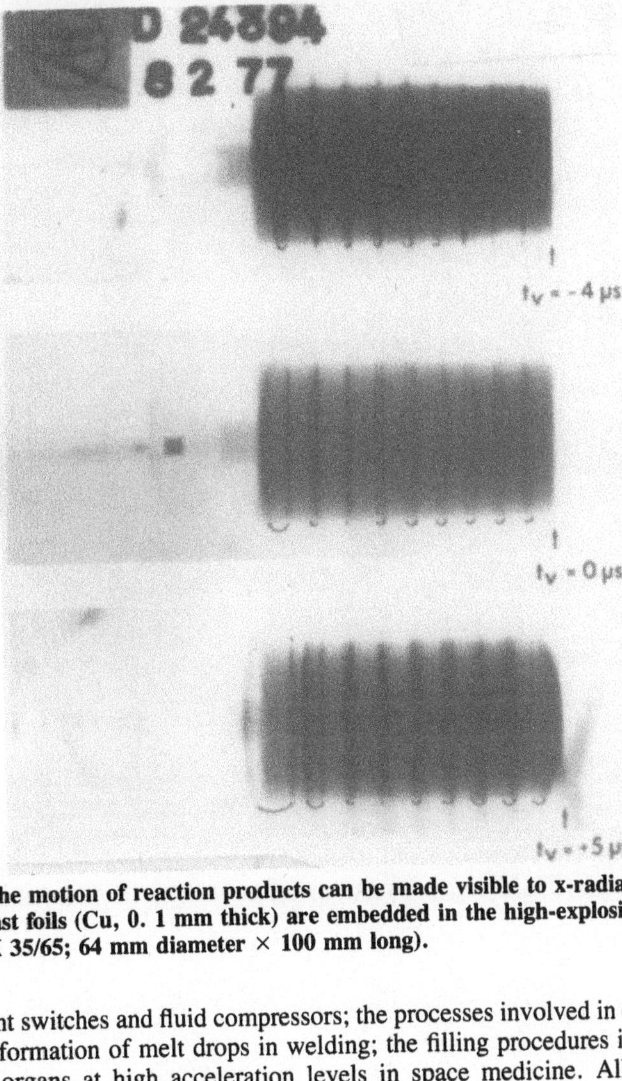

Fig. 19 The motion of reaction products can be made visible to x-radiation if thin contrast foils (Cu, 0. 1 mm thick) are embedded in the high-explosive charge (TNT/RDX 35/65; 64 mm diameter × 100 mm long).

high-current switches and fluid compressors; the processes involved in exploding wires; the formation of melt drops in welding; the filling procedures in casting; and inner organs at high acceleration levels in space medicine. All of these processes cannot easily be observed by other measurement techniques.

2. Determination of Velocities of Fragments in a Spray by Means of Only One X-Ray Flash Radiograph

As discussed previously, the velocity and position in space of a fragment can be determined by a fairly simple method using two pairs of orthogonal flash x-ray tubes. The films must be identified with fiducial marks so that the two film planes can be correlated to one another.

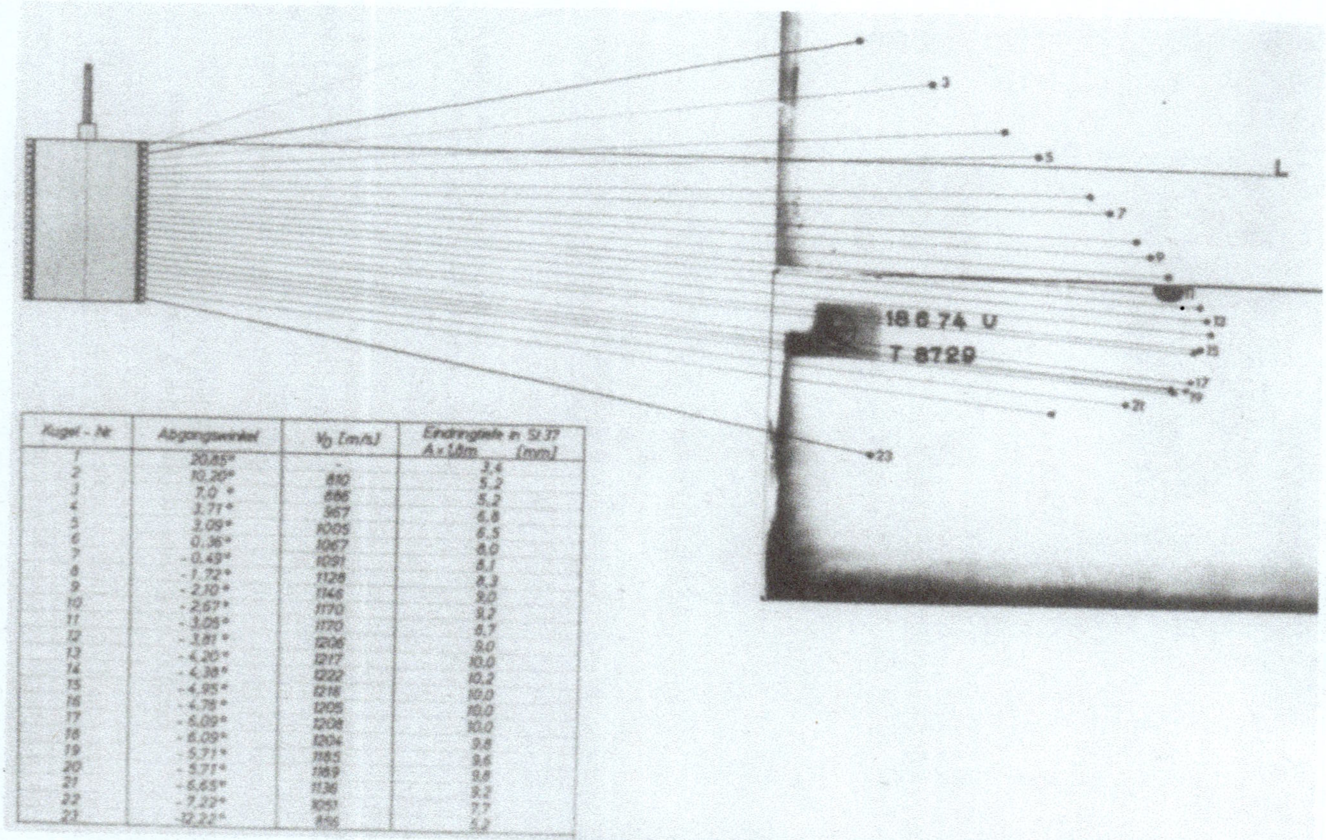

Kugel - Nr	Abgangswinkel	V_0 [m/s]	Eindringtiefe in St 37 A × 1,0m [mm]
1	20,85°	-	3,4
2	10,20°	810	5,2
3	7,0 °	886	5,2
4	3,71°	967	6,8
5	3,09°	1005	6,5
6	0,36°	1067	8,0
7	-0,49°	1091	8,1
8	-1,72°	1128	8,3
9	-2,10°	1146	9,0
10	-2,67°	1170	9,2
11	-3,05°	1170	8,7
12	-3,81°	1206	9,0
13	-4,20°	1217	10,0
14	-4,38°	1222	10,2
15	-4,95°	1216	10,0
16	-4,78°	1205	10,0
17	-6,09°	1208	10,0
18	-6,09°	1204	9,8
19	-5,71°	1185	9,6
20	-5,71°	1169	9,8
21	-6,65°	1136	9,2
22	-7,22°	1051	7,7
23	-12,22°	856	5,2

Fig. 20 **Direction and velocity of preformed fragments—in this case steel spheres of a cylindrical fragmentation charge—can be determined unambiguously by means of x-ray flash radiography with proper screening. (Kugel-Nr = sphere no.: Abganswinkel = deflection angle: Eindringtiefe in St. 37 = penetration in mild steel.)**

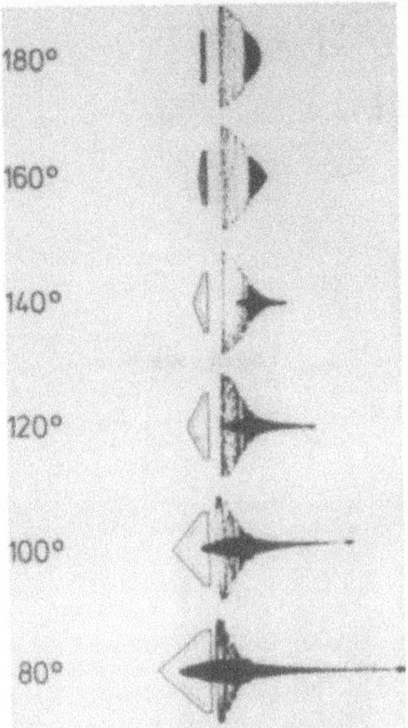

Fig. 21 Acceleration and deformation of metal can be recorded particularly well by means of x-ray flash radiography. In this picture one sees the formation of the jet and slug of copper liners at angles ranging from 180 deg to 80 deg.

A somewhat more elegant method is the stereo-sequential radiographic technique of Kronman and Gross,[6] in which four flash x-ray tubes are arranged along a line. Two tubes each, that is, a pair of tubes, cast two shadow images of the fragment at one time into a film. The position of the fragment in space can be ascertained from the mutual distance of the tubes and of the images of the fragment on the film plane.

The tubes are usually not very far from one another, and so the error in determining the spatial position is relatively great, owing to the small angles of intersection of the two reference lines. Furthermore, the correlation of the individual fragments of a fragment spray is somewhat difficult because there are four shadow images of each fragment.

This section describes another method, by which it is possible to determine the ejection angles and the true distances traveled by the fragments.[9] Therefore, the distribution of velocities of the fragments in a spray can be determined with high accuracy. This can be done with the aid of only *one* flash x-ray picture, by also using the location of fragment impacts on a target plate set up perpendicular to the fragment trajectories.

2.1 Test Setup

As an example, the velocity distribution of fragments from a so-called fragment generator shall be determined (Fig. 25). A total of 81 cube fragments with a 7.6-mm edge length and 3.5-g weight have been arranged in nine rows and

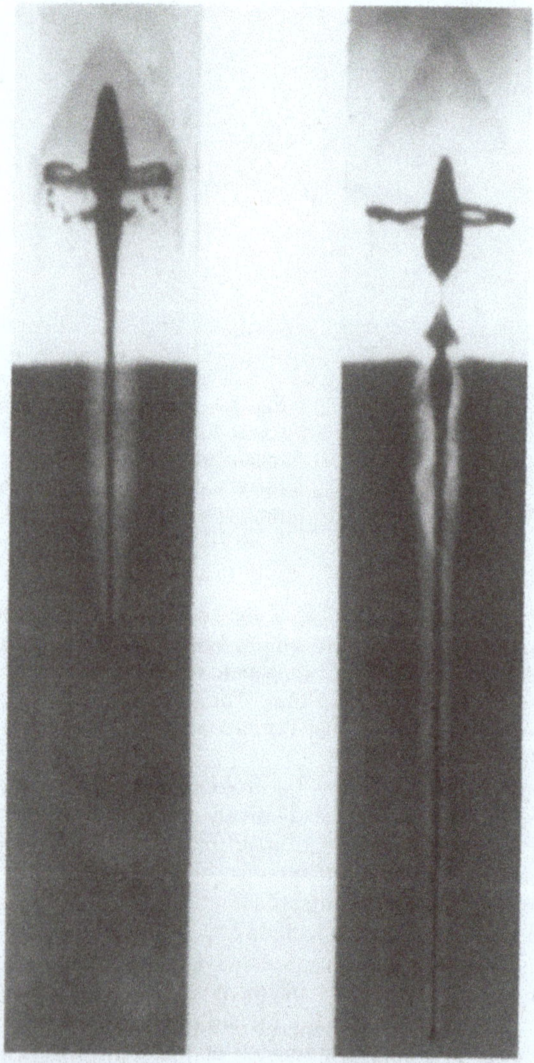

Fig. 22 Penetration of a copper jet into an aluminum block, recorded 24 and 60 μs after detonation of the shaped charge.

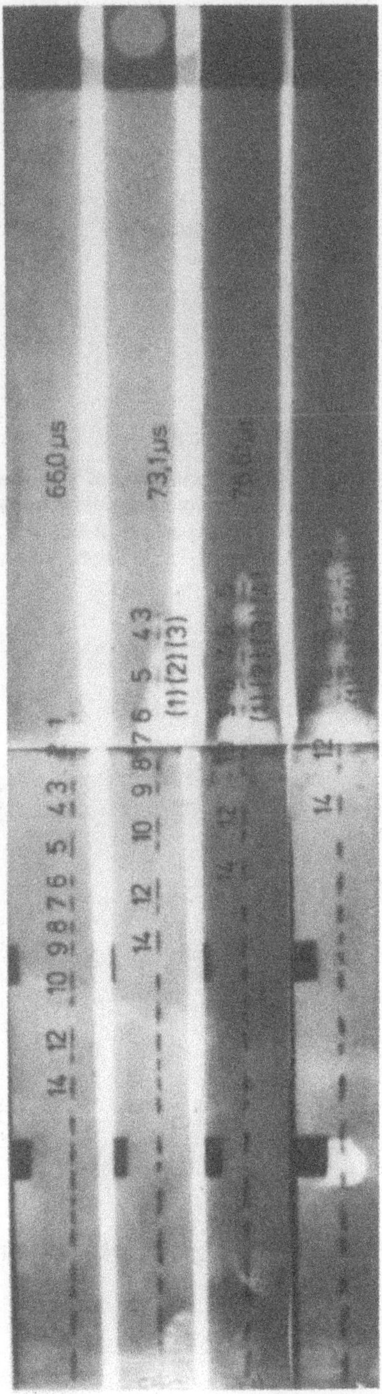

Fig. 23 Successive penetration of a particulated shaped charge jet with bubble-shaped cratering in the target, as recorded with a four-fold x-ray flash installation and with screening of the individual images (see also Fig. 8).

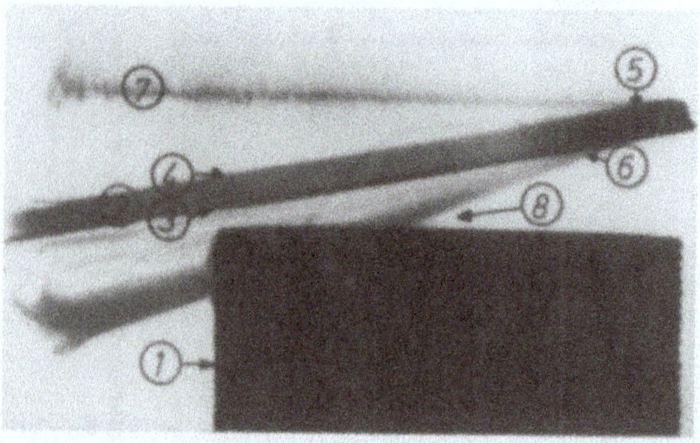

Fig. 24 Observation of jetting in explosive cladding of a copper plate to a steel block. The numbers denote the following: 1) Fe plate to be plated; 2) high explosive, still exposure; 3) Cu sheet, still exposure; 4) 0.5 mm Pb foil; 5) detonation front; 6) accelerated Cu sheet; 7) Pb accelerated upward; 8) jetting at the collapse point.

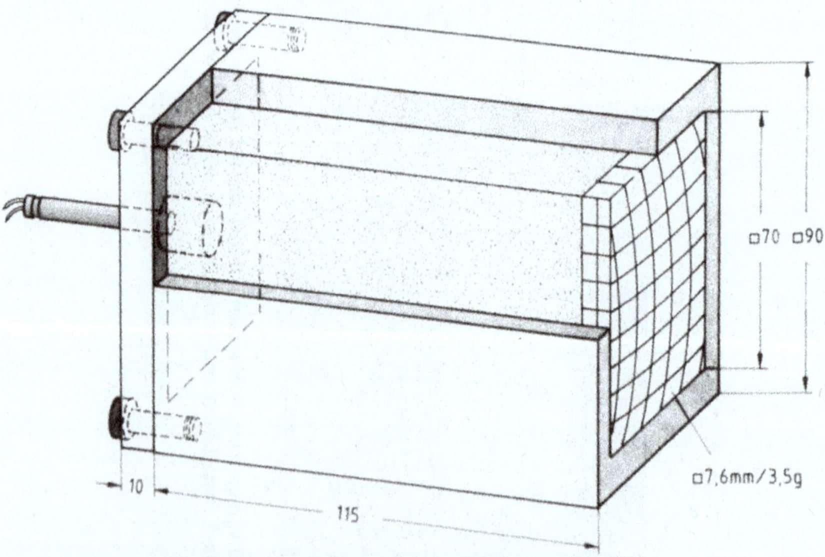

Fig. 25 Design features of a fragment generator.

Fig. 26 Front of fragment generator showing how fragments are arranged in rows and columns (columns are counted from right to left; the direction of the shot is from left to right).

nine columns. The individual fragments can be associated as an array with the individual rows *a* through *i* from top to bottom, and with the columns 1 through 9, which are to run in the direction of the shot, from left to right (Fig. 26).

This fragment generator was fired against a 2×2 m^2 mild steel (St. 37 grade) target plates 3.5 m away (Fig. 27). A flash x-ray unit producing a shadow record of the fragments on a 900×900 mm^2 x-ray film area, was triggered 640μs after the detonation of the high-explosive charge of the fragment generator. A blast shield (see Fig. 27) serves to protect the x-ray film cassettes. A photograph of the test setup is shown in Fig. 28.

2.2 Test Results

Figure 29 is the flash x-ray picture of the fragments at time 640 μs after the detonation of the high-explosive charge. The fragments have obviously traveled different distances.

On closer inspection of the individual shadows of the fragments, one can see that certain fragments in each row are focused better than the rest. The more sharply focused fragments are obviously closer to the film cassette than the ones that are imaged with a greater penumbra. In this way, the fragments can be correlated to rows that are closer to the film plane or farther from it (Fig. 30).

Quite independently of this, a relatively regular pattern of fragment impacts appears on the target plate, from which the individual impacts can be correlated unambiguously to the matrix of the 81 cube fragments (Fig. 31).

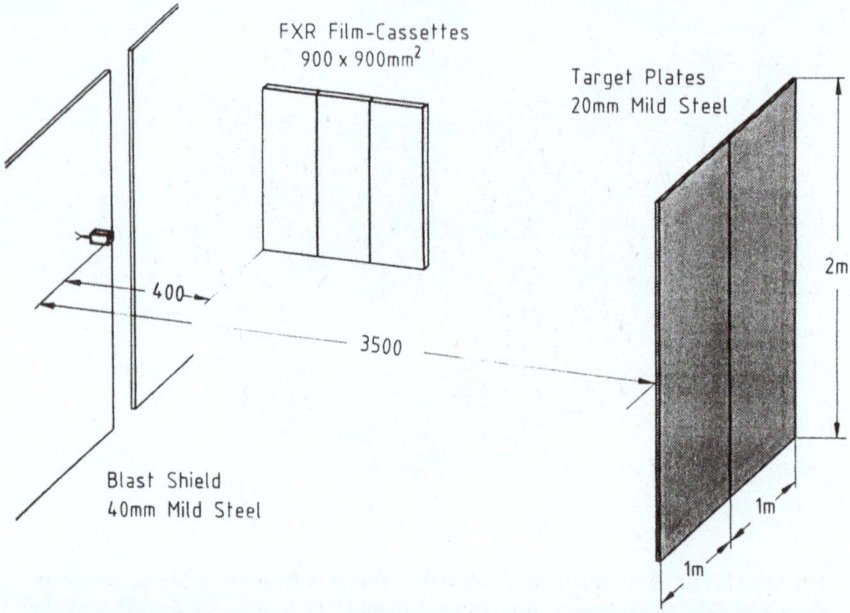

Fig. 27 Test setup for studying fragment velocities by means of x-ray flash radiographs, and of the fragment distribution and fragment performance.

2.3 Graphical Determination of the Fragment Velocities

The individual fragment rows can be recognized and correlated by the formation of groups at a uniform height on the radiograph. The individual fragments of one row can then be correlated to the individual columns, in that these fragments differ from one another in the sharpness of their images. In this manner, fragment numbers 1 and 2 are in better focus than fragment numbers 8 and 9, as can be seen particularly well on the enlarged picture of one row (Fig. 30). As explained before, this observation occurs because fragment numbers 1 and 2 are closer to the film plane than fragment numbers 8 and 9. The different sharpness of the images can thus be used to find the geometrical arrangement of the fragments on the picture as well as to correlate the fragments to their columns in each row unambiguously.

The trajectory of each fragment can be determined graphically in the plane to be analyzed, with the aid of the position of the fragments on the target plate, presenting a simple way of making corrections. For example, in Fig. 32 one sees the fragments from the corners and from the center between the flash x-ray (FXR) image and the target plate as they relate to one another.

The most expedient way to determine the true distances is to use the original flash x-ray image and to project the shadowgraphs of the fragments optically to

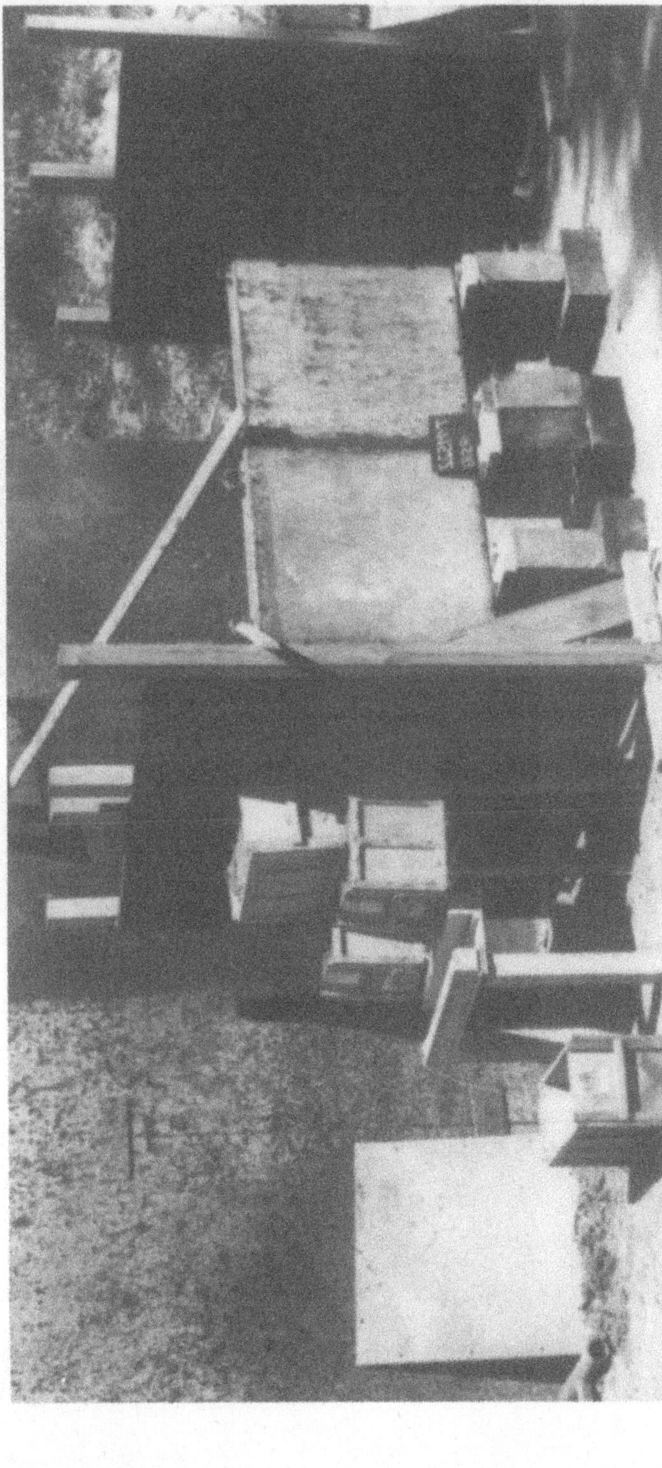

Fig. 28 Photograph of test setup showing fragment generator on the left, which, for protection of the surrounding terrain from fragments, has been shielded by three sand-filled boxes; also showing the blast shield, the protective cassette for the x-ray films, and, at the extreme right, the 2 × 2 m^2 target plate array 3.5 m away from the generator.

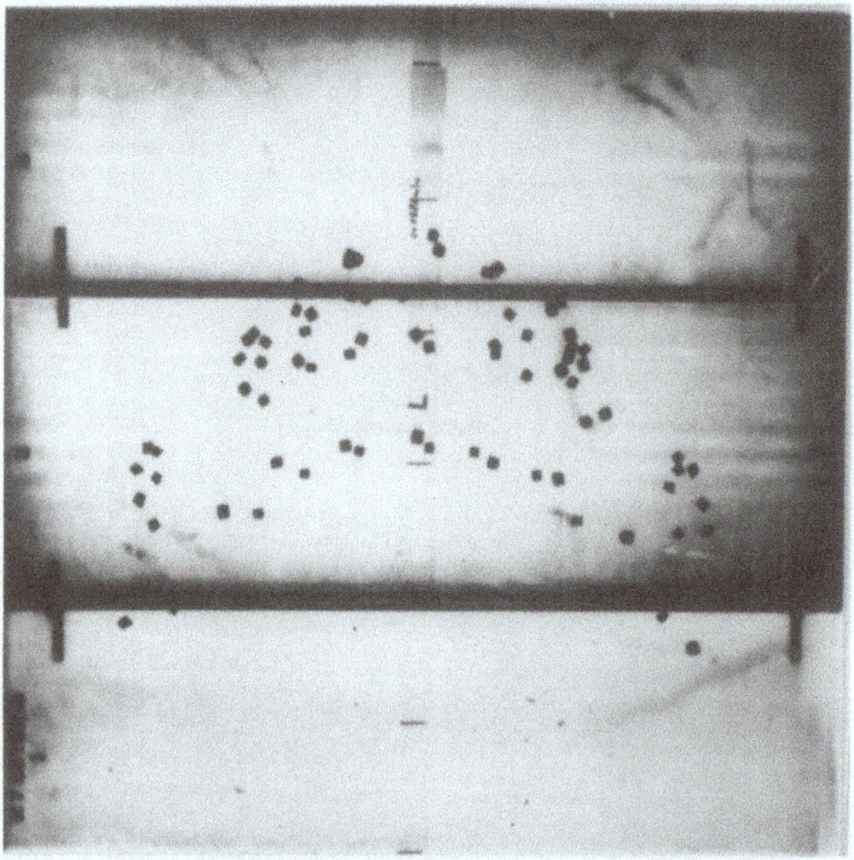

Fig. 29 X-ray flash radiograph of the fragment spray, recorded 640 μs after the detonation of the high-explosive charge. All of the 81 fragments can be identified on the original negative photograph, because this shows a wider range of densities than the positive reproductions.

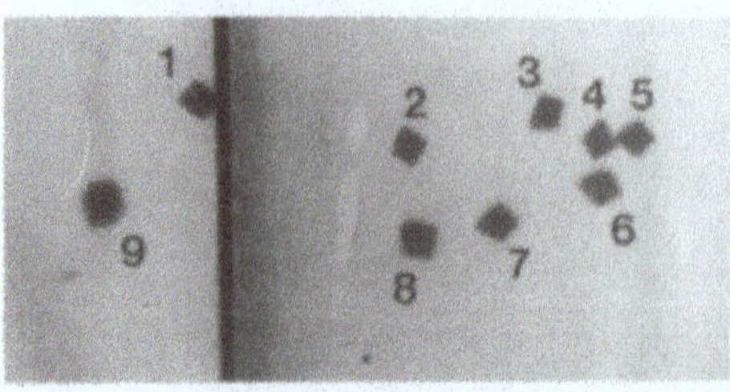

Fig. 30 Enlarged representation of the row number *i* from the flash x ray of Fig. 29.

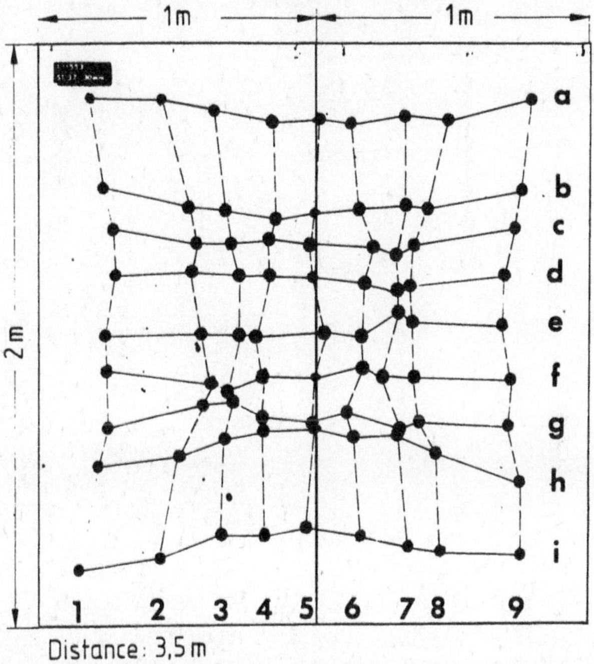

Fig. 31 Photograph of the fragment impacts on the 2 × 2 m² mild steel target area, showing the correlation of the fragment hits to the individual rows *a* through *i* and columns 1 through 9.

the film plane, which actually is vertical. From there, lines are then protracted to the focal point of the flash x-ray tube. In the same manner, parallel lines are drawn from the hits on the image of the target plate to the original positions of the fragments on the generators.

However, the location of the fragments at the time of the flash x-ray exposure can be determined in this way only for the azimuthal plane, which gives an improved determination of the distance the individual fragments have traveled in the time from detonation to flash x-ray exposure and, hence, an improved statement about their velocities. According to this representation, identical fragment velocities are located on circular arcs with the fragment generator in the center.

2.4 Determination of the Fragment Velocities by Calculation

Apart from this graphical determination of the distance traveled by the individual fragments (at least in the azimuthal plane) by means of the flash radiograph and the hit pattern on the target plate, the path in space can be calculated very precisely from the flash x-ray radiograph, the azimuth angle α and the polar angle ϵ. These angles α and ϵ of the individual fragments are determined best from the associated impacts on the target plate.

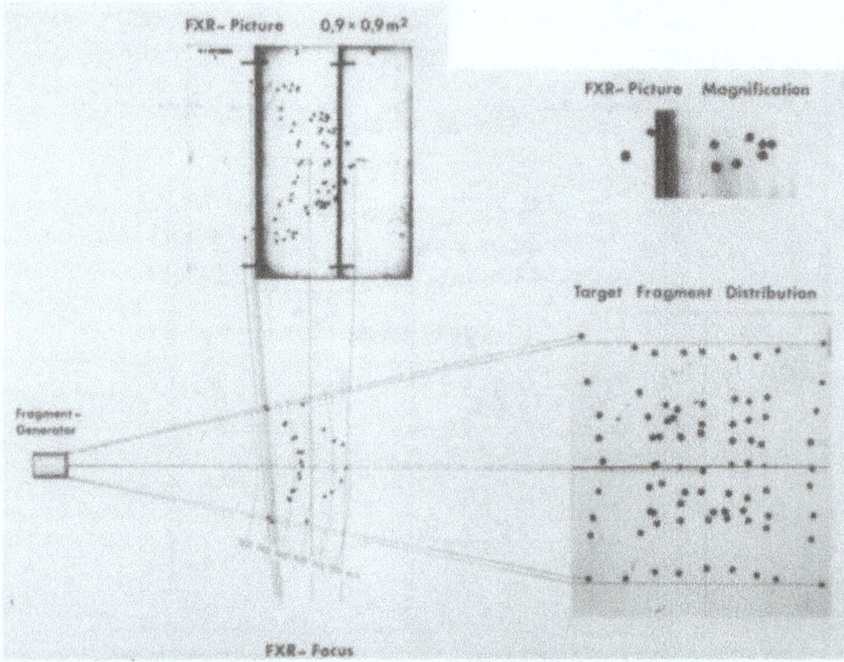

Fig. 32 Graphical analysis of the fragment velocities based on the flash x-ray radiograph and on the angular distribution as determined from the fragment hits on the target plate.

In Fig. 33 the angles and other auxiliary quantities that are necessary for the calculation are shown. The symbols used there are defined as follows:

X = horizontal distance of the fragment in the x ray, as measured from the plane of the fragment generator

R = distance from the foot of the x-ray tube to the plane of the fragment generator (The x rays are here just at an angle of 90 deg to the plane of the film. The film plane itself is arranged in parallel to the axis of the fragment generator.)

A = distance from the focal point of the x-ray tube to the film plane

B = distance from the axis of the fragment generator to the film plane, which is in parallel to it

D = path that the fragment has traveled in space in the time Δt from the detonation of the high explosive charge to flash x-ray exposure

D′ = path the fragment has traveled in time Δt in the horizontal plane, which contains the axis of the fragment generator

D″ = distance the fragment has traveled, projected to the axis of the fragment generator

The distances D of the individual fragments in space can be calculated in terms of the azimuth angle α and polar angle ϵ, both of which can be found from the

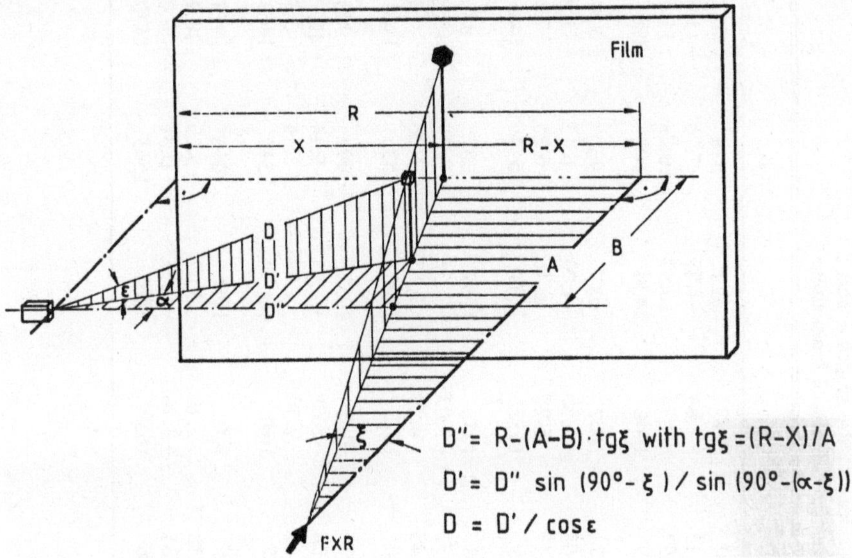

Fig. 33 Auxiliary distances and angles for the calculation of the true distances of the fragments in space.

impact on the target plate, and of the distance X on the x ray, as follows:

$$\xi = \arctan\frac{R - X}{A} \tag{1}$$

For the triangle with sides D' and D'',

$$180 = \alpha + (90 - \xi) + \Psi \tag{2a}$$

Solving for Ψ leads to

$$\Psi = 90 - (\alpha - \xi) \tag{2b}$$

The path D'' of the fragment along the charge axis can be calculated from the following formula:

$$D'' = R - (A - B) \cdot \tan \xi = R - \frac{R - X}{A} \cdot (A - B) \tag{3}$$

D' can be calculated from the triangle with D'' and the angles α and ξ by means of the law of sines.

$$D'/D'' = \sin(90 - \xi)/ \sin[90 - (\alpha - \xi)] \tag{4a}$$

$$D'/D'' = \cos \xi/ \cos(\alpha - \xi) \tag{4b}$$

Table 3 Data measured from the x-ray film for the distances X, the angles α and ϵ from the hits on the target plate, and the fragment velocities calculated from these

		1	2	3	4	5	6	7	8	9
a	$R - x$, mm	738	822	869	897	900	897	879	848	721
	α	12.19	8.40	5.60	2.38	−0.15	−1.84	−4.88	−7.76	−11.8
	ε	11.78	11.85	11.40	10.86	10.91	10.71	10.97	10.78	11.67
	v, m/s	1257	1367	1418	1445	1449	1446	1430	1409	1297
b	$R - x$, mm	835	947	987	1007	1017	1007	987	962	837
	α	11.53	6.89	4.99	2.23	0.08	−2.30	−4.90	−6.13	−11.31
	ε	6.53	6.31	6.26	5.82	6.13	6.23	6.46	6.25	7.13
	v, m/s	1377	1500	1544	1559	1569	1558	1539	1514	1405
c	$R - x$, mm	875	982	1017	1037	1068	1058	1037	987	887
	α	10.94	6.51	4.68	2.61	0.31	−3.07	−4.38	−5.37	−10.87
	ε	5.20	4.50	4.44	4.75	4.45	4.29	3.83	4.36	5.12
	v, m/s	1423	1539	1576	1594	1626	1612	1580	1535	1451
d	$R - x$, mm	897	1010	1052	1092	1097	1092	1052	997	905
	α	10.79	6.74	4.15	2.50	0.08	−2.61	−4.45	−5.14	−10.27
	ε	2.79	2.98	2.76	2.84	2.69	2.30	1.99	2.14	2.72
	v, m/s	1447	1573	1614	1657	1657	1648	1603	1542	1463
e	$R - x$, mm	902	1004	1062	1101	1115	1101	1056	1016	913
	α	11.31	6.21	4.20	3.30	−0.38	−2.46	−5.00	−5.29	−10.20
	ε	−0.38	−0.31	−0.38	−0.46	−0.31	−0.46	0.77	0.31	0.08
	v, m/s	1455	1561	1623	1668	1676	1658	1608	1563	1470

Table 3 (continued)

f	$R-x$, mm	898	998	1051	1083	1085	1082	1038	1007	888
	α	11.24	5.75	4.83	2.92	0.80	−2.54	−3.61	−5.37	−10.64
	ε	−2.26	−2.91	−3.29	−2.53	−2.61	−2.15	−2.61	−2.60	−2.79
	v, m/s	1450	1553	1016	1647	1644	1636	1587	1555	1447
g	$R-x$, mm	875	975	1016	1041	1052	1049	995	981	871
	α	11.24	6.13	4.52	2.92	0.15	−1.61	−4.53	−5.60	−10.49
	ε	−5.12	−4.05	−3.83	−4.67	−5.06	−4.53	−5.35	−5.04	−5.20
	v, m/s	1425	1528	1573	1599	1608	1602	1544	1530	1431
h	$R-x$, mm	829	929	969	1004	1017	1002	987	936	813
	α	11.68	7.34	4.99	2.84	0.08	−2.00	−4.38	−6.51	−11.09
	ε	−7.04	−6.68	−5.80	−5.44	−5.37	−5.75	−5.66	−6.62	−8.02
	v, m/s	1371	1480	1519	1556	1567	1550	1536	1487	1380
i	$R-x$, mm	732	817	862	882	892	882	847	822	702
	α	12.71	8.40	5.14	2.84	0.54	−2.38	−4.99	−6.81	−11.09
	ε	−12.12	−11.78	−10.68	−10.78	−10.42	−10.93	−11.42	−11.46	−11.08
	v, m/s	1266	1360	1405	1427	1436	1431	1399	1378	1267

$$D' = D'' \frac{\cos \epsilon}{\cos(\alpha - \epsilon)} = \left[R - \frac{R - X}{A}(A - B)\right] \frac{\cos \epsilon}{\cos(\alpha - \epsilon)} \tag{4c}$$

With the polar angle and the azimuthal distance D', the following equation is obtained for the distance D of the fragment in space:

$$D = D'/\cos \epsilon \tag{5a}$$

$$D = \left[R - \frac{R - X}{A}(A - B)\right] \cdot \frac{\cos \xi}{\cos(\alpha - \xi)} \cdot \frac{1}{\cos \epsilon} \tag{5b}$$

The 81 calculated fragment velocities, together with the distances X determined from the x rays and with the angles α and ϵ found from the target plate, are listed in Table 3. A look at the fragment velocities (Table 4) reveals that fragments from symmetrically corresponding positions have velocities that compare well with one another. Velocity increases considerably from the corner to the middle. This is clear evidence that the duration of the impulse of the detonating high-explosive charge is reduced by rare factions propagating inward from the edges.

The fragment velocities obtained this way are shown in Fig. 34 as a function of the columns, with the row number as a parameter. Comparing rows *a* and *i* (the two outermost rows), *b* and *h*, *c* and *f*, and *d* and *g* with each other, one can see that mirror-image rows give almost identical velocities. Surprisingly, the middle row shows markedly higher velocities, in particular in the middle.

3. Determination of the Response Time of Impact Fuzes for Projectiles with Flash X Ray

The response time of an impact fuze has a significant influence on the effectiveness of a projectile. When engaging aerial targets, the high-explosive projectile should not detonate until the outer surface has been penetrated. The blast effect is greatest when the projectile detonates inside the target structure, and this can widen the outer skin from inside or even tear it open. The fragments also must first penetrate the skin of the target in the case of external detonation and, therefore, lose kinetic energy and effectiveness. In the case of very flat angles of impact, the fragments do not usually penetrate the outer skin at all—they ricochet. Therefore, when a projectile with a fast fuze strikes an oblique target plate, most of the fragments will be deflected.

On the other hand, ammunition with a blast effect should have a fuze that reacts very quickly when it strikes the earth so that the projectile does not penetrate into the ground, as the fragmentation effect is practically useless there.

The response time of an impact fuze is generally a function of the following parameters: 1) impact velocity, 2) type of target material, 3) thickness of target material, and 4) angle of impact. The measuring equipment and procedure for determining the response time of projectile impact fuzes are considered next. Examples are given of typical response times of mechanical projectile fuzes that were determined using the measuring equipment and procedure. This system can be used for electrical or electronic impact fuzes.

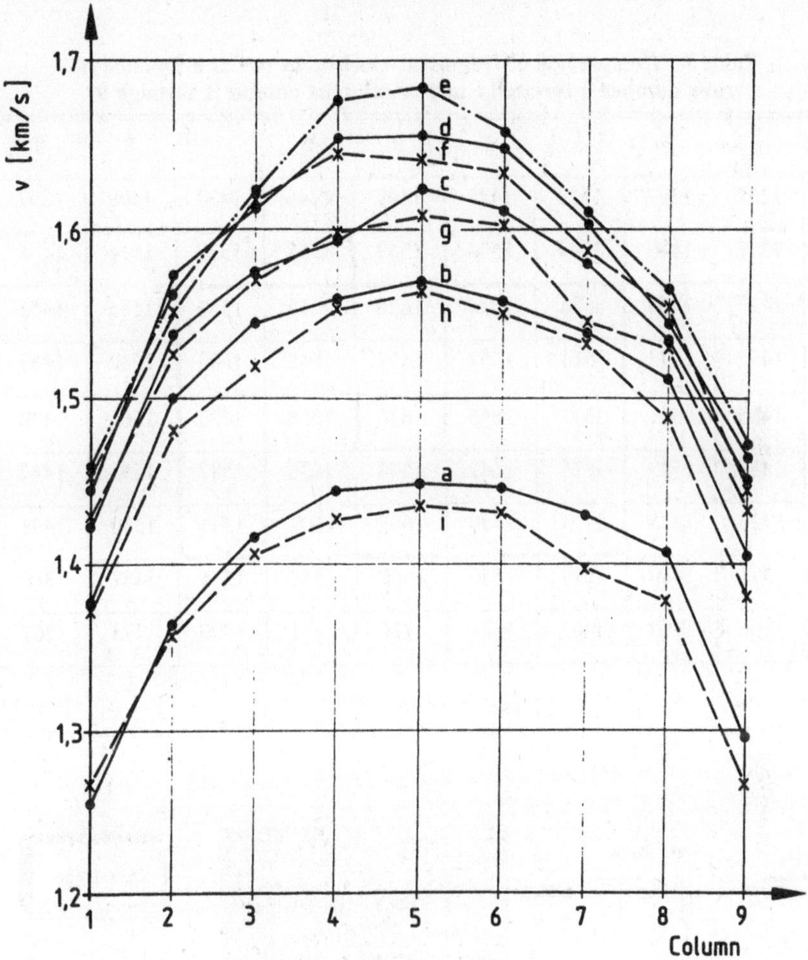

Fig. 34 Fragment velocities as a function of the columns, with the row numbers as a parameter.

3.1 Equipment and Procedures

The measuring equipment and procedure used are shown, together with a functional block diagram, in Fig. 35. The following measuring equipment is required:

- A pair of light barriers with one time counter for determining the exact impact velocity V_T of the projectile
- Short-circuit probe on the target plate with short-circuit trigger equipment for starting
- A second time counter for measuring the difference between the short circuit (impact of the projectile on the target plate) and the x-ray flash

Table 4 Comparison of fragment velocities in m/s as a function of rows number *a* through *i* and of columns number 1 through 9

	1	2	3	4	5	6	7	8	9
a	1257	1367	1418	1445	1449	1446	1430	1409	1297
b	1377	1500	1544	1559	1569	1558	1539	1514	1405
c	1423	1539	1576	1594	1626	1612	1580	1585	1451
d	1477	1573	1614	1657	1657	1648	1603	1542	1463
e	1455	1561	1623	1668	1676	1658	1608	1563	1470
f	1450	1553	1616	1647	1644	1636	1587	1555	1447
g	1425	1528	1573	1599	1608	1602	1544	1530	1431
h	1371	1480	1519	1556	1567	1550	1536	1487	1380
i	1266	1360	1405	1427	1436	1431	1399	1378	1267

$t_{FRX} = 640 \mu s$ T11553

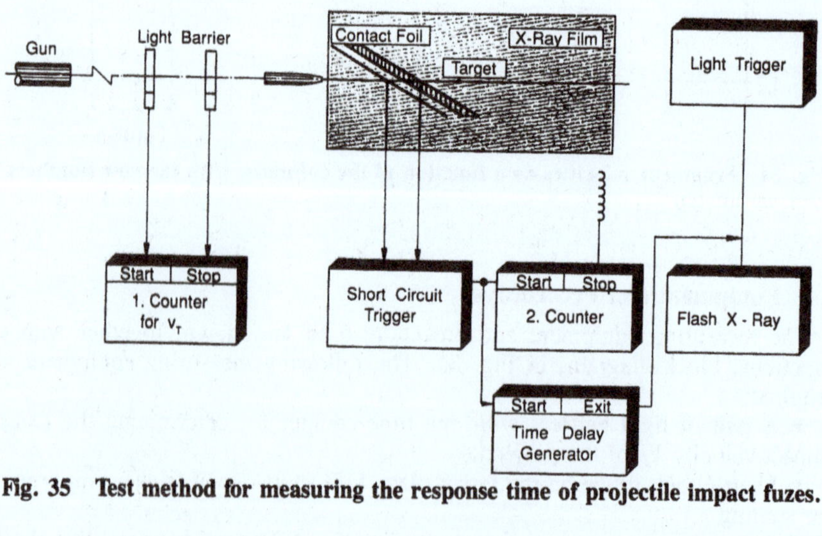

Fig. 35 Test method for measuring the response time of projectile impact fuzes.

• A time delay generator that brings about a redundant triggering of the x-ray flash equipment after a predetermined time delay
• A light trigger for triggering the x-ray flash equipment
• X-ray flash equipment for determining the reaction state of the projectile

To achieve relatively good hit accuracy on the measuring and target plates, the gun should be placed approximately 100 m away from the measuring system. At this distance the initial oscillation of the projectile will have decayed considerably. With a distance of 100 m and a $\pm 0.1\%$ hit accuracy, a minimum hit area of $\pm$ 100 mm must be chosen. This value means that relatively small plate areas can be used even when the angle of arrival is very flat. In this case the length of the plate must be multiplied by the value $1/\cos\alpha$, where α is equal to the angle between the flight path and a line normal to the surface plate.

To be able to fire at the target array at this distance at different speeds, it may be necessary to make some adjustments to the fuzes. These are particularly necessary when shots are fired at low initial velocities V_0 in order to simulate low impact velocities V_T. At such low speeds the projectile would not detonate because the fuzes would not correct function either not armed or led to the self-destruction of the projectile.

At these slower speeds it is particularly useful to employ barrels with correspondingly larger angles of twist, so that in each case the appropriate relation between terminal velocity V_T and the rotational speed of the projectile can be achieved.

In principle, the impact velocity V_T of the projectile on the target plate has to be determined exactly. It is particularly practical to use a pair of light barriers, which are separated by a distance of 2 m, and time difference measuring equipment. The target array, which usually consists of simple target plates, can be arranged at different angles to the flight path of the projectile.

In the author's experience, the arrangement of the plates proved to be very successful as the projectile was guided to the ground, where the impact could be absorbed when ricocheting or rebounding occurred. If the inclination of the plates is the reverse, a larger safety area is required in the firing direction. Ricocheting and rebounding often occur when the projectile strikes at a relatively flat angle and thicker plates are used.

A 0.1-mm thick aluminum foil or a 0.05–0.1-mm thick copper foil is fitted to the target plate. The aluminum or copper foil is insulated from the plate by a thin insulating foil. When the projectile hits the target plate, the foil is short-circuited, thus generating the signal that triggers a second time-measuring device and time delay apparatus.

The reverse side of the target plate or target array is observed using a light trigger with a large field of view. The detonating high-explosive (HE) projectile produces a flash that is received by a photodetector. A number of fast photosensitive sensors are suitable as light triggers, for example, silicon photoelements with response times of under 1 μs. The light signal triggers an x-ray flash system with a relatively short delay time (less than a few microseconds).

The reaction of the x-ray flash system is recorded by a signal derived from an antenna, and signal stops the counter. In this way the x-ray flash system records the projectile on x-ray film at the moment when an appropriate light impulse is received on the light trigger.

An x-ray flash picture is necessary for making a very exact determination of the response time of the impact fuze, as it is observed that the light emission is in no way constant with the breakup of the projectile. The light trigger signal frequently occurred at considerably different times after the detonation of the explosive charge, that is, at different phases of fragmentation. The author has not yet been able to explain this phenomenon. Figure 36 comprises three photographs of a 30-mm high-explosive incendiary shell. It is an example of the various states of the projectile at times when the light trigger reacted.

The response time of the impact fuze was defined as the time difference between the moment the short circuit occurs, caused by the projectile striking the plate, and the moment the projectile starts to break up. In order to calculate the latter, the fragmentation speed of the casing must first be determined using at least two x-ray photographs taken at different times.

With the help of the x-ray photographs taken during the tests, and by noting the distance the fragments of the casing moved, one can calculate the detonation time and the place of detonation of the projectile.

If for some reason the fuze does not function, or the flash is not sufficient to set off the light trigger, then the x-ray flash system is *forcibly* triggered by the time delay equipment, which has a fixed time delay. The time delay for this function must be set in such a way that projectiles that penetrate the plates and those that rebound from very flat target plates still appear on the x-ray film.

In the x-ray flash pictures (Fig. 37) one sees the behavior of an impact fuze of a 30-mm high-explosive incendiary shell against increasing thicknesses of the aluminum alloy plate (0.8 mm, 1.5 mm, 4 mm, and 8 mm) at an impact angle of 60 deg. The effect of the angle of inclination of a 0.8-mm-thick dural plate at 15 deg, 60 deg, and 80 deg taking a 35-mm shell fuze can be seen very clearly from the x-ray flash pictures shown in Fig. 38.

3.2 Test Results

The different designs of mechanical fuzes show to some extent quite different response behavior. For example, an impact fuze for a 20-mm shell displayed a very fast reaction time of approximately 30 μs at impact angles of 15 deg and 60 deg, and approximately 80 μs at an impact angle of 80 deg. In such cases the impact velocity was in the region of 1000 m/s. These times increase somewhat when impact velocities of 750 m/s are achieved and show relatively constant values of 80 μs at an impact velocity of 500 m/s.

In Fig. 39 is shown the analysed response time as a function of the effective target plate thickness. We take the slant thickness to be the thickness of the plate that the projectile penetrates.

From the x-ray pictures in Fig. 40, the response of the mechanical impact fuze at an impact velocity of 750 m/s against dural plates with thicknesses of 0.5 mm, 1.0 mm, 4.0 mm, and 6.0 mm and at impact angles of 15 deg, 60 deg, and 80 deg can be seen. Although the response time of this fuze varies quite considerably.

Most mechanical fuze designs show that there is a clear dependency between the response time and the slant thichkness of the target—at least as long as

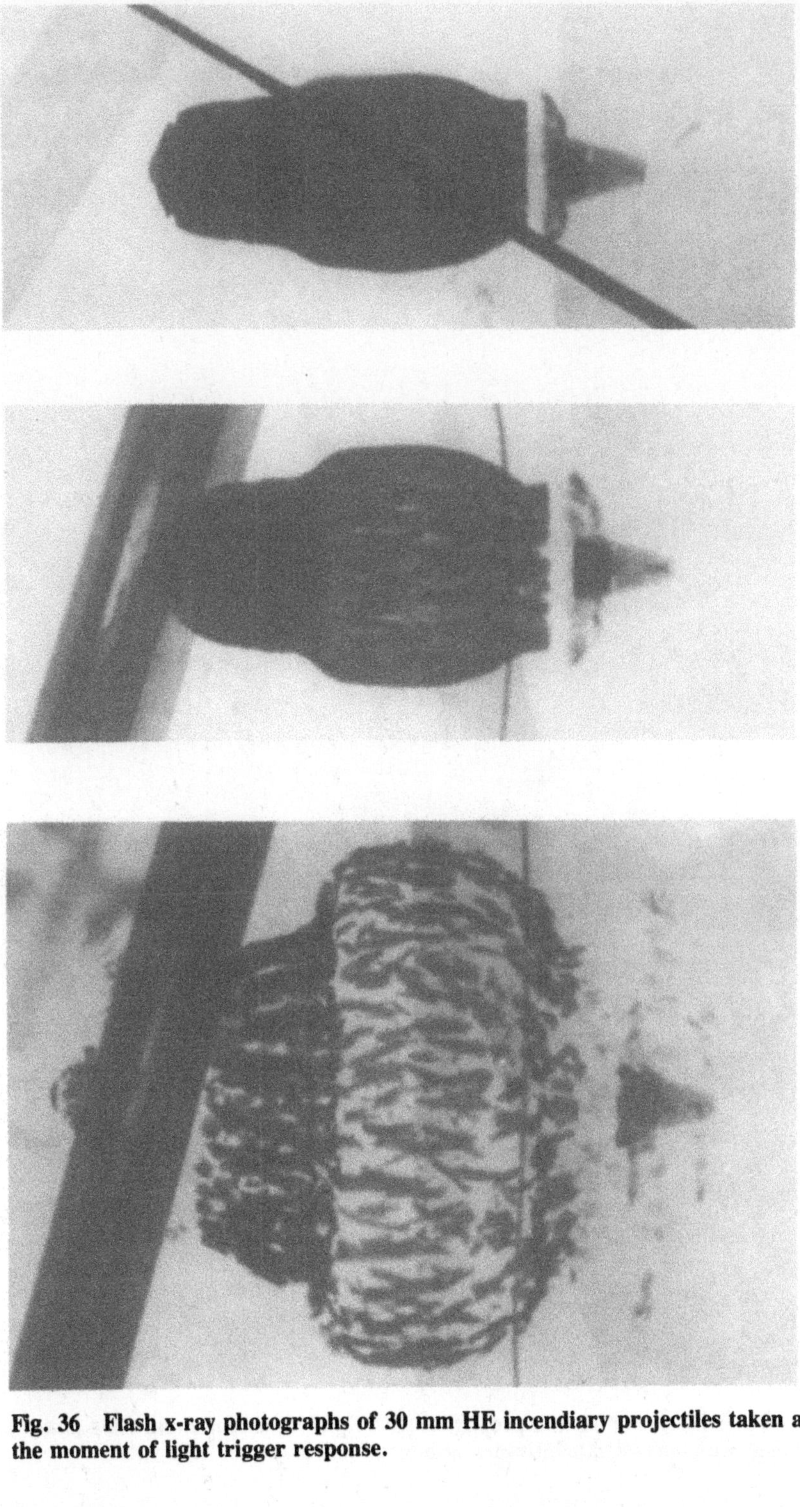

Fig. 36 Flash x-ray photographs of 30 mm HE incendiary projectiles taken at the moment of light trigger response.

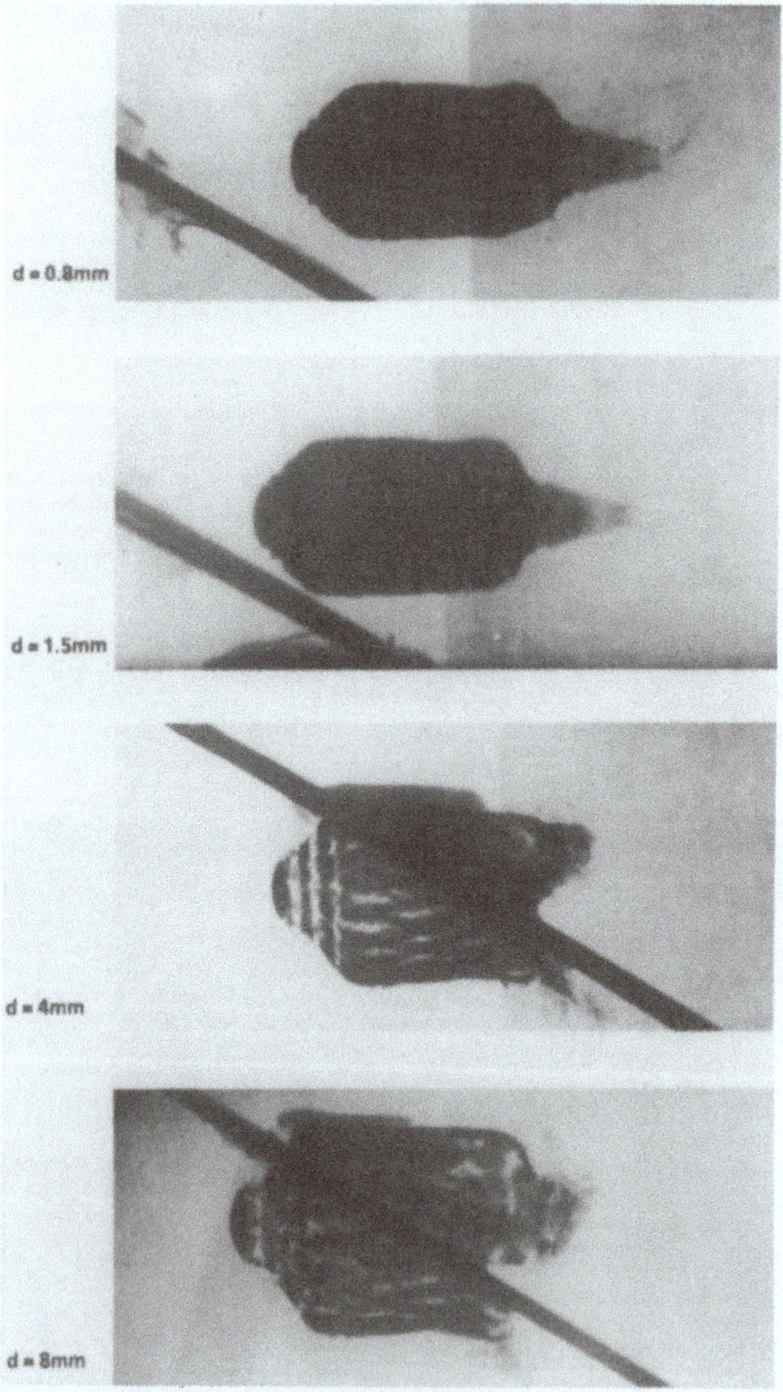

Fig. 37 Positions of 30-mm projectiles striking oblique aluminum alloy plates of 60 deg, with various thicknesses *d* at a terminal velocity of 1050 m/s.

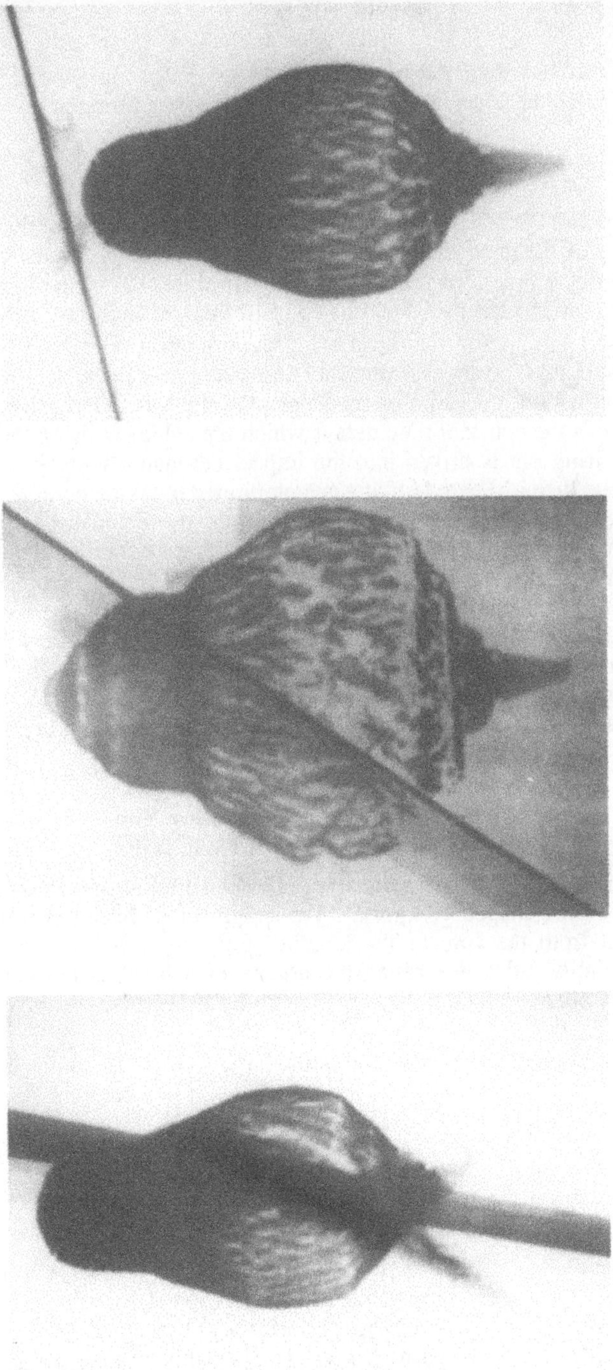

15°

60°

80°

Fig. 38 Positions of 35-mm projectiles striking aluminum alloy plates with a thickness of 0.8 mm at impact angles of 15 deg, 60 deg and 80 deg, respectively, and a terminal velocity of 1175 m/s.

certain values of the slant thicknesses are not exceeded. In general terms we were able to calculate a delay time taking the following formula:

$$t_v = B \cdot d^{1-\frac{1}{2}}$$

In this formula t_v is the time delay calculated from the x-ray picture, d' is the slant thickness of the plate, whereby $d' = d/\cos\alpha$ and B is a constant. Figure 41 shows the delay time as a function of the slant thickness of the target d' for the impact velocities 1165 m/s, 860 m/s and 610 m/s. At angles of inclination of 60° and above a certain slanting thickness d′ the time delays tend to be constant; at 80° the time delays are always constant. Above a certain thickness of the target or angle of inclination, the response mechanism of the fuze is radically different, and this explains the constant time delays which are achieved using these fuzes. Whereas the firing pin is driven into the impact detonator at steep angles, the upper cap of the fuze is sheared off at a predetermined breaking point at relatively flat angles, and the firing pin is then pushed into the detonator by a tipping mechanism. When the fuze pitches over, a relatively constant time delay occurs, which is naturally slawer at flat angles than at steep angles. On the other hand, if we take impacts which are nearly vertical or thicknesses of plates below a certain thickness, the firing pin is regularly struck into the impact detonator in such a way that after only a few tests one can exactly state the constants given above.

Similar time delays were observed for many of the designs studied, although in some cases the individual results deviate somewhat more than in this second example.

4. Determination of Particulation Time of Shaped Charge Jets with Flash X-Ray

The essential factor for the greatest possible penetration of a shaped charge is the *length* of the shaped charge jet. The length of a shaped charge jet can be determined from the sum of the lengths of the discrete particles after jet breakup. A quality index describing the quality of a *liner*, independent of its caliber, can be obtained by dividing this length by the shaped charge caliber or by the liner diameter.

To the best of our knowledge, jet particulation depends on the following parameters:

- Method of production of the liner
- Quality of the copper
- Quality of the liner inner surface
- Quality of the liner exterior surface (surface in contact with the HE)
- Adhesive layer on the liner
- Type of explosive charge, especially of the layer in contact with the liner.

Jet length is not the sole measure of liner quality. A variety of factors are significant in achieving good shaped charge performance. Particular attention should be directed to the process of particulation. The possible particle formations are sketched in Fig. 42[11]. The most desirable occurrence here is a rupture with a 100% constriction. Next best is the possibility of a transverse fracture, which occurs perpendicular to the rotational axis with a more or less pronounced con-

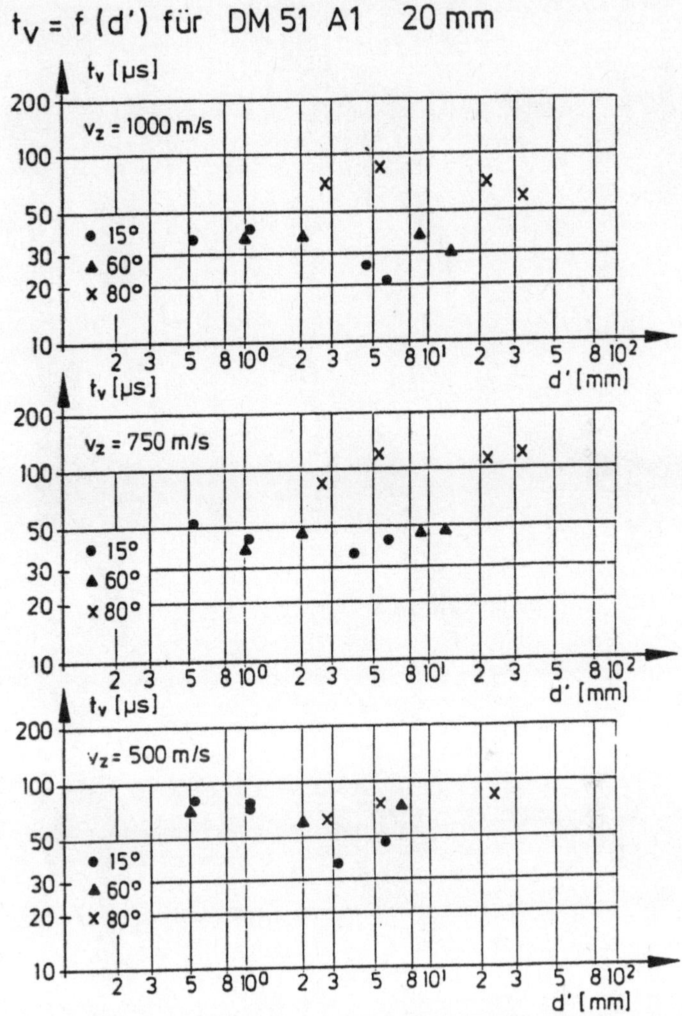

Fig. 39 Measured response times from a 20-mm projectile fuze as a function of the effective target plate thickness at three different impact velocities.

striction. However, this fracture is an indication of a brittle behavior. Least desirable are shear fractures, which normally occur under 45 deg and which generally cause off-axis forces at the ends of the particles, imparting to them a tumbling rate and drift.

4.1 Measuring Technique

The use of FXR is recommended as a means to radiograph the shaped charge jet after complete particulation. Since scaled particulation values $t_{p,s}$ of 2.5 μs/mm can occur with very good copper materials with a shaped charge of, for

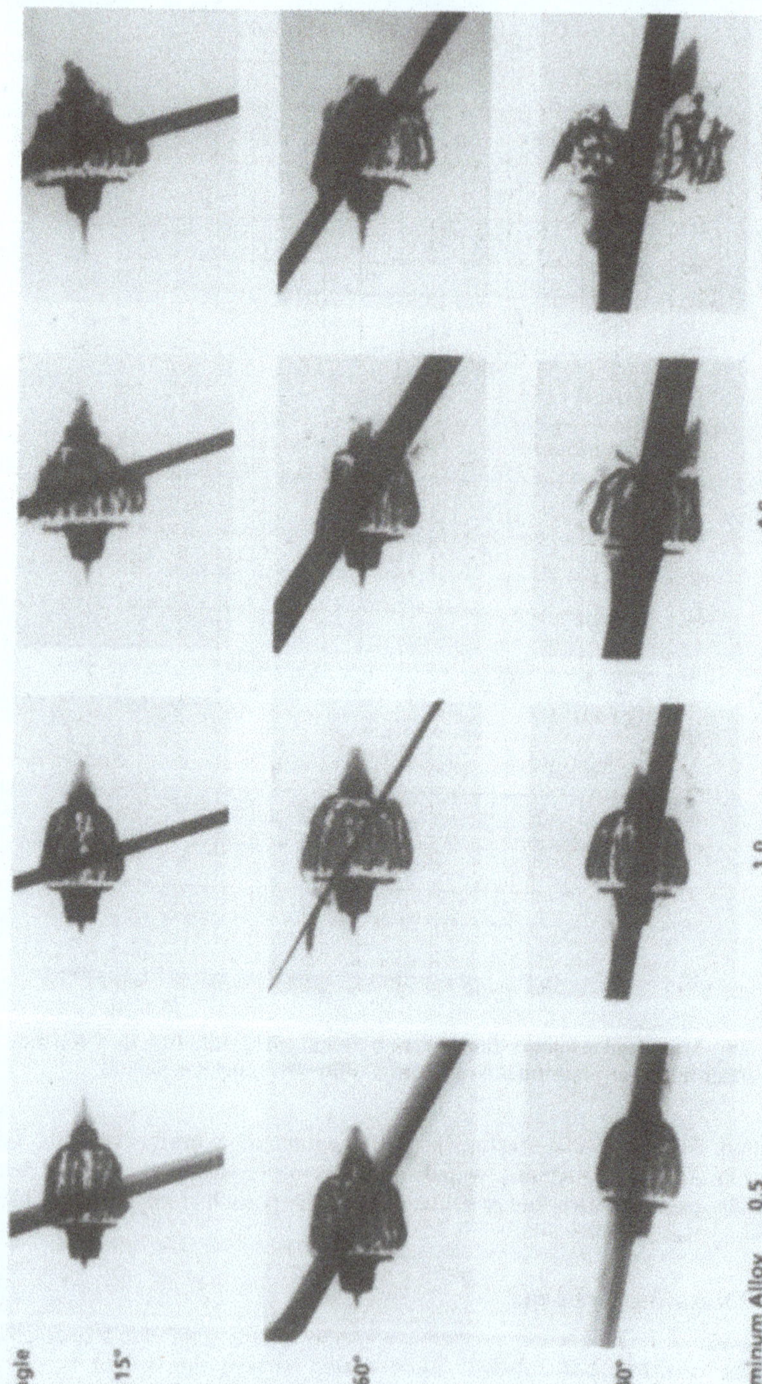

Fig. 40 Series of flash x-ray pictures of 20-mm projectiles striking aluminum alloy plates at terminal velocity of 750 m/s.

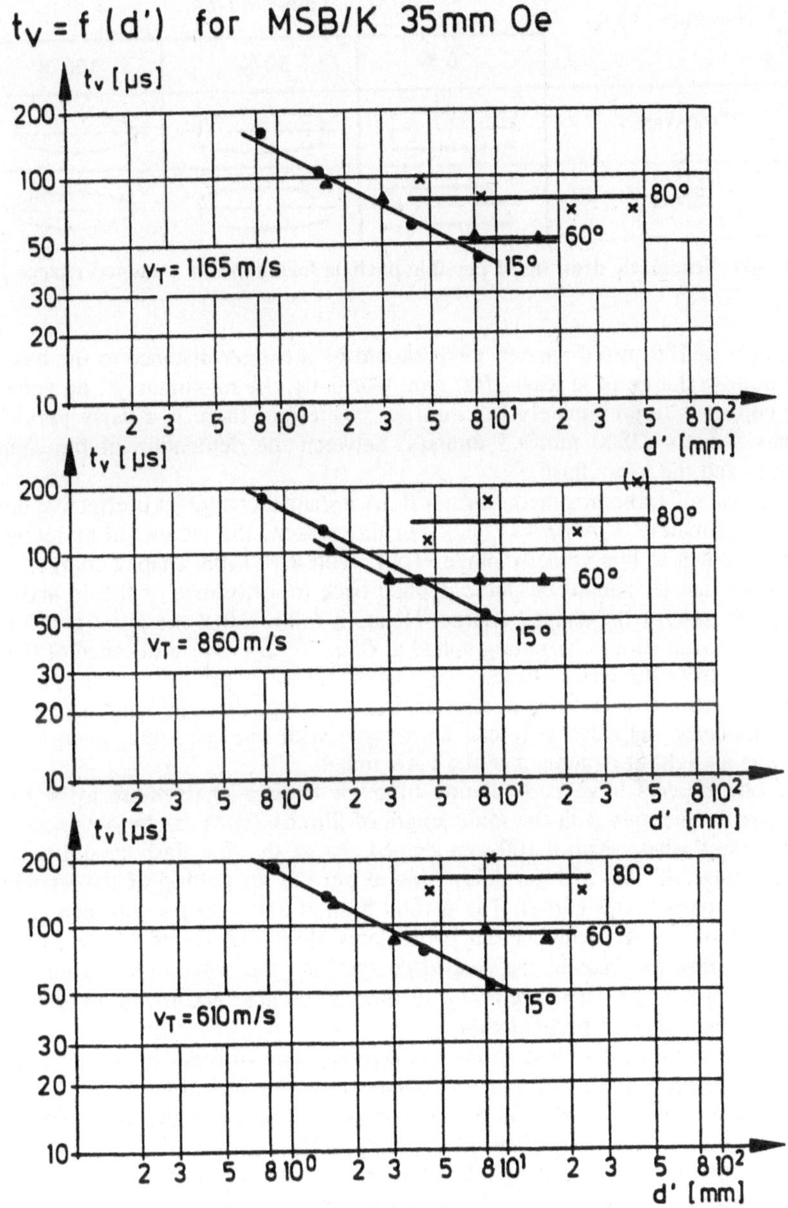

Fig. 41 Response time of a typical mechanical impact fuze (35-mm caliber) as a function of the effective target plate thickness of aluminum alloy at three terminal velocities.

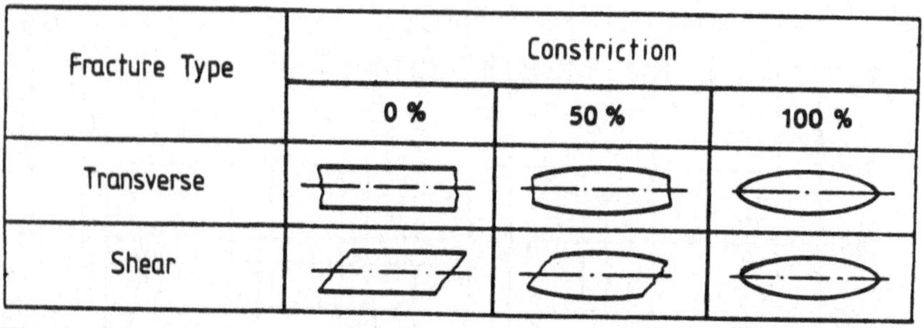

Fig. 42 Schematic drawing of possible particle formation of a shaped charge jet.

example, a 100-mm diameter, there should be a trigger distance to the base of the hollow charge of at least 2500 mm. Currently, the maximum jet tip velocity for copper is approximately 9.5 mm/μs. Therefore, there is a delay period of about 265 μs (2500 mm/9.5 mm/μs) between the detonation of the shaped charge and the x-ray flash.

Against rolled homogeneous armor RHA a shaped charge jet is effective down to approximately 3 mm/μs ($V_{j,min}$). For this reason, the jet should be recorded at least down to this velocity range. Thus, with a 100-mm shaped charge as an example, the jet should be photographed back to a distance of 0.8 m and less from the base of the shaped charge. Hence it follows that the jet of a 100-mm shaped charge should be radiographed at time 265 μs over a distance of 0.8 m to 2.5 m from the charge base.

By connecting two film cassettes each 0.9 m long, these jet lengths of a 100-mm diameter shaped charge can be x-rayed with *one* exposure, even though there is a slight enlargement in the FXR image.

If one expects large particulation times or if large charges are to be used, then useful pictures with the same length of film cassettes can be obtained with two x-ray flashes. With a 100-mm shaped charge the first flash should be shot in the range of 250–270 μs delay time to get the tip portion of the jet which generally particulates earlier. The second flash at 350–400 μs will then get the rear part of the jet, which often particulates later. The length of the film, its distance from the base of the shaped charge, and the delay times of the x-ray flashes must be adjusted according to the charge size and to the expected or experienced times of particulation.

The jet should pass close to the protective casing in order to obtain a sharp image of the jet. In this case, the enlargement factor should be below 5% and, possibly, in the order of 3%. A distance of 3 m between the x-ray tube and the film surface means that the jet should pass less than 0.15 m from the film surface, that is, approximately 0.05–0.10 m from the protecting cover plate. An example of a flash x-ray picture is given in Fig. 43.

4.2 Analysis

For an analysis, it is satisfactory to assume one virtual origin of the shaped charge jet at which all particles are generated at a single moment.[12] This as-

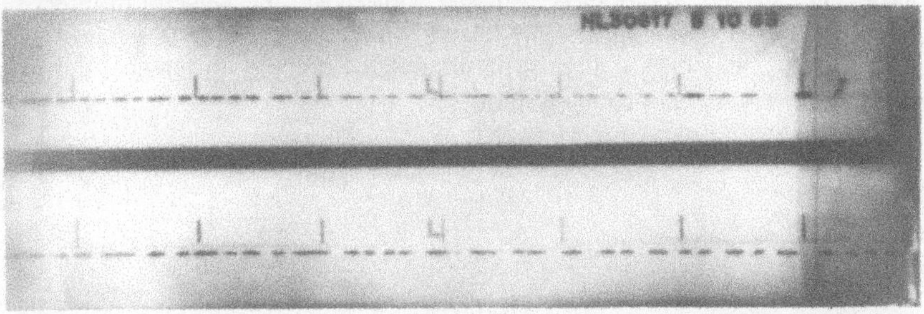

Fig. 43 Flash x-ray picture of a particulated shaped charge jet after a 250-μs (upper part) and a 350-μs (lower part) delay time between detonation and flash x-ray.

sumption is sufficient at least for a comparative evaluation. The velocities of the individual particles have to be calculated. These result from the individual trajectories from the virtual origin and the time difference, which can be ascertained from the virtual source.

Further, the individual particle lengths which correspond to the individual velocity ranges, have to be analyzed. All these can be obtained by using a digitizer and a small computer program and by entering the beginning and end points of the individual particles from the flash radiograph. The computer can then plot the individual lengths of the jet particles as a function of their velocities, as shown in of Fig. 44.

The average particulation time $\bar{t}_p$ of a shaped charge jet analyzed in this way can be interpolated quite simply from Fig. 44. To do this, the velocity differential between the jet tip velocity $v_{j,o}$ and $v_{j,min}$ (e.g., 3 mm/μs) is to be

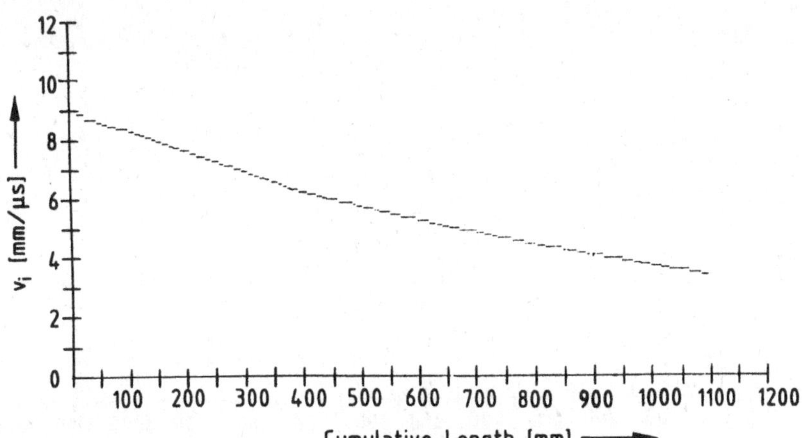

Fig. 44 Cumulative jet length of the FXR picture (Fig. 43) as a function of jet velocity.

taken, in case the whole jet is available. Otherwise, the velocity difference or range appearing on the flash x-ray image is to be used.

If the $\sum l$ = length is divided by the related velocity difference, the result will be the average particulation time:

$$\bar{t}_p = \frac{\Sigma l}{(v_{j,\phi} - v_{j,min})}, [\mu s] \tag{6}$$

The example according to Fig. 44 gives:

$$\bar{t}_p = \frac{1100 \text{ mm}}{(9 - 3.3) \text{ mm}/\mu s} = 193 \ \mu s \tag{6a}$$

Thus, $\bar{t}_p$ = 193 μs is the average particulation time of the jet in the observed velocity range.

If this evaluated average particulation time $\bar{t}_p$ is divided by the caliber (cal) of the shaped charge or by the diameter of the liner, the result will be the scaled particulation time $t_{p,s}$, which permits a comparison of the copper quality of shaped charges with different diameters:

$$\bar{t}_{p,s} = \frac{\bar{t}_p}{\text{cal}} \mu s/\text{mm} \tag{7}$$

Here, the example according to Fig. 44 gives this expression:

$$\bar{t}_{p,s} = 193 \ \mu s/96 \text{ mm} = 2.0 \ \mu s/\text{mm} \tag{7a}$$

As is clear from Fig. 44, the sum curve of the particles as a function of the jet velocity does not represent a good linear relation, but is rather made up of

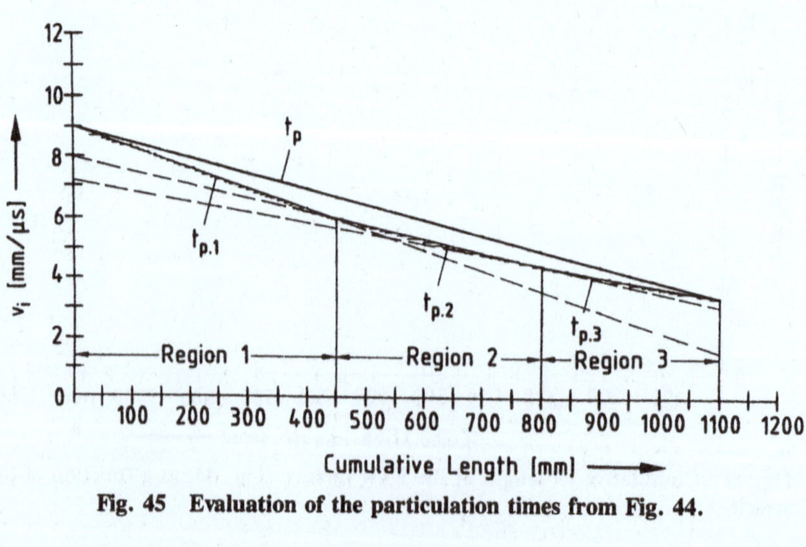

Fig. 45 Evaluation of the particulation times from Fig. 44.

three sections that, to a first approximation, can be represented by three straight lines (Fig. 45). In the upper velocity range, the straight line is steeper, and in the lower range it is flatter. This observation indicates the average particulation time is shorter in the upper region and longer in the lower region. From the three straight lines, three particulation times $t_{p,n}$ can be determined for the corresponding velocity ranges: $t_{p,1}$ for $v_{j,0}$ to $v_{j,1}$ and $t_{p,2}$ for $v_{j,1}$ to $v_{j,2}$ and so on.

Following the example of Fig. 45, we get the following particulation times $t_{p,n}$:

$$t_{p,1} = \frac{450 \text{ mm}}{(9-6) \text{ mm}/\mu\text{s}} = 150 \ \mu\text{s}$$

$$t_{p,2} = \frac{350 \text{ mm}}{(6-4.5) \text{ mm}/\mu\text{s}} = 233 \ \mu\text{s}$$

$$t_{p,3} = \frac{300 \text{ mm}}{(4.5-3.3) \text{ mm}/\mu\text{s}} = 250 \ \mu\text{s}$$

These particulation times are diagrammed as a function of the jet velocity in Fig. 46.

By comparing the particulation times $\bar{t}_p$, $\bar{t}_{p,s}$, or individual $t_{p,n}$ as a function of the jet velocity, conclusions concerning the quality of the copper can be drawn, if identical or similar types of shaped charge have been used.

4.3 Maximum Penetration Capability

Considering the theoretically possible depth of penetration of a jet, the significance of the evaluation t_p should again be stressed.

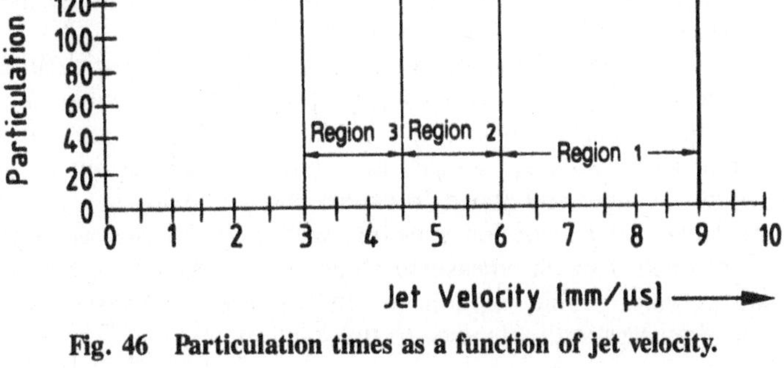

Fig. 46 Particulation times as a function of jet velocity.

The *maximum penetration capability* can be calculated from the particulation times obtained from experiment. The maximum penetration effect occurs when the jet has completely extended and thus has reached its maximum length L. All particles would then also reach the bottom of the crater and would thus contribute to the depth effect. The possible penetration of the particulated shaped charge jet is given by the following equation:

$$P = L/\gamma \tag{8}$$

where $\gamma = \sqrt{\text{Ptarget}/P\text{jet}}$.

Since the length of the particulated jet is also given by the velocity difference of the considered jet and the particulation time $\bar{t}_p$, Eq. (8) can be expressed as follows:

$$P = (v_{j,0} - v_{\text{j,min}}) \cdot \bar{t}_p/\gamma \tag{9}$$

With different particulation times for different velocity ranges of the jet, the following equation results:

$$\begin{aligned} P = {} & [(v_{\text{j},0} - v_{\text{j},1}) \cdot t_{\text{p},1} + (v_{\text{j},1} - v_{\text{j},2}) \cdot t_{\text{p},2} + \cdots \\ & + (V_{\text{j,n}-1} - v_{\text{j,min}}) \cdot t_{\text{p,n}}]/\gamma \end{aligned} \tag{10}$$

Our example of a copper jet gives the following theoretical maximum penetration in steel:

$$\begin{aligned} P &= [(9 - 6)\ \text{mm}/\mu\text{s} \cdot 150\ \mu\text{s} + (6 - 4.5)\text{mm}/\mu\text{s} \cdot 233\ \mu\text{s} \\ &\quad + (4.5 - 3)\ \text{mm}/\mu\text{s} \cdot 250\ \mu\text{s}]/\gamma \\ &= (450\ \text{mm} + 350\ \text{mm} + 375\ \text{mm}) \cdot \sqrt{8.9/7.8} \\ &= 1255\ \text{mm} \end{aligned} \tag{10a}$$

Naturally, at short range, particulation of the shaped charge jet has but a limited significance. But even with a standoff of only 1 caliber on, with a $v_{\text{j,min}}$ of 3 mm/μs and with today's particulation times of 2 μs/mm, the jet will always penetrate in a particulated condition in its end phase.

Penetration as a function of standoff distance is shown in Fig. 47, with jet velocities as a parameter and with a constant average particulation time $\bar{t}_p$ = 193 μs. In Fig. 48 penetration appears as a function of distance with the jet velocities as a parameter, for the variable particulation times in the individual velocity intervals according to Figs. 45 and 46.

A comparison of Figs. 47 and 48 shows that the variable particulation times result in penetrations that at standoffs from approximately 2 caliber to approximately 16 caliber, are larger than for a constant particulation time, with a minimum effective jet velocity $v_{\text{j,min}}$ equal to 3 mm/μs. Both results are virtually identical below 2 caliber and above 16 caliber standoff distance.

From the curves of penetration standoff, with jet velocities as a parameter (Fig. 48)—in which case the measured values of $\bar{t}_{\text{p,n}}$ as a function of the jet velocity are of considerable importance—the minimum effective jet velocity $v_{\text{j,min}}$ can be evaluated with reference to the standoff.[13]

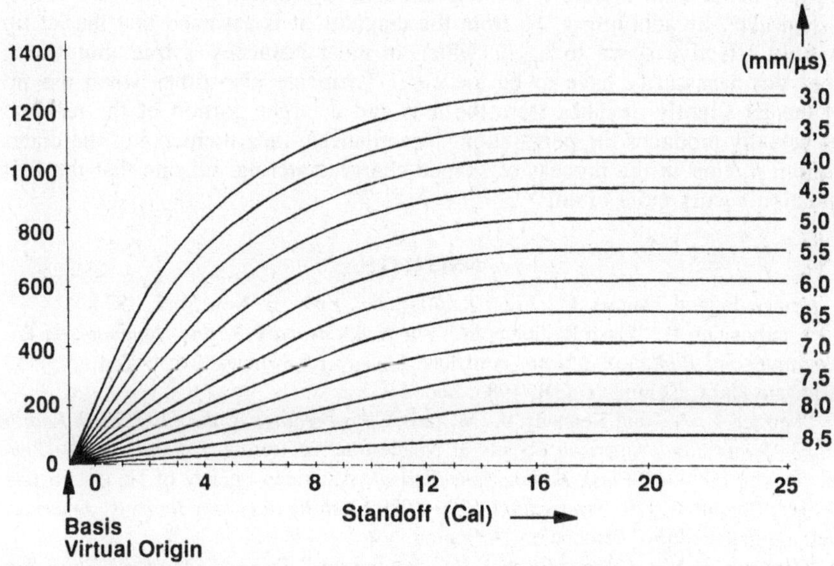

Fig. 47 Penetration vs standoff, with $v_{j,min}$ as parameter, for constant particulation times ($\bar{t}_p$ = 193 μs).

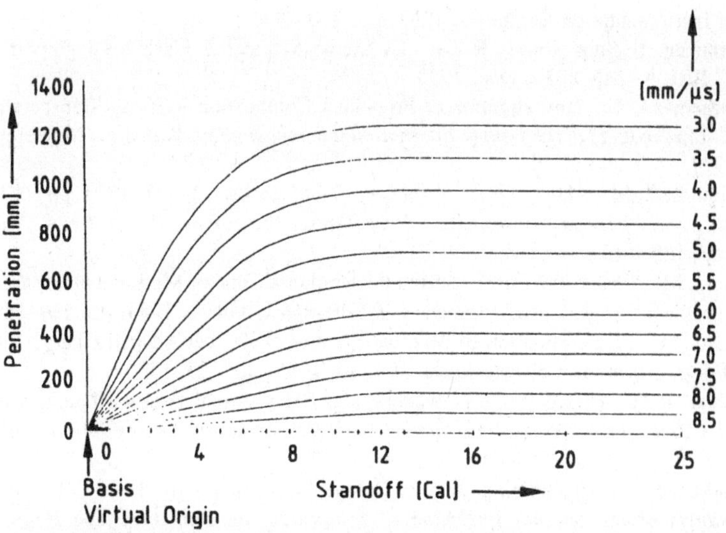

Fig. 48 Penetration vs standoff, with $v_{j,min}$ as parameter, with the particulation time as a function of jet velocity (after Fig.46).

Besides the potential performance of the shaped charge, the *efficiency* of the shaped charge can, therefore, be determined as a function of standoff.

However, by adopting $v_{j,min}$ from the diagram, it is assumed that the jet tip is fully effective down to $v_{j,min}$, which in most instances is true, but which does not necessarily have to be the case. There are also times when the tip of the jet slightly deviates from the axis and a larger portion of the residual jet actually produces the penetration. Nevertheless, measurements of the crater bottom motion in the process of shaped charge cratering indicate that the first situation occurs more often.[14]

References

[1]Jamet, F., and Thomer, G., *Flash Radiography*, Elsevier, New York, 1976.

[2]Charbonnier, F., "Flash Radiography," *The Nondestructive Testing Handbook on Radiography and Radiation Testing*, American Society for Nondestructive Testing, 4153 Arlingate Plaza, Columbus, OH, 1983, Sec. 11.

[3]Webster, E. A., and Kennedy, A. M. (eds.), *Proceedings of the 1984 Flash Radiography Symposium*, American Society of Nondestructive Testing, Columbus, OH; *Proceedings of the 1986 Flash Radiography Topical*, American Society of Nondestructive Testing, Columbus, OH; *Proceedings of the 1989 Flash Radiography Topicals*, American Defense Preparedness Association, Arlington, VA.

[4]Brewster, J. L., "Current Flash X-Ray Equipment," *Proceedings of the Flash Radiography Symposium*, edited by L. E. Bryant, American Society for Nondestructive Testing, Columbus, OH, 1976, pp. 15–22.

[5]Mattson, A., "Flash X-Ray Systems," *Proceedings of the First European Conference on Cineradiography with Photons of Particles:* edited by J. Marilleau, Society of Photo-Optical Instrumentation Engineers, 1983, pp. 270–274.

[6]Kronman, S., and Gross, R. W., "A Stereo Sequential Flash Radiographic Technique," BRL — MR 2343, Dec. 1973.

[7]Hornemann, U., "Investigation of Propellant Combustion in X-ray Transparent Gun Tubes," *Proceedings of the Fourth International Symposium on Ballistics*, Monterey, CA, Oct. 1978.

[8]Schmidt, E. M., *AIAA Journal of Spacecraft and Rockets*, Vol. 3, 1978, pp. 162–167.

[9]Held, M., "Fragment Generator," *Propellants, Explosives, Pyrotechnics*, Vol. 13, 1988, pp. 135–143.

[10]Held, M., "Procedure for Measuring the Response Time of Mechanical Impact Fuzes for Projectiles," *Propellants, Explosives, Pyrotechnics*, Vol. 9, 1984, pp. 139–146.

[11]Held, M., "Determination of Material Quality of Copper Shaped Charge Liners," *Propellants, Explosives, Pyrotechnics*, Vol. 10, 1985, pp. 125–128.

[12]Chou, P. C., Hirsch, E., and Walters, W., "The Virtual Origin Approximation of Shaped Charge Jet," *Proceedings of the Sixth International Symposium on Ballistics*, Orlando, FL, 1981, pp. 204–212.

[13]Held, M., "Characterizing Shaped Charge Performance by Stand-Off Behavior," *Proceedings of the Seventh International Symposium on Ballistics*, The Hague, The Netherlands, 1983, pp. 331–339.

[14]Held, M., "Evaluation of Shaped Charge Penetration Efficiency by Advanced Diagnostic Techniques," *Proceedings of the Sixth International Symposium on Ballistics*, Orlando, FL, 1981, pp. 480–493.

Chapter 11

High-Speed Photography

Manfred Held*
MBB Dynamics, Schrobenhausen, Germany

HIGH-speed photography plays an important role in the research and development of high-explosive warheads. Most of the applications focus upon the detonation behavior of the explosives used in the warheads. One of the most critical factors controlling the performance of precision warheads is the symmetry and smoothness of the detonation wave front. The insight into the process of detonation and, in particular, into the initiation of conventional high explosives and phenomena, which today are summarized by the term "detonics,"[1] is certainly closely related to the development of diagnostic techniques and, especially, to that of high-speed photography. The reaction and detonation velocities in the field of detonics are of the order of 1000 m/s to 9000 m/s, involving pressures of several hundreds of kilobars. Related phenomena occur at speeds of over 10,000 m/s and may involve shock pressures of more than 1 Mbar.

To obtain pictures of the detonating events with little motion blur and high time resolution, exposure times of less than 10 ns and frame rates higher than 10^6/s are desirable. Two decades ago, new discoveries and improved knowledge in the field of detonics were heavily dependent upon developments in high-speed photography. At the present time, however, special cameras are available for diagnosing detonation phenomena, and their commercial development has produced reliable and easy-to-use instruments.

Thus, the scientist working on warhead problems can now concentrate on developing measuring techniques instead of developing measuring instrumentation, and the research engineer can now concentrate his efforts on the essential problems, such as the development of measuring methods, because he no longer needs to undertake the design and manufacture of the necessary instruments. Following is a short overview of the optical instrumentation most used in the field of warhead and detonation diagnostics (see Fig. 1).

1. Radiation Sources

Light sources for recording rapid ballistic events such as warhead or detonative phenomena are being used and applied in two ways. The short exposures from

*Professor, Manager of Technology Warheads.

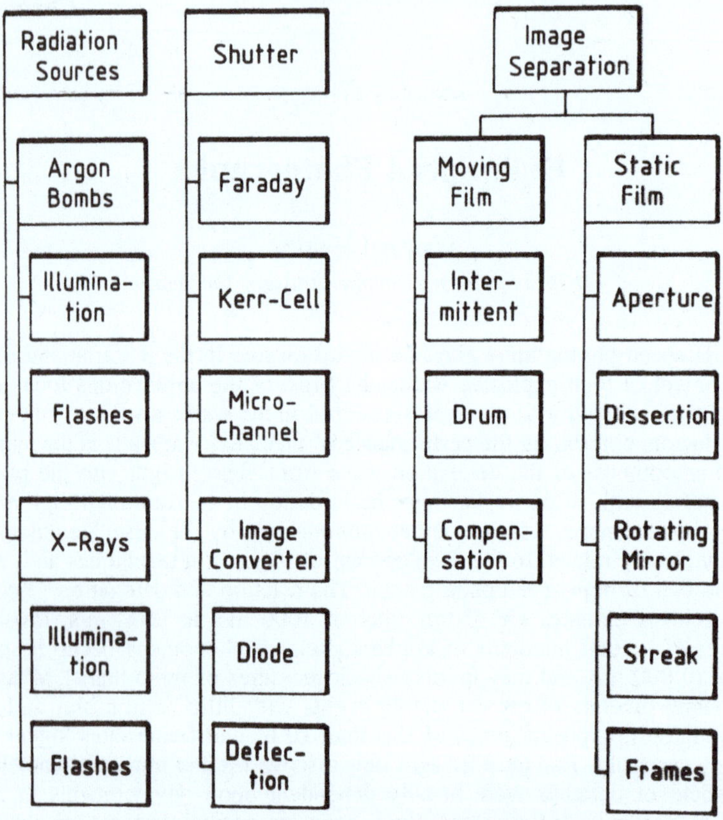

Fig. 1 Diagram of topics in this chapter.

high-speed shutters and/or the high frame rates of high-speed cameras require extremely bright illumination in order to obtain satisfactory exposures from phenomena that have little or no luminosity of their own. Up to now, the so-called "argon flash bombs" have been used to this end. They produce enough light to enable direct-light recordings with frame rates of 10^6 frames per second. Argon gas is extremely luminous when it is exposed to a detonative shock wave.

For an observation time of, say, 100 μs, an arrangement such as that shown in Fig. 2 is required, in which care is taken that a strong shock wave is fed into the argon atmosphere throughout this time. A milky-white balloon filled with argon is enough for a diffuse background (Fig. 3). A 100-g high-explosive charge is sufficient to supply the background lighting for shadow-cast records from Kerr cell or image converter shutters, which have an opening, or exposure, time of 5 ns.

Argon bombs can also be used as flash sources. This means that the light source duration is so short that fast events can be recorded on a stationary film with the camera shutter in the open position. There are several arrangements

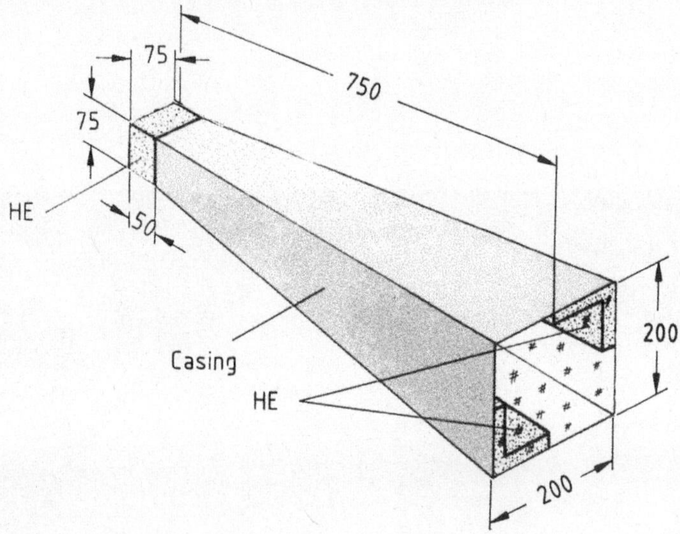

Fig. 2 Detonative shocked argon is a very bright light source; often called an argon-bomb.

and applications for this purpose. An arrangement chosen[2] is shown in Fig. 4. In contrast to other arrangements, this simple arrangement has an almost ideally rectangular light path (see Fig. 5). Several of these argon flash sources have been used for several frames with frame separation by a Hycam camera in observing an explosively formed projectile moving at a speed of 1800 m/s (see Fig. 6).

A new type of light source has certainly been found in the laser. However, because of its high cost and relatively intricate measurement arrangements, it is at present used only in special cases. As an example, a relatively simple device is a laser strobograph (see Fig. 7). This device can emit 10 laser pulses of 40 ns duration each and with intervals as short as 1 μs. This is a useful technique for examining shock wave phenomena (Fig. 8).

Without doubt, x-ray flash radiography[3] has acquired a particular significance in the whole field of warhead and detonation physics (see also Chapter 10). The fast discharge of several hundreds of thousands of volts is achieved by means of a Marx pulse generator. The x-ray flash technique also permits observations to be made of detonation waves inside a high-explosive charge itself (see Fig. 9).[4]

Extremely energetic and yet short x-ray flashes can be produced by means of linear accelerators. Figures 10 and 11 show examples of records made with a PHERMEX facility.[5] In Fig. 10, a detonative shock wave produced jets from a uranium material having triangular grooves, and these jets are just about to penetrate into a stepped aluminum block. In Fig. 11, 0.025-mm-thick tantalum foils expand together with the reaction products of a Composition B high-explosive charge that had been initiated by a plane wave generator.

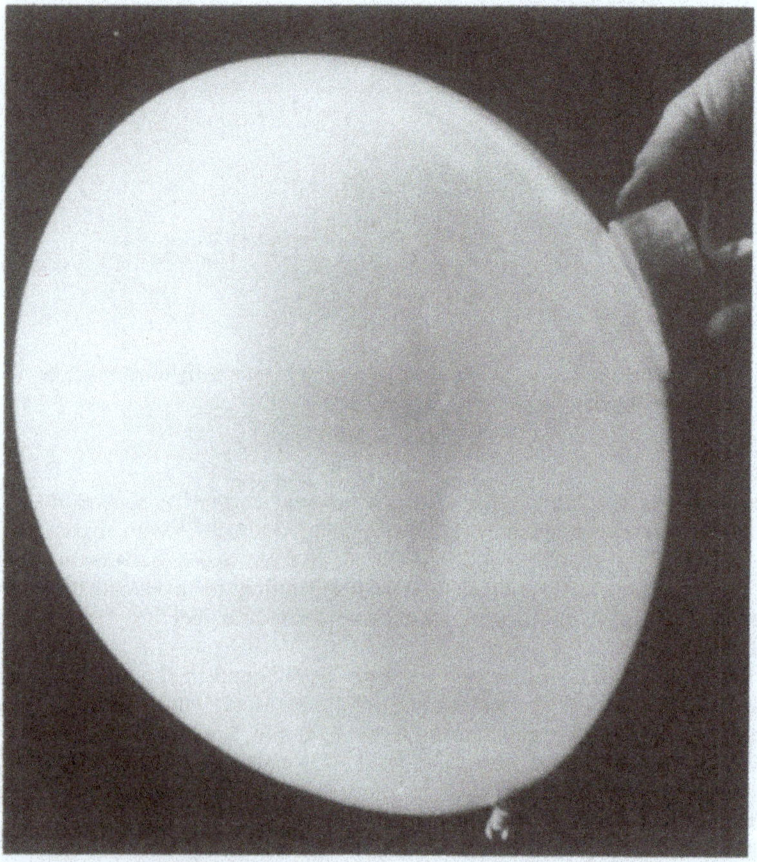

Fig. 3 Argon-filled children's balloon as a background light source, used also for 5 ns exposure time.

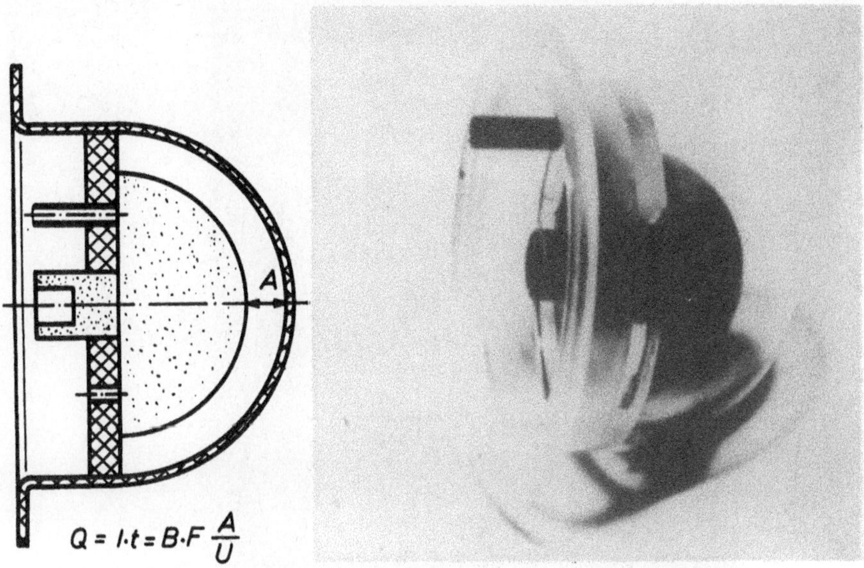

Fig. 4 Detonative argon light flash for 1 to 2 μs duration.

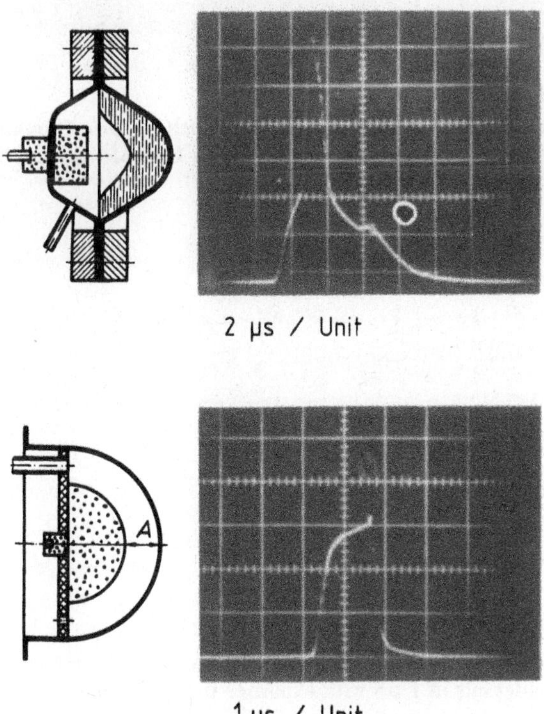

Fig. 5 Comparison of an older to a newer argon light flash design.

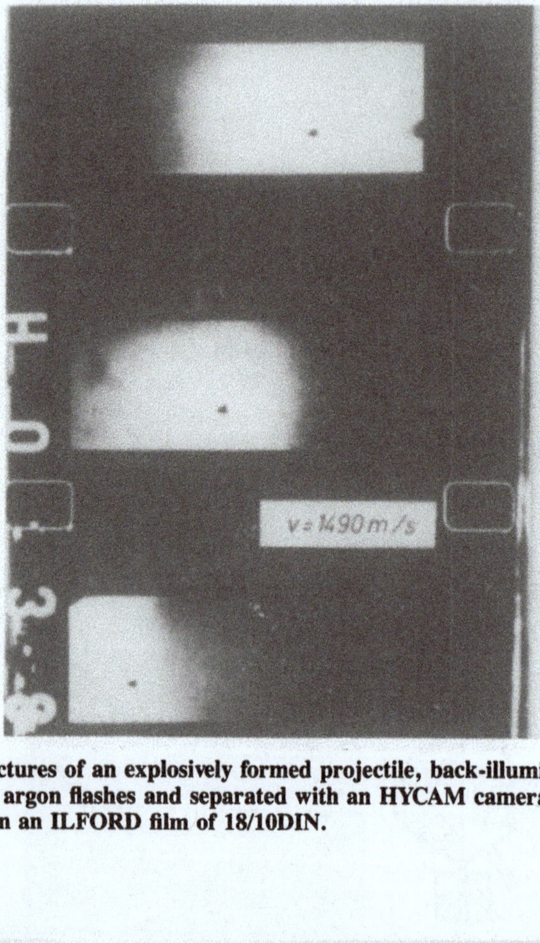

Fig. 6 Pictures of an explosively formed projectile, back-illuminated with detonative argon flashes and separated with an HYCAM camera with 10660 frames/s on an ILFORD film of 18/10DIN.

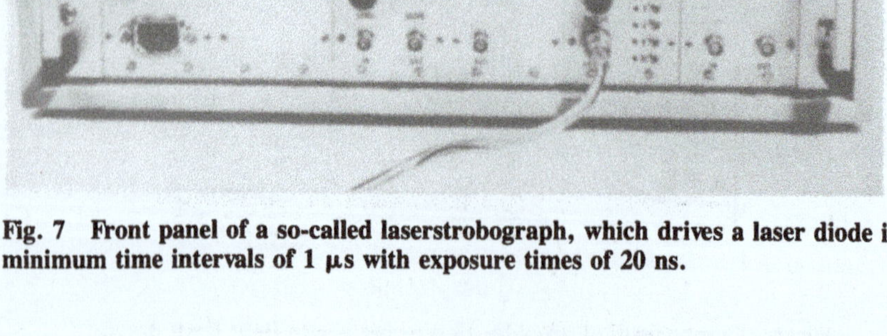

Fig. 7 Front panel of a so-called laserstrobograph, which drives a laser diode in minimum time intervals of 1 μs with exposure times of 20 ns.

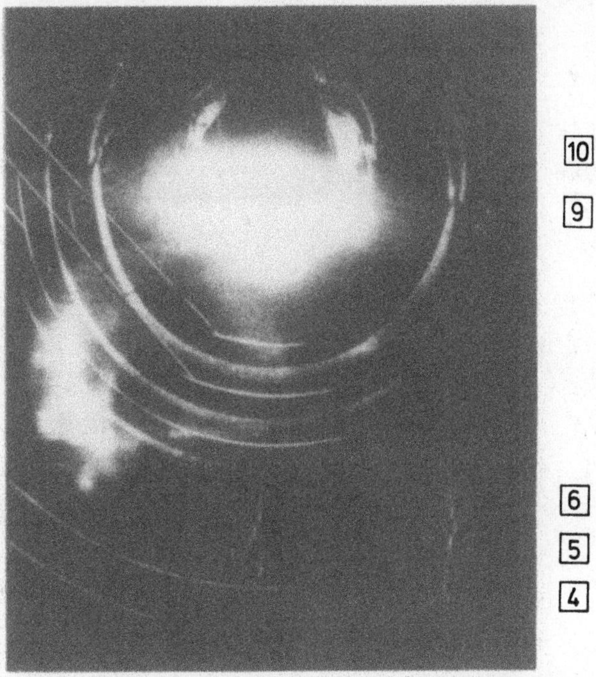

Fig. 8 Shock waves of an electric detonator DM 1225 in Plexiglas, illuminated with the laser strobograph in Schlierentechnique (darkfield).

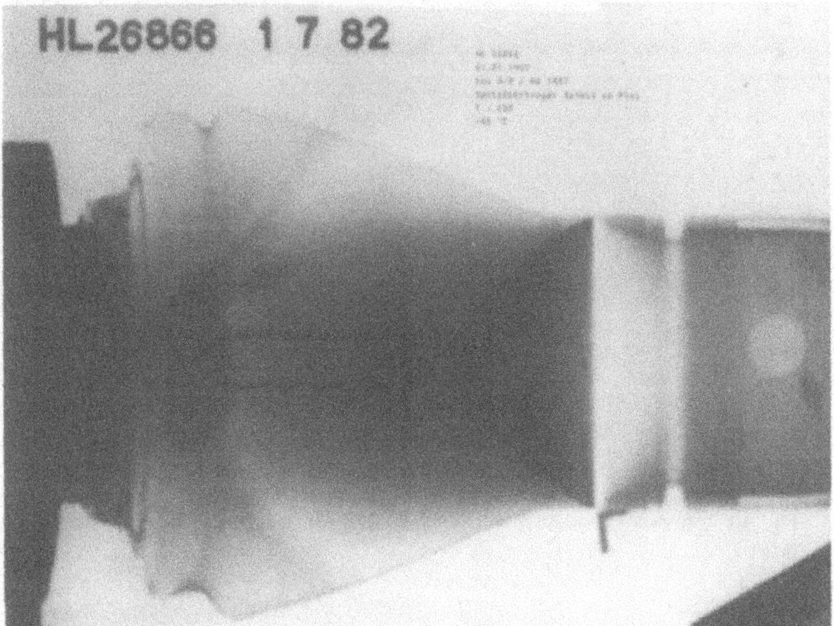

Fig. 9 In real warheads the front of the detonation wave is not easy to capture with flash x rays.[4]

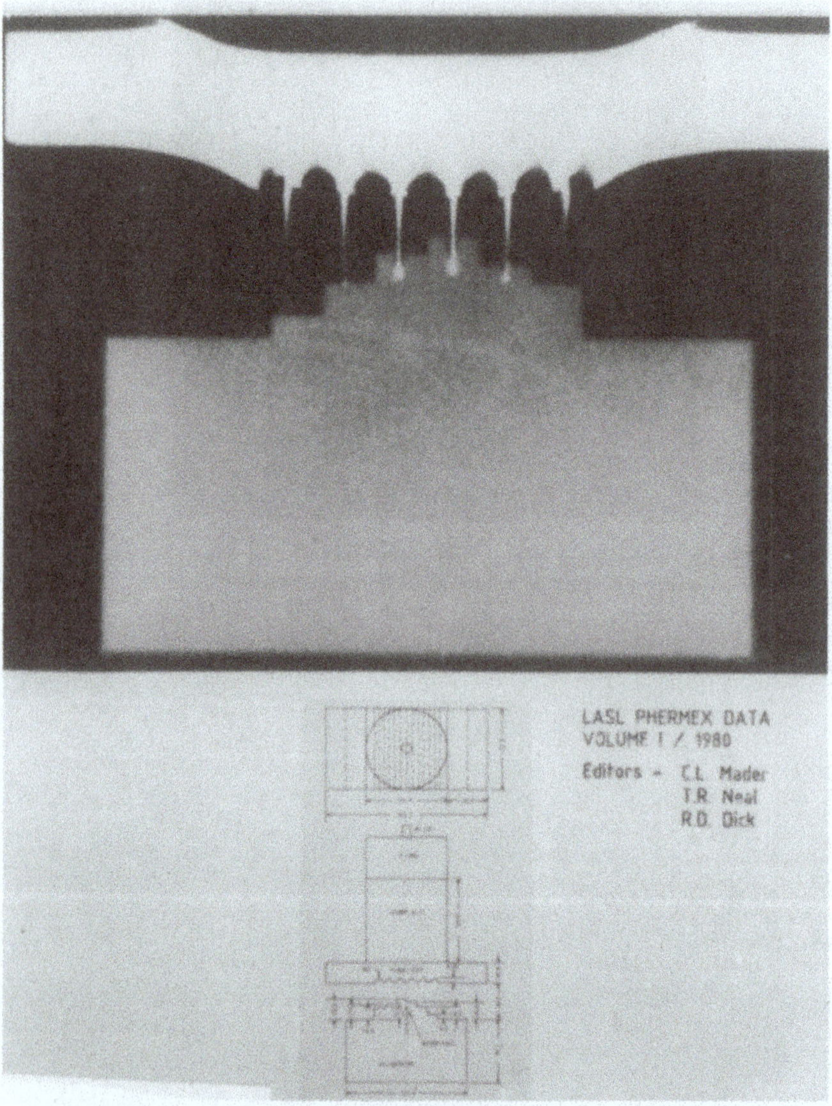

Fig. 10 Flash x-ray picture with PHERMEX of uranium jets in aluminum target.[5]

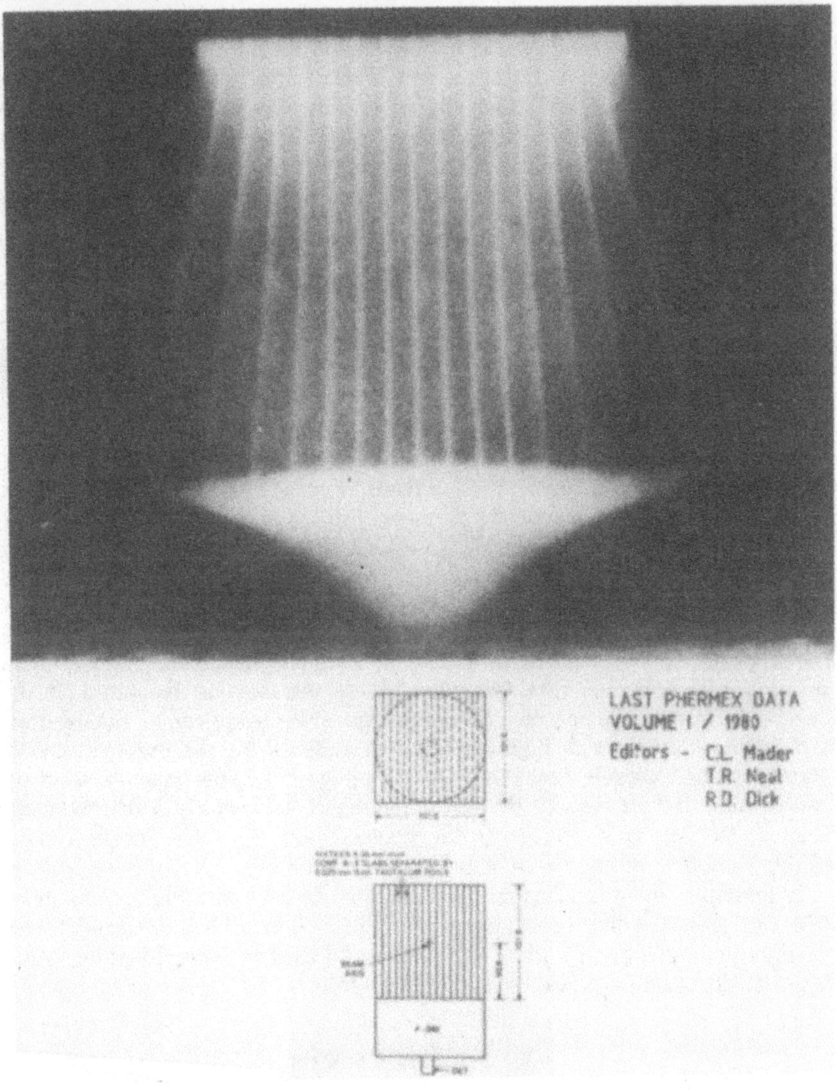

Fig. 11 Flash x-ray picture of expanding tantalum foils in a detonating HE charge taken with PHERMEX.

2. High-Speed Shutters

The Faraday shutter, which makes use of the longitudinal magneto-optical effect, has acquired only little significance in the warhead and detonation diagnostic, although it is used occasionally. This is due to the fact that its minimum shutter time is in the range of 1 μ and that, because of the polarizaton foils, its shutoff factor is not very high. Minimum shutter times of 5 ns can be achieved with the Kerr cell camera (Fig. 12), which makes use of the transverse electro-optical effect. In the 1950s and 1960s, this camera was widely used in the diagnostics of warhead and detonation phenomena (see Fig. 13). One shortcoming was that the material with by far the best Kerr constant, nitrobenzene, is toxic; the relatively low shutoff factor, together with a relatively high loss in brightness, was another drawback.

Because of the possibilities of light amplification and an improved shutoff factor, the image converter camera has now replaced both the Faraday and the Kerr cell shutters. A high resolution was obtained with the image converter diode camera (Fig. 14) through a corresponding image on a usually quite large image converter diode (Fig. 15). A very high framing rate can be obtained by means of beamsplitter systems, as shown in Fig. 16. It can be also split into 4 or even 6 arrangements.

A few examples of the high time and space resolution with shutter times of 10^{-8} s are provided in Figs. 17–19: Figure 17 shows a gas detonation[6,7] of an oxy-acetylene mixture at reduced pressure, Fig. 18 shows the determination of the roughness of the detonation front through observation of the ceasing of gap luminosity,[8,9] and Fig. 19 shows this even more clearly through a magnified image of the event on the diode.

Apart from short shutter times, the image converter camera also allows high framing rates to be obtained by means of adequate deflection systems in the special image converter tube (see Fig. 20). In this way the breakthrough of a detonation front at the end face of a high-explosive charge can be obtained as a self-luminous event, with high resolution at a rate of 5×10^6 frames/s and the corresponding exposure times of 40 ns (see Fig. 21). After magnification (the two parallel lines of Fig. 22 are only 2 mm apart) the change of the ondulation front of the detonation wave can be recorded with 20×10^6 frames/s and is visible from the ceasing of the gap luminosity (see Fig. 23).

In principle, the multichannel plate can also be used as a high-speed shutter (see Fig. 24). On this basis, a camera (Fig. 25) has been designed for the daylight photography of bullets. It can even be used for photographing people at extremely short exposure times (see Fig. 26).

3. Framing Cameras

In normal cinematography, the frames are generally separated by a moving film that is intermittently stopped for exposure. In such intermittent film transport cameras (Fig. 27), a framing rate of 500 frames/s up to a maximum of 600 frames/s can be achieved. Such a rate is by no means sufficient to resolve warhead events.

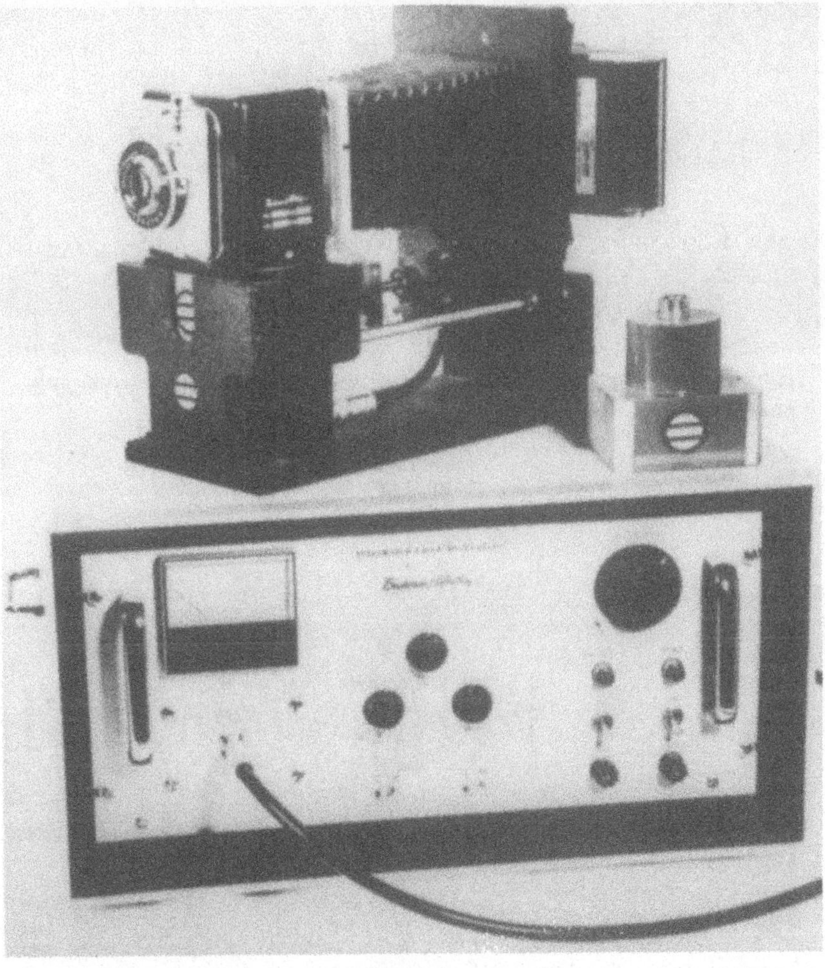

Fig. 12 Kerr cell camera with the pulser for the different exposure times (top) and the power supply for 40 KV (bottom).

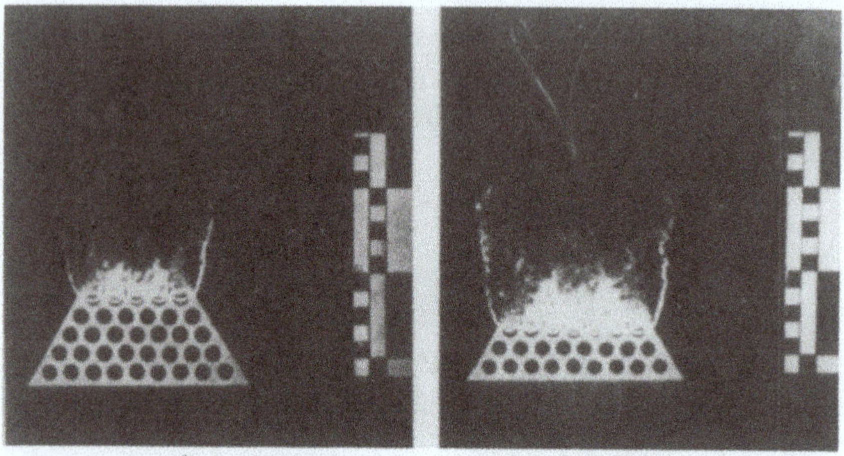

Fig. 13 Kerr cell camera pictures of the detonation front from a special line wave generator.

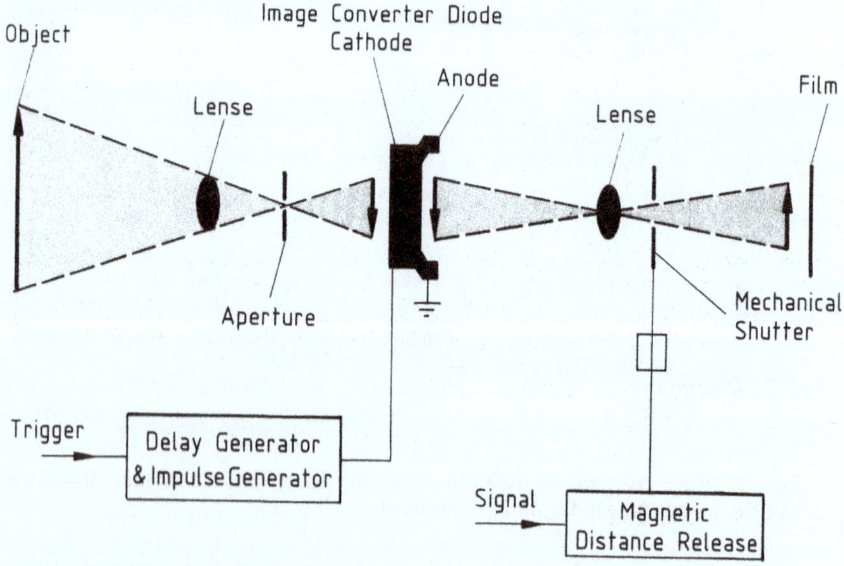

Fig. 14 Image converter camera with a biplanar diode (Beckman & Whitley XR 500).

Fig. 15 Large biplanar diode provides high-resolution image information.

Fig. 16 Parallax free pictures can be obtained with a beam splitter and two image converter shutters.

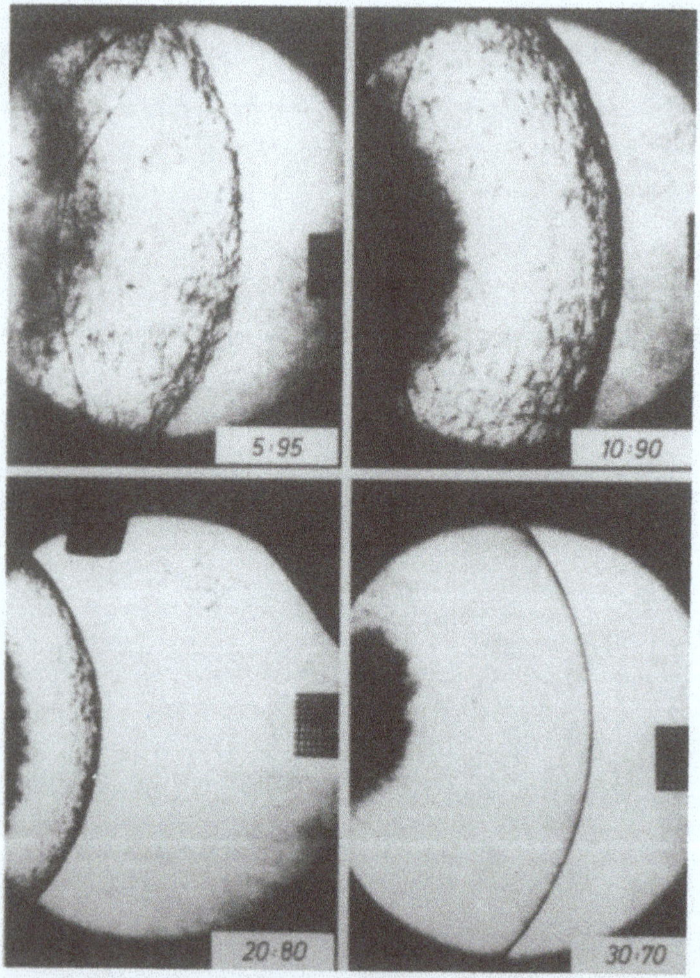

Fig. 17 High resolution of image converter cameras with biplanar diodes can be demonstrated by the spinning detonation of gaseous detonations of $C_2H_2/0_2$ mixtures at 300 Torr with 10-ns exposure time.[6,7]

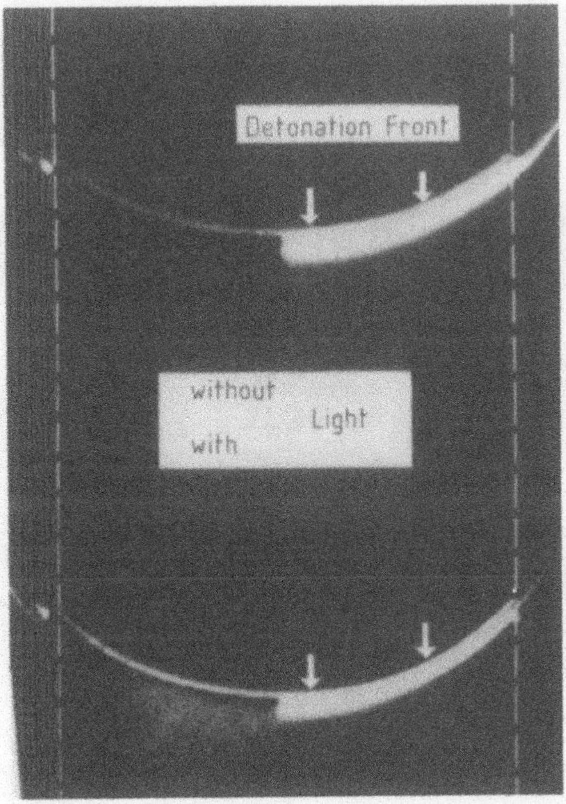

Fig. 18 Termination of the light band gives the detonation front; the upper picture is without and the lower picture is with front illumination; RDX/Wax-HE charge covered with a Plexiglas plate in 1-mm distance; 10-ns exposure time.[8,9]

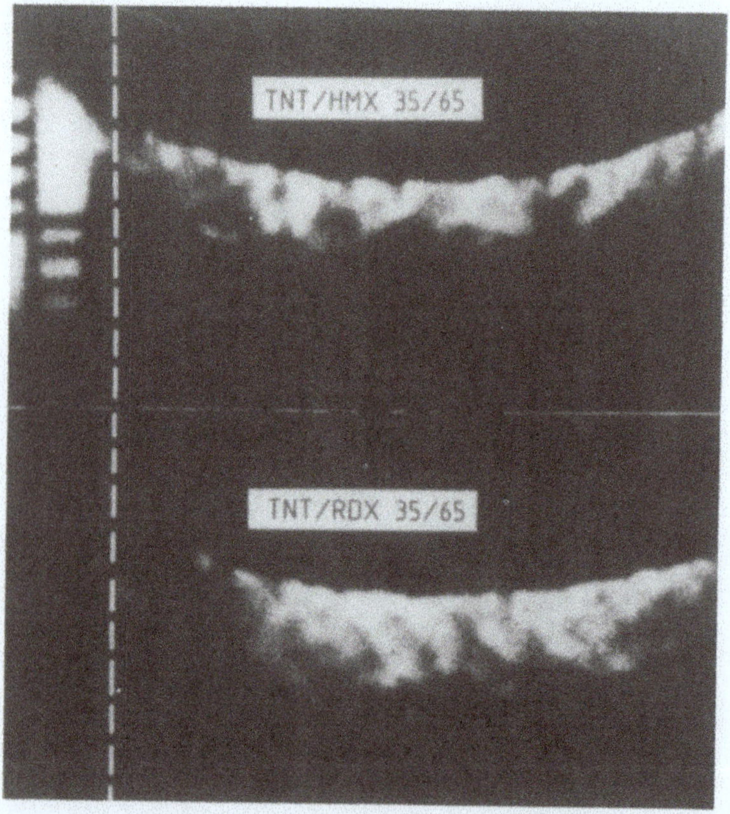

Fig. 19 Roughness of the detonation front of cast TNT/HMX and TNT/RDX 35/65; a millimeter grid is shown on the left side; 0.1-mm slit and 5-ns exposure time.[8–9]

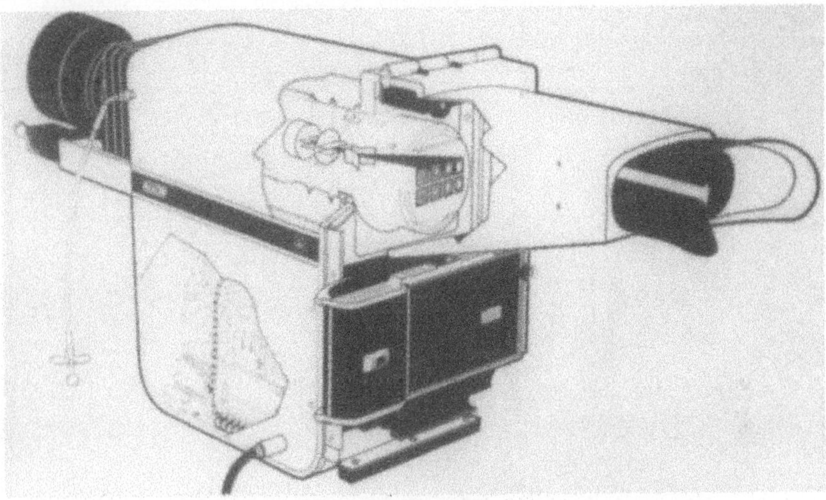

Fig. 20 Image converter camera Imacon 780 with a deflection tube.[16]

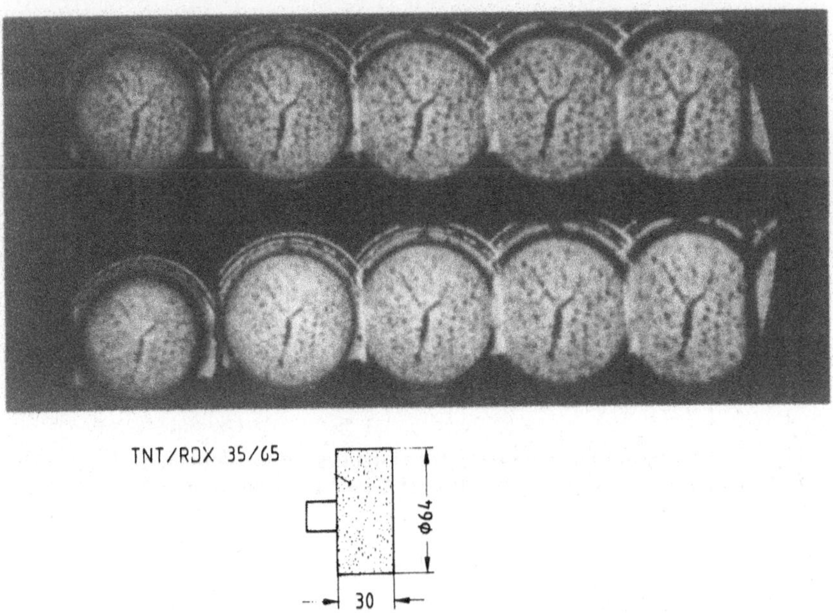

Fig. 21 Breakthrough of a detonation front at the end face of a cylindrical cast TNT/RDX 35/65 charge of 64-mm diameter and 30-mm height, recorded with Imacon 790 with 5×10^6 frames/s.

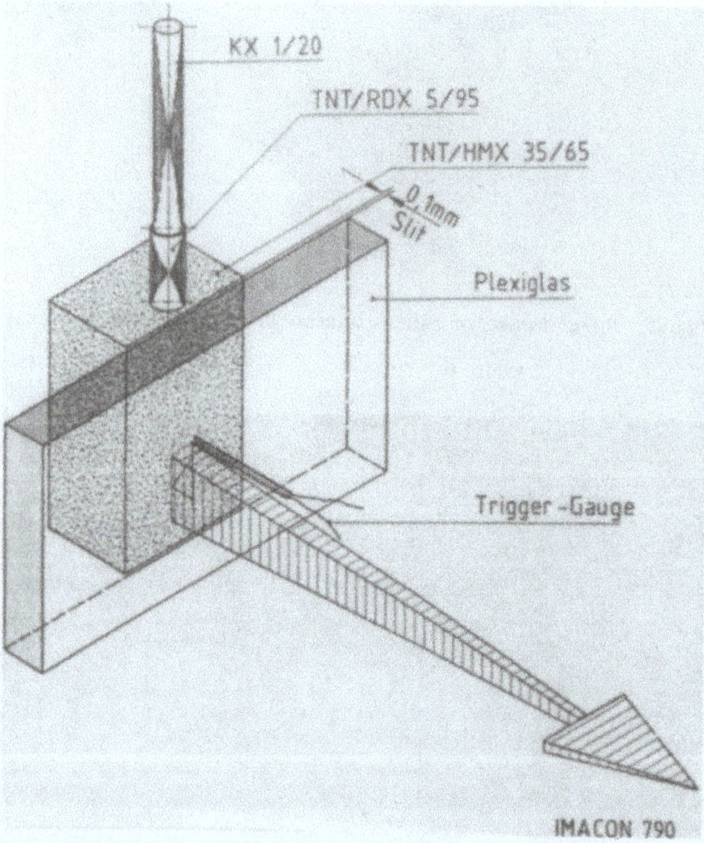

Fig. 22 Test set-up for measuring the roughness of the detonation front with high resolution; the two parallel fiducial lines have a distance of 2 mm.

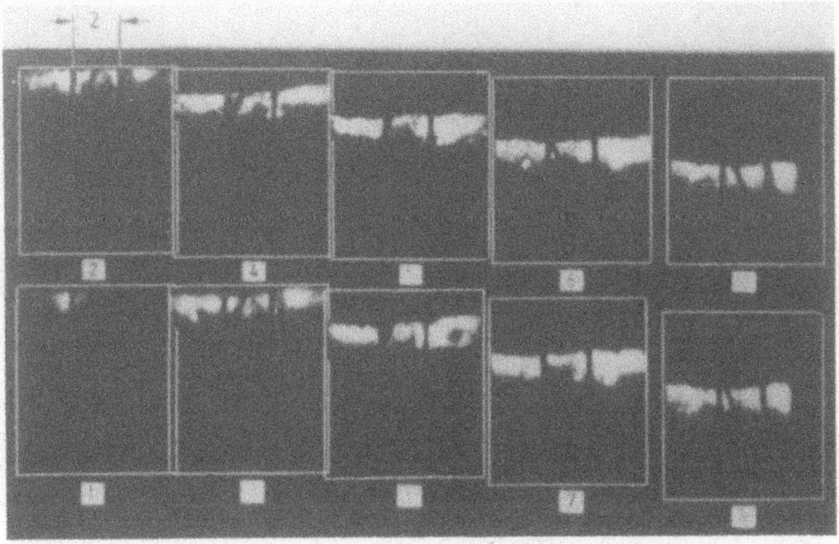

Fig. 23 20×10^6 frames/s of a cast TNT/HMX 35/65 HE charge after test set-up in Fig. 22; a wavy detonation front can be recognized.

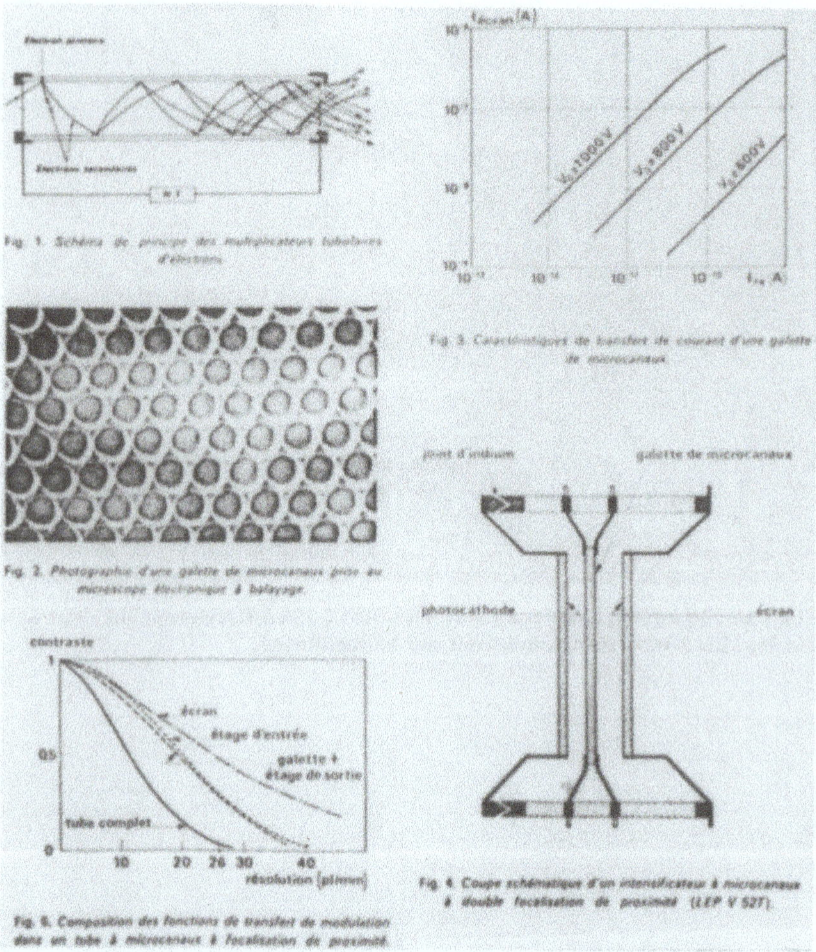

Fig. 24 Multichannel plate as an image intensifier and shutter.

Fig. 25 Available light projectile camera (ALP) from John Hadland.[16]

Fig. 26 Shortest exposure of the author (right) with John Hadland (left) and Alec Hudson (middle); exposure time $t = 0.2\ \mu s$, aperture $f = 4$, gain 5.

Fig. 27 Locam camera as a typical representative of a camera with an intermittent film transport, which operates at frame rates from 16 to 500.

In a so-called compensation camera, a rotating prism makes the picture continuously follow the film that moves with a maximum constant velocity of 80 m/s (see Fig. 28). Therefore, this type of camera is often also called a rotating-prism camera. A typical example of such a compensation camera is the Hycam, with an eight-sided compensation prism (see Fig. 29). This type of camera makes 10,000 full-format frames per second or 40,000 quarter-format frames per second; however, this is not sufficient to record warhead or detonation phenomena, except when observing explosively formed projectiles with a velocity of about 2.5 km/s, or the jet particles of shaped charges at 8 km/s more or less in the direction of their trajectory (that is, with little transverse velocity), as is done by the method known as the "objective technique."[10] This technique is mentioned here because it is a typical example of how simple equipment such as the relatively inexpensive compensation cameras can be used to obtain significant information on warhead phenomena, such as the rate of rotation of rod fragments, formation and stability of an explosively formed projectile, or the dispersion and erosion of fast shaped charge jet particles.

Rotating-mirror cameras can expose millions of frames per second. A typical representative of this type of camera is the Cordin Model 200 (Fig. 30), which apart from streak recording, permits framing records with up to 2×10^6 frames/s to be taken (see Fig. 31). The particular advantage of this feature is that apart from the streak record, which is continuous in time, there is also a spatial picture of the same event in the framing record. The latter also shows whether the desired event has been recorded on the streak record.[11] With the Cordin Model 200 camera, the event to be recorded must be controlled by the camera with an accuracy

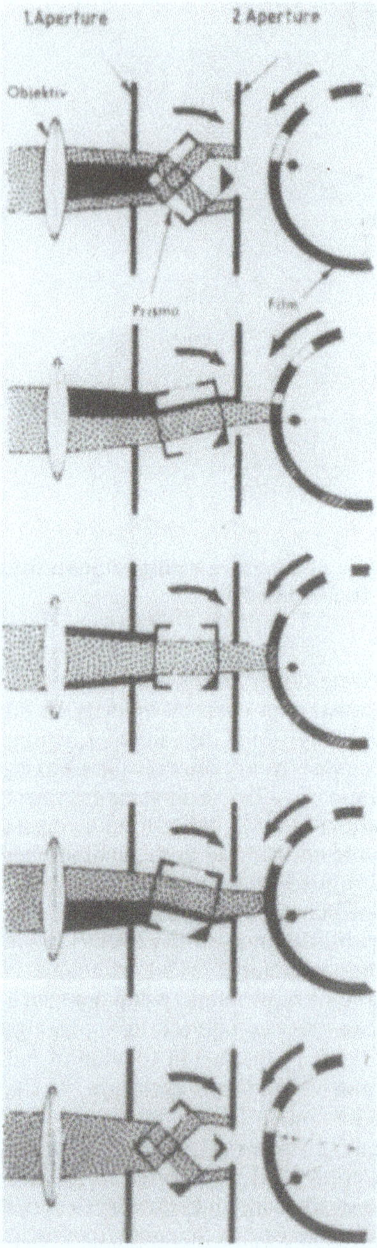

Fig. 28 Picture is "transported" with a rotating prism to the continuously moving film in the so-called compensation-camera.

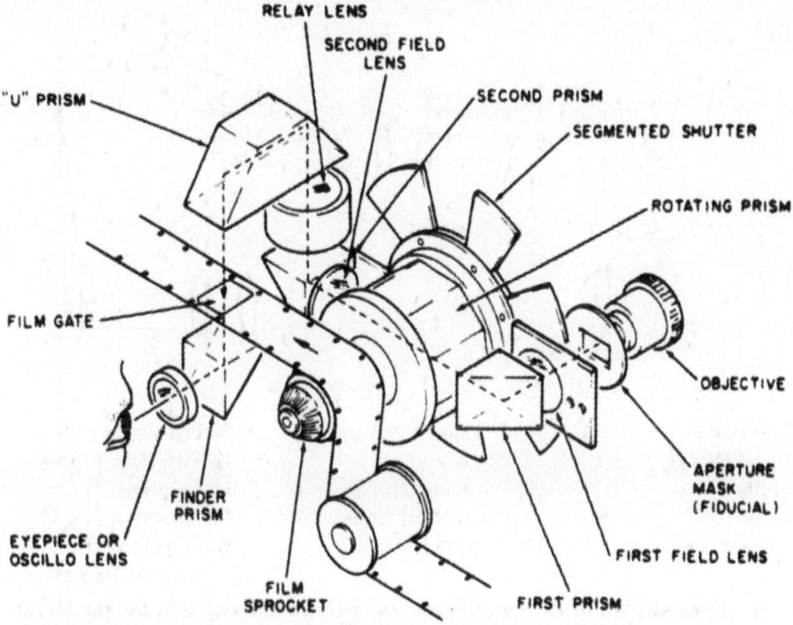

Fig. 29 Hycam camera as a typical representative of a rotating prism or compensation camera with maximum 10,000 frames/s full-frame or 40,000 frames/s 1/4 frame film height.

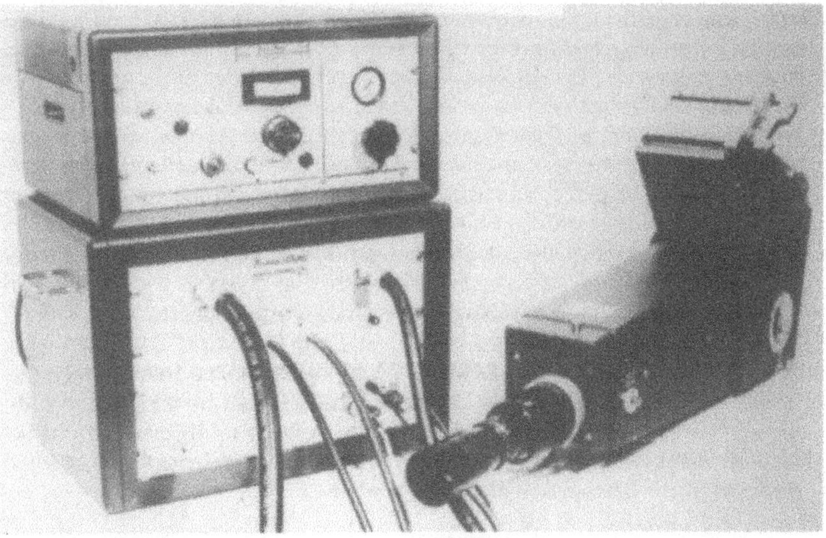

Fig. 30 Rotating-mirror camera for simultaneous streak and frame records with 14 mm/μs smearing velocity and 12 frames with 2×10^6 frames/s; Cordin Model 200.[13]

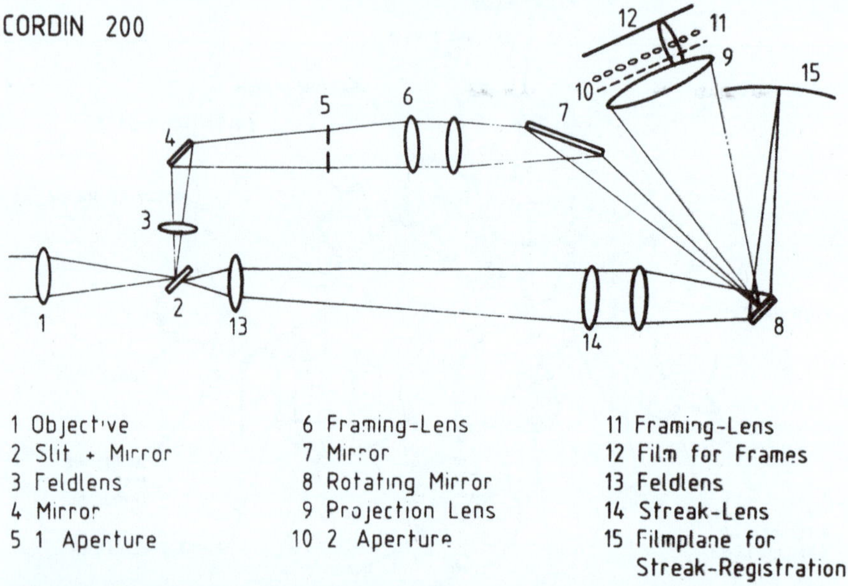

Fig. 31 Optical path of the rotating-mirror camera, Model 200, for the streak record (lower path) and frames (upper path).[13]

of microseconds in order that the relatively small recording "window" be hit. Events that are less determinate in time can be recorded with the continuously writing Cordin camera simultaneous streak and framing Model 330 (see Fig. 32). Pictures taken with a rotating-mirror camera are of such high quality that the formation of natural or of controlled fragments produced by the detonation of a high-explosive charge inside a cylinder can be traced and observed easily.

As stated above, single frames and streak records can be obtained with certain types of image converter cameras by means of a suitable deflection system. This is not to suggest that the image converter camera and the rotating-mirror camera are competing instruments in the diagnostics of warhead or detonation phenomena; an analysis of the advantages and disadvantages of these cameras will show that the instruments complement rather than compete with each other (see Table 1).[12] The specific advantages of rotating-mirror cameras are the number of frames, recording time, and, in particular, the resolution, combined with pictures and the possibility of simultaneously taking streak and framing records. The image converter camera, on the other hand, has a higher framing and recording rate and, in particular, a greater aperture. It is easier to operate and can be triggered by the event itself. The two types of cameras complement one another for problems in the diagnostics of ballistic detonics.

4. Streak Cameras

The original type of camera used for streak recording was certainly the "drum camera." In its simplest design, the film is attached to the outside of the drum,

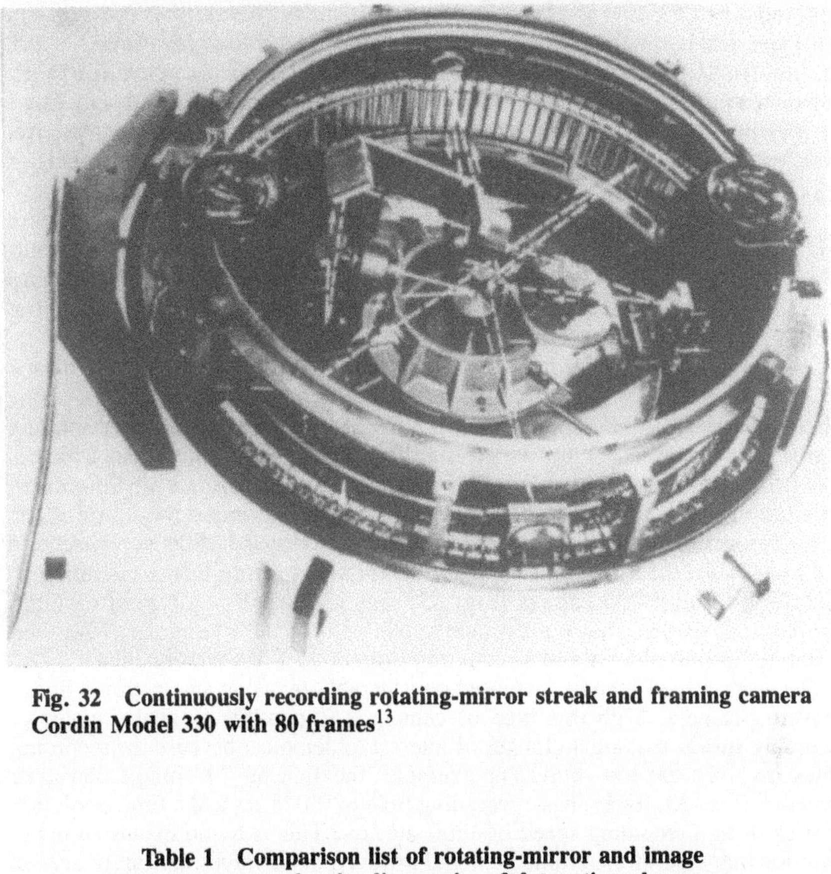

Fig. 32 Continuously recording rotating-mirror streak and framing camera Cordin Model 330 with 80 frames[13]

Table 1 Comparison list of rotating-mirror and image converter cameras for the diagnostics of detonation phenomena

	Rotating mirror	Image converter
Number of frames	+	−
Frame rate	−	+
Recording time	+	−
Writing velocity	−	+
Aperture	−	++
Resolution	+	−
Color	+	−
Triggerable	−	+
Handling	−	+
Simultaneity of streak and frames	+	−

which allows for film speeds up to 0.1 mm/μs (this limit being set by the ultimate strength of the film and the centrifugal force). Higher velocities can be achieved when the film is placed inside a high-strength drum, but this requires that the event be imaged, via a mirror arrangement, to the film inside. At 0.2 mm/μs, friction and eddies of air heat the case and, in particular, the film to such an extent as to establish another limit. With the interior of the camera evacuated it is possible to drive the speed up to the ultimate strength of the drum material. High-grade materials can endure peripheral velocities, and hence film speeds of 0.3 mm/μs are possible. However, this is the upper limit for drum cameras.

A typical drum camera is the Super Dynafax Model 318 Streak Camera (Fig. 33, top),[13] which, with a writing speed of 0.3 mm/μs, has a time resolution of 200 ns. The film is 875 mm long, and thus a recording time of 2900 μs is available at the highest writing speed. Such a time, however, is usually not needed in detonation problems.

Streak velocities of 10 mm/μs and more, which means an increase by a factor of 100 over the drum cameras, can be achieved with rotating-mirror cameras, where the light from the entrance slit is imaged to the film plane via a rotating mirror. The theoretical limit (which is called the Schardin limit[14] and which is about 0.3 ns, is determined by the width and the rotation rate of the mirror. This limit cannot be reached by a factor of about 10 because the mirror starts being deformed by the high mechanical stresses created at 5000 revolutions/s and more; also, the turbulent air around the rotating mirror impairs the image.[15] The Cordin Model 132 camera (Fig. 33, middle), which is a typical rotating-mirror streak camera, has a maximum streak velocity of 20 mm/μs. With this, a time resolution of 3 ns and a recording time of 15.5 μs are obtained.

The writing speed can be increased considerably by using an electronic image converter camera. With this type of camera it is possible to attain ranges of recording speeds that are no longer of interest to detonics, because the recording times involved are too short. For example, the Imacon 790 image converter camera[16] (Fig. 33, lower) has a recording time of 0.075 μs and a time resolution of 0.15 ns at a recording speed of 1000 mm/μs. This is by no means an upper limit for this type of electronic camera.However, such values are rarely needed in warhead or detonation diagnostics. Pictures of the three typical streak cameras are shown in Fig. 34.

Table 2 lists the maximum streak velocities v_S, the recording times t_S, the time resolutions t_R at maximum streak velocities, and the ratio Q of the three types of streak cameras mentioned above. The ratio Q, which is the ratio of t_S divided by t_R, is one measure of the quality of the particular type of camera. It should be noted, however, that the time resolution t_R cannot exceed the limit inherent in the particular type of camera. Of course, each camera type can also be operated at a slower recording rate and, hence, with a longer recording time. However, the time resolution will then be lower, if the quality ratio Q of that particular camera remains constant. The increase in writing speed and the resulting resolution of the record is at the expense of the recording time, as can be seen in Table 2.

This fact is expressed graphically in Fig. 35. The typical writing times t_S of the three fundamental types of streak cameras are shown as a function of the associated resolution t_R. It can be seen that the Imacon electronic image

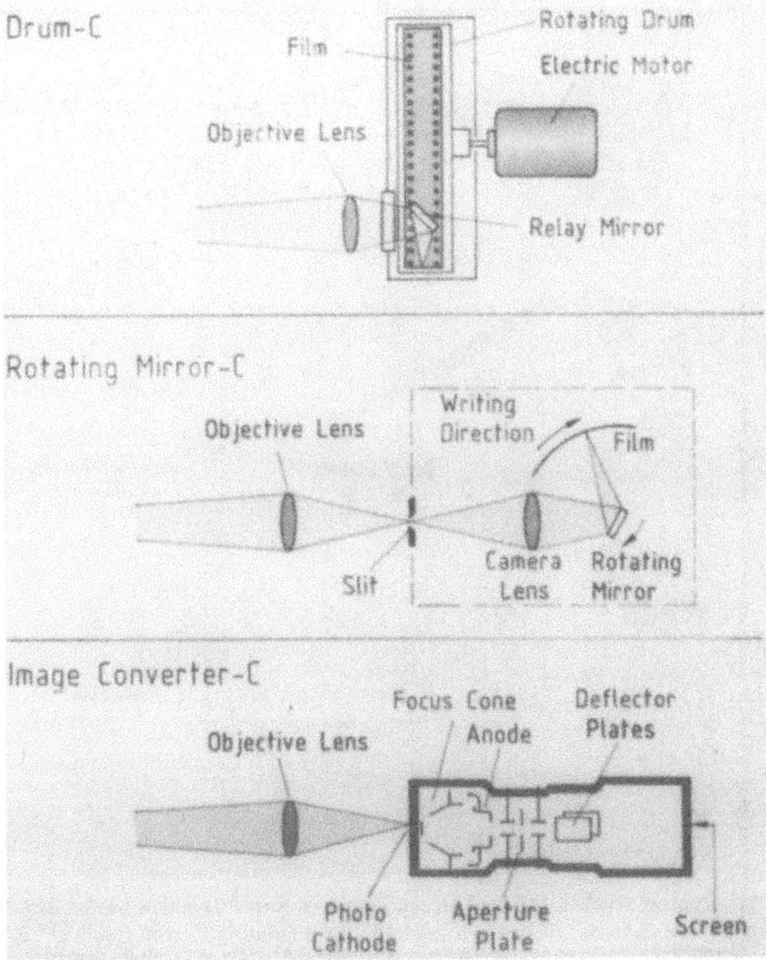

Fig. 33 Typical streak cameras: rotating drum up to 0.3 mm/μs; rotating mirror up to 20 mm/μs; and image converter camera with deflection tube up to 1000 mm/μs writing velocities.

converter camera has the highest resolution of 0.15 ns, but then only with a recording time of 0.075 μs. However, this camera can also record very slowly, up to several hundred microseconds.

The Cordin Model 132, which is a typical rotating-mirror camera, has a maximum resolution of only 3 ns, but with a writing time of 15 μs. The image converter, on the other hand, has a recording time of only about 1 μs at the same resolution. Rotating-mirror cameras are usually not very effective at low mirror speeds. Their writing velocity spans only one order of magnitude, so their maximum writing time is around 150 μs. Apart from very special cases, this recording time is sufficient for warhead processes.

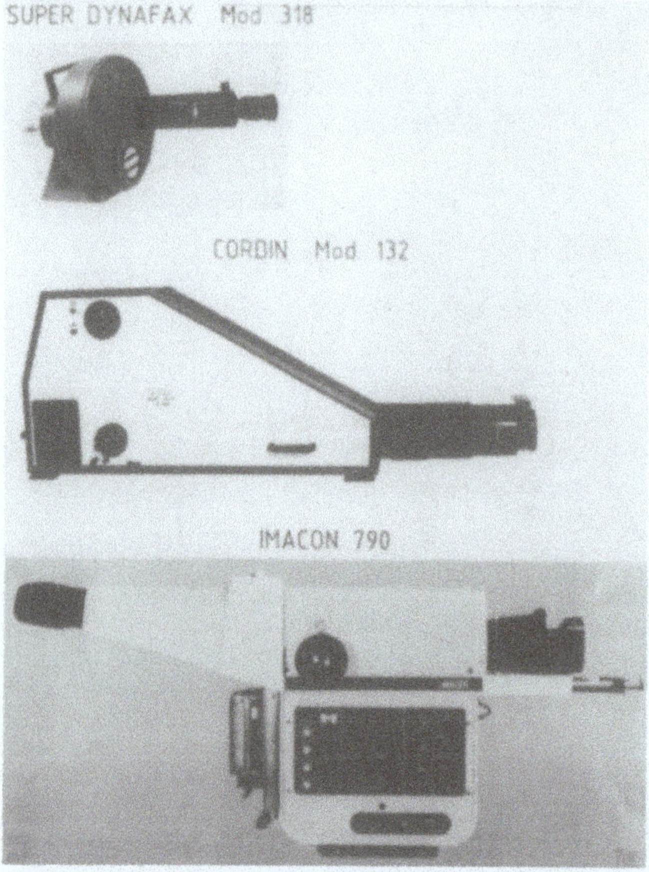

Fig. 34 Typical representative of streak cameras: Super Dynafax Model 318 for a rotating drum camera[13]; Cordin Model 132 for a rotating mirror camera[13]; and Imacon 790 for an image converter streak camera.[15]

Table 2 Maximum streak velocities, recording times, time resolutions, and quality indexes for the three typical streak camera types

	Camera Types		
	Drum	Rotating mirror	Image converter
Maximum streak rate v_S, mm/μs	0.3	20	1000
Streak time t_s at maximum streak rate, μs	2900	15	0.075
Time resolution t_R at maximum streak rate, ns	200	3	0.15
$Q = t_s/t_R \times 10^3$	14.5	5	0.5

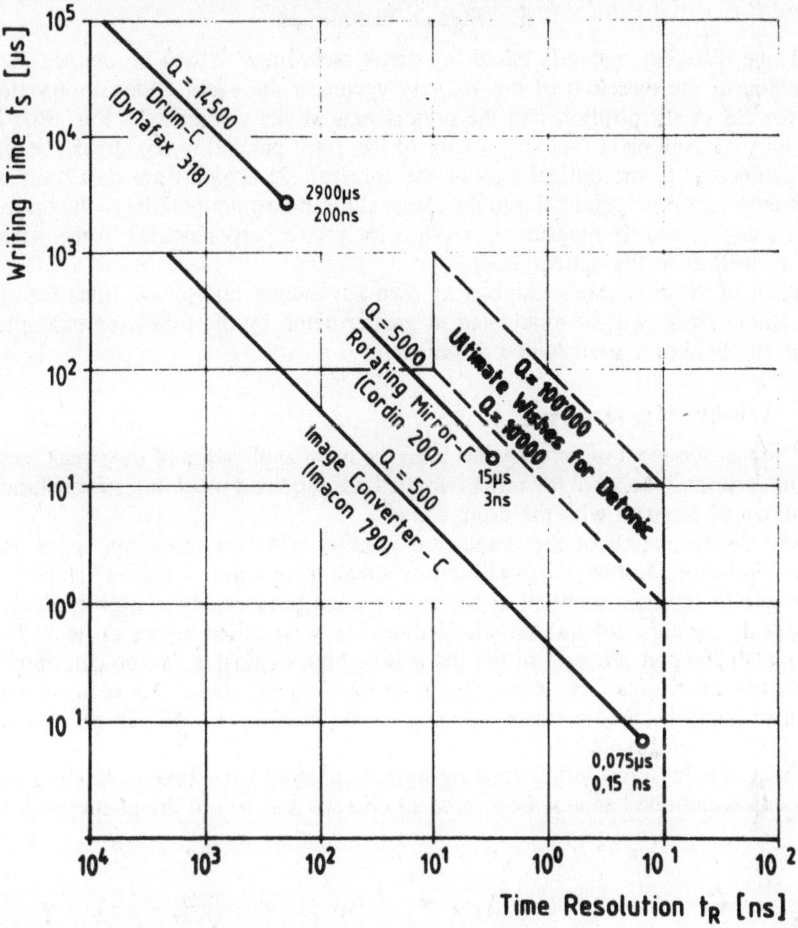

Fig. 35 Quality index Q for the different streak cameras; $Q = t_S/t_R$ with streak speed t_S and time resolution t_R.

The drum camera has a maximum time resolution of only 200 ns, but the recording time available at this speed is as long as 2900 μs. Also, the drum camera can be operated slower over two orders of magnitude, but then, of course, the resolution will be poorer.

The resolution required in shock wave and detonation physics is in the range of 10 ns to 1 ns at recording times of 10 to 100 μs, yielding Q ratios of from 10^4 to 10^5. The image converter camera meets the requirements with respect to resolution, but not with respect to the recording time. The best compromise is reached with the rotating-mirror camera. The drum camera does not have the time resolution needed in shock and detonation physics; therefore, the long recording time cannot be utilized.

5. Streak Techniques

Three different methods based on streak recording[17,18] can be defined as a function of the direction of the velocity vector of the event under observation vs the streak-slit position and the optical axis of the camera (see Fig. 36): 1) velocity measurement (velocity vector of the event parallel to the streak slit and perpendicular to the optical axis of the camera), 2) arrival time measurement (velocity vector perpendicular to the streak slit, but now antiparallel to the optical axis), and 3) profile measurement (velocity vector perpendicular to the streak slit as well as to the optical axis).

Each of these methods exhibits its own advantages and possibilities for application. These will be elucidated in greater detail by the following examples from the field of warheads and detonics.

5.1 Velocity Measurement

The measurement of velocity has been the main application of the streak technique, primarily because the other two methods require a much higher resolution than can be attained with the drum camera.

The determination of the detonation velocity with very accurate space and time resolution is often accomplished by streak techniques. Figure 37 shows an example of the measurement of the shock wave produced by a high-explosive charge in an adjacent transparent medium, in a so-called aquarium test. The Chapman-Jouguet pressure of the detonating high explosive can be determined from the shock velocity at the charge-to-medium interface. An accurate and reliable analysis can be made when reference marks are made in the middle plane of an acrylic glass block.

Since the high-resolution rotating-mirror cameras have been available, the velocity technique has also been used to observe and record the phenomena of

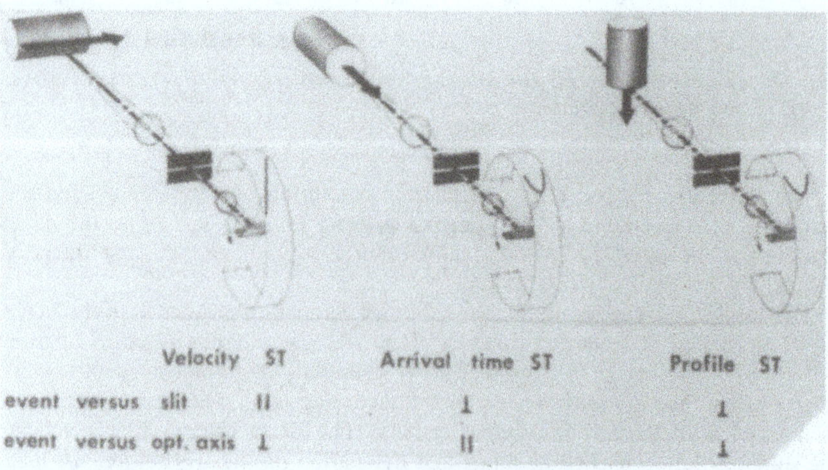

Fig. 36 Three typical streak techniques[17,18] for velocity, arrival time, and profile measurements.

Fig. 37 High time resolution is generally achieved with a rotating-mirror camera record; high space resolution can be obtained by using fiducial marks in the Plexiglas block in a so-called aquarium test, to measure the shock velocity.

buildup to detonation, yielding run-up-to detonation times and distances.

Figure 38 shows an example of transition from burning to detonation under heavy confinement, where the reaction was observed through an acrylic glass window. Apart from determining the buildup distance and the individual velocities, such as combustion and detonation, Griffith[19] has for the first time detected a retonation and has determined its velocity.

The velocity technique can also be used to determine the shock impacting on the acceptor charge in the so-called instrumented gap test. At the same time, the runup time and the buildup distance in an acceptor charge on its spontaneous transition from deflagration to detonation can be determined by the so-called transparent channel technique (Fig. 39).[20]

In fuze trains of safe-and-arm devices, the response time of the electro-explosive device or electronic detonator and the reaction times of the other elements can be recorded by means of light pipes, which pick up the luminosity at the interfaces. The ends of these light pipes are then led onto the entrance slit of the streak camera.[21] Long reaction times and considerably high dispersions of the times would then indicate an improper design of the fuze train. Several light pipes arranged in a row, with air gaps, enable the velocities inside high-explosive charges to be measured.[22]

5.2 Arrival Time Measurement

Measuring the time of arrival of a detonation front at a surface means observing the propagation of the detonation wave as it intersects with the surface. Such phase velocities are therefore often much higher than the detonation velocity itself. Hence, this also means that a particularly high resolution in time is needed, although in general there are no special requirements for the recording time. A few examples of this are given below.

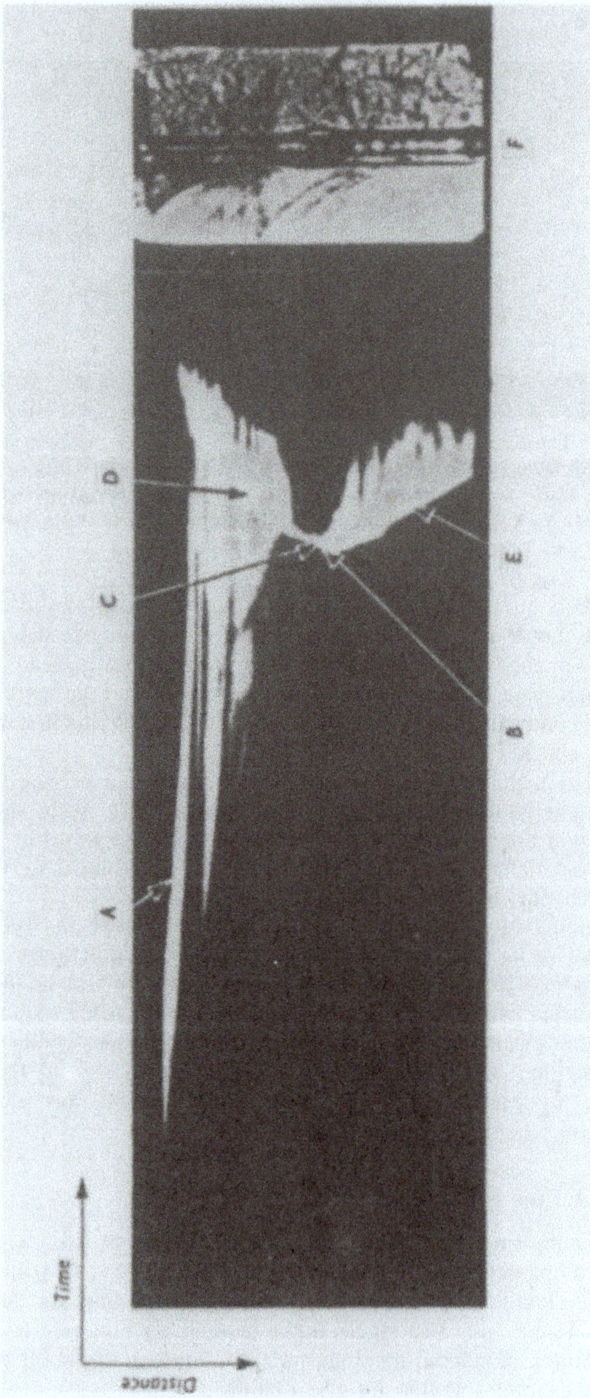

Fig. 38 First recorded retonation by Griffith.[19]

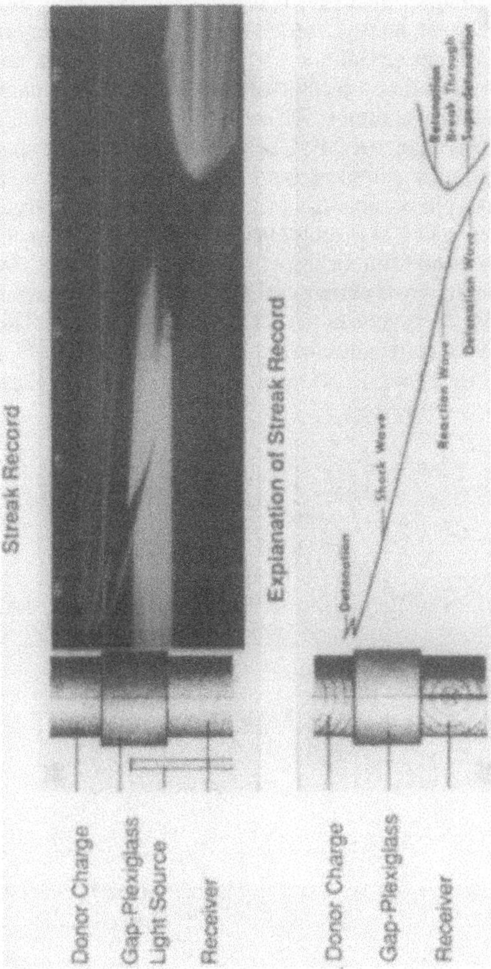

Fig. 39 Instrumented gap test to measure the buildup distance and delay times in the acceptor charge.[20]

In the corner effect,[23,24] the time of arrival of the detonation wave on the plane opposite the initiation side and the breakthrough of detonation on the perimeter of the cylindrical charge are recorded. Figure 40 shows the simple test arrangement and a typical sample record, and Fig. 41 shows a series of records where the acceptor charge diameter was increased from 25 mm to 55 mm in steps of 10 mm.[25]

In the streak technique, the event is usually observed only along a line. However, observing a larger area is often desirable. This is generally possible by means of the multistreak technique.[26] Figure 42 shows the test setup with an acrylic glass cover placed at a distance of 0.1 mm from the charge surface. The streak mask containing a number of streak slits is attached to this plate. The same figure also shows the arrival of a spherical detonation wave at a plane with five streak slits. Another sample record (Fig. 43) shows how a coaxial spherical detonation wave has swept over a multistreak mask having five concentric rings with four diagonals. This technique is widely used in specific problems of detonics, such as in determining the effectiveness of a plane wave generator.

By cleverly imaging an event and rotating part of the picture through, say, 90 deg, one can record the event on two axes simultaneously. This technique has been termed the orthogonal multistreak technique (OMST).[26] As an example, the breakthrough of a spherical detonation wave, which is slightly eccentric in one plane, is shown in Fig. 44.

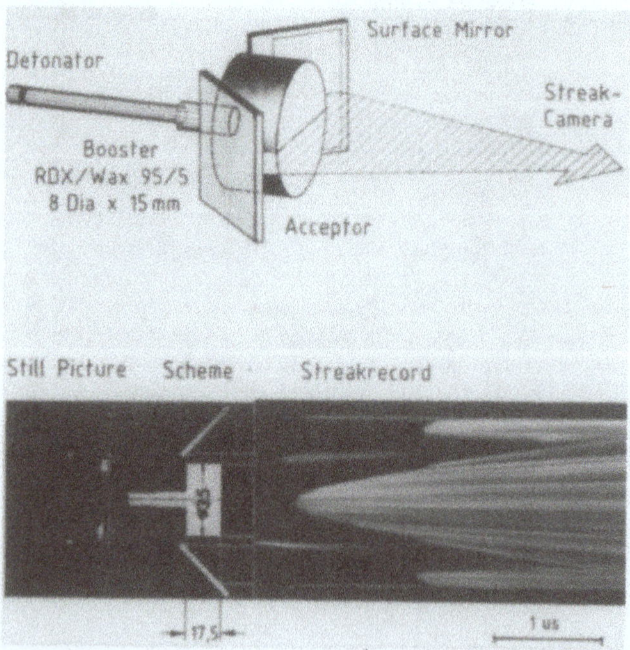

Fig. 40 Test setup for measuring the corner turning distance (CTD) in the radial test with a typical streak record.[23–25]

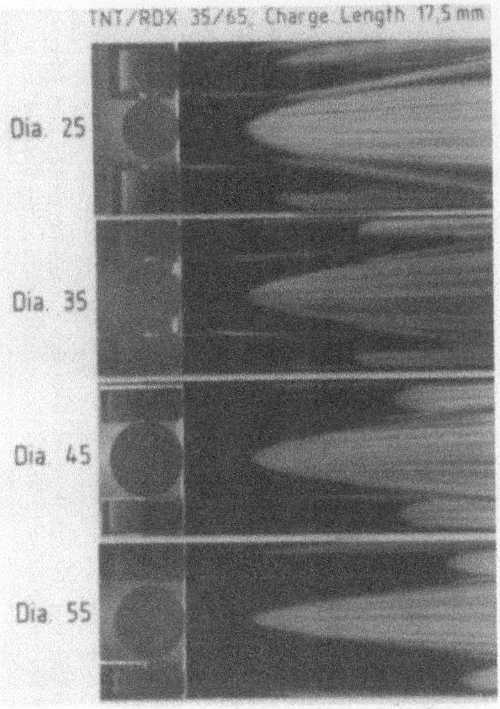

Fig. 41 The CTD is constant for different acceptor diameters, but with smaller diameters no retonation waves exist.[23,24]

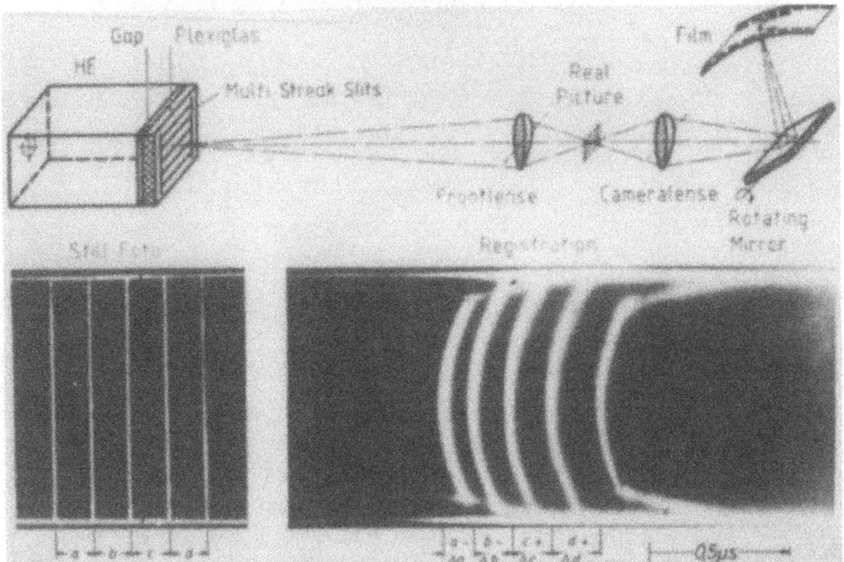

Fig. 42 Test setup for multistreak records; transparent cover with an air gap of approximately 0.1 mm distance to the front surface; and the still picture of the stream mask and the streak record of the breakthrough of a spherical detonation wave.[26]

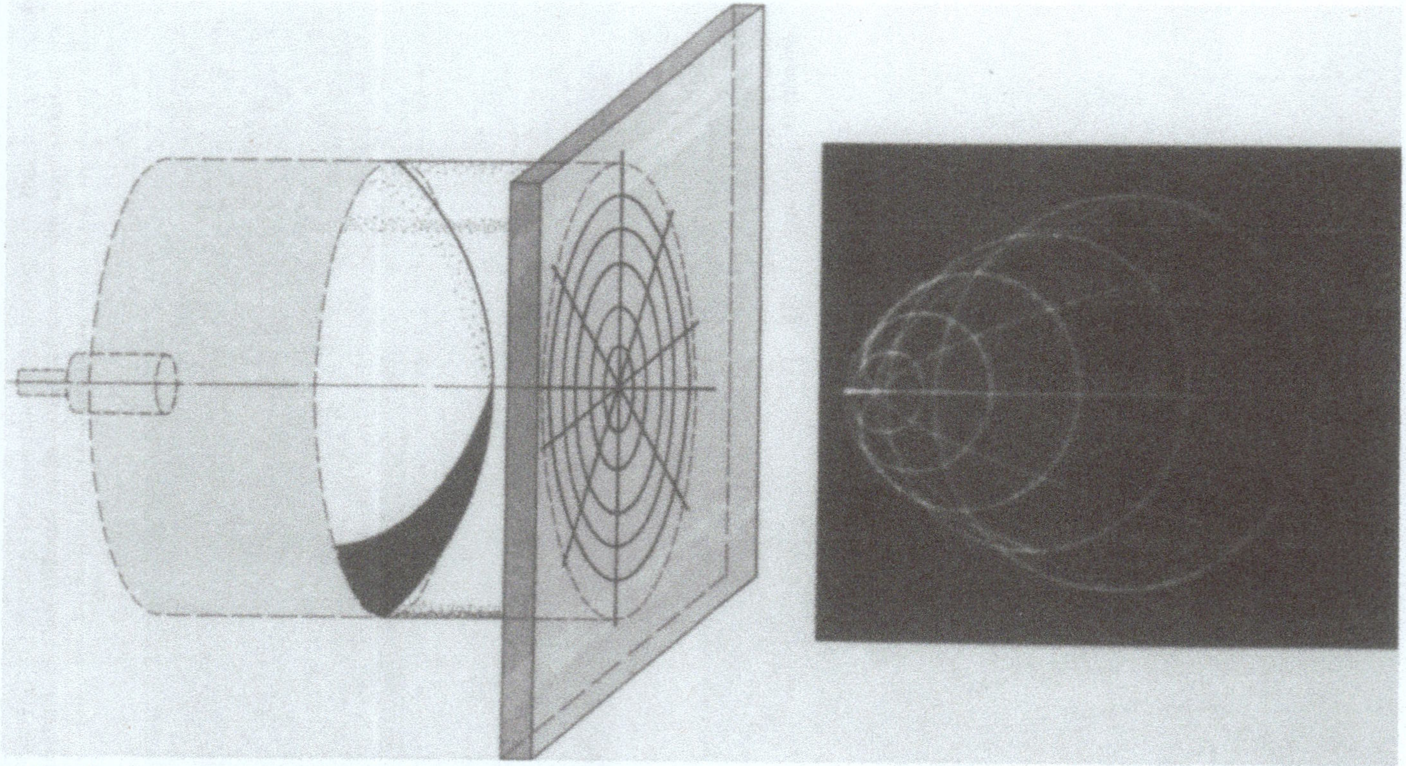

Fig. 43 Multistreak picture of a spherical detonation wave sweeping over the streak mask with five concentric circles and four diagonals. Each individual circle is covered simultaneously, from the smallest circle to the largest with increasing time by the concentric spherical detonation wave.

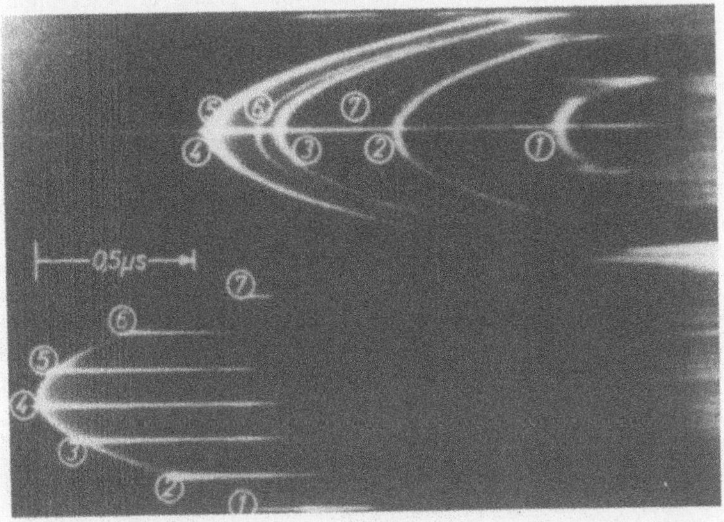

Fig. 44 Example of an orthogonal multistreak record of a spherical detonation wave, centric in the upper plane, but slightly eccentric in the lower plane.[26]

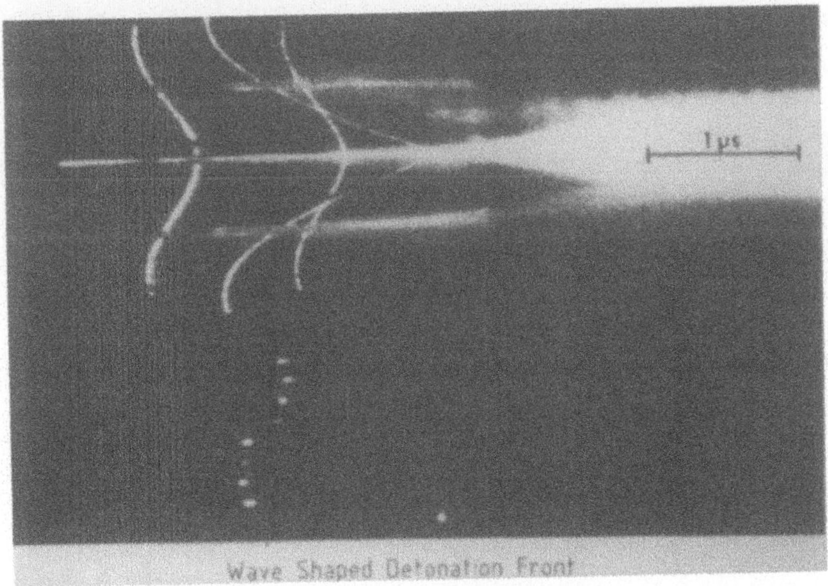

Fig. 45 The analysis of the delay times in the different lines is easier and more obvious if light fibers are also used; this technique is called time-correlated multistreak.[27]

The time differences between the individual planes of observation can be calculated from the displacement of the lines with respect to their position in this picture. Even when the scale of the reproduction is not known, these time differences can be read directly from the record when light fibers are used at the points of reference. This technique has been termed the "time correlated multistreak technique" (TMST).[27] A record of a wave-shaped detonation front, obtained by the multistreak technique, is shown in the upper part of Fig. 45, and the lower part of the same figure shows the time differences at a few selected points of this streak record via the light fibers.

When several acrylic glass covers, each with a thin air gap, are fitted to a charge, then not only the time of arrival but also the shock velocity itself can be measured in the intermediate layers. Pressure can then be calculated from these values of shock velocity. Figure 46 shows a typical test arrangement for the determination of the arrival time of a detonation front in a first gap (after this

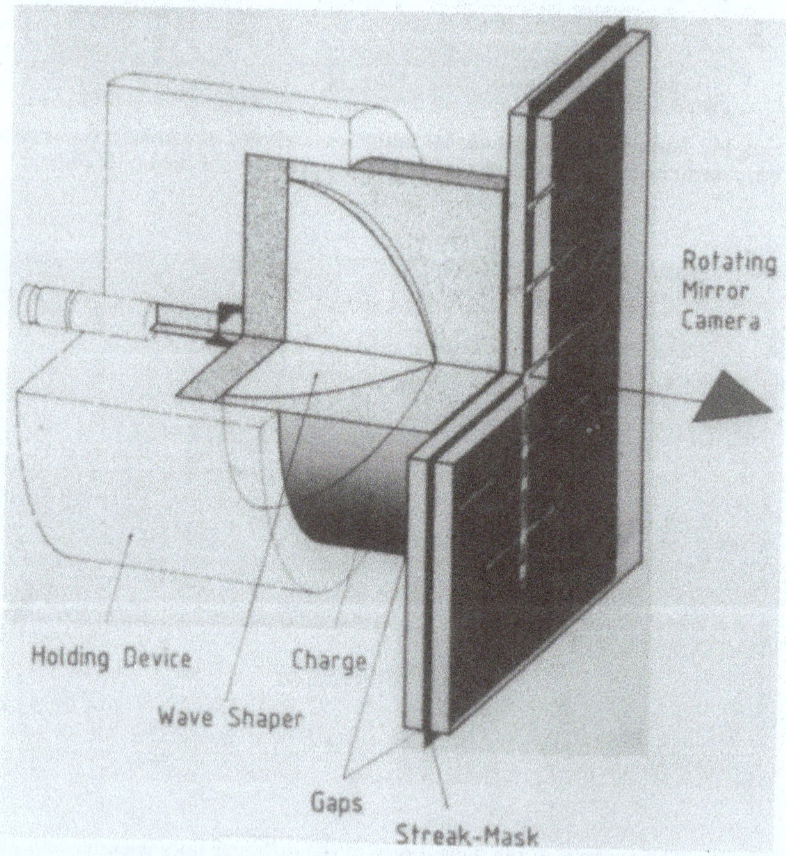

Fig. 46 Test setup for double multistreak record; this provides arrival time and indication of the pressure by measuring additionally and simultaneously the shock transit time through a given thickness of Plexiglas.

detonation has propagated around a so-called wave shaper) and the time of arrival of the shock wave in a second gap. A typical streak record is shown in Fig. 47.

The high time resolution of the rotating-mirror cameras often cannot be utilized fully with velocity measurements, because the event must be fired at a certain distance from the objective lens, owing to the blast and fragment hazard. Therefore, the image may be small compared to the entrance slit. The spatial resolution can be increased by running the event against a wedge; the angle of the wedge controls the apparent enlargement. The resolution of distance can thus be increased by a factor of 10. Very often the record trace is one of a shock wave entering water. Such a technique was used by Deal in 1958[28] to determine the Von Neumann pressure spike and the Chapman-Jouguet pressure. However, this technique requires a plane shock or detonation wave; therefore, a plane wave generator had to be used in these experiments.

A combination of the arrival time technique and of the wedge technique is advisable when no expensive plane wave generators are to be used or when processes involving a converging or diverging detonation front are to be examined.[29] Figure 48 shows the test arrangement, and Fig. 49 shows the resulting record of the time of arrival of a diverging spherical detonation wave at three planes of observation, together with the arrival times of the shock waves after the associated times of propagation in the acrylic glass wedge.

A special technique is the streak objective lens technique (SOLT), in which the event under observation proceeds at a small angle to the axis of the objective lens.[30] As an example, the determination of the strain rate of an explosively forming projectile is illustrated here. Figure 50 shows the test set up and a streak

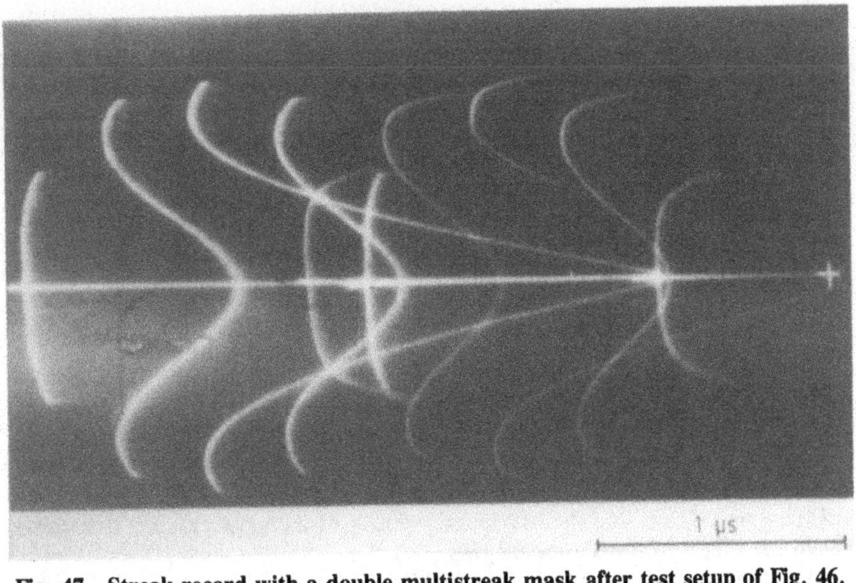

Fig. 47 Streak record with a double multistreak mask after test setup of Fig. 46.

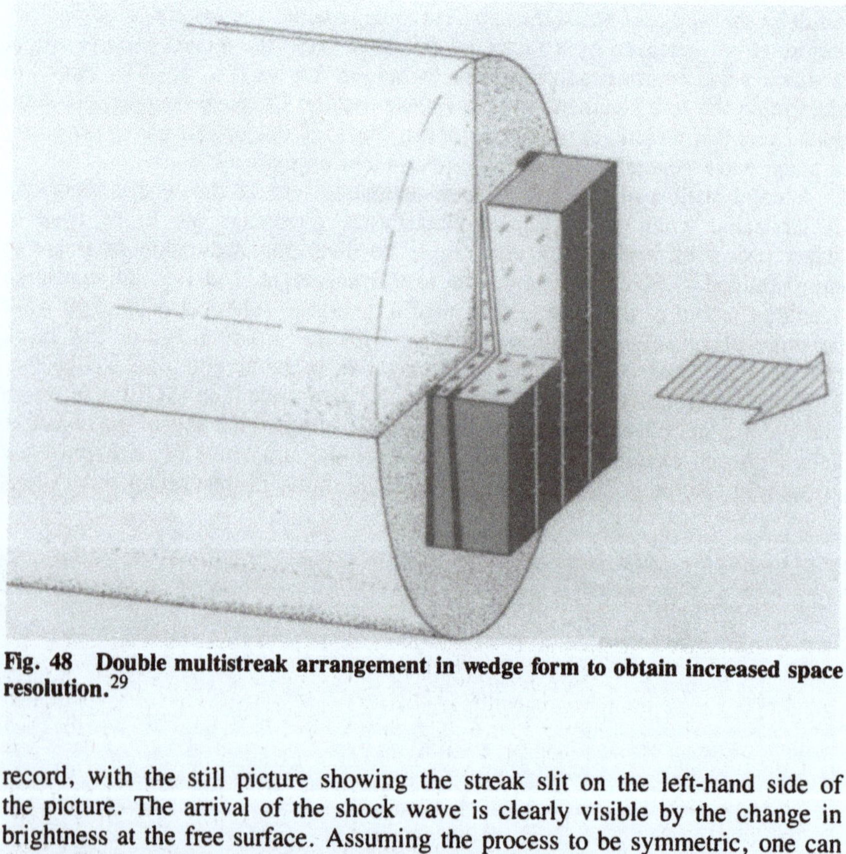

Fig. 48 Double multistreak arrangement in wedge form to obtain increased space resolution.[29]

record, with the still picture showing the streak slit on the left-hand side of the picture. The arrival of the shock wave is clearly visible by the change in brightness at the free surface. Assuming the process to be symmetric, one can determine the radial as well as the axial velocities of each individual ring element and thus calculate the strain rate.

5.3 Profile Measurement

The performance of high-explosive charges is generally tested in the so-called cylinder test by means of the profile measurement technique (PMT), in which the increase of velocity and the maximum velocity of a copper case are determined from the time-vs-distance record (Fig. 51).[31]

The multiprofile streak technique (MPST)[32,33] is a highly accurate method for determining both the detonation velocity on the periphery of the high-explosive

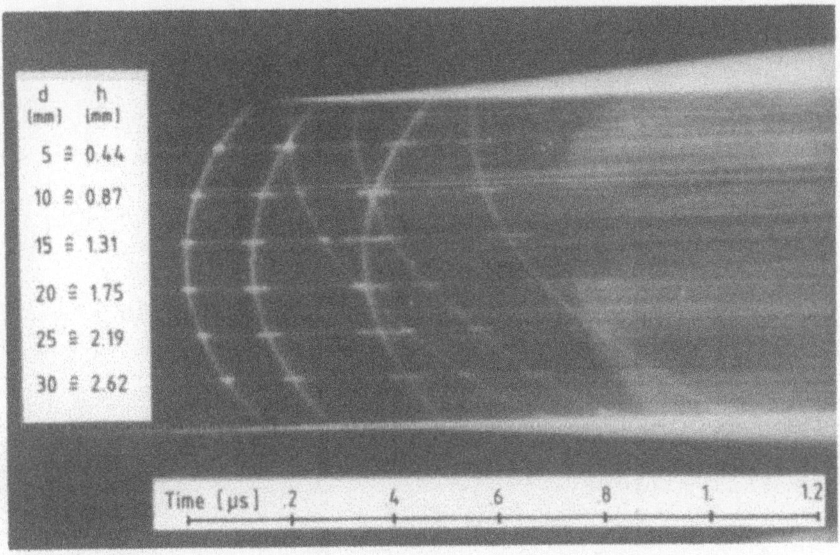

Fig. 49 Streak record of the arrival time of a detonation wave and its shock wave after the test setup Fig. 48; using a wedge angle of 5 deg the space resolution is increased by about 10 times.[29]

charge and symmetry of the detonation waves over the entire periphery at multiple lines of observation. The experimental setup is shown in Fig. 52, and two streak records are given in Fig. 53. The evaluation of the distortion of the detonation wave in the individual planes of observation is presented in Fig. 54.

The image of an object in a streak record is undistorted if the streak velocity equals the object image velocity on the film plane. This method has been termed synchro-ballistic photography or the ballistic synchro-streak technique (BaSST). It is often used with drum cameras, or even with continuously moving film cameras, which with modern film materials can have film transport velocities of up to 80 m/s.[34] In this case the velocity of the object must be adapted to the film speed by reducing the size of the image. This method can yield excellent pictures of, for example, a flying HEAT round together with the associated flow lines (Fig. 55), or of a fired APDS-FS projectile after the sabots have been discarded (Fig. 56).[35]

The synchro-streak technique (SST) (see Fig. 57)[37] is a version that has been derived from the BaSST; it allows the jet of a shaped charge to be examined, particularly with respect to deviations from its axis.

The jet can also be recorded at two subsequent distances (Fig. 58) by observing it at two subsequent planes by means of the so-called Bi-SST (see Fig. 59).[37] The velocities of the particles, and their transverse drift velocities and rotation rates, can be determined unambiguously from the time differences of the records for each of the particles at the two planes. It is also suitable for examining the

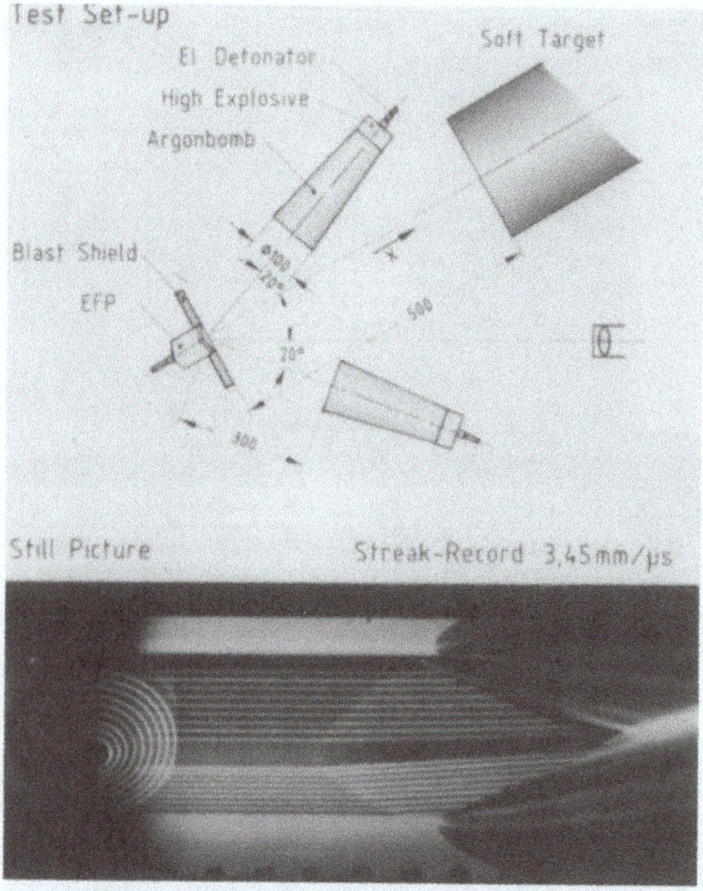

Fig. 50 Deformation of an EFP liner with regard to strain and strain rate can be observed and analyzed with the so-called streak objective lens technique; the test setup and the still picture and a streak record (bottom).[30]

shape of the particles, the process of jet breakup (i.e., particulation), the mean and individual times of particulation, and the place of particulation between two neighboring particles. Despite the fact that this is an excellent method of jet diagnostics, it is to date being used more or less only by the author of this chapter. The recording time on the film does not render the true length of the particles,because these have different velocities. The fast particles of the jet tip region appear shorter, and the slower particles from the jet end appear longer. The length of each particle follows from the time of passage multiplied by the velocity of that jet particle.

The orthogonal synchro-streak technique (OSST) makes it possible to determine the deviation of the jet in two planes, in that the jet is observed in two mutually perpendicular directions (see Fig. 60).[38] The record of a shaped charge

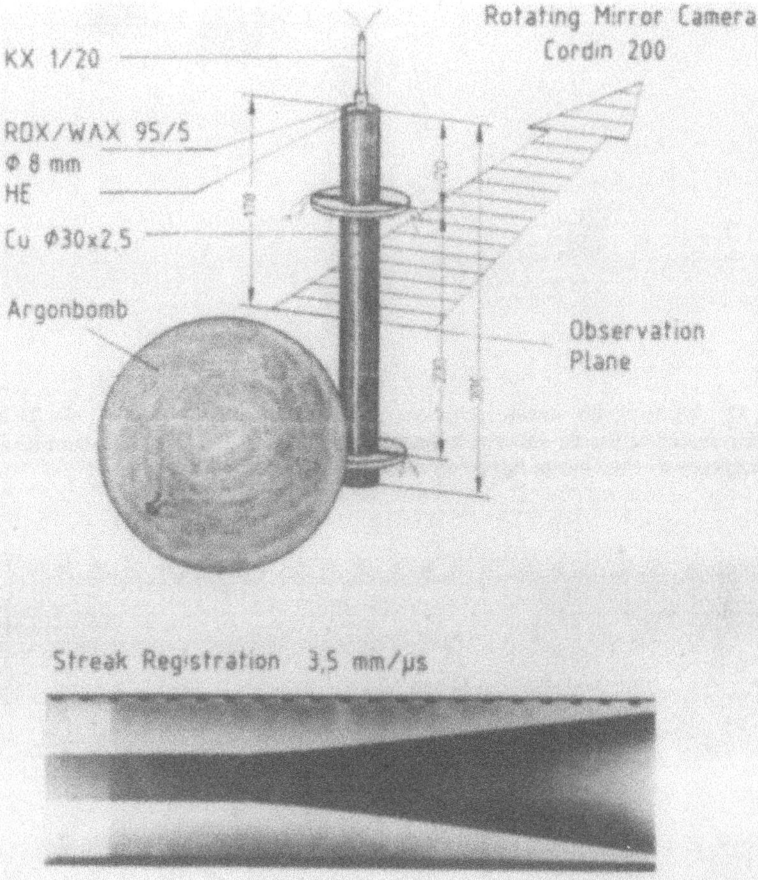

Fig. 51 Test setup for the cylinder test, together with a streak record, to get the performance and the Gurney constant of a high-explosive charge.[31]

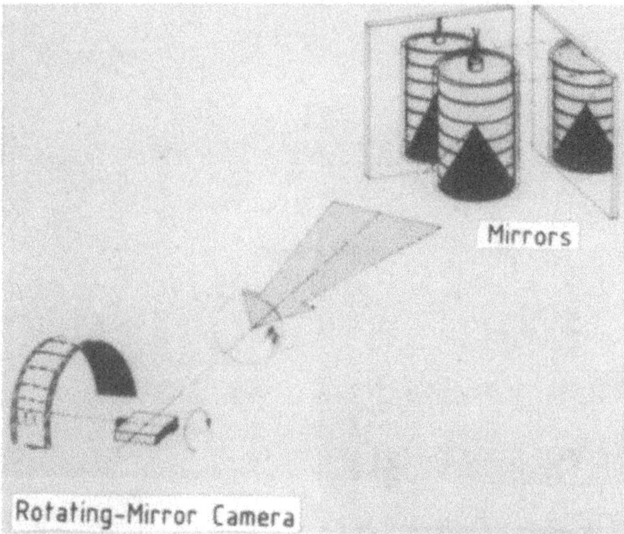

Fig. 52 Multi-profile streak technique (MPST) gives the detonation velocity and the symmetry of the detonation front in many levels without any disturbance or interference of the charge layout or event (e.g., shaped charge).[32,33]

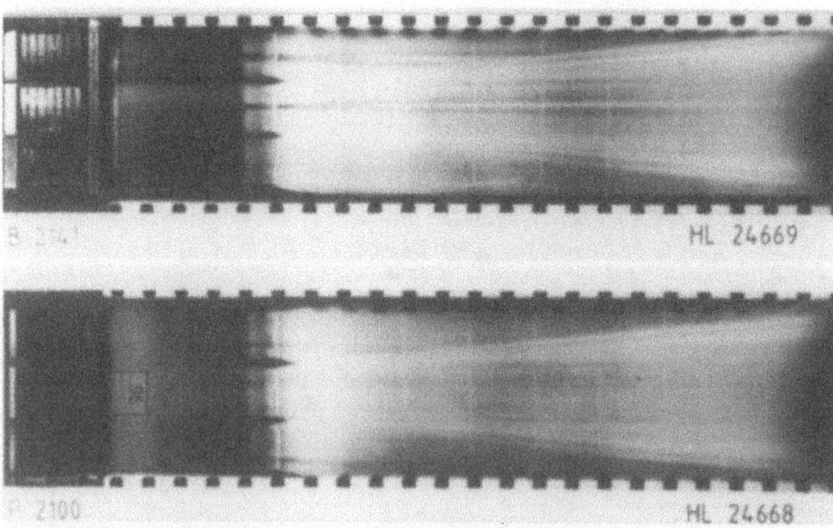

Fig. 53 Still pictures (left) and streak records (right) of the detonation of plastic bonded HE of cylindrical shaped charges with the multi-profile streak technique (Fig. 52).[31]

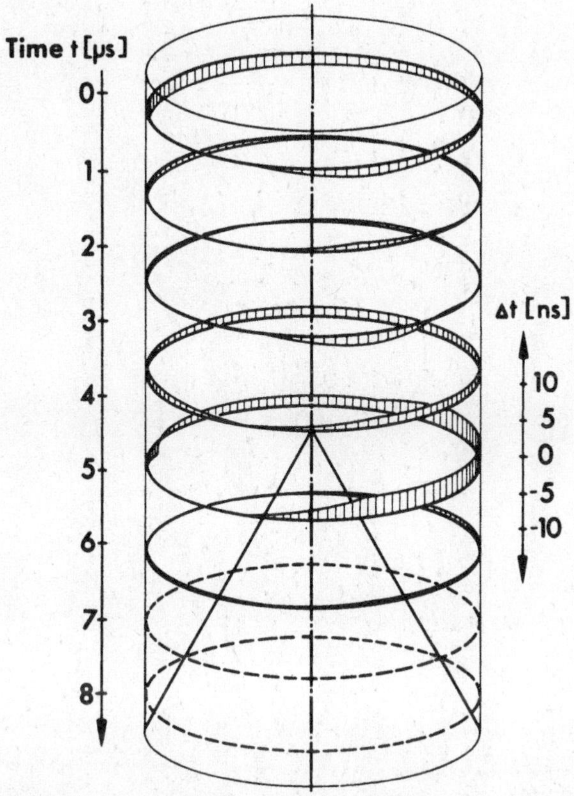

Fig. 54 Evaluation of streak record of Fig. 53 with respect to the symmetry of the detonation front in the different observed levels.

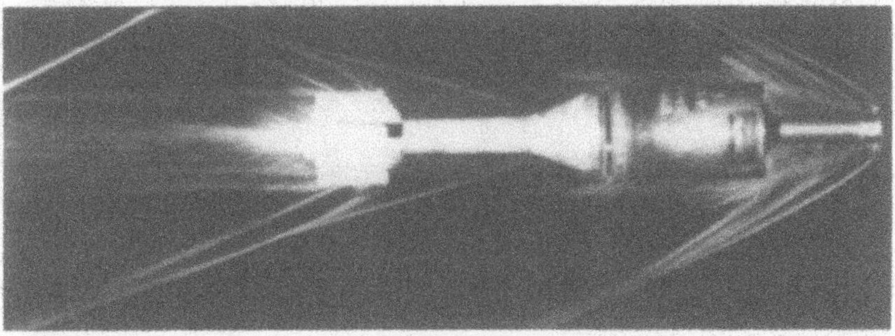

Fig. 55 Picture of a 120-mm HEAT round taken with the ballistic synchro-streak technique (BaSST).[34] (Streak record from Biele.[35])

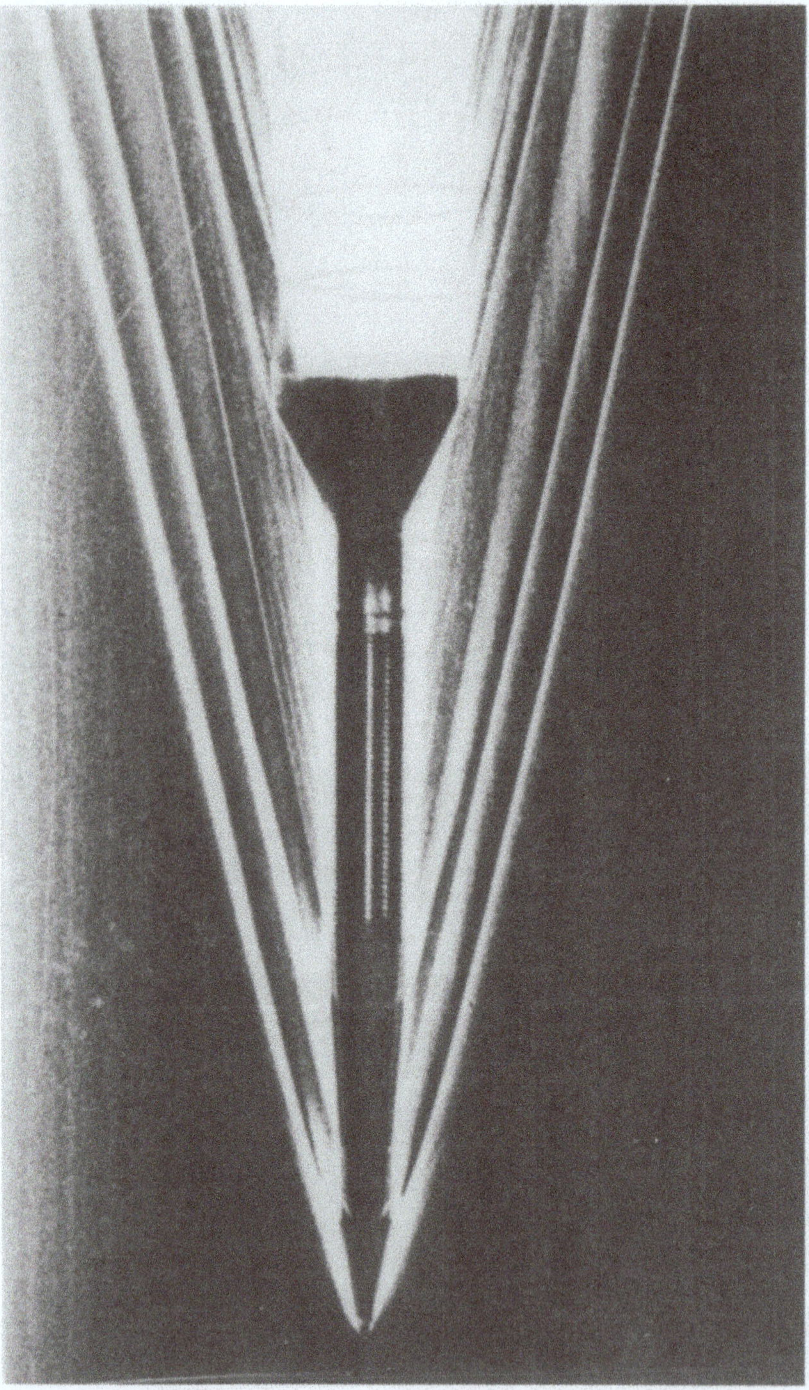

Fig. 56 Picture of an APDS-FS projectile fired from a 120-mm gun, taken with ballistic synchro-streak technique (BaSST).[34] **(Streak record from Biele.**[36]**)**

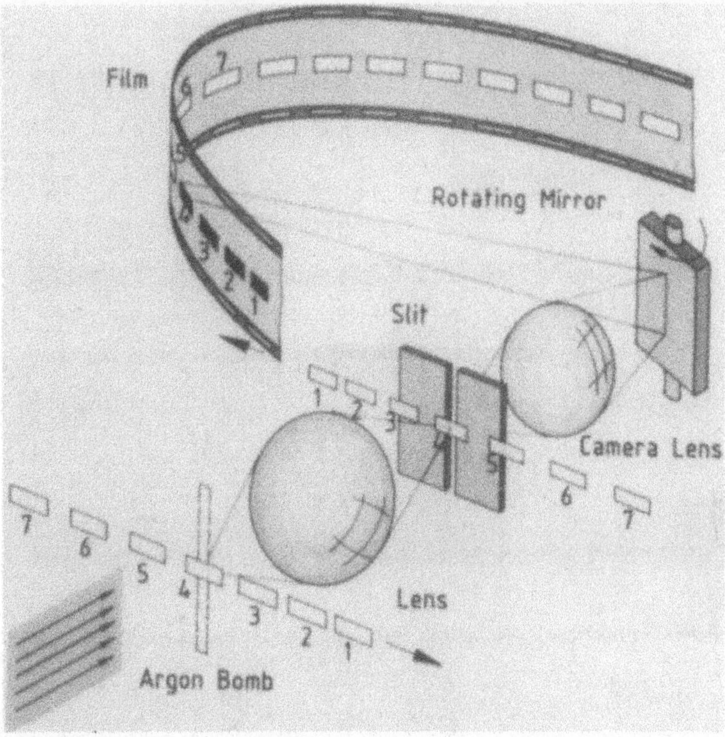

Fig. 57 Observation of a shaped charge jet with the synchro-streak technique (SST).[36,37]

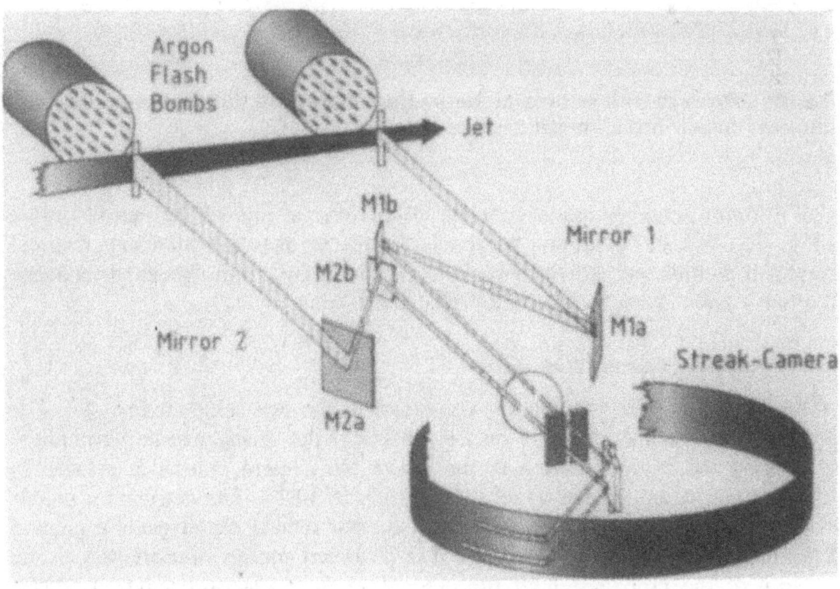

Fig. 58 Test setup for the observation of a shaped charge jet in two different distances with the synchro-streak technique (Bi-SST).[36]

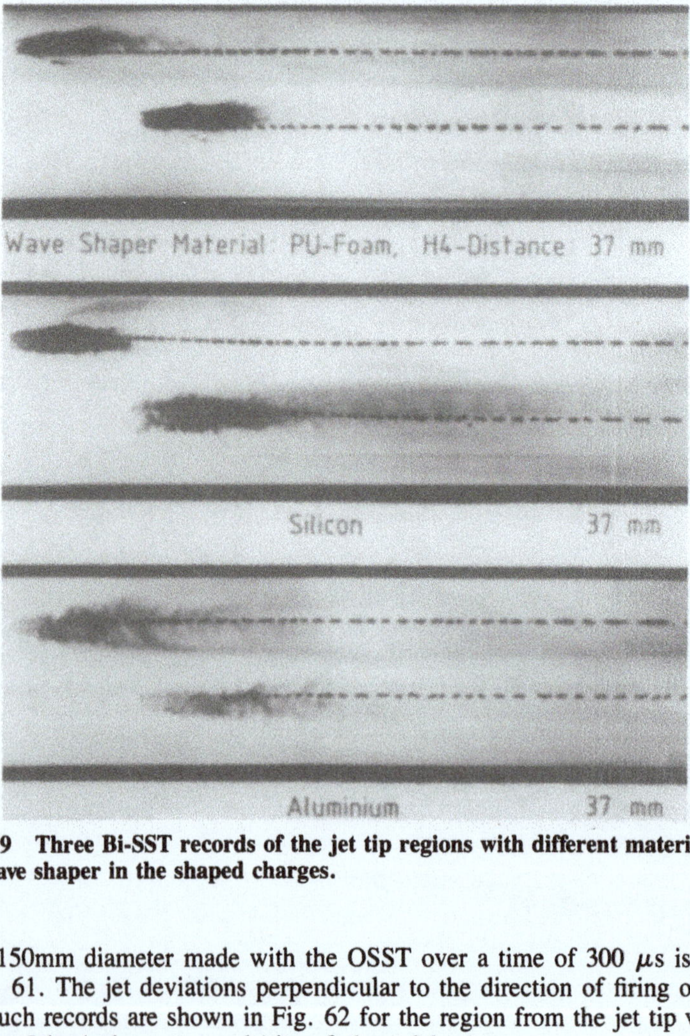

Fig. 59 Three Bi-SST records of the jet tip regions with different materials for the wave shaper in the shaped charges.

jet of 150mm diameter made with the OSST over a time of 300 μs is shown in Fig. 61. The jet deviations perpendicular to the direction of firing obtained from such records are shown in Fig. 62 for the region from the jet tip velocity of about 9 km/s down to a velocity of about 3 km/s.

5.4 Combined Streak Techniques

The three streak techniques are combined in the test setup in Fig. 63. The velocity of the detonation wave on the surface of the charge can be determined by imaging the charge surface to the slit of the camera, which is parallel to the direction of the detonation (1 corresponds to VST). The detonation profile is determined with the aid of a Dove prism that rotates the respective portion of the image through 90 deg, so that it is observed end-on, transversely to the direction of the propagation of the detonation front (2 corresponds to PST). The end face is observed via a mirror; the horizontal camera slit views the surface

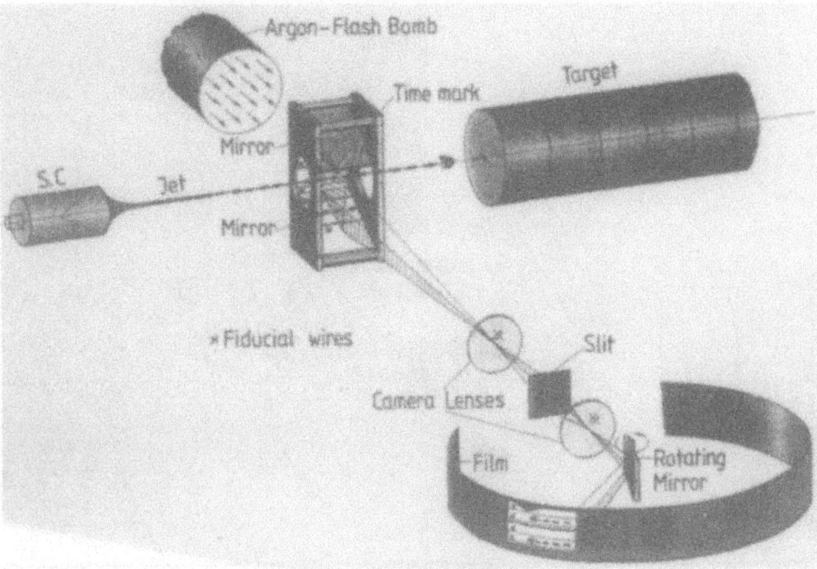

Fig. 60 Test setup for the stereoscopic observation of a shaped charge jet, called the orthogonal synchro-streak technique (OSST).[38]

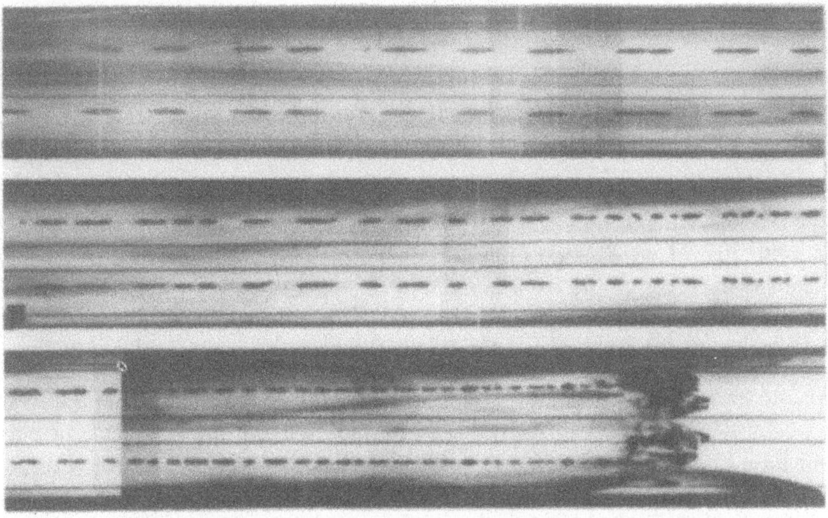

Fig. 61 Orthogonal synchro-streak record of the Cu-jet of a 150-mm shaped charge at 1800-mm distance from the basis of the liner; the cloud on top of the streak picture comes from a small detonator that is used as a time reference.

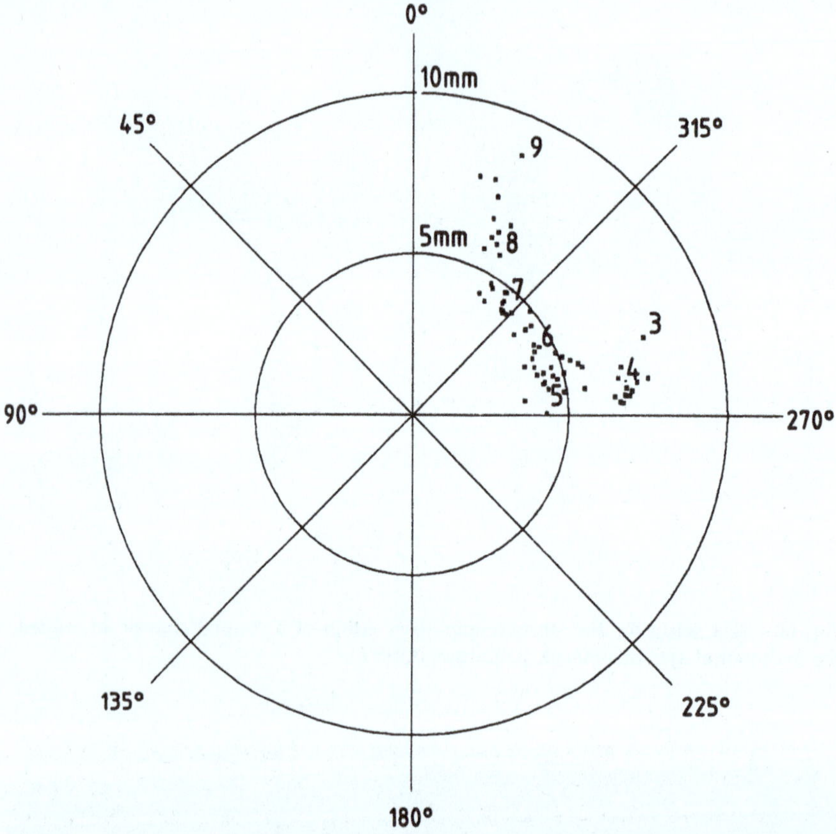

Fig. 62 Analysis of the OSST picture (Fig. 61) with regard to jet straightness in the firing direction (*z* direction); the numbers give the jet velocities.

along a diameter and thus records the time of arrival of the detonation wave along this line of observation (3 corresponds to AST).

The streak record of the measurement of detonation velocity, detonation profile, and time of arrival is given in Fig. 64; The left part of the figure shows the still picture; the right part is the streak record at 14.2 mm/μs taken with a Cordin Model 200 rotating-mirror camera at 5000 revolutions/s.

Figure 65 shows a test setup for the so-called bi-plane observation technique[20] for measuring the output of the donor charge in PST and the symmetry of the detonation wave at the output side of the acceptor charge with MST; it is used for measuring the effectiveness of a precision initiation coupler or a detonative centering device. Typical bi-plane records of such a detonative centering device, using RDX/wax/graphite 94/5/1 as transfer charges in a 6-mm brass disk with 4-, 5- and 6-mm diameters, are given in Fig. 66.[39] The analysis results of these streak records, with additional measurements from 8-, 10- and 12-mm diameters, are shown in Fig. 67.

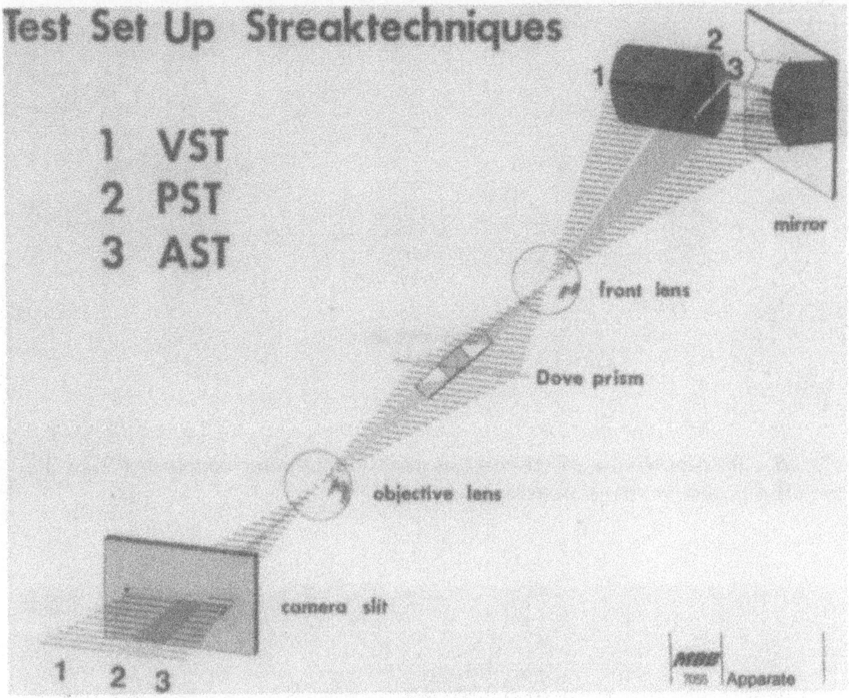

Fig. 63 Test setup for the three principal streak techniques in one firing: 1) velocity streak technique (VST), 2) profile streak technique (PST), and 3) arrival-time streak technique (AST).

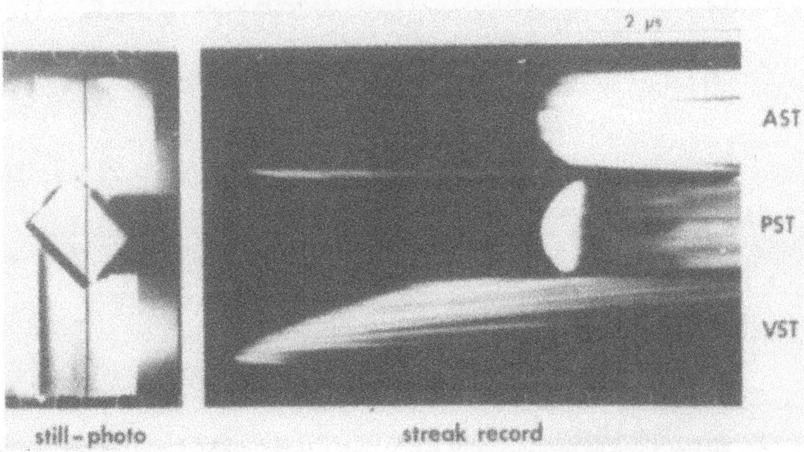

Fig. 64 Still picture and streak record of test setup of Fig. 63 starting from the bottom: 1) phase velocity of the detonation wave on the charge surface (VST), 2) profile measurement with the 90° turned picture through the Dove prism (PST), and 3) the arrive time record on the exit surface of the HE charge (AST).

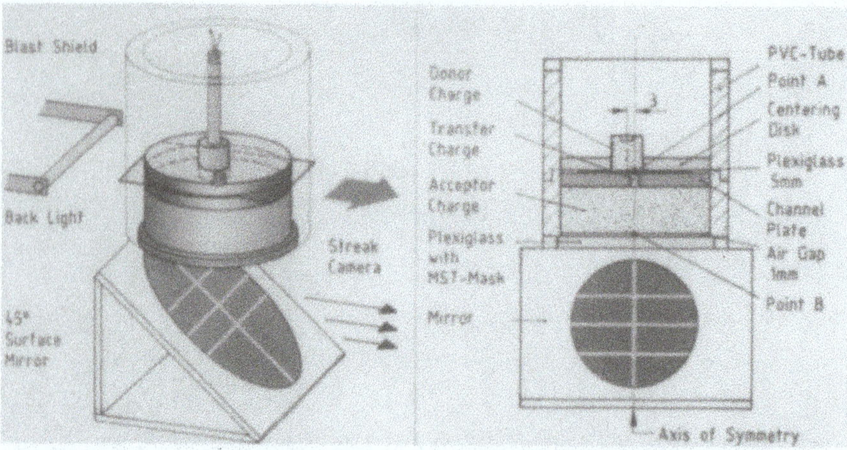

Fig. 65 To measure the effectiveness of precision initiation couplers (PICs), the so-called bi-planar streak technique is used.[20,39]

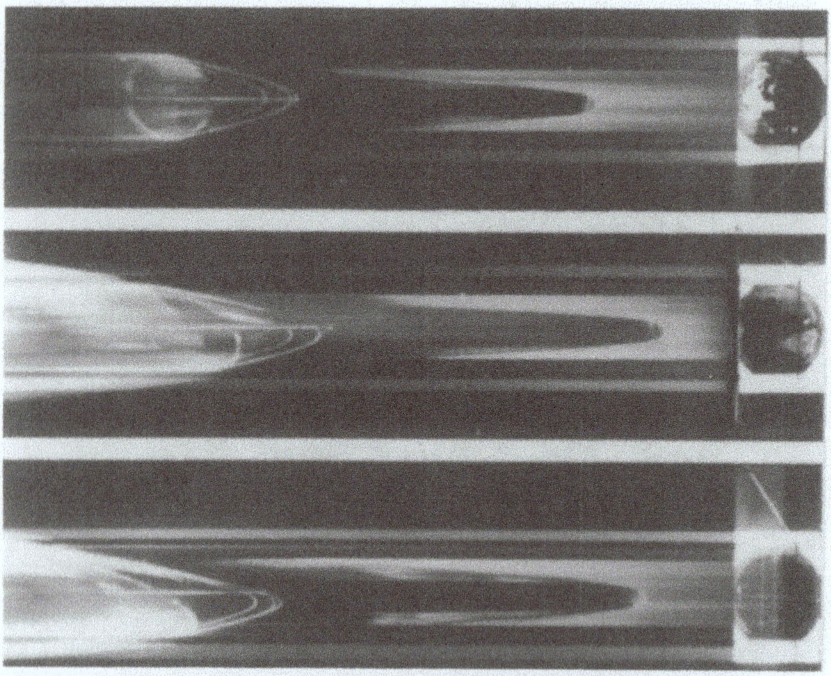

Fig. 66 Streak records of PICs with RDX/Wax/Graphite 94/5/1 as transfer charges in a 6-mm thick brass plate with 4-mm, 5-mm, and 6-mm hole diameters, from top down.[39]

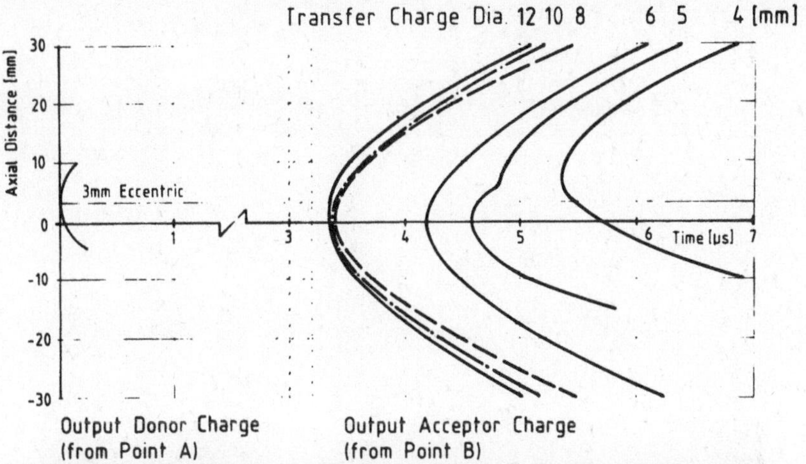

Fig. 67 Analysis of the bi-planar streak records of Fig. 66 with three additional results.[39]

The so-called open-view streak technique[40] is a combination of velocity and profile measurement. In this method, the luminous trace of high-explosive charges that are detonated in channels is recorded in such a way that the channels according to the plate design show up at various angles to the streak slit, but are all perpendicular to the optical axis (see Fig. 68). As an example, Fig. 69 shows the record taken of a plate with detonation channels for a multiple simultaneous initiation device.

6. Simultaneous Streak and Framing Records

A streak record is continuous in time and with an extremely short time of exposure, thus entailing very little motion blur. Hence, time resolution is very good for each individual camera system. Also, the resolution in space is very good along the plane, or planes, of observation. Frames, on the other hand, picture the process in space. This is useful for observing the process outside the streak plane and, in particular, for an understanding of the event or process.

The streak and framing mode recording, simultaneous and by the same camera (e.g., a Cordin Model 200 or a Cordin Model 330[13]), unites the advantages of the two methods and at the same time also indicates whether the streak record had actually been taken in the expected or predetermined planes of observation. In this manner, faulty analysis and wrong interpretation, which were frequently problems in the early days of streak recording, can be eliminated completely.[41,42]

One example is the recording by the tangent streak slit technique (TSST) of the detonation velocity of Baratol (Fig. 70) or the reaction velocity of coruscatives (Fig. 71). In the latter case, only with the help of the frames can the streak records be analyzed.[43]

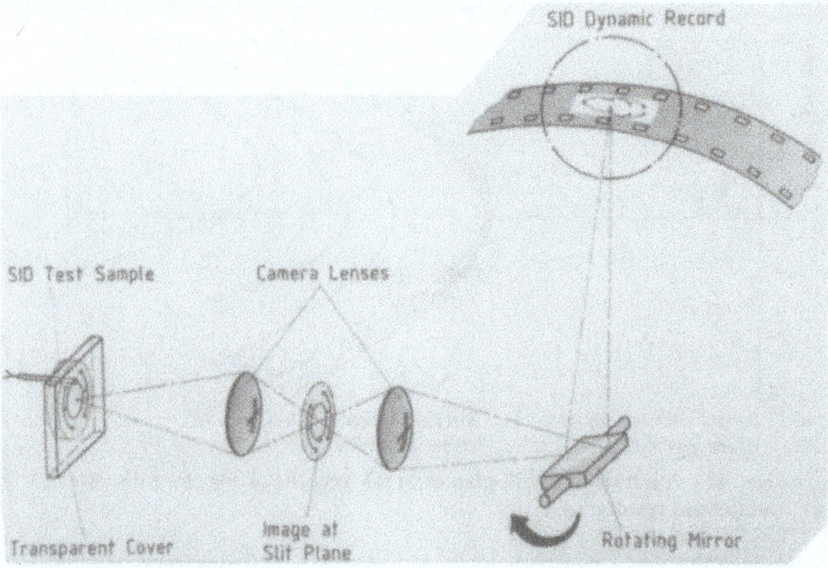

Fig. 68 Test setup for the open view streak technique (OVST), in which the streak slit in the camera is taken out.[40]

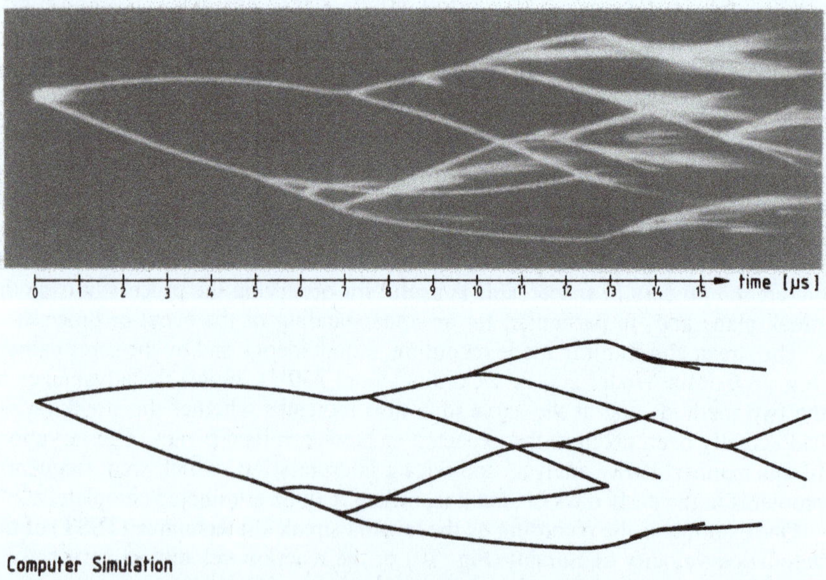

Fig. 69 Streak record of an eightfold simultaneous initiation device (SID) with the open view streak technique and the computer prediction of the traces.[40]

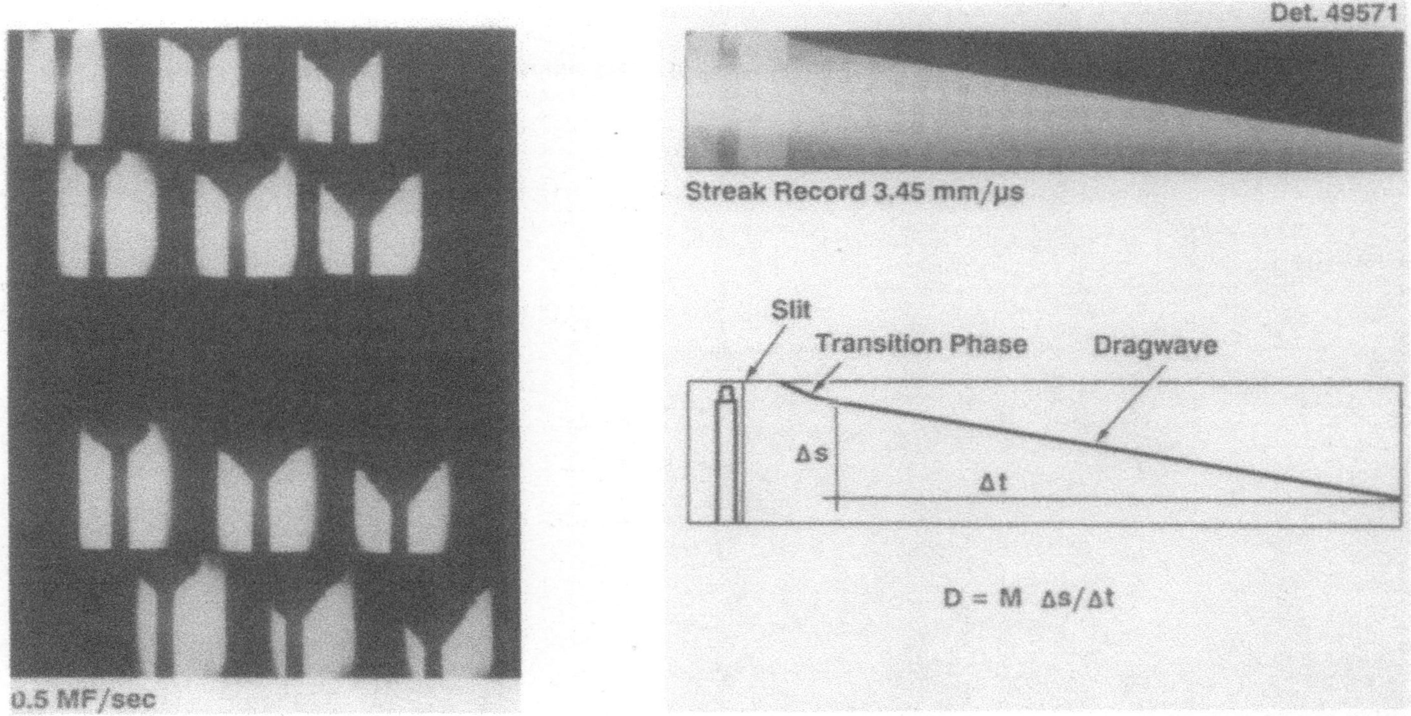

Fig. 70 **Simultaneous streak and frame records give the high time resolution from the streak picture and the understanding from the frames; the frames also give confidence in the right interpretation of the streak record, which was taken with the tangent streak slit technique (TSST)[43]; detonation velocity of a cylinder of Baratol with the density of 2.066 g/cm.[3]**

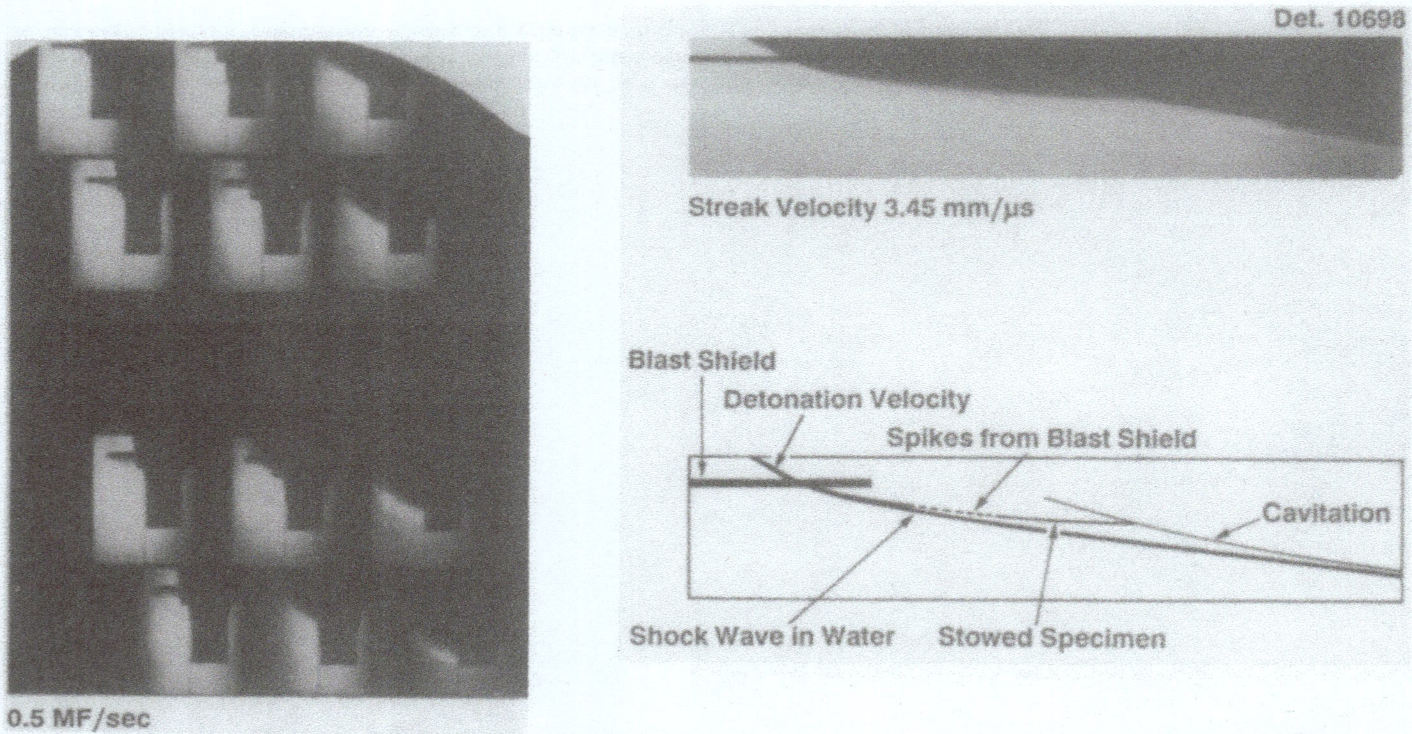

Fig. 71 Reaction behavior of a shocked coruscative Mg/Te 16/84 composition, also in TSST; without frames an interpretation of the streak record would not be easy.[43]

A second example is the testing of the initiation of acceptor charges by means of shaped charge jets with different velocities and the determination of the associated delay times and buildup distances (Figs. 72 and Fig. 73). The streak records can also only be understood and analyzed with the help of the framing pictures of these events.[44–47]

7. Computer Analysis of Streak Records

Streak techniques are not commonly used and are generally not easy to understand. Every time, one has to acquire the "feel" for these distance vs time recordings anew. Moreover, analyzing the numerous data contained in one streak record requires some effort, at least when this is done manually. The computer techniques are a considerable help in this, especially when serial measurements

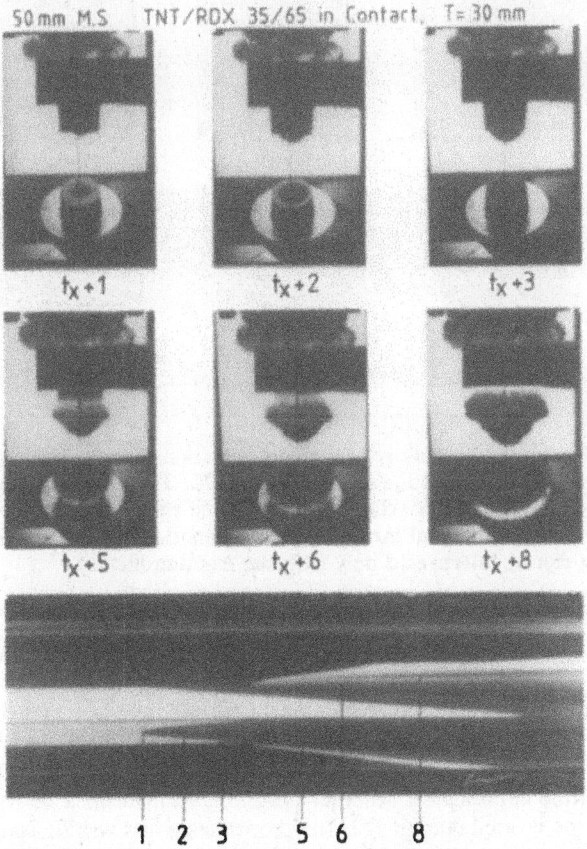

Fig. 72 Frames and streak record of the reaction behavior of a 30-mm-long acceptor charge, loaded by a shaped charge jet after perforation of a 50-mm-thick barrier.[47]

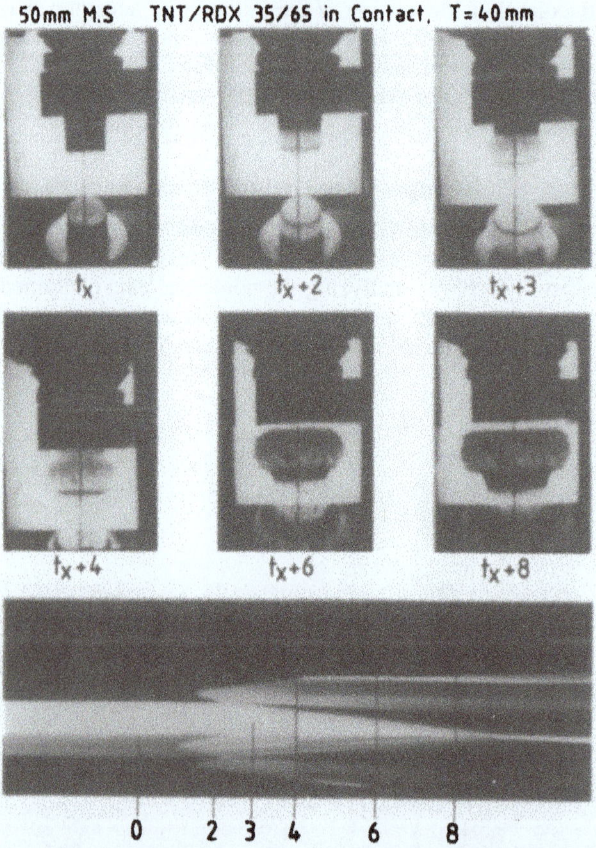

Fig. 73 Frames and streak record of the initiation of a 40-mm-long acceptor charge; the test setup was the same as in Fig. 72. The analysis of the streak record (bottom) shows the axial breakthrough of the detonation front, buildup distance in the radial breakthrough of the detonation wave, and axial expansion velocity; it can be interpreted only with the framing picture.[47]

or frequently recurring recordings or more complex recordings are to be analyzed.

The streak records are digitized by means of a scanner and a given program P1, which through a dialog terminal interfaces with the CPU, and the data are stored in memory (see Fig. 74). These data are, in general, only the information on the space and time coordinates. Another program, P2, can then be used to convert the data to the desired quantities at the proper scale and writing speed, and either to print them as figures at the line printer to present them at a graphic terminal and to make a hardcopy, or to draw them as curves on a plotter. The rather voluminous amount of data from a streak picture, such as that of a multistreak or an OSST record, can in this way be stored conveniently and quickly, and

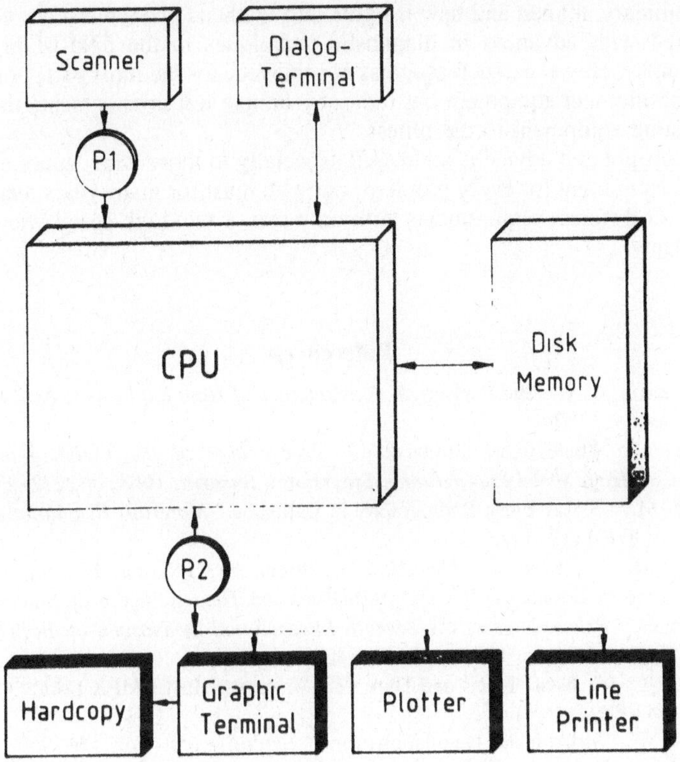

Fig. 74 Computers are extremely helpful in the quantitative analysis of the large amount of data and information from streak records that are normally not easily interpreted.

the required quantities can readily be obtained in the form of diagrams. Thus, computers and their peripherals can very much facilitate the analysis and the understanding of streak records.

8. Recommendations

In this overview chapter it is possible only to highlight the essential and to present the typical warheads and instrumentation in the field of high-speed photography.

It is absolutely impossible to base the chapter on the historical developments in order to lead to a better understanding of what has been achieved to date, or to give a detailed account of all possibilities with regard to measurement equipment and techniques. The list of references alone suggests that this is beyond the scope of this chapter. Certainly, the *Proceedings* of all the Congresses on High-Speed Photography and Photonics[48] and of the Symposia (International) on Detonation[49] are the main source for a literature search.

In summary, refined and new insights into warhead phenomena are very often correlated with advances in diagnostic techniques in the field of high-speed photography. However, such progress is achieved, for the most part, not through new measurement equipment but rather by unique test arrangements that utilize the existing equipment to the fullest.

This insight and advice is addressed especially to those who cannot afford the optimal equipment for every problem, but who must for many years meet a great number of different requirements with only one or two high-speed photographic instruments.

References

[1]Johannson, C. H., and Persson, P. A., *Detonics of High Explosives,* Academie Publishing, London, 1970.

[2]Held, M., "Selektives, Stroboskop," *Proceedings of the Eighth International Congress on High Speed Photography*, Stockholm, Sweden, 1968, pp. 232–236.

[3]Held, M., "X-Ray Flash Radiography in Ballistics," *Materials Evaluation*, Vol. 43, 1985, pp. 1104–1110, 1123.

[4]Dafelmair, X., Klee, C., Nowicki, G., Toma, K., Schmidt, H., and Wolf, A., "Investigation of Detonation Wave Perturbations and Their Influence on Shaped Charge Performance," *Proceedings of the Seventh International Symposium on Ballistics*, The Hague, The Netherlands, 1983, pp. 535–540.

[5]Mader, C. L., Neal, T. R., and Dick, R. D., "LASL PHERMEX Data," University of California Press, 1980.

[6]Held, M., "Struktur der Detonationsfront," *Explosivstoffe*, Vol. 17, 1969, pp. 241–249.

[7]Held, M., "Struktur der Detonationsfront in Gas-Luft-Gemischen," Proceedings of the Fourth Annual Internation Conference of the JCT 1973, Karlsrúhe, Germany, pp. 279–297.

[8]Held, M., "Method of Measuring the Fine Structure of Detonation Fronts in Solid Explosives," *Astronautica Acta*, Vol. 17, 1972, pp. 599–607.

[9]Held, M., and Nikowitsch, P., "Lupentechnik—Increase in Resolution of Optical Setups using Image-Converter Cameras," *Proceedings of the Eleventh International Congress on High Speed Photography*, London, UK, 1974, pp. 202–206.

[10]Held, M., and Nikowitsch, P., "Objektivtechnik," *Proceedings of the Eleventh International Congress on High Speed Photography*, 1974, pp. 65–71.

[11]Sultanoff, M., "Some Philosophical Aspects of High Speed Photographic Instrumentation," *Proceedings of the Fifth International Congress on High Speed Photography*, Washington, DC, 1962, pp. 411–416.

[12]Held, M., "Techniques in Detonation Diagnostics," *Proceedings of the Second International AVL Symposium*, Graz, Austria, 1979, C4: 1–14.

[13]Cordin (Corporation), 2230 South 3270 West, Salt Lake City, UT.

[14]Landre, J. K., "Time Resolution of Streak-Camera Equipped with Rotating Mirrors of Beryllium," *The Revue of Scientific Instruments*, Vol. 35, 1964, pp. 796–801.

[15]Landre, J. K., "Effect on Time Resolution of Ambient Gas around Rotating Mirrors," application newsletter from Beckman & Whitley, Vol. 4, No. 1, 1967, of SMPTE Vol. 75, 1966.

[16]Hadland Photonics Limited (Corporation), Newhouse Laboratories, Newhouse Road, Bovingdon, Herts HP3 OEL, UK.

[17]Held, M., "Streak Technique as a Diagnosis Method in Detonics," *Proceedings of the First International Symposium on Ballistics*, Orlando, FL, 1974, IV, pp. 177–210.

[18]Liddiard, T. P., and Drimmer, B. E., "Smear-Camera-Techniques," *Proceedings of the Fifth International Congress on High Speed Photography*, Washington, DC, 1962, pp. 322–326.

[19]Griffith, N., and Groocock, J. M., "The Burning to Detonation of Solid Explosives," *Journal of the Chemical Society*, 1960, pp. 4154–4162.

[20]Held, M., "Method of Observing Processes in the Interior of Explosives," *Proceedings of the Tenth International Congress on High Speed Photography*, Nice, France, 1972, pp. 286–291.

[21]Held, M., and Nikowitsch, P., "Neues Verfahren zur Messung der Ubertragungseigenschaften bei detonativen Zundketten," Proceedings of the Eleventh Annual International Conferences of ICT, Karlsrúhe, Germany, 1980, pp. 425–450.

[22]Held, M., and Nikowitsch, P., "Multi Stage Fibre Optic Probes," *Proceedings of Sixteenth International Congress on High Speed Photography*, Nice France, 1984, pp. 619–623.

[23]Held, M., "Der Radial-Test," Proceedings of the Nineteenth Annual International Converence of JCT, Karlsniche, Germany, 1978, V 42: 1–42.

[24]Held, M. "Corner-Turning Distance and Detonation Radius," *Propellants, Explosives, Pyrotechnics*, Vol. 14, 1989, pp. 153–161.

[25]Held, M., "The Effect of the Strength of Initiation upon the Corner Effect," *Proceedings of the International Symposium on Intense Dynamic Loading and its Effects*, *Science Press*, Beijing, China, 1986, pp. 136–143,

[26]Held, M., "Orthogonal Multi-Streak Recording Technique," *Proceedings of the Ninth International Congress on High Speed Photography*, Denver, CO, 1970, pp. 126–129.

[27]Held, M., and Nikowitsch, P., "Time-Correlated Multi-Streak," *Proceedings of the Thirteenth International Congress on High Speed Photography*, Tokyo, Japan, 1978, pp. 210–213.

[28]Deal, W. E., "Measurement of the Reflected Shock Hugoniot and Isentrope for Explosive Reaction Products," *The Physics of Fluids*, Vol. 1, 1958, pp. 523–527.

[29]Held, M., "Streak Records in Wedge Technique with High Resolution," *Proceedings of the Seventeenth International Congress on High Speed Photography and Photonics*, Pretoria, South Africa, *SPIE* Vol. 674, 1986, pp. 65–69.

[30]Blankenagel, H., and Ludwig, D., "Streak-Recordings of the Dynamic Forming of Metal Liners," *Proceedings of the Thirteenth International Congress on High Speed Photography*, 1978, pp. 802–804.

[31]Held, M., and Nikowitsch, P., "Tests fur die Bestimmung der detonativen Leistungseigenschaften von kunststoffgebundenen Sprengladungen," *Proceedings of the Thirteenth Annual International Conference of JCT*, Karlsrúhe, Germany 1982, pp. 403–425.

[32]Held, M., and Nikowitsch, P., "Multiprofile-Streak-Technique," *Proceedings of the Fifteenth International Congress on High Speed Photography*, San Diego, CA, 1982, Vol. 348, pp. 939–947.

[33]Held, M., and Nikowitsch, P., "Inverse Multi-Streak-Technique," *Proceedings of the Seventh International Symposium on International Detonation*, Anapolis, MD, 1981, pp. 751–758.

[34]Keen, S. M., "Application on the 'Slit-Aperture' Camera for Observation and Data Acquisition in Projectile Studies," *Proceedings of the Fifth International Congress on High Speed Photography*, Washington, DC, 1962, pp. 375–378.

[35]Biele, J., Erprobungsstelle 91 der Bundeswehr, 4470 Meppen, Germany.

[36]Held, M., Ludwig D., and Nikowitsch, P., "SST-Synchro-Streak-Technique," *Proceedings of the Twelveth International Congress on High Speed Photography*, Toronto, Canada, 1977, pp. 234–237.

[37]Held, M., and Nikowitsch, P., "Comparative Investigation of Flash X-Ray Pictures and SST-Records for the Diagnostic Evaluation of Shaped Charge Jets," *Proceedings of the Sixteenth International Congress on High Speed Photography*, Strasburg, France, 1984, pp. 614–618.

[38]Held, M., "The Orthogonal Synchro-Streak-Technique as a Diagnostic Tool Particularly for Shaped Charge Jets," *Propellants, Explosives, Pyrotechnics*, Vol. 11, 1986, pp. 170–175.

[39]Held, M., "Superposition of Shock Waves and Reaction Waves for the Initiation of High Explosive Charge," *Proceedings of Eighth International Symposium on Detonation*, Albuquerque, New Mexico, 1985, pp. 330–336.

[40]Held, M., and Nikowitsch, P., "Open View Streak Technique," *Proceedings of the Seventeenth International Congress on High Speed Photography and Photonics*, 1986, pp. 55–64.

[41]Sultanoff, M., "Some Philosophical Aspects of High Speed Photographic Instrumentation," *Proceedings of the Fifth International Congress on High Speed Photography*, 1962, pp. 41–416.

[42]Igel, E. A., "Analysis of Streak-Camera Measurement Errors Costs by Misorientation of the Camera Slit with Recording Direction", SANDIA Report 82–0816 (UC-37), 1982

[43]Held, M., "Tangent-Streak-Slit-Technique," *Proceedings of XIIX International Congress on High Speed Photography and Photomies*, Xiang, China, 1988, SPIE Vol. 1032, pp. 850–854

[44]Held, M., "Experiments of Initiation of Covered but Unconfined High Explosive Charges by means of Shaped Charge Jets," *Propellants, Explosives, Pyrotechnics*, Vol. 12, 1987, pp. 35–40.

[45]Held, M., "Experiments of Initiation of Covered, but Unconfined HE Charges under Different Test Conditions by Shaped Charge Jets," *Propellants, Explosives, Pyrotechnics*, Vol. 12, 1987, pp. 97–100.

[46]Held, M., "Discussion of the Experimental Findings from the Initiation of Covered with Shaped Charge Jets," *Propellant, Explosives, Pyrotechnics*, Vol. 12, 1987, pp. 167–174.

[47]Held, M., "Analysis of Shaped Charge Jet Induced Reaction of High Explosives," *Propellants, Explosives, Pyrotechnics*, Vol. 14, 1989, pp. 245–249.

[48]Proceedings of International Congresses on High Speed Photography and Photonics: 1) Washington, U.S., 1952, Published HSP, 5, SMPTE, N.Y., 1954; 2) Paris, France, 1954, Published Dunod, Paris, 1956; 3) London, England, 1956, Published Butterworth, London, 1975; 4) Koln, Germany, 1958, Published Verlag Helwich, Darmstadt, 1959; 5) Washington, U.S., 1960, Published SMPTE, N.Y., 1962; 6) The Hague, Netherlands, 1962, Published Tjeenk Willink & Zoon, Haarlan, 1963; 7) Zurich, Switzerland, 1965, Published Verlag Helwich, Darmstadt, 1967; 8) Stockholm, Sweden, Published Almqvist

& Wiksell, Stockholm, 1968; 9) Denver, U.S., 1970, Published SMPTE, New York, 1971; 10) Nice, France, 1972, Published ANRT., Paris, 1973; 11) London, England, 1974, Published Chapman & Hall, London, 1975; 12) Toronto, Canada, 1976, Published SPIE, Washington, 1977; 13) Tokyo, Japan, 1978, Published J. Soc. Prec. Eng. Tokyo, Distribution, SPIE, 1979; 14) Moscow, UDSSR, 1980, Published; 15) San Diego, U.S., 1982, Published SPIE Vol. 348, 1983; 16) Strasburg, France, 1984, Published SPIE Vol. 491, 1984; 17) Pretoria, South Africa, 1986, Published SPIE Vol.674, 1986; 18) Xiang, China, 1988, Published SPIE, Vol. 1032, 1988; 19) Cambridge, England, SPIE, Vol. 1358, 1990.

[49]Proceedings of Symposium 1–9 (Int.) on Detonation, Copies available: Naval Surface Warfare Center, White Oak, Silver Spring, MD.

Chapter 12
Part A

Special Topics in Test Methodology: Flash Radiography

Karl H. Mueller*
Los Alamos National Laboratory, Los Alamos, New Mexico 87545

1. Introduction

FLASH radiography has traditionally been used in ballistics and warhead work where it is necessary to observe the formation or flight of a projectile inside an obscuring cloud of explosive products. The x rays can easily penetrate the high-explosive (HE) gases while being totally absorbed by the warhead. The resultant image is a "stop-action" shadowgraph. Typical x-ray generators for this work are in the 100 keV to 2 MeV range with a pulse duration of ~ 50 ns and a dose at 1 m of ~ 10 to 100 mrad. Analysis of these images is generally limited to defining the position/edge of an object.

With higher-energy and higher-dose flash x-ray machines, a whole new variety of experiments (and analyses) becomes possible. Instead of x-ray shadowgraphs, we now have radiographs in which all areas of the scene are *penetrated* by the x rays. These x-ray generators are characterized by 1) energies between ~ 3 and 30 MeV, 2) doses from ~ 20 to 500 rad at 1 m, and 3) a pulse duration of ~ 30 to 200 ns. The higher x-ray energy implies a smaller mass attenuation coefficient (μ) for high Z elements and, when combined with the higher dose, adds up to more effective penetration of thick objects.

Flash radiographs obtained with penetrating x rays can be subjected to quantitative analysis. With proper calibration, the optical density in the radiograph can be converted to x-ray intensity. The physics governing the attenuation of x rays passing through material is well known and can be incorporated into a mathematical model that can be fit to the data by the method of least squares. Alternatively, one can take the data and, under certain assumptions, reconstruct the material-density cross section using tomography.

*Staff Member.

2. Mechanics

The factors influencing the choice of radiographic technique are too numerous to be dealt with here. Chapters 5 through 9 in Halmshaw[1] give a good introduction to the subject, whereas Herz[2] gives a more detailed treatment of photographic recording of x rays. Some general recommendations will be given.

Direct-recording film should be used whenever possible; despite its lower sensitivity (compared to light-screened films), the conversion from optical density to x-ray intensity is so straightforward and reliable it is always the preferred choice. For direct recording film

$$D_{\text{Total}} = D_0 + I \times S \tag{1}$$

where

$$S = \text{Unit } \Delta D/I_1 \tag{2}$$

and

$$D_0 = D_{\text{Fog}} + D_{\text{Scatter}} \tag{3}$$

Here D_{Total} is the total optical density, I is x-ray intensity, and S is the sensitivity factor for a given film type. I_1 is that x-ray dose that produces a unit change in optical density for a given film. Values for D_{Fog} and D_{Scatter} combine to form the background optical density (D_o) corresponding to zero direct x-ray intensity. The x-ray intensity at any point on the film is given by

$$I = (D_{\text{Total}} - D_o)/S \tag{4}$$

static calibration gives S and D_{Fog}, so we need only determine D_{Scatter} in the dynamic experiment to convert the film optical density values into the more meaningful x-ray intensity.

Although the x-ray sources used in flash x-ray work are bremsstrahlung radiators, we have found that they can be characterized by an average effective monochromatic energy. Thus, the attenuation of the x rays is characterized by a monochromatic mass attenuation coefficient μ_{eff}; that is,

$$I = I_{\text{Max}} e^{-\int \mu_{\text{eff}}\, dl} \tag{5}$$

where I is the x-ray intensity at any point in the radiograph, I_{Max} is the intensity of the unattenuated x-ray flux and $\int \mu_{\text{eff}}\, dl$ is integrated along a path between the x-ray source and the point of interest on the radiograph.

Combining Eq. (4) and Eq. (5) above gives

$$-\int \mu_{\text{eff}}\, dl = \ell n \frac{1}{S} \frac{D_{\text{Tot}} - D_0}{I_{\text{Max}}} \tag{6}$$

for any single given material μ_{eff}/ρ_0 = constant and is known at a given effective energy. Therefore,

$$\int \mu_{\text{eff}}\, dl = \frac{\mu_{\text{eff}}}{\rho_o} \int \rho\, dl \tag{7}$$

Without any other assumptions, the measured optical density on the film can be converted to an areal mass, or mass for each pixel, in the digitized image, that is

$$\text{Areal mass} = \int \rho\, dl \tag{8}$$

or

$$\text{Mass} = (\text{pixel area}) \int \rho\, dl \tag{9}$$

The foregoing discussion forms the basis of a very useful form of image analysis. The change from optical density to areal mass is a powerful nonlinear transformation revealing many details that are difficult to see in the original radiograph. The technique can be applied to radiographs of any object regardless of shape and can yield accurate values (5% or better) for the total mass of the object. Figure 1 shows the effect of converting the optical densities to areal mass. The details at the center of the object are more easily seen in the areal mass image.

3. Fitting to a Model

By assuming axial symmetry in the object, one can construct a mathematical model consisting of multiple cylindrical shells where the path length integrals can be explicitly evaluated in terms of the shell radii and the μ value within a shell. This model can then be least-squares fit to the radiographic data to determine the shell boundaries. An important aspect of the fitting procedure is the modeling of the radiographic system blur. This includes both a finite-x-ray-source dimension term and a contribution from blur within the detector package. The fitted radii are the locations of sharp boundaries imaged with the inherently low-resolution radiography that is a consequence of the high beam energy. The statistical uncertainty in the fitted radii may be as small as 0.1 mm even though the blur may extend over 1 or 2 mm.

4. Detector Considerations

The range of x-ray intensity that can be recorded by the detector package (film) is an important experimental consideration. Because we are dealing with exponential attenuation of the x rays, it is easy to have a situation where the flux is reduced to 0.01% of its original value. The rms noise level in a single direct-recording x-ray film is on the order of 0.1% of the maximum optical

Fig. 1 A logarithmic display [$\ell n\,(D_{\mathrm{Total}} - D_o)$] of the upper image in Fig. 3 of the section by Hanson. The gray scale in this figure is proportional to the areal mass. Detail over the whole object is more apparent here than in the original radiograph. The tomographic reconstruction [Fig. 3 of Hanson (bottom)] is yet another improvement except for the central axis.

density that can be read from the film. The effective dynamic range of the detector can be increased by one of two means. Films of different sensitivity can be incorporated into a pack, and the information from those areas exposed to less than the saturation optical density can be combined into a composite image. Alternately, many films of the same sensitivity can be exposed in a registered stack and the digitized images averaged together to reduce the rms noise level in the composite image. A combination of the two methods has proved most useful. Depending on the experiment, film packs may consist of from 4 to 20 films.

Figures 2a and 2b demonstrate the benefits of averaging 20 films. In this experiment the lead screens associated with each film absorb ~ 1.5% of the incident x-ray flux so the last film sees about 2/3 of the flux incident on the

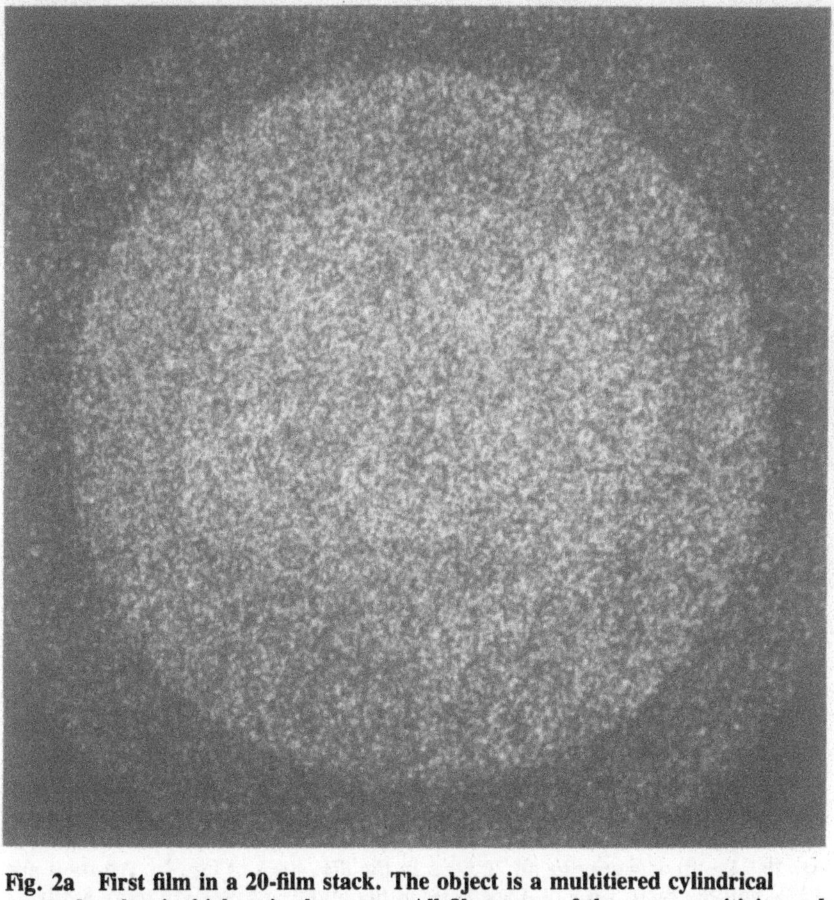

Fig. 2a First film in a 20-film stack. The object is a multitiered cylindrical stepwedge that is thickest in the center. All films were of the same sensitivity and were exposed to 24-MeV bremsstrahlung radiation.

first film. Lead screens are placed between all films to ensure statistical independence of the images, and this is borne out by the $\sim \sqrt{20}$ reduction of noise in Fig. 2b.

5. Examples

Some recent uses of flash radiography at Los Alamos have involved the formation of jets and explosively formed penetrators (EFP).[3,4] In-bore ballistics and armor penetration have also been investigated to establish the requirements for precision experiments. A few examples from the EFP experiments will serve to demonstrate the concepts outlined in the previous discussion.

Figure 3 shows a copper EFP in flight 532 μs after detonation of the device. The radiograph was made using a 2.6-MeV diode-type machine that produced

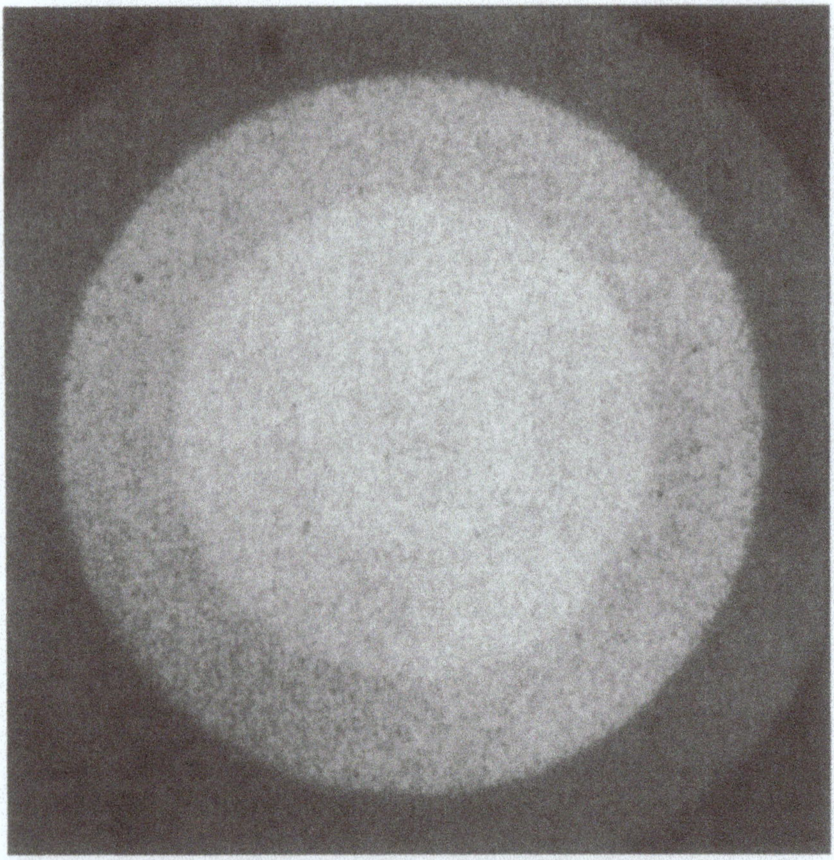

Fig. 2b Result of averaging all 20 films in the film stack of Fig. 2a. All steps are visible and the three 1.6-mm-diam holes in the outermost ring are more easily seen. A similar hole in the center of the object remains invisible. The dose at the center of the object (at the film plane) is only 50 μrad.

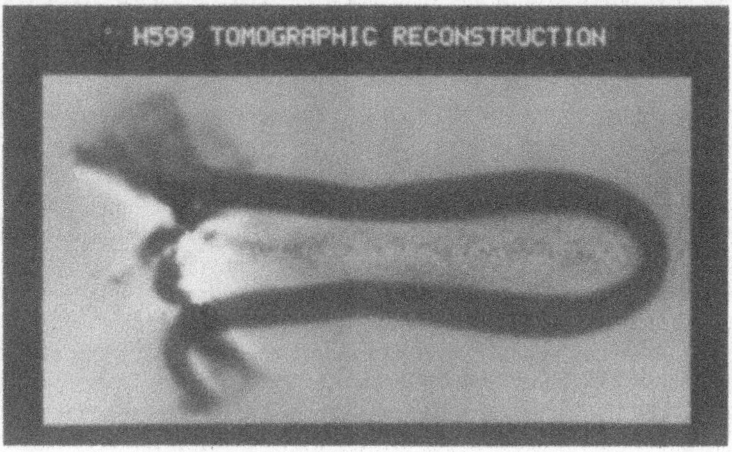

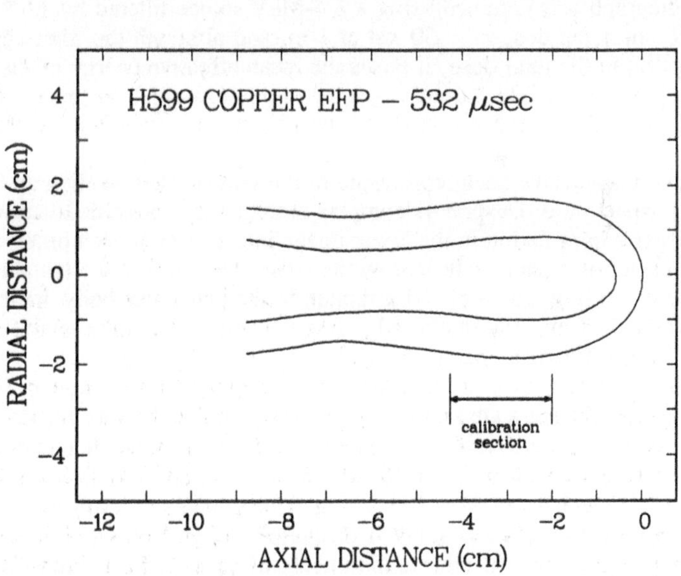

Fig. 3 Results of two types of analysis applied to a radiograph of an explosively formed penetrator. The upper image is a tomographic reconstruction; the lower figure shows the inner and outer contours of the object obtained by fitting a mathematical model to the radiographic data.

8 to 10 rad at 1 m in a 50-ns pulse. Four films were combined to form a composite image, which was then 1) tomographically reconstructed and 2) fit to a mathematical model to determine the inner and outer surface locations. Both of these procedures require the assumption of axial symmetry; for all but the tail region, this assumption is satisfactory. The tomographic reconstruction has produced a result in the tail region, but it is not physically real, and μ values are negative in some areas. For this area, there is clearly no axial symmetry, and the procedure is not valid. The fitting procedure fails to converge, and thus no results are shown for the tail.

For the symmetrical section of the device there is good convergence, and the radii are determined to ± 0.2 mm. The calibration section shown in the fitted data was used to determine $\mu_{\text{effective}}$ and $D_o + D_{\text{Scatter}}$ by assuming that the material was at normal density and the central region was air. This is necessary so that the number of free parameters in the least squares fit may be reduced. The mean effective energy of the x-ray source was ~ 0.6 MeV.

An asymmetrical tantalum penetrator radiographed by two x-ray sources (at right angles to each other and orthogonal to the penetrator motion) is shown in Fig. 4. The top view was obtained using a 450-keV (peak) portable x-ray unit that produces ~ 10 mrad at 1 m in a 30-ns pulse; there is essentially no penetration of the body of the device because of the low energy and dose. The lower radiograph was obtained using a 3.4-MeV source filtered by 12.5 mm of lead. The unfiltered dose is ~ 30 rad at 1 m, and although the filter absorbs a large fraction of the total dose, it raises the mean effective energy of the source and makes it more nearly monochromatic. In this example, the mean effective energy is 1.44 MeV and all areas of the object are penetrated by the x-ray beam.

The two images have been reproduced to the same scale revealing an approximate 2:1 aspect ratio. Despite its complex structure it is possible to convert the optical density information in the lower image into a mass at each pixel, and the resultant center of mass is indicated by the cross. The mass of the tantalum liner originally was 339 g; 262 ± 20 g remain in the penetrator body, indicating a substantial loss of material due to edge fragmentation. A similar analysis is not possible for the 450-keV image.

The final example is a static exposure of a copper test cylinder. Here all dimensions and the mass are known, and we can test the absolute accuracy of an analysis. Two films were combined and the results of tomographic reconstruction and edge fitting are shown in Fig. 5. The measured mass was 688.8 g and the reconstruction yielded a mass of 681 ± 20 g. This same mass result has also been obtained from the areal mass analysis discussed earlier. The fitted contours are uncertain to ± 0.2 mm. In other experiments using a 30-MeV bremsstrahlung source we have been able to determine surface positions to better than 0.1 mm.

In addition to the series of stepped-diameter holes in the right end of the cylinder, there are two smaller holes in the left end. The diameters are 1.6 mm and 0.8 mm, but unlike the larger holes, are too small to be seen clearly in the reconstruction or fit. Figure 6 shows the effect of transforming the optical density data to areal mass. The 1.6-mm-diam hole is clearly visible, and under optimum viewing conditions the 0.8-mm-diam hole is just perceptible.

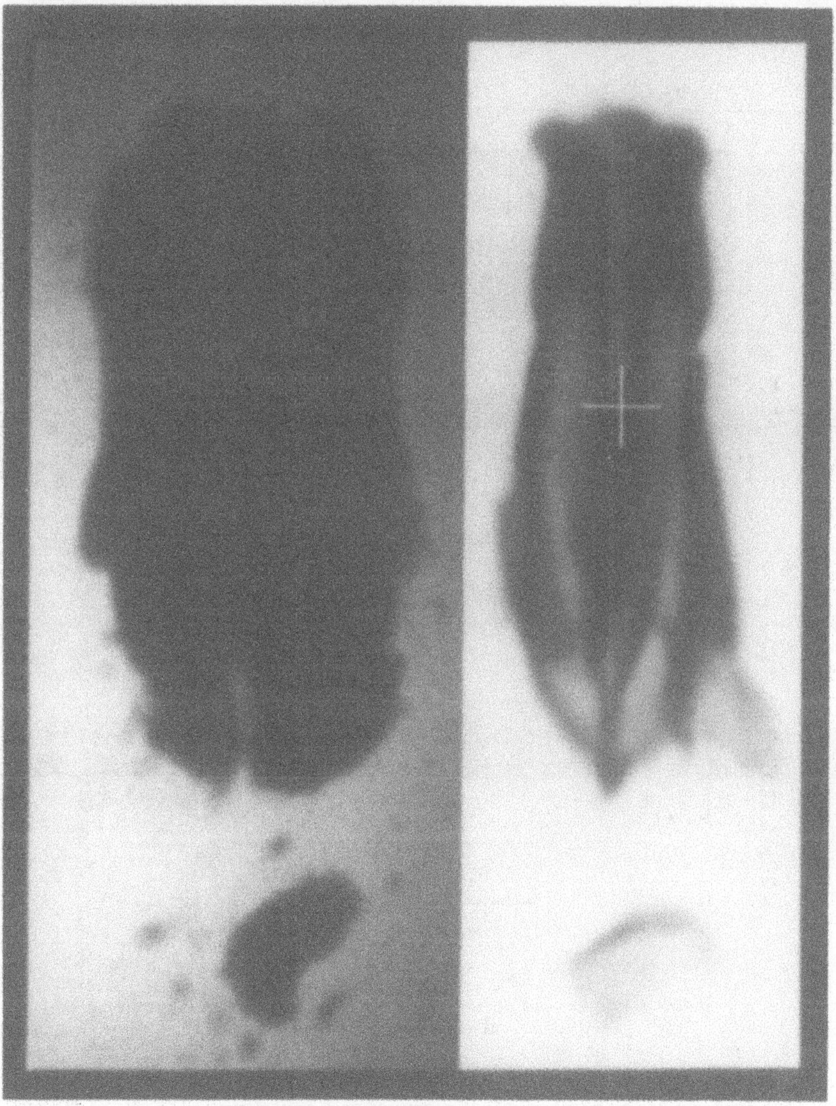

Fig. 4 Two views of an asymmetric explosively formed tantalum penetrator. The views are at a right angle to each other and perpendicular to the penetrator velocity. The left image is from a 450-keV (peak) source; the right is from a 3.4-MeV (peak) machine. Despite the greater thickness of the object, the 3.4-MeV source fully penetrates the object. The cross indicates the position of the center of mass of the main body.

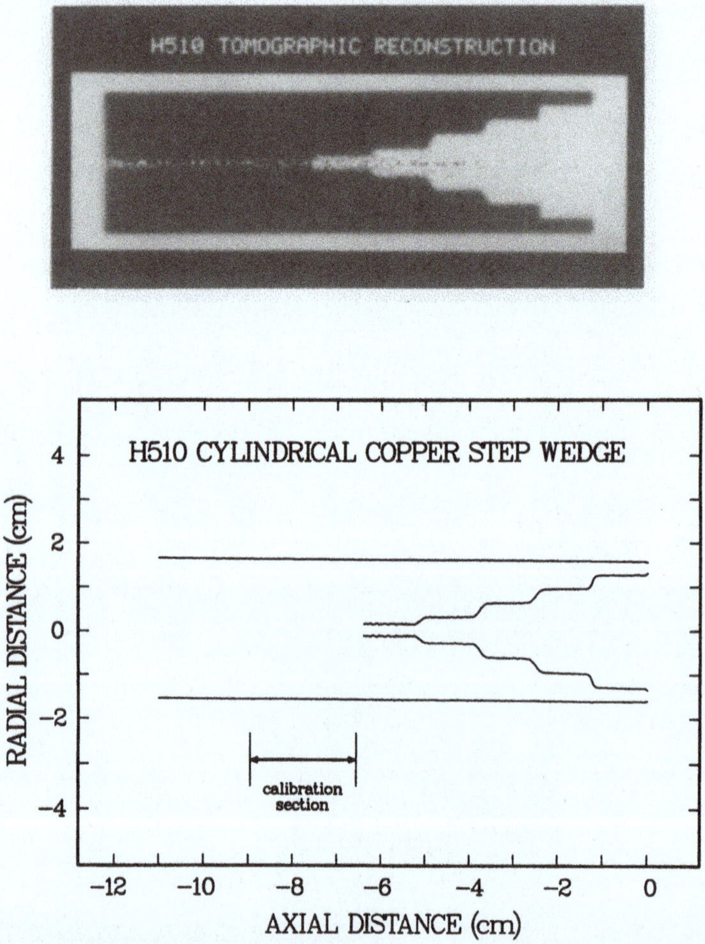

Fig. 5 Analysis of the radiograph of a copper test cylinder used in the series of explosively formed penetrator experiments.

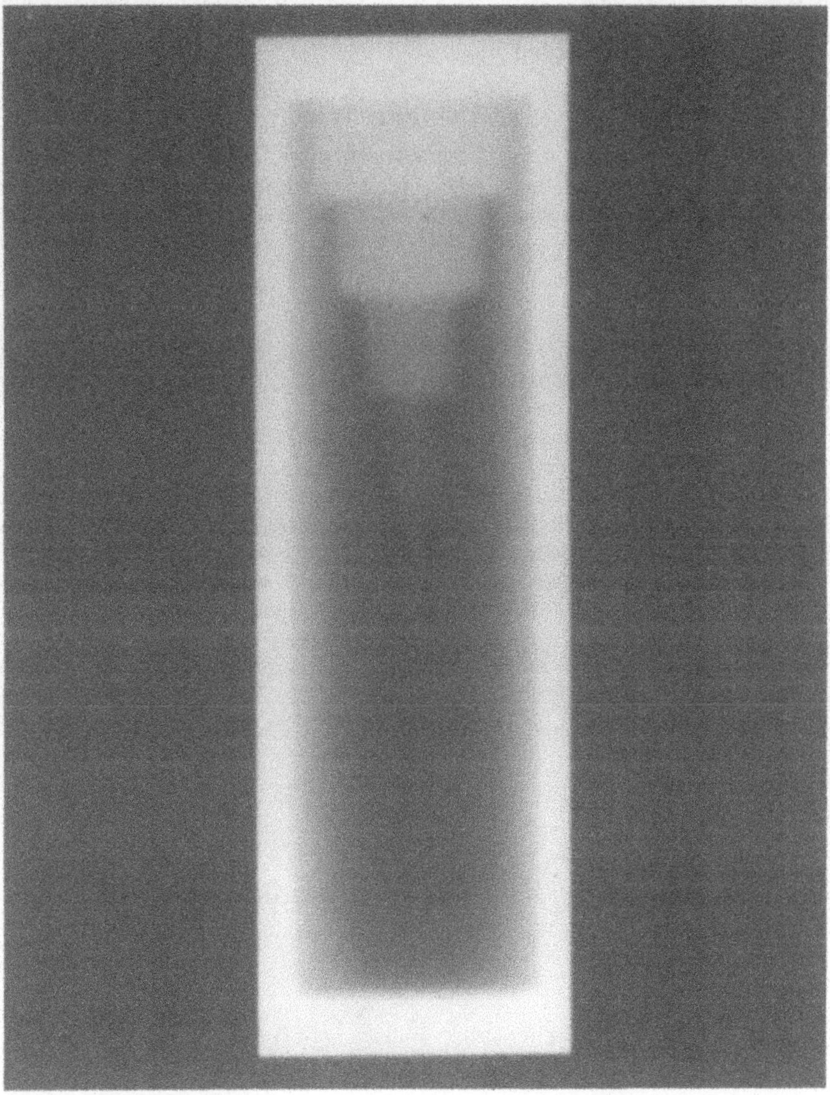

Fig. 6 A display of $\ell n(D_{\mathrm{Total}} - D_o)$ from the radiograph of Fig. 5. The absence of noise along the axis reveals the smaller holes at the left end of the object.

6. Conclusions

Flash radiography can yield quantitative results in properly designed experiments. It is possible to "weigh" objects with an accuracy of a few percent and to measure surface positions to ± 0.1 mm. This may not seem remarkable until one realizes the object may be engulfed in an explosive fireball and moving at several kilometers per second.

Acknowledgments

I would like to thank Larry Schwalbe for providing several of the figures and for his many contributions to the series of explosively formed penetrator experiments. Mike Christian was responsible for the actual firing of these shots. Allen Mathews, Ken Hanson, and Rollin Whitman provided software that made all of this work possible.

References

[1]Halmshaw, R., Ed., "Physics of Industrial Radiology," American Elsevier Publishing Co., Inc., New York, 1966, pp. 97–263.

[2]Herz, R. H., "The Photographic Action of Ionizing Radiations," Wiley-Interscience, New York, 1969, pp. 87–278.

[3]Christian, J. M., Mueller, K. H., Schwalbe, L. A., Draxler, V., and Williamson, F., "High-Energy Flash Radiography of Two Axisymmetric Copper Explosively Formed Penetrators," *International Advances in Nondestructive Testing*, Vol. 13, pp. 361–391.

[4]Schwalbe, L. A., Mueller, K. H., Jr., Christian, J. M., and DiBona, P. J., "In-Flight Radiographic Study of Two Nonaxisymmetric, Explosively Formed Penetrators," Los Alamos National Laboratory, Rept. LA-11131, March 1988.

Chapter 12
Part B

Special Topics in Test Methodology: Tomographic Reconstruction of Axially Symmetric Objects from a Single Dynamic Radiograph

K. M. Hanson*
Los Alamos National Laboratory, Los Alamos, New Mexico 87545

1. Introduction

IT is possible to reconstruct the cross section of a three-dimensional, axially symmetric object from a single radiograph, as depicted in Fig. 1. This method of analysis has been applied to many problems in many different fields in the past, including several recent applications to static and dynamic testing at Los Alamos.[1–3] In industrial radiography, where many objects have nearly circular symmetry, such an approach offers significant benefits as an image analysis tool. These benefits include improved delineation of material boundaries, enhanced display of deviations from axial symmetry (defects, for example), and estimation of the radial dependence of the attenuation coefficients of the materials. It is often possible to observe in the reconstruction subtle features of the object that are unobservable in the original radiograph. The improvements brought about by the tomographic method are due to the effective removal of overlying material from the radiograph, which allows the reconstruction to be displayed with an increase in contrast.

2. Method

First, consider the reconstruction from its projection of the radial dependence of a two-dimensional object that possesses circular symmetry. The projection of a function $f(x, y)$ of the two spatial variables x and y is defined as

$$p(R) = \int_L f(x, y)\, \mathrm{d}s \qquad (1)$$

*Staff member.

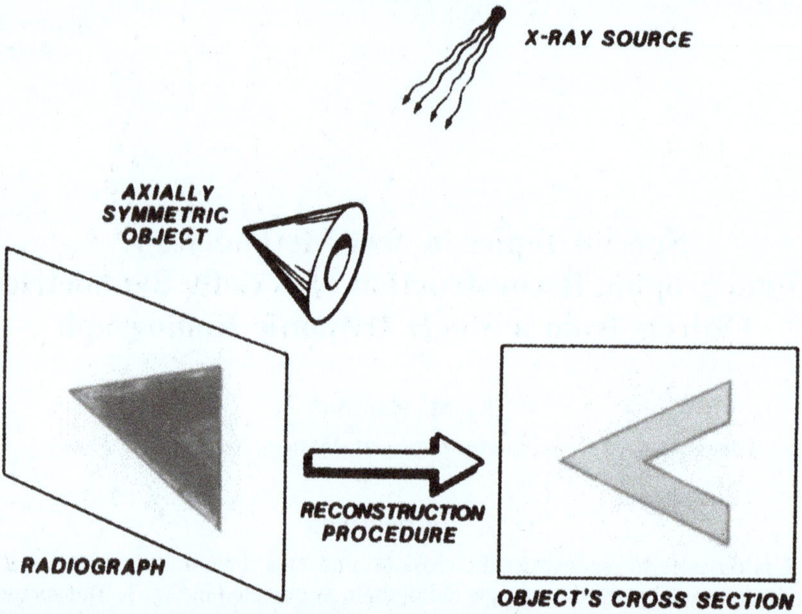

Fig. 1 Overview of the tomographic method. An object with axial symmetry is radiographed with the radiographic axis perpendicular to the axis of symmetry. The reconstruction procedure transforms the radiograph into an estimate of the cross section of the object.

where the line integral is taken along the straight line L, which is specified by its distance to the origin R. This projection is given only as a function of R because $f(x, y)$ is assumed to have circular symmetry about the origin. The solution for f given a known projection $p(R)$ was provided by Abel in 1826[4] and has been elaborated upon even in recent years.[5–8] An alternative approach avoids the mathematical difficulties that arise when one tries to evaluate the Abel inversion. In this approach, the density function $f(x, y)$ of each cross section of an object with axial symmetry is considered to be composed of a series of nested annuli, as depicted in Fig. 2. The outermost annulus, the projection of which is shown in Fig. 2, is the only annulus that contributes to the ends of the projection interval. Thus, it is possible to determine the amplitude of the outermost annulus from these projection values and consequently its contributions to each inner projection sample. The same analysis is applied to the next annulus, and so on until the center is reached. This procedure is analogous to peeling an onion by removing one layer at a time.

A minor variation on this model of nested annuli is useful in situations in which the data on the right side of the projection differ from the data on the left side. Each annulus is replaced by two annuli with tapered density profiles. The density of one of these annuli varies linearly from zero on the right side to full density on the left side; the density of the other varies linearly in the

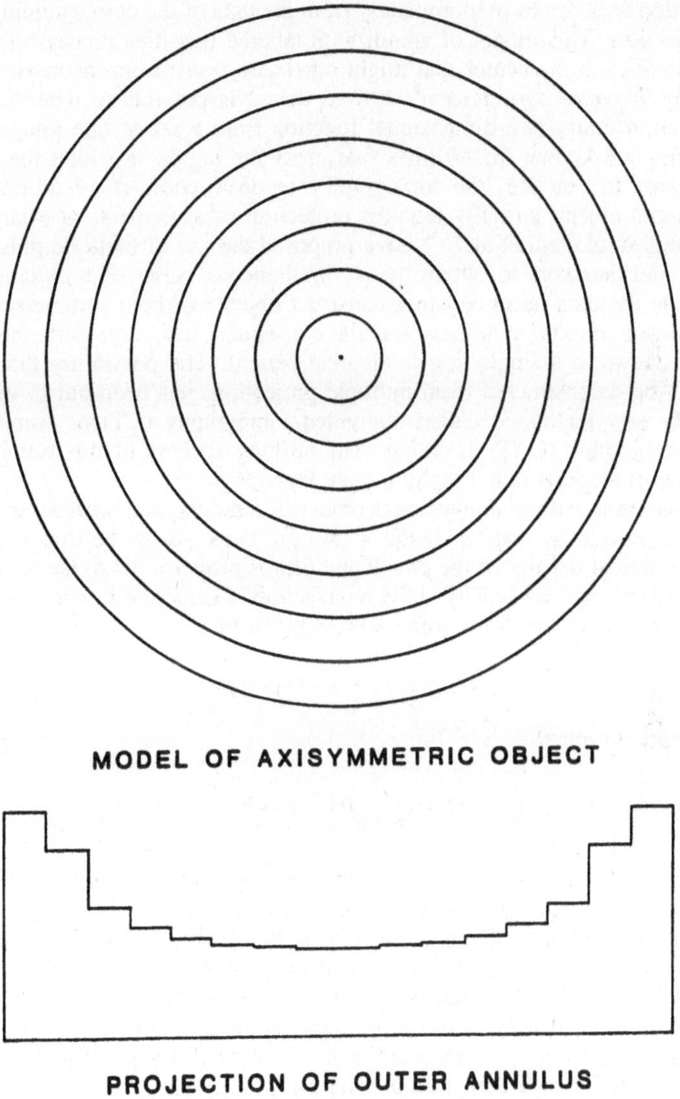

Fig. 2 Model of a two-dimensional object with circular symmetry. The model is composed of a series of annuli whose widths correspond to the width of a pixel in the digitized radiograph. Below is the projection of the outer annulus.

opposite direction. The amplitudes of the first are determined by the projection data on the left side. The amplitudes of the second are determined by the data on the right. Then the major contribution to the reconstruction on each side of the rotation axis comes predominately from the data of the corresponding side of the projection. This model of annuli with tapered densities successfully avoids discontinuities at the center that might otherwise result from inconsistent data.

Nearly 70 years ago, Radon[9] showed that it is possible to determine completely an arbitrary two-dimensional function from a set of line integrals, provided they are known for all lines that cross the region in which the function is nonzero. In practice, the tomographic reconstruction of an arbitrary two-dimensional object normally requires projection measurements at a large number of angles. Zoltani et al.[10,11] have proposed the use of multiple pulsed-x-ray sources and detectors to obtain many simultaneous views of a dynamic event to provide the data necessary to reconstruct objects without symmetry. Clearly the existence of axial symmetry and the consequent reduction in the number of required views to a single one is of great benefit. The possibility that a function may be reconstructed from multiple projections has been put to use in the relatively new technique called computed tomography (CT) or computerized axial tomographic (CAT) scanning. The striking success of this revolutionary technique in medical radiography is now legendary.

Suppose that a three-dimensional object is radiographed with an essentially monoenergetic x-ray source. If the x rays are detected by the direct-recording film, the optical density of the developed film is proportional to the x-ray intensity. The film's optical density at the intersection of each line L, which originates at the point source, with the film plane is given by

$$D_L = D_0 + D_1 \exp(-p_L) \tag{2}$$

where the path length

$$p_L = \int_L \mu(x, y, z) ds \tag{3}$$

is the line integral along line L of the object's linear attenuation coefficient distribution $\mu(x, y, z)$, evaluated at the energy of the x rays used. In Eq. (2), D_0 is the background density of the film at that point, which includes the fog value of the film and any contribution from scattered radiation; D_1 is the net density (above D_0) that would be obtained in the absence of the object. Thus, the projections needed as input for tomographic reconstruction, Eq. (1), may be obtained from radiographic measurements. Inversion of Eq. (2) yields the path length in terms of measured film density D_L:

$$p_L = -\ell n \frac{(D_L - D_0)}{D_1} \tag{4}$$

From the previous discussion, it can be seen that tomographic reconstruction based upon projections obtained in this way will yield the linear attenuation-coefficient distribution of the object.

Now consider a three-dimensional object that possesses axial or rotational symmetry. In any plane that is perpendicular to the symmetry axis, the object's cross section has circular symmetry. If the object is radiographed with a set of parallel x-ray beams, equivalent to a point source placed infinitely far from the object, all of which lie in such a plane, the path lengths obtained from applying Eq. (4) will correspond precisely to those considered in the two-dimensional example above. Thus, the radial distribution of that cross section can be reconstructed using the same technique. A line-by-line analysis carried out in this way on each line of the radiograph that is perpendicular to the symmetry axis yields the complete radial distribution of μ for the object. This line-by-line analysis is legitimate because the contributions to each line come from one, and only one, cross-sectional plane through the object, and each cross section has circular symmetry. In practice, the x-ray source cannot be placed infinitely far from the object. Indeed, often it must be placed very close. In this case, the line-by-line analysis is not strictly valid because the contributions to each line of the radiograph, with the possible exception of a single line, come from neighboring cross sections. Owing to its simplicity, the line-by-line method is employed in the present analysis even though it may not be exactly correct. The consequences of this approximation are not severe, as shown in Ref. 3.

3. Results

A test object with axial symmetry was constructed to illustrate the usefulness of the tomographic analysis technique. The object is a 70-mm-long right circular steel cylinder, 120 mm in diameter, with a 45-deg cone removed from one end to a depth of 40 mm. Four 2-mm-square grooves were machined on the flat face as well as on the inside of the cone at radii of 10, 20, 30, and 40 mm. A 2-mm-diam, 2-mm-deep hole was also drilled into the center of the flat face. This object was radiographed with a 120-Ci Co^{60} source for 15 min using Kodak AA film placed in close contact with 0.25-mm-thick lead screens, front and back. The axis of symmetry of the cylinder was perpendicular to the radiographic axis. The source diameter was about 1.5 mm. In order to closely approximate the parallel beam geometry that is assumed by the reconstruction procedure, the source was placed 4.5 m from the object, which was in turn 0.5 m from the film. The film was scanned on a Perkin-Elmer/PDS-2020 microdensitometer to produce a digital image that was 425 by 275 pixels in size with a pixel spacing of 0.4 mm. Figure 3a displays the resulting digitized radiograph. The diffuse densities in this radiograph range from about 0.35 to 1.8, the latter corresponding to the total density produced by the unattenuated x-ray beam.

When the tomographic analysis outlined above is performed on each line of the radiograph, Fig. 3a, and the reconstructed radial profiles are combined to form an image, Fig. 3b results. The entire computation time to obtain this result was 9 min of CPU time on a DEC VAX-11/785. This reconstruction closely resembles the cross section of the object itself. The value of the reconstruction in the region of the steel is very nearly the same as the tabulated value[12] of the linear attenuation coefficient for iron at 1.25 MeV, namely 0.042 mm^{-1}. The grooves, which were very difficult to observe in the original radiograph, are now easily seen. In the original radiograph, the grooves on the flat face

Fig. 3a Radiograph of an axially symmetric steel object taken with a Co^{60} source using AA film placed between two 0.25-mm-thick lead screens.

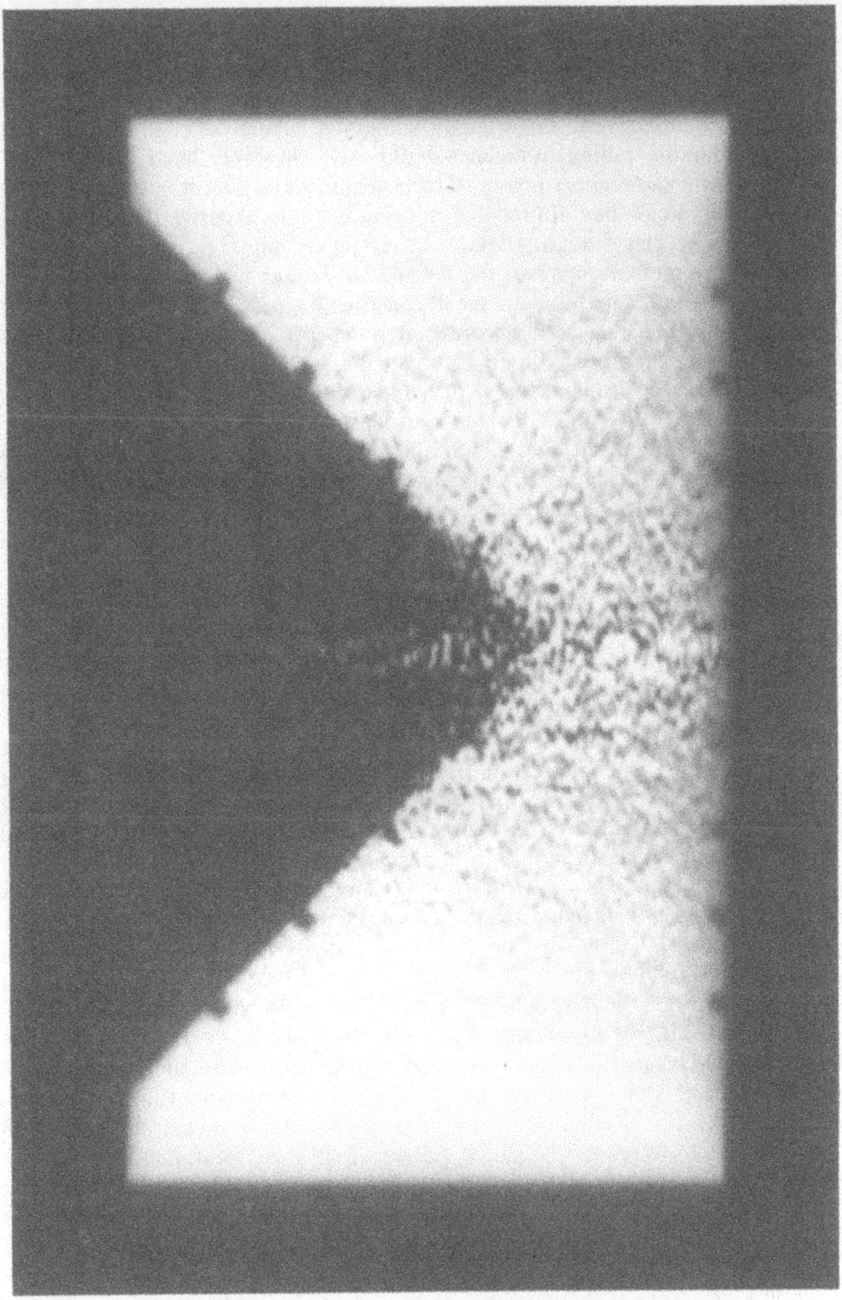

Fig. 3b Reconstruction of the cross section of the object derived from Fig. 3a. The 2- × 2-mm grooves machined in the inner cone and on the end face are much more readily visible. These grooves were nearly invisible in the original radiograph.

of the cylinder produced fluctuations in diffuse optical density ranging from 0.04 for the 10-mm-radius groove to 0.11 for the 40-mm one. These correspond to a range in contrast of luminance from 10% to 27%. Although a 10% contrast is readily observed under the best of viewing conditions,[13] it is essentially impossible to observe in the original radiograph, where the edge of the cylinder creates a rapidly falling luminance profile. On the other hand, this groove is easily seen in the reconstruction, which demonstrates that it is the noise in the original radiograph that ultimately limits one's ability to derive information from it. This is one of the major goals of image processing.

The severe enhancement of the noise near the axis of symmetry arises from a nearly singular condition in the reconstruction procedure there. This effect is due to the fact that as a ring of material with constant thickness decreases in diameter, it produces a smaller and smaller optical-density signal on the radiograph, making it increasingly difficult to observe in the presence of nearly constant film-density noise. The uncontrolled growth of noise in the reconstruction poses a problem in interpretation of the image. One method to control the noise in the reconstruction is nonlinear maximum a posteriori probability (MAP) restoration, introduced to image processing by Hunt.[14] The application of this iterative approach to the problem of reconstruction of axially symmetric objects has been reported elsewhere.[15,16]

Figures 4a and 4b show the result when the tomographic procedure is applied to the radiograph of an experimental, explosively formed copper penetrator. The void in the center of the penetrator is delineated much better in the reconstruction than in the radiograph. In dynamic radiography such as this, the reconstruction method provides a means to estimate the densities of the object in regions of mixed materials or to observe slight density fluctuations due to the presence of shock waves. The application of tomographic analysis to the radiographs of other test objects has shown that it greatly enhances the visualization of minor defects in the objects and that defects not possessing axial symmetry remain localized in the reconstruction.

Two technical points are worth making. The highly nonlinear behavior of Eq. (4) near $D = D_0$ makes the reconstruction there very sensitive to the choice of D_0. Therefore, D_0 must be known very accurately if serious systematic errors in the reconstruction are to be avoided. Slight variations in D_0 arising from variations in the scattered radiation field may be difficult to take into account.[3] Another difficulty in tomographic analysis is the assumption of a monochromatic source. Polychromatic x-ray sources are widely used. The attenuation law for these is not as simple as given in Eq. (2). The problem is exacerbated by the possible presence of many different kinds of material in industrial radiographic investigations. Whereas the slight polychromatic and beam-hardening effects present in medical CT scanners have been successfully eliminated,[17] the effects in industrial radiography can be much more significant and may be harder to cope with.

4. Conclusions

It has been shown that the extension of the Abel inversion, the reconstruction of a three-dimensional axially symmetric object from a single radiograph taken

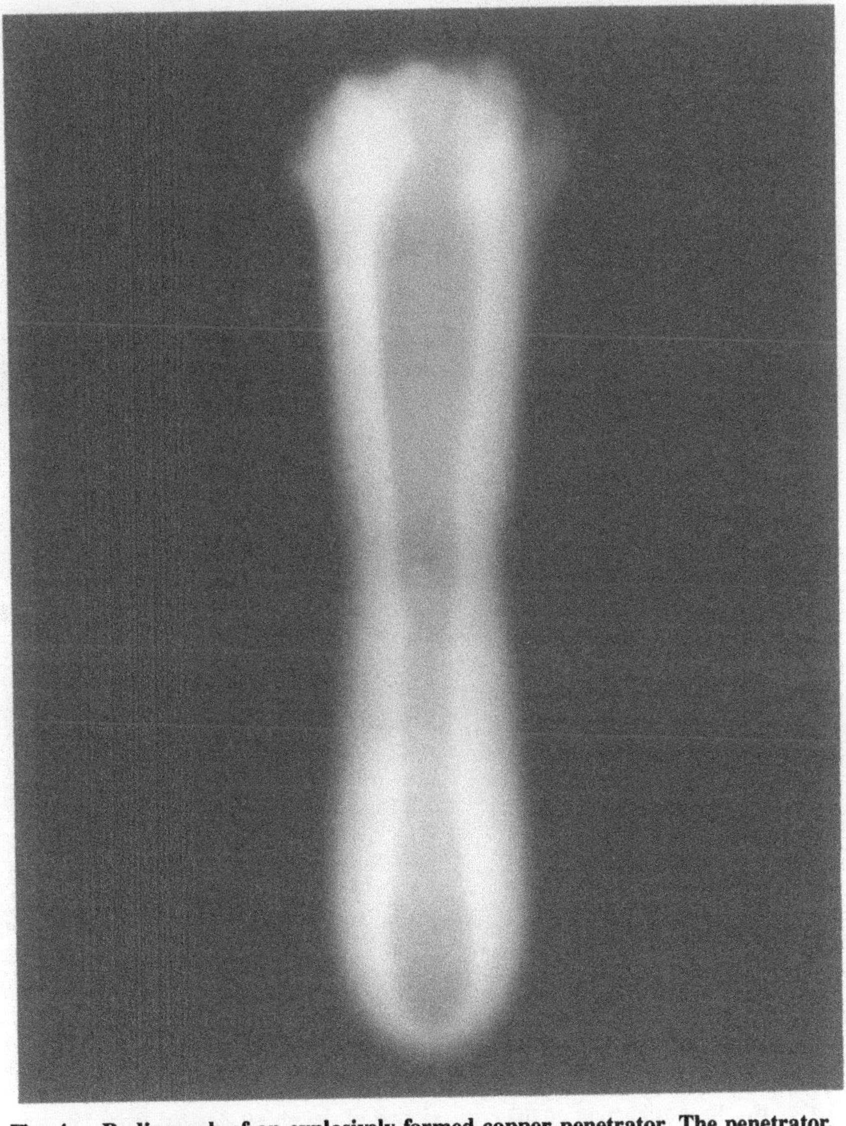

Fig. 4a Radiograph of an explosively formed copper penetrator. The penetrator, supplied by Honeywell Corp., was radiographed with a 50-ns-long pulse produced by a 0.8-MeV electron beam. Its velocity was 2.8 mm/μs.

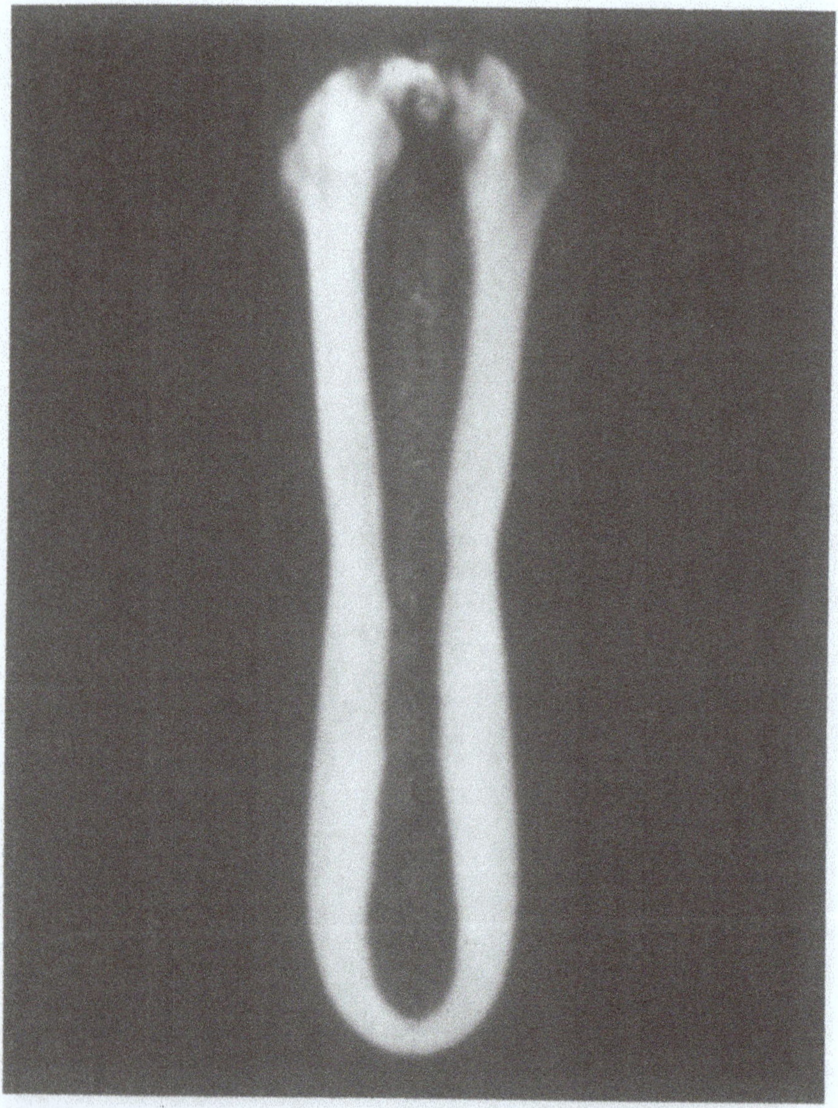

Fig. 4b Reconstruction of the cross section of the object derived from Fig. 4a. It shows the void in the middle of the penetrator much more clearly. Hence, the solution is easily realized by peeling the onion from the outside to the inside.

with the radiographic axis perpendicular to the symmetry axis, offers significant benefits as an image analysis tool. These benefits include improved delineation of material boundaries; enhanced display of minor deviations from axial symmetry, as produced by defects; and accurate estimation of the linear attenuation coefficients of the materials. Frequently, this technique allows observation of features in the object that are too subtle to be seen in the original radiograph.

Acknowledgments

The author would like to express his appreciation to George Brooks for providing the radiographs used in the static example (Fig. 3a). Rollin Whitman performed the reconstruction of this object. The reconstruction of the dynamic radiograph (Fig. 4b) was supplied by Karl Mueller, Larry Schwalbe, and Michael Christian. This work was suppoorted by the U.S. Department of Energy under contract W-7405-ENG-36.

References

[1]Fugelso, E., "Material Density Measurements from Dynamic Flash X-Ray Radiographs Using Axisymmetric Tomography," Los Alamos National Laboratory, Rept. LA-8785-MS, 1981.

[2]Berzins, G. J., McKinnon, G. C., Bates, R. H. T., and Lumpkin, A. H., "Application of Computed Tomography Techniques to Pinhole Images of Nuclear-Reactor Test Fuel Rods," *Nuclear Science and Engineering,* Vol. 77, No. 4, 1981, pp. 493–495.

[3]Hanson, K. M., "Tomographic Reconstruction of Axially Symmetric Objects from a Single Radiograph," *Proceedings of SPIE—The International Society for Optical Engineering,* Vol. 491, 1985, pp. 180–187.

[4]Abel, N. H., "Résolution d'un Probléme de Mecanique," *Journal für die Reine und Angew andte Mathematik,* Vol. 1, 1826, pp. 153–157.

[5]Porter, R. W., "Numerical Solution for Local Emission Coefficients in Axisymmetric Self-Absorbed Sources," *SIAM Review,* Vol. 6, No. 3, 1964, pp. 228–242.

[6]Corti, S., "The Problem of Abel Inversion in the Determination of Plasma Density from Phase Measurements," *Lettere Nuovo Cimento,* Vol. 29, No. 1, 1980, pp. 25–32.

[7]Deutsch, M., and Beniaminy, I., "Derivative-free Inversion of Abel's Integral Equation," *Applied Physics Letters,* Vol. 41, No. 1, 1982, pp. 27–28.

[8]Deutsch, M., and Beniaminy, I., "Inversion of Abel's Integral Equation for Experimental Data," *Applied Physics Letters,* Vol. 42, No. 3, 1983, pp. 237–239.

[9]Radon, J., "On the Determination of Functions for Their Integrals Along Certain Manifolds," *Berichte sächsische Akademieder Wissenschaften,* Vol. 29, 1917, pp. 262–277.

[10]Zoltani, C. K., White, K. J., and Kruger, R. P., "Results of Feasibility Study on Computer Assisted Tomography for Ballistic Applications," Ballistic Research Laboratory, Rept. ARBRL-TR-02513, U.S. Army Aberdeen Proving Ground, MD, 1983.

[11]Zoltani, C. K., White, K. J., and DiBianca, F. A., "The Flash X-Ray Computed Tomography Facility for Microsecond Events," *Review of Scientific Instruments,* Vol. 57, No. 4, 1986, pp. 602–611.

[12]Hubbell, J. H., Gimm, H. A., and Overbo, I., "Pair, Triplet, and Total Cross Sections for 1 MeV-100 GeV Photons in Elements Z= 1 to 100," *Journal of Physical and Chemical Reference Data,* Vol. 9, No. 4, 1980, pp. 1023–1147.

[13]Burgess, A. E., Humphrey, K., and Wagner, R. F., "Detection of Bars and Discs in Quantum Noise," *Proceedings of SPIE—The International Society for Optical Engineering,* Vol. 173, 1979, pp. 34–40.

[14]Hunt, B. R., "Bayesian Methods in Nonlinear Digital Image Restoration," *IEEE Transactions on Computers,* Vol. C-26, No. 3, 1977, pp. 219–229.

[15]Hanson, K. M., "A Bayesian Approach to Nonlinear Inversion: Abel Inversion from X-Ray Attenuation Data," *Maximum Entropy and Bayesian Methods in Applied Statistics,* edited by J. H. Justice, Cambridge University Press, Cambridge, England, 1986.

[16]Hanson, K. M., "A Bayesian Approach to Nonlinear Inversion: Abel Inversion from X-Ray Attenuation Data," *Transport Theory, Invariant Imbedding, and Integral Equations*, edited by P. Nelson, V. Faber, T. Manteuffel, D. L. Seth, and A. B. White Jr., Marcel Dekker, New York, 1989, pp. 363–378.

[17]Joseph, P. M., and Spital, R. D., "A Method for Correcting Bone Induced Artifacts in Computed Tomography Scanners," *Journal of Computer Assisted Tomography,* Vol. 2, 1978, pp. 100–108.

Chapter 12
Part C

Special Topics in Test Methodology: Electrical Pin Contactors

M. J. George*
Los Alamos National Laboratory, Los Alamos, New Mexico 87545

ELECTRICAL contactors can be used as switches to detect the arrival of a moving surface (Fig. 1a) or a shock wave within a material (Fig. 1b) at various points in space. With reasonable care, precisions of 10 ns in time and a few μm in space can be attained. Conversely, conducting foils (Fig. 1c) can be used to detect the earliest arrival of moving objects or debris at some point within defined boundaries of a surface. The success of this technique depends on the completion of a circuit when the pin or foil is contacted. Bare wire pins can be used if the circuit can be completed through the impacting material itself. If that is not possible, both circuit elements must be incorporated into each pin so that the impact will connect them. Typical examples are foil sandwiches and coaxial pins with or without a metal cap. If impact pressures are sufficiently high, then circuits can be closed through thin plastic insulators.

Figure 2 shows a typical electrical circuit for generating pin pulses. In general, $R1 \ll R2 \ll R3$. The rise time of the pulse is determined by the circuit elements associated with the pin itself and its connection to the rest of the circuit. When the pin closes, the capacitor C discharges through $R1$, and this part of the circuit determines the decay time of the pulse. If $C = 0.02\ \mu$F and $R1 = 50\ \Omega$, then the decay time will be 1 μs. Element $R2$ is a mixing resistor that allows the mixing of several pin signals or timing fiducials into the same cable to the recorder. If the cable has a characteristic impedance of 50 Ω and $R2 = 4000\ \Omega$, then the size of the pulse in the cable will be a factor of 80 less than the DC input voltage. Resistor $R3$ serves to isolate the power supply from prematurely shorted pins and makes the recharging time for the capacitor greater than the time scale of the experiment. A typical value for $R3$ might be 1 MΩ. Historically, various types of oscilloscopes have been used for recording pin pulses, but modern transient waveform digitizers also work quite well for this purpose. The DC charging

*Technical Staff Member.

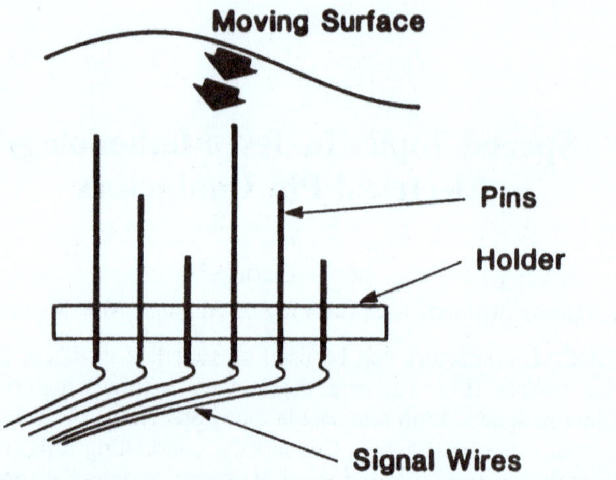

Fig. 1a Physical arrangement of pin switches to detect arrival of a free surface at specific points.

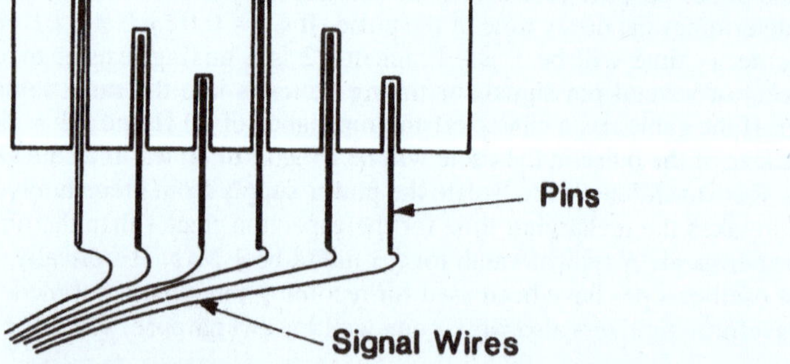

Fig. 1b Physical arrangement of pin switches to detect arrival of a shock front at points within solid material.

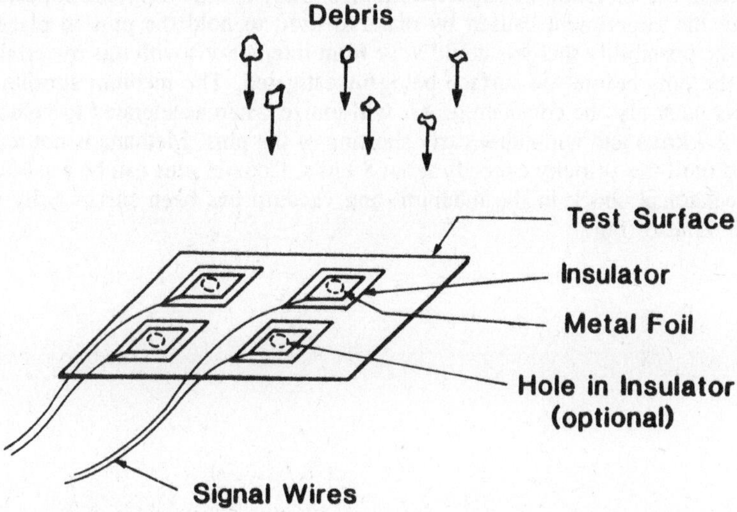

Fig. 1c Physical arrangement of foil switches to detect arrival of debris or other disturbances on elements of a surface.

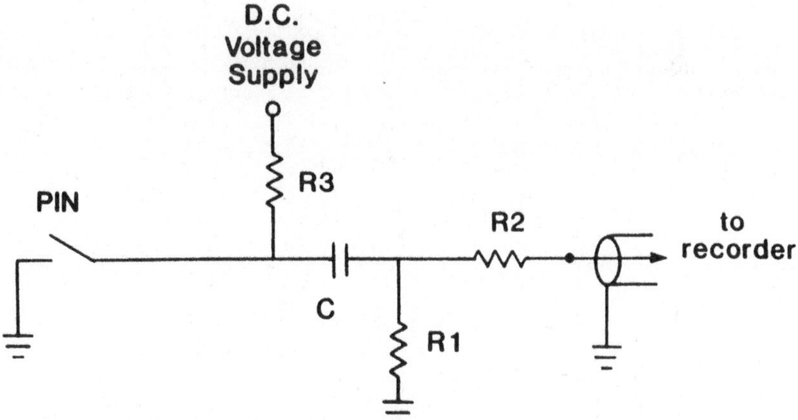

Fig. 2 Typical electrical circuit for generating pin pulses. Pulse polarity is opposite that of the DC supply voltage.

voltage must be compatible with the sensitivity of the recorder, considering the effects of other circuit elements, but should not be greater than about 300 V.

In the design of pin experiments, care should be taken to route signal wires away from the direction of expected impact. One should worry about perturbations to the experiment caused by material used to hold the pins in place, or about the possibility that jets could arise from interactions with this material and reach the pins before the surface being investigated. The medium surrounding the pins must also be considered. Air will ionize when accelerated to velocities above 2.7 km/s and will cause early shorting of the pins. Methane is not readily ionized until the velocity exceeds about 8 km/s. Coaxial pins can be sensitive to the mechanical shock in the medium, and vacuum has been successfully used for this kind of pin.

Chapter 12
Part D

Special Topics in Test Methodology: Piezoelectric Gauges

Robert R. Karpp*
Los Alamos National Laboratory, Los Alamos, New Mexico 87545

PIEZOELECTRIC pressure or stress gauges have been used to measure pressure-time histories in a variety of applications including explosively generated blast waves in air, propellant pressures in guns, combustion chamber pressures in rocket motors, and one-dimensional stress-wave profiles in solids. Many different crystals generate piezoelectric charges when strained; however, for pressure gauge applications, quartz has been used as the active element in most transducers because of its desirable electrical and mechanical properties. Crystalline quartz, either natural or reprocessed, is one of the most sensitive and stable of the piezoelectric materials.

The active element of the gauge is usually disk shaped and cut, in a particular orientation, from a single crystal. When the disk is axially compressed, an electrical potential difference develops between its two faces caused by deformation of the crystal lattice. If the crystal remains elastic, the developed voltage is proportional to the applied load, and this effect is the basis for one type of pressure gauge.

Piezoelectric pressure gauges can be constructed to operate in either of two distinct modes, the *charge mode* or the *current mode*.[1] A schematic diagram illustrating the operation of a gauge system in the charge mode is shown in Fig. 1. The dynamic pressure pulse $P(t)$ is applied to one surface of the active element through a protective buffer, and the output voltage $V(t)$, which is proportional to $P(t)$ within the operating range of the gauge, is measured across a capacitor. The response time of the gauge is limited by the time required for the active element to achieve stress equilibrium, typically hundreds of nanoseconds. Total recording times of several microseconds to many milliseconds are common, and pressures to nearly 1.0 GPa can be measured.

The result of using a charge-mode gauge to measure the blast pressure generated by the detonation of an explosive charge centrally mounted in a spherical

---

*Staff Member.

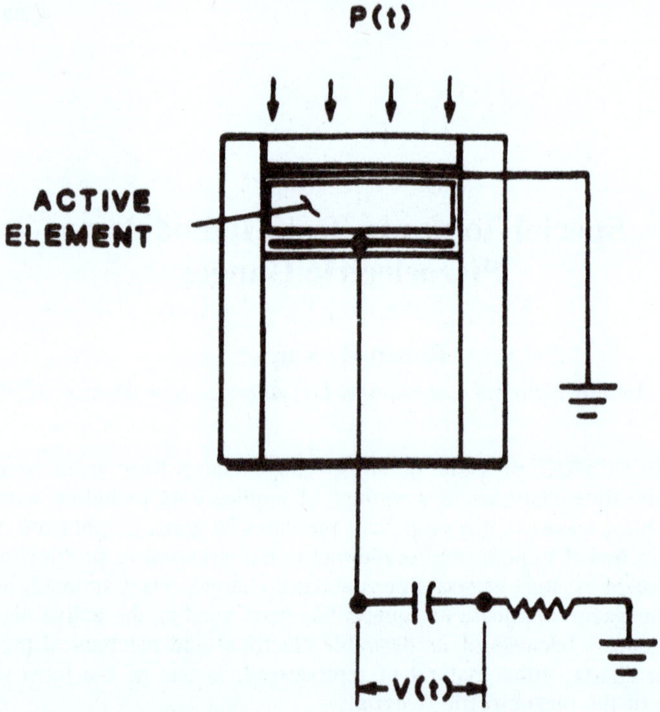

Fig. 1 Typical configuration of a charge-mode piezoelectric pressure gauge.

containment vessel is illustrated in Fig. 2. A model 109A PCB Piezotronics high-pressure transducer was mounted in the vessel wall, and the pressure applied to the wall was recorded for about a millisecond. The pressure record (upper) clearly shows the arrival of the main shock wave at the vessel wall. The main shock reflects from the vessel wall and is again reflected from the air-explosive interface, creating the double-peak structure in the primary pressure pulse. The pressure record (lower) also indicates that a substantial secondary pulse, caused by reflection of the main shock wave from the center of symmetry, arrives at the vessel wall about 200 μs after the initial pressure pulse.

As indicated in Fig. 3, many pressure gauges are constructed with two active elements connected mechanically in series but electrically in parallel to increase charge output. Also, some gauges contain built-in accelerometers that are used to subtract any signal component produced by the pressure gauge as a response to gauge acceleration. Acceleration compensation is important when the gauge is mounted in a structure that experiences significant motion.

A piezoelectric pressure gauge based upon the current mode of operation has been developed to measure pressure pulses on a submicrosecond time scale.[2] Response time of the gauge is limited only by the planarity of the wave and the frequency response of the recording system. A schematic diagram indicating the operation of the current-mode gauge is shown in Fig. 4. The pressure pulse

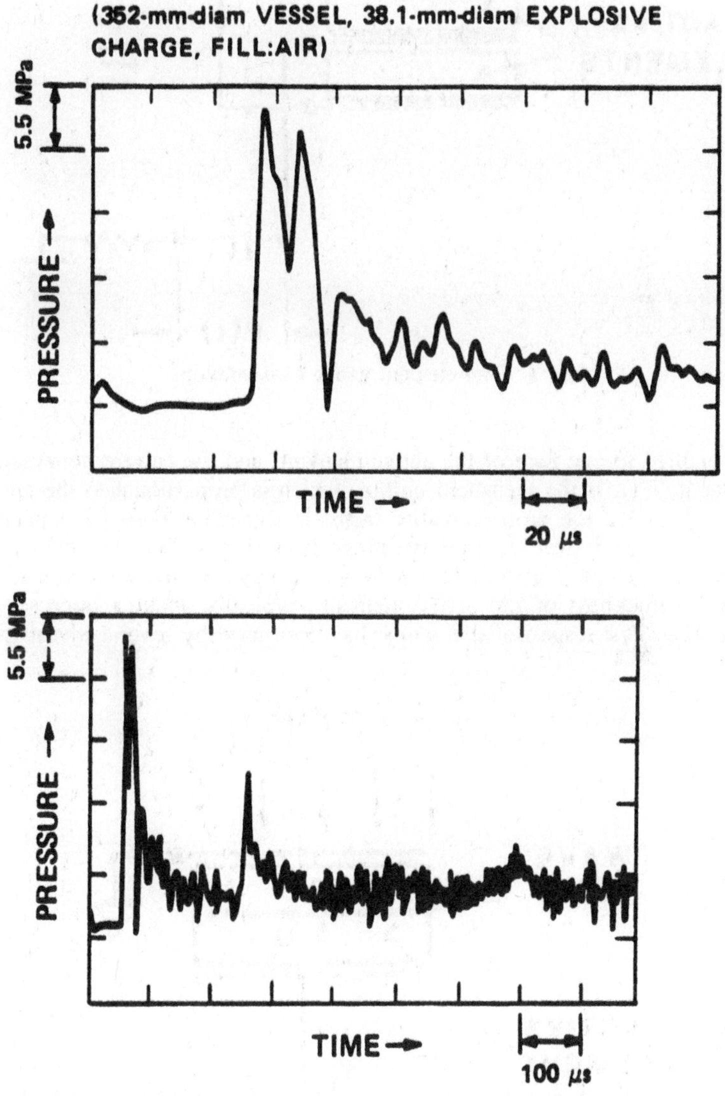

Fig. 2 Pressure pulse measured with a charge-mode piezoelectric pressure gauge.

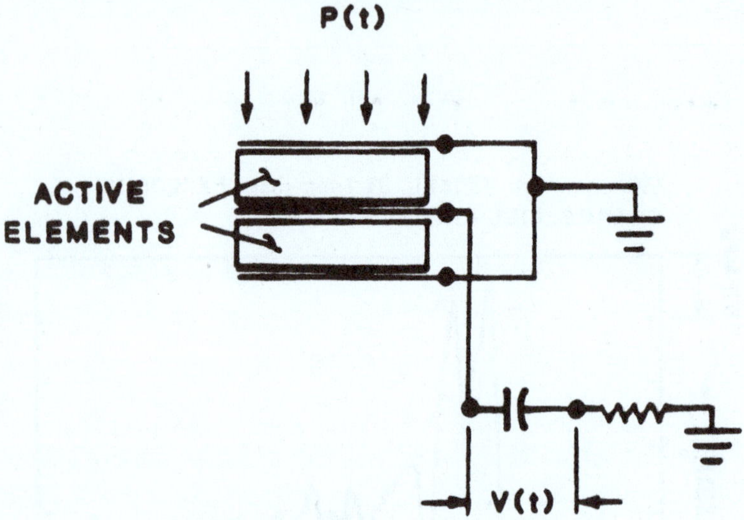

Fig. 3 Two-element gauge construction.

$P(t)$ is applied to one face of the active element, and the current generated by the element, $i(t)$, is the measured output, which is proportional to the applied pressure. Actually, the proportionality factor is somewhat pressure dependent; it varies about 21% over the pressure range from 0 to 4 GPa. In practice, the gauge recording time is usually limited to a time less than the wave transit time through the thickness of the active element—typically about a microsecond. Because of its fast response, this gauge has been used by several investigators

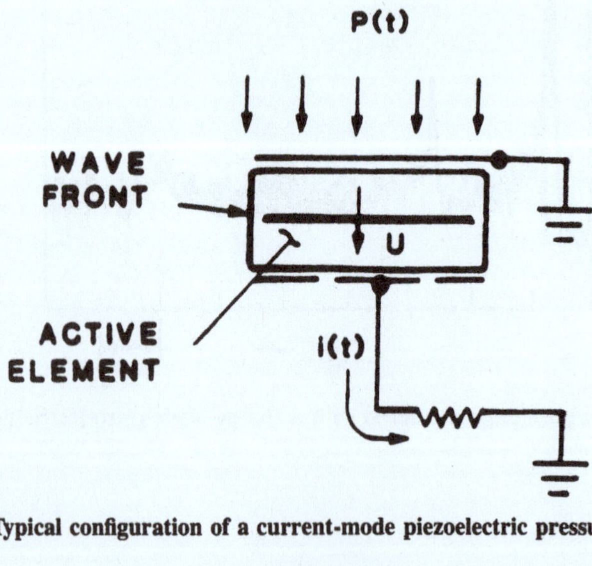

Fig. 4 Typical configuration of a current-mode piezoelectric pressure gauge.

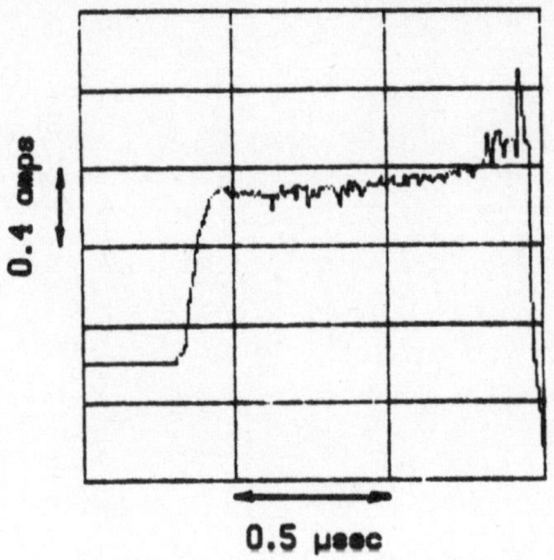

Fig. 5 Current output from a current-mode piezoelectric pressure gauge.

to measure the details of elastic-plastic wave profiles in solids. Figure 5 illustrates the current-time record obtained from a current-mode gauge, the Sandia Field Test Quartz Gage.[3] The gauge response was generated by the direct impact of an aluminum plate against the face of the gauge. This impact should produce nearly a step-shaped pressure pulse; however, the output current rises slowly after the initial jump. This current rise is a characteristic of the gauge, and some processing of the output current is required to determine the pressure history precisely. The sudden drop in current near the end of the record marks the arrival of the pressure wave at the rear surface of the gauge.

References

[1]Graham, R. A., "Lithium Niobate Stress Transducers," *Selected Papers on Piezoelectric and Impulsive "Pressure" Measurement,* edited by R. A. Graham and R. P. Reed, Sandia Rept. SAND78-1911, Dec. 1978, reprinted July 1985.

[2]Graham, R. A., Neilson, F. W., and Benedick, W. B., "Piezoelectric Current from Shock-Loaded Quartz—A Submicrosecond Stress Gauge," *Selected Papers on Piezoelectric and Impulsive "Pressure" Measurement,* edited by R. A. Graham and R. P. Reed, Sandia Rept. SAND78-1911, Dec. 1978, reprinted July 1985.

[3]Reed, R. P., "The Sandia Field Test Quartz Gage, Its Characteristics and Data Reduction," *Selected Papers on Piezoelectric and Impulsive "Pressure" Measurement,* edited by R. A. Graham and R. P. Reed, Sandia Rept. SAND78-1911, Dec. 1978, reprinted July 1985.

Chapter 12
Part E

Special Topics in Test Methodology: Piezoconductive Gauges

J. A. Morgan*
Los Alamos National Laboratory, Los Alamos, New Mexico 87545

ALMOST all materials change conductivity as the pressure (and temperature) changes. Materials that undergo phase changes frequently show large changes in conductivity. Indeed, static pressure scales were determined by abrupt changes in conductivity of selected metals, semimetals, and semiconductors. Minomura and Drickamer,[1] in a paper on silicon, germanium, and III-V compounds in 1961, reported conductivity changes of 4 to 5 orders of magnitude over very narrow pressure ranges. Shimomura et al. described the behavior of amorphous silicon and germanium in 1974[2] and amorphous InSb in 1976.[3] Elliot et al. published work on amorphous arsenic in 1977.[4] One can see from these and other papers that a set of pressure threshold probes could be made by using various materials as the active element and monitoring this sudden change in conductivity. Development of such a probe was begun several years ago at Los Alamos. Although the work has not been completed, some interesting findings will be reported here.

Figure 1 is a drawing of a piezoconductive pin. Two sizes of pins scaled from this drawing were tested. The earliest tests were made on a pin whose diameter was 6.35 mm. This size was chosen for fabrication convenience and to make transit times for shock waves through the piezoconductive material relatively long. In subsequent tests both 6.35-mm-diam and 1.0-mm-diam pins were used. Figure 2 is a schematic of the circuit used in these tests. The pin exterior was isolated from ground and used to signal arrival of a flat, stainless steel impact plate launched by a small (20-mm bore) propellant-fired gun. The velocity and tilt of the impact plate were measured independently.

Some of the results from these experiments are shown in Fig. 3. In experiment H-342 the pressure induced in the germanium element on the first shock-wave transit was 7.1 GPa. The shock wave reflected from the stainless steel center pin

---

*Staff Member.

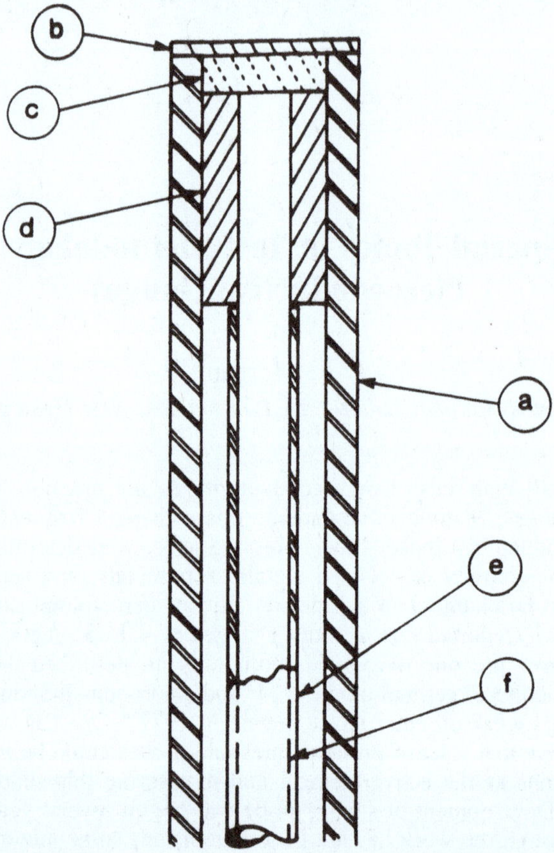

Fig. 1 General construction of a piezoconductive pin (PCP) switch. Element *a* is the stainless steel outer electrode (case); *b* is a stainless steel cap, a part of the outer electrode; *c* is the piezoconductive material (germanium in these experiments); *d* is a ceramic insulator; *e* is a plastic sleeve; and *f* is a stainless steel center conductor.

and put a stress of 9.6 GPa into the germanium element during the second transit. Since most of the circuit elements are known and the others can be reasonably estimated, it is possible to model the behavior of the pin. A model was used in which the germanium element changed from an insulator to a conductor at the shock front. Only calculations based on the germanium element becoming conductive on the reflected shock fit the experimental record. Minomura and Drickamer's static value for the transition is 12.0 to 12.5 GPa at 296-298 K. It is not uncommon, however, to observe lower transition pressures in shock experiments than in static experiments.

In experiment H-355 the pressure induced in the germanium by the shock was 14.5 GPa during the first transit and 22.1 GPa during the reflected shock transit.

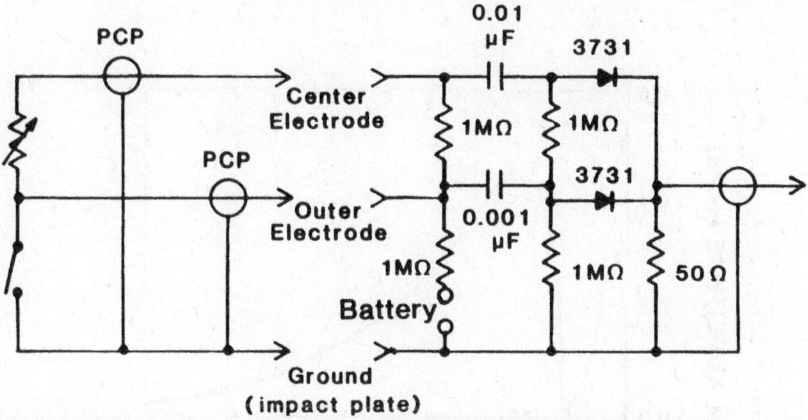

Fig. 2 Sketch of the circuit that was used for the PCP switch experiments. The left-hand side electrically represents the PCP switch. The simple switch (lower) represents the outer electrode (case) to ground circuit. When the impact plate (ground) contacts the outer electrode, a pulse is produced and the case is brought to ground potential. The variable resistor (upper) represents the change of resistance that occurs when a shock wave of sufficient amplitude passes through the germanium element. When this occurs, a second pulse is produced.

Calculations based on the germanium element becoming conductive during the first shock transit fit the observations satisfactorily.

These experiments demonstrate that a threshold pin can be used to determine if a certain pressure has been attained in an experiment. Additional work is needed to accurately determine the dynamic transition pressures for various materials considered as candidates for sensing elements. In addition, the design of the pin should be modified so that no wave is reflected back through the active element. When this is done, the presence of the second pulse would indicate that the pin threshold pressure had been achieved in the experiment; its absence would indicate that it had not.

References

[1]Minomura, S., and Drickamer, H. G., *Journal of Physics and Chemistry of Solids*, Vol. 23, 1961, p. 451.

[2]Shimomura, O., Minomura, S., Sakai, N., Asaumi, K., Tamura, K., Fukushima, J., and Endo, H., *Philosophical Magazine*, Vol. 29, 1974, p. 547.

[3]Shimomura, O., Asaumi, K., Sakai, N., and Minomura, S., *Philosophical Magazine*, Vol. 34, No. 5, 1976, p. 839.

[4]Elliot, S. R., Davis, E. A., and Pitt, G. D., *Solid State Communications*, Vol. 22, 1977, p. 481.

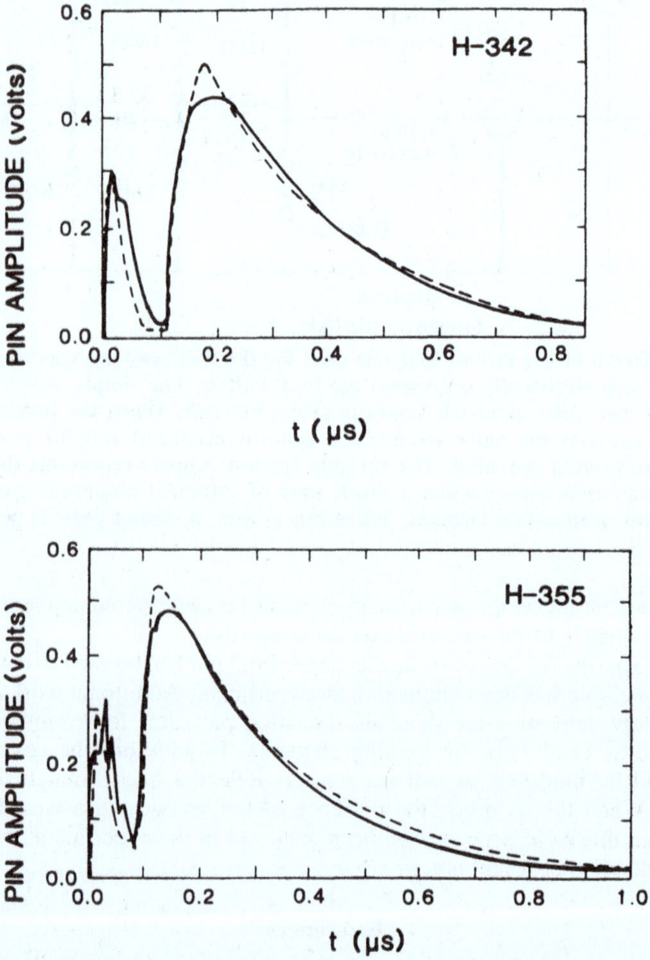

Fig. 3 Results for 1-mm-o.d. PCP switches. The smooth curves are the experimental data; the dashed curves are calculated. In the upper record, from experiment H-342, the germanium element becomes conducting only after the passage of a shock wave reflected from the stainless steel center electrode (the second shock wave). In the lower record, from experiment H-355, the germanium element becomes conducting on the first passage of a shock wave.

Chapter 12
Part F

Special Topics in Test Methodology: Manganin Gauges

G. C. Langner*
Los Alamos National Laboratory, Los Alamos, New Mexico 87545

GAUGES fabricated of manganin, a copper-manganese-nickel alloy, have been successfully used to measure dynamic stress in a variety of media. The gauges have found wide application in experiments because of their low strain sensitivity, relatively high stress sensitivity, low temperature coefficient of resistance, and a nearly linear pressure response to several hundred kilobars. The gauge configuration can be a single loop of wire or a grid of the desired size. Grids can be as small as several square millimeters in area and less than 0.01 mm thick, with all the grid lines in the same plane. Low-resistance ($<1\Omega$) or high-resistance ($> 50\Omega$) gauges can be made to suit the intended application. High-resistance gauges allow a higher voltage signal to be recorded for a given power level.

A manganin gauge operates on the principle that an applied stress causes a change in electrical resistance of the manganin. The resistance change is not measured directly but, instead, is determined by the voltage change that occurs across the arm of a Wheatstone bridge, as illustrated in Fig. 1. The gauge forms one arm of the bridge network; as many as three arms of the bridge can be gauges, if needed. Accurate measurements of gauge and lead-wire resistances are made before the bridge is balanced with the gauge in place. Resistance measurements are necessary to account for bridge nonlinearities and lead-wire contributions to the recorded voltage change. The excitation voltage applied to the bridge network is a signal of low constant amplitude ($\sim$ 2–4 Vdc) or a pulse of higher amplitude and short duration ($\sim$ 100 Vdc for $\sim$ 100 μs). If the excitation signal has excessive voltage, current, or duration, then thermal effects can alter the response of the gauge.

Gauge response is proportional to the stress on the active gauge area but is also influenced by impedance mismatches, excitation signal, electrical noise, thermal and radiation effects, and strain. In an ideal situation, the gauge responds only to stress. However, to determine the response to stress alone requires accurate

*Staff Member, Explosives Applications Group.

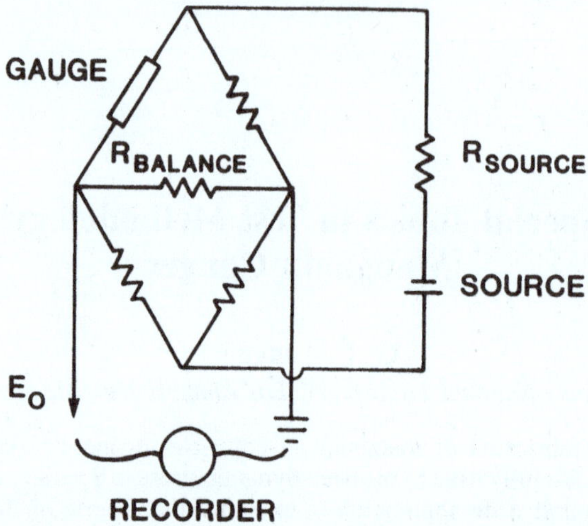

Fig. 1 Wheatstone bridge system with a manganin gauge forming one arm of the bridge.

gauge calibration and knowledge of the gauge factor, which is a relationship between stress (and strain) and resistance change. This relationship is nonlinear, and stress effects cannot be differentiated from strain effects.

An important step in limiting strain effects is to position the active area of the gauge perpendicular to the advancing stress wave. Compensation for remaining strain effects can be accomplished analytically, if a strain gauge largely unresponsive to stress can be positioned at the same location as the manganin gauge. For a grid configuration, this is accomplished by interlacing a high-stress-, low-strain-sensitivity grid of manganin with a high-strain-, low-stress-sensitivity grid of constantan, a copper-nickel alloy with constant temperature coefficient of resistance. Both gauges are then in the same plane and location. Because analytical corrections are not simple, careful design of the experiment helps to minimize strain.

Care must also be taken in connecting the gauge to power supplies and signal-recording equipment. Usually, electrical noise can be minimized by selecting proper cables and connectors and by paying attention to the details of shielding, grounding, and the routing of the gauge leads. Because the gauge is embedded in the sample being studied, a change of temperature in the host material and/or gauge can result in unequal expansion of the gauge and sample and can lead to uncertainties in the gauge response. Determining the effect of local heating is generally difficult. However, one method used for correcting this temperature effect is to connect an identical gauge to a second arm of the bridge network. The second gauge is a reference and is placed in a carefully controlled environment; it is not part of the experiment that measures the stress. Radiation fields can induce a voltage signal indirectly through thermal heating of the gauge, thus resulting in temperature- and strain-induced resistance changes in the gauge that influence its response.

Manganin gauges are very useful for measuring pressure profiles in certain materials, if used with due caution. Experimental parameters must be well defined so that gauge records can be accurately interpreted.

Chapter 12
Part G

Special Topics in Test Methodology: VISAR Interferometry

W. F. Hemsing*
Los Alamos National Laboratory, Los Alamos, New Mexico 87545

THE Velocity Interferometer System for Any Reflector (VISAR[1]) is a versatile single-pass interferometer developed to measure high velocities without the excessively high output frequencies or fringe counts of optical displacement interferometers. It has been used to acquire high-resolution measurements of surface velocities greater than 10 km/s over times ranging from several nanoseconds to milliseconds. A VISAR can provide continuous recording measurements for a single point or a line, subnanosecond response time, sensing by fiberoptic probes, computer control of hardware, and digital data acquisition.

The instrument exploits the fact that when laser light is reflected by a moving surface, the wavelength of the light is Doppler-shifted. The change in wavelength is detected by a two-beam interferometer with an optical delay in one leg. Compensating elements in the delay leg enable a VISAR to use light reflected by a diffuse surface.

The following examples illustrate the VISAR's ability to accommodate widely different experiments. Reference 1 describes a test in which a VISAR measured the velocity vs time history of a gun projectile during its acceleration from a 4-m-long barrel. The entire record produced only 1.5 cycles of interference fringe data. This can be compared to approximately 17 million fringes that would have been produced by a displacement interferometer using the same light wavelength.

The sensitivity of a VISAR can be adjusted to produce one fringe cycle for each kilometer-per-second of velocity. Using this sensitivity to measure a final velocity of 5 km/s, the VISAR would record a fringe count of five. For uniform acceleration over a time of 1 μs, this would yield a fringe frequency of 5 MHz. However, if an optical displacement interferometer were used, the fringe frequency would increase from zero to ~20 GHz during the measurement.

* Technical Staff member, Explosives Technology and Applications Division.

In these examples, the displacement interferometer signals would exceed the bandwidth or data capacity of conventional detectors and recorders, but the VISAR signals could be acquired easily.

The heart of a VISAR is its field-widened Michelson interferometer. Illustrated in Fig. 1, this interferometer contains an adjustable delay in one leg that provides its sensitivity to velocity. An eighth-wave retardation plate within the interferometer, with polarizing beamsplitters near the photodetectors, gives quadrature coding to the output signals. This important feature allows continuous velocity sensitivity at fringe extremes, makes it possible to distinguish between acceleration or deceleration, and provides a signal format that is very convenient for data reduction.

During a test, the changing Doppler shift of light scattered from an accelerating target produces a corresponding fringe shift in the interferometer. This fringe shift is proportional to velocity according to the simplified relation

$$F(t) = [U(t)2\tau]/\lambda_0$$

where $F(t)$ is the total fringe shift at time t, $U(t)$ is target velocity, τ is time delay added in one leg of the interferometer, and λ_0 is the unshifted light wavelength.

The spatial position vs time of the shifting fringes can be recorded directly by an electronic streak camera to get graphs of velocity vs time.[2,3] This special technique offers simplicity and fast time response but lacks the advantages of

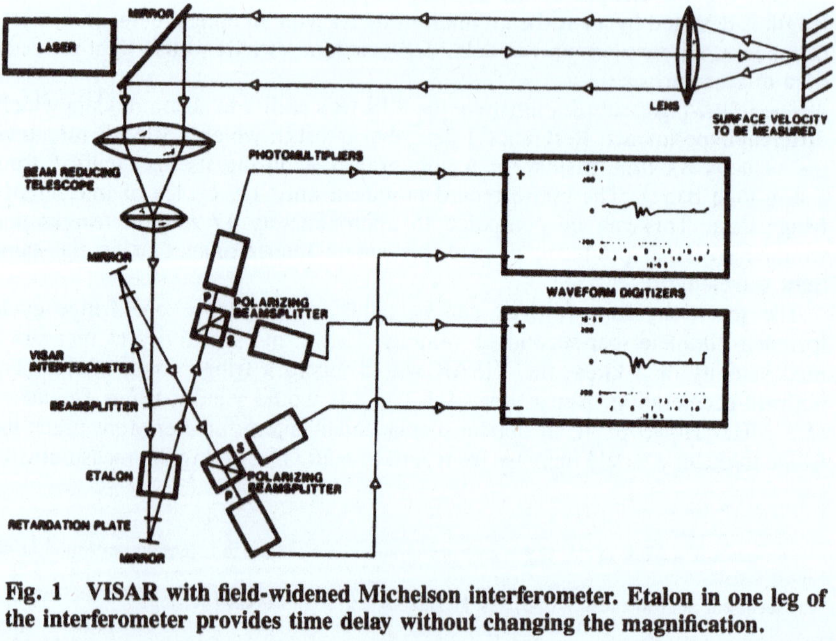

Fig. 1 VISAR with field-widened Michelson interferometer. Etalon in one leg of the interferometer provides time delay without changing the magnification.

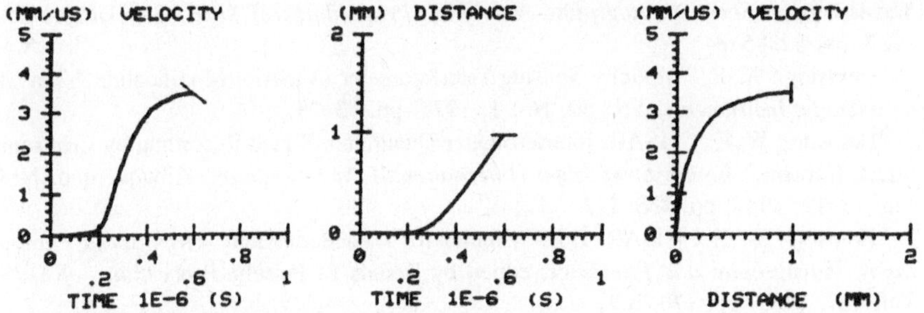

Fig. 2 Example of VISAR data plotted by a small computer several minutes after an experiment.

quadrature coding. In most VISARs, the intensities of the fringes are detected as amplitude-varying signals. The signal waveforms are analogous to $\sin[2\pi F(t)]$ and $\cos[2\pi F(t)]$ because of the quadrature coding.

The interferometer outputs can be converted to electrical signals for recording,[4] or the fringe intensities can be recorded directly by an electronic streak camera.[5] When using digital equipment, the data can be stored rapidly in a computer, and velocity information can be reduced to plots within minutes after an experiment. Figure 2 shows examples of these plots calculated using a simple algorithm.[6]

The recently developed line-imaging VISAR[7] can be a significant enhancement for some experiments. It uses a streak camera to simultaneously record the quadrature signals from many points on a surface. However, because it records pictures with their great capacity to store information, a line-imaging VISAR can capture data comparable to many conventional VISARs.

VISARs can provide subnanosecond response time, are light efficient, and can use fiberoptic probes. We have found our line-imaging VISAR extremely useful for acquiring information that simultaneously relates different points on a test object. This is most useful in experiments measuring stress-wave interactions or turbulence, where velocity gradients are important, and in tests that destroy expensive hardware—especially if test-to-test variations are important. VISARs can acquire measurements over a wide range of acceleration, velocity, and time conditions while accommodating the capabilities of the detection and recording devices. Like other optical instruments, however, a VISAR can be blinded by smoke that blocks its optical path or by the destruction of the light-reflecting surface.

References

[1]Barker, L. M., and Hollenbach, R. E., "Laser Interferometer for Measuring High Velocities of Any Reflecting Surface," *Journal of Applied Physics*, Vol. 43, No. 11, 1972, pp. 4669–4675.

[2]Parker, N. L., and Chau, H. H., "Streak Camera Recording of Interferometer Fringes," UCID-17531, Lawrence Livermore Natonal Laboratory, July 14, 1977.

[3]Bloomsquist, D. D., and Sheffield, S. A., "Optically Recording Velocity Interferometer System (ORVIS) for Subnanosecond Particle Velocity Measurements in Shock

Waves," *High Speed Photography,* Vol. 348, *Proceedings of SPIE,* San Diego, CA, 1982, pp. 523–528.

[4]Hemsing, W. F., "Velocity Sensing Interferometer (VISAR) Modification," *Review of Scientific Instruments,* Vol. 50, No. 1, 1979, pp. 73–78.

[5]Hemsing, W. F., "VISAR: Interferometer Quadrature Signal Recording by Electronic Streak Camera," *Eighth Symposium (International) on Detonation,* Albuquerque, NM, July 15–19, 1985, pp. 468–472.

[6]Hemsing, W. F., "VISAR: 2 1/2 Minutes for Data Reduction," *High Speed Photography, Videography, and Photonics,* edited by Dennis L. Paisely, *Proceedings of SPIE,* Vol. 427, 1983, pp. 199–202.

[7]Hemsing, W. F., Mathews, A. R., Warnes, R. H., George, M. J., and Whittemore, G. R., "VISAR: Line-Imaging Interferometer," *American Physical Society Topical Conference,* Williamsburg, VA, June 17–21, 1991, pp. 767–770.

Chapter 12
Part H

Special Topics in Test Methodology: Fabry-Perot Interferometry

R. H. Warnes*
Los Alamos National Laboratory, Los Alamos, New Mexico 87545

SINCE the development of the laser in 1960, many well-known optical instruments have become useful tools for making measurements on time scales that are short compared to those encountered in most laboratories and classrooms. In 1972 a velocity interferometer system for any reflector (VISAR[1]), an important variation of the Michelson interferometer, was developed in which the velocity of a moving target could be measured directly. In the VISAR, light from a moving interference fringe pattern is recorded by spatially fixed detectors, and the signals are displayed on oscilloscopes or streak cameras. The characteristics of the VISAR as it has developed over the intervening years are discussed in Chapter 12G.

Another approach to the problem of measuring the velocity of a rapidly moving object was developed in the 1970s[2] and employs a Fabry-Perot interferometer in conjunction with a streak camera to measure the position of moving fringes as a function of time. A schematic of this technique is shown in Fig. 1. Light from a laser is directed to the desired spot on a target by several mirrors and a lens placed about one focal length in front of the target. The light, after reflecting from the diffusely reflecting target, is collected and collimated by the lens. This beam is then sent to the Fabry-Perot interferometer; interference taking place during the multiple reflections between the Fabry-Perot mirrors results in a circular fringe pattern being focused by the spherical lens onto the slit plane of the streak camera. A cylindrical lens in front of the Fabry-Perot concentrates a large fraction of the light into an elongated ellipse centered on the slit. Both rotating mirror and electronic streak cameras have been used as the recording device. Considerations of the required sweep speeds and the asynchronous vs synchronous triggering of the two types of cameras should dictate which cam-

*Staff Member, Explosives Technology and Applications Division.

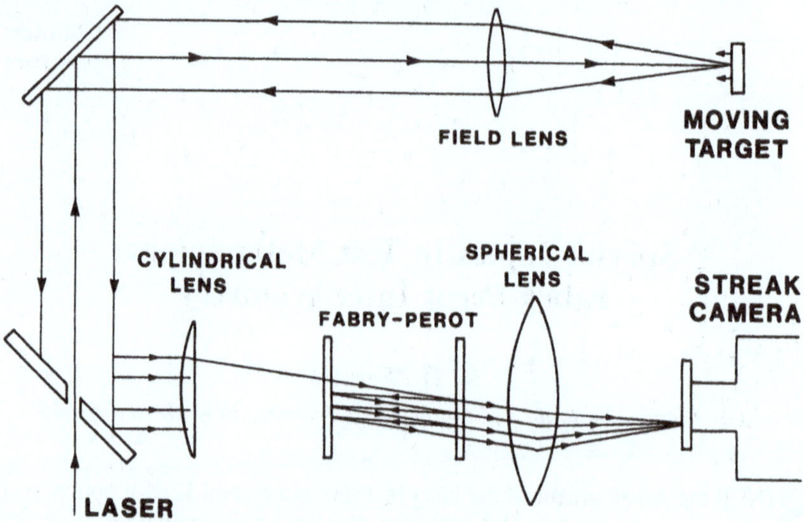

Fig. 1 Schematic of the Fabry-Perot technique.

era to use. A review article by McMillian et al.[3] discusses the details of the technique as it is frequently used.

Because the light within the Fabry-Perot undergoes multiple-beam interference, the resulting fringe pattern is very crisp, with well-defined bright fringes separated by relatively wide dark regions. When properly aligned, the major diameter of this bull's-eye pattern is placed directly on the slit of the streak camera. The result is a symmetric set of dots, which, when swept by the camera, becomes a symmetric set of lines on the recording film. When the target is motionless or is moving at a constant velocity, the frequency of the light within the Fabry-Perot is constant. Therefore, the fringe pattern will be stationary (horizontal lines in the film plane), as shown at the beginning and the end of the record in Fig. 2. When the target is accelerating, the frequency of the light within the Fabry-Perot is changing, and the fringe pattern, and therefore the dots on the slit plane of the camera, moves. On the film this results in a region of changing separation of the symmetric line pairs, as indicated by the middle portion of the record of Fig. 2. Adjustment of the sensitivity of the interferometer to velocity changes is accomplished by changing the mirror spacing, ℓ. In the velocity equation, c is the speed of light, λ is the unshifted laser wavelength, and n is an integer that allows one to handle discontinuities in the fringe record caused by large and unresolved accelerations of the target.

By measuring the separation of two adjacent line pairs in the static portion of the record and the separation of a fringe pair in the dynamic portion of the record, the velocity as a function of time may be calculated, as shown in Fig. 2. A typical record, Fig. 3, generally contains redundant fringe data. Therefore, the most precise measurement of the target velocity can usually be obtained

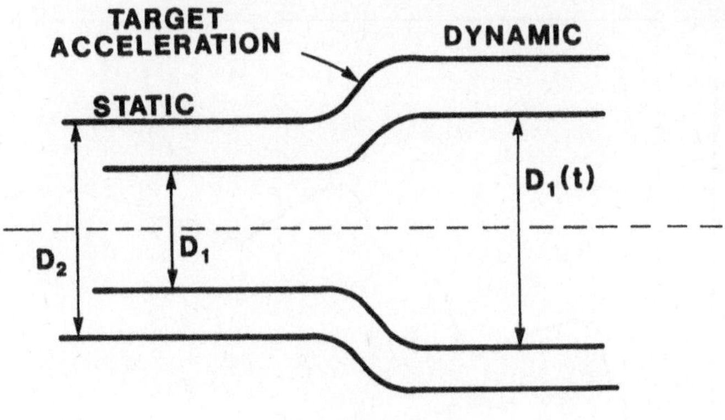

$$V(t) = \frac{a\lambda}{4\ell}\left[\frac{D_1(t)^2 - D_1^2}{D_2^2 - D_1^2}\right] \pm \frac{nc\lambda}{4\ell}$$

Fig. 2 Idealized fringe and the measurements required to calculate the velocity of a target.

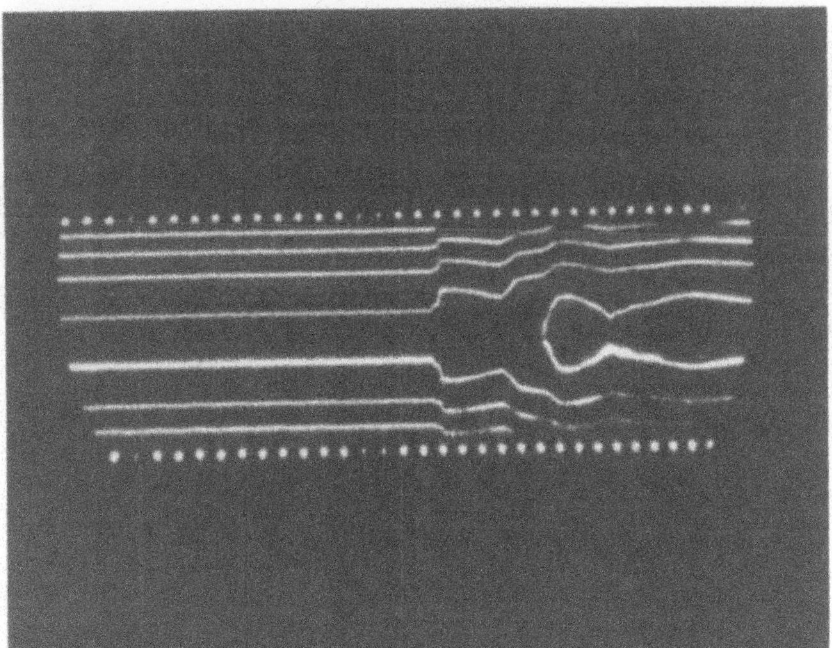

Fig. 3 Typical Fabry-Perot record. The intra-dot spacing at the top and bottom of the record is 0.2 μs and the velocity sensitivity (velocity change per fringe) is 0.501 km/s.

a)

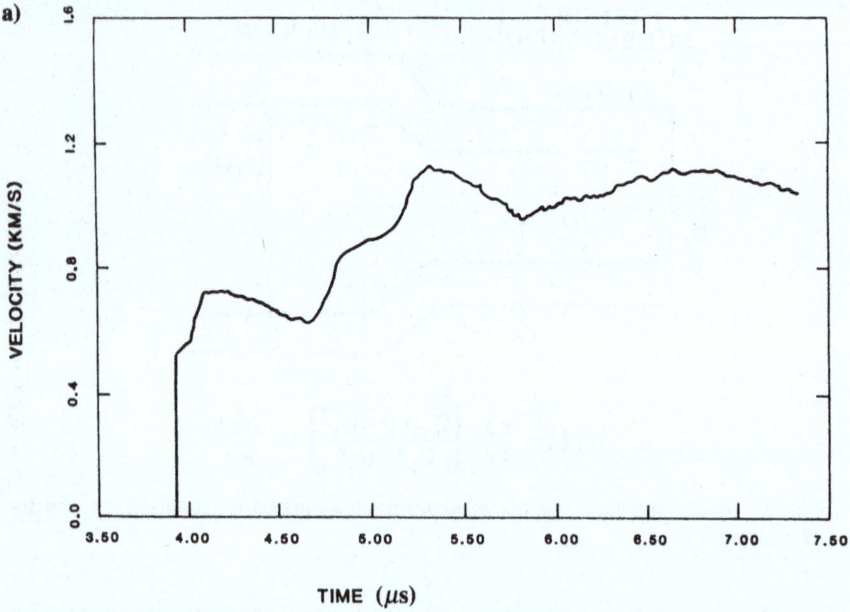

b)

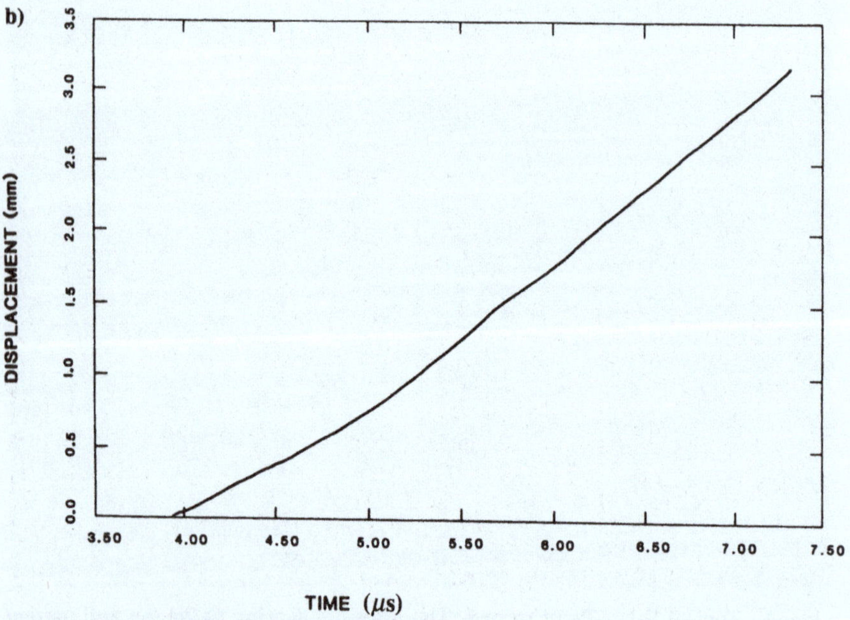

Fig. 4 Histories calculated from the record shown in Fig. 3: a) velocity-time; b) displacement-time.

by scanning the entire streak record with a microdensitometer and using all of the data, weighted appropriately, to determine a single velocity at each time step. Figure 4 is a plot of the velocity-time and displacement-time data reduced from the data of Fig. 3.

A problem common to all the optical interferometer techniques is that of maintaining adequate reflectivity of the target during the time of the measurement. Many normally reflective materials become poor reflectors while undergoing large strains or high accelerations, particularly when accompanied by large stresses, as in the case of explosively driven targets. Judicious choice of recording film and laser intensity is often sufficient for the Fabry-Perot to tolerate these large changes in target reflectivity while still maintaining good precision and time resolution. For particularly troublesome targets, a thin coating of gold, silver, platinum, or aluminum (a few tenths of a micrometer thick) may improve the reflectively enough to allow a measurement to be made.

Recent improvements in the technique have shortened the time required for analysis and increased the precision of the results. Using a CCD readout of the streak camera allows a greater range of intensities to be recorded than can be recorded with film, and the data are already digitized. More sophisticated algorithms for the weighting of redundant data have improved precision. Interactive techniques are currently being used that allow rapid analyses (<5 min) on small stand-alone computers.

For experiments that do not require time resolution better than a few nanoseconds, and particularly those with targets that cannot maintain good reflectivity for an adequate period of time, a Fabry-Perot interferometer system is a good choice for dynamic velocity measurements.

References

[1]Barker, L. M., and Hollenbach, R. E., "Laser Interferometer for Measuring High Velocities of any Reflecting Surface," *Journal of Applied Physics*, Vol. 43, No. 11, 1972, pp. 4669–4675.

[2]Durand, M., Laharrague, P., Lalle, P., Le Bihan, A., Morvan, J., and Pufols, H., "Interferometric Laser Technique for Accurate Velocity Measurement in Shock Wave Physics," *Review of Scientific Instruments*, Vol. 48, No. 3, 1977, pp. 275–278.

[3]McMillan, C. F., Goosman, D. R., Parker, N. L., Steinmetz, L. L., Chau, H. H., Huen, T., Whipkey, R. K., and Perry, S. J., "Velocimetry of Fast Surfaces Using Fabry-Perot Interferometry," *Review of Scientific Instruments*, Vol. 59, No. 1, 1988, pp. 1–20.

Chapter 12
Part I

Special Topics in Test Methodology: Axially Symmetric Magnetic Probe

J. A. Morgan*
Los Alamos National Laboratory, Los Alamos, New Mexico 87545

IN high-strain-rate or shock-wave physics it is often desirable to measure a material or particle velocity. Over the years several techniques have emerged for this purpose. Two techniques using interferometers, the VISAR[1] and the Fabry-Perot[2] (see Chaters 12G and 12H in this book), have been used extensively for these measurements. The capacitor gauge, developed by Rice,[3] was used a number of years ago in low-pressure, free-surface velocity measurements. The Dremin loop,[4] a magnetic technique, has also been used extensively. The technique described here, the axially symmetric magnetic probe (ASMP), was first described by Fritz and Morgan in 1973.[5]

Simply stated, the ASMP employs the changing inductance between a pickup coil and a moving conducting surface (linked by the magnetic field supplied by a small permanent magnet) to determine the velocity-time profile of the conducting surface. If the material between the coil and the conducting surface is neither conducting nor ferromagnetic, it has a negligible effect on the inductance and changes in the inductance between the coil and conducting surface. This insensitivity to intervening materials makes it possible to determine the velocity of interfaces between conductors and insulators—a very useful property. Several examples of experimental arrangements to exploit the advantages of the ASMP are described below.

Consider the sketch of the ASMP shown in Fig. 1. For simplicity, suppose that at some time t' the conducting surface denoted by z_f suddenly begins to move upward at a constant velocity $v(t, t')$. If the velocity is great enough or the areal conductivity ($\sigma_s = \sigma \times$ thickness) is high enough, the radial component of the magnetic field is compressed and driven through the coil; this action produces a voltage $E(t, t')$ that is proportional to the velocity $v(t, t')$. Another, and completely equivalent, way to look at the relationship between $E(t, t')$ and

*Staff Member.

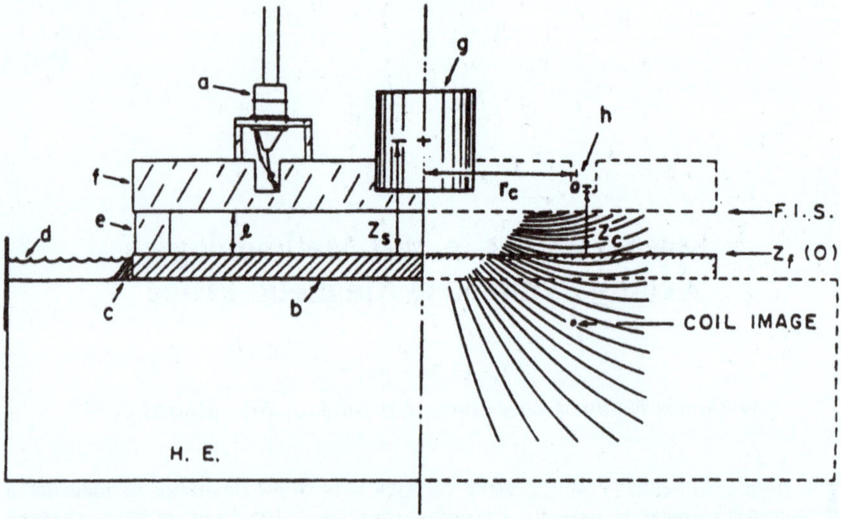

Fig. 1 Sketch of an ASMP experiment in a geometry where the sensitivity is approximately constant. Various elements and parameters in this experiment are as follows: a is the coaxial lead from the coil (note that the loop of connecting coil wire is small and is in a plane containing the magnet axis); b is the interface between the explosive and the aluminum plate; the vacuum grease barrier, c, keeps the mineral oil, d (used for suppression of blowoff from the explosive surface), from penetrating into region b; e is an insulating spacer (one of three); f is the experiment plate; g is a ceramic-5 magnet; h is the pick-up coil; z_s is the source distance; z_c and r_c define the coil position; z_f is the conductor surface; and ℓ is the initial spacing between the conductor surface and the fiducial impact surface (FIS).

$v(t, t')$ is to observe that the conducting surface suddenly becomes a magnetic mirror locking in the magnetic field at t' (again, if the velocity or areal conductivity is great enough). An image of the coil is depicted in Fig. 1. The coil image passes through the field at

$$z(t, t') = z_s - z_f(t) + z_c - z_f(t') \tag{1}$$

and r_c with velocity $v(t, t') = \mathrm{d}z(t, t')/\mathrm{d}t$. The EMF generated is then

$$E(t, t') = 2\pi r_c B(r_c, z) v(t, t'). \tag{2}$$

For the general case of a series of small velocity changes (a velocity-time profile), a set of differential images can be integrated to obtain $E(t)$. A general outline of this development may be found in Ref. 5, and a complete derivation for $E(t)$ as well as inversion of the problem to obtain $v(t)$, given $E(t)$, may be found in an article by Fritz and Morgan.[6] This report also contains a description of the method used to calibrate magnets.

Some of the advantages and disadvantages of the ASMP will be discussed below; the disadvantages will be described first. For most experiments it is most convenient to have S, the sensitivity function, be a constant [V/(km/s) = constant]. In this case the mapping of $E(t)$ to $v(t)$ is simple, and a relatively small error is involved. For an experiment to have such a value of S, $B(r_c, z)$ must be constant as z varies. This ideal is approached if the magnet is very hard (polarization is independent of magnet shape), thin, and flat. For ceramic magnetic materials (which do not complicate the situation by being conductors), a magnet with a length-to-diameter ratio of 1 is not significantly affected by changing fields during the experiment. For this aspect ratio and ceramic-5 material, the most nearly constant value of S occurs when the geometry of the experiment is about as shown in Fig. 1. The assumption is made in the analysis that the moving conducting plane is unbounded in the x-y plane. For the geometry shown in Fig. 1, this assumption is well satisfied for a conducting surface with a radius of about 1.5 r_c. These geometric considerations require the radius of the experiment to be about 4 to 8 times the value of z_s.

Another significant difficulty is that many materials undergo changes of conductivity during passage of a shock front. When this occurs, flux is trapped as the shock wave passes and the motion of the original conducting plane is masked.

A third disadvantage is related to the "reflectiveness" of the magnetic mirror. If the areal conductivity of a conducting surface in the experiment is low, the field diffuses through the conductor during the time of the experiment. This phenomenon is equivalent to the coil image receding from or lagging behind the conducting surface. The "recession velocity" for a coil image in an experiment where a thin (0.025 mm) copper foil is embedded in an insulator is about 1 km/s. Although one can usually correct for the recession velocity, in an experiment where velocity changes are not much larger than the recession velocity, the mapping of $E(t)$ to $v(t)$ can introduce large errors in the result.

Some of the disadvantages discussed above have associated advantages. For instance, the ASMP averages over a large surface, whereas the VISAR and Fabry-Perot techniques monitor very small surfaces—sometimes as small as a single grain. This averaging can have distinct advantages at low pressures or where lateral imperfections in the shock structure are found.

The following two experiments illustrate other advantages. If a plane-wave initiator were used to initiate the bottom of the high explosive of the experiment shown in Fig. 1, and if the conducting plate were aluminum, an idealized profile such as that shown in Fig. 2 would result. At a the shock wave arrives at the aluminum free surface. Between b and c a rarefaction returns to the aluminum-explosive interface and then a second compressive shock wave arrives at the aluminum free surface, d. At e the aluminum free surface collides with the fiducial impact surface (see Fig. 1). Note that the integral of $v(t)$ from a to e should equal the original distance from the aluminum free surface to the fiducial impact surface. Various other internal checks can be devised. The risetimes a to b, c to d, and e to f are a result of both the steepness of the various shock waves and the circuit parameters. To some extent the effect of circuit losses and imperfections can be reduced by analysis.

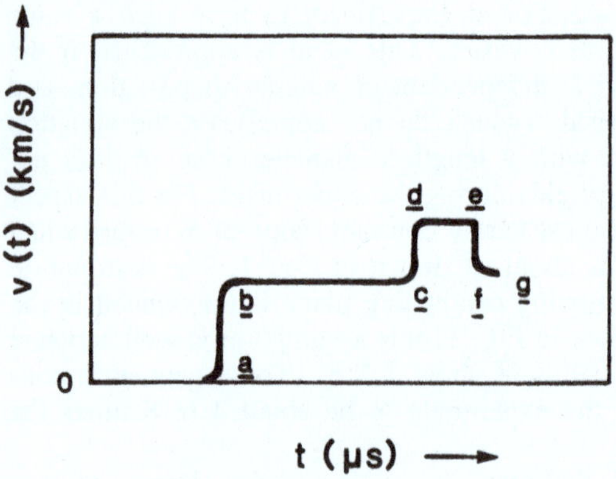

Fig. 2 Idealized $v(t)$ plot resulting from the experiment shown in Fig. 1.

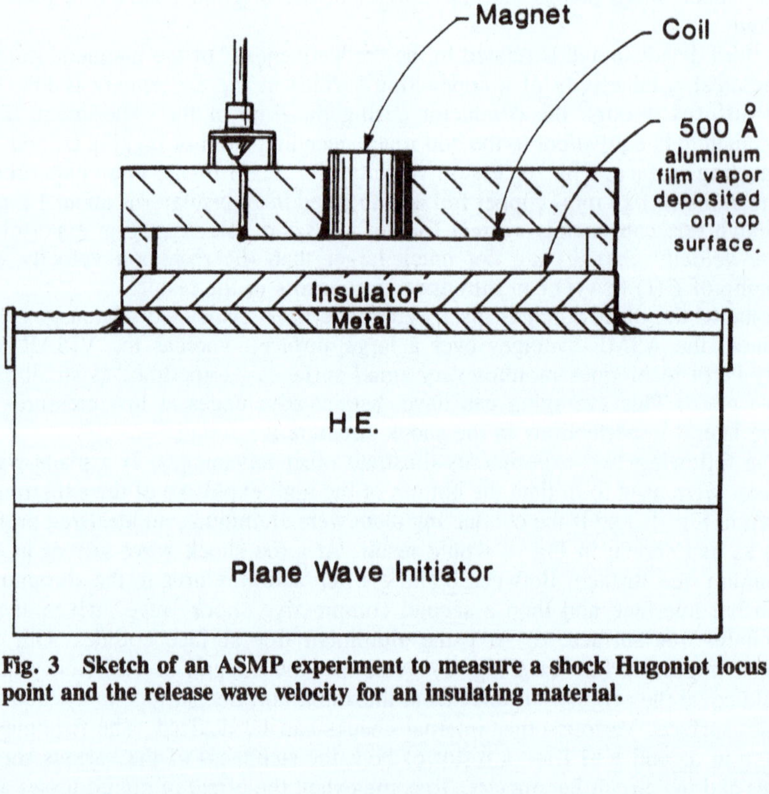

Fig. 3 Sketch of an ASMP experiment to measure a shock Hugoniot locus point and the release wave velocity for an insulating material.

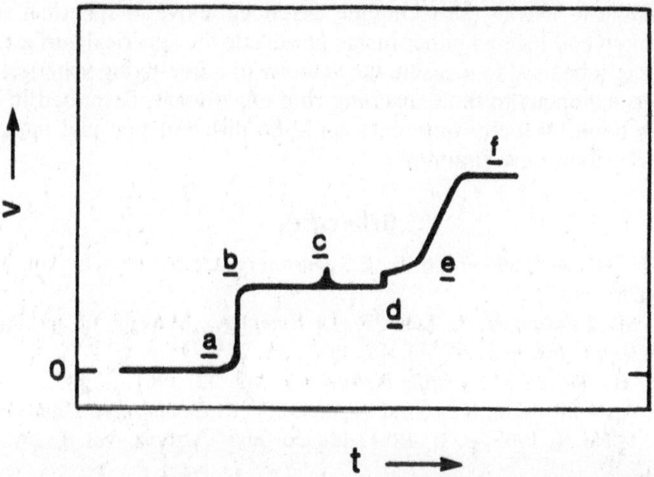

Fig. 4 Idealized *v(t)* plot resulting from the experiment shown in Fig. 3.

Consider the experiment shown in Fig. 3. In this experiment the high recession velocity of a thin conducting film is put to advantage. An idealized $v(t)$ from this experiment is shown in Fig. 4. At a the shock wave arrives at the metal-insulator interface. The constant level, b, is the velocity of the interface and hence the particle velocity in the insulator. The thin film of aluminum on the surface of the insulator has such a low areal conductivity that it does not alter the measurement until the shock wave reaches it. When the shock arrives, the thin film is accelerated and makes a small pulse on the record at c. Its recession velocity is so high that the voltage generated by this event quickly decays and makes no additional contribution to the record. The time interval from a to c is the shock transit time for the insulator from which one can calculate the shock velocity in the insulator. The motion of the interface remains undisturbed until the release wave from the insulator free surface arrives at the metal-insulator interface, d. The time interval between c and d can be used to calculate the longitudinal sound speed in the shocked insulator. As the shock strength in the experiment is increased to the melting point of the insulator, the speed of sound suddenly changes. This change is the result of melting on the insulator Hugoniot. In some cases, the technique can also be used to determine the onset of shock-induced high-pressure phase transitions in materials and to measure the sound speed in the new phase. This method of measuring sound speed has been used with stishovite,[7] a high-pressure phase of quartz.

An experiment could be designed to measure the shock-induced conductivity in a thin layer of material sandwiched between two insulators by measuring its recession velocity.

In all these considerations the paramount advantage of the ASMP is clearly that there are no leads and very little perturbation introduced into the shock flow by the technique.

It is possible to extend the technique described above to spherical surfaces[8] with the pickup coil located either inside or outside the spherical surface. Such a technique might be used to measure the velocity of a free-flying spherical surface in a manner analogous to the expanding ring experiment, described in Chapter 12K of this book. Velocity-time data could be differentiated and integrated to obtain stress, strain, and strain rate.

References

[1]Barker, L. M., and Hollenbach, R. E., *Journal of Applied Physics*, Vol. 43, 1972, pp. 4669–4675.

[2]Durand, M., Laharrague, P., Lalle, P., Le Bihan, A., Morvan, J., and Pujols, H., *Review of Scientific Instruments*, Vol. 48, 1977, pp. 275–278.

[3]Rice, M. H., *Review of Scientific Instruments*, Vol. 32, 1961, p. 449.

[4]Dremin, A. N., Penshin, S. V., and Pogorelov, V. F., *Combustion, Explosion, Shock Waves*, Vol. 1, No. 4, 1965, p. 1; also Fizika Goreniya i Vzryva, Vol. 1, No. 4, 1965, pp. 3–9.

[5]Fritz, J. N., and Morgan, J. A., *Review of Scientific Instruments*, Vol. 44, 1973, p. 215.

[6]Fritz, J. N., and Morgan, J. A., "The Use of Magnetic Fields to Measure Particle Velocity, II, General Theory, Calibration, and Utilization of the Axially-Symmetric Magnetic Probe," Los Alamos Scientific Laboratory Rept.

[7]Morgan, J. A., and Fritz, J. N., "Sound Velocity on SiO_2 Hugoniots," *High-Pressure Science and Technology*, Vol. 2, edited by K. D. Timmerhaus and M. S. Barber, Plenum Press, New York, 1979.

[8]Fritz, J. N., and Morgan, J. A., "The Axially Symmetric Magnetic Probe in Spherical Geometry," Los Alamos National Laboratory Internal Rept. (Copies are available from the author of this article.)

Chapter 12
Part J

Special Topics in Test Methodology: Freight Train Experiment

J. A. Morgan*
Los Alamos National Laboratory, Los Alamos, New Mexico 87545
and
A. Eugene Carden†
University of Alabama, Tuscaloosa, Alabama 35486

THE freight train experiment (FTE)[1] is an experiment to measure high-strain-rate, high-strain constitutive relations for materials. It is similar to the split Hopkinson bar[2-4] or dynamic-forging[5,6] experiments performed by many experimenters. In this experiment the incident bar, or tup, falls or is propelled onto the sample, which rests on the pressure bar (see Fig. 1).

In the FTE, stress, measured under the usual assumption of one-dimensional stress in the sample, is deduced from the strain response of the 25-mm-diam maraging steel pressure bar. Load-cell measurements are made at 14 and 28 diameters from the specimen. Semiconductor strain gauges, 1.5 mm long and with a gauge factor of about 100, were arranged in a Wheatstone bridge configuration so as to cancel out bending components. The instrumental rise time of the bridge and associated amplifiers was about 300 ns.

The assumption of one-dimensional stress in the sample implies that the sample remains cylindrical. To confirm this during the experiment, pictures of the sample are taken at from three to five predetermined strains. If in any of these pictures the sample is seen to be acylindrical, data at this and larger strains are discarded. Each of the cameras for these pictures is a 40-mm-diam, microchannel-plate electronic camera equipped with a lens system to provide 5× to 10× magnification. Each has an exposure time of about 20 ns.

The height of the sample is measured continuously with a Michelson interferometer. The strain is computed from the original height and the interferometer

---

*Staff Member.

†Professor, Department of Engineering Mechanics.

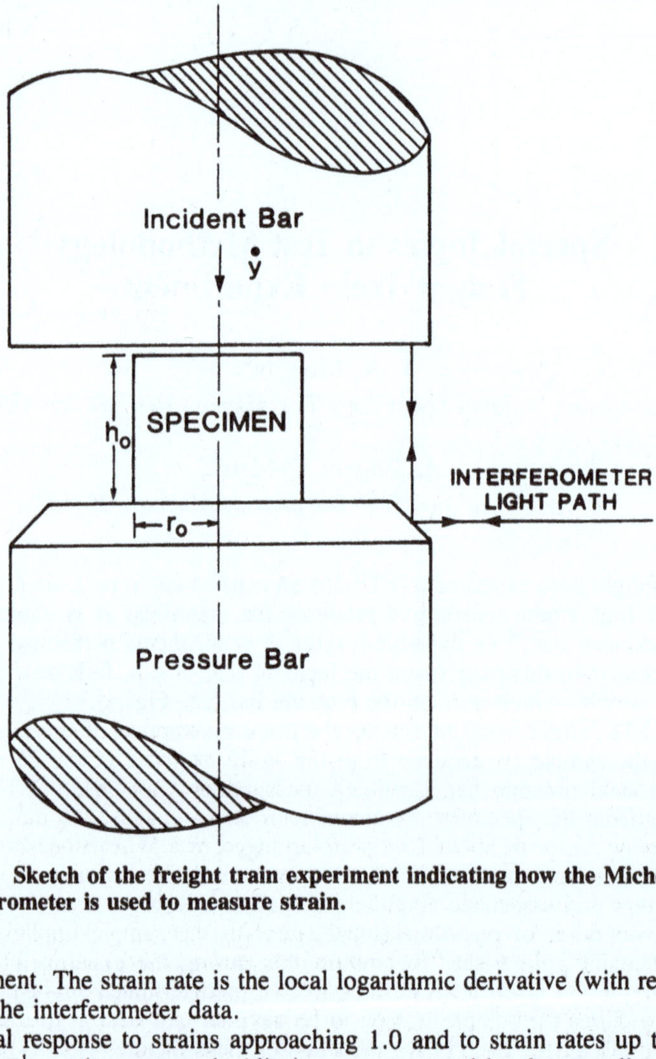

Fig. 1 Sketch of the freight train experiment indicating how the Michelson interferometer is used to measure strain.

measurement. The strain rate is the local logarithmic derivative (with respect to time) of the interferometer data.

Material response to strains approaching 1.0 and to strain rates up to about 2×10^5 s^{-1} can be measured. Aliasing errors caused by the sampling rate of the digitizers (2×10^8 samples/s) limit the maximum strain rates that can be recorded on these experiments. Faster digitizers with sampling rates of 10^9s^{-1} will be used when they become available.

In an experiment the interferometer data, the two load-cell records, and timing information are recorded on four digitizer channels. These data are transferred to a large computer and processed. Pockhammer-Chree corrections for dispersal of the stress signal in the pressure bar are made. Options in the processing allow for digital noise filtering and corrections for sample/bar friction and sample inertia.

Precision in the strain and the strain-rate data is better than 1%. Accuracy in the strain measurement depends on knowing when the incident bar contacted the

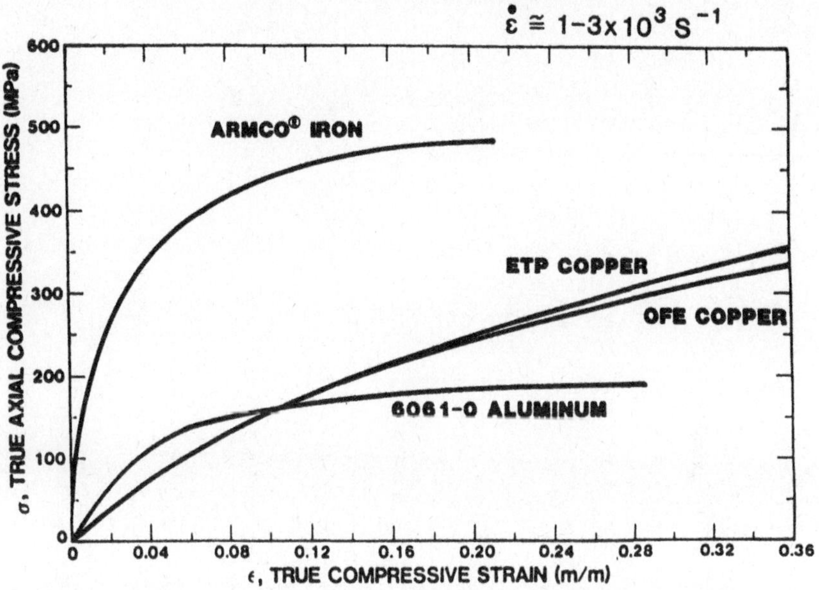

Fig. 2 Illustrative sampling of data from the freight train experiment.

sample. The strain accuracy is 1–2% depending on specimen size. The strain-rate accuracy is better than 1%. Accuracy in the stress measurement depends on the sample area in contact with the bars; this, of course, varies with initial specimen size and changes during the experiment. For samples down to 1 mm o.d. a satisfactory signal-to-noise ratio is achieved and accuracy is about 2–3%. For smaller samples a more sensitive load cell is required.

A sampling of data from the FTE is shown in Fig. 2.

References

[1]Morgan, J. A., and Carden, E. A., "Metallurgical Applications of Shock-Wave and High-Strain-Rate Phenomena," *EXPLOMET '85*, Portland, OR, 1985, edited by L. E. Murr, K. P. Staudhammer, and M. A. Meyers, Marcel Dekker, New York, 1986, p. 553.

[2]Bertholf, L. D., and Karnes, C. H., *Journal of the Mechanics and Physics of Solids*, Vol. 23, 1975, p. 1.

[3]Davies, E. D., and Hunter, S. C., *Journal of the Mechanics and Physics of Solids*, Vol. 11, 1963, p. 155.

[4]Lindhold, U. S., *Journal of Mechanics and Physics of Solids*, Vol. 12, 1964, p. 317.

[5]Samanta, S. K., *Journal of Mechanics and Physics of Solids*, Vol. 19, 1971, p. 117.

[6]Gorham, D. A., Mentha, S. N., and Cox, O. P., *Proceedings of the Eighth International Conference* on *HERF*, ASME, 1984, p. 207.

Chapter 12
Part K

Special Topics in Test Methodology: Freely Expanding Ring Test

Robert R. Karpp* and Richard H. Warnes*
Los Alamos National Laboratory, Los Alamos, New Mexico 87545

THE freely expanding ring test is a dynamic testing technique for measuring the flow stress of a material that is rapidly deforming in a uniaxial stress state. The test consists of monitoring the motion of a thin circular ring of test material that is placed in a state of uniform radial expansion, as indicated in Fig. 1. Because external forces are absent, the ring decelerates under the action of internal hoop stresses. Newton's law of motion applied to an element of the ring yields the relationship

$$\sigma = -\rho r \ddot{r} \tag{1}$$

where σ is the hoop stress, ρ is the density of the ring material (which is usually assumed constant), and r and $\ddot{r}$ are the current ring radius and acceleration. The true strain ϵ and strain rate $\dot{\epsilon}$ are given by

$$\epsilon = \ell n \frac{r}{r_0} \tag{2}$$

and

$$\dot{\epsilon} = \frac{\dot{r}}{r} \tag{3}$$

where r_0 is the initial ring radius and $\dot{r}$ is the radial velocity of the ring. If the ring is expanding in the plastic state, the hoop stress is the yield stress or flow stress of the material at the strain and strain rate indicated by Eqs. (2) and (3). Thus, if the ring motion is measured, $r = r(t)$, the flow stress, strain,

*Staff Member.

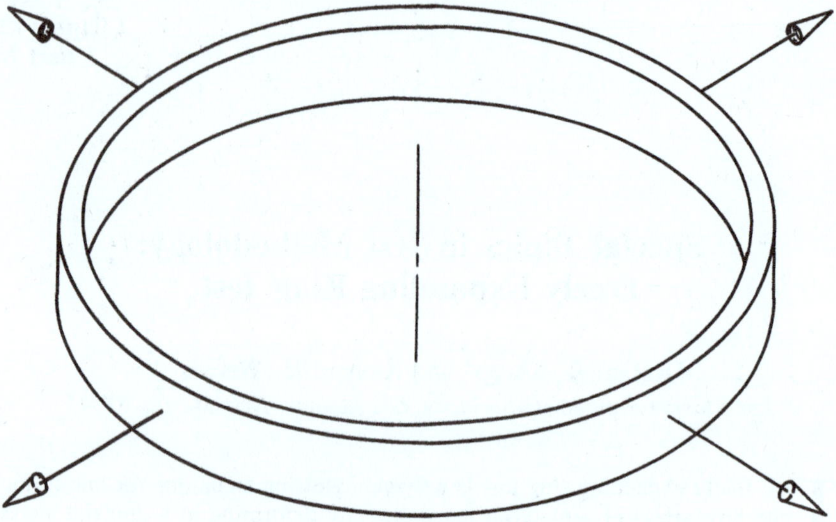

Fig. 1 Circular ring in a state of uniform radial expansion.

and strain rate can be determined from Eqs. (1), (2), and (3). As indicated by Eq. (3), extremely high strain rates can be achieved by using small rings propelled at high radial velocities. Strain rates of 10^4 s^{-1} and higher can be achieved.

The expanding ring test was introduced by Johnson et al.[1] over three decades ago. To propel the ring, they used an explosive driving system similar to the one illustrated in Fig. 2. Detonation of the explosive drives a shock wave across the high-strength steel driver propelling the test ring radially outward. The time-dependent radius of the ring was measured with a streak camera. Although the technique produced stress, strain, and strain-rate information, significant data

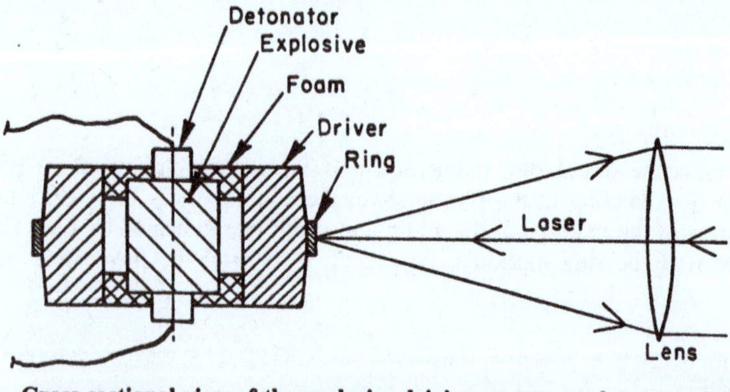

Fig. 2 Cross-sectional view of the explosive driving system used to propel the test ring.[3]

reduction difficulties arose because double differentiation of the experimental radius-time data, an inherently noisy process, had to be performed to determine the acceleration required in Eq. (1). By using the same experimental procedure, Hoggat and Recht[2] refined the technique by fitting a parabola to the experimental radius-time data. The parabola was then analytically differentiated twice to obtain the acceleration. This procedure improved the results but constrained the acceleration to be constant over the useful data window.

Warnes et al.[3] recently overcame the double differentiation problem by directly measuring the ring velocity with a laser velocity interferometer system (see Fig. 3). The stresses, strains, and strain rates were then obtained from Eqs. (1), (2), and (3) by integrating and differentiating the velocity-time data only once. Additional improvements,[4,5] including the use of three rings to reduce the time necessary for the test ring (central ring) to achieve one-dimensional motion, have resulted in the test configuration illustrated in Fig. 4. A typical velocity-time curve obtained from this test configuration is shown in Fig. 5. Two stress-strain curves obtained from the expanding ring test at an average strain rate of $4\times10^3\ s^{-1}$ are shown in Fig. 6. Also shown in this graph are two curves obtained from the well-known split Hopkinson bar test. One Hopkinson bar specimen was prestrained 5% before testing, and the other specimen was not. Figure 6 shows that the Hopkinson bar results using the prestrained specimen agree very well with results of the expanding ring test. This agreement suggests that shock hardening may have occurred during the launch process of the expanding ring test. Therefore, this effect should be considered when data from the explosively driven expanding ring test are analyzed. Figure 7 compares results from the expanding ring test with results from the Hopkinson bar test for a wide range of strain rates.

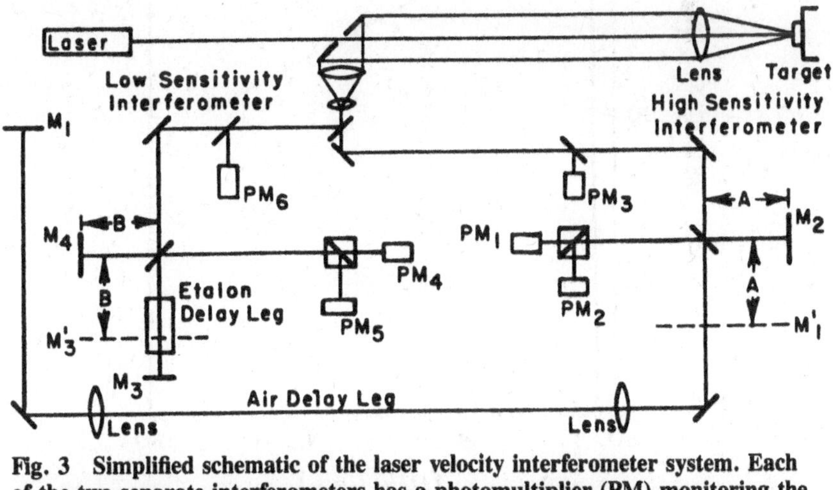

Fig. 3 Simplified schematic of the laser velocity interferometer system. Each of the two separate interferometers has a photomultiplier (PM) monitoring the return light and two PMs recording the interference information. Additional optics (not shown) permit quadrature recording of the interference information, enabling the sign of the target acceleration to be determined.[3]

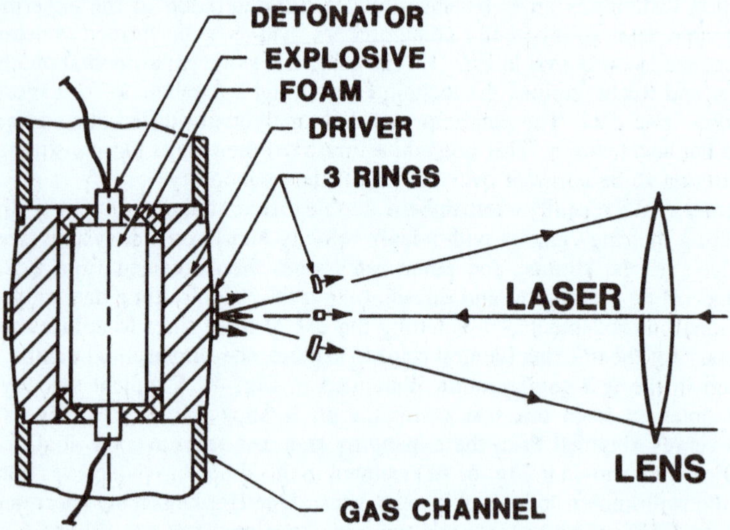

Fig. 4 Recent version of the test configuration showing the three rings before and after launch from the driver. Typical ring diameters range from 2 inches down to 1/2 in. (Ref. 5).

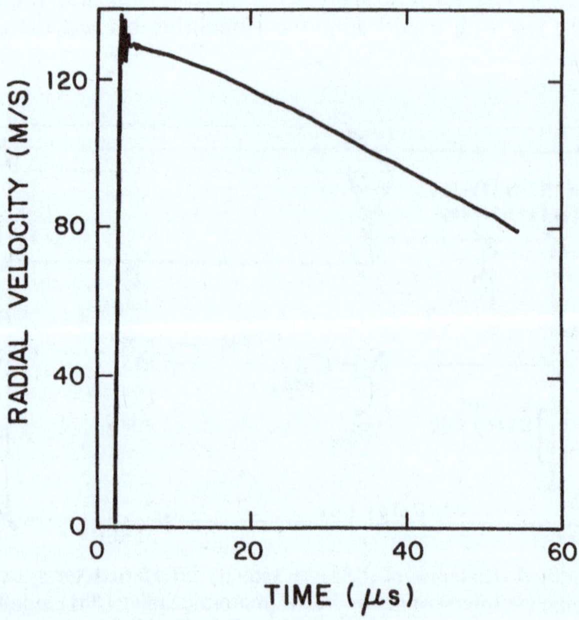

Fig. 5 Typical velocity-time curve obtained from the driver configuration shown in Fig. 4.[5]

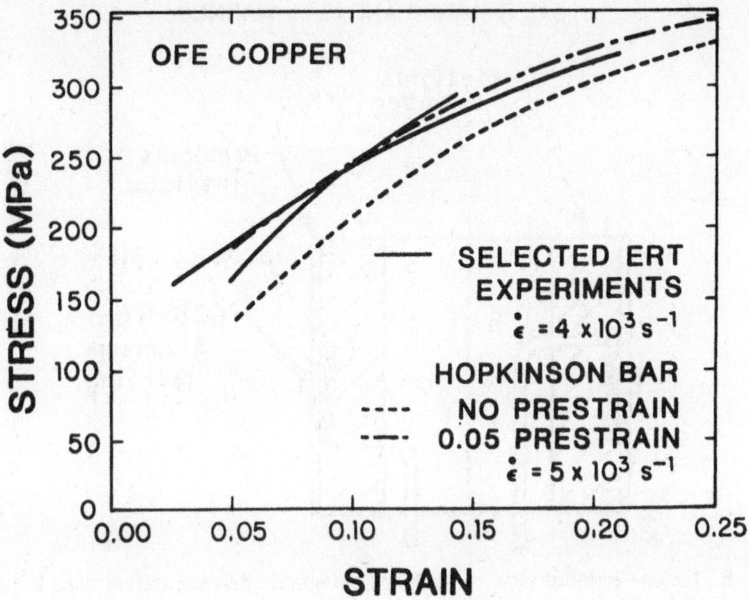

Fig. 6 Stress-strain curves obtained from the expanding ring test compared with results from the Hopkinson bar test. The test material was fully annealed OFE copper.[5]

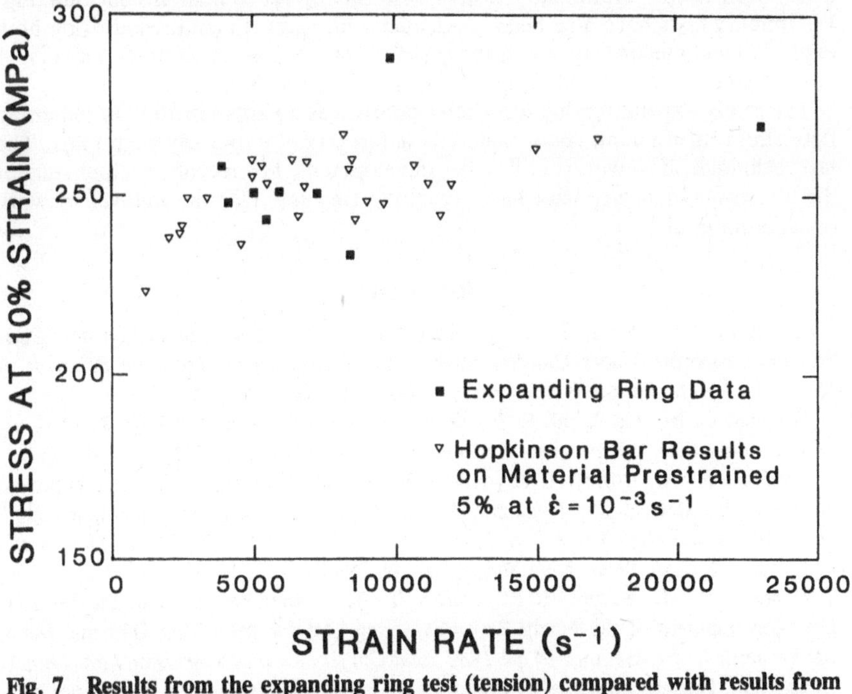

Fig. 7 Results from the expanding ring test (tension) compared with results from the split Hopkinson bar test (compression). The test material was fully annealed OFE copper. The Hopkinson bar samples were prestrained in compression to 5%.[5]

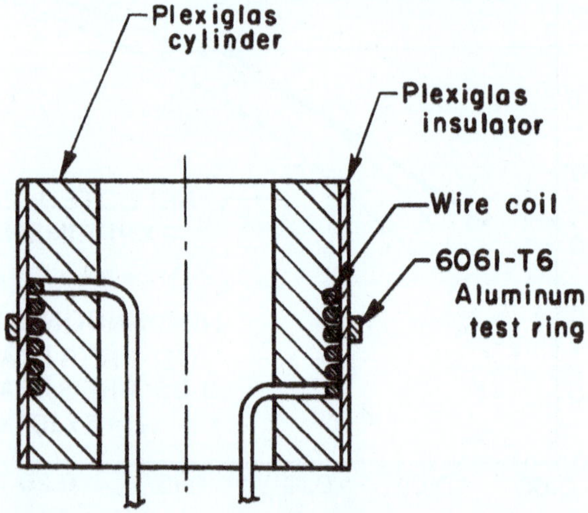

Fig. 8 Cross-sectional view of the electromagnetic driving system used to propel the test ring. [4]

To eliminate shock hardening, the electromagnetic driver system shown in Fig. 8 has been used.[4] Preliminary results with this driver system are encouraging. Preliminary tests have also been conducted with rings driven inward using both explosive and electromagnetic drive systems. These data are limited to relatively modest strains since buckling instabilities eventually develop.

The freely expanding ring test shows promise as a simple method of obtaining flow stress information about materials deforming at extremely high rates. The test technique, however, is still under development. For recent developments of the electromagnetically launched expanding ring test, see the innovative work of Gourdin et al.[6]

References

[1]Johnson, P. C., Stein, B. A., and Davis, R. S., "Measurement of Dynamic Plastic Flow Properties Under Uniform Stress," *Symposium on the Dynamic Behavior of Materials*, ASTM Special Publication No. 336, 1963, pp. 195–205.

[2]Hoggat, C. R., and Recht, R. F., "Stress-Strain Data Obtained at High Rates Using an Expanding Ring," *Experimental Mechanics*, Vol. 9, No. 10, 1969, pp. 441–448.

[3]Warnes, R. H., Duffey, T. A., Karpp, R. R., and Carden, A. E., "An Improved Technique for Determining Dynamic Material Properties Using the Expanding Ring," *Shock Waves and High-Strain-Rate Phenomena in Metals*, edited by M. A. Meyers and L. E. Murr, Plenum Press, New York, 1981, pp. 23–36.

[4]Warnes, R. H., Karpp, R. R., Duffey, T. A., Carden, A. E., and Jacobson, J. D., "Development of the Freely Expanding Ring Test for Measuring Dynamic Material Properties," *Proceedings of the 1982 Joint Conference on Experimental Mechanics*, Oahu-Maui, Hawaii, May 23–28, 1982; SESA, Brookfield Center, CT, Part II, pp. 656–661.

[5]Warnes, R. H., Karpp, R. R., and Follansbee, P. S., "The Freely Expanding Ring Test—A Test to Determine Material Strength at High Strain Rates," *Journal of Engineering Materials and Technology*, Vol. 108, No. 4, Oct. 1986, pp. 335–339.
[6]Gourdin, W. H., Weinland, S. L., and Boling, R. M., "Development of the Electromagnetically Launched Expanding Ring as a High-Strain-Rate Test Technique," *Review of Scientific Instruments*, Vol. 60, No. 3, 1989, pp. 427–432.

Author Index

PROGRESS IN ASTRONAUTICS AND AERONAUTICS SERIES VOLUMES

***1. Solid Propellant Rocket Research** (1960)
Martin Summerfield
Princeton University

***2. Liquid Rockets and Propellants** (1960)
Loren E. Bollinger
Ohio State University
Martin Goldsmith
The Rand Corp.
Alexis W. Lemmon Jr.
Battelle Memorial Institute

***3. Energy Conversion for Space Power** (1961)
Nathan W. Snyder
Institute for Defense Analyses

***4. Space Power Systems** (1961)
Nathan W. Snyder
Institute for Defense Analyses

***5. Electrostatic Propulsion** (1961)
David B. Langmuir
Space Technology Laboratories, Inc.
Ernst Stuhlinger
NASA George C. Marshall Space Flight Center
J.M. Sellen Jr.
Space Technology Laboratories, Inc.

***6. Detonation and Two-Phase Flow** (1962)
S.S. Penner
California Institute of Technology
F.A. Williams
Harvard University

***7. Hypersonic Flow Research** (1962)
Frederick R. Riddell
AVCO Corp.

***8. Guidance and Control** (1962)
Robert E. Roberson, Consultant
James S. Farrior
Lockheed Missiles and Space Co.

***9. Electric Propulsion Development** (1963)
Ernst Stuhlinger
NASA George C. Marshall Space Flight Center

***10. Technology of Lunar Exploration** (1963)
Clifford I. Cummings
Harold R. Lawrence
Jet Propulsion Laboratory

***11. Power Systems for Space Flight** (1963)
Morris A. Zipkin
Russell N. Edwards
General Electric Co.

***12. Ionization in High-Temperature Gases** (1963)
Kurt E. Shuler, Editor
National Bureau of Standards
John B. Fenn, Associate Editor
Princeton University

***13. Guidance and Control—II** (1964)
Robert C. Langford
General Precision Inc.
Charles J. Mundo
Institute of Naval Studies

***14. Celestial Mechanics and Astrodynamics** (1964)
Victor G. Szebehely
Yale University Observatory

***15. Heterogeneous Combustion** (1964)
Hans G. Wolfhard
Institute for Defense Analyses
Irvin Glassman
Princeton University
Leon Green Jr.
Air Force Systems Command

***16. Space Power Systems Engineering** (1966)
George C. Szego
Institute for Defense Analyses
J. Edward Taylor
TRW Inc.

***17. Methods in Astrodynamics and Celestial Mechanics** (1966)
Raynor L. Duncombe
U.S. Naval Observatory
Victor G. Szebehely
Yale University Observatory

***18. Thermophysics and Temperature Control of Spacecraft and Entry Vehicles** (1966)
Gerhard B. Heller
NASA George C. Marshall Space Flight Center

*Out of print.

***19. Communication Satellite Systems Technology** (1966)
Richard B. Marsten
Radio Corporation of America

***20. Thermophysics of Spacecraft and Planetary Bodies: Radiation Properties of Solids and the Electromagnetic Radiation Environment in Space** (1967)
Gerhard B. Heller
NASA George C. Marshall Space Flight Center

***21. Thermal Design Principles of Spacecraft and Entry Bodies** (1969)
Jerry T. Bevans
TRW Systems

***22. Stratospheric Circulation** (1969)
Willis L. Webb
Atmospheric Sciences Laboratory, White Sands, and University of Texas at El Paso

***23. Thermophysics: Applications to Thermal Design of Spacecraft** (1970)
Jerry T. Bevans
TRW Systems

24. Heat Transfer and Spacecraft Thermal Control (1971)
John W. Lucas
Jet Propulsion Laboratory

25. Communication Satellites for the 70's: Technology (1971)
Nathaniel E. Feldman
The Rand Corp.
Charles M. Kelly
The Aerospace Corp.

26. Communication Satellites for the 70's: Systems (1971)
Nathaniel E. Feldman
The Rand Corp.
Charles M. Kelly
The Aerospace Corp.

27. Thermospheric Circulation (1972)
Willis L. Webb
Atmospheric Sciences Laboratory, White Sands, and University of Texas at El Paso

28. Thermal Characteristics of the Moon (1972)
John W. Lucas
Jet Propulsion Laboratory

***29. Fundamentals of Spacecraft Thermal Design** (1972)
John W. Lucas
Jet Propulsion Laboratory

30. Solar Activity Observations and Predictions (1972)
Patrick S. McIntosh
Murray Dryer
Environmental Research Laboratories, National Oceanic and Atmospheric Administration

31. Thermal Control and Radiation (1973)
Chang-Lin Tien
University of California at Berkeley

32. Communications Satellite Systems (1974)
P.L. Bargellini
COMSAT Laboratories

33. Communications Satellite Technology (1974)
P.L. Bargellini
COMSAT Laboratories

***34. Instrumentation for Airbreathing Propulsion** (1974)
Allen E. Fuhs
Naval Postgraduate School
Marshall Kingery
Arnold Engineering Development Center

35. Thermophysics and Spacecraft Thermal Control (1974)
Robert G. Hering
University of Iowa

36. Thermal Pollution Analysis (1975)
Joseph A. Schetz
Virginia Polytechnic Institute
ISBN 0-915928-00-0

37. Aeroacoustics: Jet and Combustion Noise; Duct Acoustics (1975)
Henry T. Nagamatsu, Editor
General Electric Research and Development Center
Jack V. O'Keefe, Associate Editor
The Boeing Co.
Ira R. Schwartz, Associate Editor
NASA Ames Research Center
ISBN 0-915928-01-9

38. Aeroacoustics: Fan, STOL, and Boundary Layer Noise; Sonic Boom; Aeroacoustics Instrumentation (1975)
Henry T. Nagamatsu, Editor
General Electric Research and Development Center
Jack V. O'Keefe, Associate Editor
The Boeing Co.
Ira R. Schwartz, Associate Editor
NASA Ames Research Center
ISBN 0-915928-02-7

39. Heat Transfer with Thermal Control Applications (1975)
M. Michael Yovanovich
University of Waterloo
ISBN 0-915928-03-5

***40. Aerodynamics of Base Combustion** (1976)
S.N.B. Murthy, Editor
J.R. Osborn,
Associate Editor
Purdue University
A.W. Barrows
J.R. Ward,
Associate Editors
Ballistics Research Laboratories
ISBN 0-915928-04-3

41. Communications Satellite Developments: Systems (1976)
Gilbert E. LaVean
Defense Communications Agency
William G. Schmidt
CML Satellite Corp.
ISBN 0-915928-05-1

42. Communications Satellite Developments: Technology (1976)
William G. Schmidt
CML Satellite Corp.
Gilbert E. LaVean
Defense Communications Agency
ISBN 0-915928-06-X

***43. Aeroacoustics: Jet Noise, Combustion and Core Engine Noise** (1976)
Ira R. Schwartz, Editor
NASA Ames Research Center
Henry T. Nagamatsu,
Associate Editor
General Electric Research and Development Center
Warren C. Strahle,
Associate Editor
Georgia Institute of Technology
ISBN 0-915928-07-8

***44. Aeroacoustics: Fan Noise and Control; Duct Acoustics; Rotor Noise** (1976)
Ira R. Schwartz, Editor
NASA Ames Research Center
Henry T. Nagamatsu,
Associate Editor
General Electric Research and Development Center
Warren C. Strahle,
Associate Editor
Georgia Institute of Technology
ISBN 0-915928-08-6

***45. Aeroacoustics: STOL Noise; Airframe and Airfoil Noise** (1976)
Ira R. Schwartz, Editor
NASA Ames Research Center
Henry T. Nagamatsu,
Associate Editor
General Electric Research and Development Center
Warren C. Strahle,
Associate Editor
Georgia Institute of Technology
ISBN 0-915928-09-4

***46. Aeroacoustics: Acoustic Wave Propagation; Aircraft Noise Prediction; Aeroacoustic Instrumentation** (1976)
Ira R. Schwartz, Editor
NASA Ames Research Center
Henry T. Nagamatsu,
Associate Editor
General Electric Research and Development Center
Warren C. Strahle,
Associate Editor
Georgia Institute of Technology
ISBN 0-915928-10-8

47. Spacecraft Charging by Magnetospheric Plasmas (1976)
Alan Rosen
TRW Inc.
ISBN 0-915928-11-6

48. Scientific Investigations on the Skylab Satellite (1976)
Marion I. Kent
Ernst Stuhlinger
NASA George C. Marshall Space Flight Center
Shi-Tsan Wu
University of Alabama
ISBN 0-915928-12-4

49. Radiative Transfer and Thermal Control (1976)
Allie M. Smith
ARO Inc.
ISBN 0-915928-13-2

50. Exploration of the Outer Solar System (1976)
Eugene W. Greenstadt
TRW Inc.
Murray Dryer
National Oceanic and Atmospheric Administration
Devrie S. Intriligator
University of Southern California
ISBN 0-915928-14-0

51. Rarefied Gas Dynamics, Parts I and II (two volumes) (1977)
J. Leith Potter
ARO Inc.
ISBN 0-915928-15-9

52. Materials Sciences in Space with Application to Space Processing (1977)
Leo Steg
General Electric Co.
ISBN 0-915928-16-7

53. Experimental Diagnostics in Gas Phase Combustion Systems (1977)
Ben T. Zinn, Editor
Georgia Institute of Technology
Craig T. Bowman, Associate Editor
Stanford University
Daniel L. Hartley, Associate Editor
Sandia Laboratories
Edward W. Price, Associate Editor
Georgia Institute of Technology
James G. Skifstad, Associate Editor
Purdue University
ISBN 0-015928-18-3

54. Satellite Communications: Future Systems (1977)
David Jarett
TRW Inc.
ISBN 0-915928-18-3

55. Satellite Communications: Advanced Technologies (1977)
David Jarett
TRW Inc.
ISBN 0-915928-19-1

56. Thermophysics of Spacecraft and Outer Planet Entry Probes (1977)
Allie M. Smith
ARO Inc.
ISBN 0-915928-20-5

57. Space-Based Manufacturing from Nonterrestrial Materials (1977)
Gerard K. O'Neill, Editor
Brian O'Leary, Assistant Editor
Princeton University
ISBN 0-915928-21-3

58. Turbulent Combustion (1978)
Lawrence A. Kennedy
State University of New York at Buffalo
ISBN 0-915928-22-1

59. Aerodynamic Heating and Thermal Protection Systems (1978)
Leroy S. Fletcher
University of Virginia
ISBN 0-915928-23-X

60. Heat Transfer and Thermal Control Systems (1978)
Leroy S. Fletcher
University of Virginia
ISBN 0-915928-24-8

61. Radiation Energy Conversion in Space (1978)
Kenneth W. Billman
NASA Ames Research Center
ISBN 0-915928-26-4

62. Alternative Hydrocarbon Fuels: Combustion and Chemical Kinetics (1978)
Craig T. Bowman
Stanford University
Jorgen Birkeland
Department of Energy
ISBN 0-915928-25-6

63. Experimental Diagnostics in Combustion of Solids (1978)
Thomas L. Boggs
Naval Weapons Center
Ben T. Zinn
Georgia Institute of Technology
ISBN 0-915928-28-0

64. Outer Planet Entry Heating and Thermal Protection (1979)
Raymond Viskanta
Purdue University
ISBN 0-915928-29-9

65. Thermophysics and Thermal Control (1979)
Raymond Viskanta
Purdue University
ISBN 0-915928-30-2

66. Interior Ballistics of Guns (1979)
Herman Krier
University of Illinois at Urbana-Champaign
Martin Summerfield
New York University
ISBN 0-915928-32-9

***67. Remote Sensing of Earth from Space: Role of "Smart Sensors"**(1979)
Roger A. Breckenridge
NASA Langley Research Center
ISBN 0-915928-33-7

68. Injection and Mixing in Turbulent Flow (1980)
Joseph A. Schetz
Virginia Polytechnic Institute and State University
ISBN 0-915928-35-3

69. Entry Heating and Thermal Protection (1980)
Walter B. Olstad
NASA Headquarters
ISBN 0-915928-38-8

70. Heat Transfer, Thermal Control, and Heat Pipes (1980)
Walter B. Olstad
NASA Headquarters
ISBN 0-915928-39-6

***71. Space Systems and Their Interactions with Earth's Space Environment** (1980)
Henry B. Garrett
Charles P. Pike
Hanscom Air Force Base
ISBN 0-915928-41-8

72. Viscous Flow Drag Reduction (1980)
Gary R. Hough
Vought Advanced Technology Center
ISBN 0-915928-44-2

73. Combustion Experiments in a Zero-Gravity Laboratory (1981)
Thomas H. Cochran
NASA Lewis Research Center
ISBN 0-915928-48-5

74. Rarefied Gas Dynamics, Parts I and II (two volumes) (1981)
Sam S. Fisher
University of Virginia
ISBN 0-915928-51-5

75. Gasdynamics of Detonations and Explosions (1981)
J.R. Bowen
University of Wisconsin at Madison
N. Manson
Université de Poitiers
A.K. Oppenheim
University of California at Berkeley
R.I. Soloukhin
Institute of Heat and Mass Transfer, BSSR Academy of Sciences
ISBN 0-915928-46-9

76. Combustion in Reactive Systems (1981)
J.R. Bowen
University of Wisconsin at Madison
N. Manson
Université de Poitiers
A.K. Oppenheim
University of California at Berkeley
R.I. Soloukhin
Institute of Heat and Mass Transfer, BSSR Academy of Sciences
ISBN 0-915928-47-7

77. Aerothermodynamics and Planetary Entry (1981)
A.L. Crosbie
University of Missouri-Rolla
ISBN 0-915928-52-3

78. Heat Transfer and Thermal Control (1981)
A.L. Crosbie
University of Missouri-Rolla
ISBN 0-915928-53-1

79. Electric Propulsion and Its Applications to Space Missions (1981)
Robert C. Finke
NASA Lewis Research Center
ISBN 0-915928-55-8

80. Aero-Optical Phenomena (1982)
Keith G. Gilbert
Leonard J. Otten
Air Force Weapons Laboratory
ISBN 0-915928-60-4

81. Transonic Aerodynamics (1982)
David Nixon
Nielsen Engineering & Research, Inc.
ISBN 0-915928-65-5

82. Thermophysics of Atmospheric Entry (1982)
T.E. Horton
University of Mississippi
ISBN 0-915928-66-3

83. Spacecraft Radiative Transfer and Temperature Control (1982)
T.E. Horton
University of Mississippi
ISBN 0-915928-67-1

84. Liquid-Metal Flows and Magnetohydrodynamics (1983)
H. Branover
Ben-Gurion University of the Negev
P.S. Lykoudis
Purdue University
A. Yakhot
Ben-Gurion University of the Negev
ISBN 0-915928-70-1

85. Entry Vehicle Heating and Thermal Protection Systems: Space Shuttle, Solar Starprobe, Jupiter Galileo Probe (1983)
Paul E. Bauer
McDonnell Douglas Astronautics Co.
Howard E. Collicott
The Boeing Co.
ISBN 0-915928-74-4

86. Spacecraft Thermal Control, Design, and Operation (1983)
Howard E. Collicott
The Boeing Co.
Paul E. Bauer
McDonnell Douglas Astronautics Co.
ISBN 0-915928-75-2

87. Shock Waves, Explosions, and Detonations (1983)
J.R. Bowen
University of Washington
N. Manson
Université de Poitiers
A.K. Oppenheim
University of California at Berkeley
R.I. Soloukhin
Institute of Heat and Mass Transfer, BSSR Academy of Sciences
ISBN 0-915928-76-0

88. Flames, Lasers, and Reactive Systems (1983)
J.R. Bowen
University of Washington
N. Manson
Université de Poitiers
A.K. Oppenheim
University of California at Berkeley
R.I. Soloukhin
Institute of Heat and Mass Transfer, BSSR Academy of Sciences
ISBN 0-915928-77-9

89. Orbit-Raising and Maneuvering Propulsion: Research Status and Needs (1984)
Leonard H. Caveny
Air Force Office of Scientific Research
ISBN 0-915928-82-5

90. Fundamentals of Solid-Propellant Combustion (1984)
Kenneth K. Kuo
Pennsylvania State University
Martin Summerfield
Princeton Combustion Research Laboratories, Inc.
ISBN 0-915928-84-1

91. Spacecraft Contamination: Sources and Prevention (1984)
J.A. Roux
University of Mississippi
T.D. McCay
NASA Marshall Space Flight Center
ISBN 0-915928-85-X

92. Combustion Diagnostics by Nonintrusive Methods (1984)
T.D. McCay
NASA Marshall Space Flight Center
J.A. Roux
University of Mississippi
ISBN 0-915928-86-8

93. The INTELSAT Global Satellite System (1984)
Joel Alper
COMSAT Corp.
Joseph Pelton
INTELSAT
ISBN 0-915928-90-6

94. Dynamics of Shock Waves, Explosions, and Detonations (1984)
J.R. Bowen
University of Washington
N. Manson
Université de Poitiers
A.K. Oppenheim
University of California at Berkely
R.I. Soloukhin
Institute of Heat and Mass Transfer, BSSR Academy of Sciences
ISBN 0-915928-91-4

95. Dynamics of Flames and Reactive Systems (1984)
J.R. Bowen
University of Washington
N. Manson
Université de Poitiers
A.K. Oppenheim
University of California at Bereley
R.I. Soloukhin
Institute of Heat and Mass Transfer, BSSR Academy of Sciences
ISBN 0-915928-92-2

96. Thermal Design of Aeroassisted Orbital Transfer Vehicles (1985)
H.F. Nelson
University of Missouri-Rolla
ISBN 0-915928-94-9

97. Monitoring Earth's Ocean, Land, and Atmosphere from Space — Sensors, Systems, and Applications (1985)
Abraham Schnapf
Aerospace Systems Engineering
ISBN 0-915928-98-1

98. Thrust and Drag: Its Prediction and Verification (1985)
Eugene E. Covert
Massachusetts Institute of Technology
C.R. James
Vought Corp.
William F. Kimzey
Sverdrup Technology AEDC Group
George K. Richey
U.S. Air Force
Eugene C. Rooney
U.S. Navy Department of Defense
ISBN 0-930403-00-2

99. Space Stations and Space Platforms — Concepts, Design, Infrastructure, and Uses (1985)
Ivan Bekey
Daniel Herman
NASA Headquarters
ISBN 0-930403-01-0

100. Single- and Multi-Phase Flows in an Electromagnetic Field: Energy, Metallurgical, and Solar Applications (1985)
Herman Branover
Ben-Gurion University of the Negev
Paul S. Lykoudis
Purdue University
Michael Mond
Ben-Gurion University of the Negev
ISBN 0-930403-04-5

101. MHD Energy Conversion: Physiotechnical Problems (1986)
V.A. Kirillin
A.E. Sheyndlin
Soviet Academy of Sciences
ISBN 0-930403-05-3

102. Numerical Methods for Engine-Airframe Integration (1986)
S.N.B. Murthy
Purdue University
Gerald C. Paynter
Boeing Airplane Co.
ISBN 0-930403-09-6

103. Thermophysical Aspects of Re-Entry Flows (1986)
James N. Moss
NASA Langley Research Center
Carl D. Scott
NASA Johnson Space Center
ISBN 0-930403-10-X

104. Tactical Missile Aerodynamics (1986)
M.J. Hemsch
PRC Kentron, Inc.
J.N. Nielsen
NASA Ames Research Center
ISBN 0-930403-13-4

105. Dynamics of Reactive Systems Part I: Flames and Configurations; Part II: Modeling and Heterogeneous Combustion (1986)
J.R. Bowen
University of Washington
J.-C. Leyer
Université de Poitiers
R.I. Soloukhin
Institute of Heat and Mass Transfer, BSSR Academy of Sciences
ISBN 0-930403-14-2

106. Dynamics of Explosions (1986)
J.R. Bowen
University of Washington
J.-C. Leyer
Université de Poitiers
R.I. Soloukhin
Institute of Heat and Mass Transfer, BSSR Academy of Sciences
ISBN 0-930403-15-0

107. Spacecraft Dielectric Material Properties and Spacecraft Charging (1986)
A.R. Frederickson
U.S. Air Force Rome Air Development Center
D.B. Cotts
SRI International
J.A. Wall
U.S. Air Force Rome Air Development Center
F.L. Bouquet
Jet Propulsion Laboratory, California Institute of Technology
ISBN 0-930403-17-7

108. Opportunities for Academic Research in a Low-Gravity Environment (1986)
George A. Hazelrigg
National Science Foundation
Joseph M. Reynolds
Louisiana State University
ISBN 0-930403-18-5

109. Gun Propulsion Technology (1988)
Ludwig Stiefel
U.S. Army Armament Research, Development and Engineering Center
ISBN 0-930403-20-7

110. Commercial Opportunities in Space (1988)
F. Shahrokhi
K.E. Harwell
University of Tennessee Space Institute
C.C. Chao
National Cheng Kung University
ISBN 0-930403-39-8

111. Liquid-Metal Flows: Magnetohydrodynamics and Applications (1988)
Herman Branover, Michael Mond, and Yeshajahu Unger
Ben-Gurion University of the Negev
ISBN 0-930403-43-6

112. Current Trends in Turbulence Research (1988)
Herman Branover, Michael Mond, and Yeshajahu Unger
Ben-Gurion University of the Negev
ISBN 0-930403-44-4

113. Dynamics of Reactive Systems Part I: Flames; Part II: Heterogeneous Combustion and Applications (1988)
A.L. Kuhl
R & D Associates
J.R. Bowen
University of Washington
J.-C. Leyer
Université de Poitiers
A. Borisov
USSR Academy of Sciences
ISBN 0-930403-46-0

114. Dynamics of Explosions (1988)
A.L. Kuhl
R & D Associates
J.R. Bowen
University of Washington
J.-C. Leyer
Université de Poitiers
A. Borisov
USSR Academy of Sciences
ISBN 0-930403-47-9

115. Machine Intelligence and Autonomy for Aerospace (1988)
E. Heer
Heer Associates, Inc.
H. Lum
NASA Ames Research Center
ISBN 0-930403-48-7

116. Rarefied Gas Dynamics: Space-Related Studies (1989)
E.P. Muntz
University of Southern California
D.P. Weaver
U.S. Air Force Astronautics Laboratory (AFSC)
D.H. Campbell
University of Dayton Research Institute
ISBN 0-930403-53-3

117. Rarefied Gas Dynamics: Physical Phenomena (1989)
E.P. Muntz
University of Southern California
D.P. Weaver
U.S. Air Force Astronautics Laboratory (AFSC)
D. Campbell
University of Dayton Research Institute
ISBN 0-930403-54-1

118. Rarefied Gas Dynamics: Theoretical and Computational Techniques (1989)
E.P. Muntz
University of Southern California
D.P. Weaver
U.S. Air Force Astronautics Laboratory (AFSC)
D.H. Campbell
University of Dayton Research Institute
ISBN 0-930403-55-X

119. Test and Evaluation of the Tactical Missile (1989)
Emil J. Eichblatt Jr.
Pacific Missile Test Center
ISBN 0-930403-56-8

120. Unsteady Transonic Aerodynamics (1989)
David Nixon
Nielsen Engineering & Research, Inc.
ISBN 0-930403-52-5

121. Orbital Debris from Upper-Stage Breakup (1989)
Joseph P. Loftus Jr.
NASA Johnson Space Center
ISBN 0-930403-58-4

122. Thermal-Hydraulics for Space Power, Propulsion and Thermal Management System Design (1989)
William J. Krotiuk
General Electric Co.
ISBN 0-930403-64-9

123. Viscous Drag Reduction in Boundary Layers (1990)
Dennis M. Bushnell
Jerry N. Hefner
NASA Langley Research Center
ISBN 0-930403-66-5

124. Tactical and Strategic Missile Guidance (1990)
Paul Zarchan
Charles Stark Draper Laboratory, Inc.
ISBN 0-930403-68-1

125. Applied Computational Aerodynamics (1990)
P.A. Henne
Douglas Aircraft Company
ISBN 0-930403-69-X

126. Space Commercialization: Launch Vehicles and Programs (1990)
F. Shahrokhi
University of Tennessee Space Institute
J.S. Greenberg
Princeton Synergetics Inc.
T. Al-Saud
Ministry of Defense and Aviation Kingdom of Saudi Arabia
ISBN 0-930403-75-4

127. Space Commercialization: Platforms and Processing (1990)
F. Shahrokhi
University of Tennessee Space Institute
G. Hazelrigg
National Science Foundation
R. Bayuzick
Vanderbilt University
ISBN 0-930403-76-2

128. Space Commercialization: Satellite Technology (1990)
F. Shahrokhi
University of Tennessee Space Institute
N. Jasentuliyana
United Nations
N. Tarabzouni
King Abulaziz City for Science and Technology
ISBN 0-930403-77-0

129. Mechanics and Control of Large Flexible Structures (1990)
John L. Junkins
Texas A&M University
ISBN 0-930403-73-8

130. Low-Gravity Fluid Dynamics and Transport Phenomena (1990)
Jean N. Koster
Robert L. Sani
University of Colorado at Boulder
ISBN 0-930403-74-6

131. Dynamics of Deflagrations and Reactive Systems: Flames (1991)
A. L. Kuhl
Lawrence Livermore National Laboratory
J.-C. Leyer
Université de Poitiers
A. A. Borisov
USSR Academy of Sciences
W. A. Sirignano
University of California
ISBN 0-930403-95-9

132. Dynamics of Deflagrations and Reactive Systems: Heterogeneous Combustion (1991)
A. L. Kuhl
Lawrence Livermore National Laboratory
J.-C. Leyer
Université de Poitiers
A. A. Borisov
USSR Academy of Sciences
W. A. Sirignano
University of California
ISBN 0-930403-96-7

133. Dynamics of Detonations and Explosions: Detonations (1991)
A. L. Kuhl
Lawrence Livermore National Laboratory
J.-C. Leyer
Université de Poitiers
A. A. Borisov
USSR Academy of Sciences
W. A. Sirignano
University of California
ISBN 0-930403-97-5

134. Dynamics of Detonations and Explosions: Explosion Phenomena (1991)
A. L. Kuhl
Lawrence Livermore National Laboratory
J.-C. Leyer
Université de Poitiers
A. A. Borisov
USSR Academy of Sciences
W. A. Sirignano
University of California
ISBN 0-930403-98-3

135. Numerical Approaches to Combustion Modeling (1991)
Elaine S. Oran
Jay P. Boris
Naval Research Laboratory
ISBN 1-56347-004-7

136. Aerospace Software Engineering (1991)
Christine Anderson
U.S. Air Force Wright Laboratory
Merlin Dorfman
Lockheed Missiles & Space Company, Inc.
ISBN 1-56346-005-5

137. High-Speed Flight Propulsion Systems (1991)
S. N. B. Murthy
Purdue University
E. T. Curran
Wright Laboratory
ISBN 1-56347-011-X

138. Propagation of Intensive Laser Radiation in Clouds (1992)
O. A. Volkovitsky
Yu. S. Sedunov
L. P. Semenov
Institute of Experimental Meteorology
ISBN 1-56347-020-9

139. Gun Muzzle Blast and Flash (1992)
Günter Klingenberg
Fraunhofer-Institut für Kurzzeitdynamik, Ernst-Mach-Institut (EMI)
Joseph M. Heimerl
U.S. Army Ballistic Research Laboratory (BRL)
ISBN 1-56347-012-8

140. Thermal Structures and Materials for High-Speed Flight (1992)
Earl A. Thornton
University of Virginia
ISBN 1-56347-017-9

141. Tactical Missile Aerodynamics: General Topics (1992)
Michael J. Hemsch
Lockheed Engineering & Sciences Company
ISBN 1-56347-015-2

142. Tactical Missile Aerodynamics: Prediction Methodology (1992)
Michael R. Mendenhall
Nielsen Engineering & Research, Inc.
ISBN 1-56347-016-0

143. Nonsteady Burning and Combustion Stability of Solid Propellants (1992)
Luigi De Luca
Politecnico di Milano
Edward W. Price
Georgia Institute of Technology
Martin Summerfield
Princeton Combustion Research Laboratories, Inc.
ISBN 1-56347-014-4

144. Space Economics (1992)
Joel S. Greenberg
Princeton Synergetics, Inc.
Henry R. Hertzfeld
HRH Associates
ISBN 1-56347-042-X

145. Mars: Past, Present, and Future (1992)
E. Brian Pritchard
NASA Langley Research Center
ISBN 1-56347-043-8

146. Computational Nonlinear Mechanics in Aerospace Engineering (1992)
Satya N. Atluri
Georgia Institute of Technology
ISBN 1-56347-044-6

147. Modern Engineering for Design of Liquid-Propellant Rocket Engines (1992)
Dieter K. Huzel
David H. Huang
ISBN 1-56347-013-6

148. Metallurgical Technologies, Energy Conversion, and Magnetohydrodynamic Flows (1993)
Herman Branover
Yeshajahu Unger
Ben-Gurion University of the Negev
ISBN 1-56347-019-5

149. Advances in Turbulence Studies (1993)
Herman Branover
Yeshajahu Unger
Ben-Gurion University of the Negev
ISBN 1-56347-018-7

150. Structural Optimization: Status and Promise (1993)
Manohar P. Kamat
Georgia Institute of Technology
ISBN 1-56347-056-X

151. Dynamics of Gaseous Combustion (1993)
A. L. Kuhl
Lawrence Livermore National Laboratory
J.-C. Leyer
Université de Poitiers
A. A. Borisov
Russian Academy of Sciences
W. A. Sirignano
University of California
ISBN 1-56347-060-8

152. Dynamics of Heterogeneous Combustion and Reacting Systems (1993)
A. L. Kuhl
Lawrence Livermore National Laboratory
J.-C. Leyer
Université de Poitiers
A. A. Borisov
Russian Academy of Sciences
W. A. Sirignano
University of California
ISBN 1-56347-058-6

153. Dynamic Aspects of Detonations (1993)
A. L. Kuhl
Lawrence Livermore National Laboratory
J.-C. Leyer
Université de Poitiers
A. A. Borisov
Russian Academy of Sciences
W. A. Sirignano
University of California
ISBN 1-56347-057-8

154. Dynamic Aspects of Explosion Phenomena (1993)
A. L. Kuhl
Lawrence Livermore National Laboratory
J.-C. Leyer
Université de Poitiers
A. A. Borisov
Russian Academy of Sciences
W. A. Sirignano
University of California
ISBN 1-56347-059-4

155. Tactical Missile Warheads (1993)
Joseph Carleone
Aerojet General Corporation
ISBN 1-56347-067-5

(Other Volumes are planned.)